List of Selected Tables for Analysis and Design

SIXTH EDITION

Modern Control Systems

Richard C. Dorf
University of California, Davis

ADDISON-WESLEY PUBLISHING COMPANY

Reading, Massachusetts•Menlo Park, California•New York
Don Mills, Ontario•Wokingham, England•Amsterdam•Bonn
Sydney•Singapore•Tokyo•Madrid•San Juan•Milan•Paris

This book is in the Addison-Wesley Series in Electrical and Computer Engineering: Control Engineering
Consulting Editor: Karl J. Åström

Adaptive Control, 09720
Karl J. Åström and Björn Wittenmark

Introduction to Robotics, Second Edition, 09528
John J. Craig

Modern Control Systems, Sixth Edition, 51713
Richard C. Dorf

Digital Control of Dynamic Systems, Second Edition, 11938
Gene F. Franklin, J. David Powell, and Michael L. Workman

Computer Control of Machines and Processes, 10645
John G. Bollinger and Neil A. Duffie

Feedback Control of Dynamic Systems, Second Edition, 50862
Gene F. Franklin, J. David Powell, and Abbas Emami-Naeini

Adaptive Control of Mechanical Manipulators, 10490
John J. Craig

Reprinted with corrections March, 1993

ISBN 0-201-54343-5

1 2 3 4 5 6 7 8 9 10 MA 959493

DEDICATION

Man cannot inherit the past, he has to recreate it.

TO

MY STUDENTS as they seek to create

PREFACE

The National Academy of Engineering identified in 1990 the ten outstanding engineering achievements of the preceding twenty-five years. These feats included five accomplishments made possible by utilizing modern control engineering: the Apollo lunar landing, satellites, computer-aided manufacturing, computerized axial tomography, and the jumbo jet.

Automation and robotics are critical ingredients in the world's efforts toward an improved standard of living for all. Automation, the automatic operation of processes, and robotics, which includes the manipulator, controller, and associated devices, are all critical to effective operation of our plants, factories, and institutions. The most important and productive approach to learning is for each of us to rediscover and recreate anew the answers and methods of the past. Thus the ideal is to present the student with a series of problems and questions and point to some of the answers that have been obtained over the past decades. The traditional method—to confront the student not with the problem but with the finished solution—is to deprive the student of all excitement, to shut off the creative impulse, to reduce the adventure of humankind to a dusty heap of theorems. The issue, then, is to present some of the unanswered and important problems which we continue to confront. For it may be asserted that what we have truly learned and understood we discovered ourselves.

The purpose of this book is to present the structure of feedback control theory and to provide a sequence of exciting discoveries as we proceed through the text and problems. If this book is able to assist the student in discovering feedback control system theory and practice, it will have succeeded.

The book is organized around the concepts of control system theory as they have been developed in the frequency- and time-domain. A real attempt has been made to make the selection of topics, as well as the systems discussed in the examples and problems, modern in the best sense. Therefore this book includes a discussion of sensitivity, performance indices, state variables, robotics, and computer control systems, to name a few. However, a valiant attempt has been made to retain the classical topics of control theory that have proved to be so very useful in practice.

Written in an integrated form, the text should be read from the first to the last chapter. However, it is not necessary to include all the sections of a given chapter in any given course, and there appears to be quite a large number of combinations of sequences of the sections for study. The book is designed for an introductory undergraduate course in control systems for engineering students. There is very little demarcation between electrical, mechanical, chemical, and industrial engineering in control system practice; therefore this text is written without any conscious bias toward one discipline. Thus it is hoped that this book will be equally useful for all engineering disciplines and, perhaps, will assist in illustrating the unity of control engineering. The problems and examples are chosen from all fields, and the examples of the sociological, biological, ecological, and economic control systems are intended to provide the reader with an awareness of the general applicability of control theory to many facets of life.

The book is primarily concerned with linear, constant parameter control systems. This is a deliberate limitation because I believe that for an introduction to control systems, it is wisest initially to consider linear systems. Nevertheless, several nonlinear systems are introduced and discussed where appropriate.

Chapter 1 provides an introduction to and basic history of control theory. Chapter 2 is concerned with developing mathematical models of these systems. With the models available, the text describes the characteristics of feedback control systems in Chapter 3 and illustrates why feedback is introduced in a control system. Chapter 4 examines the performance of control systems, and Chapter 5 investigates the stability of feedback systems. Chapter 6 is concerned with the s-plane representation of the characteristic equation of a system and the root locus. Chapters 7 and 8 treat the frequency response of a system and the investigation of stability using the Nyquist criterion. Chapter 9 develops the time-domain concepts in terms of the state variables of a system. Chapter 10 describes and develops several approaches to designing and compensating a control system. Chapter 11 discusses computer control systems, robust systems, and robotics. Finally, Chapter 12 introduces and illustrates the all-important topic of engineering design.

This book is suitable for an introductory course in control systems. In its first five editions, the text has been used in senior-level courses for engineering students at more than 400 colleges and universities. Also, it has been used in courses for engineering graduate students with no previous background in control system theory.

The text presumes a reasonable familiarity with the Laplace transformation and transfer functions as developed in a first course in linear system analysis or network analysis. These concepts are discussed in Chapter 2 and are used to develop mathematical models for control system components. Answers to selected exercises are provided along with the exercises. Answers to selected problems are provided at the end of the book.

The sixth edition has incorporated several important developments in the field of control systems, with particular reference to robots and robust systems. In addition, a valuable feature is the exercises immediately preceding the problems. The purpose of these exercises is to permit students to utilize readily the concepts and methods introduced in each chapter in the solution of relatively straightforward exercises before attempting the more complex problems. The sixth edition expands the emphasis on design and incorporates a design example and several design problems in each chapter.

The sixth edition uses the computer program the Control System Design Program (CSDP) to assist in the solution of selected examples throughout the book. The student should first understand and use the tools and concepts before proceeding to utilize computer solutions. Nevertheless, the computer-aided analysis of the CSDP can be an invaluable aid in solving complex problems.

This edition also provides a preview as well as a summary of the terms and concepts for each chapter. There is an expanded Chapter 11, which discusses the useful proportional-integral-derivative (PID) controller and robust control systems. An enhanced Chapter 12 encompassing the all-important topic of design of real-world, complex control systems completes this edition.

This material has been developed with the assistance of many individuals to whom I wish to express my sincere appreciation. Finally, I can only partially acknowledge the encouragement and patience of my wife, Joy, who helped to make this book possible.

Davis, California R.C.D.

CONTENTS

5 The Stability of Linear Feedback Systems 207

6 The Root Locus Method 235

7 Frequency Response Methods 301

10 The Design and Compensation of Feedback Control Systems 487

11 Robust Control Systems 573

CHAPTER 1

Introduction to Control Systems

Preview

A system, consisting of interconnected components, is built to achieve a desired purpose. The performance of this system can be examined, and methods for controlling its performance can be proposed. It is the purpose of this chapter to describe the general approach to designing and building a control system.

In order to understand the purpose of a control system, it is useful to examine some examples of control systems through the course of history. Even these early systems incorporated the idea of feedback, which we will discuss throughout this book.

Modern control engineering practice includes the use of control strategies for aircraft, rapid transit, the artificial heart, and steel making, among others. We will examine these very interesting applications of control engineering.

1.1 Introduction

Engineering is concerned with understanding and controlling the materials and forces of nature for the benefit of humankind. Control system engineers are concerned with understanding and controlling segments of their environment, often called *systems,* in order to provide useful economic products for society. The twin goals of understanding and control are complementary because, in order to be controlled more effectively, the systems under control must be understood and modeled. Furthermore, control engineering must often consider the control of poorly understood systems such as chemical process systems. The present challenge to control engineers is the modeling and control of modern, complex, interrelated systems such as traffic-control systems, chemical processes, and robotic systems. However, simultaneously, the fortunate engineer has the opportunity to control many very useful and interesting industrial automation systems. Perhaps the most characteristic quality of control engineering is the opportunity to control machines, and industrial and economic processes for the benefit of society.

Control engineering is based on the foundations of feedback theory and linear system analysis, and it integrates the concepts of network theory and communication theory. Therefore control engineering is not limited to any engineering discipline but is equally applicable for aeronautical, chemical, mechanical, environmental, civil, and electrical engineering. For example, quite often a control system includes electrical, mechanical, and chemical components. Furthermore, as the understanding of the dynamics of business, social, and political systems increases, the ability to control these systems will increase also.

A *control system* is an interconnection of components forming a system configuration that will provide a desired system response. The basis for analysis of a system is the foundation provided by linear system theory, which assumes a cause-effect relationship for the components of a system. Therefore a component or *process* to be controlled can be represented by a block as shown in Fig. 1.1. The input-output relationship represents the cause and effect relationship of the process, which in turn represents a processing of the input signal to provide an output signal variable, often with a power amplification. An *open-loop* control system utilizes a controller or control actuator in order to obtain the desired response, as shown in Fig. 1.2.

In contrast to an open-loop control system, a closed-loop control system utilizes an additional measure of the actual output in order to compare the actual output with the desired output response. The measure of the output is called the *feedback signal.* A simple *closed-loop feedback control system* is shown in Fig. 1.3.

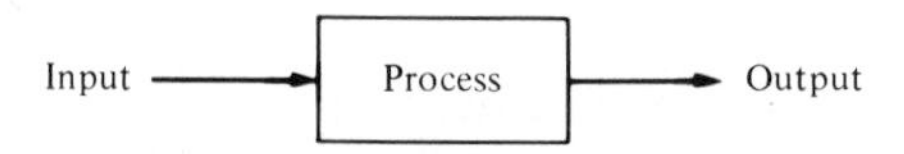

Figure 1.1. Process to be controlled.

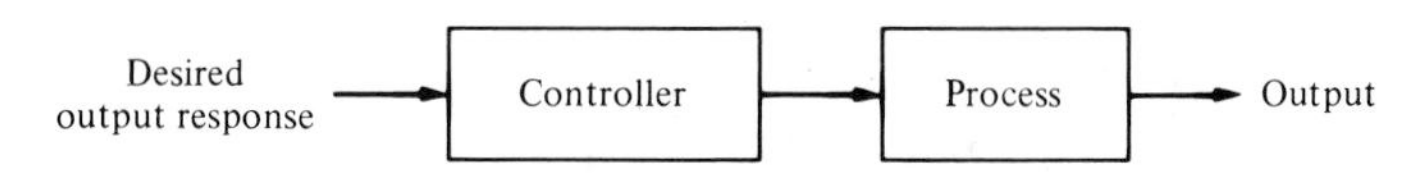

Figure 1.2. Open-loop control system.

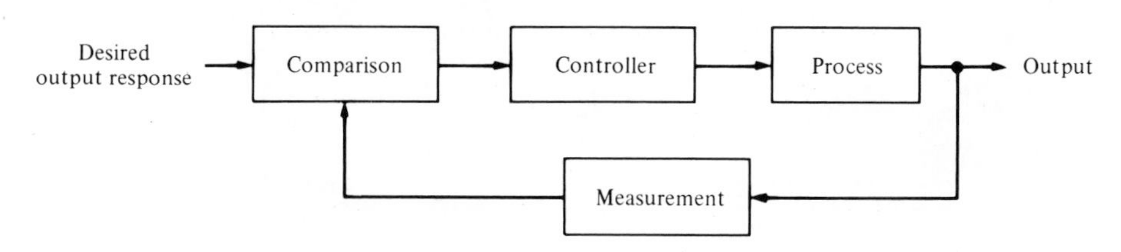

Figure 1.3. Closed-loop feedback control system.

A standard definition of a feedback control system is as follows: A feedback control system is a control system that tends to maintain a prescribed relationship of one system variable to another by comparing functions of these variables and using the difference as a means of control.

A feedback control system often uses a function of a prescribed relationship between the output and reference input to control the process. Often the difference between the output of the process under control and the reference input is amplified and used to control the process so that the difference is continually reduced. The feedback concept has been the foundation for control system analysis and design.

Due to the increasing complexity of the system under control and the interest in achieving optimum performance, the importance of control system engineering has grown in this decade. Furthermore, as the systems become more complex, the interrelationship of many controlled variables must be considered in the control scheme. A block diagram depicting a *multivariable control system* is shown in Fig. 1.4. A humorous example of a closed-loop feedback system is shown in Fig. 1.5.

A common example of an open-loop control system is an electric toaster in the kitchen. An example of a closed-loop control system is a person steering an automobile (assuming his or her eyes are open).

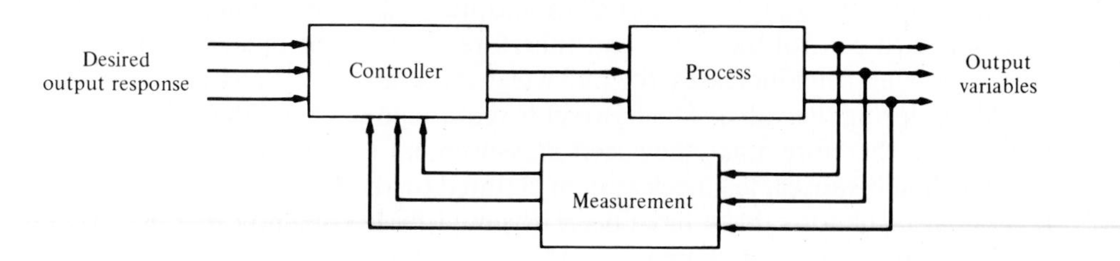

Figure 1.4. Multivariable control system.

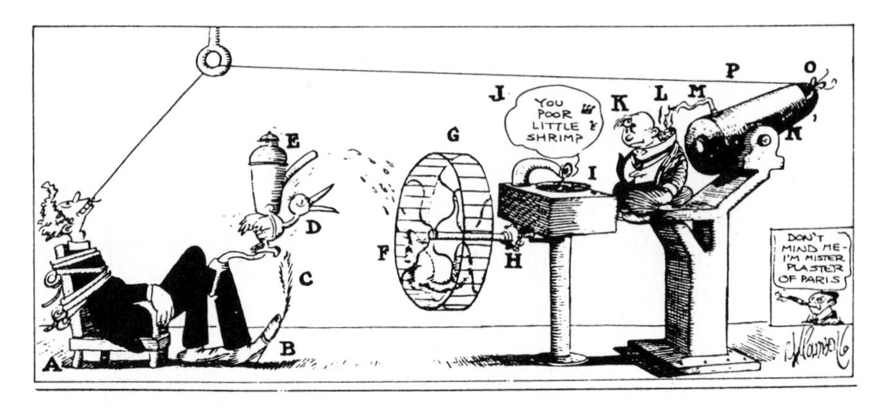

Figure 1.5. Rube Goldberg's elaborate creations were almost all closed-loop feedback systems. Goldberg called this simply, "Be Your Own Dentist." (© Rube Goldberg, permission granted by King Features Syndicate, Inc., 1979.)

1.2 History of Automatic Control

The use of feedback in order to control a system has had a fascinating history. The first applications of feedback control rest in the development of float regulator mechanisms in Greece in the period 300 to 1 B.C. [1, 2]. The water clock of Ktesibios used a float regulator (refer to Problem 1.11). An oil lamp devised by Philon in approximately 250 B.C. used a float regulator in an oil lamp for maintaining a constant level of fuel oil. Heron of Alexandria, who lived in the first century A.D., published a book entitled *Pneumatica,* which outlined several forms of water-level mechanisms using float regulators [1].

The first feedback system to be invented in modern Europe was the temperature regulator of Cornelis Drebbel (1572–1633) of Holland [1]. Dennis Papin [1647–1712] invented the first pressure regulator for steam boilers in 1681. Papin's pressure regulator was a form of safety regulator similar to a pressure-cooker valve.

The first automatic feedback controller used in an industrial process is generally agreed to be James Watt's *flyball governor* developed in 1769 for controlling the speed of a steam engine [1, 2]. The all-mechanical device, shown in Fig. 1.6, measured the speed of the output shaft and utilized the movement of the flyball with speed to control the valve and therefore the amount of steam entering the engine. As the speed increases, the ball weights rise and move away from the shaft axis thus closing the valve. The flyweights require power from the engine in order to turn and therefore make the speed measurement less accurate.

The first historical feedback system claimed by the Soviet Union is the water-level float regulator said to have been invented by I. Polzunov in 1765 [4]. The level regulator system is shown in Fig. 1.7. The float detects the water level and controls the valve that covers the water inlet in the boiler.

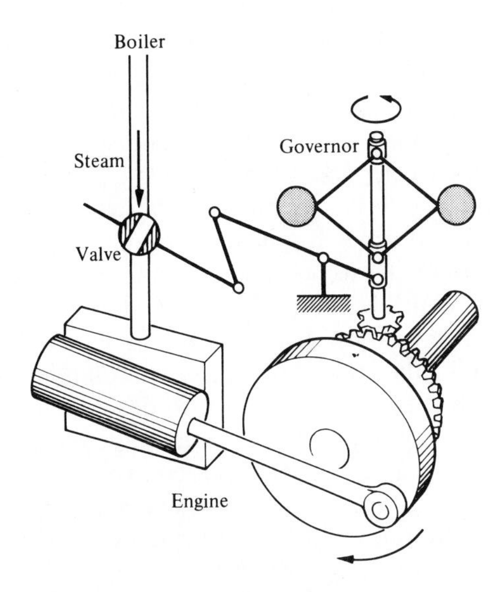

Figure 1.6. Watt flyball governor.

The period preceding 1868 was characterized by the development of automatic control systems by intuitive invention. Efforts to increase the accuracy of the control system led to slower attenuation of the transient oscillations and even to unstable systems. It then became imperative to develop a theory of automatic control. J. C. Maxwell formulated a mathematical theory related to control theory using a differential equation model of a governor [5]. Maxwell's study was concerned with the effect various system parameters had on the system performance. During the same period, I. A. Vyshnegradskii formulated a mathematical theory of regulators [6].

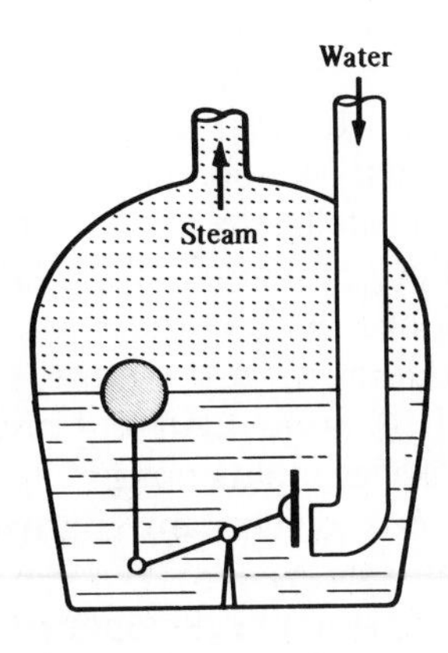

Figure 1.7. Water-level float regulator.

Prior to World War II, control theory and practice developed in the United States and Western Europe in a different manner than in the U.S.S.R. and Eastern Europe. A main impetus for the use of feedback in the United States was the development of the telephone system and electronic feedback amplifiers by Bode, Nyquist, and Black at Bell Telephone Laboratories [7, 8, 9, 10, 12]. The frequency domain was used primarily to describe the operation of the feedback amplifiers in terms of bandwidth and other frequency variables. In contrast, the eminent mathematicians and applied mechanicians in the Soviet Union inspired and dominated the field of control theory. Therefore, the Russian theory tended to utilize a time-domain formulation using differential equations.

A large impetus to the theory and practice of automatic control occurred during World War II when it became necessary to design and construct automatic airplane pilots, gun-positioning systems, radar antenna control systems, and other military systems based on the feedback control approach. The complexity and expected performance of these military systems necessitated an extension of the available control techniques and fostered interest in control systems and the development of new insights and methods. Prior to 1940, for most cases, the design of control systems was an art involving a trial-and-error approach. During the 1940s, mathematical and analytical methods increased in number and utility, and control engineering became an engineering discipline in its own right [10, 11, 12].

Frequency-domain techniques continued to dominate the field of control following World War II with the increased use of the Laplace transform and the complex frequency plane. During the 1950s, the emphasis in control engineering theory was on the development and use of the s-plane methods and, particularly, the root locus approach. Furthermore, during the 1980s, the utilization of digital computers for control components became routine. These new controlling elements possessed an ability to calculate rapidly and accurately that was formerly not available to the control engineer. There are now over two hundred thousand digital process control computers installed in the United States [13, 44]. These computers are employed especially for process control systems in which many variables are measured and controlled simultaneously by the computer.

With the advent of Sputnik and the space age, another new impetus was imparted to control engineering. It became necessary to design complex, highly accurate control systems for missiles and space probes. Furthermore, the necessity to minimize the weight of satellites and to control them very accurately has spawned the important field of optimal control. Due to these requirements, the time-domain methods developed by Liapunov, Minorsky, and others have met with great interest in the last decade. Furthermore, new theories of optimal control have been developed by L. S. Pontryagin in the Soviet Union and R. Bellman in the United States. It now is clear that control engineering must consider both the time-domain and the frequency-domain approaches simultaneously in the analysis and design of control systems.

A selected history of control system development is summarized in Table 1.1.

Table 1.1. Selected Historical Developments of Control Systems

1769	James Watt's steam engine and governor developed. The Watt steam engine is often used to mark the beginning of the Industrial Revolution in Great Britain. During the industrial revolution, great strides were made in the development of mechanization, a technology preceding automation.
1800	Eli Whitney's concept of interchangeable parts manufacturing demonstrated in the production of muskets. Whitney's development is often considered as the beginning of mass production.
1868	J. C. Maxwell formulates a mathematical model for a governor control of a steam engine.
1913	Henry Ford's mechanized assembly machine introduced for automobile production.
1927	H. W. Bode analyzes feedback amplifiers.
1932	H. Nyquist develops a method for analyzing the stability of systems.
1952	Numerical control (NC) developed at Massachusetts Institute of Technology for control of machine-tool axes.
1954	George Devol develops "programmed article transfer," considered to be the first industrial robot design.
1960	First Unimate robot introduced, based on Devol's designs. Unimate installed in 1961 for tending die-casting machines.
1980	Robust control system design widely studied.
1990	Export-oriented manufacturing companies emphasize automation.

1.3 Two Examples of Engineering Creativity

Harold S. Black graduated from Worcester Polytechnic Institute in 1921 and joined Bell Laboratories of American Telegraph and Telephone (AT&T). In 1921, the major task confronting Bell Labs was the improvement of the telephone system and the design of improved signal amplifiers. Black was assigned the task of linearizing, stabilizing, and improving the amplifiers that were used in tandem to carry conversations over distances of several thousand miles.

In a recent article, Black reports [8]:

> Then came the morning of Tuesday, August 2, 1927, when the concept of the negative feedback amplifier came to me in a flash while I was crossing the Hudson River on the Lackawanna Ferry, on my way to work. For more than 50 years I have pondered how and why the idea came, and I can't say any more today than I could that morning. All I know is that after several years of hard work on the problem, I suddenly realized that if I fed the amplifier output back to the input, in reverse phase, and kept the device from oscillating (singing, as we called it then), I would have exactly what I wanted: a means of canceling out the distortion in the output. I opened my morning newspaper and on a page of *The New York Times* I sketched a simple canonical diagram of a negative feedback amplifier plus

> the equations for the amplification with feedback. I signed the sketch, and 20 minutes later, when I reached the laboratory at 463 West Street, it was witnessed, understood, and signed by the late Earl C. Blessing.
>
> I envisioned this circuit as leading to extremely linear amplifiers (40 to 50 db of negative feedback), but an important question is: How did I know I could avoid self-oscillations over very wide frequency bands when many people doubted such circuits would be stable? My confidence stemmed from work that I had done two years earlier on certain novel oscillator circuits and three years earlier in designing the terminal circuits, including the filters, and developing the mathematics for a carrier telephone system for short toll circuits.

Another example of the discovery of an engineering solution to a control system problem was that of the creation of a gun director by David B. Parkinson of Bell Telephone Laboratories. In the spring of 1940, Parkinson was a 29-year-old engineer intent on improving the automatic level recorder, an instrument that used strip-chart paper to plot the record of a voltage. A critical component was a small potentiometer that was used to control the pen of the recorder through an actuator.

Parkinson had a dream about an antiaircraft gun that was successfully felling airplanes. Parkinson described the situation [58]:

> After three or four shots one of the men in the crew smiled at me and beckoned me to come closer to the gun. When I drew near he pointed to the exposed end of the left trunnion. Mounted there was the control potentiometer of my level recorder!

The next morning Parkinson realized the significance of his dream:

> If my potentiometer could control the pen on the recorder, something similar could, with suitable engineering, control an antiaircraft gun.

After considerable effort, an engineering model was delivered for testing to the U.S. Army on December 1, 1941. Production models were available by early 1943 and eventually 3000 gun controllers were delivered. Input to the controller was provided by radar and the gun was aimed by taking the data of the airplane's present position and calculating the target's future position.

1.4 Control Engineering Practice

Control engineering is concerned with the analysis and design of goal-oriented systems. Therefore the mechanization of goal-oriented policies has grown into a hierarchy of goal-oriented control systems. Modern control theory is concerned with systems with the self-organizing, adaptive, robust, learning, and optimum qualities. This interest has aroused even greater excitement among control engineers.

The control of an industrial process (manufacturing, production, and so on) by automatic rather than human means is often called *automation.* Automation is prevalent in the chemical, electric power, paper, automobile, and steel industries, among others. The concept of automation is central to our industrial society. Automatic machines are used to increase the production of a plant per worker in order to offset rising wages and inflationary costs. Thus industries are concerned with the productivity per worker of their plant. *Productivity* is defined as the ratio of physical output to physical input. In this case we are referring to labor productivity, which is real output per hour of work. In a study conducted by the U.S. Commerce Department it was determined that labor productivity grew at an average annual rate of 2.8% from 1948 to 1990 [13]. In order to continue these productivity gains, expenditures for factory automation in the United States are expected to increase from $5.0 billion in 1988 to $12.0 billion in 1994 [26]. Worldwide expenditures for process control and manufacturing plant control are expected to grow from $12.0 billion in 1988 to $28.0 billion in 1994 [26]. The U.S. manufacturers currently supply approximately one-half of worldwide control equipment.

The transformation of the U.S. labor force in the country's brief history follows the progressive mechanization of work that attended the evolution of the agrarian republic into an industrial world power. In 1820 more than 70% of the labor force worked on the farm. By 1900 fewer than 40% were engaged in agriculture. Today, fewer than 5% work in agriculture [15].

In 1925 some 588,000 people—about 1.3% of the nation's labor force—were needed to mine 520 million tons of bituminous coal and lignite, almost all of it from underground. By 1980 production was up to 774 million tons, but the work force had been reduced to 208,000. Furthermore, only 136,000 of that number were employed in underground mining operations. The highly mechanized and highly productive surface mines, with just 72,000 workers, produced 482 million tons, or 62% of the total [27].

The easing of human labor by technology, a process that began in prehistory, is entering a new stage. The acceleration in the pace of technological innovation inaugurated by the Industrial Revolution has until recently resulted mainly in the displacement of human muscle power from the tasks of production. The current revolution in computer technology is causing an equally momentous social change: the expansion of information gathering and information processing as computers extend the reach of the human brain [56].

The decline in the work week in the United States is illustrated by Fig. 1.8.

Control systems are used to achieve (1) increased productivity and (2) improved performance of a device or system. Automation is used to improve productivity and obtain high quality products. Automation is the automatic operation or control of a process, device, or system. We utilize automatic control of machines and processes in order to produce a product within specified tolerances [28].

The term *automation* first became popular in the automobile industry. Transfer lines were coupled with automatic machine tools to create long machinery

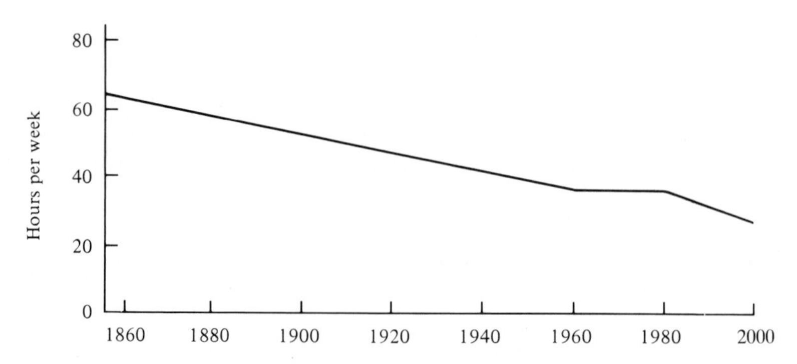

Figure 1.8. The work week in U.S. manufacturing industries was shortened from 67 hours in 1860 to about 38 hours in 1990.

lines that could produce engine parts, such as the cylinder block, virtually without operator intervention. In body-parts manufacturing, automatic-feed mechanisms were coupled with high-speed stamping presses to increase productivity in sheet-metal forming. In many other areas where designs were relatively stable, such as radiator production, entire automated lines replaced manual operations.

With the demand for flexible, custom production emerging in the 1990s, a need for flexible automation and robots is growing [31].

There are about 150,000 control engineers in the United States and also in Japan, and over 100,000 control engineers in the Soviet Union. In the United States alone, the control industry does a business of over $40 billion per year! The theory, practice, and application of automatic control is a large, exciting, and extremely useful engineering discipline. One can readily understand the motivation for a study of modern control systems.

1.5 Examples of Modern Control Systems

Feedback control is a fundamental fact of modern industry and society. Driving an automobile is a pleasant task when the auto responds rapidly to the driver's commands. Many cars have power steering and brakes, which utilize hydraulic amplifiers for amplification of the force to the brakes or the steering wheel. A simple block diagram of an automobile steering control system is shown in Fig. 1.9(a). The desired course is compared with a measurement of the actual course in order to generate a measure of the error as shown in Fig. 1.9(b). This measurement is obtained by visual and tactile (body movement) feedback. There is an additional feedback from the feel of the steering wheel by the hand (sensor). This feedback system is a familiar version of the steering control system in an ocean liner or the flight controls in a large airplane. All these systems operate in a closed-loop sequence as shown in Fig. 1.10. The actual and desired outputs are compared, and a measure of the difference is used to drive the power amplifier. The

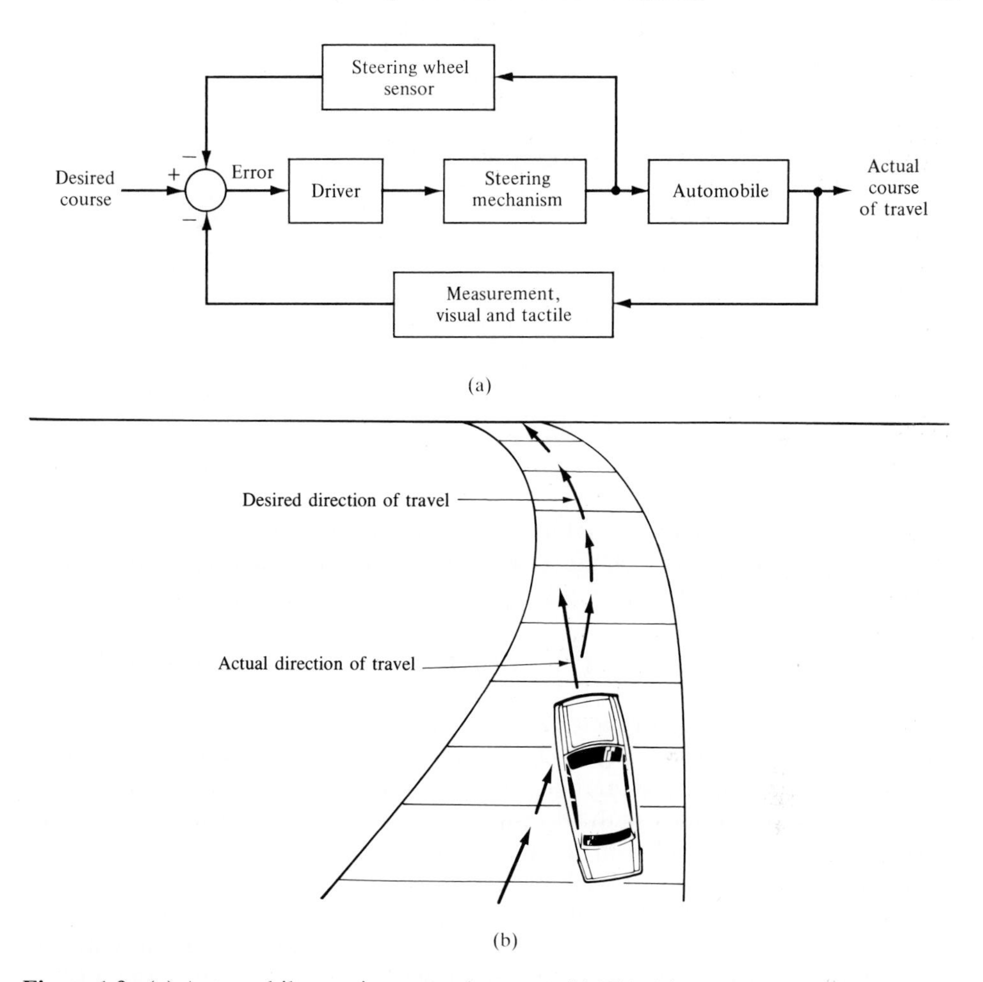

Figure 1.9. (a) Automobile steering control system. (b) The driver uses the difference between the actual and desired direction of travel to generate a controlled adjustment of the steering wheel.

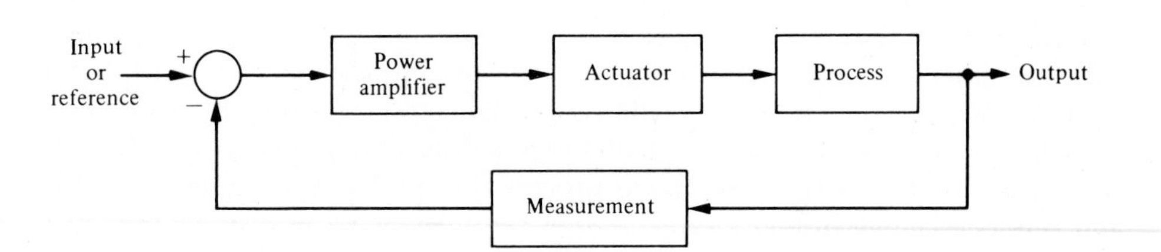

Figure 1.10. Basic closed-loop control system.

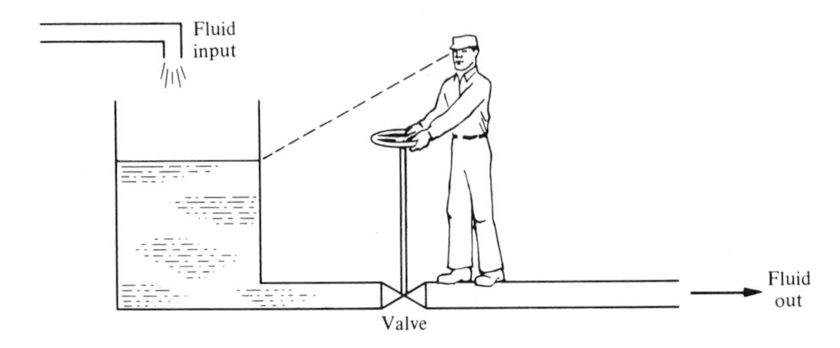

Figure 1.11. A manual control system for regulating the level of fluid in a tank by adjusting the output valve. The operator views the level of fluid through a port in the side of the tank.

power amplifier causes the actuator to modulate the process in order to reduce the error. The sequence is such that if the ship, for instance, is heading incorrectly to the right, the rudder is actuated in order to direct the ship to the left. The system shown in Fig. 1.10 is a *negative feedback* control system, because the output is subtracted from the input and the difference is used as the input signal to the power amplifier.

A basic manually controlled closed-loop system for regulating the level of fluid in a tank is shown in Fig. 1.11. The input is a reference level of fluid that the operator is instructed to maintain. (This reference is memorized by the operator.) The power amplifier is the operator and the sensor is visual. The operator compares the actual level with the desired level and opens or closes the valve (actuator) to maintain the desired level.

Other familiar control systems have the same basic elements as the system shown in Fig. 1.10. A refrigerator has a temperature setting or desired temperature, a thermostat to measure the actual temperature and the error, and a compressor motor for power amplification. Other examples in the home are the oven, furnace, and water heater. In industry, there are speed controls, process temperature and pressure controls, position, thickness, composition, and quality controls among many others [25, 29, 30].

In its modern usage, automation can be defined as a technology that uses programmed commands to operate a given process, combined with feedback of information to determine that the commands have been properly executed. Automation is often used for processes that were previously operated by humans. When automated, the process can operate without human assistance or interference. In fact, most automated systems are capable of performing their functions with greater accuracy and precision, and in less time, than humans are able to do.

An example of a semiautomated process that incorporates human workers and robots is shown in Fig. 1.12.

A robot is a computer controlled machine and is a technology closely associated with automation. Industrial robotics can be defined as a particular field of

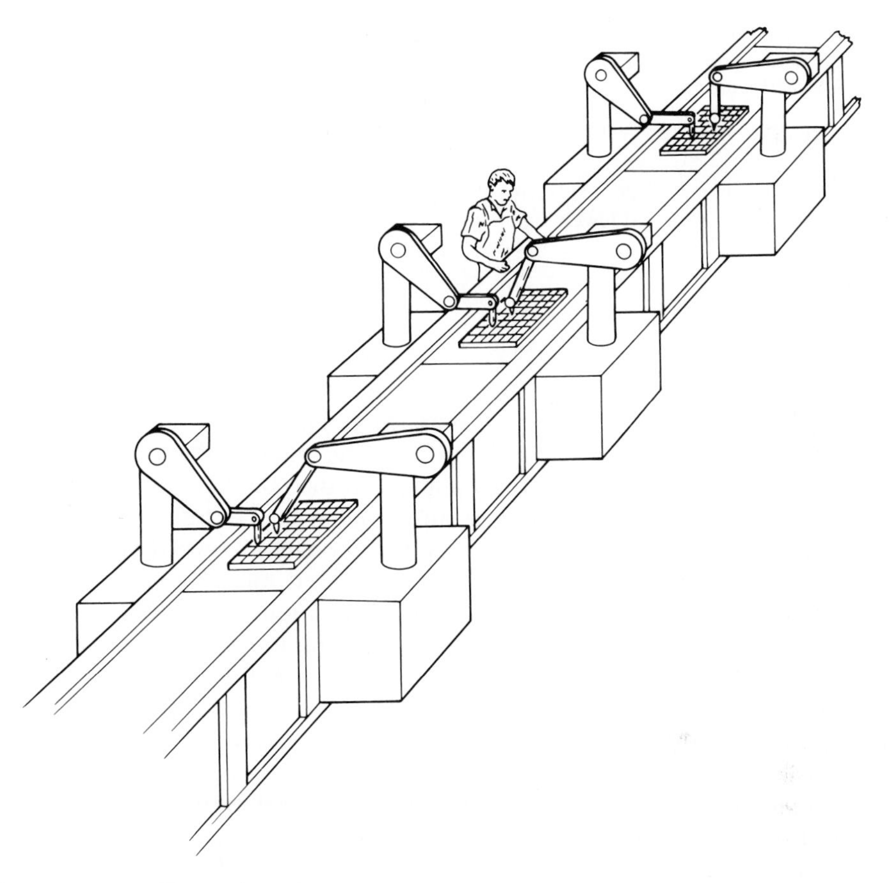

Figure 1.12. An illustration of a typical hybrid (human-robot) workstation. This illustration is based on the on-line operation phase of a production line where operator and intelligent robot cooperation must occur.

automation in which the automated machine (i.e., the robot) is designed to substitute for human labor [18, 19]. To do this, robots possess certain human-like characteristics. Today, the most common human-like characteristic is a mechanical manipulator that is patterned somewhat after the human arm and wrist. We recognize that the automatic machine is well suited to some tasks, as noted in Table 1.2, while other tasks are best carried out by humans.

Another very important application of control technology is in the control of the modern automobile [20, 59]. Control systems for suspension, steering, and engine control are being introduced. Many new autos have a four-wheel-steering system, as well as an antiskid control system.

There has been considerable discussion recently concerning the gap between practice and theory in control engineering. However, it is natural that theory precedes the applications in many fields of control engineering. Nonetheless, it is

Table 1.2. Task Difficulty: Human Versus Automatic Machine

Tasks Difficult for a Machine	Tasks Difficult for a Human
Inspect seedlings in a nursery.	Inspect a system in a hot, toxic environment.
Drive a vehicle through rugged terrain.	Repetitively assemble a clock.
Identify the most expensive jewels on a tray of jewels.	Land an airliner at night, in bad weather.

interesting to note that in the electric power industry, the largest industry in the United States, the gap is relatively insignificant. The electric power industry is primarily interested in energy conversion, control, and distribution. It is critical that computer control be increasingly applied to the power industry in order to improve the efficient use of energy resources. Also, the control of power *plants* for minimum waste emission has become increasingly important. The modern, large-capacity plants, which exceed several hundred megawatts, require automatic control systems that account for the interrelationship of the process variables and the optimum power production. It is common to have as many as 90 or more manipulated variables under coordinated control. A simplified model showing several of the important control variables of a large boiler-generator system is shown in Fig. 1.13. This is an example of the importance of measuring many variables, such as pressure and oxygen, in order to provide information to the computer for control calculations. It is estimated that more than two hundred thousand computer control systems have been installed in the United States [44, 50]. The diagram of a computer control system is shown in Fig. 1.14. The electric power industry has utilized the modern aspects of control engineering for significant and interesting applications. It appears that in the process industry, the factor that maintains the applications gap is the lack of instrumentation to measure all the important process variables, including the quality and composition of the product. As these instruments become available, the applications of modern control theory to industrial systems should increase measurably.

Another important industry, the metallurgical industry, has had considerable success in automatically controlling its processes. In fact, in many cases, the control applications are beyond the theory. For example, a hot-strip steel mill, which involves a $100-million investment, is controlled for temperature, strip width, thickness, and quality.

Rapidly rising energy costs coupled with threats of energy curtailment are resulting in new efforts for efficient automatic energy management. Computer controls are used to control energy use in industry and to stabilize and connect loads evenly to gain fuel economy [30].

There has been considerable interest recently in applying the feedback control concepts to automatic warehousing and inventory control. Furthermore, automatic control of agricultural systems (farms) is meeting increased interest. Auto-

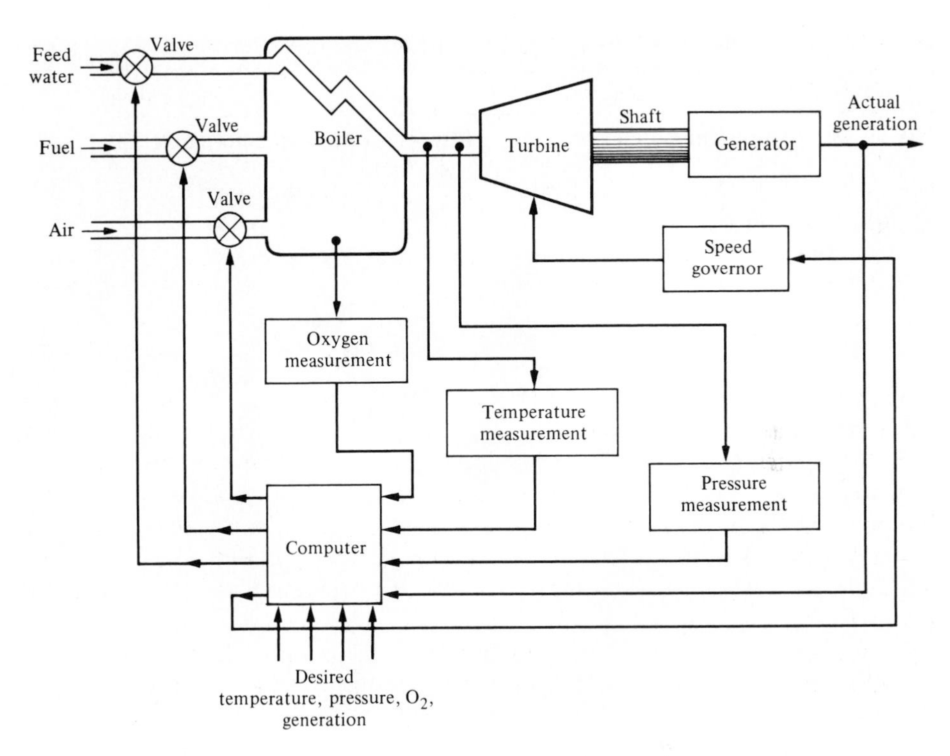

Figure 1.13. Coordinated control system for a boiler-generator.

matically controlled silos and tractors have been developed and tested. Automatic control of wind turbine generators, solar heating and cooling, and automobile engine performance are important modern examples [20, 32].

Also, there have been many applications of control system theory to biomedical experimentation, diagnosis, prosthetics, and biological control systems [24, 38]. The control systems under consideration range from the cellular level to the central nervous system, and include temperature regulation and neurological, respiratory, and cardiovascular control. Most physiological control systems are closed-loop systems. However, we find not one controller but rather control loop

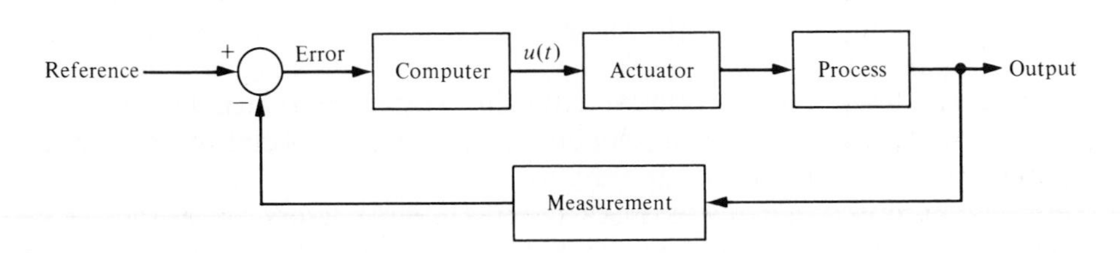

Figure 1.14. A computer control system.

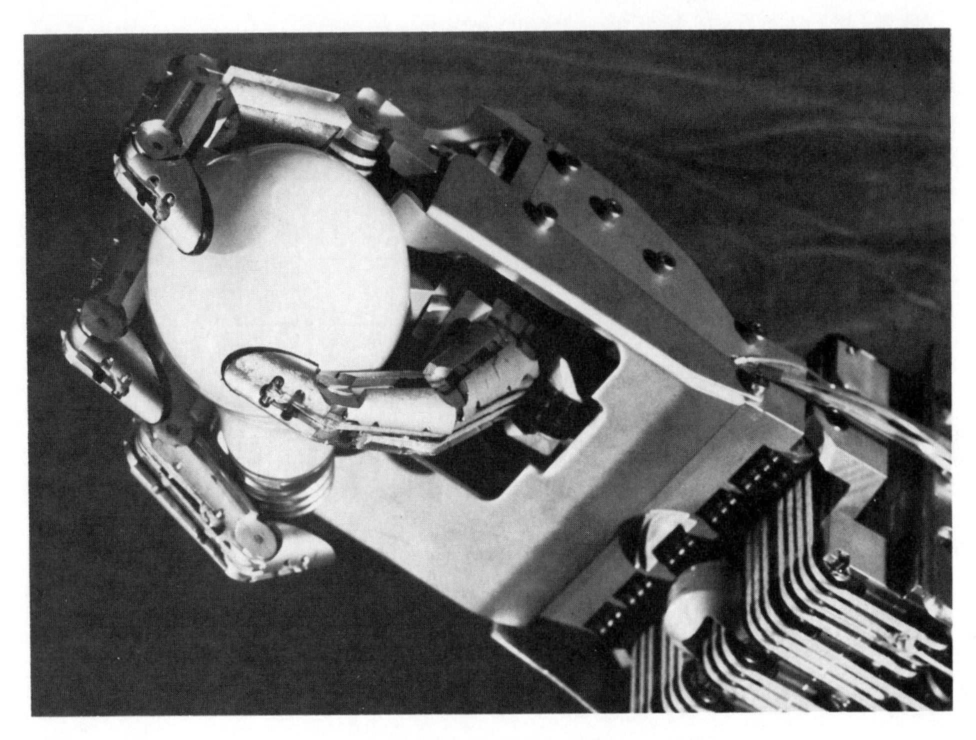

Figure 1.15. The Utah/MIT Dextrous Robotic Hand: A dextrous robotic hand having 18 degrees of freedom, developed as a research tool by the Center for Engineering Design at the University of Utah and the Artificial Intelligence Laboratory at M.I.T. It is controlled by five Motorola 68000 microprocessors and actuated by 36 high-performance electropneumatic actuators via high-strength polymeric tendons. The hand has three fingers and a thumb. It uses touch sensors and tendons for control. (Photograph by Michael Milochik. Courtesy of University of Utah.)

within control loop, forming a hierarchy of systems. The modeling of the structure of biological processes confronts the analyst with a high-order model and a complex structure. Prosthetic devices that aid the 46 million handicapped individuals in the United States are designed to provide automatically controlled aids to the disabled [19, 22, 24, 36]. An artificial hand that uses force feedback signals and is controlled by the amputee's bioelectric control signals, which are called electromyographic signals, is shown in Fig. 1.15. The Jarvik artificial heart is shown in Fig. 1.16.

Finally, it has become of interest and value to attempt to model the feedback processes prevalent in the social, economic, and political spheres. This approach is undeveloped at present but appears to have a reasonable future. Society, of course, is comprised of many feedback systems and regulatory bodies such as the Interstate Commerce Commission and the Federal Reserve Board, which are controllers exerting the necessary forces on society in order to maintain a desired

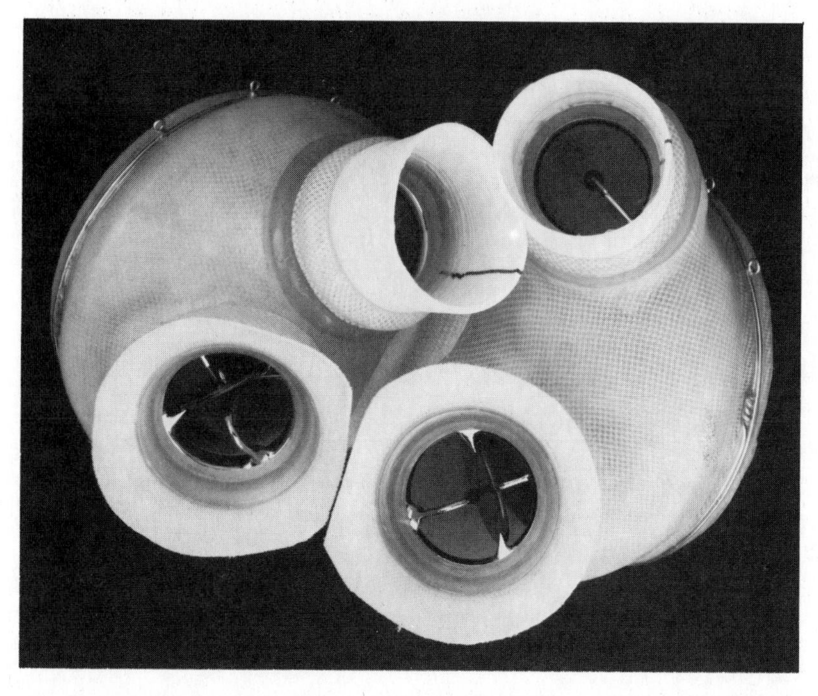

Figure 1.16. The Jarvik-7 artificial heart. William Schroeder received such an artificial heart on November 25, 1984, in Louisville, Kentucky. This system is driven by an external air compressor that moves the heart diaphragms. This is an advanced control system for the human blood-flow system. (Courtesy of Symbion, Inc.) [47]

output. A simple lumped model of the national income feedback control system is shown in Fig. 1.17. This type of model helps the analyst to understand the effects of government control—granted its existence—and the dynamic effects of government spending. Of course, many other loops not shown also exist, since, theoretically, government spending cannot exceed the tax collected without a deficit, which is itself a control loop containing the Internal Revenue Service and the Congress. Of course, in a communist country the loop due to consumers is deemphasized and the government control is emphasized. In that case, the measurement block must be accurate and must respond rapidly; both are very difficult characteristics to realize from a bureaucratic system. This type of political or social feedback model, while usually nonrigorous, does impart information and understanding.

Feedback control systems are used extensively in industrial applications. An industrial robot is shown in Fig. 1.18. Thousands of industrial robots are currently in use. Manipulators can pick up objects weighing hundreds of pounds and position them with an accuracy of one-tenth of an inch or better [28]. A mobile automaton capable of avoiding objects and traveling through a room or industrial plant is shown in Fig. 1.19.

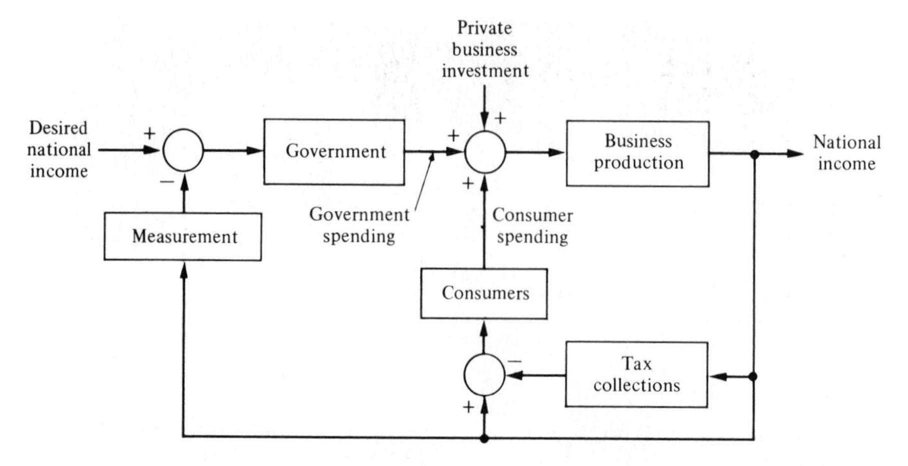

Figure 1.17. A feedback control system model of the economy.

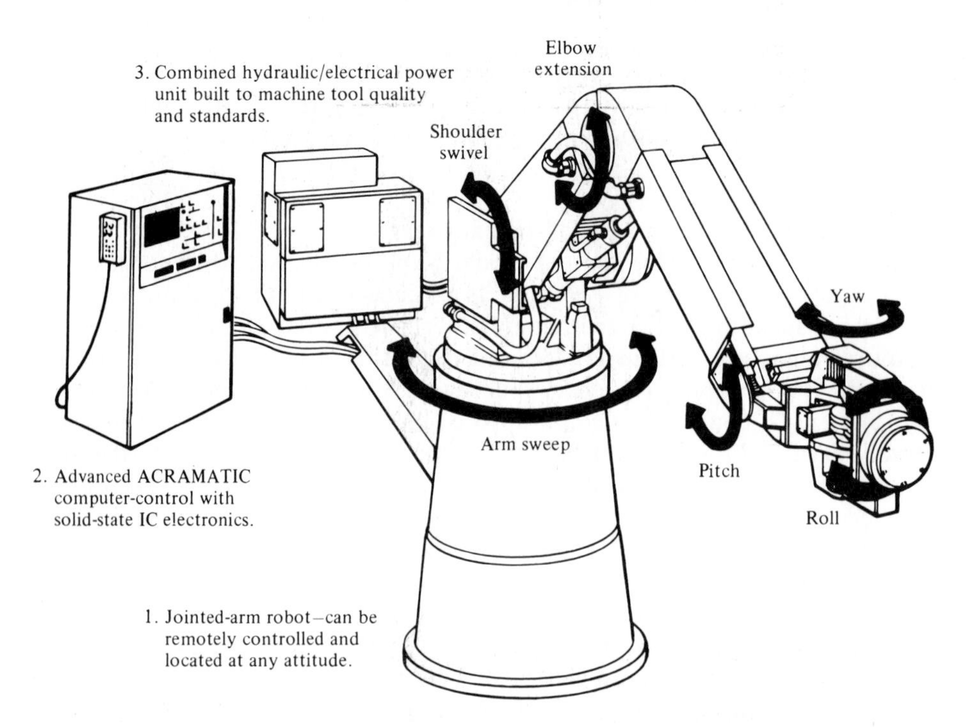

Figure 1.18. The T^3 industrial robot. (Courtesy of Cincinnati Milacron.)

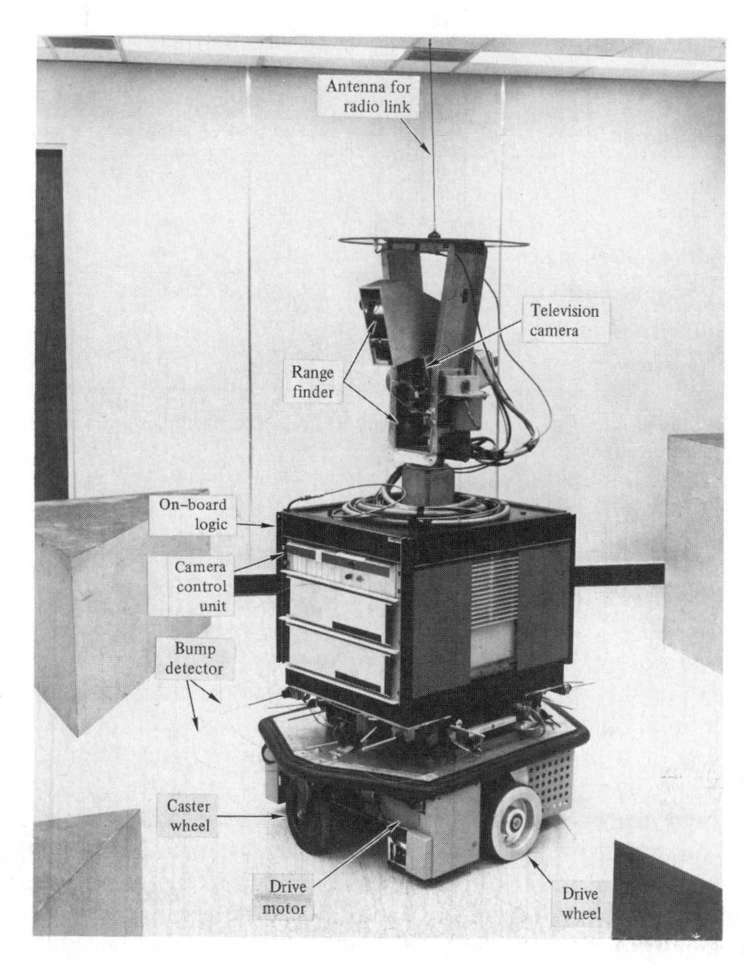

Figure 1.19. The automaton vehicle used by SRI in the application of artificial intelligence principles to the development of integrated robot systems. The vehicle is propelled by electric motors and carries a television camera and optical range finder in the movable "head." The vehicle responds to commands from a computer. The sensors are the bump detector, the TV camera, and the range finder [36]. (Courtesy of Stanford Research Institute.)

The potential future application of feedback control systems and models appears to be unlimited. Estimates of the U.S. markets for several control systems applications are given in Table 1.3. It appears that the theory and practice of modern control systems have a bright and important future and certainly justify the study of modern automatic control system theory and application. In the next chapter, we shall study the system models further to obtain quantitative mathematical models useful for engineering analysis and design.

Table 1.3. Applications of Control Systems

Application	(Millions of dollars)* 1972	1973	1976	1980	1990
Motor controls (speed, position)	90.3	100.5	112	150	250
Numerical controls	43.4	47.3	76	100	170
Thickness controls (steel, paper)	45.4	57.8	99	180	240
Process controls (oil, chemical)	318.5	357.2	449	700	2000
Pollution monitoring and control	14.0	17.0	26	75	300
Nuclear reactor control	9.3	11.1	19	25	60

*U.S. market estimates for several control system applications. The examples given in parentheses are not all-inclusive of the applications.

1.6 Control Engineering Design

Design is a purposeful activity in which a designer has in mind an idea about a desired outcome. It is the process of originating systems and predicting how these systems will fulfill objectives. Engineering design is the process of producing a set of descriptions of a system that satisfy a set of performance requirements and constraints.

The design process can be considered to incorporate three phases: analysis, synthesis, and evaluation. The first task is to diagnose, define, and prepare—that is, to understand the problem and produce an explicit statement of goals. The second task involves finding plausible solutions. The third task concerns judging the validity of solutions relative to the goals and selecting among alternatives. A cycle is implied in which the solution is revised and improved by reexamining the analysis. These three phases form the basis of a framework for planning, organizing, and evolving design projects.

Engineering design is the process of converting an idea or market need into the detailed information from which a product, process, or system can be made. A four-phase process of engineering design is summarized in Fig. 1.20. It is this process that we utilize in a design example for each chapter.

1.7 Design Example: Insulin Delivery Control System

Control systems have been utilized in the biomedical field to create implanted automatic drug-delivery systems to patients [54, 55]. Automatic systems can be used to regulate blood pressure, blood sugar level, and heart rate. A common application of control engineering is in the field of open-loop system drug delivery

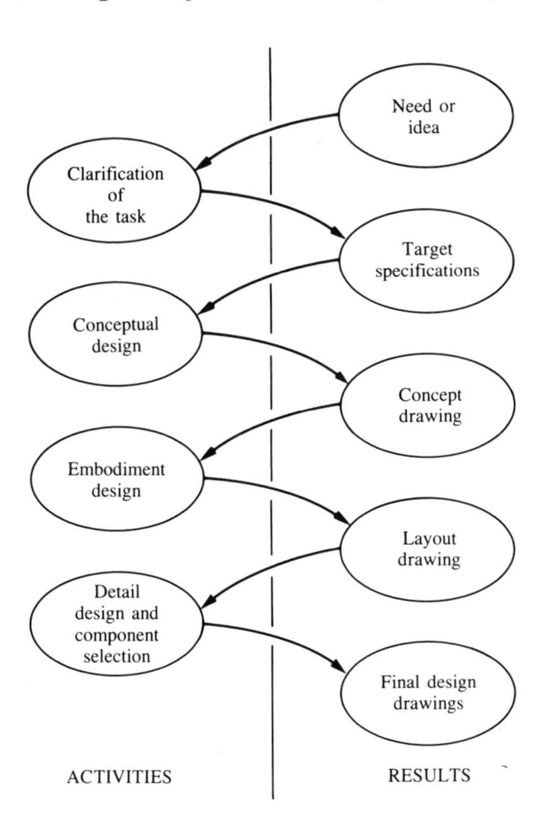

Figure 1.20. The design process.

in which mathematical models of the dose-effect relationship of the drugs are used. A drug-delivery system implanted in the body uses an open-loop system since miniaturized glucose sensors are not yet available. The best solutions rely on individually programmable pocket-sized insulin pumps that can deliver insulin according to a preset time history. More complicated systems will use closed-loop control for the measured blood glucose levels.

The goal of this design is (1) to design the block diagram of an open-loop control system and (2) to design a closed-loop control system to regulate the blood sugar concentration of a diabetic. The blood glucose and insulin concentrations are shown in Fig. 1.21 for a normal person. The designed system must provide the insulin from a reservoir implanted within the person.

An open-loop system would use a preprogrammed signal generator and miniature motor pump to regulate the insulin delivery rate as shown in Fig. 1.22(a). The feedback control system would use a sensor to measure the actual glucose level and compare that level with the desired level, thus turning the motor pump on when it is required as shown in Fig. 1.22(b).

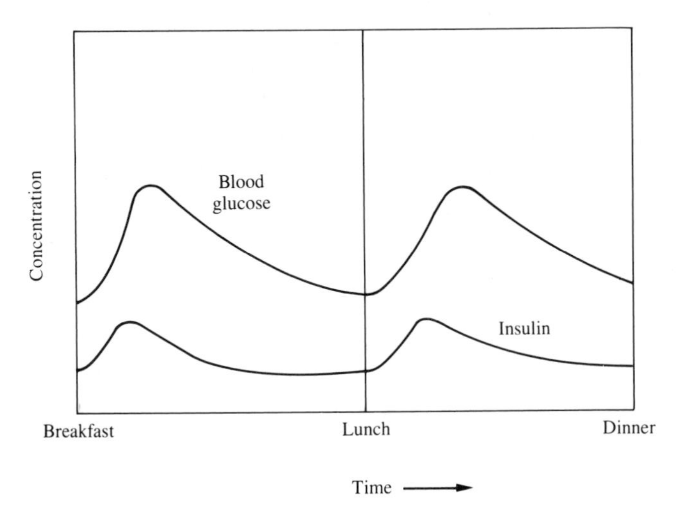

Figure 1.21. The blood glucose and insulin levels for a healthy person.

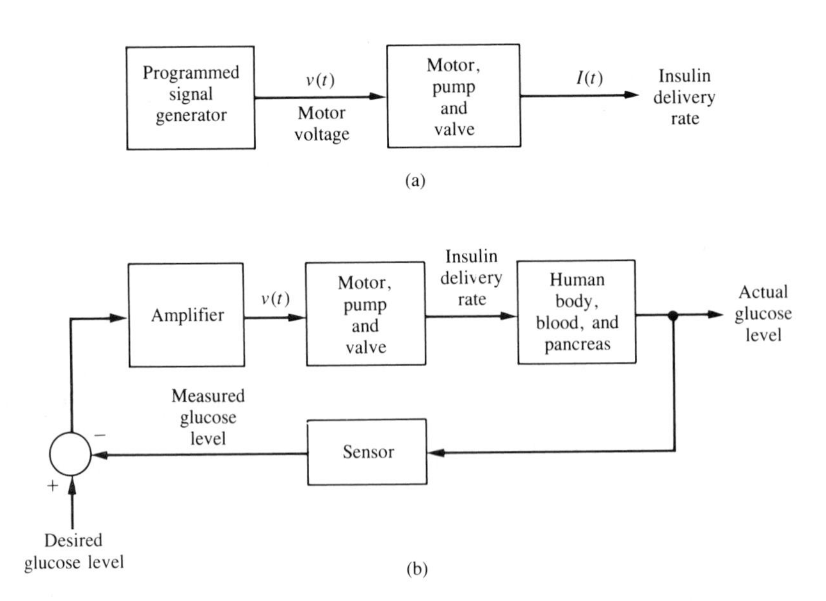

Figure 1.22. (a) Open-loop control and (b) closed-loop control of blood glucose.

Exercises

(Exercises are straightforward applications of the concepts of the chapter.)

E1.1. A precise optical signal source can control the output power level to within 1% [51]. A laser is controlled by an input current to yield the power output. A microprocessor controls the input current to the laser. The microprocessor compares the desired power level with a measured signal proportional to the laser power output obtained from a sensor. Show the block diagram to represent this closed-loop control system.

E1.2. An automobile driver uses a control system to maintain the speed of the car at a prescribed level. Draw a block diagram to illustrate this feedback system.

E1.3. The number of robots sold annually for automation of U.S. industry could grow to 20,000 by 1992. Visit a local fast-food restaurant and discuss whether a robot could replace one or more of the workers.

E1.4. An autofocus camera will adjust the distance of the lens from the film by using a beam of infrared or ultrasound to determine the distance to the subject. Draw a block diagram of this open-loop control system and briefly explain its operation.

Problems

(Problems require extending the concepts of this chapter to new situations.)

P1.1. Draw a schematic block diagram of a home heating system. Identify the function of each element of the thermostatically controlled heating system.

P1.2. Control systems have, in the past, used a human operator as part of a closed-loop control system. Draw the block diagram of the valve control system shown in Fig. P1.2.

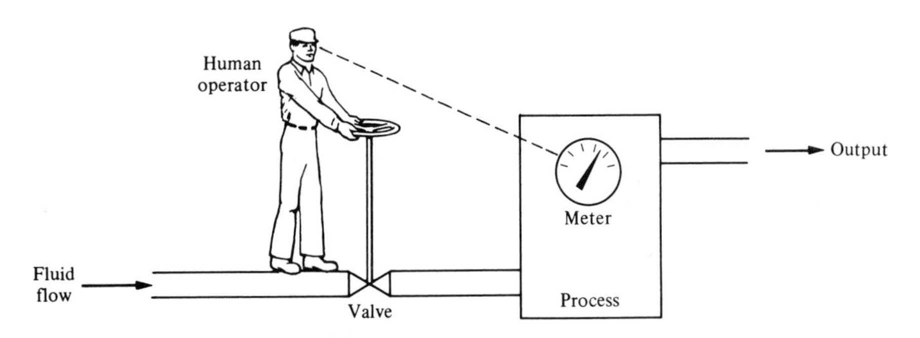

Figure P1.2. Fluid flow control.

P1.3. In a chemical process control system, it is valuable to control the chemical composition of the product. In order to control the composition, a measurement of the composition may be obtained by using an infrared stream analyzer as shown in Fig. P1.3. The valve on the additive stream may be controlled. Complete the control feedback loop and draw a block diagram describing the operation of the control loop.

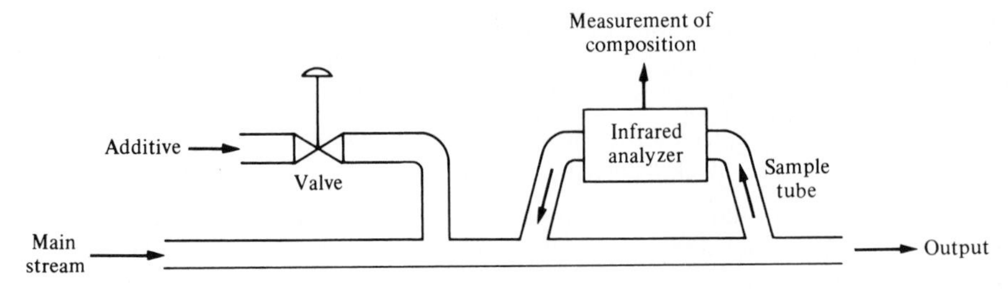

Figure P1.3.

P1.4. The accurate control of a nuclear reactor is important for power system generators. Assuming the number of neutrons present is proportional to the power level, an ionization chamber is used to measure the power level. The current, i, is proportional to the power level. The position of the graphite control rods moderates the power level. Complete the control system of the nuclear reactor shown in Fig. P1.4 and draw the block diagram describing the operation of the control loop.

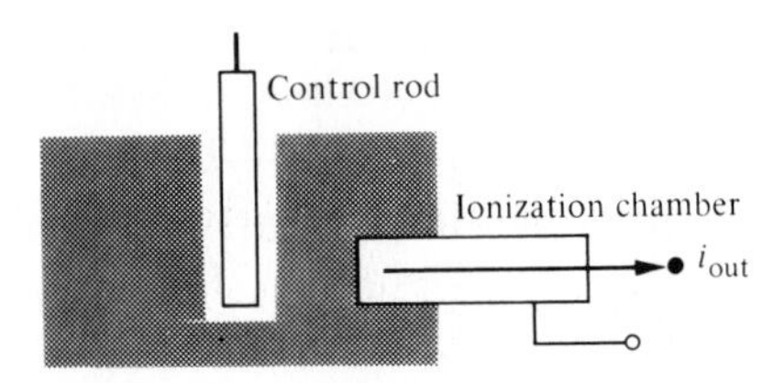

Figure P1.4. Nuclear reactor control.

P1.5. A light-seeking control system, used to track the sun, is shown in Fig. P1.5. The output shaft, driven by the motor through a worm reduction gear, has a bracket attached on which are mounted two photocells. Complete the closed-loop system in order that the system follows the light source.

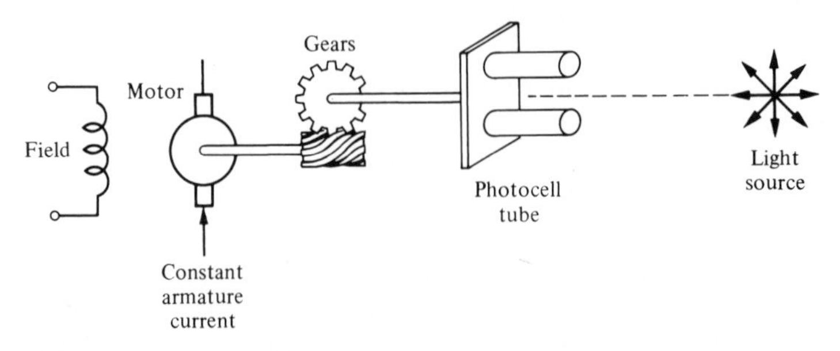

Figure P1.5. A photocell is mounted in each tube. The light reaching each cell is equal only when the light source is exactly in the middle as shown.

P1.6. Feedback systems are not always negative feedback systems in nature. Economic inflation, which is evidenced by continually rising prices, is a *positive feedback* system. A positive feedback control system, as shown in Fig. P1.6, *adds* the feedback signal to the input signal, and the resulting signal is used as the input to the process. A simple model of the price-wage inflationary spiral is shown in Fig. P1.6. Add additional feedback loops, such as legislative control or control of the tax rate, in order to stabilize the system. It is assumed that an increase in workers' salaries, after some time delay, results in an increase in prices. Under what conditions could prices be stabilized by falsifying or delaying the availability of cost-of-living data? How would a national wage and price economic guideline program affect the feedback system?

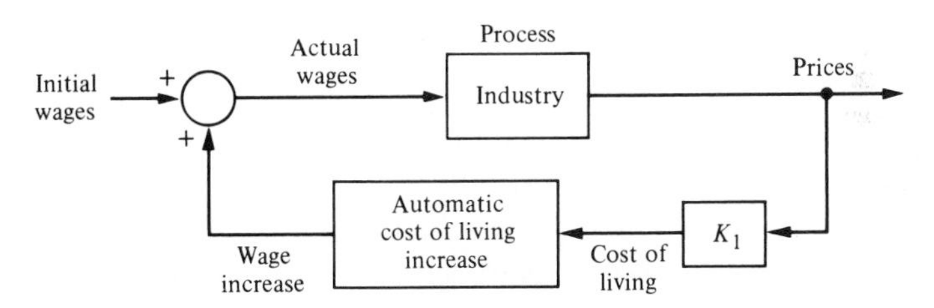

Figure P1.6. Positive feedback.

P1.7. The story is told about the sergeant who stopped at the jewelry store every morning at nine o'clock and compared and reset his watch with the chronometer in the window. Finally, one day the sergeant went into the store and complimented the owner on the accuracy of the chronometer.

"Is it set according to time signals from Arlington?" asked the sergeant.

"No," said the owner, "I set it by the five o'clock (P.M.) cannon fired from the fort. Tell me, Sergeant, why do you stop every day and check your watch?"

The sergeant replied, "I'm the gunner at the fort!"

Is the feedback prevalent in this case positive or negative? The jeweler's chronometer loses one minute each 24-hour period and the sergeant's watch loses one minute during each eight hours. What is the net time error of the cannon at the fort after 15 days?

P1.8. The student–teacher learning process is inherently a feedback process intended to reduce the system error to a minimum. The desired output is the knowledge being studied and the student may be considered the process. With the aid of Fig. 1.3, construct a feedback model of the learning process and identify each block of the system.

P1.9. Models of physiological control systems are valuable aids to the medical profession. A model of the heart-rate control system is shown in Fig. P1.9 [24, 38]. This model includes the processing of the nerve signals by the brain. The heart-rate control system is, in fact, a multivariable system, and the variables x, y, w, v, z, and u are vector variables. In other words, the variable x represents many heart variables $x_1, x_2, \ldots, x_n$. Examine the model of the heart-rate control system and add or delete the blocks, if necessary. Determine a control-system model of one of the following physiological control systems:

1. Respiratory control system
2. Adrenalin control system

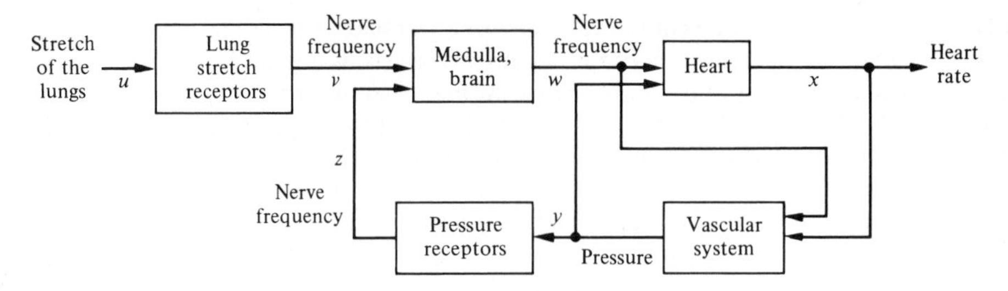

Figure P1.9. Heart rate control.

3. Human arm control system
4. Eye control system
5. Pancreas and the blood-sugar-level control system
6. Circulatory system

P1.10. The role of air traffic control systems is increasing as airplane traffic increases at busy airports. A tragic example of a traffic control mishap was the collision of the Pacific Southwest Airways 727 and a privately owned Cessna at San Diego airport in October 1978 [25]. Engineers are developing flight control systems, air traffic control systems, and collision avoidance systems [41]. Investigate these and other systems designed to improve air traffic safety; select one and draw a simple block diagram of its operation.

P1.11. Automatic control of water level using a float level was used in the Middle East for a water clock [1, 11]. The water clock, shown in Fig. P1.11, was used from sometime before Christ until the 17th century. Discuss the operation of the water clock and establish how the float provides a feedback control which maintains the accuracy of the clock.

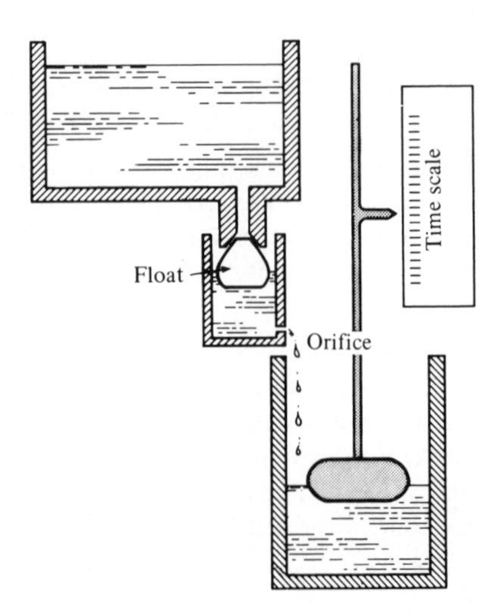

Figure P1.11. (From Newton, Gould, and Kaiser, *Analytical Design of Linear Feedback Controls.* Wiley, New York, 1957, with permission.)

P1.12. An automatic turning gear for windmills was invented by Meikle in about 1750 [1, 11]. The fantail gear shown in Fig. P1.12 automatically turns the windmill into the wind. The fantail windmill at right angle to the mainsail is used to turn the turret. The gear ratio is of the order of 3000 to 1. Discuss the operation of the windmill and establish the feedback operation that maintains the main sails into the wind.

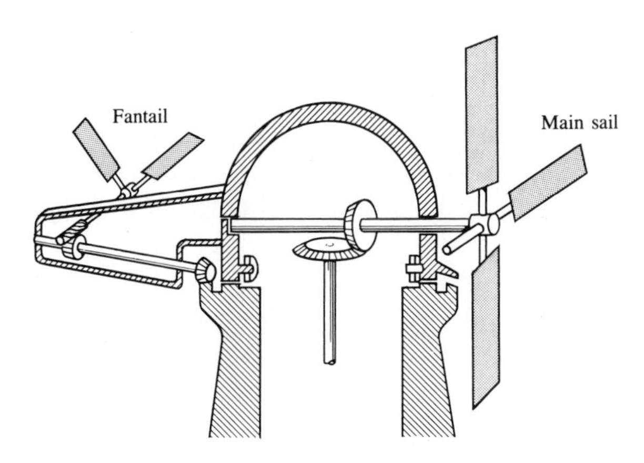

Figure P1.12. (From Newton, Gould, and Kaiser, *Analytical Design of Linear Feedback Controls.* Wiley, New York, 1957, with permission.)

P1.13. A common example of a two-input control system is a home shower with separate valves for hot and cold water. The objective is to obtain (1) a desired temperature of the shower water and (2) a desired flow of water. Draw a block diagram of the closed-loop control system. Would you be willing to take a shower under open-loop control by another person?

P1.14. Adam Smith (1723–1790) discussed the issue of free competition between the participants of an economy in his book *Wealth of Nations.* It may be said that Smith employed social feedback mechanisms to explain his theories [39]. Smith suggests that (1) the available workers as a whole compare the various possible employments and enter that one which offers the greatest rewards; and (2) in any employment the rewards diminish as the number of competing workers rises. Let r = total of rewards averaged over all trades; c = total of rewards in a particular trade; q = influx of workers into the specific trade. Draw a feedback loop to represent this system.

P1.15. Small computers are used in automobiles to control emissions and obtain improved gas mileage. A computer controlled transmission and engine that automatically adjusts itself to the road and driving conditions could improve automobile performance up to 30%. Sketch a block diagram of such a system for an automobile.

P1.16. All humans have experienced a fever associated with an illness. A fever is related to the changing of the control input in your body's thermostat. This thermostat, within the brain, normally regulates your temperature near 98°F in spite of external temperatures ranging from 0 to 100°F or more. For a fever, the input or desired temperature is increased. Even to many scientists it often comes as a surprise to learn that fever does not indicate something wrong with body temperature control but rather well contrived regulation at an elevated level of desired input. Sketch a block diagram of the temperature control system and explain how aspirin will lower a fever.

P1.17. An outfielder for a baseball team uses feedback to judge a fly ball [34]. Determine the method used by the fielder to judge where the ball will land so he can be in the right spot to catch it.

P1.18. A cutaway view of a commonly used pressure regulator is shown in Fig. P1.18. The desired pressure is set by turning a calibrated screw. This compresses the spring and sets up a force that opposes the upward motion of the diaphragm. The bottom side of the diaphragm is exposed to the water pressure that is to be controlled. Thus the motion of the diaphragm is an indication of the pressure difference between the desired and the actual pressures. It acts like a comparator. The valve is connected to the diaphragm and moves according to the pressure difference until it reaches a position in which the difference is zero. Sketch a block diagram showing the control system with the output pressure as the regulated variable.

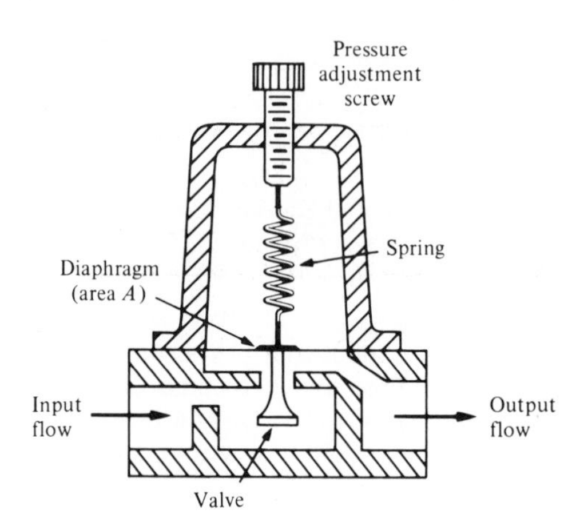

Figure P1.18. Pressure regulator.

P1.19. A horizontal airfoil creates vertical lift, and an upright wing can produce a horizontal force to push a boat. In England John Walker has used this idea to develop an automatically controlled wingsail to power a personal sailing ship or a large cargo boat, as shown in Fig. P1.19, on next page [60]. The controls are required to turn the sails at the appropriate angle to maximize the force. Furthermore, the ship will not heel over due to the force. Examine the dynamics of the system and draw a closed-loop feedback control system block diagram.

P1.20. Ichiro Masaki of General Motors has patented a system that automatically adjusts a car's speed to keep a safe distance from vehicles in front. Using a video camera, the system detects and stores a reference image of the car in front. It then compares this image with a stream of incoming live images as the two cars move down the highway and calculates the distance. Masaki suggests that the system could control steering as well as speed, allowing drivers to lock on to the car ahead and get a "computerized tow." Draw a block diagram for the control system.

Figure P1.19. Wingsail ship.

Design Problems 

(Design problems emphasize the design task.)

DP1.1. The road and vehicle noise that invades an automobile's cabin hastens occupant fatigue. Design the block diagram of an "antinoise" system that will eliminate the effect of unwanted noises. Indicate the device within each block.

DP1.2. Many cars are fitted with cruise control that, at the press of a button, automatically maintains a set speed. In this way the driver can cruise at a speed limit or economic speed without continually checking the speedometer. Design a feedback control in block diagram form for a cruise control system.

DP1.3. As part of the automation of a dairy farm, the automation of cow milking is under study [57]. Design a milking machine that can milk cows four or five times a day at the cow's demand. Show a block diagram and indicate the devices in each block.

DP1.4. A large braced robot arm for welding large structures is shown in Fig. DP1.4. Draw the block diagram of a closed-loop feedback control system for accurately controlling the location of the weld tip.

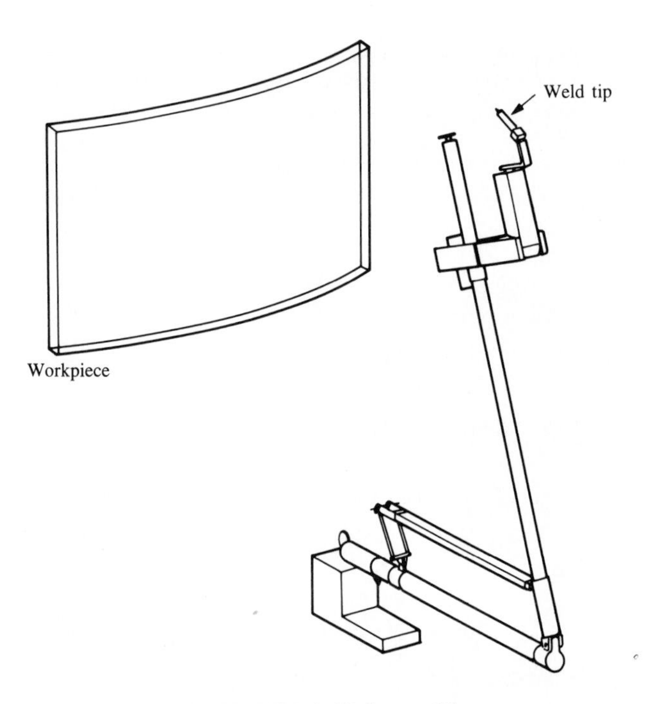

Figure DP1.4. Robot welder.

DP1.5. Vehicle traction control, which includes antiskid braking and antispin acceleration, can enhance vehicle performance and handling. The objective of this control is to maximize tire traction by preventing locked brakes as well as tire spinning during acceleration. Wheel slip, the difference between the vehicle speed and the wheel speed, is chosen as the controlled variable because of its strong influence on the tractive force between the tire and the road [59]. The adhesion coefficient, between the wheel and the road, reaches a maximum at a low slip. Develop a block diagram model of one wheel of a traction control system.

DP1.6. The front-page engineering story of 1990 was the plight of the Hubble space telescope. Launched on April 24, 1990, into a 610-kilometer-high orbit, the $2.5 billion U.S. spacecraft was to have been the most precise observatory ever built. But once the telescope was in orbit, the nightmare began. First, a loop of cable on one of the two high-gain communications antennas turned out to be improperly tied to the antenna's mast, limiting the time during which the antenna could transmit data back to earth. Next, the solar panels vibrated for three to six minutes each time the spacecraft passed into or out of sunlight, throwing off the sensors fixing the telescope on a guide star and disrupting the observation in progress. Worst of all, the telescope's 2.4-meter-diameter (94.5-inch) main mirror was found to have been ground and polished to the wrong shape, thus rendering it unable to focus an image.

To compensate for the aberration in the main mirror over the long term, NASA is modifying the curvature or "prescription" of the lenses of an already planned second wide-field and planetary camera scheduled to replace the current one during a routine maintenance mission in 1993. The most challenging problem now is damping the jitter that vibrates the spacecraft each time it passes into or out of the earth's shadow. The worst vibration has a period of about 10 seconds, or a frequency of 0.1 hertz.

Design a feedback system that will reduce the vibrations of the Hubble space telescope.

Terms and Concepts

Automation The control of an industrial process by automatic means.

Closed-loop feedback control system A system that uses a measurement of the output and compares it with the desired output.

Control system An interconnection of components forming a system configuration that will provide a desired response.

Design The process of conceiving the form and parts of a system to achieve a purpose.

Feedback signal A measure of the output of the system used for feedback to control the system.

Flyball governor A mechanical device for controlling the speed of a steam engine.

Multivariable control system A system with more than one input variable and more than one output variable.

Negative feedback The output signal is fedback so that it subtracts from the input signal.

Open-loop control system A system that utilizes a device to control the process without using feedback. Thus the output has no effect upon the signal to the process.

Plant See Process.

Positive feedback The output signal is fedback so that it adds to the input signal.

Process The device, plant, or system under control.

Productivity The ratio of physical output to physical input of an industrial process.

System An interconnection of elements and devices for a desired purpose.

References

1. O. Mayr, *The Origins of Feedback Control,* M.I.T. Press, Cambridge, Mass., 1970.
2. O. Mayr, "The Origins of Feedback Control," *Scientific American,* **223,** 4, October 1970; pp. 110–18.
3. O. Mayr, *Feedback Mechanisms in the Historical Collections of the National Museum of History and Technology,* Smithsonian Institution Press, Washington, D.C., 1971.
4. E. P. Popov, *The Dynamics of Automatic Control Systems,* Gostekhizdat, Moscow, 1956; Addison-Wesley, Reading, Mass., 1962.

5. J. C. Maxwell, "On Governors," *Proc. of the Royal Society of London,* **16,** 1868; in *Selected Papers on Mathematical Trends in Control Theory.* Dover, New York, 1964; pp. 270–83.
6. I. A. Vyshnegradskii, "On Controllers of Direct Action," *Izv. SPB Tekhnolog. Inst.,* 1877.
7. H. W. Bode, "Feedback—The History of an Idea," in *Selected Papers on Mathematical Trends in Control Theory.* Dover, New York, 1964; pp. 106–23.
8. H. S. Black, "Inventing the Negative Feedback Amplifier," *IEEE Spectrum,* December 1977; pp. 55–60.
9. J. E. Brittain, *Turning Points in American Electrical History,* IEEE Press, New York, 1977, Sect. II-E.
10. G. J. Thaler, *Automatic Control Systems,* West Publishing, St. Paul, Minn., 1989.
11. G. Newton, L. Gould, and J. Kaiser, *Analytical Design of Linear Feedback Controls,* John Wiley & Sons, New York, 1957.
12. M. D. Fagen, *A History of Engineering and Science on the Bell Systems,* Bell Telephone Laboratories, 1978; Ch. 3.
13. B. W. Niebel, *Modern Manufacturing Process Engineering,* McGraw-Hill, New York, 1989.
14. J. Gertber, *Low Cost Automation,* IFAC Conference, Pergamon Press, New York, 1990.
15. W. D. Rasmussen, "The Mechanization of Agriculture," *Scientific American,* September 1982; pp. 26–37.
16. R. Shonreshi, "Optically Driven Learning Control for Industrial Manipulators," *IEEE Control Systems,* October 1989; pp. 21–26.
17. W. B. Arthur, "Positive Feedbacks in the Economy," *Scientific American,* February 1990; pp. 92–99.
18. S. Derby, "Mechatronics for Robots," *Mechanical Engineering,* July 1990; pp. 40–42.
19. S. T. Venkataraman, *Dextrous Robot Hands,* Springer-Verlag, New York, 1990.
20. J. Krueger, "Developments in Automotive Electronics," *Automotive Engineering,* September 1990; pp. 37–44.
21. "High Speed Trains," *The Economist,* September 30, 1989; pp. 52–53.
22. L. A. Geddes, *Principles of Applied Biomedical Instrumentation,* John Wiley & Sons, New York, 1989.
23. B. W. Mar and O. A. Bakken, "Applying Classical Control Theory to Energy-Economics Modeling," *Management Science,* January 1984; pp. 81–91.
24. U. Zwiener et al., "Non-Invasive System Analysis of the Arterial Pressure Control by Digital Modelling," *Proceed. of the IFAC World Congress,* July 1987; pp. 50–56.
25. P. K. Shih, "Structures for Hypervelocity Flight," *Aerospace America,* May 1989; pp. 14–18.
26. M. L. Dertouzos, *Made in America,* M.I.T. Press, Cambridge, Mass., 1989.
27. R. C. Dorf, *The Encyclopedia of Robotics,* John Wiley & Sons, New York; 1988.
28. R. C. Dorf, *Robotics and Automated Manufacturing,* Reston Publishing, Reston, Virginia, 1983.
29. "Industrial U.S. Market Report," *Electronics,* January 6, 1990; pp. 90–94.
30. K. S. Betts, "Process Control Takes Command," *Mechanical Engineering,* July 1990; pp. 64–68.
31. S. Ashley, "A Mosaic for Machine Tools," *Mechanical Engineering,* September 1990; pp. 38–43.

32. P. M. Moretti and L. V. Divone, "Modern Windmills," *Scientific American,* June 1986; pp. 110–18.
33. M. Drela and J. S. Langford, "Human Powered Flight," *Scientific American,* November 1985; pp. 114–51.
34. P. J. Brancazio, "Science and the Game of Baseball," *Science Digest,* July 1984; pp. 66–70.
35. H. Chestnut, "Applying Adaptive Control Principles to Resolving International Conflicts," *Proceed. of the IFAC World Congress,* July 1987; pp. 149–59.
36. R. C. Dorf, *Introduction to Computers and Computer Science,* Boyd and Fraser, San Francisco, 3d ed., 1982; Chs. 13, 14.
37. J. D. Erickson, "Manned Spacecraft Automation and Robotics," *Proceed. of the IEEE,* March 1987; pp. 417–27.
38. R. C. Dorf and J. Unmack, "A Time-Domain Model of the Heart Rate Control System," *Proceed. of the San Diego Symposium for Biomedical Engineering,* 1965; pp. 43–47.
39. O. Mayr, "Adam Smith and the Concept of the Feedback System," *Technology and Culture,* **12,** 1, January 1971; pp. 1–22.
40. C. F. Lorenzo, "An Intelligent Control System for Rocket Engines," *Proceed. of the American Automatic Control Conference,* 1990; pp. 974–82.
41. S. Winchester, "Leviathans of the Sky," *Atlantic Monthly,* October 1990; pp. 107–17.
42. H. Tamura, *Large-Scale Systems Control,* Marcel Dekker, New York, 1990.
43. R. T. Howe et al., "Silicon Micromechanics: Sensors and Actuators on a Chip," *IEEE Spectrum,* July 1990; pp. 29–35.
44. P. Cleaveland, "Programmable Controllers: What's Ahead," *Instruments and Control Systems,* March 1990; pp. 37–58.
45. "Some Industry Views on Control Use," *IEEE Control Systems,* December 1987; pp. 20–24.
46. G. J. Blickley, "Where Control Dollars Are Spent," *Control Engineering,* August 1987; p. 13.
47. K. Fitzgerald, "The Artificial Heart," *IEEE Spectrum,* November 1989; p. 22.
48. J. D. Ryder and D. G. Fink, *Engineers and Electrons,* IEEE Press, Piscataway, N.J., 1984.
49. D. G. Johnson, *Programmable Controllers for Factory Automation,* Marcel Dekker, New York, 1987.
50. H. Van Dyke Parunak, "Focus on Intelligent Control," *Inter. J. of Integrated Manufacturing,* February 1990; pp. 1–5.
51. C. W. De Silva, *Control Sensors and Actuators,* Prentice Hall, Englewood Cliffs, N.J., 1989.
52. A. Goldsmith, "Autofocus Cameras," *Popular Science,* March 1988; pp. 70–72.
53. P. Dorato, "Robust Control: A Historical Review," *IEEE Control Systems,* April 1987; pp. 44–46.
54. G. W. Neat, "Expert Adaptive Control for Drug Delivery Systems," *IEEE Control Systems,* June 1989; pp. 20–23.
55. S. S. Hacisalihzade, "Control Engineering and Therapeutic Drug Delivery," *IEEE Control Systems,* June 1989; pp. 44–46.
56. R. Kurzweil, *The Age of Intelligent Machines,* M.I.T. Press, Cambridge, Mass., 1990.
57. C. Klomp et al., "Development of an Autonomous Cow-Milking Robot Control System," *IEEE Control Systems,* October 1990; pp. 11–19.

58. G. Zorpette, "Parkinson's Gun Director," *IEEE Spectrum,* April 1989; p. 43.
59. H. S. Tan and M. Tomizuka, "Controller Design for Robust Vehicle Traction," *IEEE Control Systems,* April 1990; pp. 107–13.
60. D. Scott, "Wingsail Trimaran," *Popular Science,* June 1990, pp. 116–17.
61. T. S. Perry, "Improving Air Traffic Control System," *IEEE Spectrum,* February 1991, pp. 22–36.

CHAPTER 2

Mathematical Models of Systems

Preview

In order to analyze and design control systems, we use quantitative mathematical models of these systems. We will consider a wide range of systems including mechanical, electrical, and fluid. We will first describe the dynamic behavior of these systems using differential equations. In order to use the Laplace transform we will develop the means of obtaining a linear model of components of a system. Then we will be able to unite the differential equations describing the system and obtain the Laplace transform of these equations.

We will then proceed to obtain the input-output relationship for components and subsystems in the form of a transfer function. Then the set of transfer functions representing the interconnected components will be represented by a block diagram model or a signal-flow graph. Using various analytical methods, we will be able to obtain the equations for selected outputs of a control system as they are regulated by selected inputs of the system.

2.1 Introduction

In order to understand and control complex systems, one must obtain quantitative *mathematical models* of these systems. Therefore it is necessary to analyze the relationships between the system variables and to obtain a mathematical model. Because the systems under consideration are dynamic in nature, the descriptive equations are usually *differential equations.* Furthermore, if these equations can be *linearized,* then the *Laplace transform* can be utilized in order to simplify the method of solution. In practice, the complexity of systems and the ignorance of all the relevant factors necessitate the introduction of *assumptions* concerning the system operation. Therefore we shall often find it useful to consider the physical system, delineate some necessary assumptions, and linearize the system. Then, by using the physical laws describing the linear equivalent system, we can obtain a set of linear differential equations. Finally, utilizing mathematical tools, such as the Laplace transform, we obtain a solution describing the operation of the system. In summary, the approach to dynamic system problems can be listed as follows:

1. Define the system and its components.
2. Formulate the mathematical model and list the necessary assumptions.
3. Write the differential equations describing the model.
4. Solve the equations for the desired output variables.
5. Examine the solutions and the assumptions.
6. Reanalyze or design.

2.2 Differential Equations of Physical Systems

The differential equations describing the dynamic performance of a physical system are obtained by utilizing the physical laws of the process [1, 2, 3]. This approach applies equally well to mechanical [1], electrical [3], fluid, and thermodynamic systems [4]. A summary of the variables of dynamic systems is given in Table 2.1 [5]. We prefer to use the International System of units (SI) in contrast to the British System of units. The International System of units is given in Table 2.2. The conversion of other systems of units to SI units is facilitated by Table

Table 2.1. Summary of Through- and Across-Variables for Physical Systems

System	Variable Through Element	Integrated Through Variable	Variable Across Element	Integrated Across Variable
Electrical	Current, i	Charge, q	Voltage difference, v_{21}	Flux linkage, λ_{21}
Mechanical translational	Force, F	Translational momentum, P	Velocity difference, v_{21}	Displacement difference, y_{21}
Mechanical rotational	Torque, T	Angular momentum, h	Angular velocity difference, ω_{21}	Angular displacement difference, θ_{21}
Fluid	Fluid volumetric rate of flow, Q	Volume, V	Pressure difference, P_{21}	Pressure momentum, γ_{21}
Thermal	Heat flow rate, q	Heat energy, H	Temperature difference, τ_{21}	

Table 2.2. The International System of Units (SI)

	Unit	Symbol
Basic Units		
Length	meter	m
Mass	kilogram	kg
Time	second	s
Temperature	kelvin	K
Electric current	ampere	A
Derived Units		
Velocity	meters per second	m/s
Area	square meter	m^2
Force	newton	$N = kgm/s^2$
Torque	kilogram-meter	kgm
Pressure	pascal	Pa
Energy	joule	J = Nm
Power	watt	W = J/s

Table 2.3. Conversion Factors for Converting to SI Units

From	Multiply by	To Obtain
Length		
inches	25.4	millimeters
feet	30.48	centimeters
Speed		
miles per hour	0.4470	meters per second
Mass		
pounds	0.4536	kilograms
Force		
pounds-force	4.448	newtons
Torque		
foot-pounds	0.1383	kilogram-meters
Power		
horsepower	746	watts
Energy		
British thermal unit	1055	joules
kilowatt-hour	3.6×10^6	joules

2.3. A summary of the describing equations for lumped, linear, dynamic elements is given in Table 2.4 [5]. The equations in Table 2.4 are idealized descriptions and only approximate the actual conditions (for example, when a linear, lumped approximation is used for a distributed element).

■ Nomenclature

- *Through-variable:* F = force, T = torque, i = current, Q = fluid volumetric flow rate, q = heat flow rate.
- *Across-variable:* v = translational velocity, ω = angular velocity, v = voltage, P = pressure, τ = temperature.
- *Inductive storage:* L = inductance, $1/k$ = reciprocal translational or rotational stiffness, I = fluid inertance.
- *Capacitive storage:* C = capacitance, M = mass, J = moment of inertia, C_f = fluid capacitance, C_t = thermal capacitance.
- *Energy dissipators:* R = resistance, f = viscous friction, R_f = fluid resistance, R_t = thermal resistance.

The symbol $v(t)$ is used for both voltage in electrical circuits and velocity in translational mechanical systems, and is distinguished within the context of each differential equation. For mechanical systems, one utilizes Newton's laws, and for

Table 2.4. Summary of Describing Differential Equations for Ideal Elements

Type of Element	Physical Element	Describing Equation	Energy E or Power $\mathcal{P}$	Symbol
	Electrical inductance	$v_{21} = L\dfrac{di}{dt}$	$E = \dfrac{1}{2}Li^2$	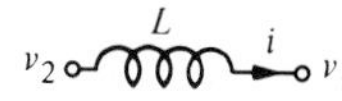 v_2 L i v_1
Inductive storage	Translational spring	$v_{21} = \dfrac{1}{K}\dfrac{dF}{dt}$	$E = \dfrac{1}{2}\dfrac{F^2}{K}$	 v_2 K v_1 F
	Rotational spring	$\omega_{21} = \dfrac{1}{K}\dfrac{dT}{dt}$	$E = \dfrac{1}{2}\dfrac{T^2}{K}$	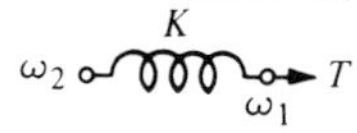 ω_2 K ω_1 T
	Fluid inertia	$P_{21} = I\dfrac{dQ}{dt}$	$E = \dfrac{1}{2}IQ^2$	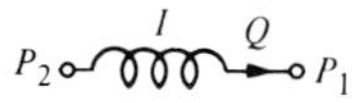 P_2 I Q P_1
	Electrical capacitance	$i = C\dfrac{dv_{21}}{dt}$	$E = \dfrac{1}{2}Cv_{21}^2$	 v_2 i C v_1
	Translational mass	$F = M\dfrac{dv_2}{dt}$	$E = \dfrac{1}{2}Mv_2^2$	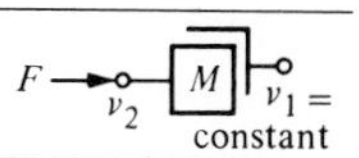 F v_2 M $v_1 =$ constant
Capacitive storage	Rotational mass	$T = J\dfrac{d\omega_2}{dt}$	$E = \dfrac{1}{2}J\omega_2^2$	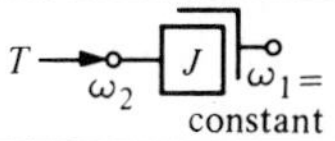 T ω_2 J $\omega_1 =$ constant
	Fluid capacitance	$Q = C_f\dfrac{dP_{21}}{dt}$	$E = \dfrac{1}{2}C_fP_{21}^2$	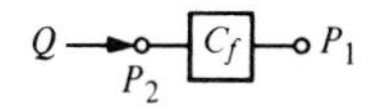 Q P_2 C_f P_1
	Thermal capacitance	$q = C_t\dfrac{d\mathcal{T}_2}{dt}$	$E = C_t\mathcal{T}_2$	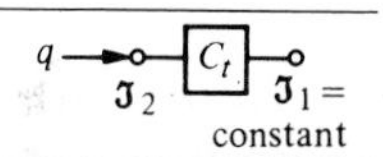 q $\mathfrak{T}_2$ C_t $\mathfrak{T}_1 =$ constant
	Electrical resistance	$i = \dfrac{1}{R}v_{21}$	$\mathcal{P} = \dfrac{1}{R}v_{21}^2$	 v_2 R i v_1
	Translational damper	$F = fv_{21}$	$\mathcal{P} = fv_{21}^2$	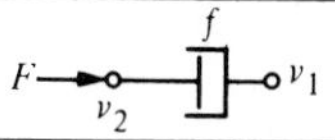 F v_2 f v_1
Energy dissipators	Rotational damper	$T = f\omega_{21}$	$\mathcal{P} = f\omega_{21}^2$	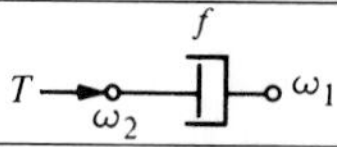 T ω_2 f ω_1
	Fluid resistance	$Q = \dfrac{1}{R_f}P_{21}$	$\mathcal{P} = \dfrac{1}{R_f}P_{21}^2$	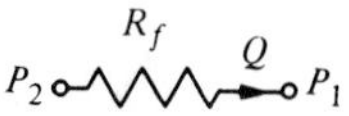 P_2 R_f Q P_1
	Thermal resistance	$q = \dfrac{1}{R_t}\mathcal{T}_{21}$	$\mathcal{P} = \dfrac{1}{R_t}\mathcal{T}_{21}$	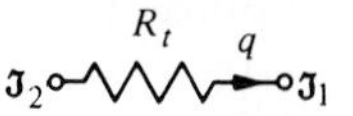 $\mathfrak{T}_2$ R_t q $\mathfrak{T}_1$

electrical systems Kirchhoff's voltage laws. For example, the simple spring-mass-damper mechanical system shown in Fig. 2.1 is described by Newton's second law of motion. (This system could represent, for example, an automobile shock absorber.) Therefore, we obtain

$$M\frac{d^2y(t)}{dt^2} + f\frac{dy(t)}{dt} + Ky(t) = r(t), \tag{2.1}$$

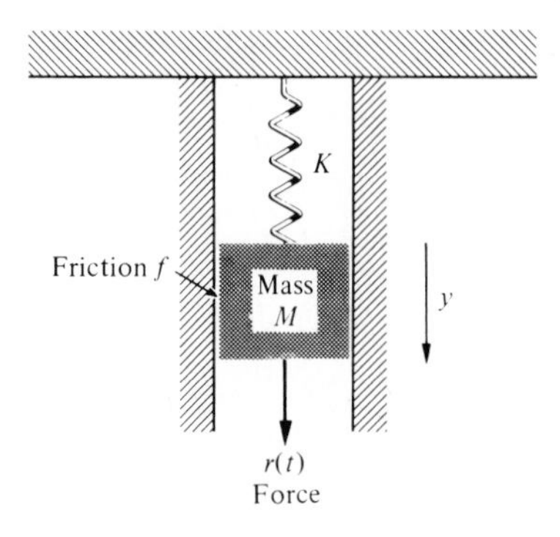

Figure 2.1. Spring-mass-damper system.

where K is the spring constant of the ideal spring and f is the friction constant. Equation 2.1 is a linear constant coefficient differential equation of second order.

Alternatively, one may describe the electrical RLC circuit of Fig. 2.2 by utilizing Kirchhoff's current law. Then we obtain the following integrodifferential equation:

$$\frac{v(t)}{R} + C\frac{dv(t)}{dt} + \frac{1}{L}\int_0^t v(t)\,dt = r(t). \tag{2.2}$$

The solution of the differential equation describing the process may be obtained by classical methods such as the use of integrating factors and the method of undetermined coefficients [1]. For example, when the mass is displaced initially a distance $y(t) = y(0)$ and released, the dynamic response of an *underdamped* system is represented by an equation of the form

$$y(t) = K_1 e^{-\alpha_1 t} \sin(\beta_1 t + \theta_1). \tag{2.3}$$

A similar solution is obtained for the voltage of the RLC circuit when the circuit is subjected to a constant current $r(t) = I$. Then the voltage is

$$v(t) = K_2 e^{-\alpha_2 t} \cos(\beta_2 t + \theta_2). \tag{2.4}$$

A voltage curve typical of an underdamped RLC circuit is shown in Fig. 2.3.

In order to further reveal the close similarity between the differential equa-

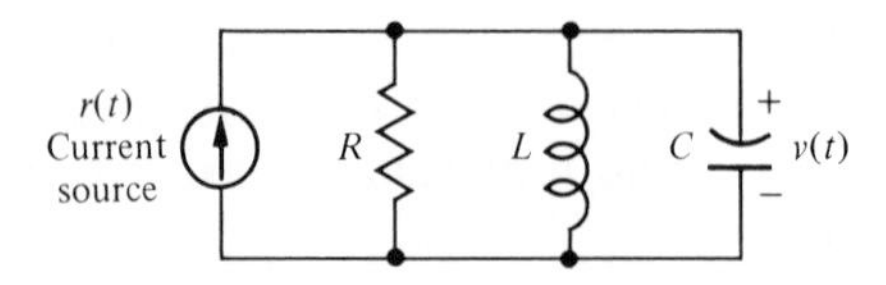

Figure 2.2. RLC circuit.

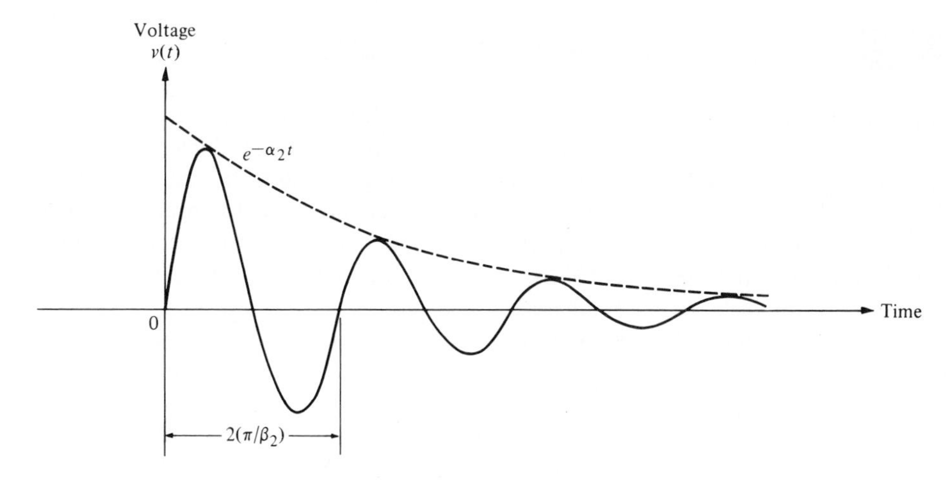

Figure 2.3. Typical voltage curve for underdamped *RLC* circuit.

tions for the mechanical and electrical systems, we shall rewrite Eq. (2.1) in terms of velocity,

$$v(t) = \frac{dy(t)}{dt}.$$

Then we have

$$M\frac{dv(t)}{dt} + fv(t) + K\int_0^t v(t)\,dt = r(t). \tag{2.5}$$

One immediately notes the equivalence of Eqs. (2.5) and (2.2) where velocity $v(t)$ and voltage $v(t)$ are equivalent variables, usually called *analogous* variables, and the systems are analogous systems. Therefore the solution for velocity is similar to Eq. (2.4) and the curve for an underdamped system is shown in Fig. 2.3. The concept of analogous systems is a very useful and powerful technique for system modeling. The voltage-velocity analogy, often called the force–current analogy, is a natural analogy because it relates the analogous through- and across-variables of the electrical and mechanical systems. However, another analogy that relates the velocity and current variables is often used and is called the force-voltage analogy.

Analogous systems with similar solutions exist for electrical, mechanical, thermal, and fluid systems. The existence of analogous systems and solutions provides the analyst with the ability to extend the solution of one system to all analogous systems with the same describing differential equations. Therefore what one learns about the analysis and design of electrical systems is immediately extended to an understanding of fluid, thermal, and mechanical systems.

2.3 Linear Approximations of Physical Systems

A great majority of physical systems are linear within some range of the variables. However, all systems ultimately become nonlinear as the variables are increased without limit. For example, the spring-mass-damper system of Fig. 2.1 is linear and described by Eq. (2.1) so long as the mass is subjected to small deflections $y(t)$. However, if $y(t)$ were continually increased, eventually the spring would be overextended and break. Therefore the question of linearity and the range of applicability must be considered for each system.

A system is defined as linear in terms of the system excitation and response. In the case of the electrical network, the excitation is the input current $r(t)$ and the response is the voltage $v(t)$. In general, a *necessary condition* for a linear system can be determined in terms of an excitation $x(t)$ and a response $y(t)$. When the system at rest is subjected to an excitation $x_1(t)$, it provides a response $y_1(t)$. Furthermore, when the system is subjected to an excitation $x_2(t)$, it provides a corresponding response $y_2(t)$. For a linear system, it is *necessary* that the excitation $x_1(t) + x_2(t)$ result in a response $y_1(t) + y_2(t)$. This is usually called the *principle of superposition.*

Furthermore, it is necessary that the magnitude scale factor is preserved in a linear system. Again, consider a system with an input x which results in an output y. Then it is necessary that the response of a linear system to a constant multiple β of an input x is equal to the response to the input multiplied by the same constant so that the output is equal to βy. This is called the property of *homogeneity.* A system is linear if and only if the properties of superposition and homogeneity are satisfied.

A system characterized by the relation $y = x^2$ is not linear because the superposition property is not satisfied. A system represented by the relation $y = mx + b$ is not linear because it does not satisfy the homogeneity property. However, this device may be considered linear about an operating point x_0, y_0 for small changes Δx and Δy. When $x = x_0 + \Delta x$ and $y = y_0 + \Delta y$, we have

$$y = mx + b$$

or

$$y_0 + \Delta y = mx_0 + m\,\Delta x + b$$

and therefore $\Delta y = m\,\Delta x$, which satisfies the necessary conditions.

The linearity of many mechanical and electrical elements can be assumed over a reasonably large range of the variables [7]. This is not usually the case for thermal and fluid elements, which are more frequently nonlinear in character. Fortunately, however, one can often linearize nonlinear elements assuming small-signal conditions. This is the normal approach used to obtain a linear equivalent circuit for electronic circuits and transistors. Consider a general element with an excitation (through-) variable $x(t)$ and a response (across-) variable $y(t)$. Several examples of dynamic system variables are given in Table 2.1. The relationship of the two variables is written as

$$y(t) = g(x(t)), \tag{2.6}$$

where $g(x(t))$ indicates $y(t)$ is a function of $x(t)$. The relationship might be shown graphically, as in Fig. 2.4. The normal operating point is designated by x_0. Because the curve (function) is continuous over the range of interest, a *Taylor series* expansion about the operating point may be utilized. Then we have

$$y = g(x) = g(x_0) + \left.\frac{dg}{dx}\right|_{x=x_0} \frac{(x - x_0)}{1!} + \left.\frac{d^2g}{dx^2}\right|_{x=x_0} \frac{(x - x_0)^2}{2!} + \cdots \tag{2.7}$$

The slope at the operating point,

$$\left.\frac{dg}{dx}\right|_{x=x_0},$$

is a good approximation to the curve over a small range of $(x - x_0)$, the deviation from the operating point. Then, as a reasonable approximation, Eq. (2.7) becomes

$$y = g(x_0) + \left.\frac{dg}{dx}\right|_{x=x_0} (x - x_0) = y_0 + m(x - x_0), \tag{2.8}$$

where m is the slope at the operating point. Finally, Eq. (2.8) can be rewritten as the linear equation

$$(y - y_0) = m(x - x_0)$$

or

$$\Delta y = m\,\Delta x. \tag{2.9}$$

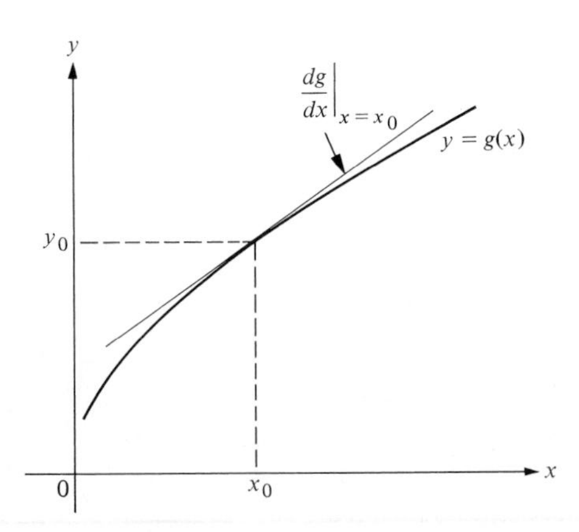

Figure 2.4. A graphical representation of a nonlinear element.

This *linear approximation* is as accurate as the assumption of small signals is applicable to the specific problem.

If the dependent variable y depends upon several excitation variables, x_1, x_2, . . . , x_n, then the functional relationship is written as

$$y = g(x_1, x_2, \ldots, x_n). \tag{2.10}$$

The Taylor series expansion about the operating point $x_{1_0}, x_{2_0}, \ldots, x_{n_0}$ is useful for a linear approximation to the nonlinear function. When the higher-order terms are neglected, the linear approximation is written as

$$y = g(x_{1_0}, x_{2_0}, \ldots, x_{n_0}) + \left.\frac{\partial g}{\partial x_1}\right|_{x=x_0}(x_1 - x_{1_0}) + \left.\frac{\partial g}{\partial x_2}\right|_{x=x_0}(x_2 - x_{2_0}) + \cdots + \left.\frac{\partial g}{\partial x_n}\right|_{x=x_0}(x_n - x_{n_0}), \tag{2.11}$$

where x_0 is the operating point. An example will clearly illustrate the utility of this method.

■ Example 2.1 Pendulum oscillator model

Consider the pendulum oscillator shown in Fig. 2.5(a). The torque on the mass is

$$T = MgL \sin \theta, \tag{2.12}$$

where g is the gravity constant. The equilibrium condition for the mass is $\theta_0 = 0°$. The nonlinear relation between T and θ is shown graphically in Fig. 2.5(b). The first derivative evaluated at equilibrium provides the linear approximation, which is

$$\begin{aligned} T &= MgL \left.\frac{\partial \sin \theta}{\partial \theta}\right|_{\theta=\theta_0}(\theta - \theta_0) \\ &= MgL(\cos 0°)(\theta - 0°) \\ &= MgL\theta. \end{aligned} \tag{2.13}$$

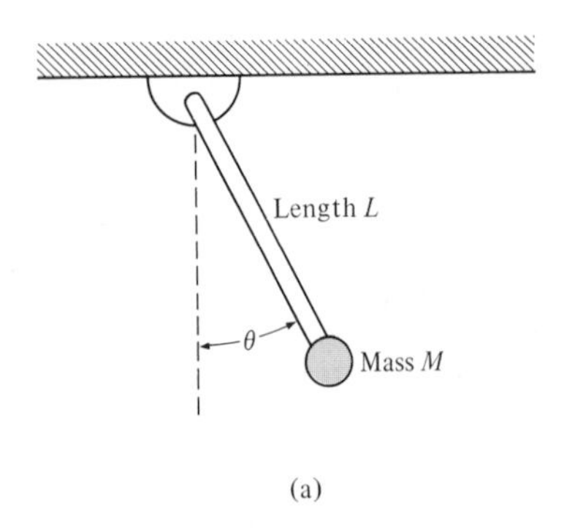

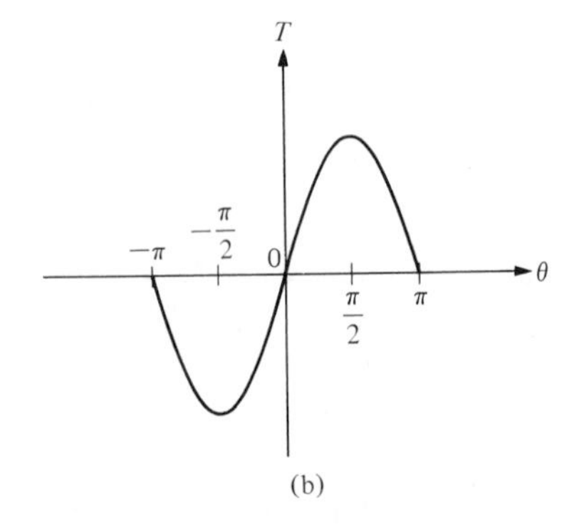

Figure 2.5. Pendulum oscillator.

This approximation is reasonably accurate for $-\frac{\pi}{4} \le \theta \le \frac{\pi}{4}$. For example, the response of the linear model for the swing through $\pm 30°$ is within 2% of the actual nonlinear pendulum response.

2.4 The Laplace Transform

The ability to obtain linear approximations of physical systems allows the analyst to consider the use of the *Laplace transformation.* The Laplace transform method substitutes the relatively easily solved algebraic equations for the more difficult differential equations [1, 3]. The time response solution is obtained by the following operations:

1. Obtain the differential equations.
2. Obtain the Laplace transformation of the differential equations.
3. Solve the resulting algebraic transform of the variable of interest.

The Laplace transform exists for linear differential equations for which the transformation integral converges. Therefore, in order that $f(t)$ be transformable, it is sufficient that

$$\int_0^\infty |f(t)|\, e^{-\sigma_1 t}\, dt < \infty$$

for some real, positive σ_1 [1]. If the magnitude of $f(t)$ is $|f(t)| < Me^{\alpha t}$ for all positive t, the integral will converge for $\sigma_1 > \alpha$. The region of convergence is therefore given by $\infty > \sigma_1 > \alpha$, and σ_1 is known as the abscissa of absolute convergence. Signals that are physically possible always have a Laplace transform. The Laplace transformation for a function of time, $f(t)$, is

$$F(s) = \int_0^\infty f(t)e^{-st}\, dt = \mathcal{L}\{f(t)\}. \tag{2.14}$$

The *inverse Laplace transform* is written as

$$f(t) = \frac{1}{2\pi j}\int_{\sigma - j\infty}^{\sigma + j\infty} F(s)e^{+st}\, ds. \tag{2.15}$$

The transformation integrals have been used to derive tables of Laplace transforms that are ordinarily used for the great majority of problems. A table of Laplace transforms is provided in Appendix A. Some important Laplace transform pairs are given in Table 2.5.

Alternatively, the Laplace variable s can be considered to be the differential operator so that

$$s \equiv \frac{d}{dt}. \tag{2.16}$$

Then we also have the integral operator

$$\frac{1}{s} \equiv \int_{0+}^{t} dt. \tag{2.17}$$

The inverse Laplace transformation is usually obtained by using the Heaviside partial fraction expansion. This approach is particularly useful for systems analysis and design, because the effect of each characteristic root or eigenvalue may be clearly observed.

In order to illustrate the usefulness of the Laplace transformation and the steps involved in the system analysis, reconsider the spring-mass-damper system described by Eq. (2.1), which is

$$M\frac{d^2y}{dt^2} + f\frac{dy}{dt} + Ky = r(t). \tag{2.18}$$

We wish to obtain the response, y, as a function of time. The Laplace transform of Eq. (2.18) is

$$M\left(s^2Y(s) - sy(0^+) - \frac{dy(0^+)}{dt}\right) + f(sY(s) - y(0^+)) + KY(s) = R(s). \tag{2.19}$$

Table 2.5. Important Laplace Transform Pairs

$f(t)$	$F(s)$
Step function, $u(t)$	$\frac{1}{s}$
e^{-at}	$\frac{1}{s+a}$
$\sin \omega t$	$\frac{\omega}{s^2+\omega^2}$
$\cos \omega t$	$\frac{s}{s^2+\omega^2}$
$e^{-at}f(t)$	$F(s+a)$
t^n	$\frac{n!}{s^{n+1}}$
$f^{(k)}(t) = \frac{d^kf(t)}{dt^k}$	$s^kF(s) - s^{k-1}f(0^+) - s^{k-2}f'(0^+) - \cdots - f^{(k-1)}(0^+)$
$\int_{-\infty}^{t} f(t)dt$	$\frac{F(s)}{s} + \frac{\int_{-\infty}^{0} f\,dt}{s}$
Impulse function $\delta(t)$	1

When

$$r(t) = 0, \qquad y(0^+) = y_0, \qquad \left.\frac{dy}{dt}\right|_{t=0+} = 0,$$

we have

$$Ms^2Y(s) - Msy_0 + fsY(s) - fy_0 + KY(s) = 0. \tag{2.20}$$

Solving for $Y(s)$, we obtain

$$Y(s) = \frac{(Ms + f)y_0}{Ms^2 + fs + K} = \frac{p(s)}{q(s)}. \tag{2.21}$$

The denominator polynomial $q(s)$, when set equal to zero, is called the *characteristic equation,* because the roots of this equation determine the character of the time response. The roots of this characteristic equation are also called the *poles* or *singularities* of the system. The roots of the numerator polynomial $p(s)$ are called the *zeros* of the system; for example, $s = -f/M$. Poles and zeros are critical frequencies. At the poles the function $Y(s)$ becomes infinite; while at the zeros, the function becomes zero. The complex frequency *s-plane* plot of the poles and zeros graphically portrays the character of the natural transient response of the system.

For a specific case, consider the system when $K/M = 2$ and $f/M = 3$. Then Eq. (2.21) becomes

$$Y(s) = \frac{(s + 3)y_0}{(s + 1)(s + 2)}. \tag{2.22}$$

The poles and zeros of $Y(s)$ are shown on the s-plane in Fig. 2.6.

Expanding Eq. (2.22) in a partial fraction expansion, we obtain

$$Y(s) = \frac{k_1}{s + 1} + \frac{k_2}{s + 2}, \tag{2.23}$$

where k_1 and k_2 are the coefficients of the expansion. The coefficients k_i, are called *residues* and are evaluated by multiplying through by the denominator factor of

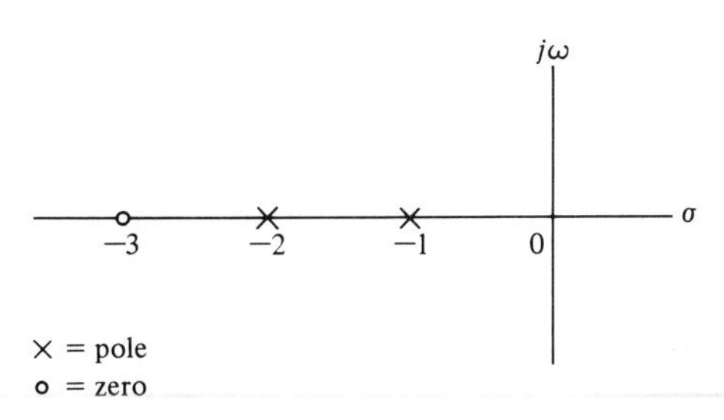

Figure 2.6. An s-plane pole and zero plot.

Eq. (2.22) corresponding to k_i and setting s equal to the root. Evaluating k_1 when $y_0 = 1$, we have

$$k_1 = \left.\frac{(s - s_1)p(s)}{q(s)}\right|_{s=s_1} = \left.\frac{(s + 1)(s + 3)}{(s + 1)(s + 2)}\right|_{s_1=-1} = 2 \tag{2.24}$$

and $k_2 = -1$. Alternatively, the residues of $Y(s)$ at the respective poles may be evaluated graphically on the s-plane plot, since Eq. (2.24) may be written as

$$k_1 = \left.\frac{s + 3}{s + 2}\right|_{s=s_1=-1} = \left.\frac{s_1 + 3}{s_1 + 2}\right|_{s_1=-1} = 2. \tag{2.25}$$

The graphical representation of Eq. (2.25) is shown in Fig. 2.7. The graphical method of evaluating the residues is particularly valuable when the order of the characteristic equation is high and several poles are complex conjugate pairs.

The inverse Laplace transform of Eq. (2.22) is then

$$y(t) = \mathcal{L}^{-1}\left\{\frac{2}{s + 1}\right\} + \mathcal{L}^{-1}\left\{\frac{-1}{s + 2}\right\}. \tag{2.26}$$

Using Table 2.5, we find that

$$y(t) = 2e^{-t} - 1e^{-2t}. \tag{2.27}$$

Finally, it is usually desired to determine the *steady-state* or *final value* of the response of $y(t)$. For example, the final or steady-state rest position of the spring-mass-damper system should be calculated. The final value can be determined from the relation

$$\lim_{t\to\infty} y(t) = \lim_{s\to 0} sY(s), \tag{2.28}$$

where a simple pole of $Y(s)$ at the origin is permitted, but poles on the imaginary axis and in the right half-plane and higher-order poles at the origin are excluded.

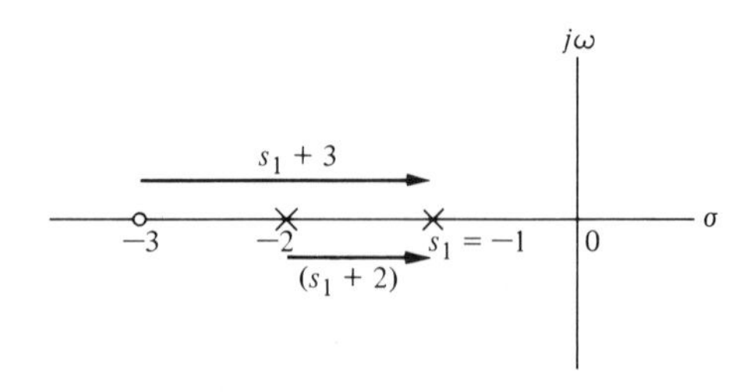

Figure 2.7. Graphical evaluation of the residues.

Therefore, for the specific case of the spring, mass, and damper, we find that

$$\lim_{t\to\infty} y(t) = \lim_{s\to 0} sY(s) = 0. \tag{2.29}$$

Hence the final position for the mass is the normal equilibrium position $y = 0$.

In order to illustrate clearly the salient points of the Laplace transform method, let us reconsider the mass-spring-damper system for the underdamped case. The equation for $Y(s)$ may be written as

$$\begin{aligned} Y(s) &= \frac{(s + f/M)(y_0)}{(s^2 + (f/M)s + K/M)} \\ &= \frac{(s + 2\zeta\omega_n)(y_0)}{s^2 + 2\zeta\omega_n s + \omega_n^2}, \end{aligned} \tag{2.30}$$

where ζ is the dimensionless *damping ratio* and ω_n is the *natural frequency* of the system. The roots of the characteristic equation are

$$s_1, s_2 = -\zeta\omega_n \pm \omega_n\sqrt{\zeta^2 - 1}, \tag{2.31}$$

where, in this case $\omega_n = \sqrt{K/M}$ and $\zeta = f/(2\sqrt{KM})$. When $\zeta > 1$, the roots are real; and when $\zeta < 1$, the roots are complex and conjugates. When $\zeta = 1$, the roots are repeated and real and the condition is called *critical damping.*

When $\zeta < 1$, the response is underdamped and

$$s_{1,2} = -\zeta\omega_n \pm j\omega_n\sqrt{1 - \zeta^2}. \tag{2.32}$$

The s-plane plot of the poles and zeros of $Y(s)$ is shown in Fig. 2.8, where $\theta = \cos^{-1}\zeta$. As ζ varies with ω_n constant, the complex conjugate roots follow a circular locus as shown in Fig. 2.9. The transient response is increasingly oscillatory as the roots approach the imaginary axis when ζ approaches zero.

The inverse Laplace transform can be evaluated using the graphical residue evaluation. The partial fraction expansion of Eq. (2.30) is

$$Y(s) = \frac{k_1}{(s - s_1)} + \frac{k_2}{(s - s_2)}. \tag{2.33}$$

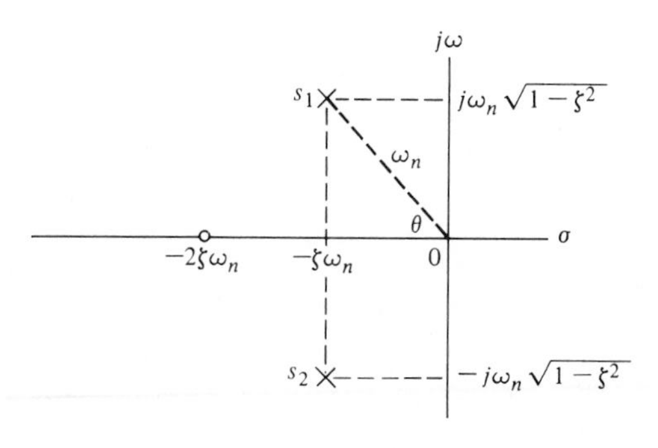

Figure 2.8. An s-plane plot of the poles and zeros of $Y(s)$.

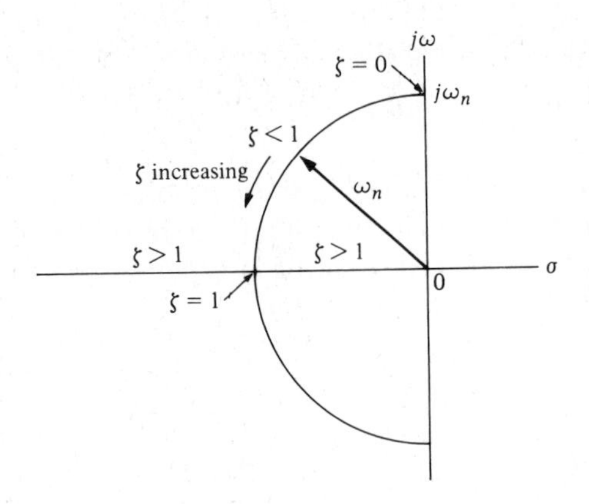

Figure 2.9. The locus of roots as ζ varies with ω_n constant.

Since s_2 is the complex conjugate of s_1, the residue k_2 is the complex conjugate of k_1 so that we obtain

$$Y(s) = \frac{k_1}{(s - s_1)} + \frac{k_1^*}{(s - s_1^*)},$$

where the asterisk indicates the conjugate relation. The residue k_1 is evaluated from Fig. 2.10 as

$$k_1 = \frac{(y_0)(s_1 + 2\zeta\omega_n)}{(s_1 - s_1^*)} = \frac{(y_0)M_1 e^{j\theta}}{M_2 e^{j\pi/2}}, \tag{2.34}$$

where M_1 is the magnitude of $(s_1 + 2\zeta\omega_n)$ and M_2 is the magnitude of $(s_1 - s_1^*)$.† In this case, we obtain

$$k_1 = \frac{(y_0)(\omega_n e^{j\theta})}{(2\omega_n \sqrt{1 - \zeta^2}\, e^{j\pi/2})} = \frac{(y_0)}{2\sqrt{1 - \zeta^2}\, e^{j(\pi/2 - \theta)}}, \tag{2.35}$$

where $\theta = \cos^{-1} \zeta$. Therefore,

$$k_2 = \frac{(y_0)}{2\sqrt{1 - \zeta^2}} e^{j(\pi/2 - \theta)}. \tag{2.36}$$

†A review of complex numbers appears in Appendix E.

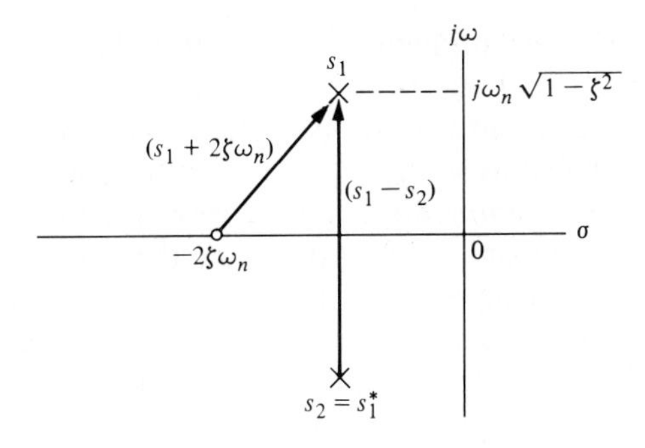

Figure 2.10. Evaluation of the residue k_1.

Finally, we find that

$$
\begin{aligned}
y(t) &= k_1 e^{s_1 t} + k_2 e^{s_2 t} \\
&= \frac{y_0}{2\sqrt{1-\zeta^2}} \left(e^{j(\theta-\pi/2)} e^{-\zeta\omega_n t} e^{j\omega_n\sqrt{1-\zeta^2}t} + e^{j(\pi/2-\theta)} e^{-\zeta\omega_n t} e^{-j\omega_n\sqrt{1-\zeta^2}t}\right) \qquad (2.37) \\
&= \frac{y_0}{\sqrt{1-\zeta^2}} e^{-\zeta\omega_n t} \sin\left(\omega_n\sqrt{1-\zeta^2}\, t + \theta\right).
\end{aligned}
$$

The transient response of the overdamped ($\zeta > 1$) and underdamped ($\zeta < 1$) cases are shown in Fig. 2.11. The transient response occurs when $\zeta < 1$ exhibits an

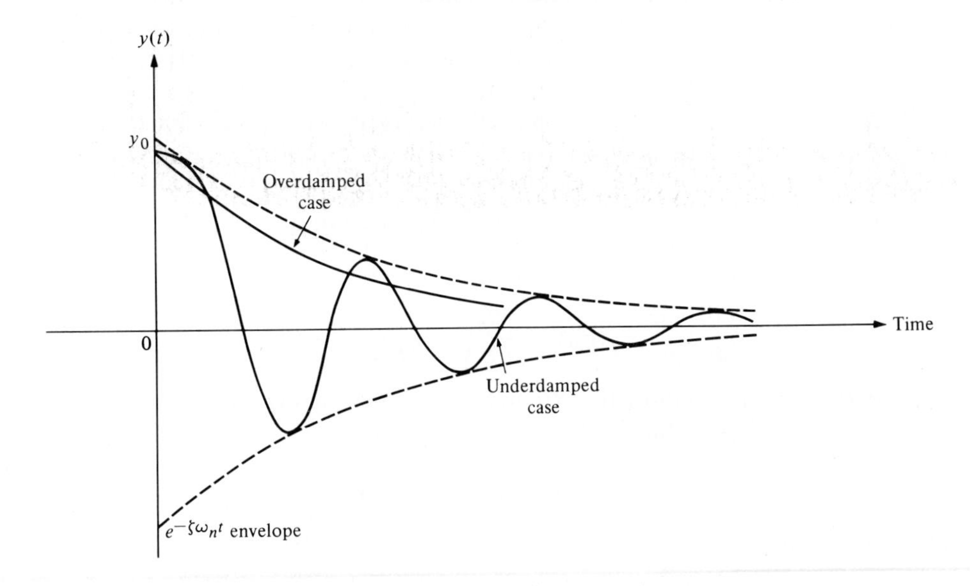

Figure 2.11. Response of the spring-mass-damper system.

oscillation in which the amplitude decreases with time, and it is called a *damped oscillation.*

The direct and clear relationship between the s-plane location of the poles and the form of the transient response is readily interpreted from the s-plane pole-zero plots. Furthermore, the magnitude of the response of each root, represented by the residue, is clearly visualized by examining the graphical residues on the s-plane. The Laplace transformation and the s-plane approach is a very useful technique for system analysis and design where emphasis is placed on the transient and steady-state performance. In fact, because the study of control systems is concerned primarily with the transient and steady-state performance of dynamic systems, we have real cause to appreciate the value of the Laplace transform techniques.

2.5 The Transfer Function of Linear Systems

The *transfer function* of a linear system is defined as the ratio of the Laplace transform of the output variable to the Laplace transform of the input variable, with all initial conditions assumed to be zero. The transfer function of a system (or element) represents the relationship describing the dynamics of the system under consideration.

A transfer function may only be defined for a linear, stationary (constant parameter) system. A nonstationary system, often called a time-varying system, has one or more time-varying parameters, and the Laplace transformation may not be utilized. Furthermore, a transfer function is an input–output description of the behavior of a system. Thus the transfer function description does not include any information concerning the internal structure of the system and its behavior.

The transfer function of the spring-mass-damper system is obtained from the original describing equation, Eq. (2.19), rewritten with zero initial conditions as follows:

$$Ms^2Y(s) + fsY(s) + KY(s) = R(s). \tag{2.38}$$

Then the transfer function is

$$\frac{\text{Output}}{\text{Input}} = G(s) = \frac{Y(s)}{R(s)} = \frac{1}{Ms^2 + fs + K}. \tag{2.39}$$

The transfer function of the RC network shown in Fig. 2.12 is obtained by writing the Kirchhoff voltage equation, which yields

$$V_1(s) = \left(R + \frac{1}{Cs}\right) I(s). \tag{2.40}$$

The output voltage is

$$V_2(s) = I(s)\left(\frac{1}{Cs}\right). \tag{2.41}$$

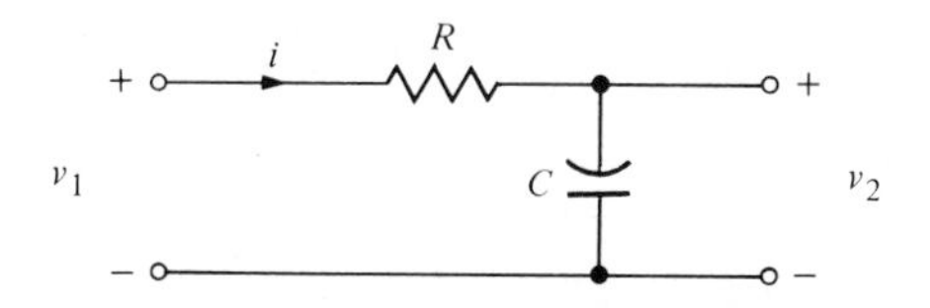

Figure 2.12. An *RC* network.

Therefore, solving Eq. (2.40) for $I(s)$ and substituting in Eq. (2.41), we have

$$V_2(s) = \frac{(1/Cs)V_1(s)}{R + 1/Cs}. \tag{2.42}$$

Then the transfer function is obtained as the ratio $V_2(s)/V_1(s)$, which is

$$\begin{aligned} G(s) &= \frac{V_2(s)}{V_1(s)} = \frac{1}{RCs + 1} \\ &= \frac{1}{\tau s + 1} \\ &= \frac{(1/\tau)}{s + 1/\tau}, \end{aligned} \tag{2.43}$$

where $\tau = RC$, the *time constant* of the network. Equation (2.43) could be immediately obtained if one observes that the circuit is a voltage divider, where

$$\frac{V_2(s)}{V_1(s)} = \frac{Z_2(s)}{Z_1(s) + Z_2(s)} \tag{2.44}$$

and $Z_1(s) = R$, $Z_2 = 1/Cs$.

A multiloop electrical circuit or an analogous multiple mass mechanical system results in a set of simultaneous equations in the Laplace variable. It is usually more convenient to solve the simultaneous equations by using matrices and determinants [1, 3, 16]. An introduction to matrices and determinants is provided in Appendix C [45].

■ Example 2.2 Transfer function of system

Consider the mechanical system shown in Fig. 2.13(a) and its electrical circuit analog shown in Fig. 2.13(b). The electrical circuit analog is a force-current analog as outlined in Table 2.1. The velocities, $v_1(t)$ and $v_2(t)$, of the mechanical system are directly analogous to the node voltage $v_1(t)$ and $v_2(t)$ of the electrical circuit. The simultaneous equations, assuming the initial conditions are zero, are

$$M_1 s V_1(s) + (f_1 + f_2)V_1(s) - f_1 V_2(s) = R(s), \tag{2.45}$$

$$M_2 s V_2(s) + f_1(V_2(s) - V_1(s)) + K\frac{V_2(s)}{s} = 0. \tag{2.46}$$

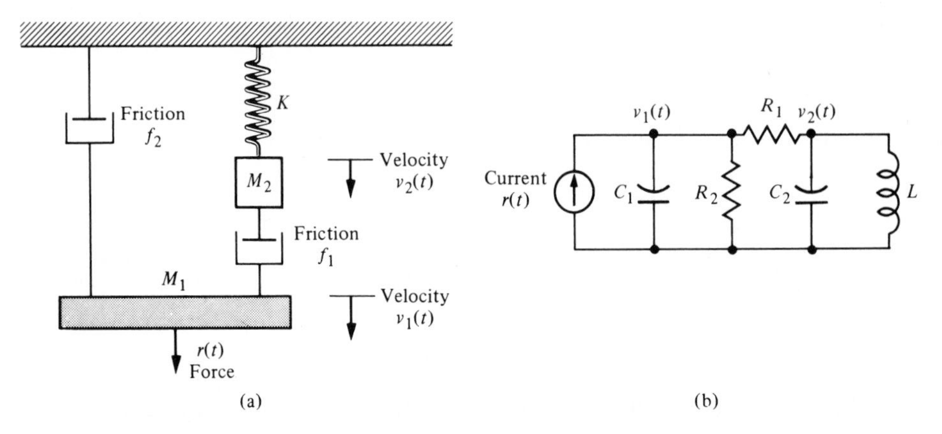

Figure 2.13. (a) Two-mass mechanical system. (b) Two-node electric circuit analog $C_1 = M_1$, $C_2 = M_2$, $L = 1/K$, $R_1 = 1/f_1$, $R_2 = 1/f_2$.

Rearranging Eqs. (2.45) and (2.46) we obtain

$$(M_1 s + (f_1 + f_2))V_1(s) + (-f_1)V_2(s) = R(s), \tag{2.47}$$

$$(-f_1)V_1(s) + \left(M_2 s + f_1 + \frac{K}{s}\right) V_2(s) = 0, \tag{2.48}$$

or, in matrix form, we have

$$\begin{bmatrix} (M_1 s + f_1 + f_2) & (-f_1) \\ (-f_1) & \left(M_2 s + f_1 + \dfrac{K}{s}\right) \end{bmatrix} \begin{bmatrix} V_1(s) \\ V_2(s) \end{bmatrix} = \begin{bmatrix} R(s) \\ 0 \end{bmatrix}. \tag{2.49}$$

Assuming the velocity of M_1 is the output variable, we solve for $V_1(s)$ by matrix inversion or Cramer's rule to obtain [1, 3]

$$V_1(s) = \frac{(M_2 s + f_1 + (K/s))R(s)}{(M_1 s + f_1 + f_2)(M_2 s + f_1 + (K/s)) - f_1^2}. \tag{2.50}$$

Then the transfer function of the mechanical (or electrical) system is

$$\begin{aligned} G(s) = \frac{V_1(s)}{R(s)} &= \frac{(M_2 s + f_1 + (K/s))}{(M_1 s + f_1 + f_2)(M_2 s + f_1 + (K/s)) - f_1^2} \\ &= \frac{(M_2 s^2 + f_1 s + K)}{(M_1 s + f_1 + f_2)(M_2 s^2 + f_1 s + K) - f_1^2 s}. \end{aligned} \tag{2.51}$$

If the transfer function in terms of the position $x_1(t)$ is desired, then we have

$$\frac{X_1(s)}{R(s)} = \frac{V_1(s)}{sR(s)} = \frac{G(s)}{s}. \tag{2.52}$$

As an example, let us obtain the transfer function of an important electrical control component, the *dc motor* [7].

■ Example 2.3 Transfer function of dc motor

The *dc motor* is a power actuator device that delivers energy to a load as shown in Fig. 2.14(a) and a sketch of a dc motor is shown in Fig. 2.14(b). A cutaway view of a pancake dc motor is given in Fig. 2.15. The dc motor converts direct current (dc) electrical energy into rotational mechanical energy. A major fraction of the torque generated in the rotor (armature) of the motor is available to drive an external load. Because of features such as high torque, speed controllability over a wide range, portability, well-behaved speed-torque characteristics, and adaptability to various types of control methods, dc motors are still widely used in numerous control applications including robotic manipulators, tape transport mechanisms, disk drives, machine tools, and servovalve actuators.

The transfer function of the dc motor will be developed for a linear approximation to an actual motor, and second-order effects, such as hysteresis and the voltage drop across the brushes, will be neglected. The input voltage may be applied to the field or armature terminals. The air-gap flux of the motor is proportional to the field current, provided the field is unsaturated, so that

$$\phi = K_f i_f. \tag{2.53}$$

The torque developed by the motor is assumed to be related linearly to ϕ and the armature current as follows:

$$T_m = K_1 \phi i_a(t) = K_1 K_f i_f(t) i_a(t). \tag{2.54}$$

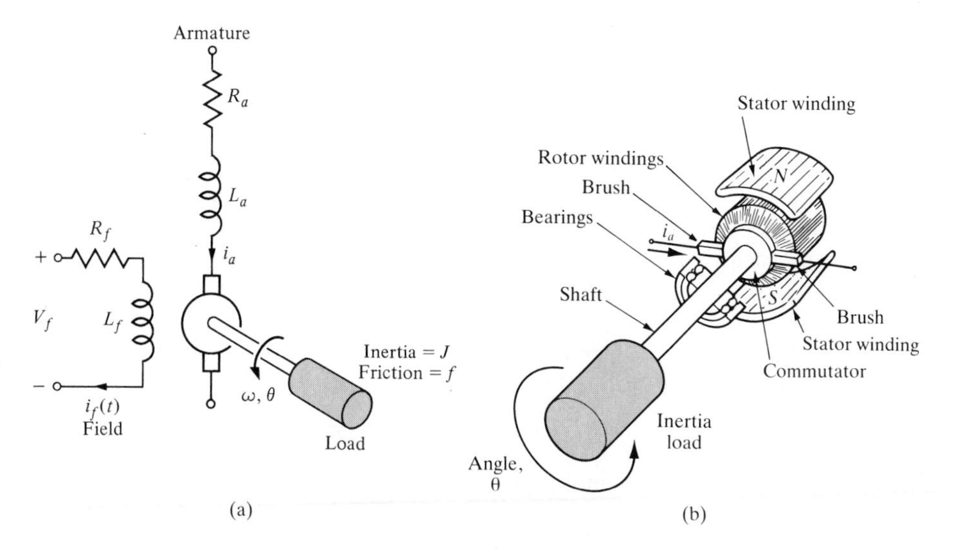

Figure 2.14. A dc motor. (a) Wiring diagram. (b) Sketch.

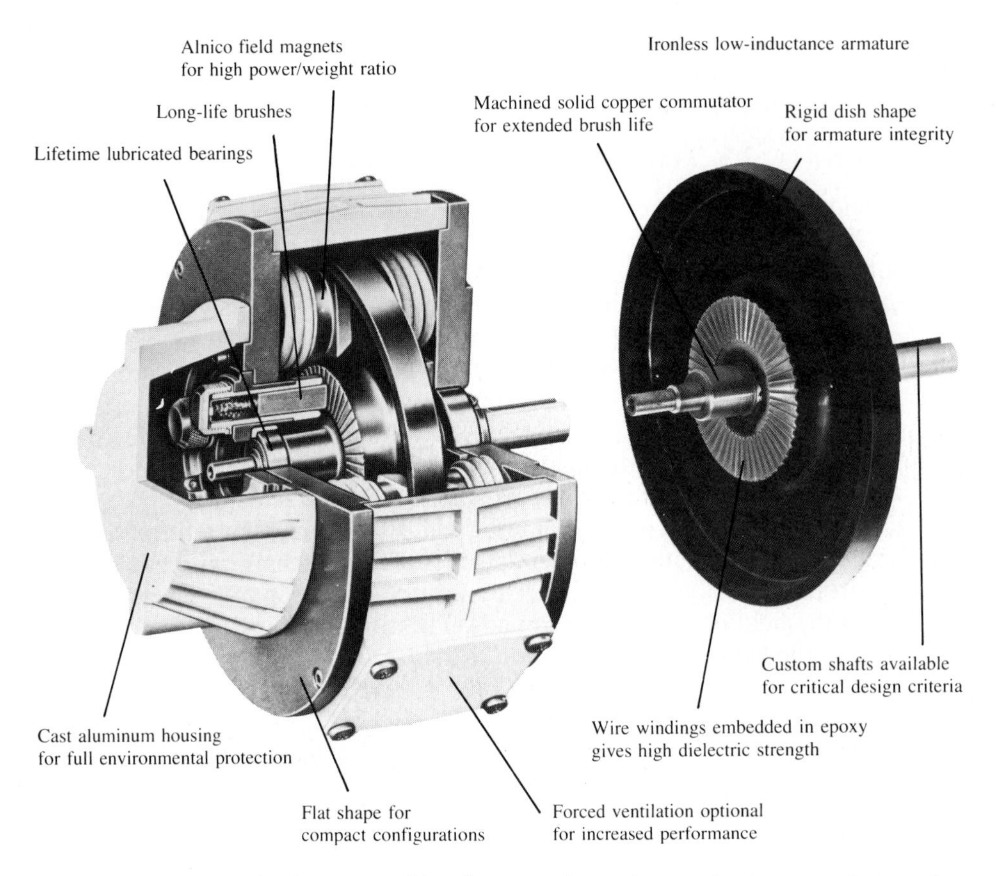

Figure 2.15. A pancake dc motor with a flat wound armature and a permanent magnet rotor. These motors are capable of providing high torque with a low rotor inertia. A typical mechanical time constant is in the range of 15 ms. (Courtesy of Mavilor Motors.)

It is clear from Eq. (2.54) that in order to have a linear element one current must be maintained constant while the other current becomes the input current. First, we shall consider the field current controlled motor, which provides a substantial power amplification. Then we have in Laplace transform notation

$$T_m(s) = (K_1 K_f I_a) I_f(s) = K_m I_f(s), \tag{2.55}$$

where $i_a = I_a$ is a constant armature current and K_m is defined as the motor constant. The field current is related to the field voltage as

$$V_f(s) = (R_f + L_f s) I_f(s). \tag{2.56}$$

The motor torque $T_m(s)$ is equal to the torque delivered to the load. This relation may be expressed as

$$T_m(s) = T_L(s) + T_d(s), \tag{2.57}$$

where $T_L(s)$ is the load torque and $T_d(s)$ is the disturbance torque, which is often negligible. However, the disturbance torque often must be considered in systems subjected to external forces such as antenna wind-gust forces. The load torque for rotating inertia as shown in Fig. 2.14 is written as

$$T_L(s) = Js^2\theta(s) + fs\theta(s). \tag{2.58}$$

Rearranging Eqs. (2.55), (2.56), and (2.57), we have

$$T_L(s) = T_m(s) - T_d(s), \tag{2.59}$$

$$T_m(s) = K_m I_f(s), \tag{2.60}$$

$$I_f(s) = \frac{V_f(s)}{R_f + L_f s}. \tag{2.61}$$

Therefore the transfer function of the motor-load combination is

$$\begin{aligned}\frac{\theta(s)}{V_f(s)} &= \frac{K_m}{s(Js + f)(L_f s + R_f)} \\ &= \frac{K_m/JL_f}{s(s + f/J)(s + R_f/L_f)}.\end{aligned} \tag{2.62}$$

The block diagram model of the field controlled dc motor is shown in Fig. 2.16. Alternatively, the transfer function may be written in terms of the time constants of the motor as

$$\frac{\theta(s)}{V_f(s)} = G(s) = \frac{K_m/fR_f}{s(\tau_f s + 1)(\tau_L s + 1)}, \tag{2.63}$$

where $\tau_f = L_f/R_f$ and $\tau_L = J/f$. Typically, one finds that $\tau_L > \tau_f$ and often the field time constant may be neglected.

The *armature controlled dc motor* utilizes a constant field current, and therefore the motor torque is

$$T_m(s) = (K_1 K_f I_f) I_a(s) = K_m I_a(s). \tag{2.64}$$

The armature current is related to the input voltage applied to the armature as

$$V_a(s) = (R_a + L_a s) I_a(s) + V_b(s), \tag{2.65}$$

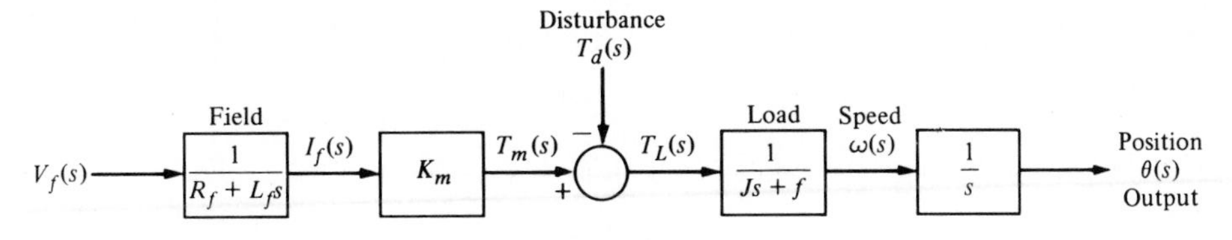

Figure 2.16. Block diagram model of field controlled dc motor.

where $V_b(s)$ is the back electromotive-force voltage proportional to the motor speed. Therefore we have

$$V_b(s) = K_b\omega(s), \tag{2.66}$$

and the armature current is

$$I_a(s) = \frac{V_a(s) - K_b\omega(s)}{(R_a + L_a s)}. \tag{2.67}$$

Equations (2.58) and (2.59) represent the load torque so that

$$T_L(s) = Js^2\theta(s) + fs\theta(s) = T_m(s) - T_d(s). \tag{2.68}$$

The relations for the armature controlled dc motor are shown schematically in Fig. 2.17. Using Eqs. (2.64), (2.67), and (2.68), or, alternatively, the block diagram, we obtain the transfer function

$$G(s) = \frac{\theta(s)}{V_a(s)} = \frac{K_m}{s[(R_a + L_a s)(Js + f) + K_b K_m]} = \frac{K_m}{s(s^2 + 2\zeta\omega_n s + \omega_n^2)}. \tag{2.69}$$

However, for many dc motors, the time constant of the armature, $\tau_a = L_a/R_a$, is negligible, and therefore

$$G(s) = \frac{\theta(s)}{V_a(s)} = \frac{K_m}{s[R_a(Js + f) + K_b K_m]} = \frac{[K_m/(R_a f + K_b K_m)]}{s(\tau_1 s + 1)}, \tag{2.70}$$

where the equivalent time constant $\tau_1 = R_a J/(R_a f + K_b K_m)$.

It is of interest to note that K_m is equal to K_b. This equality may be shown by considering the steady-state motor operation and the power balance when the rotor resistance is neglected. The power input to the rotor is $(K_b\omega)i_a$ and the power delivered to the shaft is $T\omega$. In the steady-state condition, the power input is equal to the power delivered to the shaft so that $(K_b\omega)i_a = T\omega$; and since $T = K_m i_a$ (Eq. 2.64), we find that $K_b = K_m$.

Electric motors are used for moving loads when a rapid response is not

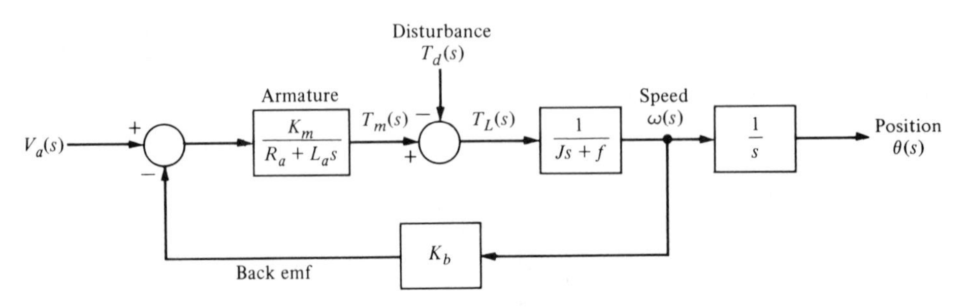

Figure 2.17. Armature controlled dc motor.

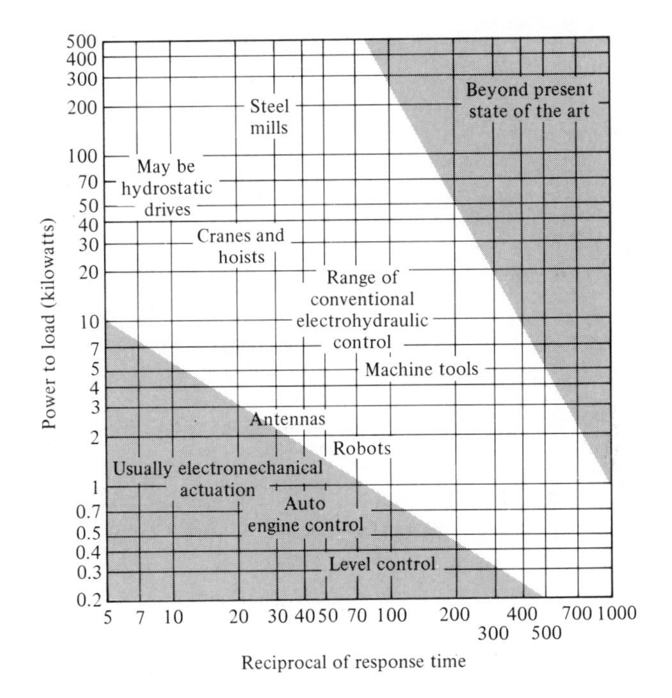

Figure 2.18. Range of control response time and power to load for electromechanical and electrohydraulic devices.

required and for relatively low power requirements. Actuators that operate as a result of hydraulic pressure are used for large loads. Figure 2.18 shows the usual ranges of use for electromechanical drives as contrasted to electrohydraulic drives. Typical applications are also shown on the figure.

■ Example 2.4 Transfer function of hydraulic actuator

A useful actuator for the linear positioning of a mass is the hydraulic actuator shown in Table 2.6, entry 9 [10]. The hydraulic actuator is capable of providing a large power amplification. It will be assumed that the hydraulic fluid is available from a constant pressure source and that the compressibility of the fluid is negligible. A downward input displacement, x, moves the control valve, and thus fluid passes into the upper part of the cylinder and the piston is forced downward. A small, low-power displacement of $x(t)$ causes a larger, high-power displacement, $y(t)$. The volumetric fluid flow rate Q is related to the input displacement $x(t)$ and the differential pressure across the piston as $Q = g(x, P)$. Using the Taylor series linearization as in Eq. (2.11), we have

$$\begin{aligned} Q &= \left(\frac{\partial g}{\partial x}\right)_{x_0 P_0} x + \left(\frac{\partial g}{\partial P}\right)_{P_0 x_0} P \\ &= k_x x - k_P P \end{aligned} \tag{2.71}$$

Table 2.6. Transfer Functions of Dynamic Elements and Networks

Element or System	G(s)
1. Integrating circuit	
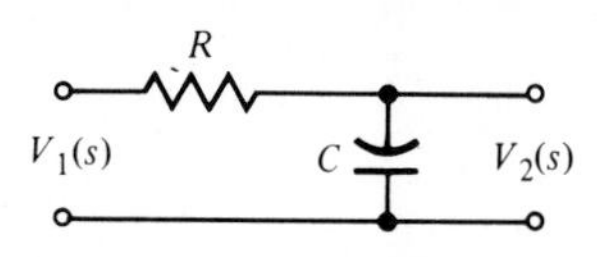	$\dfrac{V_2(s)}{V_1(s)} = \dfrac{1}{RCs + 1}$
2. Differentiating circuit	
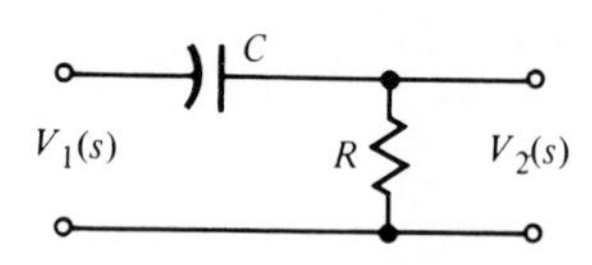	$\dfrac{V_2(s)}{V_1(s)} = \dfrac{RCs}{RCs + 1}$
3. Differentiating circuit	
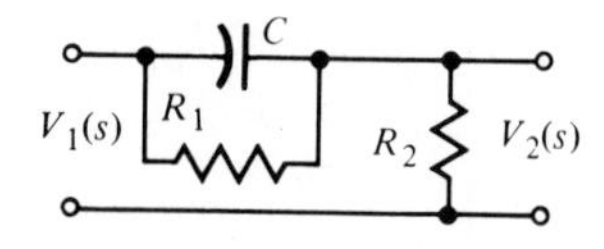	$\dfrac{V_2(s)}{V_1(s)} = \dfrac{s + 1/R_1C}{s + (R_1 + R_2)/R_1R_2C}$
4. Lead-lag filter circuit	
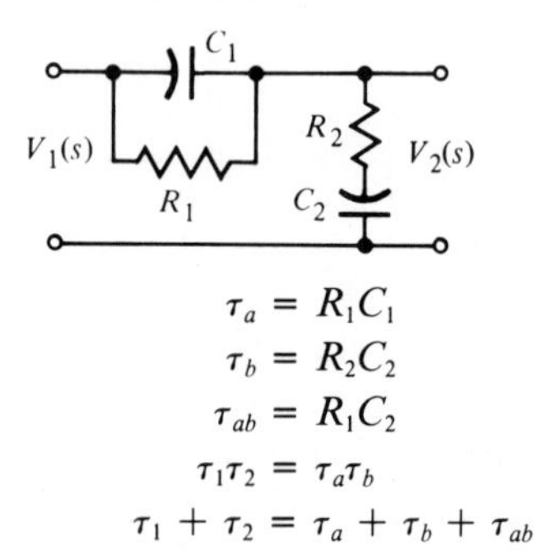 $\tau_a = R_1C_1$ $\tau_b = R_2C_2$ $\tau_{ab} = R_1C_2$ $\tau_1\tau_2 = \tau_a\tau_b$ $\tau_1 + \tau_2 = \tau_a + \tau_b + \tau_{ab}$	$\dfrac{V_2(s)}{V_1(s)} = \dfrac{(1 + s\tau_a)(1 + s\tau_b)}{\tau_a\tau_b s^2 + (\tau_a + \tau_b + \tau_{ab})s + 1}$ $= \dfrac{(1 + s\tau_a)(1 + s\tau_b)}{(1 + s\tau_1)(1 + s\tau_2)}$
5. dc-motor, field controlled	
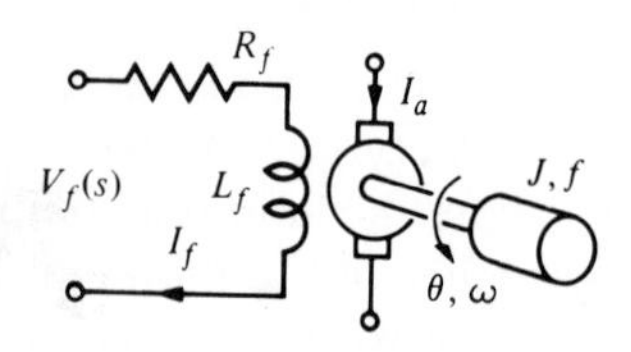	$\dfrac{\theta(s)}{V_f(s)} = \dfrac{K_m}{s(Js + f)(L_f s + R_f)}$
6. dc-motor, armature controlled	
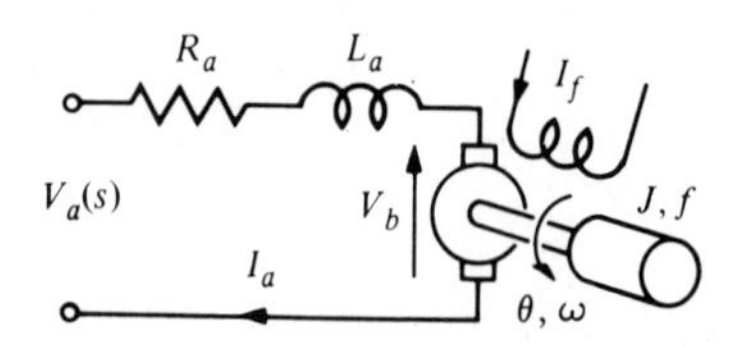	$\dfrac{\theta(s)}{V_a(s)} = \dfrac{K_m}{s[(R_a + L_a s)(Js + f) + K_bK_m]}$

Table 2.6.—*Continued*

Element or System	$G(s)$
7. ac-motor, two-phase control field 	$\dfrac{\theta(s)}{V_c(s)} = \dfrac{K_m}{s(\tau s + 1)}$ $\tau = J/(f - m)$ m = slope of linearized torque-speed curve (normally negative)
8. Amplidyne 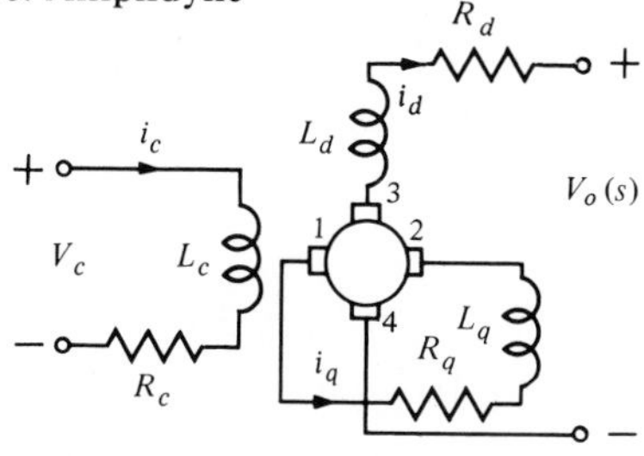	$\dfrac{V_o(s)}{V_c(s)} = \dfrac{(K/R_cR_q)}{(s\tau_c + 1)(S\tau_q + 1)}$ $\tau_c = L_c/R_c, \quad \tau_q = L_q/R_q$ For the unloaded case, $i_d \simeq 0$, $\tau_c \simeq \tau_q$, $0.05 \text{ sec} < \tau_c < 0.5 \text{ sec}$ $V_{12} = V_q$, $V_{34} = V_d$
9. Hydraulic actuator 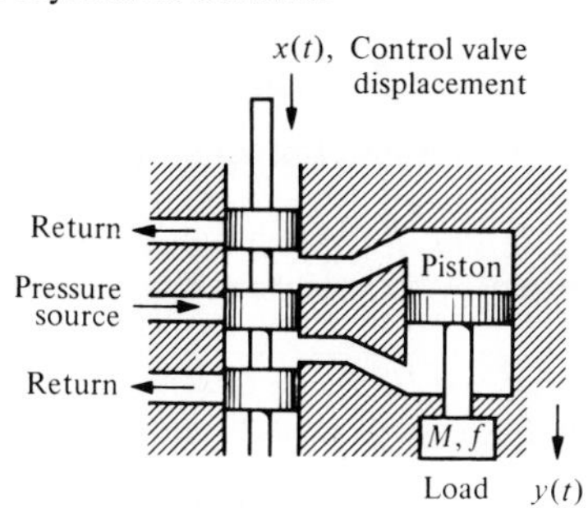	$\dfrac{Y(s)}{X(s)} = \dfrac{K}{s(Ms + B)}$ $K = \dfrac{Ak_x}{k_p}, \quad B = \left(f + \dfrac{A^2}{k_p}\right)$ $k_x = \left.\dfrac{\partial g}{\partial x}\right\|_{x0}, \quad k_p = \left.\dfrac{\partial g}{\partial P}\right\|_{P0},$ $g = g(x, P)$ = flow A = area of piston
10. Gear train 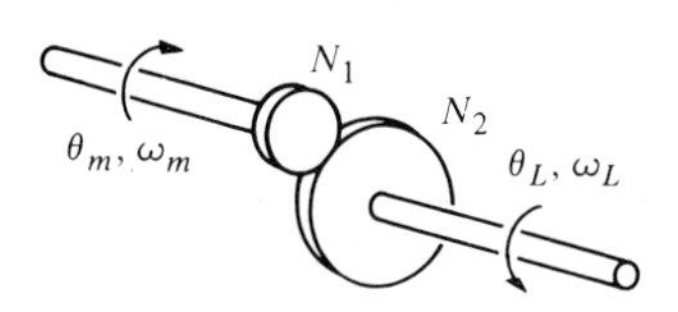	Gear ratio $= n = \dfrac{N_1}{N_2}$ $N_2\theta_L = N_1\theta_m, \quad \theta_L = n\theta_m$ $\omega_L = n\omega_m$
11. Potentiometer 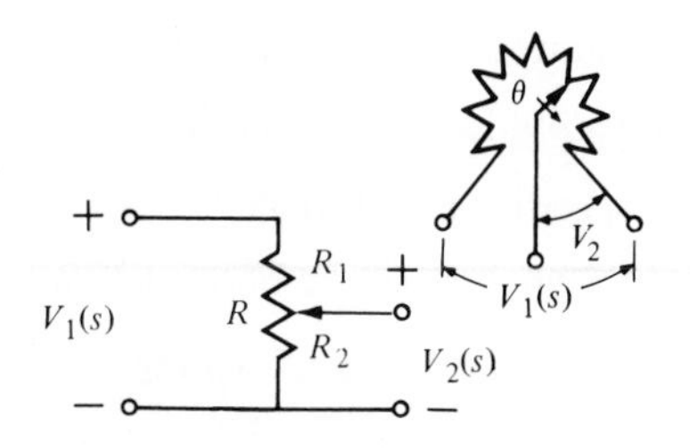	$\dfrac{V_2(s)}{V_1(s)} = \dfrac{R_2}{R} = \dfrac{R_2}{R_1 + R_2}$ $\dfrac{R_2}{R} = \dfrac{\theta}{\theta_{max}}$

Continued

Table 2.6.—*Continued*

Element or System	G(s)
12. Potentiometer error detector bridge 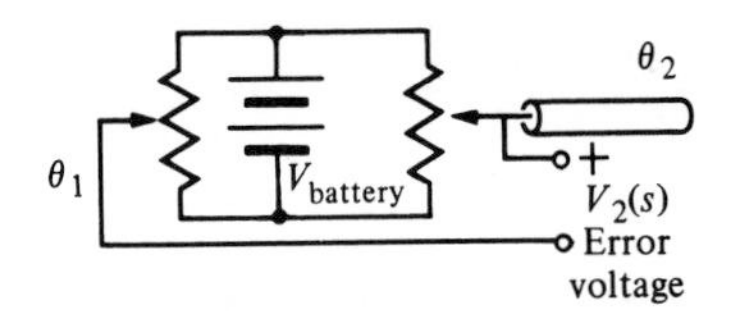	$V_2(s) = k_s(\theta_1(s) - \theta_2(s))$ $V_2(s) = k_s\theta_{error}(s)$ $k_s = \dfrac{V_{battery}}{\theta_{max}}$
13. Tachometer 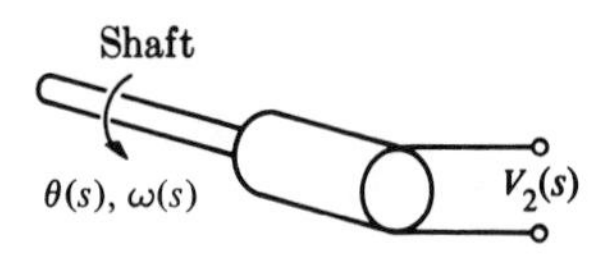	$V_2(s) = K_t\omega(s) = K_ts\theta(s)$; K_t = constant
14. dc-amplifier 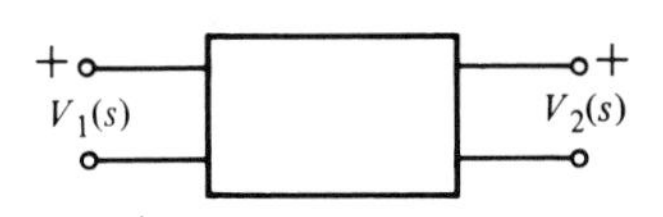	$\dfrac{V_2(s)}{V_1(s)} = \dfrac{k_a}{s\tau + 1}$ R_o = output resistance C_o = output capacitance $\tau = R_oC_o$, $\tau \ll 1$ and is often negligible for servomechanism amplifier
15. Accelerometer 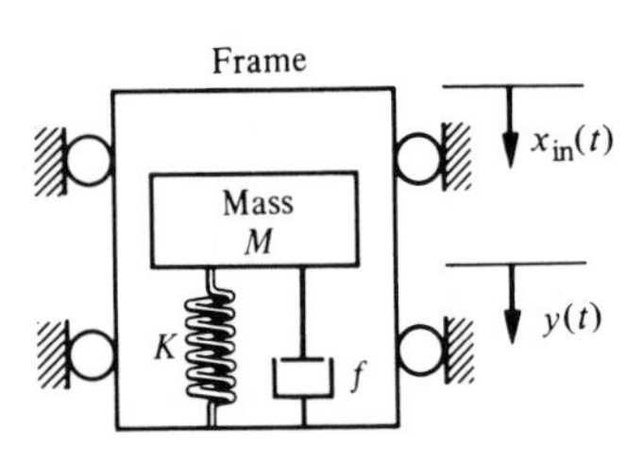	$x_o(t) = y(t) - x_{in}(t)$, $\dfrac{X_o(s)}{X_{in}(s)} = \dfrac{-s^2}{s^2 + (f/M)s + K/M}$ For low-frequency oscillations, where $\omega < \omega_n$, $\dfrac{X_o(j\omega)}{X_{in}(j\omega)} \simeq \dfrac{\omega^2}{K/M}$
16. Thermal heating system 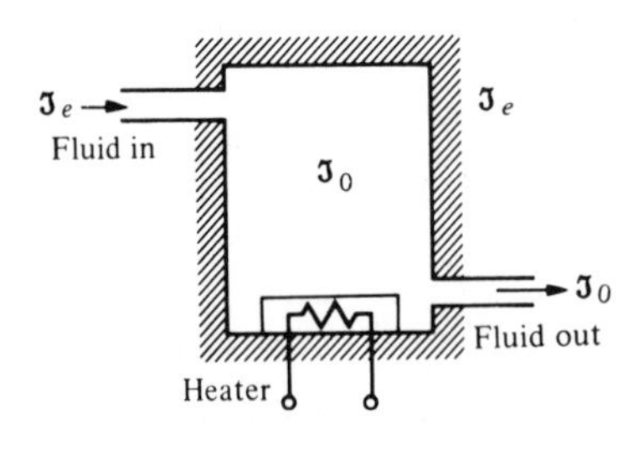	$\dfrac{\tau(s)}{q(s)} = \dfrac{1}{C_ts + (QS + 1/R)}$, where $\tau = \tau_o - \tau_e$ = temperature difference due to thermal process C_t = thermal capacitance Q = fluid flow rate = constant S = specific heat of water R_t = thermal resistance of insulation $q(s)$ = rate of heat flow of heating element

where $g = g(x, P)$ and (x_0, P_0) is the operating point. The force developed by the actuator piston is equal to the area of the piston, A, multiplied by the pressure, P. This force is applied to the mass, and therefore we have

$$AP = M\frac{d^2y}{dt^2} + f\frac{dy}{dt}. \tag{2.72}$$

Thus, substituting Eq. (2.71) into Eq. (2.72), we obtain

$$\frac{A}{k_P}(k_x x - Q) = M\frac{d^2y}{dt^2} + f\frac{dy}{dt}. \tag{2.73}$$

Furthermore, the volumetric fluid flow is related to the piston movement as

$$Q = A\frac{dy}{dt}. \tag{2.74}$$

Then, substituting Eq. (2.74) into Eq. (2.73) and rearranging, we have

$$\frac{Ak_x}{k_P}x = M\frac{d^2y}{dt^2} + \left(f + \frac{A^2}{k_P}\right)\frac{dy}{dt}. \tag{2.75}$$

Therefore, using the Laplace transformation, we have the transfer function

$$\frac{Y(s)}{X(s)} = \frac{K}{s(Ms + B)}, \tag{2.76}$$

where

$$K = \frac{Ak_x}{k_P} \quad \text{and} \quad B = \left(f + \frac{A^2}{k_P}\right).$$

Note that the transfer function of the hydraulic actuator is similar to that of the electric motor. Also, for an actuator operating at high pressure levels and requiring a rapid response of the load, the effect of the compressibility of the fluid must be accounted for [4, 5].

The SI units of the variables are given in Table B.1 in Appendix B. Also a complete set of conversion factors for the British system of units are given in Table B.2.

The transfer function concept and approach is very important because it provides the analyst and designer with a useful mathematical model of the system elements. We shall find the transfer function to be a continually valuable aid in the attempt to model dynamic systems. The approach is particularly useful since the s-plane poles and zeros of the transfer function represent the transient response of the system. The transfer functions of several dynamic elements are given in Table 2.6.

2.6 Block Diagram Models

The dynamic systems that comprise automatic control systems are represented mathematically by a set of simultaneous differential equations. As we have noted in the previous sections, the introduction of the Laplace transformation reduces the problem to the solution of a set of linear algebraic equations. Since control systems are concerned with the control of specific variables, the interrelationship of the controlled variables to the controlling variables is required. This relationship is typically represented by the transfer function of the subsystem relating the input and output variables. Therefore one can correctly assume that the transfer function is an important relation for control engineering.

The importance of the cause and effect relationship of the transfer function is evidenced by the interest in representing the relationship of system variables by diagrammatic means. The *block diagram* representation of the systems relationships is prevalent in control system engineering. Block diagrams consist of *unidirectional,* operational blocks that represent the transfer function of the variables of interest. A block diagram of a field controlled dc motor and load is shown in Fig. 2.19. The relationship between the displacement $\theta(s)$ and the input voltage $V_f(s)$ is clearly portrayed by the block diagram.

In order to represent a system with several variables under control, an interconnection of blocks is utilized. For example, the system shown in Fig. 2.20 has two input variables and two output variables [6]. Using transfer function relations, we can write the simultaneous equations for the output variables as

$$C_1(s) = G_{11}(s)R_1(s) + G_{12}(s)R_2(s), \tag{2.77}$$

$$C_2(s) = G_{21}(s)R_1(s) + G_{22}(s)R_2(s), \tag{2.78}$$

where $G_{ij}(s)$ is the transfer function relating the ith output variable to the jth input variable. The block diagram representing this set of equations is shown in Fig. 2.21. In general, for J inputs and I outputs, we write the simultaneous equation in matrix form as

$$\begin{bmatrix} C_1(s) \\ C_2(s) \\ \vdots \\ C_I(s) \end{bmatrix} = \begin{bmatrix} G_{11}(s) \cdots G_{1J}(s) \\ G_{21}(s) \cdots G_{2J}(s) \\ \vdots \qquad \vdots \\ G_{I1}(s) \cdots G_{IJ}(s) \end{bmatrix} \begin{bmatrix} R_1(s) \\ R_2(s) \\ \vdots \\ R_J(s) \end{bmatrix} \tag{2.79}$$

or, simply,

$$\mathbf{C} = \mathbf{G}\mathbf{R}. \tag{2.80}$$

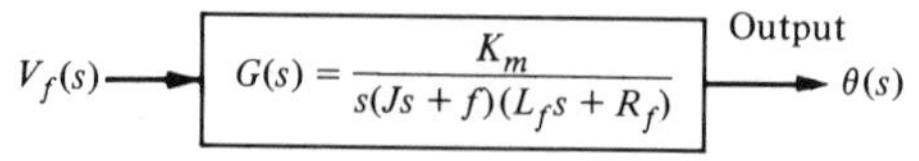

$V_f(s)$ → $G(s) = \dfrac{K_m}{s(Js + f)(L_f s + R_f)}$ → Output $\theta(s)$

Figure 2.19. Block diagram of dc motor.

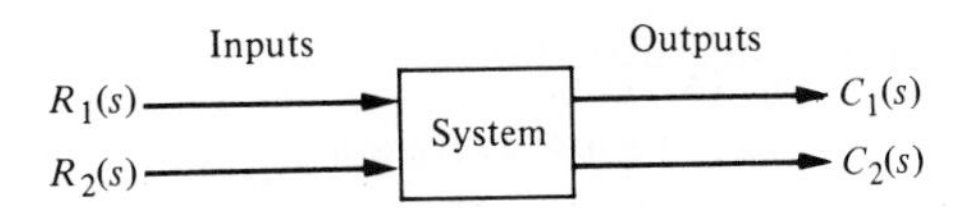

Figure 2.20. General block representation of two-input, two-output system.

Here the **C** and **R** matrices are column matrices containing the I output and the J input variables, respectively, and **G** is an I by J transfer function matrix. The matrix representation of the interrelationship of many variables is particularly valuable for complex multivariable control systems. An introduction to matrix algebra is provided in Appendix C for those unfamiliar with matrix algebra or who would find a review helpful [2].

The block diagram representation of a given system may often be reduced by block diagram reduction techniques to a simplified block diagram with fewer blocks than the original diagram. Since the transfer functions represent linear systems, the multiplication is commutative. Therefore, as in Table 2.7, entry 1, we have

$$X_3(s) = G_1(s)G_2(s)X_2(s) = G_2(s)G_1(s)X_1(s).$$

When two blocks are connected in cascade as in entry 1 of Table 2.7 we assume that

$$X_3(s) = G_2(s)G_1(s)X_1(s)$$

holds true. This assumes that when the first block is connected to the second block, loading of the first block is negligible. Loading and interaction between interconnected components or systems may occur. If loading of interconnected devices does occur, the engineer must account for this change in the transfer function and use the corrected transfer function in subsequent calculations.

Block diagram transformations and reduction techniques are derived by considering the algebra of the diagram variables. For example, consider the block

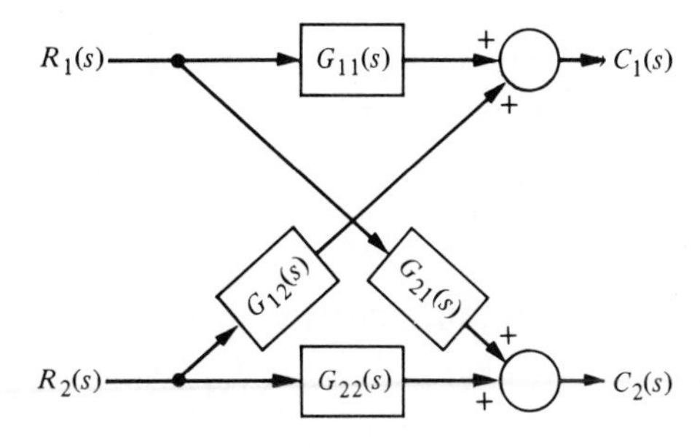

Figure 2.21. Block diagram of interconnected system.

Table 2.7. Block Diagram Transformations

Transformation	Original Diagram	Equivalent Diagram
1. Combining blocks in cascade	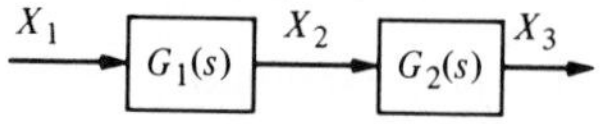 X_1, $G_1(s)$, X_2, $G_2(s)$, X_3	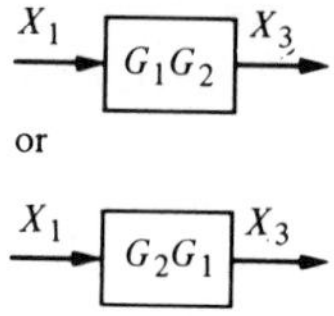 X_1, G_1G_2, X_3 or X_1, G_2G_1, X_3
2. Moving a summing point behind a block	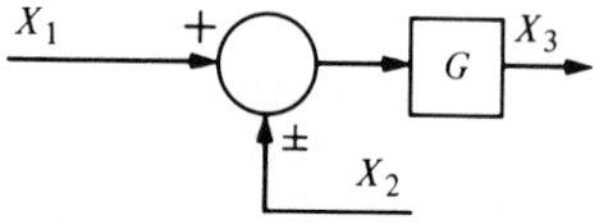 X_1, +, ±, X_2, G, X_3	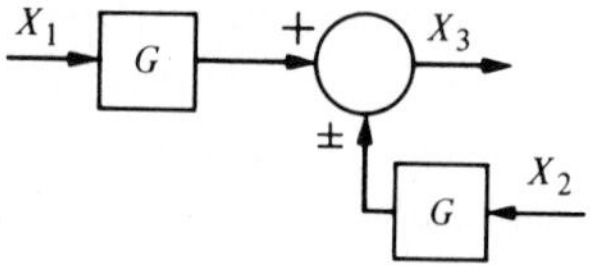 X_1, G, +, ±, X_3, G, X_2
3. Moving a pickoff point ahead of a block	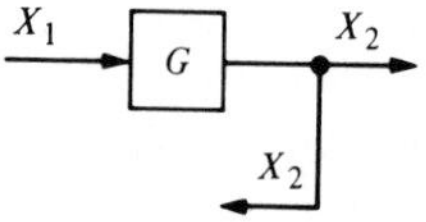 X_1, G, X_2, X_2	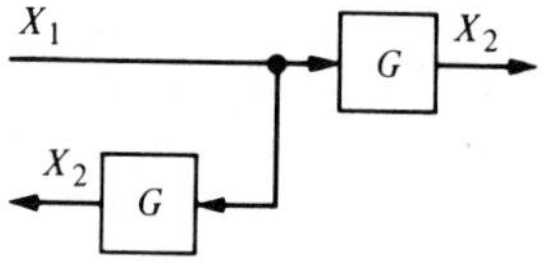 X_1, G, X_2, X_2, G
4. Moving a pickoff point behind a block	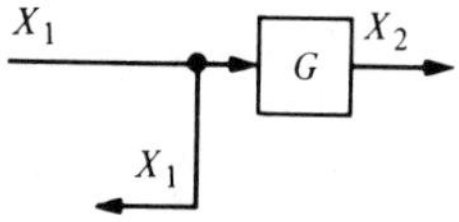 X_1, G, X_2, X_1	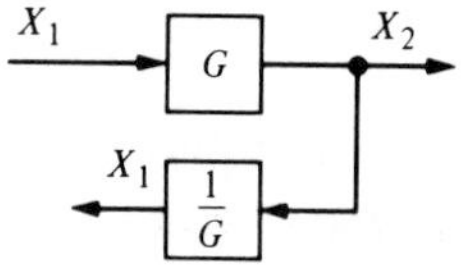 X_1, G, X_2, X_1, $\frac{1}{G}$
5. Moving a summing point ahead of a block	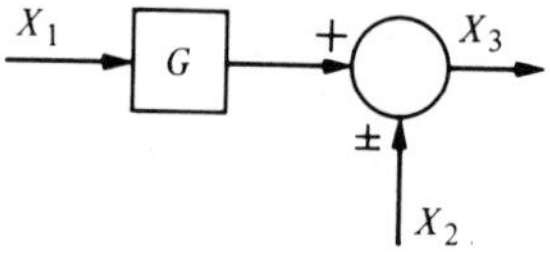 X_1, G, +, ±, X_3, X_2	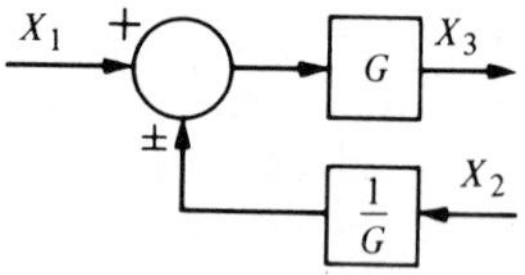 X_1, +, ±, G, X_3, $\frac{1}{G}$, X_2
6. Eliminating a feedback loop	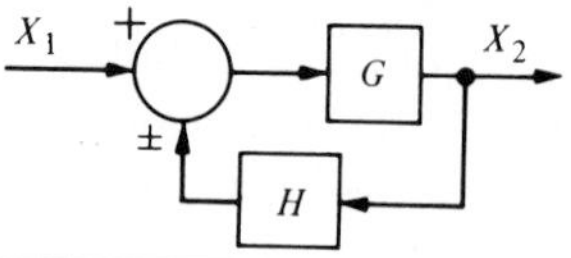 X_1, +, ±, G, X_2, H	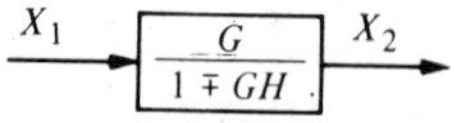 X_1, $\frac{G}{1 \mp GH}$, X_2

diagram shown in Fig. 2.22. This negative feedback control system is described by the equation for the actuating signal

$$\begin{aligned} E_a(s) &= R(s) - B(s) \\ &= R(s) - H(s)C(s). \end{aligned} \tag{2.81}$$

Because the output is related to the actuating signal by $G(s)$, we have

$$C(s) = G(s)E_a(s), \tag{2.82}$$

and therefore

$$C(s) = G(s)(R(s) - H(s)C(s)). \tag{2.83}$$

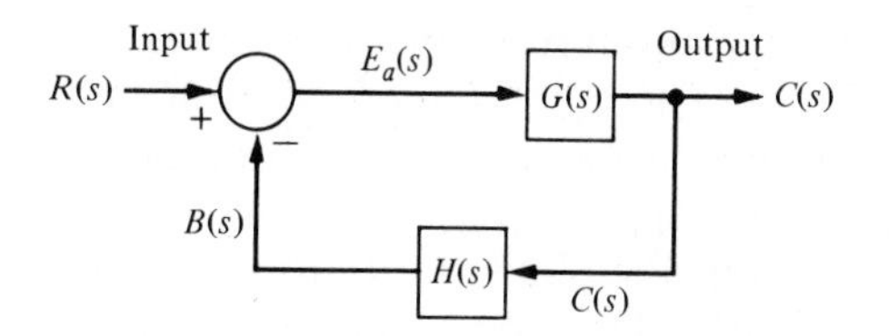

Figure 2.22. Negative feedback control system.

Solving for $C(s)$, we obtain

$$C(s)(1 + G(s)H(s)) = G(s)R(s). \tag{2.84}$$

Therefore the transfer function relating the output $C(s)$ to the input $R(s)$ is

$$\frac{C(s)}{R(s)} = \frac{G(s)}{1 + G(s)H(s)}. \tag{2.85}$$

This *closed-loop transfer function* is particularly important because it represents many of the existing practical control systems.

The reduction of the block diagram shown in Fig. 2.22 to a single block representation is one example of several useful block diagram reductions. These diagram transformations are given in Table 2.7. All the transformations in Table 2.7 can be derived by simple algebraic manipulation of the equations representing the blocks. System analysis by the method of block diagram reduction has the advantage of affording a better understanding of the contribution of each component element than is possible to obtain by the manipulation of equations. The utility of the block diagram transformations will be illustrated by an example of a block diagram reduction.

■ Example 2.5 Block diagram reduction

The block diagram of a multiple-loop feedback control system is shown in Fig. 2.23. It is interesting to note that the feedback signal $H_1(s)C(s)$ is a positive feedback signal and the loop $G_3(s)G_4(s)H_1(s)$ is called a *positive feedback loop*. The

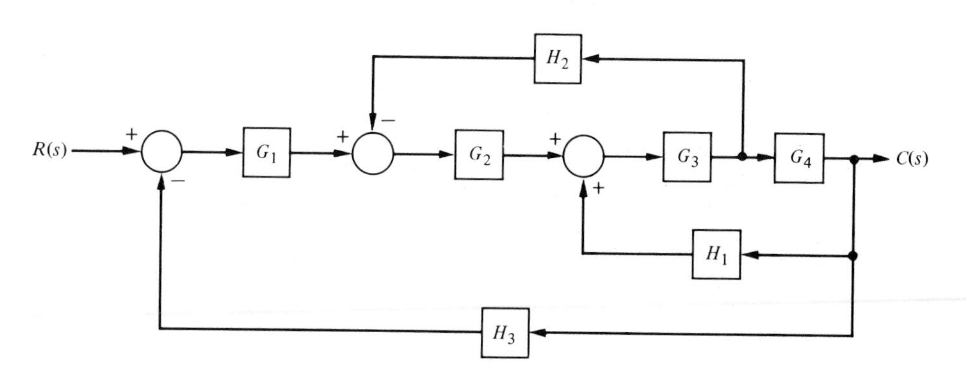

Figure 2.23. Multiple-loop feedback control system.

block diagram reduction procedure is based on the utilization of rule 6, which eliminates feedback loops. Therefore, the other tranformations are used in order to transform the diagram to a form ready for eliminating feedback loops. First, in order to eliminate the loop $G_3G_4H_1$, we move H_2 behind block G_4 by using rule 4, and therefore obtain Fig. 2.24(a). Eliminating the loop $G_3G_4H_1$ by using rule 6, we obtain Fig. 2.24(b). Then, eliminating the inner loop containing H_2/G_4, we obtain Fig. 2.24(c). Finally, by reducing the loop containing H_3 we obtain the closed-loop system transfer function as shown in Fig. 2.24(d). It is worthwhile to examine the form of the numerator and denominator of this closed-loop transfer function. We note that the numerator is composed of the cascade transfer function of the feedforward elements connecting the input $R(s)$ and the output $C(s)$. The denominator is comprised of 1 minus the sum of each loop transfer function. The sign of the loop $G_3G_4H_1$ is plus because it is a positive feedback loop, whereas

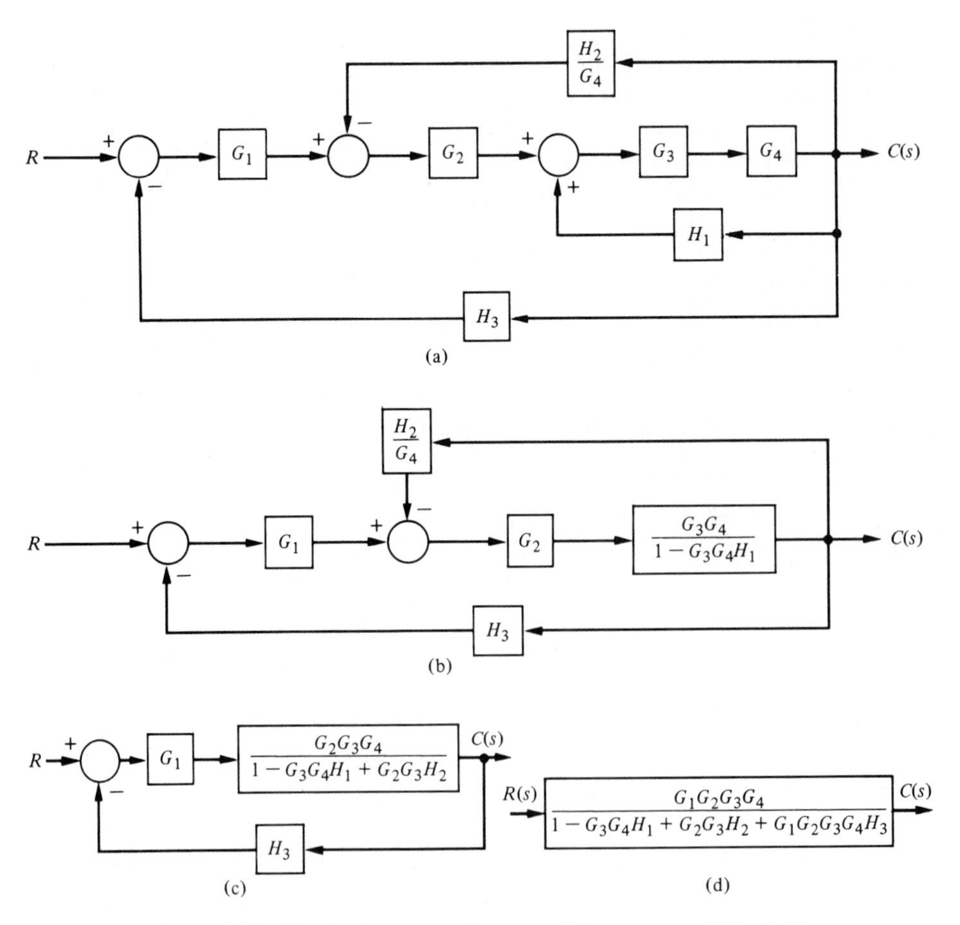

Figure 2.24. Block diagram reduction of the system of Fig. 2.23.

the loops $G_1G_2G_3G_4H_3$ and $G_2G_3H_2$ are negative feedback loops. The denominator can be rewritten as

$$q(s) = 1 - (+G_3G_4H_1 - G_2G_3H_2 - G_1G_2G_3G_4H_3) \quad (2.86)$$

in order to illustrate this point. This form of the numerator and denominator is quite close to the general form for multiple-loop feedback systems, as we shall find in the following section.

The block diagram representation of feedback control systems is a valuable and widely used approach. The block diagram provides the analyst with a graphical representation of the interrelationships of controlled and input variables. Furthermore, the designer can readily visualize the possibilities for adding blocks to the existing system block diagram in order to alter and improve the system performance. The transition from the block diagram method to a method utilizing a line path representation instead of a block representation is readily accomplished and is presented in the following section.

2.7 Signal-Flow Graph Models

Block diagrams are adequate for the representation of the interrelationships of controlled and input variables. However, for a system with reasonably complex interrelationships, the block diagram reduction procedure is cumbersome and often quite difficult to complete. An alternative method for determining the relationship between system variables has been developed by Mason and is based on a representation of the system by line segments [4]. The advantage of the line path method, called the signal-flow graph method, is the availability of a flow graph gain formula, which provides the relation between system variables without requiring any reduction procedure or manipulation of the flow graph.

The transition from a block diagram representation to a directed line segment representation is easy to accomplish by reconsidering the systems of the previous section. A *signal-flow graph* is a diagram consisting of nodes that are connected by several directed branches and is a graphical representation of a set of linear relations. Signal-flow graphs are particularly useful for feedback control systems because feedback theory is primarily concerned with the flow and processing of signals in systems. The basic element of a signal flow graph is a unidirectional path segment called a *branch,* which relates the dependency of an input and an output variable in a manner equivalent to a block of a block diagram. Therefore, the branch relating the output of a dc motor, $\theta(s)$, to the field voltage, $V_f(s)$, is similar to the block diagram of Fig. 2.19 and is shown in Fig. 2.25. The input and output points or junctions are called *nodes.* Similarly, the signal-flow graph rep-

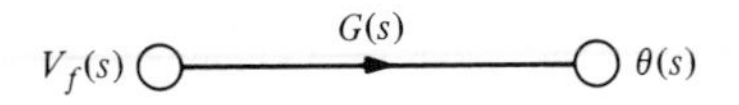

Figure 2.25. Signal-flow graph of the dc motor.

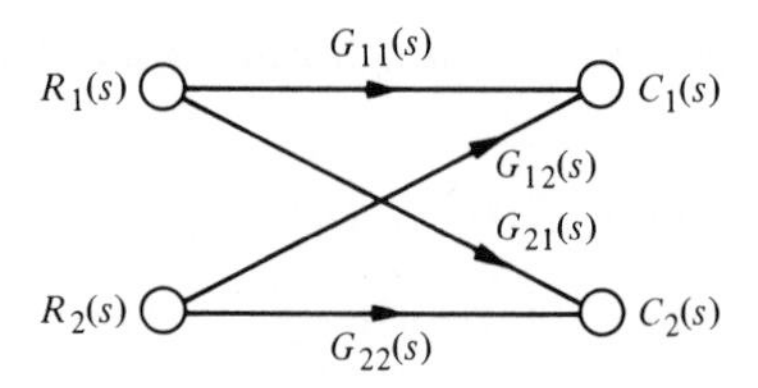

Figure 2.26. Signal-flow graph of interconnected system.

resenting Eqs. (2.77) and (2.78) and Fig. 2.21 is shown in Fig. 2.26. The relation between each variable is written next to the directional arrow. All branches leaving a node pass the nodal signal to the output node of each branch (unidirectionally). All branches entering a node summate as a total signal at the node. A *path* is a branch or a continuous sequence of branches that can be traversed from one signal (node) to another signal (node). A *loop* is a closed path that originates and terminates on the same node, and along the path no node is met twice. Therefore, reconsidering Fig. 2.26, we obtain

$$C_1(s) = G_{11}(s)R_1(s) + G_{12}(s)R_2(s), \tag{2.87}$$

$$C_2(s) = G_{21}(s)R_1(s) + G_{22}(s)R_2(s). \tag{2.88}$$

The flow graph is simply a pictorial method of writing a system of algebraic equations so as to indicate the interdependencies of the variables. As another example, consider the following set of simultaneous algebraic equations:

$$a_{11}x_1 + a_{12}x_2 + r_1 = x_1 \tag{2.89}$$

$$a_{21}x_1 + a_{22}x_2 + r_2 = x_2. \tag{2.90}$$

The two input variables are r_1 and r_2, and the output variables are x_1 and x_2. A signal-flow graph representing Eqs. (2.89) and (2.90) is shown in Fig. 2.27. Equations (2.89) and (2.90) may be rewritten as

$$x_1(1 - a_{11}) + x_2(-a_{12}) = r_1, \tag{2.91}$$

$$x_1(-a_{21}) + x_2(1 - a_{22}) = r_2. \tag{2.92}$$

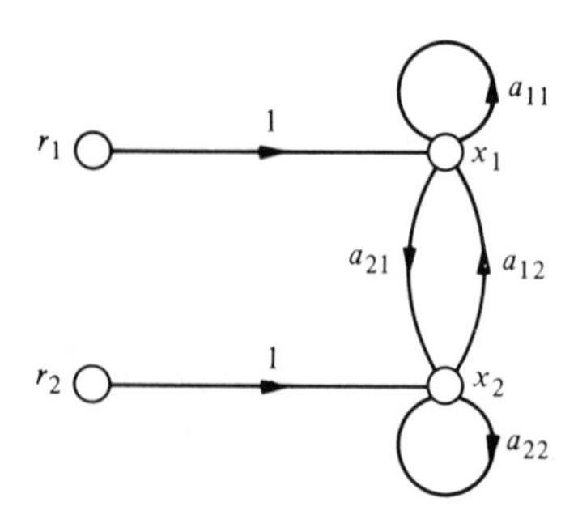

Figure 2.27. Signal-flow graph of two algebraic equations.

The simultaneous solution of Eqs. (2.91) and (2.92) using Cramer's rule results in the solutions

$$x_1 = \frac{(1 - a_{22})r_1 + a_{12}r_2}{(1 - a_{11})(1 - a_{22}) - a_{12}a_{21}} = \frac{(1 - a_{22})}{\Delta} r_1 + \frac{a_{12}}{\Delta} r_2, \tag{2.93}$$

$$x_2 = \frac{(1 - a_{11})r_2 + a_{21}r_1}{(1 - a_{11})(1 - a_{22}) - a_{12}a_{21}} = \frac{(1 - a_{11})}{\Delta} r_2 + \frac{a_{21}}{\Delta} r_1. \tag{2.94}$$

The denominator of the solution is the determinant Δ of the set of equations and is rewritten as

$$\begin{aligned} \Delta &= (1 - a_{11})(1 - a_{22}) - a_{12}a_{21} \\ &= 1 - a_{11} - a_{22} + a_{11}a_{22} - a_{12}a_{21}. \end{aligned} \tag{2.95}$$

In this case, the denominator is equal to 1 minus each self-loop a_{11}, a_{22}, and $a_{12}a_{21}$, plus the product of the two nontouching loops a_{11} and a_{22}.

The numerator for x_1 with the input r_1 is 1 times $(1 - a_{22})$, which is the value of Δ not touching the path 1 from r_1 to x_1. Therefore the numerator from r_2 to x_1 is simply a_{12} because the path through a_{12} touches all the loops. The numerator for x_2 is symmetrical to that of x_1.

In general, the linear dependence T_{ij} between the independent variable x_i (often called the input variable) and a dependent variable x_j is given by Mason's *loop rule* [8, 11]:

$$T_{ij} = \frac{\Sigma_k P_{ij_k} \Delta_{ij_k}}{\Delta}, \tag{2.96}$$

where P_{ij_k} = kth path from variable x_i to variable x_j,
Δ = determinant of the graph,
Δ_{ij_k} = cofactor of the path P_{ij_k},

and the summation is taken over all possible k paths from x_i to x_j. The cofactor Δ_{ij_k} is the determinant with the loops touching the kth path removed. The determinant Δ is

$$\Delta = 1 - \sum_{n=1}^{N} L_n + \sum_{m=1, q=1}^{M,Q} L_m L_q - \sum L_r L_s L_t + \cdots, \tag{2.97}$$

where L_q equals the value of the qth loop transmittance. Therefore the rule for evaluating Δ in terms of loops $L_1, L_2, L_3, \ldots, L_N$ is

$\Delta = 1 -$ (sum of all different loop gains)
$+$ (sum of the gain products of all combinations of 2 nontouching loops)
$-$ (sum of the gain products of all combinations of 3 nontouching loops)
$+ \cdots$.

Two loops are nontouching if they do not have any common nodes.

The gain formula is often used to relate the output variable $C(s)$ to the input variable $R(s)$ and is given in somewhat simplified form as

$$T = \frac{\Sigma_k P_k \Delta_k}{\Delta}, \tag{2.98}$$

where $T(s) = C(s)/R(s)$. The path gain or transmittance P_k (or P_{ij_k}) is defined as the continuous succession of branches that are traversed in the direction of the arrows and with no node encountered more than once. A loop is defined as a closed path in which no node is encountered more than once per traversal.

Several examples will illustrate the utility and ease of this method. Although the gain equation (2.96) appears to be formidable, one must remember that it represents a summation process, not a complicated solution process.

■ Example 2.6: Transfer Function of Interacting System

A two-path signal-flow graph is shown in Fig. 2.28. An example of a control system with multiple signal paths is the multi-legged robot shown in Fig. 2.29. The paths connecting the input $R(s)$ and output $C(s)$ are

$$\text{path 1: } P_1 = G_1G_2G_3G_4$$

and

$$\text{path 2: } P_2 = G_5G_6G_7G_8.$$

There are four self-loops:

$$L_1 = G_2H_2, \quad L_2 = H_3G_3, \quad L_3 = G_6H_6, \quad L_4 = G_7H_7.$$

Loops L_1 and L_2 do not touch L_3 and L_4. Therefore, the determinant is

$$\Delta = 1 - (L_1 + L_2 + L_3 + L_4) + (L_1L_3 + L_1L_4 + L_2L_3 + L_2L_4). \tag{2.99}$$

The cofactor of the determinant along path 1 is evaluated by removing the loops that touch path 1 from Δ. Therefore we have

$$L_1 = L_2 = 0 \quad \text{and} \quad \Delta_1 = 1 - (L_3 + L_4).$$

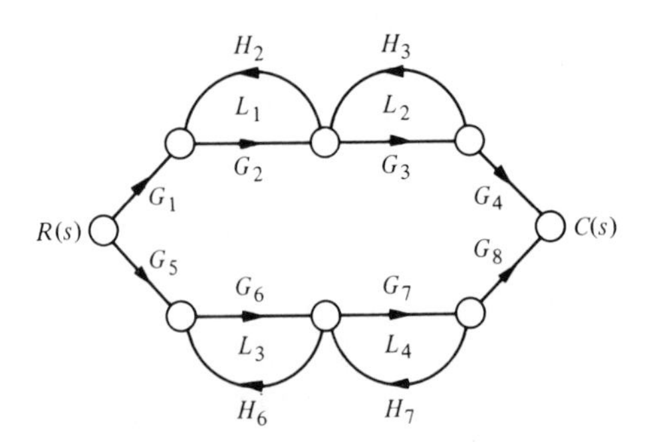

Figure 2.28. Two-path interacting system.

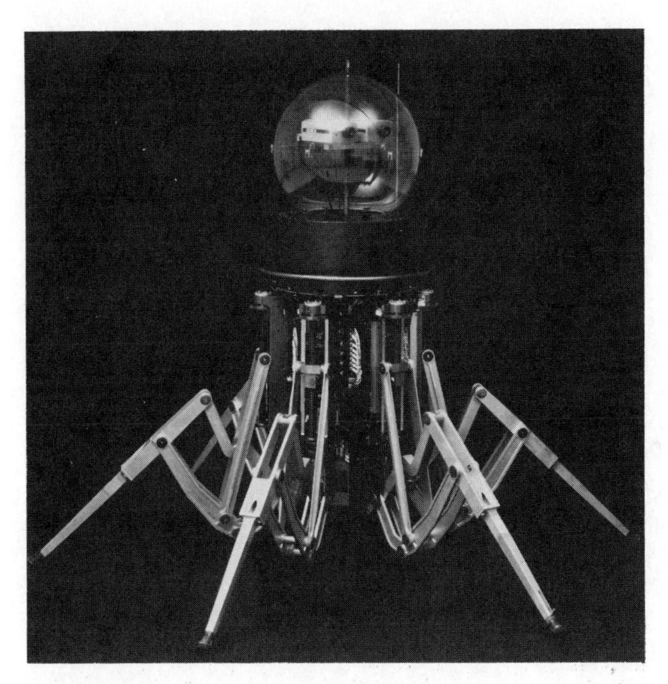

Figure 2.29. This six-legged robot can climb into a truck or lift and maneuver at the same time. It walks over rough terrain and is a forerunner of robots for many uses. (Courtesy of Odetics, Inc., Anaheim, Calif.)

Similarly, the cofactor for path 2 is

$$\Delta_2 = 1 - (L_1 + L_2).$$

Therefore the transfer function of the system is

$$\frac{C(s)}{R(s)} = T(s) = \frac{P_1\Delta_1 + P_2\Delta_2}{\Delta}$$
$$= \frac{G_1G_2G_3G_4(1 - L_3 - L_4) + G_5G_6G_7G_8(1 - L_1 - L_2)}{1 - L_1 - L_2 - L_3 - L_4 + L_1L_3 + L_1L_4 + L_2L_3 + L_2L_4}. \tag{2.100}$$

The signal-flow graph gain formula provides a reasonably straightforward approach for the evaluation of complicated systems. In order to compare the method with block diagram reduction, which is really not much more difficult, let us reconsider the complex system of Example 2.4.

■ Example 2.7: Transfer Function of Multiple Loop System

A multiple-loop feedback system is shown in Fig. 2.23 in block diagram form. There is no reason to redraw the diagram in signal-flow graph form, and so we shall proceed as usual by using the signal-flow gain formula, Eq. (2.98). There is one forward path $P_1 = G_1G_2G_3G_4$. The feedback loops are

$$L_1 = -G_2G_3H_2, \quad L_2 = G_3G_4H_1, \quad L_3 = -G_1G_2G_3G_4H_3. \tag{2.101}$$

All the loops have common nodes and therefore are all touching. Furthermore, the path P_1 touches all the loops and hence $\Delta_1 = 1$. Thus the closed-loop transfer function is

$$\begin{aligned} T(s) = \frac{C(s)}{R(s)} &= \frac{P_1\Delta_1}{1 - L_1 - L_2 - L_3} \\ &= \frac{G_1G_2G_3G_4}{1 + G_2G_3H_2 - G_3G_4H_1 + G_1G_2G_3G_4H_3}. \end{aligned} \tag{2.102}$$

■ Example 2.8: Transfer Function of Complex System

Finally, we shall consider a reasonably complex system that would be difficult to reduce by block diagram techniques. A system with several feedback loops and feedforward paths is shown in Fig. 2.30. The forward paths are

$$P_1 = G_1G_2G_3G_4G_5G_6, \quad P_2 = G_1G_2G_7G_6, \quad P_3 = G_1G_2G_3G_4G_8.$$

The feedback loops are

$$L_1 = -G_2G_3G_4G_5H_2, \quad L_2 = -G_5G_6H_1, \quad L_3 = -G_8H_1, \quad L_4 = -G_7H_2G_2,$$

$$L_5 = -G_4H_4, \quad L_6 = -G_1G_2G_3G_4G_5G_6H_3, \quad L_7 = -G_1G_2G_7G_6H_3,$$

$$L_8 = -G_1G_2G_3G_4G_8H_3.$$

Loop L_5 does not touch loop L_4 and loop L_7; loop L_3 does not touch loop L_4; and all other loops touch. Therefore the determinant is

$$\Delta = 1 - (L_1 + L_2 + L_3 + L_4 + L_5 + L_6 + L_7 + L_8) + (L_5L_7 + L_5L_4 + L_3L_4). \tag{2.103}$$

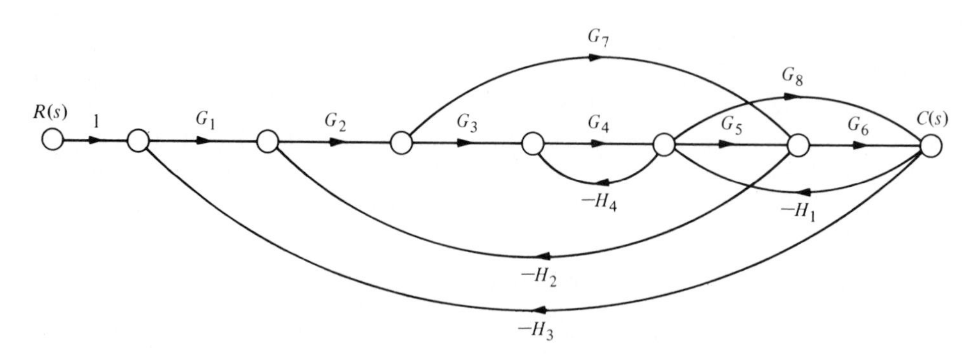

Figure 2.30. Multiple-loop system.

The cofactors are

$$\Delta_1 = \Delta_3 = 1 \quad \text{and} \quad \Delta_2 = 1 - L_5 = 1 + G_4H_4.$$

Finally, the transfer function is then

$$T(s) = \frac{C(s)}{R(s)} = \frac{P_1 + P_2\Delta_2 + P_3}{\Delta} \tag{2.104}$$

Signal-flow graphs and the signal-flow gain formula may be used profitably for the analysis of feedback control systems, analog computer diagrams, electronic amplifier circuits, statistical systems, mechanical systems, among many other examples.

2.8 Computer Analysis of Control Systems

When a model is available for a component or system, a computer can be utilized to investigate the behavior of the system. A computer model of a system in a mathematical form suitable for demonstrating the system's behavior may be utilized to investigate various designs of a planned system without actually building the system itself. A *computer simulation* uses a model and the actual conditions of the system being modeled and actual input commands to which the system will be subjected.

A system may be simulated using analog or digital computers. An electronic analog computer is used to establish a model of a system, using the analogy between the voltage of the electronic amplifier and the variable of the system being modeled [1, 7, 9, 14, 16]. An electronic analog computer usually has available the mathematical functions of integration, multiplication by a constant, multiplication of two variables, and the summation of several variables, among others. These functions are often sufficient to develop a simulation model of a system. The analog simulation model of a second-order system is shown in Fig. 2.31 for the system with negative unity feedback and a plant transfer function

$$G(s) = \frac{C(s)}{E(s)} = \frac{K}{s(s + p)}. \tag{2.105}$$

The differential equation necessary to yield the simulation of the plant is obtained by cross-multiplying in Eq. (2.105) to yield

$$s(s + p)C(s) = KE(s). \tag{2.106}$$

Since $sC(s)$ is the derivative of $c(t)$ in the s-domain we have:

$$\frac{d^2c(t)}{dt^2} = -p\frac{dc(t)}{dt} + Ke(t). \tag{2.107}$$

This equation is represented on the analog diagram by the integration in the center of the diagram with the output dc/dt. This analog simulation arrangement

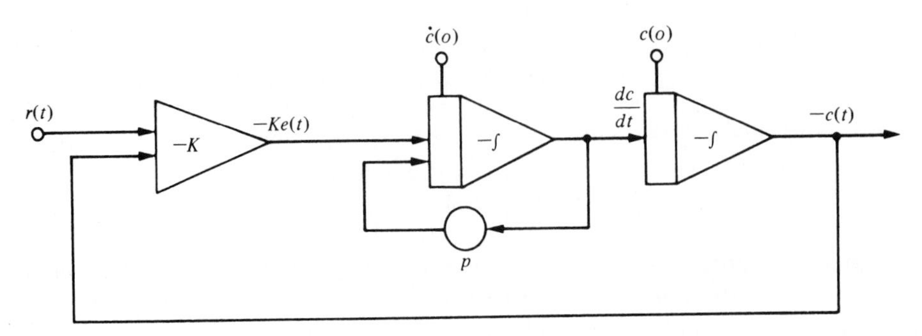

Figure 2.31. An analog computer simulation model for a second-order system with negative unity feedback.

may be realized physically on an electronic analog computer in order to yield an output recording of the simulated response of the system. The parameters K and p may be varied in order to ascertain the effect of the parameter change.

Simulation models may also be utilized with digital computers. A computer simulation may be developed in common computer language such as Pascal or BASIC or in a language specifically developed for simulation [15, 16]. Two widely used languages for the simulation of systems operating in continuous time are CSMP and ACSL. CSMP is an acronym for Continuous System Modeling Program and is an IBM product available for most IBM computers. A graphic feature is also available with CSMP. CSMP provides up to 42 functions such as integration and random number generation and including many nonlinear functions. A portion of a CSMP computer program for simulating the second-order control system of Eq. (2.105) is shown in Fig. 2.32. In this simulation, COUT = $c(t)$ and CDOT = dc/dt. The function REALPL simulates the function with one real pole, p. This simulation will yield an output printed at spacings of one-half second over a 20-second interval.

Digital simulation software like CSMP has been available for mainframe computers for over a decade. These programs run on either mainframe computers or minicomputers and often the only graphical output available is in the form of

```
DYNAMIC
                ERROR = RIN − COUT
                CONTL = GAIN*ERROR
                CDOT = REALPL (0.0, P, CONTL)
                COUT = INTGRL (0,0, CDOT)
PARAMETER P = 1.5, RIN = 1.0
PARAMETER GAIN = (1.0, 5.0, 10.0)
      TIMER FINTIM = 20.0, OUTDEL = 0.5
      PRINT COUT, CDOT
```

Figure 2.32. A portion of a digital computer program written in the simulation language CSMP is shown for a second-order unity feedback system.

a line printer plot. All of these programs have the disadvantage of being available on computer systems that are not typically perceived by student and faculty as being user friendly. Programs for use on more user-friendly systems have recently become available. Several simulation programs are now available for a personal computer similar to those available on mainframe computers [16]. These programs are written to be interactive and are considerably more user friendly than the batch-mode mainframe simulators in use today. They offer user help, error detection, and medium resolution graphics for presentation of system responses.

Assuming that a model and the simulation are reliably accurate, the advantages of computer simulation are [42]:

1. System performance can be observed under all conceivable conditions.
2. Results of field-system performance can be extrapolated with a simulation model for prediction purposes.
3. Decisions concerning future systems presently in a conceptual stage can be examined.
4. Trials of systems under test can be accomplished in a much reduced period of time.
5. Simulation results can be obtained at lower cost than real experimentation.

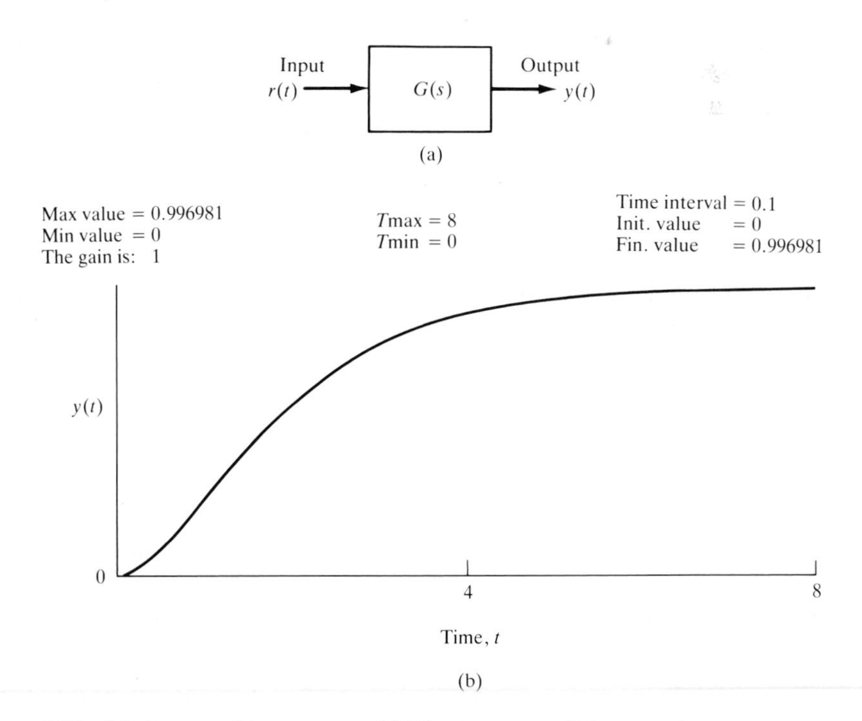

Figure 2.33. (a) An open-loop system. (b) The response of the open-loop system to a unit step input when $K = 1$. The response asympotically approaches a final value of 1.0.

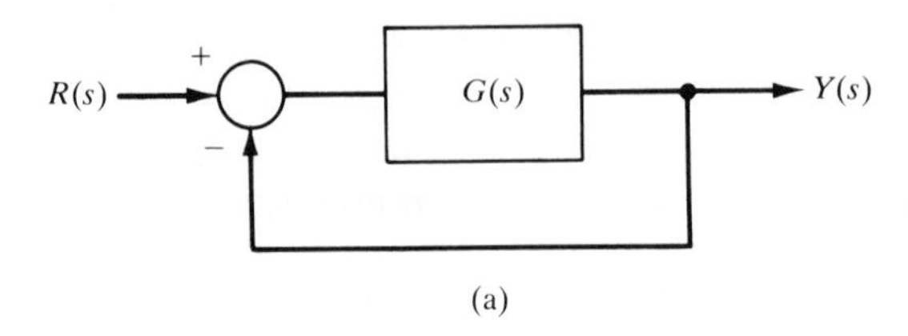

(a)

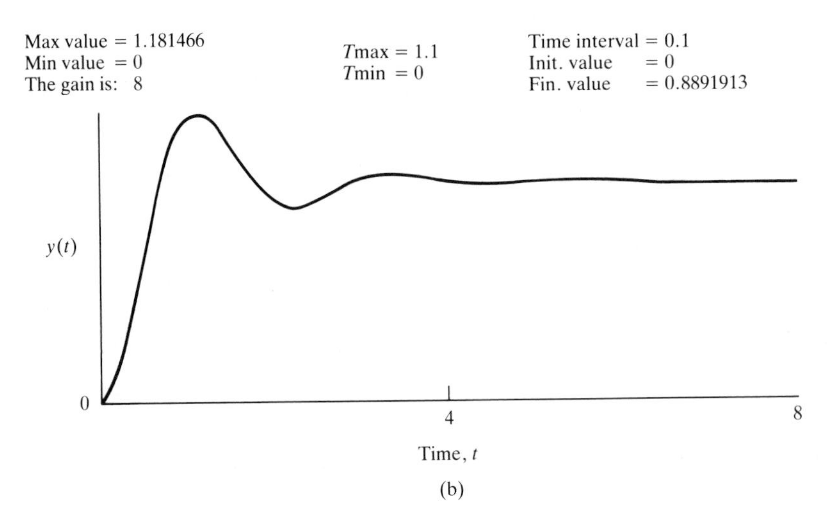

(b)

Figure 2.34. (a) A closed-loop system. (b) The response of the closed-loop system to a unit step input when $K = 8$.

6. Study of hypothetical situations can be achieved even when the hypothetical situation would be unrealizable in actual life at the present time.
7. Computer modeling and simulation is often the only feasible or safe technique to analyze and evaluate a system.

The Control System Design Program (CSDP) permits the analyst to determine the response of control systems for various input signals using the IBM PC or a compatible computer. For example, the response of the open-loop control system shown in Fig. 2.33(a) is shown in Fig. 2.33(b) for a unit step input when

$$G(s) = \frac{K}{(s+1)^2} \tag{2.108}$$

and $K = 1$.

Utilizing CSDP one can proceed to readily obtain the response of the closed-loop system shown in Fig. 2.34(a) when $G(s)$ is that of Eq. (2.108) and $K = 8$. The response of this closed-loop system to a unit step input is shown in Fig. 2.34(b).

2.9 Design Examples

■ Example 2.9 Electric traction motor control

A majority of modern trains and local transit vehicles utilize electric traction motors. The electric motor drive for a railway vehicle is shown in block diagram form in Fig. 2.35(a) incorporating the necessary control of the velocity of the vehi-

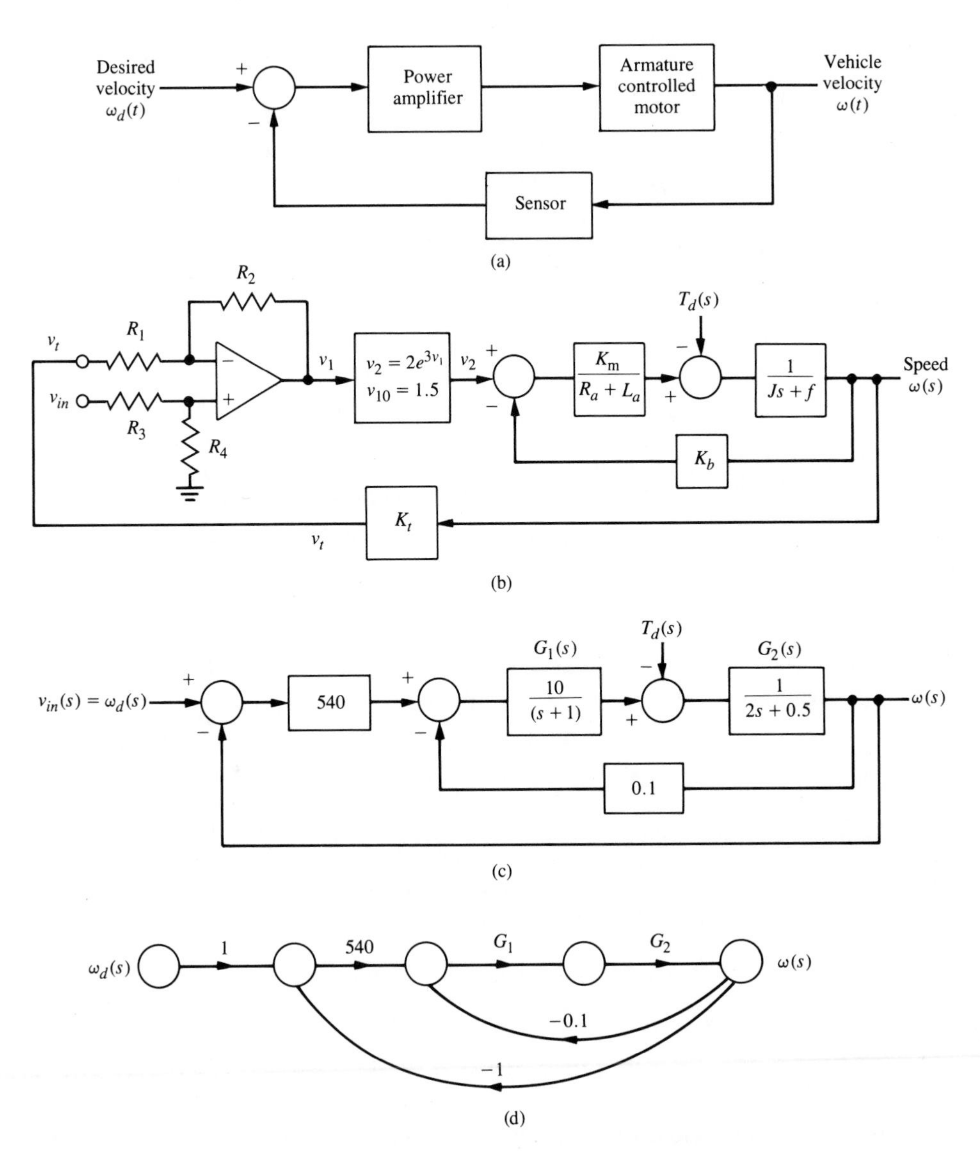

Figure 2.35. Speed control of an electric traction motor.

cle. The goal of the design is to obtain a system model and the closed-loop transfer function of the system, $\omega(s)/\omega_d(s)$.

The first step is to describe the transfer function of each block. We propose the use of a tachometer to generate a voltage proportional to velocity and to connect that voltage, v_t, to one input of a difference amplifier, as shown in Fig. 2.35(b). The power amplifier is nonlinear and can be approximately represented by $v_2 = 2e^{3v_1}$, an exponential function with a normal operating point, $v_{10} = 1.5V$. Using the technique in Section 2.3, we then obtain a linear model

$$\begin{aligned} v_2 &= 2\left[\frac{dg(v_1)}{dv_1}\bigg|_{v_{10}}\right]\Delta v_1 \\ &= 2[3e^{3v}_{10}]\,\Delta v_1 \\ &= 2[270]\,\Delta v_1 \\ &= 540\,\Delta v_1. \end{aligned} \tag{2.109}$$

Then, discarding the delta notation and writing the transfer function, we have

$$V_2(s) = 540V_1(s).$$

The transfer function of the differential amplifier is

$$v_1 = \frac{1 + R_2/R_1}{1 + R_3/R_4} v_{\text{in}} - \frac{R_2}{R_1} v_t\,. \tag{2.110}$$

We wish to obtain an input control that sets $\omega_d(t) = v_{\text{in}}$ where the units of ω_d are rad/s and the units of v_{in} are volts. Then, when $v_{\text{in}} = 10\ V$, the steady-state speed is $\omega = 10$ rad/s. We note that $v_t = K_t\omega_d$, in steady state and we expect, in balance, the steady-state output, v_1, to be

$$v_1 = \frac{1 + R_2/R_1}{1 + R_3/R_4} v_{\text{in}} - \left(\frac{R_2}{R_1}\right) K_t(v_{\text{in}}). \tag{2.111}$$

When the system is in balance $v_1 = 0$ and when $K_t = 0.1$, we have

$$\frac{1 + R_2/R_1}{1 + R_3/R_4} = \left(\frac{R_2}{R_1}\right) K_t = 1.$$

This relation can be achieved when

$$R_2/R_1 = 10 \qquad \text{and} \qquad R_3/R_4 = 10.$$

Table 2.8. Parameters of a Large dc Motor

$K_m = 10$	$J = 2$
$R_a = 1$	$f = 0.5$
$L_a = 1$	$K_b = 0.1$

The parameters of the motor and load are given in Table 2.8. The overall system is shown in Figure 2.35(c). Using Mason's signal flow rule with the signal flow diagram of Figure 2.35(d), we have

$$
\begin{aligned}
\frac{\omega(s)}{\omega_d(s)} &= \frac{540G_1(s)G_2(s)}{1 + 0.1G_1G_2 + 540G_1G_2} \\
&= \frac{540G_1G_2}{1 + 540.1G_1G_2} \\
&= \frac{5400}{(s + 1)(2s + 0.5) + 5401} \qquad (2.112) \\
&= \frac{5400}{2s^2 + 2.5s + 5401.5} \\
&= \frac{2700}{s^2 + 1.25s + 2700.75}.
\end{aligned}
$$

Since the characteristic equation is second order, we note that $\omega_n = 52$ and $\zeta = 0.012$, and we expect the response of the system to be highly oscillatory.

■ Example 2.10 Mechanical accelerometer

A mechanical accelerometer is used to measure the acceleration of a levitated test sled, as shown in Fig. 2.36. The test sled is magnetically levitated above a guide rail a small distance δ. The accelerometer provides a measurement of the accel-

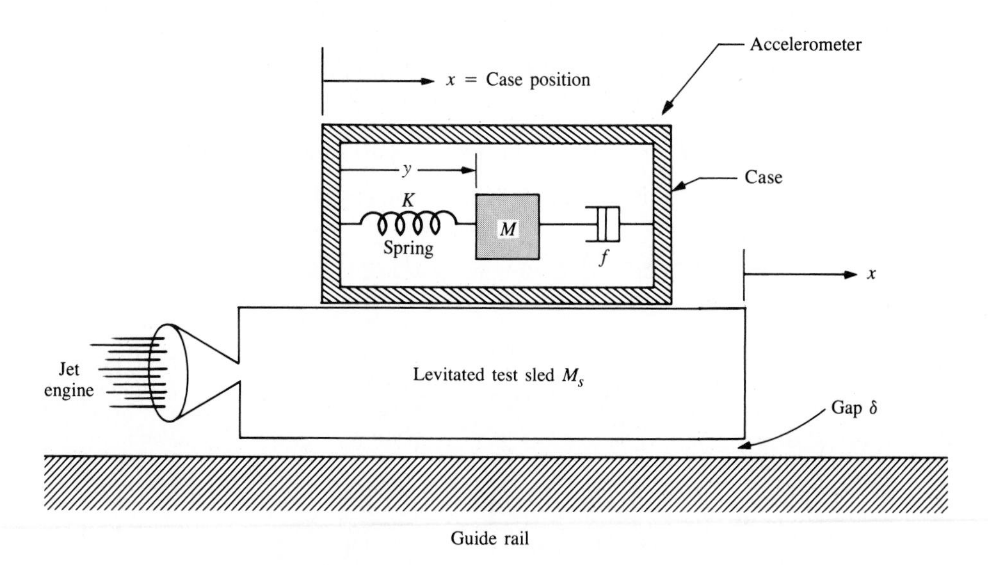

Figure 2.36. An accelerometer mounted on a jet engine test sled.

eration $a(t)$ of the sled since the position y of the mass M, with respect to the accelerometer case, is proportional to the acceleration of the case (and the sled). The goal is to design an accelerometer with an appropriate dynamic responsiveness. We wish to design an accelerometer with an acceptable time for the desired measurement characteristic, $y(t) = qa(t)$, to be attained (q is a constant).

The sum of the forces acting on the mass is

$$-f\frac{dy}{dt} - Ky = M\frac{d^2}{dt^2}(y - x)$$

or

$$M\frac{d^2y}{dt^2} + f\frac{dy}{dt} + Ky = M\frac{d^2x}{dt^2}. \tag{2.113}$$

Since

$$M_s\frac{d^2x}{dt^2} = F(t),$$

the engine force, we have

$$M\ddot{y} + f\dot{y} + Ky = \frac{M}{M_s}F(t)$$

or

$$\ddot{y} + \frac{f}{M}\dot{y} + \frac{K}{M}y = \frac{F(t)}{M_s}. \tag{2.114}$$

We select the coefficients where $f/M = 3$, $K/M = 2$, $F(t)/M_s = Q(t)$, and we consider the initial conditions $y(0) = -1$ and $\dot{y}(0) = 2$. We then obtain the Laplace transform equation, when the force and thus $Q(t)$ is a step function, as follows:

$$(s^2Y(s) - sy(0) - \dot{y}(0)) + 3\,(sY(s) - y(0)) + 2Y(s) = Q(s). \tag{2.115}$$

Since $Q(s) = P/s$, where P is the magnitude of the step function, we obtain

$$(s^2Y(s) + s - 2) + 3(sY(s) + 1) + 2Y(s) = \frac{P}{s}$$

or

$$(s^2 + 3s + 2)Y(s) = \frac{-(s^2 + s - P)}{s}. \tag{2.116}$$

Thus the output transform is

$$Y(s) = \frac{-(s^2 + s - P)}{s(s^2 + 3s + 2)} = \frac{-(s^2 + s - P)}{s(s + 1)(s + 2)}. \tag{2.117}$$

Expanding in partial fraction form,

$$Y(s) = \frac{k_1}{s} + \frac{k_2}{s+1} + \frac{k_3}{s+2}. \quad (2.118)$$

We then have

$$k_1 = \left.\frac{-(s^2 + s - P)}{(s+1)(s+2)}\right|_{s=0} = \frac{P}{2}. \quad (2.119)$$

Similarly, $k_2 = -P$ and $k_3 = \dfrac{P-2}{2}$.

Thus

$$Y(s) = \frac{P}{2s} - \frac{P}{s+1} + \frac{(P-2)}{2(s+2)}. \quad (2.220)$$

Therefore the output measurement is

$$y(t) = \tfrac{1}{2}\,[P - 2Pe^{-t} + (P-2)e^{-2t}]\,,\ t \geqslant 0$$

A plot of $y(t)$ is shown in Fig. 2.37 for $P = 3$. We can see that $y(t)$ is proportional to the magnitude of the force, and thus the acceleration, after four seconds. If this period is excessively long, we must increase the spring constant, K, and the fric-

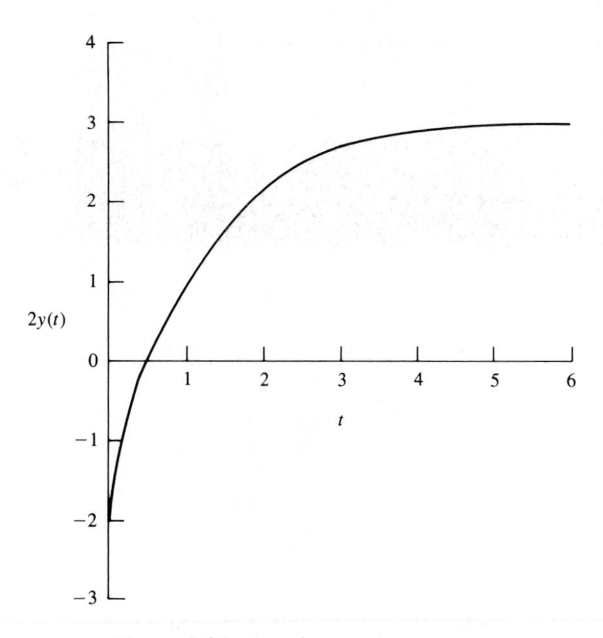

Figure 2.37. Accelerometer response.

tion, f, while reducing the mass, M. If we are able to select the components so that $f/M = 12$ and $K/M = 32$, the accelerometer will attain the proportional response in one second. (It is left to the reader to show this.)

2.10 Summary

In this chapter, we have been concerned with quantitative mathematical models of control components and systems. The differential equations describing the dynamic performance of physical systems were utilized to construct a mathematical model. The physical systems under consideration included mechanical, electrical, fluid, and thermodynamic systems. A linear approximation using a Taylor series expansion about the operating point was utilized to obtain a small-signal linear approximation for nonlinear control components. Then, with the approximation of a linear system, one may utilize the Laplace transformation and its related input–output relationship, the transfer function. The transfer function approach to linear systems allows the analyst to determine the response of the system to various input signals in terms of the location of the poles and zeros of the transfer function. Using transfer function notations, block diagram models of systems of interconnected components were developed. The block functions were

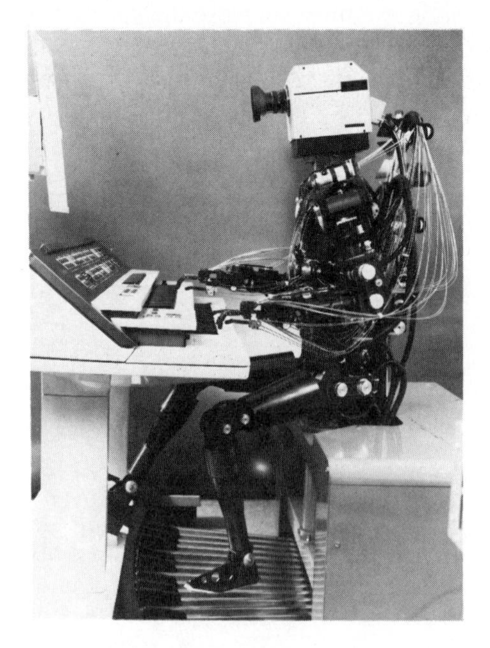

Figure 2.38. A keyboard instrument playing robot with the capability to see, hear, and speak. It can operate an electronic organ with its hands and feet. Equipped with arms and legs, it has a total of 50 joints. It has fingers capable of striking keys at a rate of 15 strokes per second. (Courtesy of Sumitomo Electric.)

obtained. Additionally, an alternative use of transfer function models in signal-flow graph form was investigated. The signal-flow graph gain formula was investigated. The signal-flow graph gain formula was found to be useful for obtaining the relationship between system variables in a complex feedback system. The advantage of the signal-flow graph method was the availability of Masons's flow graph gain formula, which provides the relation between system variables without requiring any reduction or manipulation of the flow graph. Thus in Chapter 2 we have obtained a useful mathematical model for feedback control systems by developing the concept of a transfer function of a linear system and the relationship among system variables using block diagram and signal-flow graph models. Finally, we considered the utility of the computer simulation of linear and nonlinear systems in order to determine the response of a system for several conditions of the system parameters and the environment. An example of a complex system which requires all the approaches discussed in Chapter 2 is shown in Fig. 2.38.

Exercises

(Exercises are straightforward applications of the concepts of the chapter.)

E2.1. A unity, negative feedback system has a nonlinear function $c = f(e) = e^2$ as shown in the Fig. E2.1. For an input r in the range of zero to four calculate and plot the open-loop and closed-loop output versus input and show that the feedback system results in a more linear relationship.

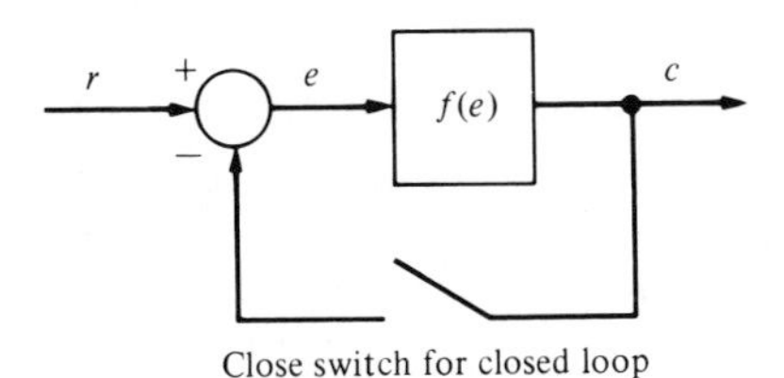

Figure E2.1. Open and closed loop.

E2.2. A thermistor has a response to temperature represented by

$$R = R_o e^{-0.1T},$$

where $R_o = 10{,}000\ \Omega$, R = resistance, and T = temperature in degrees Celsius. Find the linear model for the thermistor operating at $T = 20°\text{C}$ and for a small range of variation of temperature.

Answer: $\Delta R = -135\ \Delta T$

E2.3. The force versus displacement for a spring is shown in Fig. E2.3 for the spring-mass-damper system of Fig. 2.1. Graphically find the spring constant for the equilibrium point of $y = 0.5$ cm and a range of operation of ± 1.5 cm.

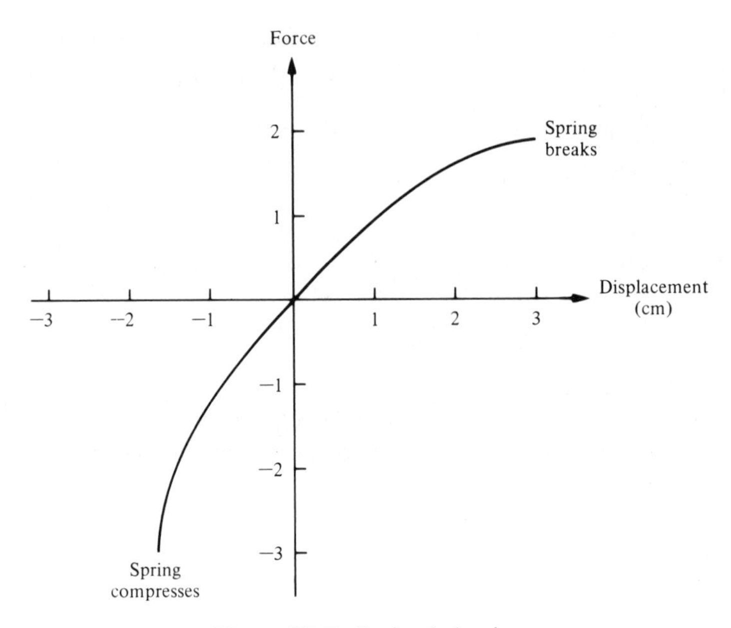

Figure E2.3. Spring behavior.

E2.4. A laser jet printer uses a laser beam to print copy rapidly for a computer [18, 19]. The laser is positioned by a control input, $r(t)$, so that we have

$$Y(s) = \frac{500(s + 100)}{s^2 + 60s + 500} R(s).$$

(a) If $r(t)$ is a unit step input, find the output $y(t)$. (b) What is the final value of $y(t)$?

E2.5. A switching circuit is used to convert one level of DC voltage to an output DC voltage. The switching circuit is shown in Fig. E2.5(a), on next page [27]. The filter circuit to filter out the high frequencies is shown in Fig. E2.5(b). Calculate the transfer function $V_2(s)/V_1(s)$.

E2.6. A nonlinear device is represented by the function

$$\begin{aligned} y &= f(x) \\ &= x^{1/2} \end{aligned}$$

where the operating point for the input x is $x_o = 1/2$. Determine the linear approximation in the form of Eq. 2.9.

Answer: $\Delta y = \Delta x/\sqrt{2}$

E2.7. A lamp's intensity stays constant when monitored by an optotransistor-controlled feedback loop. When the voltage drops, the lamp's output also drops, and optotransistor Q_1 draws less current. As a result, a power transistor conducts more heavily and charges a capacitor more rapidly [8]. The capacitor voltage controls the lamp voltage directly. A flow diagram of the system is shown in the Fig. E2.7, on page 88. Find the closed-loop transfer

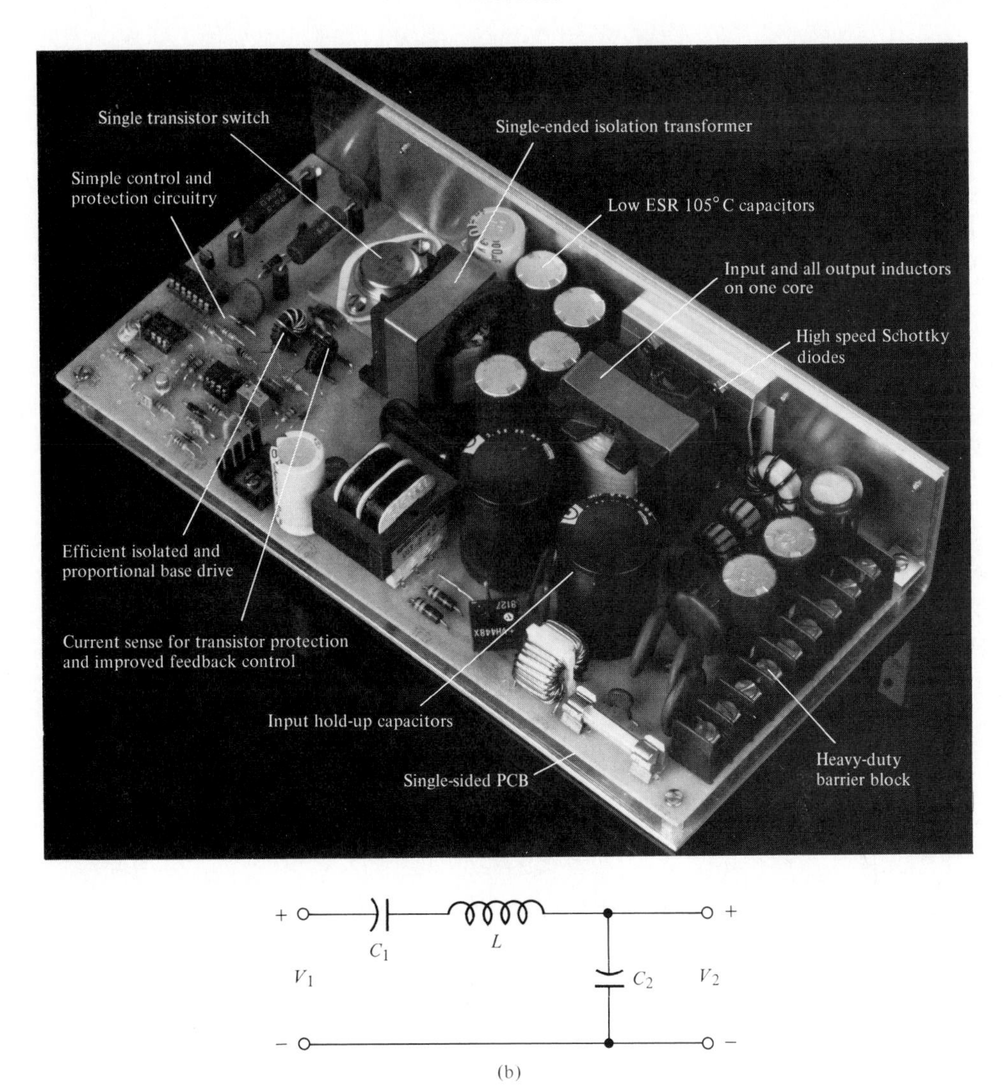

Figure E2.5. (a) Prototype circuit for switcher. (b) Filter circuit for 150 w switcher (idealized). (Photo courtesy of Ron Burian and TESLAco.)

function, $I(s)/R(s)$ where $I(s)$ is in the lamp intensity and $R(s)$ is the command or desired level of light.

E2.8. A four-wheel antilock automobile braking system uses electronic feedback to control automatically the brake force on each wheel [24]. A simplified flow graph of a brake control system is shown in Fig. E2.8, on next page where $F_f(s)$ and $F_R(s)$ are the braking force of the front and rear wheels, respectively, and $R(s)$ is the desired automobile response on an icy road. Find $F_f(s)/R(s)$.

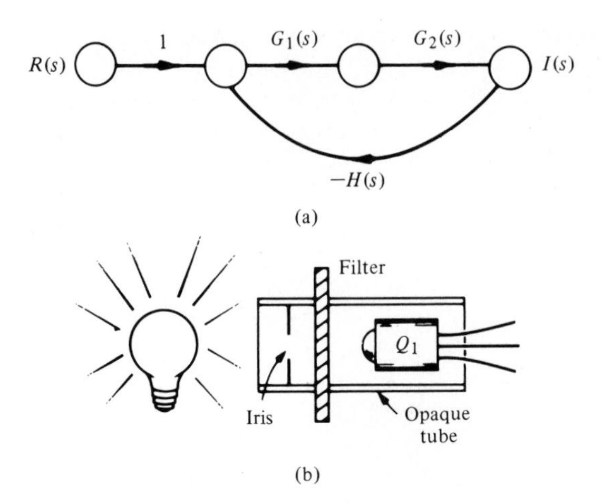

Figure E2.7. Lamp controller.

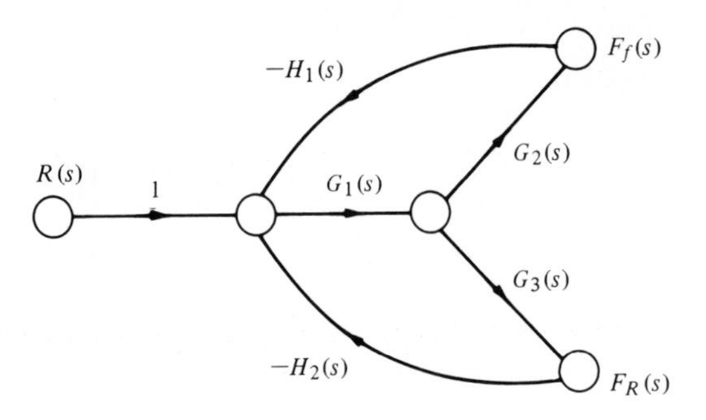

Figure E2.8. Brake control system.

E2.9. A control engineer, N. Minorsky, designed an innovative ship steering system in the 1930s for the U.S. Navy. The system is represented by the signal-flow graph shown in Fig. E2.9 where $C(s)$ is the ship's course, $R(s)$ is the desired course, and $A(s)$ is the rudder angle [41]. Find the transfer function $C(s)/R(s)$.

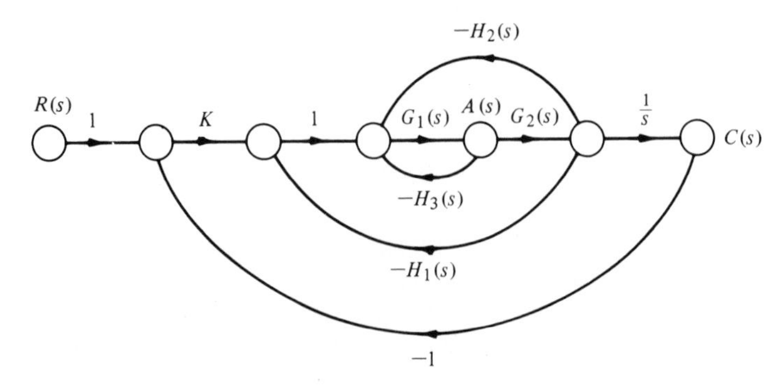

Figure E2.9. Ship steering system.

E2.10. Off-road vehicles experience many disturbance inputs as they traverse over rough roads. An active suspension system can be controlled by a sensor which looks "ahead" at the road conditions. An example of a simple suspension system that can accommodate the bumps is shown in Fig. E2.10. Find the appropriate gain K_1 so that the vehicle does not bounce when the desired deflection is $R(s) = 0$.

Answer: $K_1K_2 = 1$

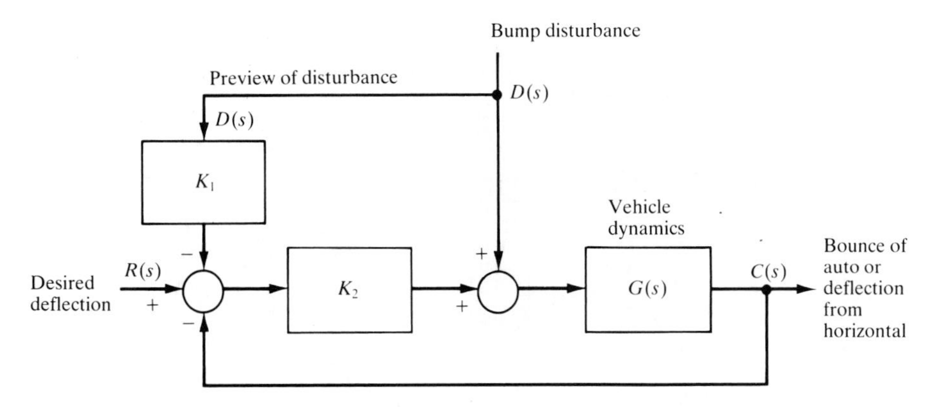

Figure E2.10. Active suspension system.

E2.11. A spring exhibits a force versus displacement characteristic as shown in Fig. E2.11. For small deviations from the operating point, find the spring constant when x_o is (a) -1.4, (b) 0, (c) 3.5.

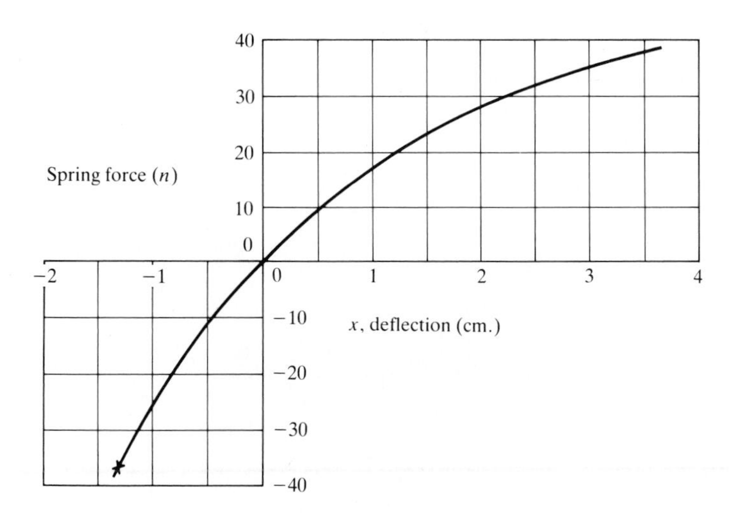

Figure E2.11. Spring characteristic.

E2.12. One of the most potentially beneficial applications of automotive control systems is the active control of the suspension system. One feedback control system uses a shock absorber consisting of a cylinder filled with a compressible fluid that provides both spring and damping forces [24]. The cylinder has a plunger activated by a gear motor, a displacement-measuring sensor, and a piston. Spring force is generated by piston displacement, which compresses the fluid. During piston displacement, the pressure imbalance across the piston is used to control damping. The plunger varies the internal volume of the cylinder. This feedback system is shown in Fig. E2.12. Develop a linear model for this device using a block diagram model.

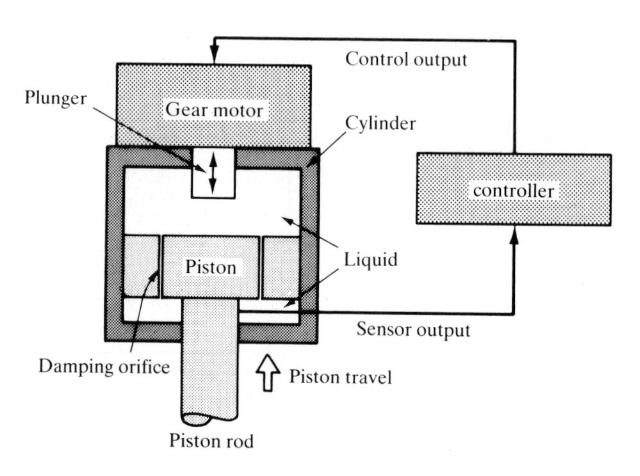

Figure E2.12. Shock absorber.

E2.13. Find the transfer function

$$\frac{C_1(s)}{R_2(s)}$$

for the multivariable system in Fig. E2.13.

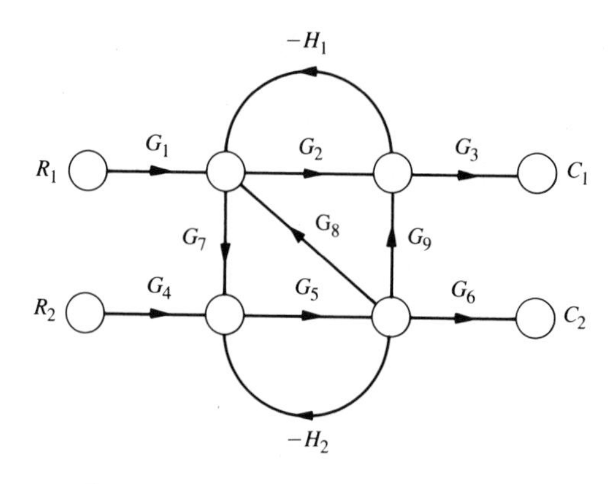

Figure E2.13. Multivariable system.

E2.14. Obtain the differential equations in terms of i_1 and i_2 for the circuit in Fig. E2.14.

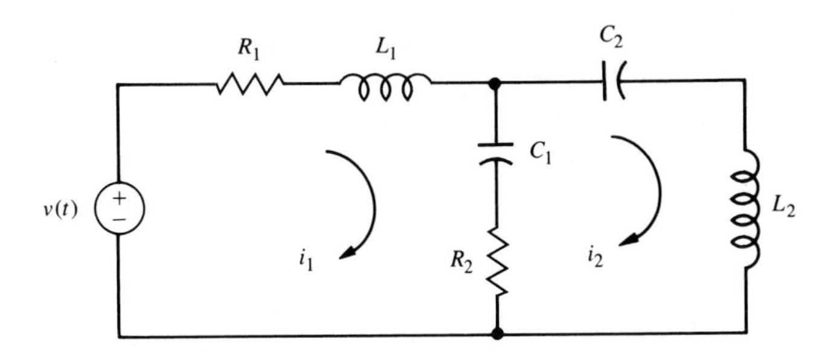

Figure E2.14. Electric circuit.

E2.15. The position control system for a spacecraft platform is governed by the following equations:

$$\frac{d^2p}{dt^2} + \frac{dp}{dt} + 4p = \theta$$

$$v_1 = r - p$$

$$\frac{d\theta}{dt} = 0.4v_2$$

$$v_2 = 7v_1.$$

The variables involved are as follows:

$r(t)$ = desired platform position

$p(t)$ = actual platform position

$v_1(t)$ = amplifier input voltage

$v_2(t)$ = amplifier output voltage

$\theta(t)$ = motor shaft position

Draw a signal flow diagram of the system, identifying the component parts and their transmittances, then determine the system transfer function $P(s)/R(s)$.

E2.16. A spring used in an auto shock absorber develops a force, f, represented by the relation

$$f = kx^3,$$

where x is the displacement of the spring. Determine a linear model for the spring when $x_0 = 1$.

E2.17. The output, y, and input, x, of a device are related by

$$y = x + 0.5x^3.$$

(a) Find the values of the output for steady-state operation at the two operating points $x_0 = 1$ and $x_0 = 2$. (b) Obtain a linearized model for both operating points and compare them.

E2.18. The transfer function of a system is

$$\frac{C(s)}{R(s)} = \frac{10(s+2)}{s^2+8s+15}$$

Determine $c(t)$ when $r(t)$ is a unit step input.

Answer: $c(t) = 1.33 + 1.67e^{-3t} - 3e^{-5t},\ t \geq 0$

Problems

(Problems require an extension of the concepts of the chapter to new situations.)

P2.1. An electric circuit is shown in Fig. P2.1. Obtain a set of simultaneous integrodifferential equations representing the network.

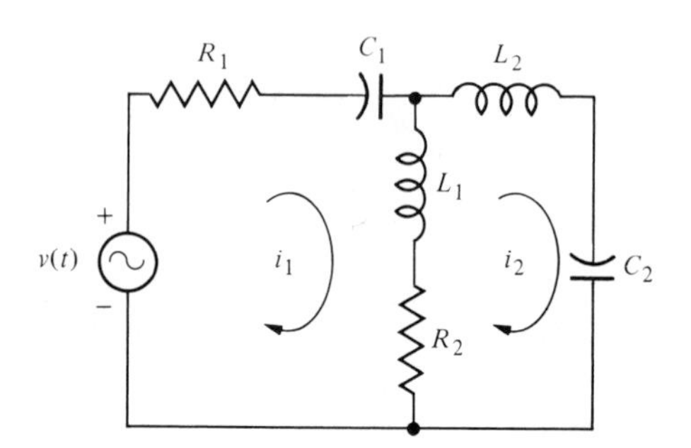

Figure P2.1. Electric circuit.

P2.2. A dynamic vibration absorber is shown in Fig. P2.2. This system is representative of many situations involving the vibration of machines containing unbalanced components. The parameters M_2 and k_{12} may be chosen so that the main mass M_1 does not vibrate when $F(t) = a \sin \omega_0 t$. (a) Obtain the differential equations describing the system. (b) Draw the analogous electrical circuit based on the force-current analogy.

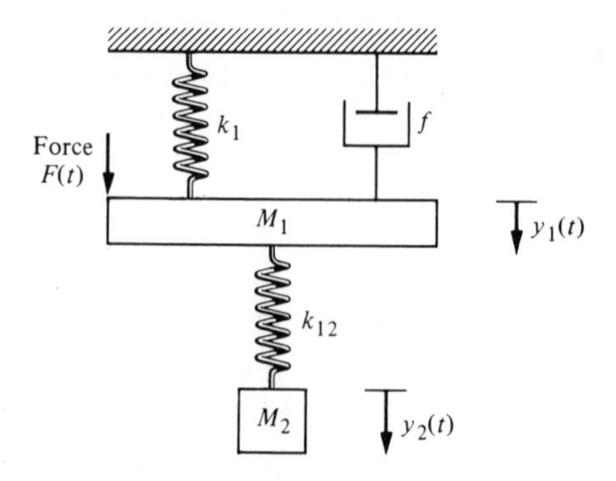

Figure P2.2. Vibration absorber.

P2.3. A coupled spring-mass system is shown in Fig. P2.3. The masses and springs are assumed to be equal. (a) Obtain the differential equations describing the system. (b) Draw an analogous electrical circuit based on the force–current analogy.

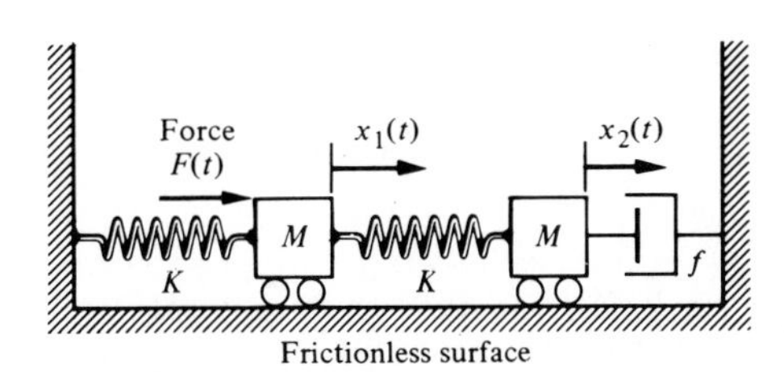

Figure P2.3. Two mass system.

P2.4. A nonlinear amplifier can be described by the following characteristic:

$$v_o(t) = \begin{cases} v_{in}^2 & v_{in} \geqslant 0 \\ -v_{in}^2 & v_{in} < 0 \end{cases}.$$

The amplifier will be operated over a range for v_{in} of ± 0.5 volts at the operating point. Describe the amplifier by a linear approximation (a) when the operating point is $v_{in} = 0$; and (b) when the operating point is $v_{in} = 1$ volt. Obtain a sketch of the nonlinear function and the approximation for each case.

P2.5. Fluid flowing through an orifice can be represented by the nonlinear equation

$$Q = K(P_1 - P_2)^{1/2},$$

where the variables are shown in Fig. P2.5 and K is a constant. (a) Determine a linear approximation for the fluid flow equation. (b) What happens to the approximation obtained in (a) if the operating point is $P_1 - P_2 = 0$?

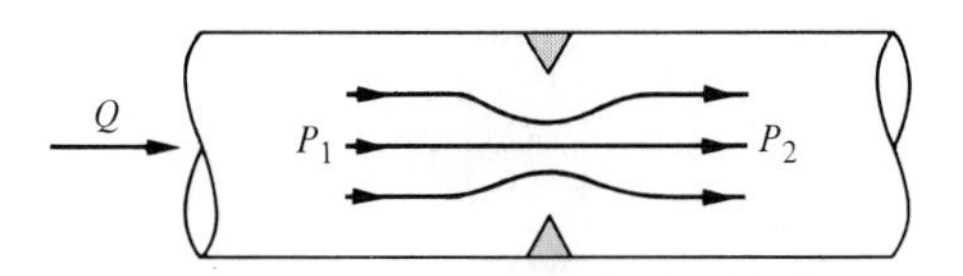

Figure P2.5. Flow through an orifice.

P2.6. Using the Laplace transformation, obtain the current $I_2(s)$ of Problem 2.1. Assume that all the initial currents are zero, the initial voltage across capacitor C_1 is zero, $v(t)$ is zero, and the initial voltage across C_2 is 10 volts.

P2.7. Obtain the transfer function of the differentiating circuit shown in Fig. P2.7.

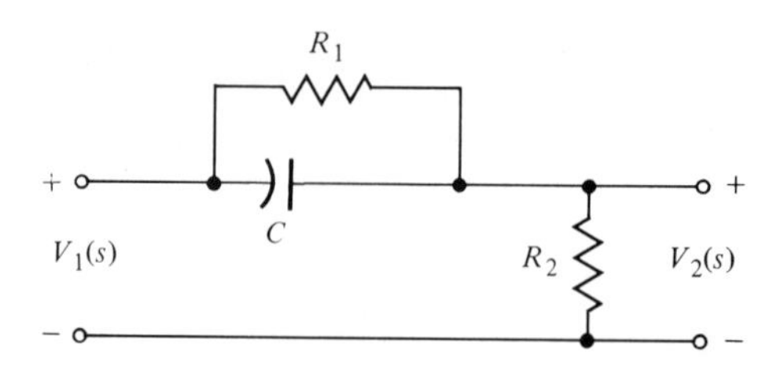

Figure P2.7.

P2.8. A bridged-T network is often used in ac control systems as a filter network. The circuit of one bridged-T network is shown in Fig. P2.8. Show that the transfer function of the network is

$$\frac{V_{out}(s)}{V_{in}(s)} = \frac{1 + 2R_1Cs + R_1R_2C^2s^2}{1 + (2R_1 + R_2)Cs + R_1R_2C^2s^2}.$$

Draw the pole-zero diagram when $R_1 = 0.5$, $R_2 = 1$, and $C = 0.5$.

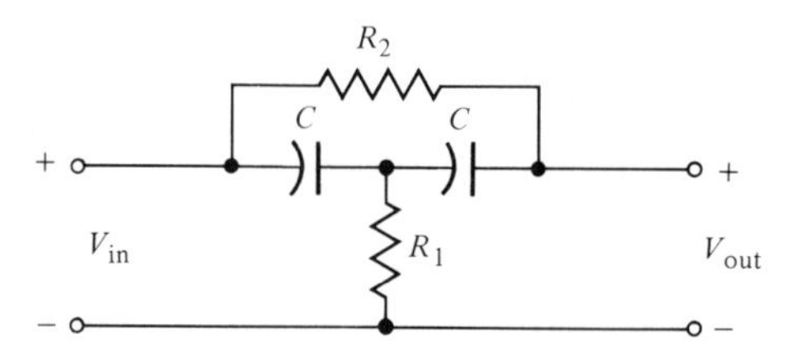

Figure P2.8. Bridged-T network.

P2.9. Determine the transfer function $X_1(s)/F(s)$ for the coupled spring-mass system of Problem 2.3. Draw the s-plane pole-zero diagram for low damping when $M = 1$, $f/K = 1$ and

$$\zeta = \frac{1}{2}\frac{f}{\sqrt{KM}} = 0.2.$$

P2.10. Determine the transfer function $Y_1(s)/F(s)$ for the vibration absorber system of Problem 2.2. Determine the necessary parameters M_2 and k_{12} so that the mass M_1 does not vibrate when $F(t) = a \sin \omega_0 t$.

P2.11. For electromechanical systems that require large power amplification, rotary amplifiers are often used. An amplidyne is a power amplifying rotary amplifier. An amplidyne and a servomotor are shown in Fig. P2.11. Obtain the transfer function $\theta(s)/V_c(s)$ and draw the block diagram of the system.

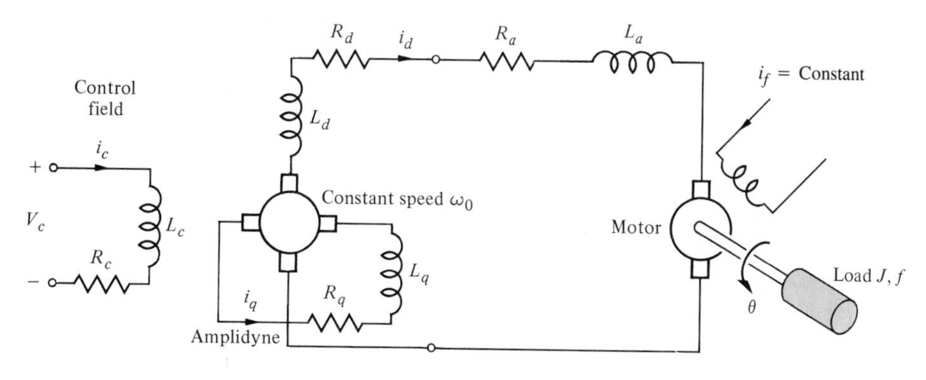

Figure P2.11. Amplidyne and motor.

P2.12. An electromechanical open-loop control system is shown in Fig. P2.12. The generator, driven at a constant speed, provides the field voltage for the motor. The motor has an inertia J_m and bearing friction f_m. Obtain the transfer function $\theta_L(s)/V_f(s)$ and draw a block diagram of the system. The generator voltage can be assumed to be proportional to the field current.

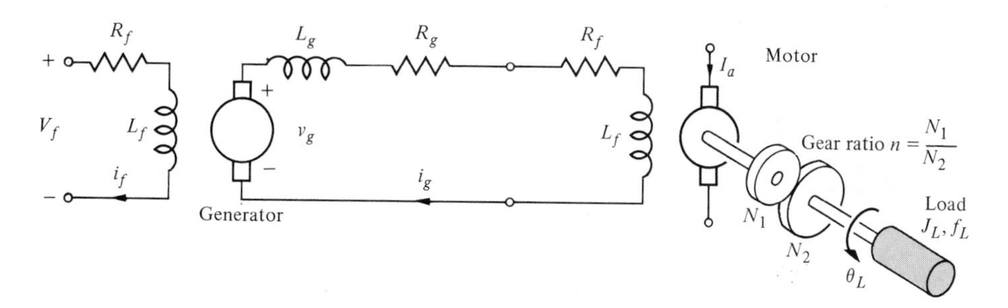

Figure P2.12. Motor and generator.

P2.13. A fluid flow system is shown in Fig. P2.13, where an incompressible fluid is flowing into an open tank. One may assume that the change in outflow ΔQ_2 is proportional to the change in head ΔH. At steady-state $Q_1 = Q_2$ and $Q_2 = kH^{1/2}$. Using a linear approximation, obtain the transfer function of the tank, $\Delta Q_2(s)/\Delta Q_1(s)$.

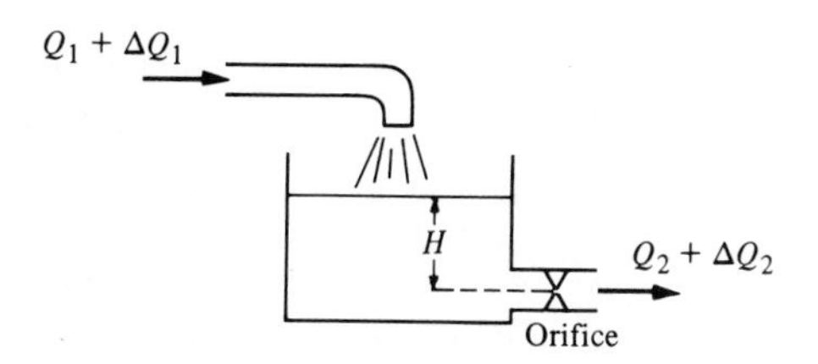

Figure P2.13. Fluid flow system.

P2.14. A rotating load is connected to a field-controlled dc electric motor through a gear system. The motor is assumed to be linear. A test results in the output load reaching a speed of 1 rad/sec within ½ sec when a constant 100 v is applied to the motor terminals. The output steady-state speed is 2 rad/sec. Determine the transfer function of the motor, $\theta(s)/V_f(s)$ in rad/V. The inductance of the field may be assumed to be negligible (see Fig. 2.16). Also, note that the application of 100 V to the motor terminals is a step input of 100 V in magnitude.

P2.15. As the complexity of interconnected power systems grows, the potential for interactive dynamic phenomena increases. Mechanical analogies can be used to explain complex interactive dynamics. Consider the weight–rubber band analogy shown in Fig. P2.15. In this weight-rubber band analogy, the weights represent the rotating mass of the turbine generators and the rubber bands are analogous to the inductance of transmission lines. Any disturbance—a "pull" on a weight—causes oscillations to be set up in all synchronous machines in the system. Determine a set of differential equations to describe this system.

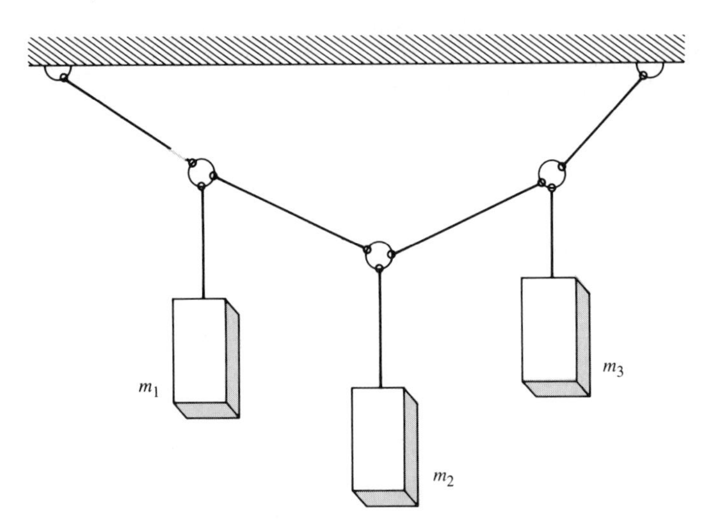

Figure P2.15. Three mass analogy.

P2.16. Obtain a signal-flow graph to represent the following set of algebraic equations where x_1 and x_2 are to be considered the dependent variables and 8 and 13 are the inputs:

$$x_1 + 2x_2 = 8,$$

$$2x_1 + 3x_2 = 13.$$

Determine the value of each dependent variable by using the gain formula. After solving for x_1 by Mason's formula, verify the solution by using Cramer's rule.

P2.17. A mechanical system is shown in Fig. P2.17, which is subjected to a known displacement $x_3(t)$ with respect to the reference. (a) Determine the two independent equations of motion. (b) Obtain the equations of motion in terms of the Laplace transform assuming that the initial conditions are zero. (c) Draw a signal-flow graph representing the system of equations. (d) Obtain the relationship between $X_1(s)$ and $X_3(s)$, $T_{13}(s)$, by using Mason's gain formula. Compare the work necessary to obtain $T_{13}(s)$ by matrix methods to that using Mason's gain formula.

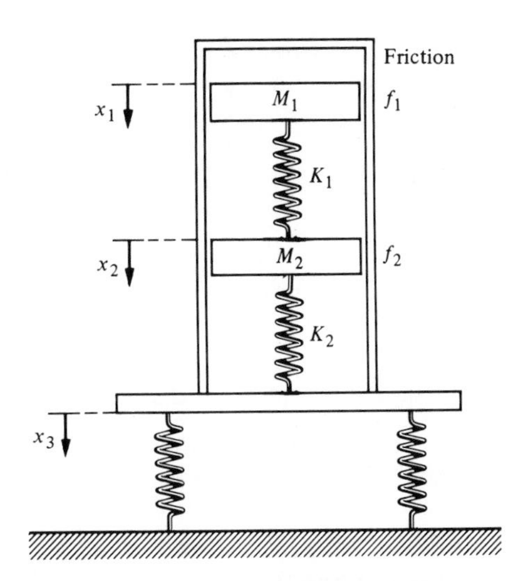

Figure P2.17. Mechanical system.

P2.18. An *LC* ladder network is shown in Fig. P2.18. One may write the equations describing the network as follows:

$$I_1 = (V_1 - V_a)Y_1, \qquad V_a = (I_1 - I_a)Z_2,$$

$$I_a = (V_a - V_2)Y_3, \qquad V_2 = I_aZ_4.$$

Construct a flow graph from the equations and determine the transfer function $V_2(s)/V_1(s)$.

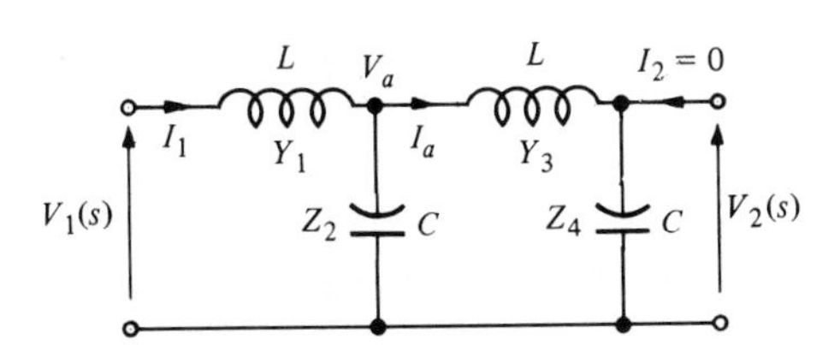

Figure P2.18. LC Ladder network.

P2.19. The basic noninverting operational amplifier is shown in Fig. P2.19(a), on next page and the signal-flow representation of the equations of the circuit is shown in Fig. P2.19(b) [8]. (a) Write the voltage equations and verify the representations in the flow graph. (b) Using the signal-flow graph, calculate the gain of the amplifier and verify that $T(s) = (R_1 + R_f)/R_1$ when $A \gg 10^3$.

P2.20. The source follower amplifier provides lower output impedence and essentially unity gain. The circuit diagram is shown in Fig. P2.20(a), and the small signal model is shown in Fig. P2.20(b). This circuit uses an FET and provides a gain of approximately unity. Assume that $R_2 \gg R_1$ for biasing purposes and that $R_g \gg R_2$. (a) Solve for the amplifier gain. (b) Solve for the gain when $g_m = 2000$ μmhos and $R_s = 10$ Kohms where $R_s = R_1 + R_2$. (c) Sketch a signal-flow diagram that represents the circuit equations.

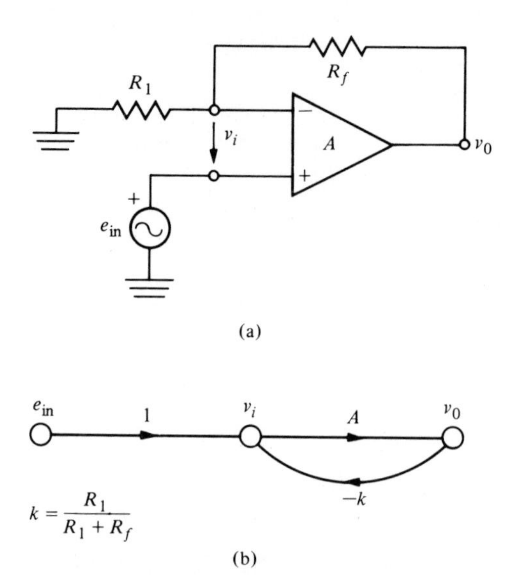

Figure P2.19. Source follower.

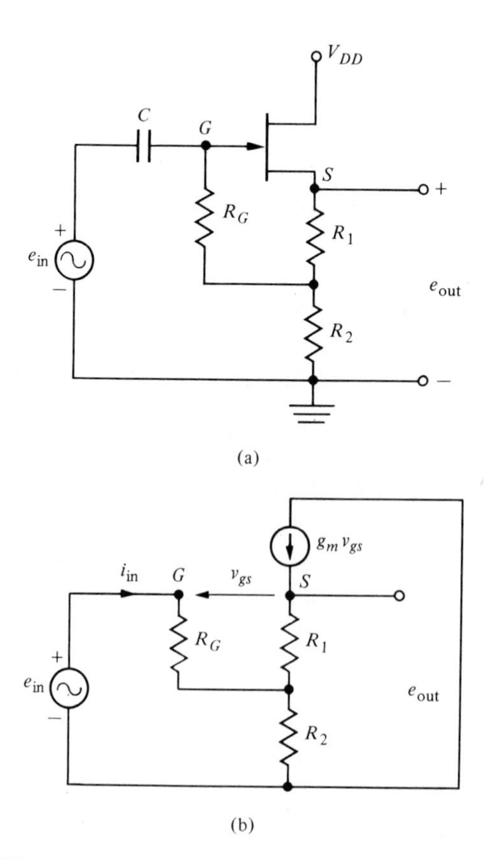

Figure P2.20. The source follower or common drain amplifier using an FET.

P2.21. A hydraulic servomechanism with mechanical feedback is shown in Fig. P2.21 [17]. The power piston has an area equal to A. When the valve is moved a small amount Δz, then the oil will flow through to the cylinder at a rate $p \cdot \Delta z$, where p is the port coefficient. The input oil pressure is assumed to be constant. (a) Determine the closed-loop signal-flow graph for this mechanical system. (b) Obtain the closed-loop transfer function $Y(s)/X(s)$.

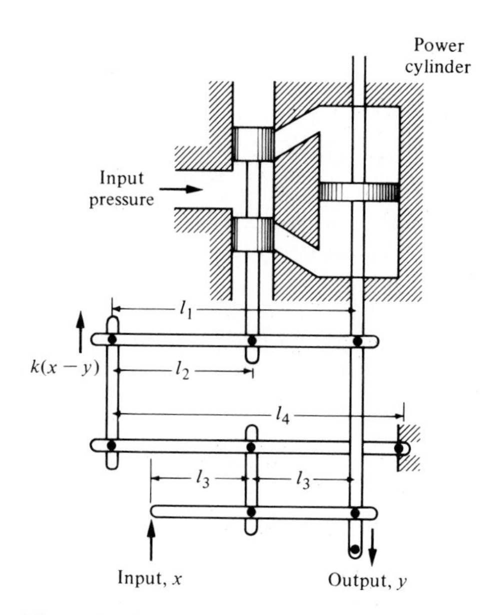

Figure P2.21. Hydraulic servomechanism.

P2.22. Figure P2.22 shows two pendulums suspended from frictionless pivots and connected at their midpoints by a spring [1]. Assume that each pendulum can be represented by a mass M at the end of a massless bar of length L. Also assume that the displacement is small and linear approximations can be used for sin θ and cos θ. The spring located in the middle of the bars is unstretched when $\theta_1 = \theta_2$. The input force is represented by $f(t)$, which influences the left-hand bar only. (a) Obtain the equations of motion and draw a signal-flow diagram for them. (b) Determine the transfer function $T(s) = \theta_1(s)/F(s)$. (c) Draw the location of the poles and zeros of $T(s)$ on the s-plane.

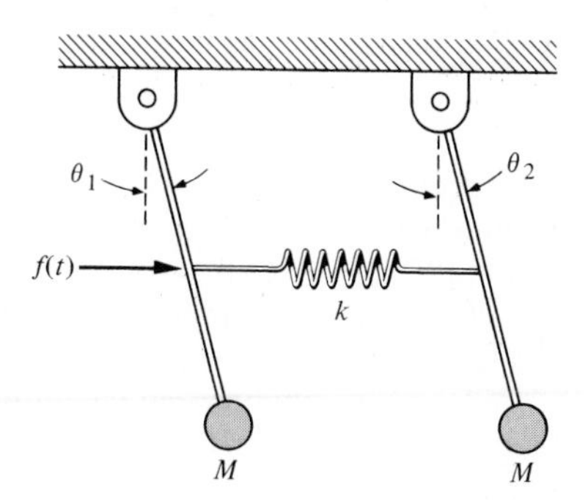

Figure P2.22. The bars are each of length L and the spring is located at $L/2$.

P2.23. The small-signal equivalent circuit of a common-emitter transistor amplifier is shown in Fig. P2.23. The transistor amplifier includes a feedback resistor R_f. Obtain a signal-flow graph model of the feedback amplifier and determine the input-output ratio v_{ce}/v_{in}.

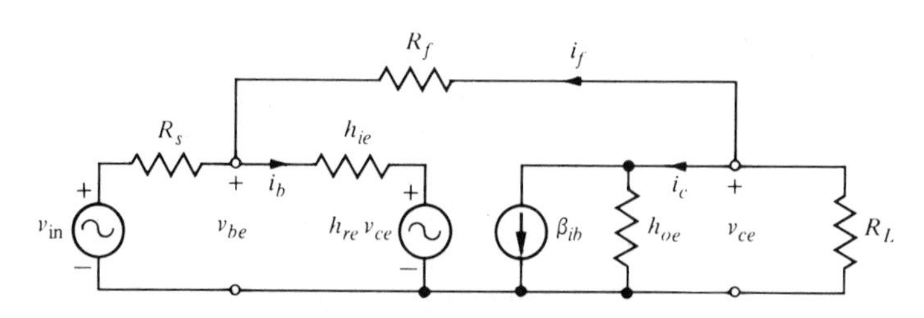

Figure P2.23. CE amplifier.

P2.24. A two-transistor series voltage feedback amplifier is shown in Fig. P2.24(a). This ac equivalent circuit neglects the bias resistors and the shunt capacitors. A signal-flow graph representing the circuit is shown in Fig. P2.24(b). This flow graph neglects the effect of h_{re}, which is usually an accurate approximation, and assumes that $(R_2 + R_L) \gg R_1$. (a) Determine the voltage gain, $e_{\text{out}}/e_{\text{in}}$. (b) Determine the current gain, i_{c_2}/i_{b_1}. (c) Determine the input impedance, e_{in}/i_{b_1}.

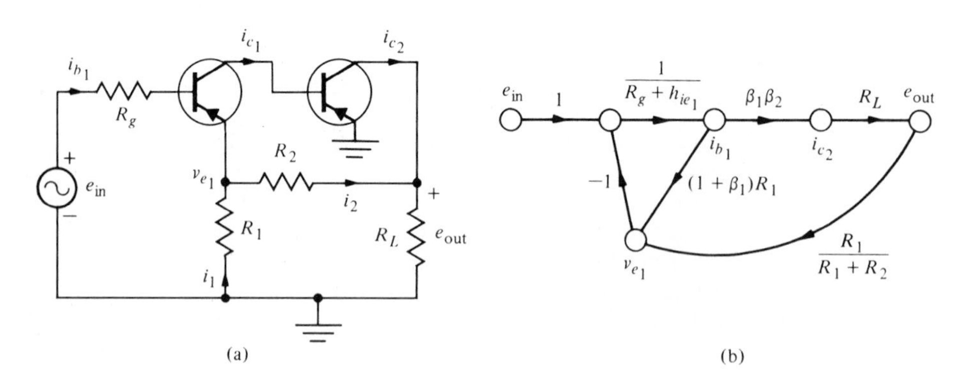

Figure P2.24. Feedback amplifier.

P2.25. Often overlooked is the fact that H. S. Black, who is noted for developing a negative feedback amplifier in 1927, three years earlier had invented a circuit design technique known as feedforward correction [22, 23]. Recent experiments have shown that this technique offers the potential for yielding excellent amplifier stabilization. Black's amplifier is shown in Fig. P2.25(a) in the form recorded in 1924. The signal-flow graph is shown in Fig. P2.25(b). Determine the transfer function between the output $C(s)$ and the input $R(s)$ and between the output and the disturbance $D(s)$. $G(s)$ is used for the amplifier represented by μ in Fig. P2.25(a).

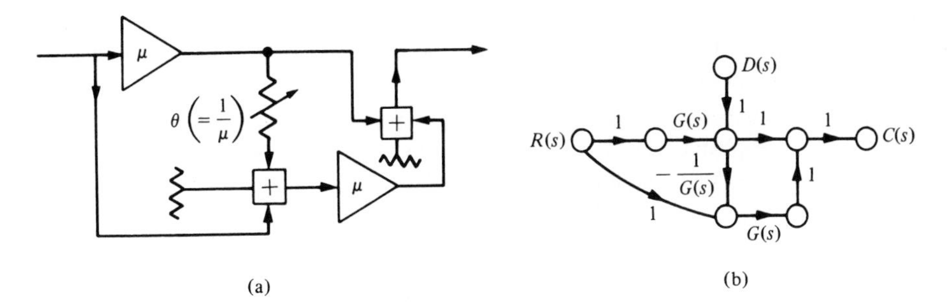

Figure P2.25. H. S. Black's amplifier.

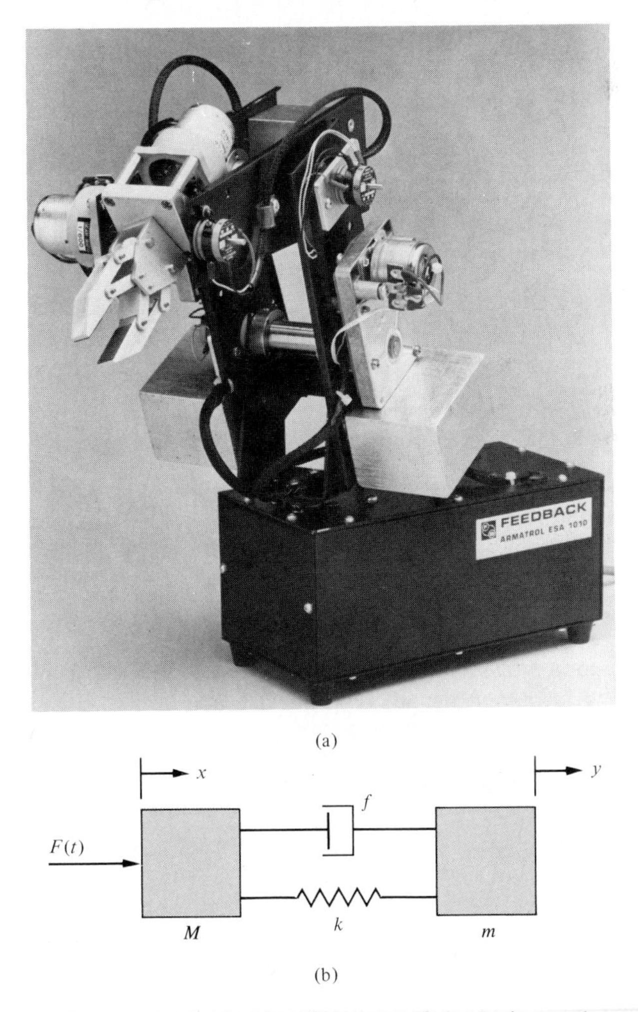

Figure P2.26. (a) The Armatrol robot has four axes of revolution and a two-fingered gripper with feedback from all axes. (Courtesy of Feedback, Inc.) (b) The spring-mass-damper model.

P2.26. A robot such as the one shown in Fig. P2.26(a) on the previous page includes significant flexibility in the arm members with a heavy load in the gripper [32]. A two-mass model of the robot is shown in Fig. P2.26(b). Find the transfer function $Y(s)/F(s)$.

P2.27. Magnetic levitation trains provide a high speed, very low friction alternative to steel wheels on steel rails. The train floats on an air gap as shown in Fig. P2.27 [29]. The levitation force F_L is controlled by the coil current i in the levitation coils and may be approximated by

$$F_L = k\frac{i^2}{z^2},$$

where z is the air gap. This force is opposed by the downward force $F = mg$. Determine the linearized relationship between the air gap z and the controlling current near the equilibrium condition.

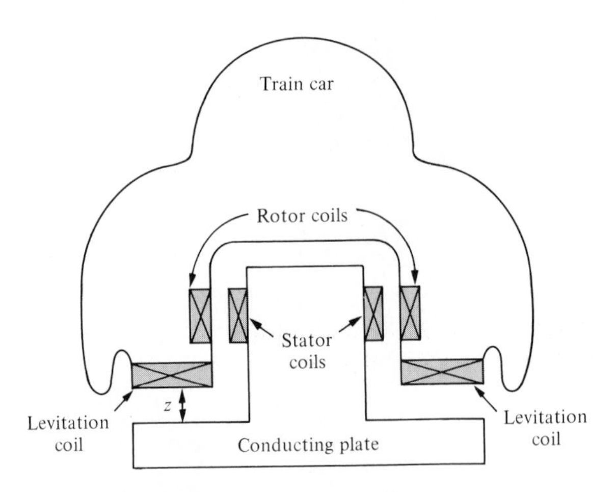

Figure P2.27. Cutaway view of train.

P2.28. A multiple-loop model of an urban ecological system might include the following variables: number of people in the city (P), modernization (M), migration into the city (C), sanitation facilities (S), number of diseases (D), bacteria/area (B), and amount of garbage/ area (G), where the symbol for the variable is given in the parentheses. The following causal loops are hypothesized:

1. $P \to G \to B \to D \to P$
2. $P \to M \to C \to P$
3. $P \to M \to S \to D \to P$
4. $P \to M \to S \to B \to D \to P$

Draw a signal-flow graph for these causal relationships, using appropriate gain symbols. Indicate whether you believe each gain transmission is positive or negative. For example, the causal link S to B is negative because improved sanitation facilities lead to reduced bacteria/area. Which of the four loops are positive feedback loops and which are negative feedback loops?

P2.29. We desire to balance a rolling ball on a tilting beam as shown in Fig. P2.29. We will assume the motor input current i controls the torque with negligible friction. Assume the beam may be balanced near the horizontal ($\phi = 0$); therefore we have a small deviation of ϕ. Find the transfer function $X(s)/I(s)$ and draw a block diagram illustrating the transfer function showing $\phi(s)$, $X(s)$ and $I(s)$.

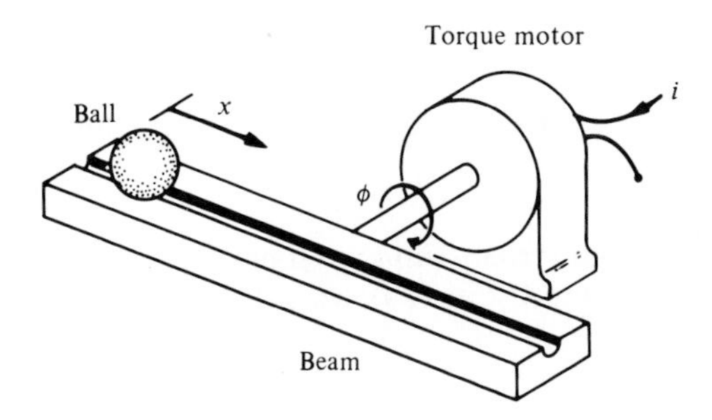

Figure P2.29. Tilting beam and ball.

P2.30. The measurement or sensor element in a feedback system is important to the accuracy of the system [37]. The dynamic response of the sensor is often important. Most sensor elements possess a transfer function

$$H(s) = \frac{k}{\tau s + 1}.$$

Investigate several sensor elements available today and determine the accuracy available and the time constant of the sensor. Consider two of the following sensors: (1) linear position, (2) temperature with a thermistor, (3) strain measurement, (4) pressure.

P2.31. A cable reel control system uses a tachometer to measure the speed of the cable as it leaves the reel. The output of the tachometer is used to control the motor speed of the reel as the cable is unwound off the reel. The system is shown in Fig. P2.31. The radius of the reel, R, is 4 meters when full and 2 meters when empty. The moment of inertia of the reel is $I = 18.5R^4 - 221$. The rate of change of the radius is

$$\frac{dR}{dt} = \frac{-D^2\dot{\omega}}{2\pi W},$$

where W = width of the reel and D = diameter of the cable. The actual speed of the cable is $v(t) = R\dot{\omega}$. The desired cable speed is 50 m/sec. Develop a digital computer simulation of this system and obtain the response of the speed over 20 seconds for the three values of gain $K = 0.5$, 1.0, and 1.5. The reel angular velocity $\dot{\omega} = d\theta/dt$ is equal to $1/I$ times the integral of the torque. Note that the inertia changes with time as the reel is unwound. However, an equation for I within the simulation will account for this change.

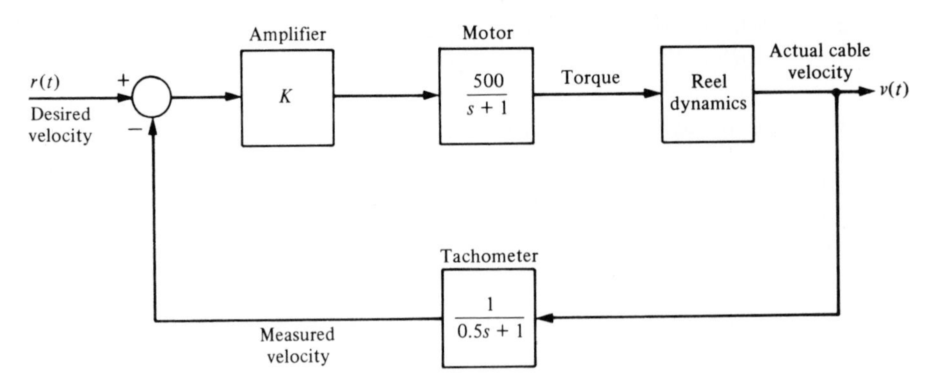

Figure P2.31. Cable reel control system.

P2.32. An interacting control system with two inputs and two outputs is shown in Fig. P2.32. Solve for $C_1(s)/R_1(s)$ and $C_2(s)/R_1(s)$, when $R_2 = 0$.

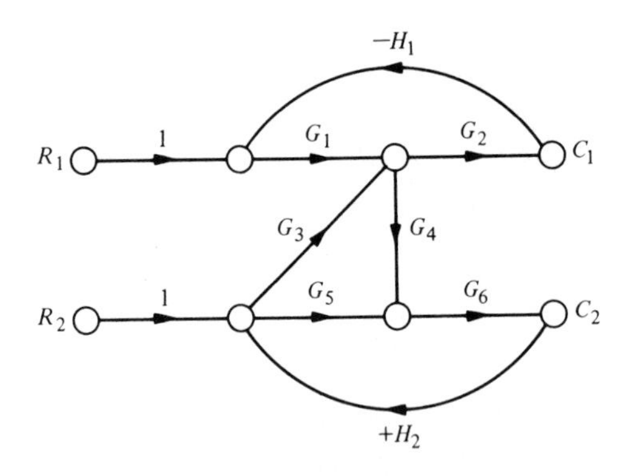

Figure P2.32.

P2.33. The system shown in Fig. P2.33(a), on the next page consists of two electric motors which are coupled by a continuous flexible belt. The belt also passes over a swinging arm which is instrumented to allow measurement of the belt speed and tension. The basic control problem is to regulate the belt speed and tension by varying the motor torques.

An example of a practical system similar to that shown occurs in textile fiber manufacturing processes when yarn is wound from one spool to another at high speed. Between the two spools the yarn is processed in some way which may require the yarn speed and tension to be controlled to within defined limits. A model of the system is shown in Fig. P2.33(b). Find $C_2(s)/R_1(s)$. Determine a relationship for the system that will make C_2 independent of R_1.

P2.34. Find the transfer function for $C(s)/R(s)$ for the idle speed control system for a fuel injected engine as shown in Fig. P2.34, on page 106.

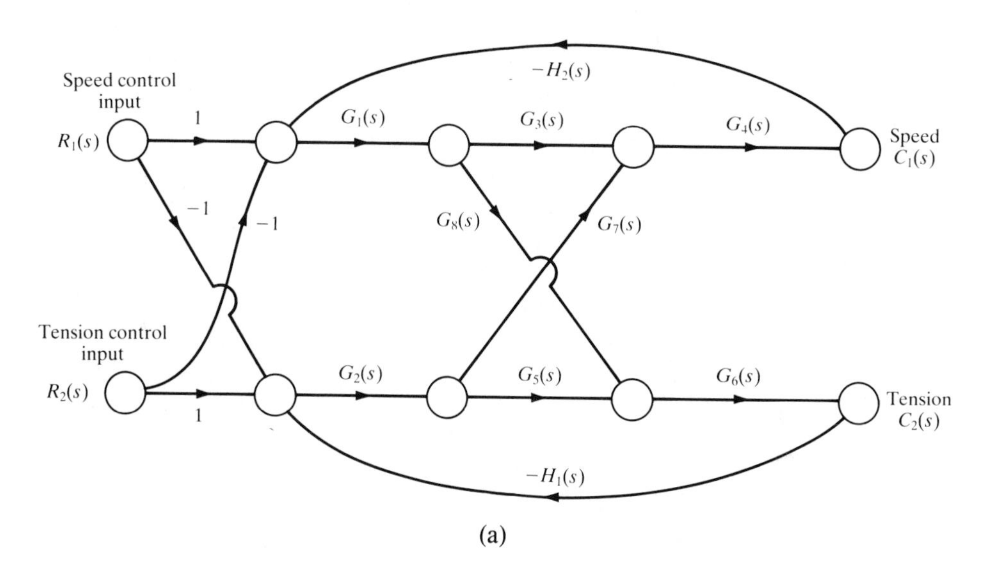

(a)

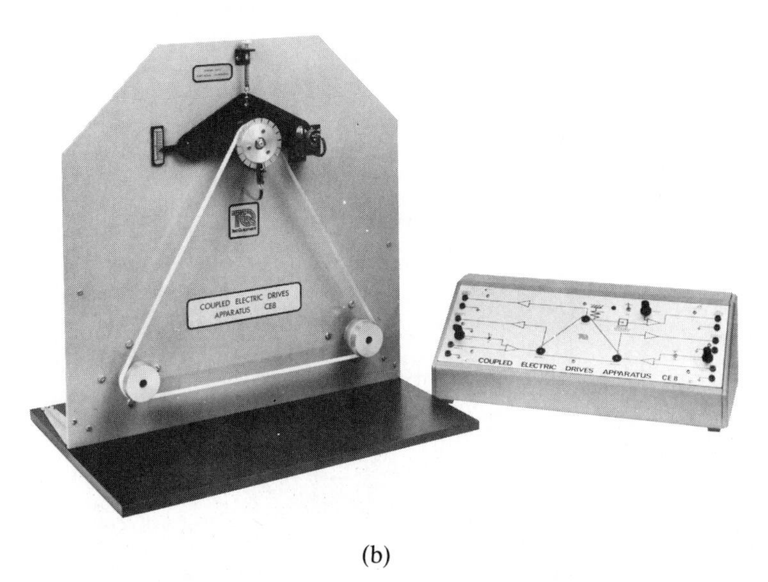

(b)

Figure P2.33. (a) A model of the coupled motor drives. (b) The coupled motor drives. Courtesy of TecQuipment Inc.

P2.35. The suspension system for one wheel of an old-fashioned pick-up truck is illustrated in Fig. P2.35. The mass of the vehicle is m_1 and the mass of the wheel is m_2. The suspension spring has a spring constant k_1 and the tire has a spring constant k_2. The damping constant of the shock absorber is f. Obtain the transfer function $Y_1(s)/X(s)$, which represents the vehicle response to bumps in the road.

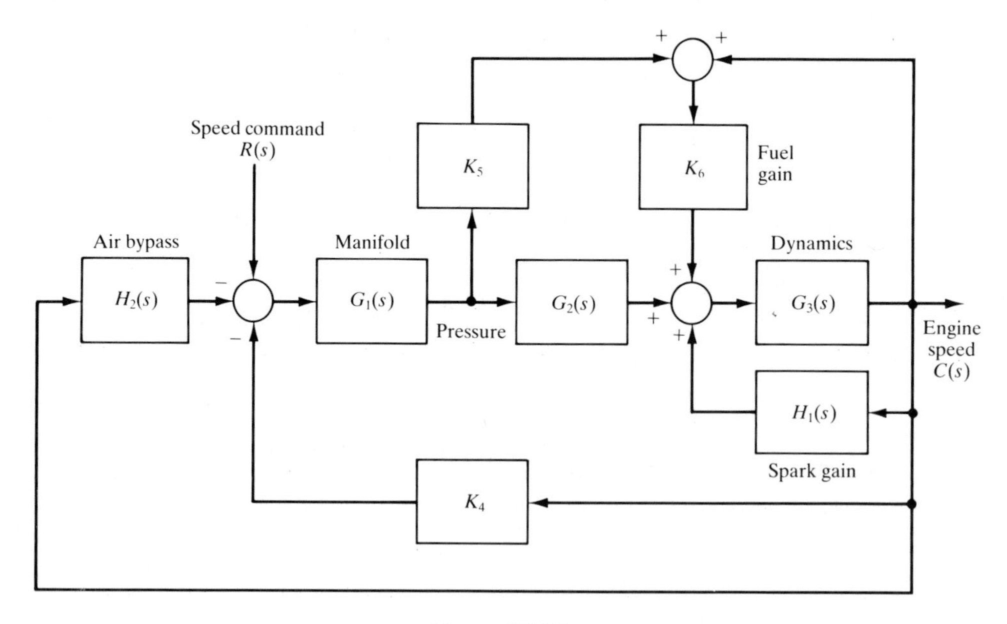

Figure P2.34.

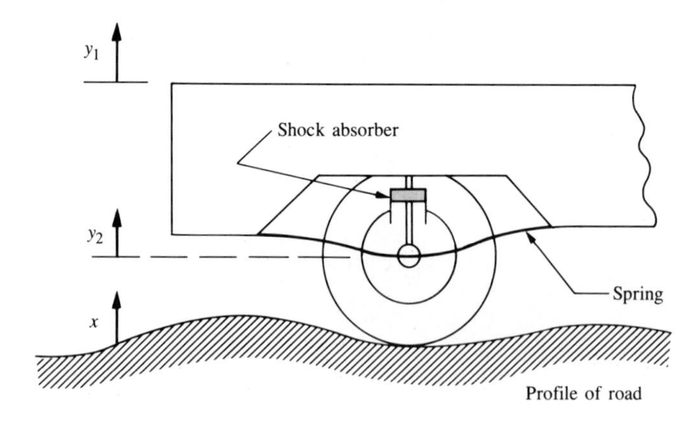

Figure P2.35. Pick-up truck suspension.

P2.36. A feedback control system has the structure shown in Fig. P2.36. Determine the closed-loop transfer function $C(s)/R(s)$ (a) by block diagram manipulation and (b) by using a signal flow graph and Mason's formula. (c) Select the gains K_1 and K_2 so that the closed-loop response to a step input is critically damped with two equal roots at $s = -10$. (d) Plot the critically damped response for a unit step input. What is the time required for the step response to reach 90% of its final value?

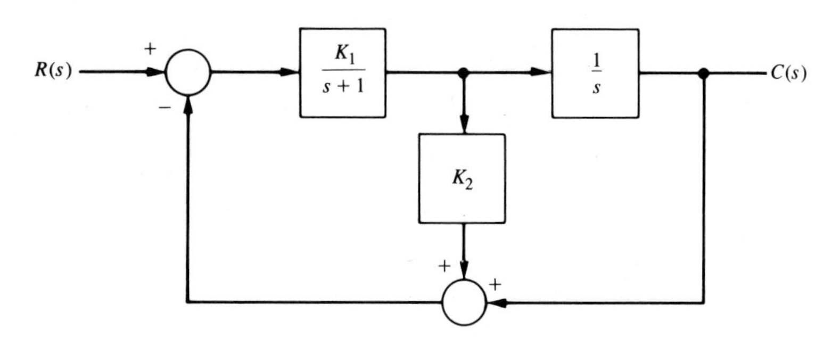

Figure P2.36.

P2.37. A system is represented by Fig. P2.37. (a) Determine the partial fraction expansion and $c(t)$ for a ramp input, $r(t) = t, t \geq 0$. (b) Obtain a plot of $c(t)$ for part (a) and find $c(t)$ for $t = 1.5s$. (c) Determine the impulse response of the system $c(t)$ for $t \geq 0$. (d) Obtain a plot of $c(t)$ for part (c) and find $c(t)$ for $t = 1.5s$.

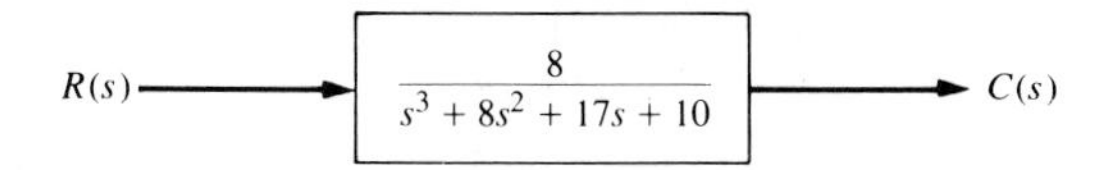

Figure P2.37.

P2.38. A two-mass system is shown in Fig. P2.38 with an input force $u(t)$. When $m_1 = m_2 = 1$ and $K_1 = K_2 = 1$, find the set of differential equations describing the system.

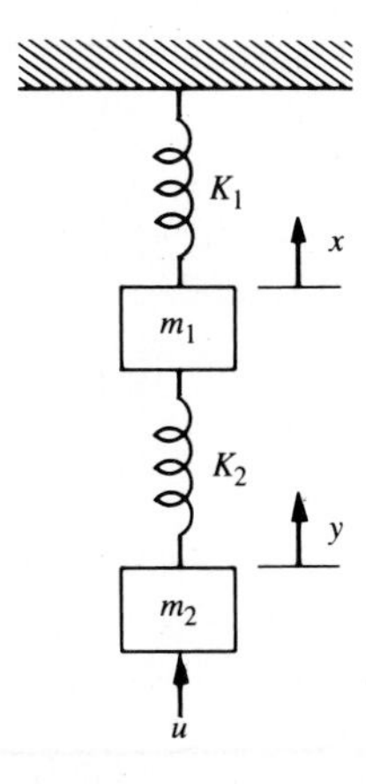

Figure P2.38. Two mass sytem.

P2.39. A winding oscillator consists of two steel spheres on each end of a long slender rod, as shown in Fig. P2.39. The rod is hung on a thin wire that can be twisted many revolutions without breaking. The device will be wound up 4000 degrees. How long will it take until the motion decays to a swing of only 10 degrees? Assume that the damping constant for the sphere in air is $2 \times 10^{-4}\ N \cdot m/\text{rad}$ and each sphere has a mass of 1 kg.

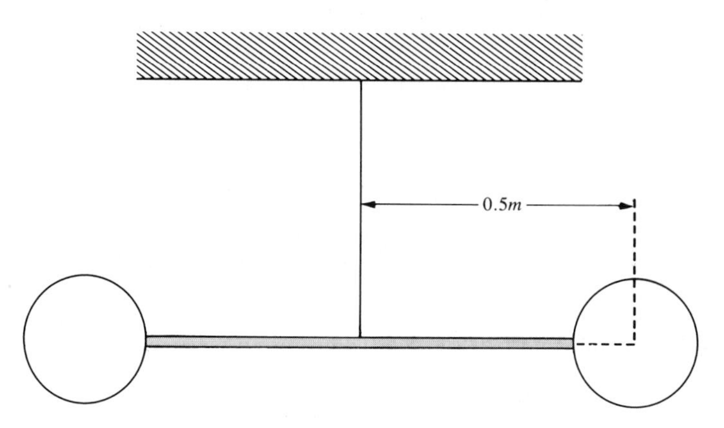

Figure P2.39. Winding oscillator.

P2.40. A damping device is used to reduce the undesired vibrations of machines. A viscous fluid, such as a heavy oil, is placed between the wheels, as shown in Fig. P2.40. When vibration becomes excessive, the relative motion of the two wheels creates damping. When the device is rotating without vibration, there is no relative motion and no damping occurs. Find $\theta_1(s)$ and $\theta_2(s)$. Assume that the shaft has a spring constant K and that f is the damping constant of the fluid. The load torque is T.

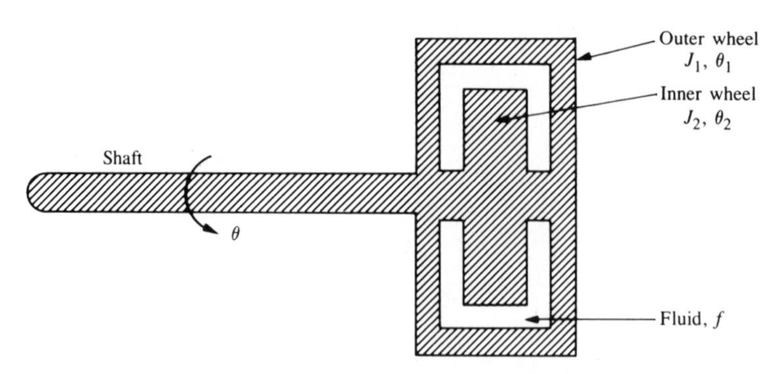

Figure P2.40. Cutaway view of damping device.

P2.41. For the circuit of Fig. P2.41, determine the transform of the output voltage $V_0(s)$. Assume that the circuit is in steady state when $t < 0$. Assume that the switch moves instantaneously from contact 1 to contact 2 at $t = 0$.

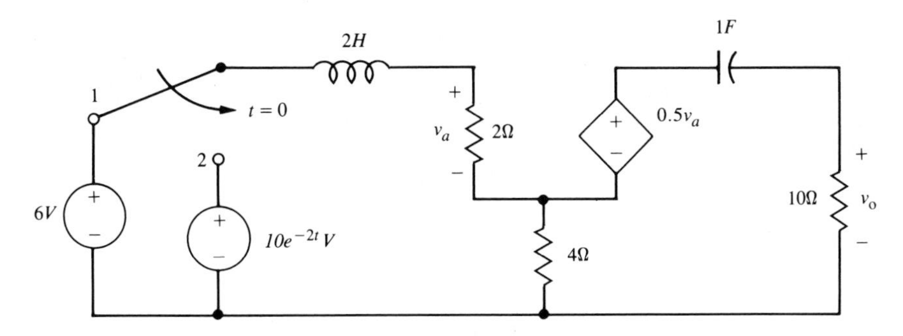

Figure P2.41.

P2.42. The lateral control of a rocket with a gimbaled engine is shown in Fig. P2.42. The lateral deviation from the desired trajectory is h, and the forward rocket speed is V. The control torque of the engine is T_c, and the disturbance torque is T_d. Derive the describing equations of a linear model of the system and draw the block diagram with the appropriate transfer functions.

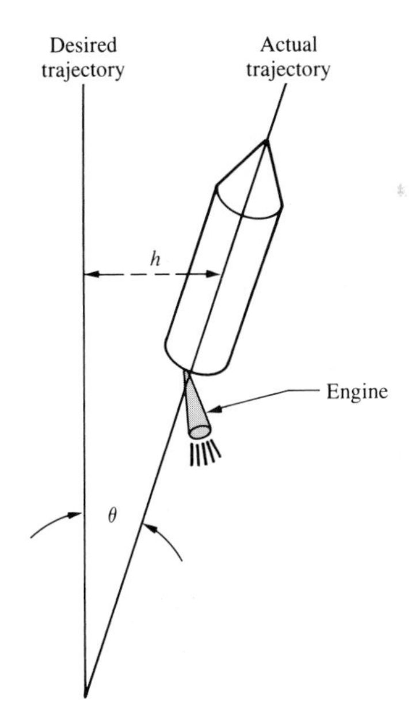

Figure P2.42. Rocket with gimbaled engine.

P2.43. In many applications, such as reading product codes in supermarkets and in printing and manufacturing, an optical scanner is utilized to read codes, as shown in Fig. P2.43. As the mirror rotates, a friction force is developed that is proportional to its angular speed. The friction constant is equal to 0.05 $N \cdot s$/rad, and the moment of inertia is equal to 0.1 kg-m^2. The output variable is the velocity, $\omega(t)$. (a) Obtain the differential equation for the motor. (b) Find the response of the system when the input motor torque is a unit step and the initial velocity at $t = 0$ is equal to 1.

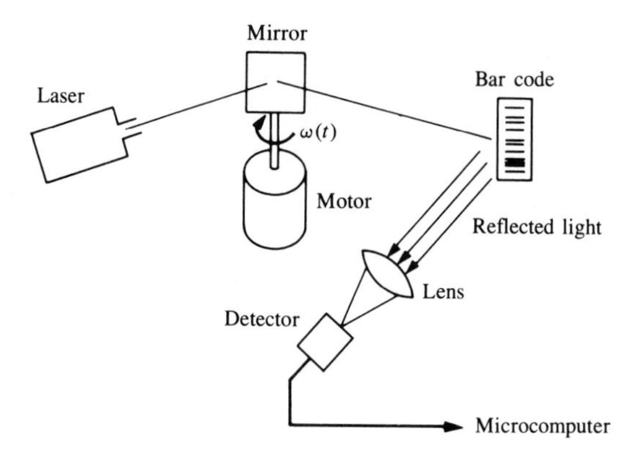

Figure P2.43. Optical scanner.

P2.44. An ideal set of gears is shown as item 10 of Table 2.6. Neglect the inertia and friction of the gears and assume that the work done by one gear is equal to that of the other. Derive the relationships given in item 10 of Table 2.6. Also determine the relationship between the torques T_m and T_L.

P2.45. An ideal set of gears is connected to a load inertia with friction coefficient f as shown in Fig. P2.45. Obtain the torque equation, in terms of θ_1, at the input side of the gears where the input is a torque T. The gear ratio is $n = N_1/N_2$.

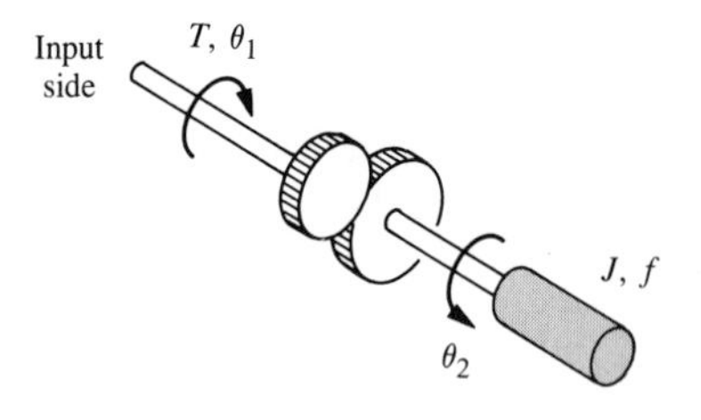

Figure P2.45.

Design Problems ✦

(Design problems emphasize the design task.)

DP2.1. A control system is shown in Fig. DP2.1. The transfer functions $G_2(s)$ and $H_2(s)$ are fixed. Determine the transfer functions $G_1(s)$ and $H_1(s)$ so that the closed-loop transfer function $C(s)/R(s)$ is exactly equal to 1.

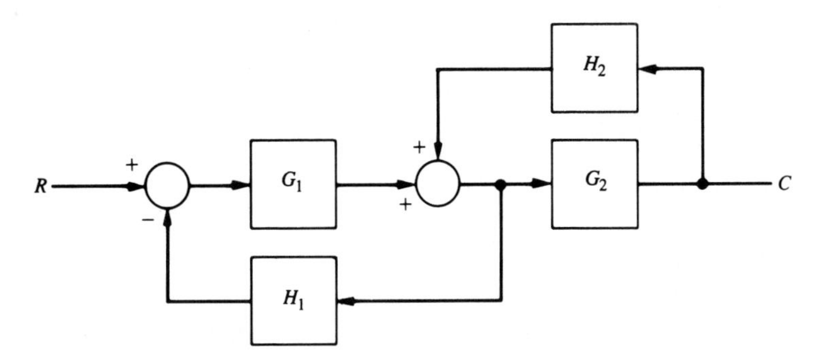

Figure DP2.1. Selection of transfer functions.

DP2.2. The television beam circuit of a television set is represented by the model in Fig. DP2.2. Select the unknown conductance, G, so that the voltage, v, is 24 V. Each conductance is given in siemens (S).

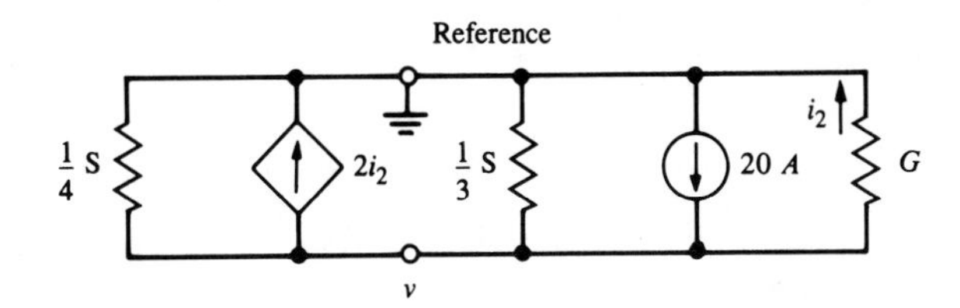

Figure DP2.2. Television beam circuit.

DP2.3. An input $r(t) = e^{-t}u(t)$ is applied to a black box with a transfer function $G(s)$. The resulting output response, when the initial conditions are zero, is

$$c(t) = 2 - 3e^{-t} + e^{-2t} \cos 2t, \; t \geq 0.$$

Determine $G(s)$ for this system.

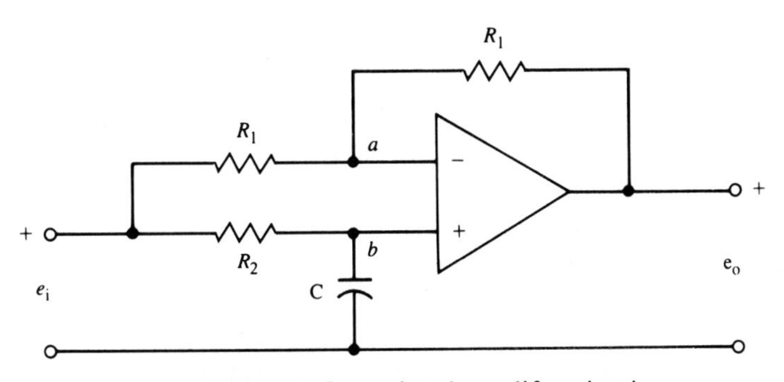

Figure DP2.4. Operational amplifier circuit.

DP2.4. An operational amplifier circuit that can serve as a filter circuit is shown in Fig. DP2.4. (a) Determine the transfer function of the circuit, assuming an ideal op amp. Find $e_o(t)$ when the input is $e_I(t) = At$, $t \geq 0$.

Terms and Concepts

Actuator The device that causes the process to provide the output. The device that provides the motive power to the process.

Block diagrams Unidirectional, operational blocks that represent the transfer functions of the elements of the system.

Characteristic equation The relation formed by equating to zero the denominator of a transfer function.

Critical damping The case where damping is on the boundary between underdamped and overdamped.

Damped oscillation An oscillation in which the amplitude decreases with time.

Damping ratio A measure of damping. A dimensionless number for the second-order characteristic equation.

dc motor An electric actuator that uses an input voltage as a control variable.

Laplace transform A transformation of a function $f(t)$ from the time domain into the complex frequency domain yielding $F(s)$.

Linear approximation An approximate model that results in a linear relationship between the output and the input of the device.

Mason loop rule A rule that enables the user to obtain a transfer function by tracing paths and loops within a system.

Mathematical models Descriptions of the behavior of a system using mathematics.

Signal-flow graph A diagram that consists of nodes connected by several directed branches and that is a graphical representation of a set of linear relations.

Simulation A model of a system that is used to investigate the behavior of a system by utilizing actual input signals.

Transfer function The ratio of the Laplace transform of the output variable to the Laplace transform of the input variable.

References

1. R. C. Rosenberg and D. C. Karnopp, *Introduction to Physical System Design,* McGraw-Hill, New York, 1986.
2. R. C. Dorf, *Introduction to Electric Circuits,* John Wiley & Sons, New York, 1989.
3. J. W. Nilsson, *Electric Circuits,* 3rd ed., Addison-Wesley, Reading, Mass., 1990.
4. E. Kamen, *Introduction to Signals and Systems,* 2nd ed., Macmillan, New York, 1990.

5. F. Raven, *Automatic Control Engineering,* 2nd ed., McGraw-Hill, New York, 1990.
6. R. C. Dorf, *The Encyclopedia of Robotics,* John Wiley & Sons, New York, 1988.
7. S. Ashley, "A Mosaic for Machine Tools," *Mechanical Engineering,* September 1990; pp. 38–43.
8. D. L. Schilling, *Electronic Circuits,* 3rd ed., McGraw-Hill, New York, 1989.
9. B. C. Kuo, *Automatic Control Systems,* 3rd ed., Prentice Hall, Englewood Cliffs, N.J., 1990.
10. J. L. Shearer, *Dynamic Modeling and Control of Engineering Systems,* Macmillan, New York, 1990.
11. S. Franco, *Design with Operational Amplifiers,* McGraw-Hill, New York, 1988.
12. R. J. Smith and R. C. Dorf, *Circuits, Devices and Systems,* 5th ed., John Wiley & Sons, New York, 1991.
13. S. Derby, "Mechatronics for Robots," *Mechanical Engineering,* July 1990; pp. 40–44.
14. H. M. Paynter, "The Differential Analyzer," *IEEE Control Systems,* December 1989; pp. 3–8.
15. J. J. Ovaska, "Modular Control System Simulator," *Simulation,* October 1989; pp. 181–85.
16. R. C. Dorf, *Introduction to Computers and Computer Science,* 3rd ed., Boyd and Fraser, San Francisco, 1982.
17. A. Wolfe, "Micromachine," *Mechanical Engineering,* September 1990; pp. 49–53.
18. P. C. Krause, *Electromechanical Motion Devices,* McGraw-Hill, New York, 1989.
19. A. Dane, "Hubble: Heartbreak and Hope," *Popular Mechanics,* October 1990; pp. 130–31.
20. T. Yoshikawa, *Foundation of Robotics,* M.I.T. Press, Cambridge, Mass., 1990.
21. J. G. Bollinger, *Computer Control of Machines and Processes,* Addison-Wesley, Reading, Mass., 1989.
22. R. K. Jurgen, "Feedforward Correction: A Late-Blooming Design," *IEEE Spectrum,* April 1972; pp. 41–43.
23. H. S. Black, "Stabilized Feed-Back Amplifiers," *Electrical Engineering,* **53,** January 1934; pp. 114–20. Also in *Turning Points in American History,* J. E. Brittain, ed., IEEE Press, New York, 1977; pp. 359–61.
24. D. H. McMahon, "Vehicle Modeling for Automated Highway Systems," *Proceed. of the American Control Conference,* 1990; pp. 297–303.
25. D. E. Johnson, J. L. Hilburn, and J. R. Johnson, *Basic Electric Circuits,* 2nd ed., Prentice Hall, Englewood Cliffs, N.J., 1990.
26. R. C. Dorf, *Energy, Resources, and Policy,* Addison-Wesley, Reading, Mass., 1978.
27. S. Cuk and J. F. Brewer, "Low-Noise, Low Cost 150w Off-Line Switcher," *Power Conversion International,* April 1983; pp. 40–58.
28. G. J. Thaler, *Automatic Control Systems,* West Publishing, St. Paul, Minn., 1990.
29. "The Flying Train Takes Off," *U.S. News and World Report,* July 23, 1990; p. 52.
30. H. Oman, "Accelerating Hypersonic Airplanes with Ground-Power," *IEEE AES Magazine,* April 1990; pp. 9–14.
31. J. Murphy, *Power Control of AC Motors,* Pergamon Press, New York, 1990.
32. S. C. Jacobsen, "Control Strategies for Tendon-Driven Manipulators," *IEEE Control Systems,* February 1990; pp. 23–28.
33. Z. Wang, "Design and Characterization of a Linear Motion Piezoelectric Micropositioner," *IEEE Control Systems,* February 1990; pp. 10–15.
34. A. Marchant, *Optical Recording,* Addison-Wesley, Reading, Mass., 1990.
35. J. B. Shung et al., "Feedback Control and Simulation of a Wheelchair," *Trans. of the*

American Society of Mechanical Engineers, J. of Dynamic Systems, June 1983; pp. 96–100.

36. L. V. Merritt, "The Space Station Rotary Joint Motor Controller," *Motion,* February 1990; pp. 14–21.
37. C. W. DeSilva, *Control Sensors and Actuators,* Prentice Hall, Englewood Cliffs, N.J., 1990.
38. D. M. Auslander, *Real-Time Software for Control,* Prentice Hall, Englewood Cliffs, N.J., 1990.
39. R. E. Klein, "Using Bicycles to Teach System Dynamics," *IEEE Control Systems,* April 1989; pp. 4–8.
40. R. Kurzweil, *The Age of Intelligent Machines,* M.I.T. Press, Cambridge, Mass., 1990.
41. S. Bennett, "Nicholas Minorsky and the Automatic Steering of Ships," *IEEE Control Systems,* November 1984; pp. 10–15.
42. G. Jackson, "Software for Control Systems Design," *Mechanical Engineering,* July 1990; pp. 44–45.
43. K. Andersen, "Artificial Neural Networks Applied to Arc Welding Control," *IEEE Trans. Industry Applications,* October 1990; pp. 824–28.
44. "Hubble's Legacy," *Scientific American,* June 1990; pp. 18–19.
45. R. C. Houts, *Signal Analysis in Linear Systems,* W. B. Saunders, Philadelphia, 1991.

CHAPTER 3

Feedback Control System Characteristics

Preview

With the mathematical models obtained in Chapter 2 we are able to develop analytical tools for describing the characteristics of a feedback control system. In this chapter, we will develop the concepts of the system error signal. This signal is used to control the process, and our ultimate goal is to reduce the error to the smallest feasible amount.

We also develop the concept of the sensitivity of a system to a parameter change, since it is desirable to minimize the effects of unwanted parameter variation. We also will describe the transient performance of a feedback system and show how this performance can be readily improved.

We wish to reduce the effect of unwanted input signals, called disturbances, on the output signal. We will show how we may design a control system to reduce the impact of disturbance signals. Of course, the benefits of a control system come with an attendant cost. We will demonstrate how the cost of using feedback in a control system is associated with the selection of the feedback sensor device.

3.1 Open- and Closed-Loop Control Systems

Now that we are able to obtain mathematical models of the components of control systems, we shall examine the characteristics of control systems. A control system was defined in Section 1.1 as an interconnection of components forming a system configuration that will provide a desired system response. Because a desired system response is known, a signal proportional to the error between the desired and the actual response is generated. The utilization of this signal to control the process results in a closed-loop sequence of operations that is called a feedback system. This closed-loop sequence of operations is shown in Fig. 3.1. The introduction of feedback in order to improve the control system is often necessary. It is interesting that this is also the case for systems in nature, such as biological and physiological systems, and feedback is inherent in these systems. For example, the human heart-rate control system is a feedback control system.

In order to illustrate the characteristics and advantages of introducing feedback, we shall consider a simple, single-loop feedback system. Although many control systems are not single-loop in character, a single-loop system is illustrative. A thorough comprehension of the benefits of feedback can best be obtained from the single-loop system and then extended to multiloop systems.

An open-loop control system is shown in Fig. 3.2. For contrast, a closed-loop,

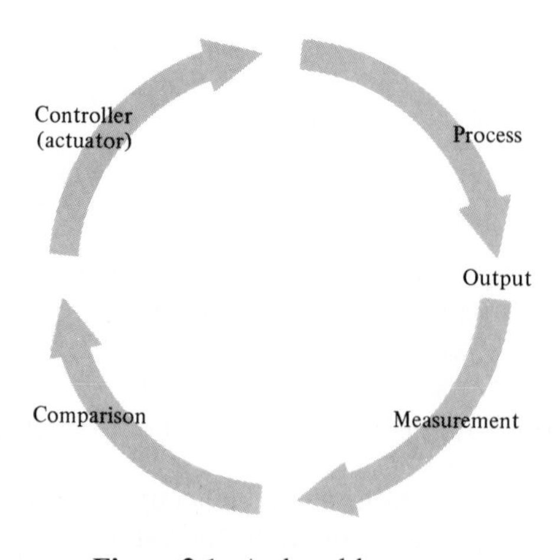

Figure 3.1. A closed-loop system.

Figure 3.2. An open-loop system.

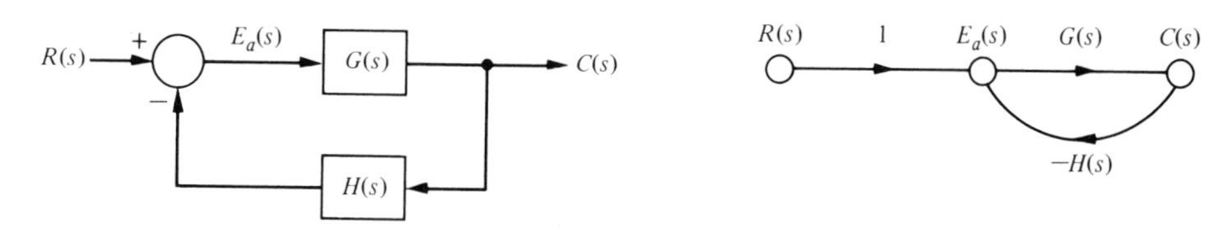

Figure 3.3. A closed-loop control system.

negative feedback control system is shown in Fig. 3.3. The two control systems are shown in both block diagram and signal-flow graph form, although signal-flow graphs will be used predominantly for subsequent diagrams. The prime difference between the open- and closed-loop systems is the generation and utilization of the *error signal.* The closed-loop system, when operating correctly, operates so that the error will be reduced to a minimum value. The signal $E_a(s)$ is a measure of the error of the system and is equal to the error $E(s) = R(s) - C(s)$ when $H(s) = 1$. The output of the open-loop system is

$$C(s) = G(s)R(s). \tag{3.1}$$

The output of the closed-loop system is

$$C(s) = G(s)E_a(s) = G(s)(R(s) - H(s)C(s)),$$

and therefore

$$C(s) = \frac{G(s)}{1 + GH(s)} R(s). \tag{3.2}$$

The actuating error signal is

$$E_a(s) = \frac{1}{1+GH(s)} R(s). \tag{3.3}$$

It is clear that in order to reduce the error, the magnitude of $1 + GH(s)$ must be greater than one over the range of s under consideration.

3.2 Sensitivity of Control Systems to Parameter Variations

A process, represented by the transfer function $G(s)$, whatever its nature, is subject to a changing environment, aging, ignorance of the exact values of the process parameters, and other natural factors that affect a control process. In the open-loop system, all these errors and changes result in a changing and inaccurate output. However, a closed-loop system senses the change in the output due to the process changes and attempts to correct the output. The *sensitivity* of a control system to parameter variations is of prime importance. A primary advantage of a closed-loop feedback control system is its ability to reduce the system's sensitivity [1, 2, 25].

For the closed-loop case, if $GH(s) \gg 1$ for all complex frequencies of interest, then from Eq. (3.2) we obtain

$$C(s) \cong \frac{1}{H(s)} R(s). \tag{3.4}$$

Then, the output is affected only by $H(s)$, which may be a constant. If $H(s) = 1$, we have the desired result; that is, the output is equal to the input. However, before we use this approach for all control systems, we must note that the requirement that $G(s)H(s) \gg 1$ may cause the system response to be highly oscillatory and even unstable. But the fact that as we increase the magnitude of the loop transfer function $G(s)H(s)$, we reduce the effect of $G(s)$ on the output, is an exceedingly useful concept. Therefore, the *first advantage* of a feedback system is that the effect of the *variation of the parameters* of the process, $G(s)$, is reduced.

In order to illustrate the effect of parameter variations, let us consider a change in the process so that the new process is $G(s) + \Delta G(s)$. Then, in the open-loop case, the change in the transform of the output is

$$\Delta C(s) = \Delta G(s) R(s). \tag{3.5}$$

In the closed-loop system, we have

$$C(s) + \Delta C(s) = \frac{G(s) + \Delta G(s)}{1 + (G(s) + \Delta G(s))H(s)} R(s). \tag{3.6}$$

Then the change in the output is

$$\Delta C(s) = \frac{\Delta G(s)}{(1 + GH(s) + \Delta GH(s))(1 + GH(s))} R(s). \tag{3.7}$$

When $GH(s) \gg \Delta GH(s)$, as is often the case, we have

$$\Delta C(s) = \frac{\Delta G(s)}{(1 + GH(s))^2} R(s). \tag{3.8}$$

Examining Eq. (3.8), we note that the change in the output of the closed-loop system is reduced by the factor $1 + GH(s)$, which is usually much greater than one over the range of complex frequencies of interest. The factor $1 + GH(s)$ plays a very important role in the characteristics of feedback control systems.

The *system sensitivity* is defined as the ratio of the percentage change in the system transfer function to the percentage change of the process transfer function. The system transfer function is

$$T(s) = \frac{C(s)}{R(s)}, \tag{3.9}$$

and therefore the sensitivity is defined as

$$S = \frac{\Delta T(s)/T(s)}{\Delta G(s)/G(s)}. \tag{3.10}$$

In the limit, for small incremental changes, Eq. (3.10) becomes

$$S = \frac{\partial T/T}{\partial G/G} = \frac{\partial \ln T}{\partial \ln G}. \tag{3.11}$$

Clearly, from Eq. (3.5), the sensitivity of the open-loop system is equal to one. The sensitivity of the closed-loop is readily obtained by using Eq. (3.11). The system transfer function of the closed-loop system is

$$T(s) = \frac{G}{1 + GH}. \tag{3.12}$$

Therefore the sensitivity of the feedback system is

$$S = \frac{\partial T}{\partial G} \cdot \frac{G}{T} = \frac{1}{(1 + GH)^2} \cdot \frac{G}{G/(1 + GH)} = \frac{1}{1 + GH(s)}. \tag{3.13}$$

Again we find that the sensitivity of the system may be reduced below that of the open-loop system by increasing $GH(s)$ over the frequency range of interest.

The sensitivity of the feedback system to changes in the feedback element $H(s)$ is

$$S_H^T = \frac{\partial T}{\partial H} \cdot \frac{H}{T} = \left(\frac{G}{1 + GH}\right)^2 \cdot \frac{-H}{G/(1 + GH)} = \frac{-GH}{1 + GH}. \tag{3.14}$$

When GH is large, the sensitivity approaches unity and the changes in $H(s)$ directly affect the output response. Therefore it is important to use feedback components that will not vary with environmental changes or that can be maintained constant.

Very often the transfer function of the system $T(s)$ is a fraction of the form [1]:

$$T(s,\alpha) = \frac{N(s,\alpha)}{D(s,\alpha)} \tag{3.15}$$

where α is a parameter that may be subject to variation due to the environment. Then we may obtain the sensitivity to α by rewriting Eq. (3.11) as:

$$\begin{aligned} S_\alpha^T &= \frac{\partial \ln T}{\partial \ln \alpha} = \left.\frac{\partial \ln N}{\partial \ln \alpha}\right|_{\alpha_0} - \left.\frac{\partial \ln D}{\partial \ln \alpha}\right|_{\alpha_0} \\ &= S_\alpha^N - S_\alpha^D, \end{aligned} \tag{3.16}$$

where α_0 is the nominal value of the parameter.

The ability to reduce the effect of the variation of parameters of a control system by adding a feedback loop is an important advantage of feedback control systems. To obtain highly accurate open-loop systems, the components of the open-loop $G(s)$ must be selected carefully in order to meet the exact specifications. However, a closed-loop system allows $G(s)$ to be less accurately specified because the sensitivity to changes or errors in $G(s)$ is reduced by the loop gain $1 + GH(s)$. This benefit of closed-loop systems is a profound advantage for the electronic amplifiers of the communication industry. A simple example will illustrate the value of feedback for reducing sensitivity.

■ Example 3.1 Inverting Amplifier

The integrated circuit operational amplifier can be fabricated on a single chip of silicon and sold for less than a dollar. As a result, IC operational amplifiers (op amps) are widely used. The model symbol of an op amp is shown in Fig. 3.4(a). We can assume that the gain A is at least 10^4. The basic inverting amplifier circuit is shown in Fig. 3.4(b). Because of the high input impedance of the op amp, the amplifier input current is negligibly small. At node n we may write the current equation as

$$\frac{e_{\text{in}} - v_n}{R_1} + \frac{v_o - v_n}{R_f} = 0. \tag{3.17}$$

Because the gain of the amplifier is A, $v_o = Av_n$ and therefore

$$v_n = \frac{v_o}{A}, \tag{3.18}$$

and we may substitute Eq. (3.18) into Eq. (3.17), obtaining

$$\frac{e_{\text{in}}}{R_1} - \frac{v_o}{AR_1} + \frac{v_o}{R_f} - \frac{v_o}{AR_f} = 0. \tag{3.19}$$

Solving for the output voltage, we have

$$v_o = \frac{A(R_f/R_1)e_{\text{in}}}{(R_f/R_1) - A}. \tag{3.20}$$

Alternatively, we may rewrite Eq. (3.20) as follows:

$$\frac{v_o}{e_{\text{in}}} = \frac{A}{1 - A(R_1/R_f)} = \frac{A}{1 - Ak}, \tag{3.21}$$

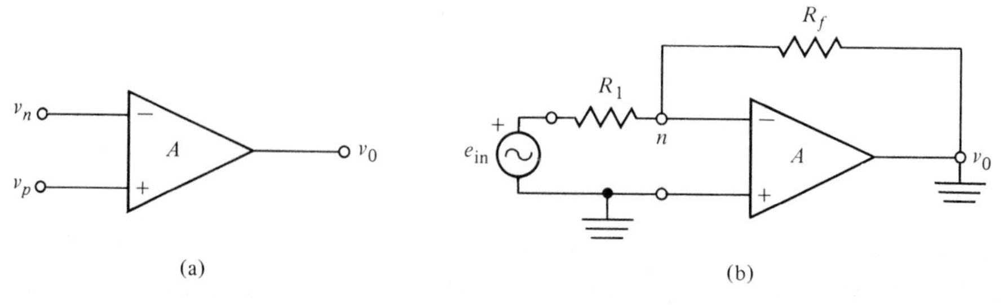

Figure 3.4. (a) An operational amplifier model symbol. (b) Inverting amplifier circuit.

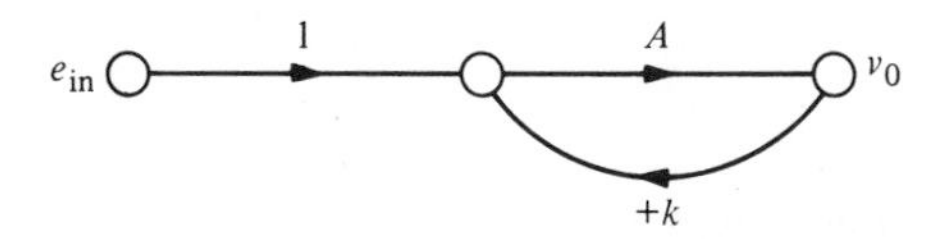

Figure 3.5. Signal-flow graph of inverting amplifier.

where $k = R_1/R_f$. The signal-flow graph representation of the inverting amplifier is shown in Fig. 3.5. Note that when $A \gg 1$ we have

$$\frac{v_o}{e_{in}} = -\frac{R_f}{R_1}. \tag{3.22}$$

The feedback factor in the diagram is $H(s) = k$ and the open-loop transfer function is $G(s) = A$.

The op amp is subject to variations in the amplification A. The sensitivity of the open loop is unity. The sensitivity of the closed-loop amplifier is

$$S_A^T = \frac{\partial T/T}{\partial A/A} = \frac{1}{1 - GH} = \frac{1}{1 - AK}. \tag{3.23}$$

If $A = 10^4$ and $k = 0.1$, we have

$$S_A^T = \frac{1}{1 - 10^3}, \tag{3.24}$$

or the magnitude of the sensitivity is approximately equal to 0.001, which is one-thousandth of the magnitude of the open loop sensitivity. The sensitivity due to changes in the feedback resistance R_f (or the factor k) is

$$S_k^T = \frac{GH}{1 - GH} = \frac{Ak}{1 - Ak}, \tag{3.25}$$

and the sensitivity to k is approximately equal to 1.

We shall return to the concept of sensitivity in subsequent chapters to emphasize the importance of sensitivity in the design and analysis of control systems.

3.3 Control of the Transient Response of Control Systems

One of the most important characteristics of control systems is their transient response. The *transient response* is the response of a system as a function of time. Because the purpose of control systems is to provide a desired response, the transient response of control systems often must be adjusted until it is satisfactory. If an open-loop control system does not provide a satisfactory response, then the process, $G(s)$, must be replaced with a suitable process. By contrast, a closed-loop system can often be adjusted to yield the desired response by adjusting the feedback loop parameters. It should be noted that it is often possible to alter the response of an open-loop system by inserting a suitable cascade filter, $G_1(s)$,

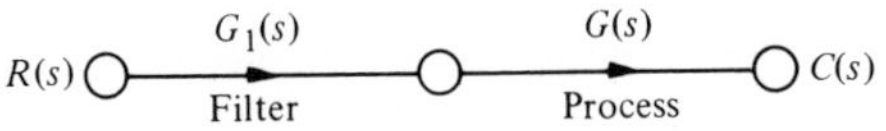

$G_1(s)$ $G(s)$

$R(s)$ $C(s)$

Filter Process

Figure 3.6. Cascade filter open-loop system.

preceding the process, $G(s)$, as shown in Fig. 3.6. Then it is necessary to design the cascade transfer function $G_1(s)G(s)$ so that the resulting transfer function provides the desired transient response.

In order to make this concept more readily comprehensible, let us consider a specific control system, which may be operated in an open- or closed-loop manner. A speed control system, which is shown in Fig. 3.7, is often used in industrial processes to move materials and products. Several important speed control systems are used in steel mills for rolling the steel sheets and moving the steel through the mill. The transfer function of the open-loop system was obtained in Eq. (2.70) and for $\omega(s)/V_a(s)$ we have

$$\frac{\omega(s)}{V_a(s)} = G(s) = \frac{K_1}{(\tau_1 s + 1)}, \tag{3.26}$$

where

$$K_1 = \frac{K_m}{(R_a f + K_b K_m)} \quad \text{and} \quad \tau_1 = \frac{R_a J}{(R_a f + K_b K_m)}.$$

In the case of a steel mill, the inertia of the rolls is quite large and a large armature controlled motor is required. If the steel rolls are subjected to a step command for a speed change of

$$V_a(s) = \frac{k_2 E}{s}, \tag{3.27}$$

the output response is

$$\omega(s) = G(s)V_a(s). \tag{3.28}$$

The transient speed change is then

$$\omega(t) = K_1(k_2 E)(1 - e^{-t/\tau_1}). \tag{3.29}$$

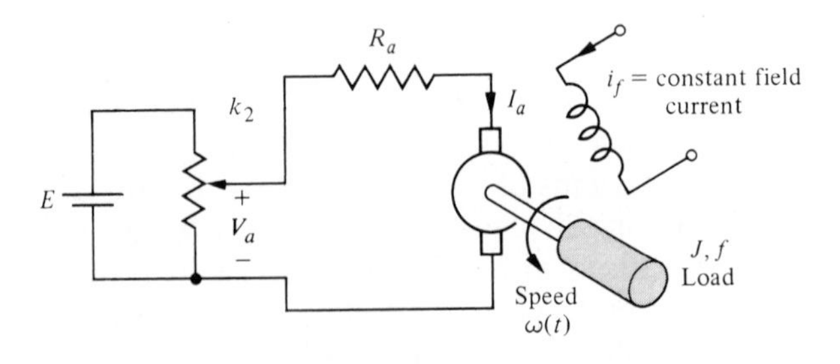

Figure 3.7. Open-loop speed control system.

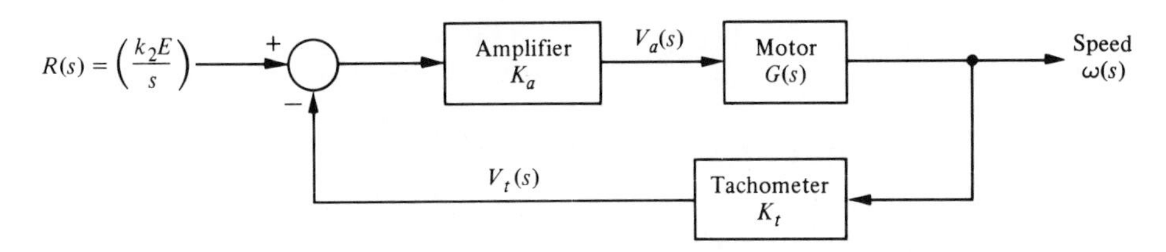

Figure 3.8. Closed-loop speed control system.

If this overdamped transient response is too slow, there is little choice but to choose another motor with a different time constant τ_1, if possible. However, because τ_1 is dominated by the load inertia, little hope for much alteration of the transient response remains.

A closed-loop speed control system is easily obtained by utilizing a tachometer to generate a voltage proportional to the speed, as shown in Fig. 3.8. This voltage is subtracted from the potentiometer voltage and amplified as shown in Fig. 3.8. A practical transistor amplifier circuit for accomplishing this feedback in low power applications is shown in Fig. 3.9 [3]. The closed-loop transfer function is

$$\begin{aligned}\frac{\omega(s)}{R(s)} &= \frac{K_aG(s)}{1 + K_a K_tG(s)}\\ &= \frac{K_aK_1}{\tau_1 s + 1 + K_a K_t K_1}\\ &= \frac{K_aK_1/\tau_1}{s + [(1 + K_a K_t K_1)/\tau_1]}\,.\end{aligned} \tag{3.30}$$

The amplifier gain K_a may be adjusted to meet the required transient response specifications. Also, the tachometer gain constant K_t may be varied, if necessary.

The transient response to a step change in the input command is then

$$\omega(t) = \frac{K_aK_1}{(1 + K_a K_t K_1)}(k_2E)(1 - e^{-pt}), \tag{3.31}$$

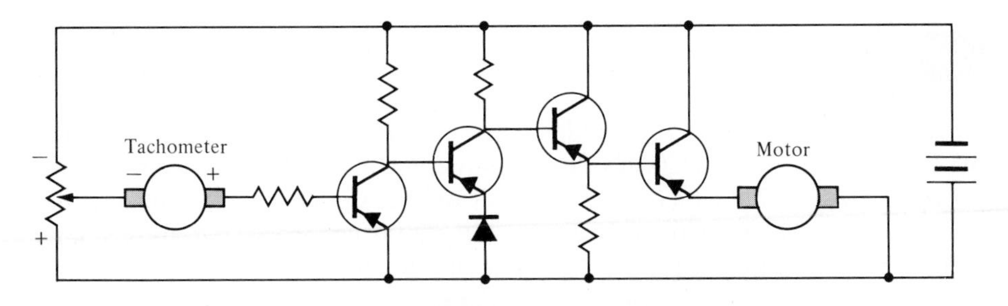

Figure 3.9. Transistorized speed control system.

where $p = (1 + K_a K_t K_1)/\tau_1$. Because the load inertia is assumed to be very large, we alter the response by increasing K_a, and we have the approximate response

$$\omega(t) \simeq \frac{1}{K_t}(k_2 E)\left[1 - exp\left(\frac{-(K_a K_t K_1)t}{\tau_1}\right)\right]. \tag{3.32}$$

For a typical application, the open-loop pole might be $1/\tau_1 = 0.10$, whereas the closed-loop pole could be at least $(K_a K_t K_1)/\tau_1 = 10$, a factor of 100 in the improvement of the speed of response. It should be noted that in order to attain the gain $K_a K_t K_1$, the amplifier gain K_a must be reasonably large, and the armature voltage signal to the motor and its associated torque signal will be larger for the closed-loop than for the open-loop operation. Therefore a larger motor will be required in order to avoid saturation of the motor.

Also, while we are considering this speed control system, it will be worthwhile to determine the sensitivity of the open- and closed-loop systems. As before, the sensitivity of the open-loop system to a variation in the motor constant or the potentiometer constant k_2 is unity. The sensitivity of the closed-loop system to a variation in K_m is

$$\begin{aligned} S^T_{K_m} &= \frac{1}{1 + GH(s)} \\ &= \frac{1}{1 + K_a K_t G(s)} \\ &= \frac{[s + (1/\tau_1)]}{[s + (K_a K_t K_1 + 1)/\tau_1]}. \end{aligned} \tag{3.33}$$

Using the typical values given in the previous paragraph, we have

$$S^T_{K_m} = \frac{(s + 0.10)}{(s + 10)}. \tag{3.34}$$

We find that the sensitivity is a function of s and must be evaluated for various values of frequency. This type of frequency analysis is straightforward but will be deferred until a later chapter. However, it is clearly seen that at a specific frequency, for example, $s = j\omega = j1$, the magnitude of the sensitivity is approximately $|S^T_{K_m}| \cong 0.1$.

3.4 Disturbance Signals in a Feedback Control System

The third most important effect of feedback in a control system is the control and partial elimination of the effect of disturbance signals. A *disturbance signal* is an unwanted signal which affects the system's output signal. Many control systems are subject to extraneous disturbance signals that cause the system to provide an inaccurate output. Electronic amplifiers have inherent noise generated within the integrated circuits or transistors; radar antennas are subjected to wind gusts; and

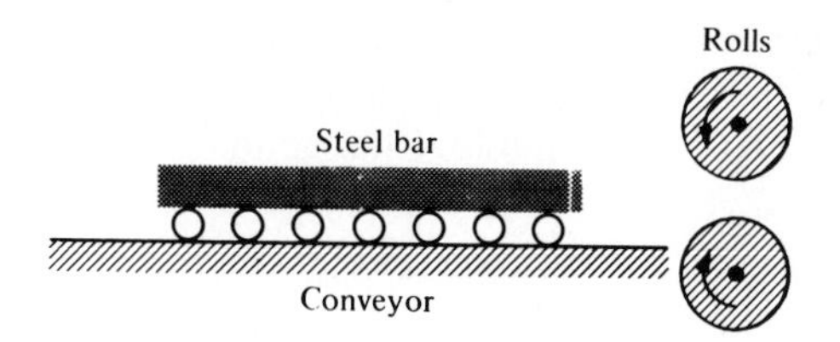

Figure 3.10. Steel rolling mill.

many systems generate unwanted distortion signals due to nonlinear elements. Feedback systems have the beneficial aspect that the effect of distortion, noise, and unwanted disturbances can be effectively reduced.

As a specific example of a system with an unwanted disturbance, let us reconsider the speed control system for a steel rolling mill. Rolls passing steel through are subject to large load changes or disturbances. As a steel bar approaches the rolls (see Fig. 3.10), the rolls turn unloaded. However, when the bar engages in the rolls, the load on the rolls increases immediately to a large value. This loading effect can be approximated by a step change of disturbance torque as shown in Fig. 3.11. Alternatively, we might examine the speed–torque curves of a typical motor as shown in Fig. 3.12.

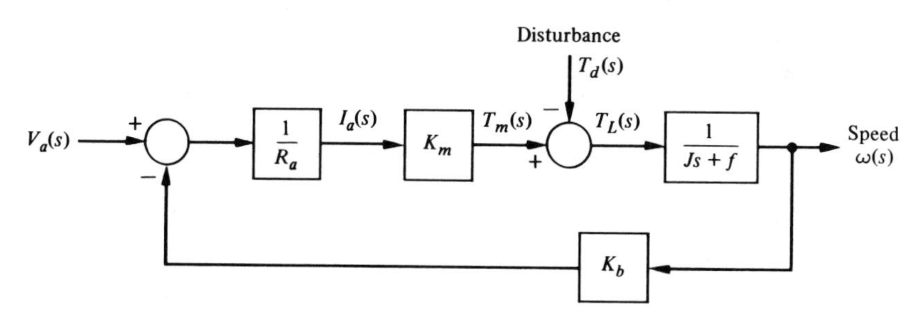

Figure 3.11. Open-loop speed control system.

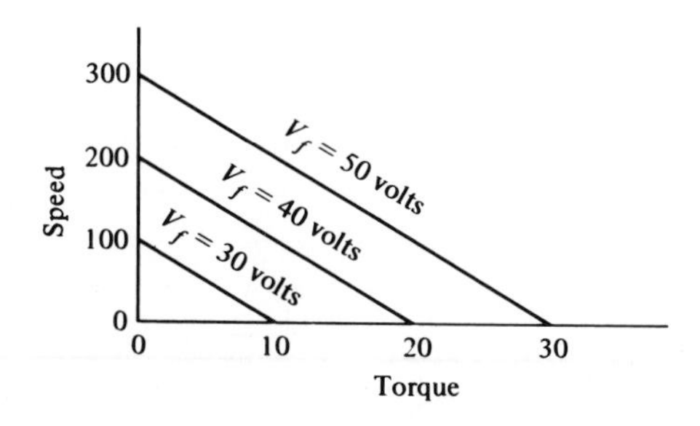

Figure 3.12. Motor speed–torque curves.

The transfer function model of an armature controlled dc-motor with a load torque disturbance was determined in Example 2.3 and is shown in Fig. 3.11, where it is assumed that L_a is negligible. Change in speed due to load disturbance is

$$\omega(s) = \left(\frac{-1}{Js + f + (K_m K_b / R_a)}\right) T_d(s). \tag{3.35}$$

The steady-state error in speed due to the load torque $T_d(s) = D/s$ is found by using the final-value theorem. Therefore, for the open-loop system, we have

$$\begin{aligned}\lim_{t\to\infty} \omega(t) &= \lim_{s\to 0} s\omega(s) = \lim_{s\to 0} s\left(\frac{-1}{Js + f + (K_m K_b / R_a)}\right)\left(\frac{D}{s}\right) \\ &= \frac{-D}{f + (K_m K_b / R_a)}.\end{aligned} \tag{3.36}$$

The closed-loop speed control system is shown in block diagram form in Fig. 3.13. The closed-loop system is shown in the more general signal-flow graph form in Fig. 3.14. The output, $\omega(s)$, of the closed-loop system of Fig. 3.14 can be obtained by utilizing the signal-flow gain formula and is

$$\omega(s) = \frac{-G_2(s)}{1 + G_1(s)G_2(s)H(s)} T_d(s). \tag{3.37}$$

Then, if $G_1G_2H(s)$ is much greater than one over the range of s, we obtain the approximate result

$$\omega(s) \simeq \frac{-1}{G_1(s)H(s)} T_d(s). \tag{3.38}$$

Therefore if $G_1(s)$ is made sufficiently large, the effect of the disturbance can be decreased by closed-loop feedback.

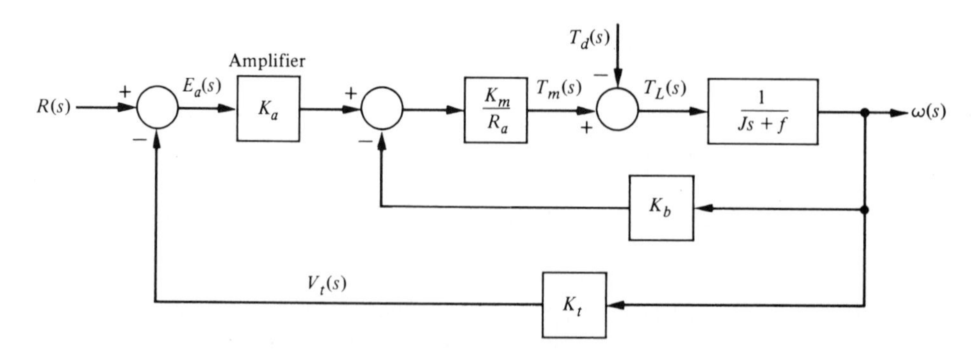

Figure 3.13. Closed-loop speed tachometer control system.

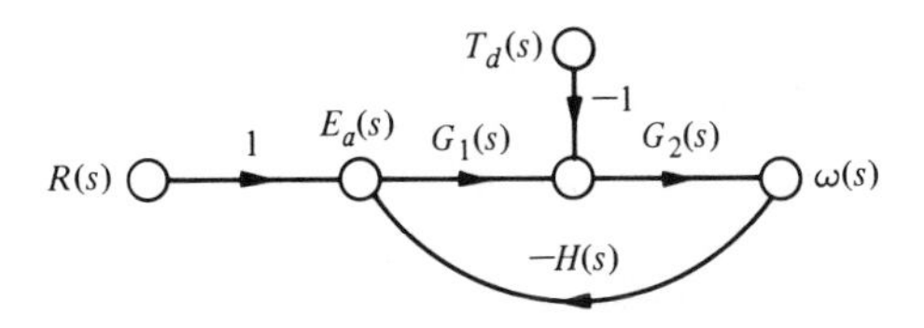

Figure 3.14. Signal-flow graph of closed-loop system.

The output for the speed control system of Fig. 3.13 due to the load disturbance, when the input $R(s) = 0$, may be obtained by using Mason's formula as

$$\begin{aligned}\omega(s) &= \frac{-[1/(Js + f)]}{1 + (K_t K_a K_m/R_a)[1/(Js + f)] + (K_m K_b/R_a)[1/(Js + f)]} T_d(s) \\ &= \frac{-1}{Js + f + (K_m/R_a)(K_t K_a + K_b)} T_d(s).\end{aligned} \tag{3.39}$$

Again, the steady-state output is obtained by utilizing the final-value theorem, and we have

$$\begin{aligned}\lim_{t\to\infty} \omega(t) &= \lim_{s\to 0} (s\omega(s)) \\ &= \frac{-1}{f + (K_m/R_a)(K_t K_a + K_b)} D;\end{aligned} \tag{3.40}$$

when the amplifier gain is sufficiently high, we have

$$\omega(\infty) \simeq \frac{-R_a}{K_a K_m K_t} D. \tag{3.41}$$

The ratio of closed-loop to open-loop steady-state speed output due to an undesirable disturbance is

$$\frac{\omega_c(\infty)}{\omega_0(\infty)} = \frac{R_a f + K_m K_b}{K_a K_m K_t}, \tag{3.42}$$

and is usually less than 0.02.

This advantage of a feedback speed control system can also be illustrated by considering the speed–torque curves for the closed-loop system. The closed-loop system speed–torque curves are shown in Fig. 3.15. Clearly, the improvement of the feedback system is evidenced by the almost horizontal curves, which indicate that the speed is almost independent of the load torque.

In general, a primary reason for introducing feedback is the ability to alleviate the effects of disturbances and noise signals occurring within the feedback loop. A noise signal that is prevalent in many systems is the noise generated by the measurement sensor. This disturbance or noise, $N(s) = T_d(s)$, can be represented

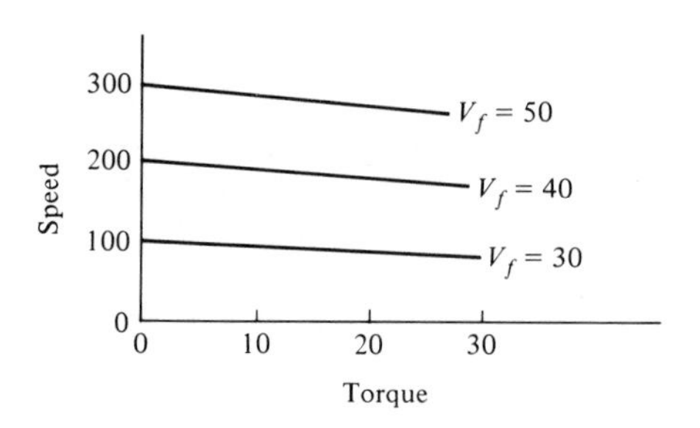

Figure 3.15. The closed-loop system speed–torque curves.

as shown in Fig. 3.16. The effect of the noise on the output is

$$C(s) = \frac{-G_1G_2H_2(s)}{1 + G_1G_2H_1H_2(s)} N(s), \tag{3.43}$$

which is approximately

$$C(s) \cong -\frac{1}{H_1(s)} N(s). \tag{3.44}$$

Clearly, the designer must obtain a maximum value of $H_1(s)$, which is equivalent to maximizing the signal-to-noise ratio of the measurement sensor. This necessity is equivalent to requiring that the feedback elements $H(s)$ be well designed and operated with minimum noise, drift, and parameter variation. This is equivalent to the requirement determined from the sensitivity function, Eq. (3.14), which showed that $S_H^T \cong 1$. Therefore one must be aware of the necessity of assuring the quality and constancy of the feedback sensors and elements. This is usually possible because the feedback elements operate at low power levels and can be well designed at reasonable cost.

The equivalency of sensitivity S_G^T and the response of the closed-loop system to a disturbance input can be illustrated by considering Fig. 3.14. The sensitivity of the system to G_2 is

$$S_{G_2}^T = \frac{1}{1 + G_1G_2H(s)} \cong \frac{1}{G_1G_2H(s)}. \tag{3.45}$$

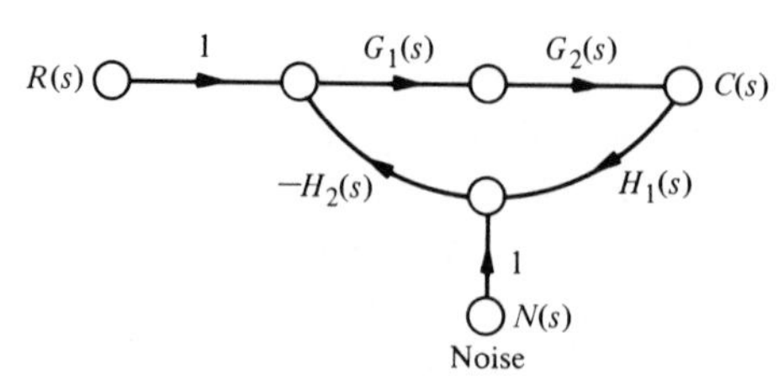

Figure 3.16. Closed-loop control system with measurement noise.

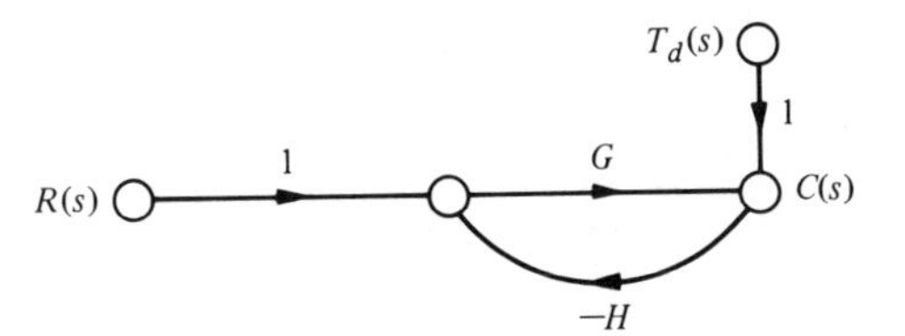

Figure 3.17. Closed-loop control system with output noise.

The effect of the disturbance on the output is

$$\frac{C(s)}{T_d(s)} = \frac{-G_2(s)}{1 + G_1G_2H(s)} \cong \frac{-1}{G_1H(s)}. \tag{3.46}$$

In both cases, we found that the undesired effects could be alleviated by increasing $G_1(s) = K_a$, the amplifier gain. The utilization of feedback in control systems is primarily to reduce the sensitivity of the system to parameter variations and the effect of disturbance inputs. It is noteworthy that the effort to reduce the effects of parameter variations or disturbances is equivalent and we have the fortunate circumstance that they reduce simultaneously. As a final illustration of this fact, we note that for the system shown in Fig. 3.17, the effect of the noise or disturbance on the output is

$$\frac{C(s)}{T_d(s)} = \frac{1}{1 + GH(s)}, \tag{3.47}$$

which is identically equal to the sensitivity S_G^T.

Quite often, noise is present at the input to the control system. For example, the signal at the input to the system might be $r(t) + n(t)$, where $r(t)$ is the desired system response and $n(t)$ is the noise signal. The feedback control system, in this case, will simply process the noise as well as the input signal $r(t)$ and will not be able to improve the signal-noise ratio, which is present at the input to the system. However, if the frequency spectrums of the noise and input signals are of a different character, the output signal-noise ratio can be maximized, often by simply designing a closed-loop system transfer function which has a low-pass frequency response.

3.5 Steady-State Error

A feedback control system is valuable because it provides the engineer with the ability to adjust the transient response. In addition, as we have seen, the sensitivity of the system and the effect of disturbances can be reduced significantly. However, as a further requirement, one must examine and compare the final steady-state error for an open-loop and a closed-loop system. The *steady-state error* is the error after the transient response has decayed leaving only the continuous response.

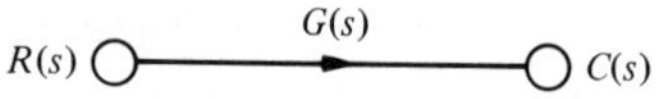

Figure 3.18. Open-loop control system.

The error of the open-loop system shown in Fig. 3.18 is

$$\begin{aligned} E_0(s) &= R(s) - C(s) \\ &= (1 - G(s))R(s). \end{aligned} \tag{3.48}$$

The error of the closed-loop system, $E_c(s)$, shown in Fig. 3.19, when $H(s) = 1$, is

$$E_c(s) = \frac{1}{1 + G(s)} R(s). \tag{3.49}$$

In order to calculate the steady-state error, we utilize the final-value theorem, which is

$$\lim_{t \to \infty} e(t) = \lim_{s \to 0} sE(s). \tag{3.50}$$

Therefore, using a unit step input as a comparable input, we obtain for the open-loop system

$$\begin{aligned} e_0(\infty) &= \lim_{s \to 0} s(1 - G(s)) \left(\frac{1}{s}\right) \\ &= \lim_{s \to 0} (1 - G(s)) \\ &= 1 - G(0). \end{aligned} \tag{3.51}$$

For the closed-loop system, when $H(s) = 1$, we have

$$\begin{aligned} e_c(\infty) &= \lim_{s \to 0} s\left(\frac{1}{1 + G(s)}\right)\left(\frac{1}{s}\right) \\ &= \frac{1}{1 + G(0)}. \end{aligned} \tag{3.52}$$

The value of $G(s)$ when $s = 0$ is often called the dc-gain and is normally greater than one. Therefore, the open-loop system will usually have a steady-state error of significant magnitude. By contrast, the closed-loop system with a reasonably large dc-loop gain $GH(0)$ will have a small steady-state error.

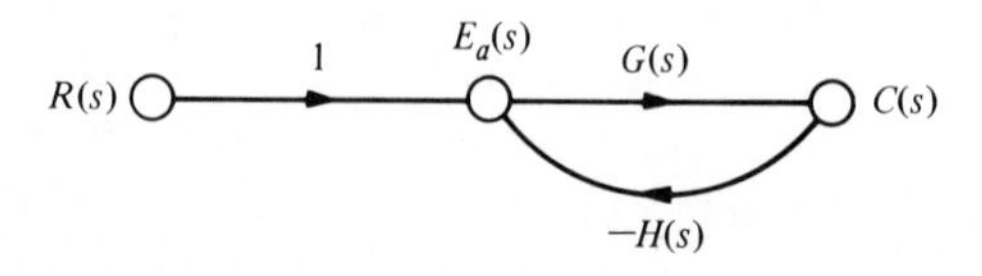

Figure 3.19. Closed-loop control system.

Upon examination of Eq. (3.51), one notes that the open-loop control system can possess a zero steady-state error by simply adjusting and calibrating the dc-gain, $G(0)$, of the system, so that $G(0) = 1$. Therefore, one may logically ask, what is the advantage of the closed-loop system in this case? Again, we return to the concept of the sensitivity of the system to parameter changes as our answer to this question. In the open-loop system, one may calibrate the system so that $G(0) = 1$, but during the operation of the system it is inevitable that the parameters of $G(s)$ will change due to environmental changes and the dc-gain of the system will no longer be equal to one. However, because it is an open-loop system the steady-state error will remain other than zero until the system is maintained and recalibrated. By contrast, the closed-loop feedback system continually monitors the steady-state error and provides an actuating signal in order to reduce the steady-state error. Thus we find that it is the sensitivity of the system to parameter drift, environmental effects, and calibration errors that encourages the introduction of negative feedback. An example of a ingenious feedback control system is shown in Fig. 3.20.

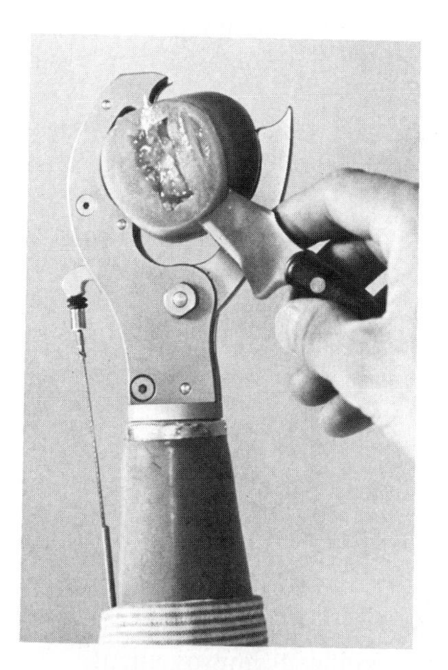

Figure 3.20. The Grip II is a prosthesis artificial hand that is cable operated. It can be used to operate an automobile manual shift, drive a nail, slice a tomato, and other normal tasks requiring two hands. It is based on a "pull to close" cable action and has a gripping force ranging from 0 to 110 pounds. The hand provides the movement of a thumb and forefinger and closes when effort is exerted on the cable by the person's back muscles. A person's vision system provides feedback, but the person does lack the normal sense of touch most of us use to grasp an object lightly. (Courtesy of Therapeutic Recreation Systems, Inc.)

The advantage of the closed-loop system in reducing the steady-state error of the system resulting from parameter changes and calibration errors may be illustrated by an example. Let us consider a system with a process transfer function

$$G(s) = \frac{K}{\tau s + 1}, \tag{3.53}$$

which would represent a thermal control process, a voltage regulator, or a water-level control process. For a specific setting of the desired input variable, which may be represented by the normalized unit step input function, we have $R(s) = 1/s$. Then, the steady-state error of the open-loop system is, as in Eq. (3.51),

$$e_0(\infty) = 1 - G(0) = 1 - K \tag{3.54}$$

when a consistent set of dimensional units are utilized for $R(s)$ and K. The steady-state error for the closed-loop system, with unity-feedback, is

$$e_c(\infty) = \frac{1}{1 + G(0)} = \frac{1}{1 + K}. \tag{3.55}$$

For the open-loop system, one would calibrate the system so that $K = 1$ and the steady-state error is zero. For the closed-loop system, one would set a large gain K, for example, $K = 100$. Then the closed-loop system steady-state error is $e_c(\infty) = 1/101$.

If the calibration of the gain setting drifts or changes in some way by $\Delta K/K = 0.1$, a 10% change, the open-loop steady-state error is $\Delta e_0(\infty) = 0.1$ and the percent change from the calibrated setting is

$$\frac{\Delta e_0(\infty)}{|r(t)|} = \frac{0.10}{1}, \tag{3.56}$$

or 10%. By contrast, the steady-state error of the closed-loop system, with $\Delta K/K = 0.1$, is $e_c(\infty) = 1/91$ if the gain decreases. Thus the change in the

$$\Delta e_c(\infty) = \frac{1}{101} - \frac{1}{91}, \tag{3.57}$$

and the relative change is

$$\frac{\Delta e_c(\infty)}{|r(t)|} = 0.0011, \tag{3.58}$$

or 0.11%. This is indeed a significant improvement.

3.6 The Cost of Feedback

The addition of feedback to a control system results in the advantages outlined in the previous sections. However, it is natural that these advantages have an attendant cost. The cost of feedback is first manifested in the increased number of *components* and the *complexity* of the system. In order to add the feedback, it is necessary to consider several feedback components, of which the measurement component (sensor) is the key component. The sensor is often the most expensive component in a control system. Furthermore, the sensor introduces noise and inaccuracies into the system.

The second cost of feedback is the *loss of gain*. For example, in a single-loop system, the open-loop gain is $G(s)$ and is reduced to $G(s)/(1 + G(s))$ in a unity negative feedback system. The reduction in closed-loop gain is $1/(1 + G(s))$, which is exactly the factor that reduces the sensitivity of the system to parameter variations and disturbances. Usually, we have open-loop gain to spare, and we are more than willing to trade it for increased control of the system response.

However, we should note that it is the gain of the input-output transmittance that is reduced. The control system does possess a substantial power gain, which is fully utilized in the closed-loop system.

Finally, a cost of feedback is the introduction of the possibility of *instability*. While the open-loop system is stable, the closed-loop system may not be always stable. The question of the stability of a closed-loop system is deferred until Chapter 5, where it can be treated more completely.

The addition of feedback to dynamic systems results in several additional problems for the designer. However, for most cases, the advantages far outweigh the disadvantages, and a feedback system is utilized. Therefore it is necessary to consider the additional complexity and the problem of stability when designing a control system. One complex control system is shown in Fig. 3.21.

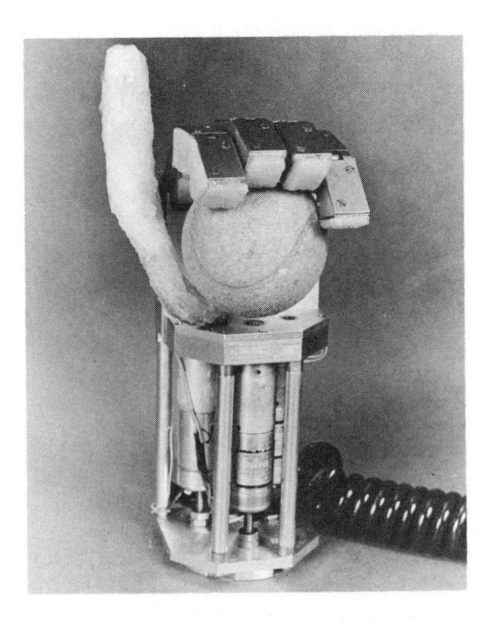

Figure 3.21. The Belgrade-USC hand [9]. This dextrous hand using sensory integration has five fingers but only two motors. There are no tendons; rather a mechanical linkage allows all the fingers to close as far as possible, thus making the hand automatically adapt to the shape of the grasped object.

It has become clear that it is desired that the output of the system $C(s)$ equal the input $R(s)$. However, upon reflection, one might ask, "Why not simply set the transfer function $G(s) = C(s)/R(s)$ equal to 1?" (See Fig. 3.2.) The answer to this question becomes apparent once we recall that the process (or plant) $G(s)$ was necessary in order to provide the desired output; that is, the transfer function $G(s)$ represents a real process and possesses dynamics that cannot be neglected. If we set $G(s)$ equal to 1, we imply that the output is directly connected to the input. However, one must recall that a specific output, such as temperature, shaft rotation, or engine speed, is desired, whereas the input might be a potentiometer setting or a voltage. The process $G(s)$ is necessary in order to provide the physical process between $R(s)$ and $C(s)$. Therefore, a transfer function $G(s) = 1$ is unrealizable, and we must settle for a practical transfer function.

3.7 Design Example: English Channel Boring Machines

The construction of the tunnel under the English Channel from France to Great Britain began in December 1987. The first connection of the boring tunnels from each country was achieved in November 1990. The tunnel is 23.5 miles long and is bored 200 feet below sea level. The tunnel, when completed in 1992, will accommodate 500 train trips daily and may have a total cost of over $14 billion. This construction will be a critical link between Europe and Great Britain, making it possible for a train to reach Paris from London in three hours.

The machines operating from both ends of the Channel are boring toward the middle. In order to link up in the middle of the Channel with the necessary accuracy, a laser guidance system keeps the machines precisely aligned. A model of the boring machine control is shown in Fig. 3.22, where $C(s)$ is the actual angle of direction of travel of the boring machine and $R(s)$ is the desired angle. The effect of load on the machine is represented by the disturbance $D(s)$.

The design objective is to select the gain K so that the response to input angle changes is desirable while maintaining minimal error due to the disturbance. Using Mason's rule, the output due to the two inputs is

$$
\begin{aligned}
C(s) &= T(s)R(s) + T_d(s)D(s) \\
&= \frac{KG(s)}{1 + KG(s)} R(s) + \frac{G(s)}{1 + KG(s)} D(s) \qquad (3.59) \\
&= \frac{K}{s^2 + 12s + K} R(s) + \frac{1}{s^2 + 12s + K} D(s).
\end{aligned}
$$

Thus, to reduce the effect of the disturbance, we wish to set the gain greater than 10. When we select $K = 100$ and let $d(t) = 0$, we have the step response for a unit step input $r(t)$, as shown in Fig. 3.23(a). Letting the input $r(t) = 0$ and determining the response to the unit step disturbance, we obtain $c(t)$, as shown in Fig. 3.23(b). Clearly, the effect of the disturbance is quite small. If we set the gain K

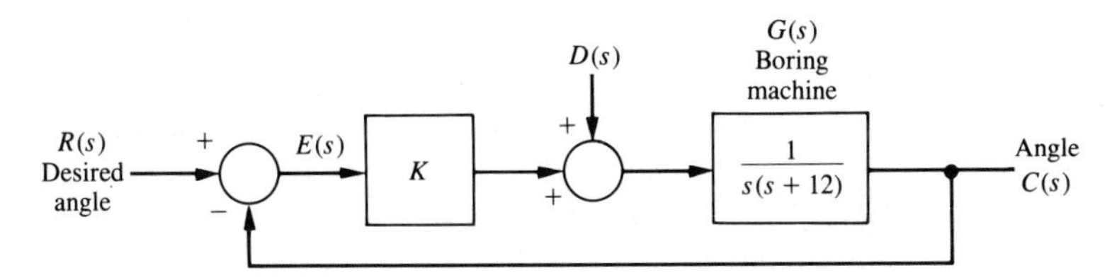

Figure 3.22. A block diagram model of a boring machine control system.

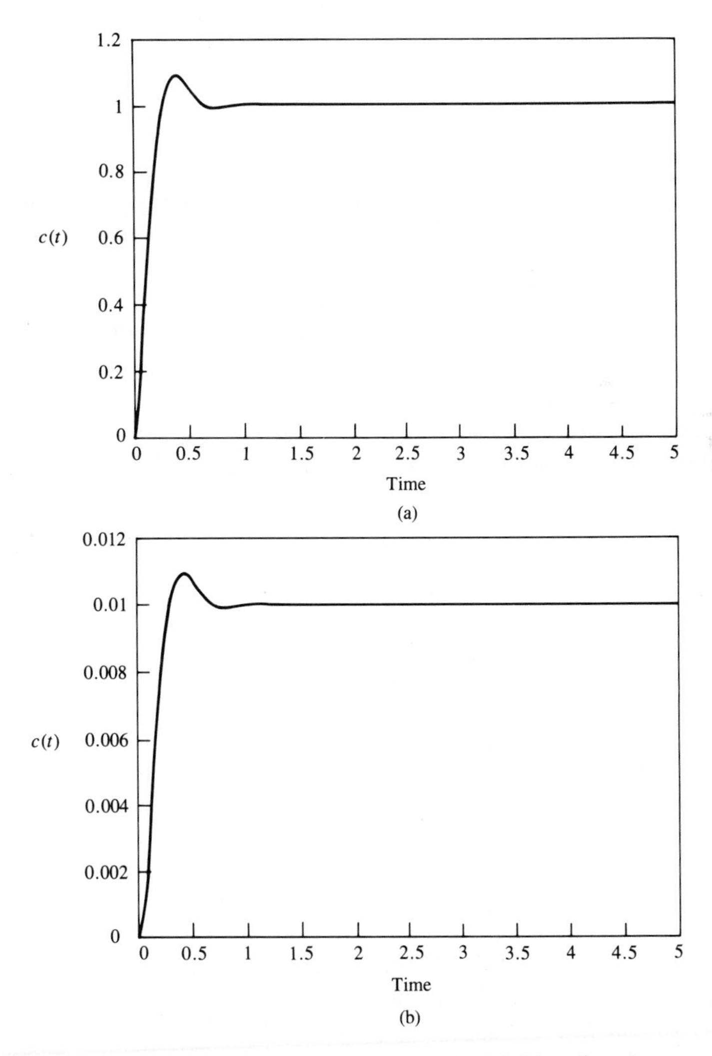

Figure 3.23. The response $c(t)$ to (a) an input step $r(t)$ and (b) a disturbance step input $d(t)$ for $K = 100$.

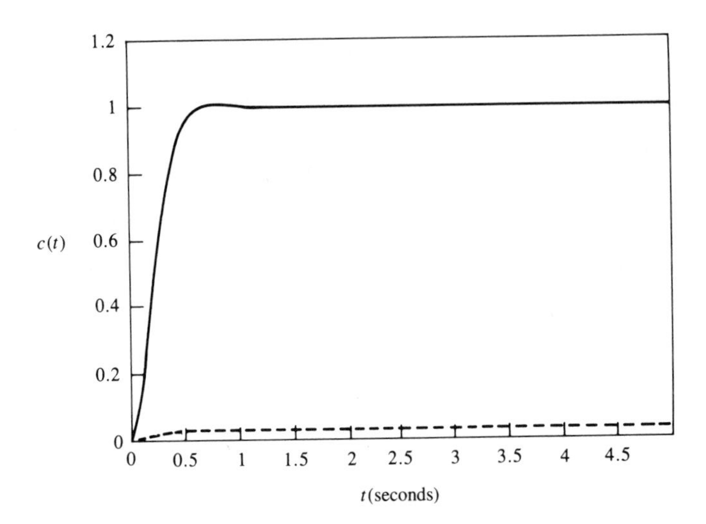

Figure 3.24. The response $c(t)$ for a step input (solid line) and for a step disturbance (dashed line) for $K = 50$.

equal to 50, the responses of $c(t)$ due to a unit step input $r(t)$ and $d(t)$ are displayed together in Fig. 3.24. Since the overshoot of the response is small (less than 2%) and the steady state is attained in one second, we would prefer $K = 50$. The results are summarized in Table 3.1.

The steady-state error of the system to a unit step input $R(s) = 1/s$ is

$$\begin{aligned} \lim_{t \to \infty} e(t) &= \lim_{s \to 0} s \frac{1}{1 + KG(s)} \left(\frac{1}{s}\right) \\ &= 0. \end{aligned} \tag{3.60}$$

The steady-state value of $c(t)$ when the disturbance is a unit step, $D(s) = 1/s$, and the desired value is $r(t) = 0$ is

$$\begin{aligned} \lim_{t \to \infty} c(t) &= \lim_{s \to 0} s \left[\frac{G(s)}{1 + KG(s)}\right] D(s) \\ &= \lim_{s \to 0} \left[\frac{1}{s(s + 12) + K}\right] \\ &= \frac{1}{K}. \end{aligned} \tag{3.61}$$

Thus the steady-state value is 0.01 and 0.02 for $K = 100$ and 50, respectively.

Table 3.1. Response of the Boring System for Two Gains

Gain K	Overshoot of response to $r(t)$ = step	Time for response to $r(t)$ = step to reach steady state	Steady-state response $c(t)$ for $d(t)$ = step with $r(t) = 0$	Steady-state error of response to $r(t)$ = step with $d(t) = 0$
100	10%	1.5s	0.01	0
50	1%	1.0s	0.02	0

Finally, we examine the sensitivity of the system to a change in the process $G(s)$ using Eq. (3.13). Then,

$$S = \frac{1}{1 + KG(s)} = \frac{s(s+12)}{s(s+12)+K}. \tag{3.62}$$

For low frequencies ($|s| < 4$), the sensitivity can be approximated by

$$S \simeq \frac{12s}{K}, \tag{3.63}$$

where $K \geq 50$. Thus the sensitivity of the system is reduced by increasing the gain K. In this case we choose K, which reasonably satisfies all our design requirements, thus selecting $K = 50$ for a reasonable design compromise.

3.8 Summary

The fundamental reasons for using feedback, despite its cost and additional complexity, are as follows:

1. Decrease in the sensitivity of the system to variations in the parameters of the process $G(s)$
2. Ease of control and adjustment of the transient response of the system
3. Improvement in the rejection of the disturbance and noise signals within the system
4. Improvement in the reduction of the steady-state error of the system

The benefits of feedback can be illustrated by considering the system shown in Fig. 3.25(a). This system can be considered for several values of gain, K. Table 3.2 summarizes the results of the system operated as an open loop system (disconnect the feedback path) and for several values of gain, K with the feedback

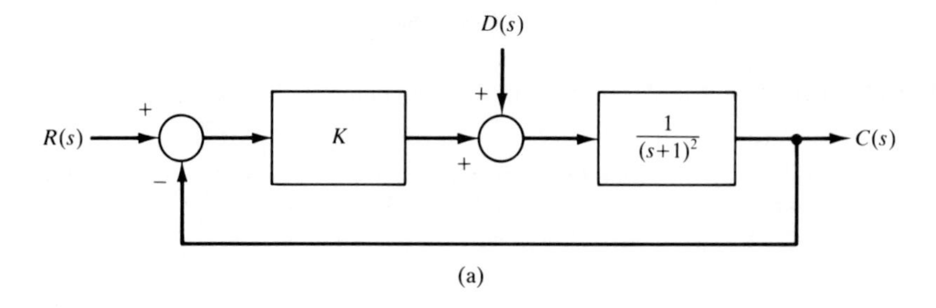

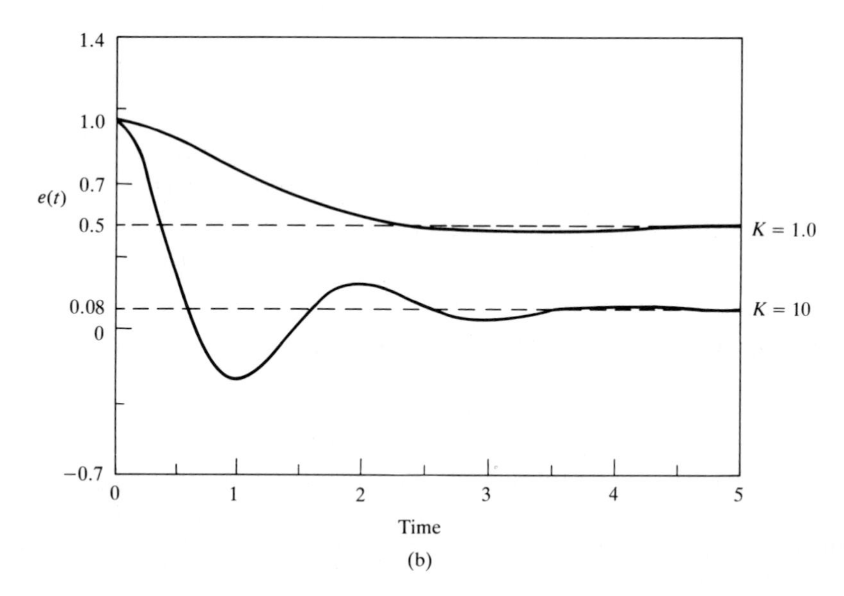

Figure 3.25. (a) A single-loop feedback control system. (b) The error response for a unit step disturbance.

Table 3.2.

	Open Loop	**Closed Loop**		
	$K = 1$	$K = 1$	$K = 8$	$K = 10$
Rise time (seconds) (10% to 90% of final value)	3.35	1.52	0.45	0.38
Percent overshoot (%)	0	4.31	33	40
Final value of $c(t)$ due to a disturbance, $D(s) = \frac{1}{s}$	1.0	0.50	0.11	0.09
Percent steady-state error for unit step input	0	50%	11%	9%
Percent change in steady-state error due to 10% decrease in K	10%	5.3%	1.2%	0.9%

connected. It is clear that the rise time of the system and sensitivity of the system is reduced as the gain is increased. Also, the feedback system demonstrates excellent reduction of the steady state error as the gain is increased. Finally, Fig. 3.25(b) shows the response for a unit step disturbance (when $R(s) = 0$) and shows how a larger gain will reduce the effect of the disturbance.

Feedback control systems possess many beneficial characteristics, and it is not surprising that one finds a multitude of feedback control systems in industry, government, and nature.

Exercises

E3.1. A closed-loop system is used to track the sun in order to obtain maximum power from a photovoltaic array. The tracking system may be represented by Fig. 3.3 with $H(s) = 1$ and

$$G(s) = \frac{100}{3s + 1}.$$

(a) Calculate the sensitivity of this system. (b) Calculate the time constant of the closed-loop system response.

Answers: $s = (3s + 1)/(3s + 100)$; $\tau_c = 3/101$

E3.2. A digital audio system is designed to minimize the effect of disturbances and noise as shown in Fig. E3.2. As an approximation, we may represent $G(s) = K_2$. (a) Calculate the sensitivity of the system due to K_2. (b) Calculate the effect of the disturbance on V_{out}. (c) What value would you select for K_1 to minimize the effect of the disturbance?

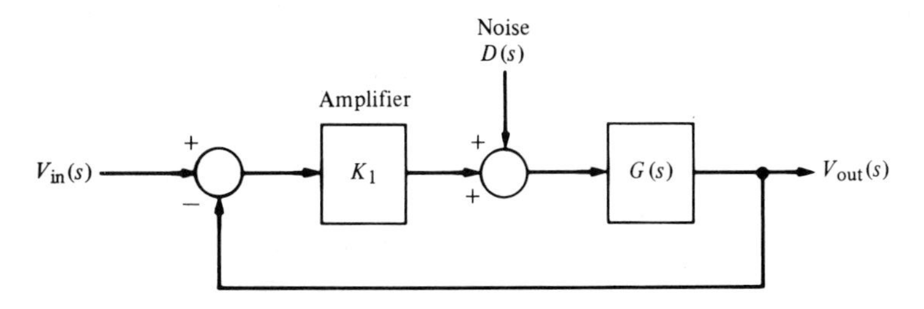

Figure E3.2. Digital audio system.

E3.3. A robot arm and camera could be used to pick fruit as shown in Fig. E3.3(a). The camera is used to close the feedback loop to a microcomputer, which controls the arm [14]. The process is

$$G(s) = \frac{K}{(s + 2)^2}.$$

(a) Calculate the expected steady-state error of the gripper for a step command A as a function of K. (b) Name a possible disturbance signal for this system.

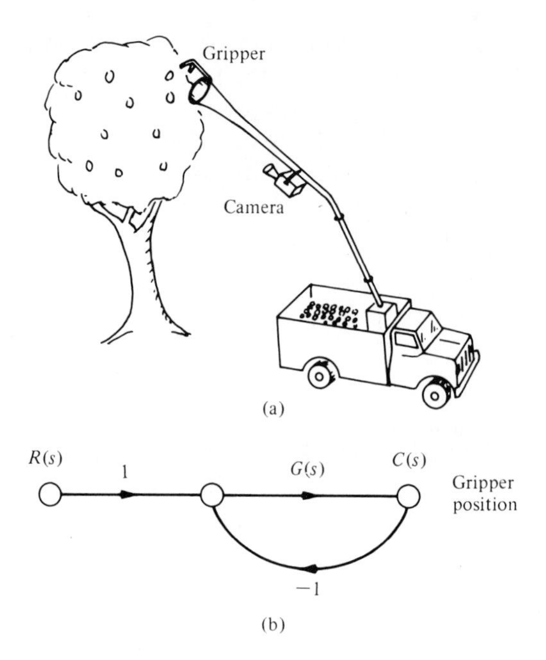

Figure E3.3. Robot fruit picker.

E3.4. A magnetic disk drive requires a motor to position a read/write head over tracks of data on a spinning disk as shown in Fig. E3.4, on the next page. The motor and head may be represented by

$$G(s) = \frac{10}{s(\tau s + 1)},$$

where $\tau = 0.001$ sec. The controller takes the difference of the actual and desired positions and generates an error. This error is multiplied by an amplifier K. (a) What is the steady-state position error for a step change in the desired input? (b) Calculate the required K in order to yield a steady-state error of 1 mm for a ramp input of 10 cm/sec.

Answers: $e_{ss} = 0$; $K = 10$

E3.5. Most people have experienced an out of focus slide projector. A projector with an automatic focus adjusts for variations in slide position and temperature disturbances [19]. Draw the block diagram of an autofocus system and describe how the system works. An unfocused slide projection is a visual example of steady-state error.

E3.6. Four-wheel drive autos are popular in regions where winter road conditions are often slippery due to snow and ice. A four-wheel drive with antilock brakes uses a sensor to keep each wheel rotating to maintain traction. One system is shown in Fig. E3.6. Find the closed-loop response of this system as it attempts to maintain a constant speed of the wheel. Use a computer program to determine the response when $R(s) = A/s$.

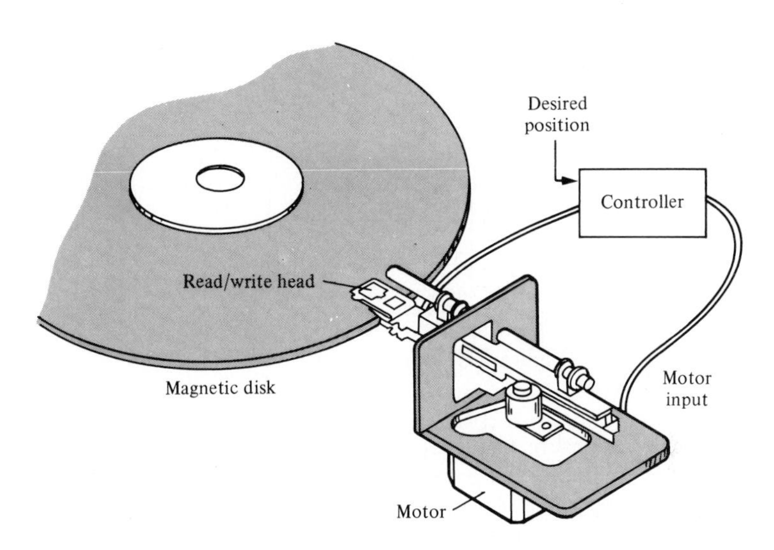

Figure E3.4. Disk drive control.

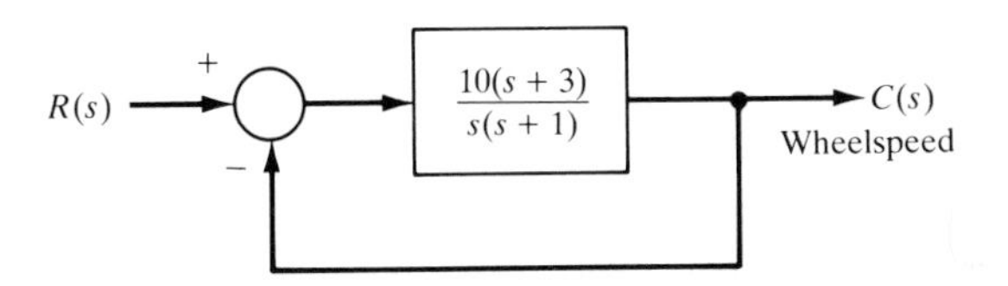

Figure E3.6. Four-wheel drive auto.

Problems

P3.1. The open-loop transfer function of a fluid-flow system obtained in Problem 2.13 can be written as

$$G(s) = \frac{\Delta Q_2(s)}{\Delta Q_1(s)} = \frac{1}{\tau s + 1},$$

where $\tau = RC$, R is a constant equivalent to the resistance offered by the orifice so that $1/R = \frac{1}{2}kH_0^{-1/2}$, and $C =$ the cross-sectional area of the tank. Since $\Delta H = R\Delta Q_2$, we have for the transfer function relating the head to the input change:

$$G_1(s) = \frac{\Delta H(s)}{\Delta Q_1(s)} = \frac{R}{RCs + 1}.$$

For a closed-loop feedback system, a float level sensor and valve may be used as shown in Fig. P3.1 [11]. Assuming the float is a negligible mass, the valve is controlled so that a reduction in the flow rate, ΔQ_1, is proportional to an increase in head, ΔH, or $\Delta Q_1 = -K\,\Delta H$. Draw a closed-loop flow graph or block diagram. Determine and compare the open-loop and closed-loop system for (a) sensitivity to changes in the equivalent coefficient R and the feedback coefficient K; (b) ability to reduce the effects of a disturbance in the level $\delta H(s)$; and (c) steady-state error of the level (head) for a step change of the input $\Delta Q_1(s)$.

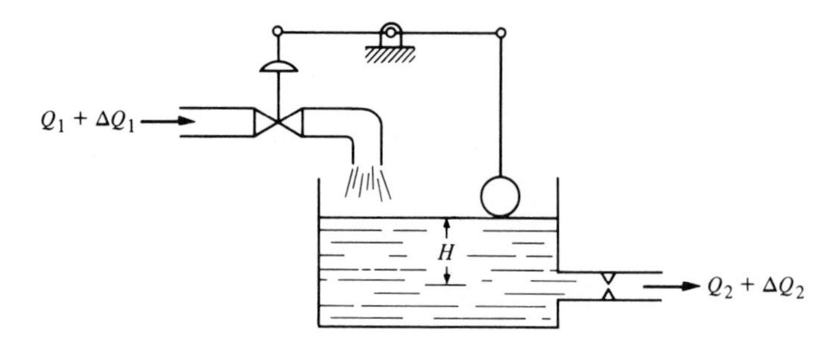

Figure P3.1. Tank level control.

P3.2. It is important to ensure passenger comfort on ships by stabilizing the ship's oscillations due to waves [5, 6]. Most ship stabilization systems use fins or hydrofoils projecting into the water in order to generate a stabilization torque on the ship. A simple diagram of a ship stabilization system is shown in Fig. P3.2. The rolling motion of a ship can be regarded as an oscillating pendulum with a deviation from the vertical of θ degrees and a typical period of 3 sec. The transfer function of a typical ship is

$$G(s) = \frac{\omega_n^2}{s^2 + 2\zeta\omega_n s + \omega_n^2},$$

where $\omega_n = 2\pi/T = 2$, $T = 3.14$ sec, and $\zeta = 0.10$. With this low damping factor ζ, the oscillations continue for several cycles and the rolling amplitude can reach 18° for the expected amplitude of waves in a normal sea. Determine and compare the open-loop and closed-loop system for (a) sensitivity to changes in the actuator constant K_a and the roll sensor K_1; and (b) the ability to reduce the effects of the disturbance of the waves. Note that the desired roll $\theta_d(s)$ is zero degrees.

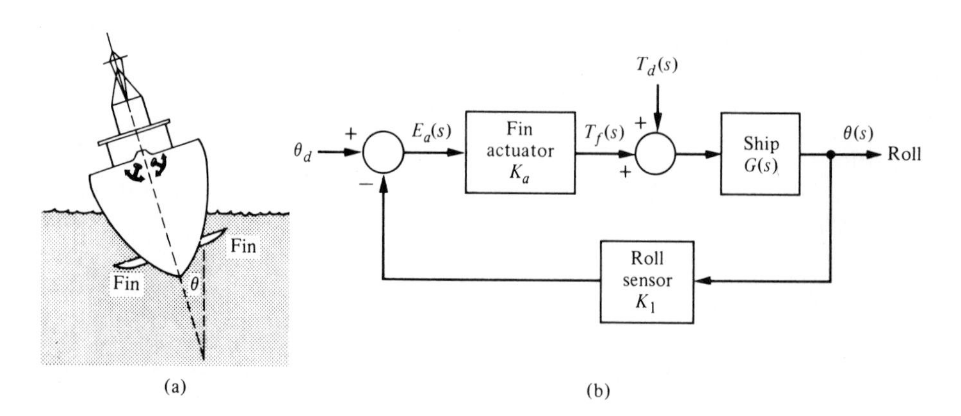

Figure P3.2. Ship stabilization system.

P3.3. One of the most important variables that must be controlled in industrial and chemical systems is temperature. A simple representation of a thermal control system is shown in Fig. P3.3 [20]. The temperature $\mathcal{T}$ of the process is controlled by the heater with a resistance R. An approximate representation of the dynamics linearly relates the heat loss from the process to the temperature difference $(\mathcal{T} - \mathcal{T}_e)$. This relation holds if the temperature difference is relatively small and the energy storage of the heater and the vessel walls is negligible. Also, it is assumed that the voltage connected to the heater e_h, is proportional to e_{desired} or $e_h = kE_b = k_a E_b e(t)$, where k_a is the constant of the actuator. Then the linearized open-loop response of the system is

$$\mathcal{T}(s) = \frac{(k_1 k_a E_b)}{\tau s + 1} E(s) + \frac{\mathcal{T}_e(s)}{\tau s + 1},$$

where

$\tau = MC/\rho A$,
M = mass in tank,
A = surface area of tank,
ρ = heat transfer constant,
C = specific heat constant,
k_1 = a dimensionality constant,
e_{th} = output voltage of thermocouple.

Determine and compare the open-loop and closed-loop systems for (a) sensitivity to changes in the constant $K = k_1 k_a E_b$; (b) ability to reduce the effects of a step disturbance in the environmental temperature $\Delta\mathcal{T}_e(s)$; and (c) the steady-state error of the temperature controller for a step change in the input, e_{desired}.

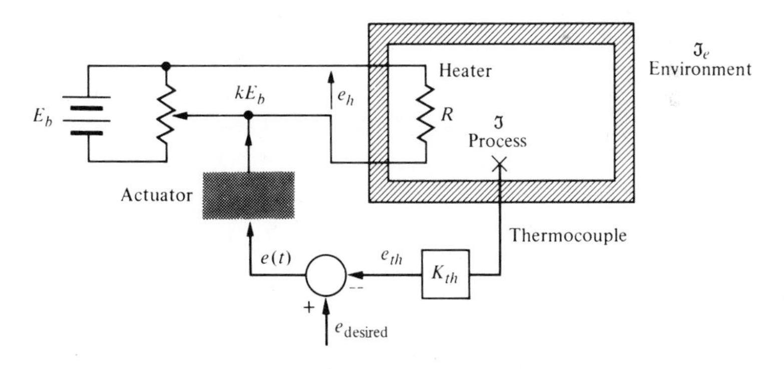

Figure P3.3. Temperature control.

P3.4. A control system has two forward signal paths as shown in Fig. P3.4. (a) Determine the overall transfer function $T(s) = C(s)/R(s)$. (b) Calculate the sensitivity S_G^T using Eq. (3.16). (c) Does the sensitivity depend on $U(s)$ or $M(s)$?

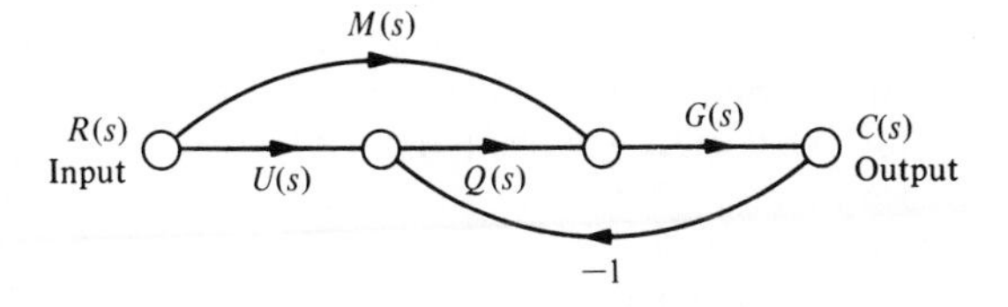

Figure P3.4. Two path system.

P3.5. Large microwave antennas have become increasingly important for missile tracking, radio astronomy, and satellite tracking. A large antenna, for example, with a diameter of 60 ft, is subject to large wind gust torques. A proposed antenna is required to have an error of less than 0.20° in a 35 mph wind. Experiments show that this wind force exerts a maximum disturbance at the antenna of 200,000 lb-ft at 35 mph, or equivalent to 20 volts at the input, $T_d(s)$, to the amplidyne. Also, one problem of driving large antennas is the form of the system transfer function that possesses a structural resonance. The antenna servosystem is shown in Fig. P3.5. The transfer function of the antenna, drive motor, and amplidyne is approximated by

$$G(s) = \frac{\omega_n^2}{s^2 + 2\zeta\omega_n s + \omega_n^2},$$

where $\zeta = 0.6$ and $\omega_n = 10$. The transfer function of the magnetic amplifier is approximately

$$G_1(s) = \frac{k_a}{\tau s + 1},$$

where $\tau = 0.20$ sec. (a) Determine the sensitivity of the system to a change of the parameter k_a. (b) The system is subjected to a disturbance $T_d(s) = 15/s$. Determine the required magnitude of k_a in order to maintain the steady-state error of the system less than 0.20° when the input $R(s)$ is zero. (c) Determine the error of the system when subjected to a disturbance $T_d(s) = 15/s$ when it is operating as an open-loop system ($k_s = 0$) with $R(s) = 0$.

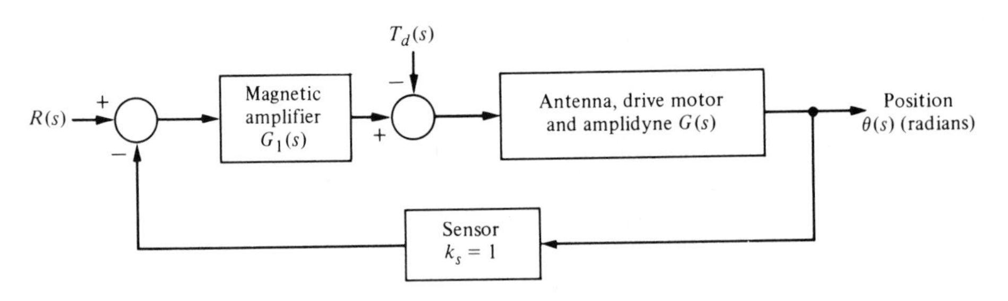

Figure P3.5. Antenna control system.

P3.6. An automatic speed control system will be necessary for passenger cars traveling on the automatic highways of the future. A model of a feedback speed control system for a standard vehicle is shown in Fig. P3.6. The load disturbance due to a percent grade, $\Delta D(s)$, is shown also. The engine gain K_e varies within the range of 10 to 1,000 for various models of automobiles. The engine time constant, τ_e, is 20 sec. (a) Determine the sensitivity of the system to changes in the engine gain K_e. (b) Determine the effect of the load torque on the speed. (c) Determine the constant percent grade $\Delta D(s) = \Delta d/s$ for which the engine stalls in terms of the gain factors. Note that since the grade is constant, the steady-state solution is sufficient. Assume that $R(s) = 30/s$ km/hr and that $K_e K_1 \gg 1$. When $(K_g/K_1) = 2$, what percent grade Δd would cause the automobile to stall?

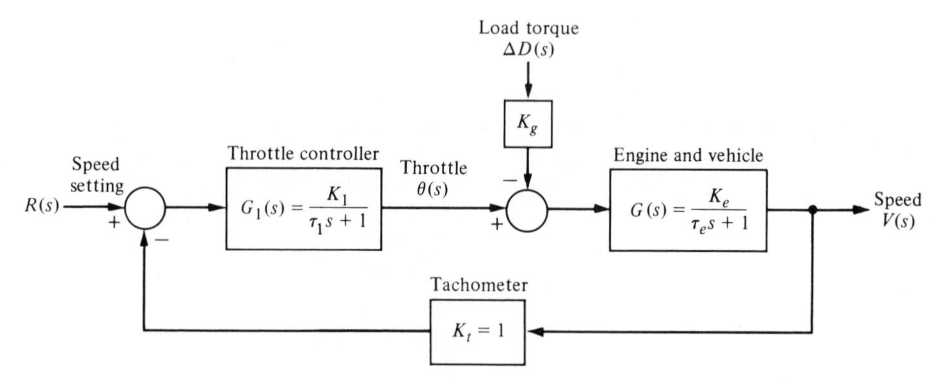

Figure P3.6. Automobile speed control.

P3.7. A robot using an electro-hydraulic actuator is shown in Fig. P3.7(a), on the next page. The EHA 1050 is a microprocessor-controlled hydraulic robot incorporating five axes and a two-fingered gripper that can be programmed by either a remote teaching pendant or an optional external computer. The robot can store eight program sequences in its non-volatile memory, each consisting of up to 64 preprogrammed arm positions. The closed-loop system provides continuous position feedback from all axes including gripper, thus allowing object recognition and maximum grip control.

The robot's two-speed system automatically switches from normal to low speed as the arm approaches programmed preset coordinates, thus enabling more precise positioning. The ARMDRAULIC 1052 robot has a maximum payload of five pounds fully retracted, and three pounds at full extension.

The system will be deflected by the load carried in the gripper. Thus the system may be represented by Fig. P3.7(b) where the load torque is D/s. Assume $R(s) = 0$ at the index position. (a) What is the effect of $T_L(s)$ on $C(s)$? (b) Determine the sensitivity of the closed loop to K_2. (c) What is the steady-state error when $R(s) = 1/s$ and $T_L(s) = 0$?

P3.8. Extreme temperature changes result in many failures of electronic circuits [18]. Temperature control feedback systems reduce the change of temperature by using a heater to overcome outdoor low temperatures. A block diagram of one system is shown in Fig. P3.8, on the next page. The effect of a drop in environmental temperature is a step decrease in $D(s)$. The actual temperature of the electronic circuit is $C(s)$. The dynamics of the electronic circuit temperature change are represented by

$$G(s) = \frac{324}{s^2 + 20s + 324}.$$

(a) Determine the sensitivity of the system to K. (b) Obtain the effect of the disturbance $D(s)$ on the output $C(s)$.

P3.9. A useful unidirectional sensing device is the photoemitter sensor [6]. A light source is sensitive to the emitter current flowing and alters the resistance of the photosensor. Both the light source and the photoconductor are packaged in a single four-terminal device. This device provides a large gain and total isolation. A feedback circuit utilizing this device is

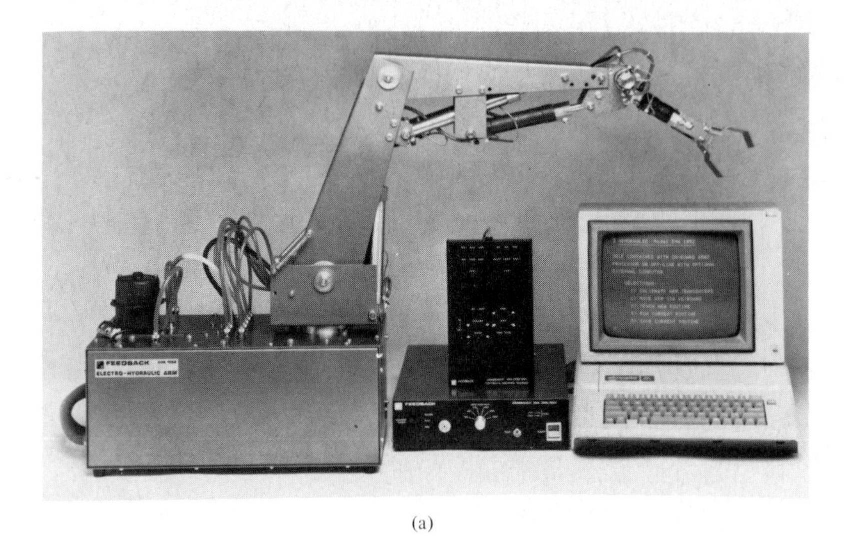

(a)

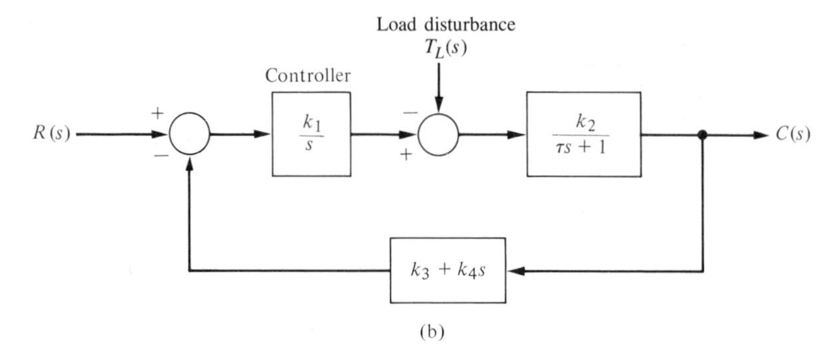

(b)

Figure P3.7. Robot control. (Courtesy of Feedback, Inc.)

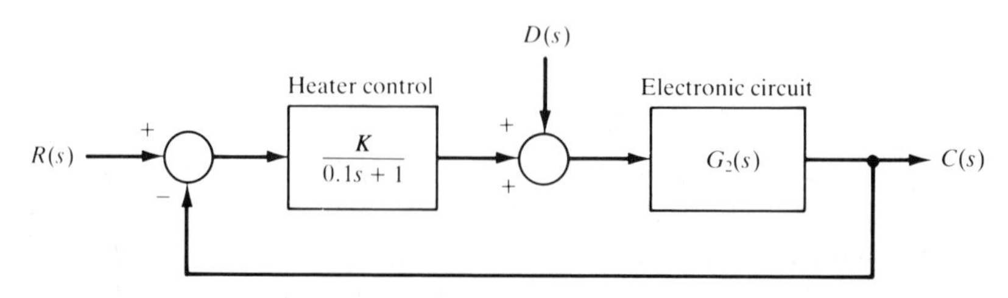

Figure P3.8. Temperature control.

shown in Fig. P3.9(a), and the nonlinear resistance-current characteristic is shown in Fig. 3.9(b) for the Raytheon CK1116. The resistance curve can be represented by the equation

$$\log_{10} R = \frac{0.175}{(i - 0.005)^{1/2}}$$

where i is the lamp current. The normal operating point is obtained when $e_{out} = 35$ V, and $e_{in} = 2.0$ V. (a) Determine the closed-loop transfer function of the system. (b) Determine the sensitivity of the system to changes in the gain K.

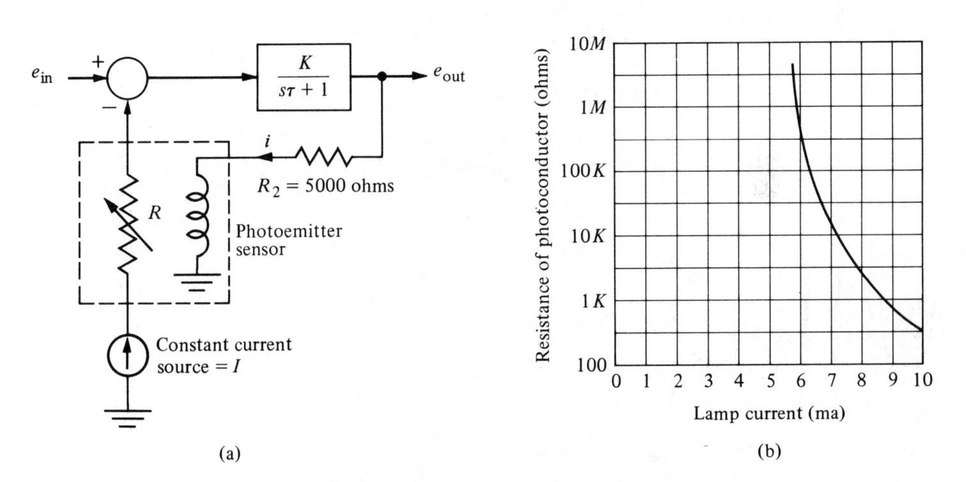

Figure P3.9. Photosensor system.

P3.10. For a paper processing plant, it is important to maintain a constant tension on the continuous sheet of paper between the windoff and windup rolls. The tension varies as the widths of the rolls change, and an adjustment in the take-up motor speed is necessary, as shown in Fig. P3.10. If the windup motor speed is uncontrolled, as the paper transfers from the windoff roll to the windup roll, the velocity v_0 decreases and the tension of the paper drops [10, 15]. The three-roller and spring combination provides a measure of the tension of the paper. The spring force is equal to $k_1 y$ and linear differential transformer, rectifier, and amplifier may be represented by $e_0 = -k_2 y$. Therefore the measure of the tension is described by the relation $2T(s) = k_1 y$, where y is the deviation from the equilibrium condition and $T(s)$ is the vertical component of the deviation in tension from the equilibrium condition. The time constant of the motor is $\tau = L_a/R_a$ and the linear velocity of the windup roll is twice the angular velocity of the motor; that is, $v_0(t) = 2\omega_0(t)$. The equation of the motor is then

$$E_0(s) = \frac{1}{K_m}[\tau s\omega_0(s) + \omega_0(s)] + k_3\,\Delta T(s),$$

where ΔT = a tension disturbance. (a) Draw the closed-loop block diagram for the system, including the disturbance $\Delta T(s)$. (b) Add the effect of a disturbance in the windoff roll velocity $\Delta V_1(s)$ to the block diagram. (c) Determine the sensitivity of the system to the motor constant K_m. (d) Determine the steady-state error in the tension when a step disturbance in the input velocity $\Delta V_1(s) = A/s$ occurs.

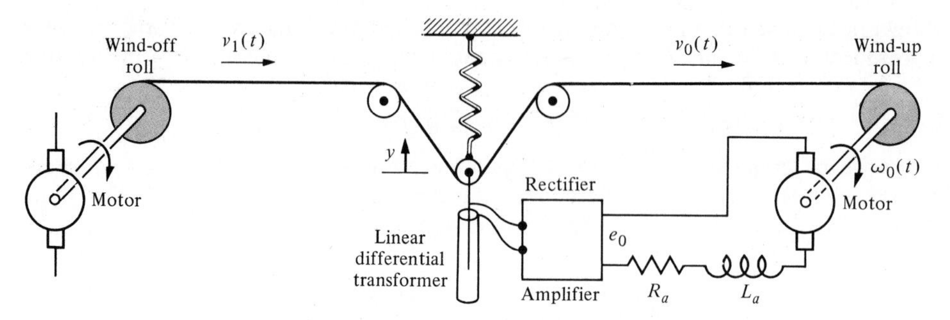

Figure P3.10. Paper tension control.

P3.11. One important objective of the paper-making process is to maintain uniform consistency of the stock output as it progresses to drying and rolling. A diagram of the thick stock consistency dilution control system is shown in Fig. P3.11(a). The amount of water added determines the consistency. The signal-flow diagram of the system is shown in Fig. P3.11(b). Let $H(s) = 1$ and

$$G_c(s) = \frac{K}{(10s + 1)},$$

$$G(s) = \frac{1}{(2s + 1)}.$$

Determine (a) the closed-loop transfer function $T(s) = C(s)/R(s)$, (b) the sensitivity S_K^T, and (c) the steady-state error for a step change in the desired consistency $R(s) = A/s$. (d) Calculate the value of K required for an allowable steady-state error of 1%.

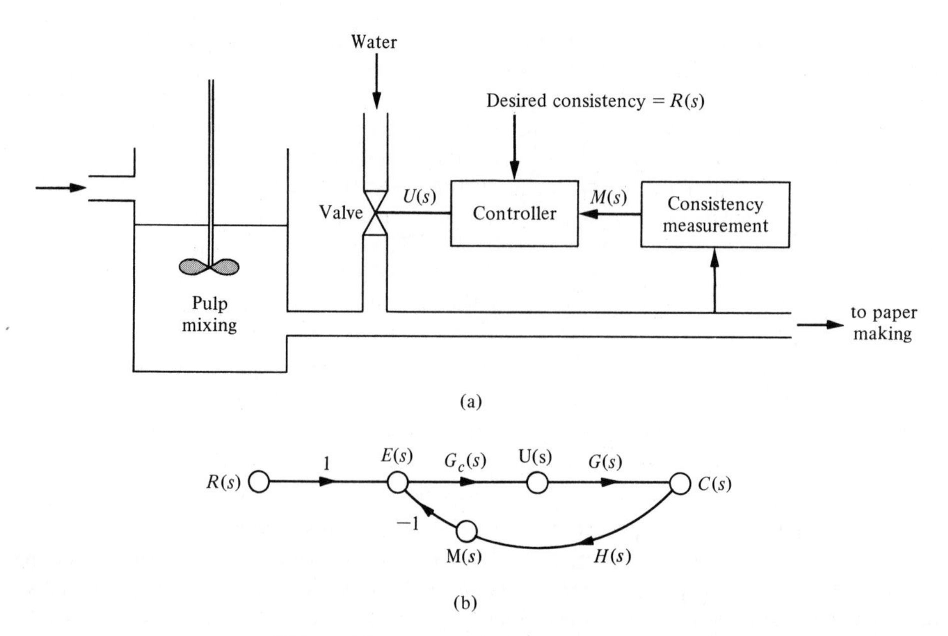

Figure P3.11. Paper making control.

P3.12. Two feedback systems are shown in signal-flow diagram form in Fig. P3.12(a) and (b). (a) Evaluate the closed-loop transfer functions T_1 and T_2 for each system. (b) Show that $T_1 = T_2 = 100$ when $K_1 = K_2 = 100$. (c) Compare the sensitivities of the two systems with respect to the parameter K_1 for the nominal values of $K_1 = K_2 = 100$.

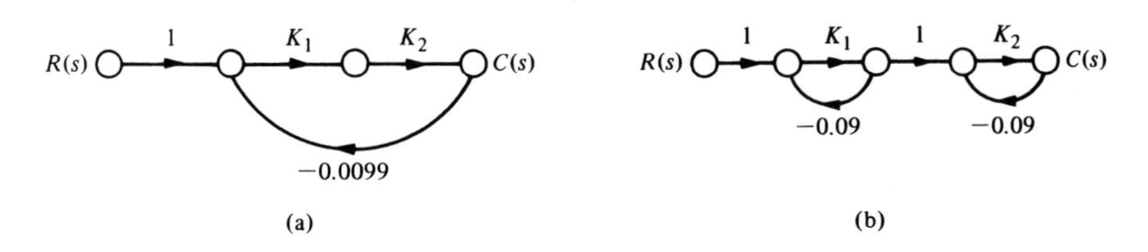

Figure P3.12.

P3.13. One form of closed-loop transfer function is

$$T(s) = \frac{G_1(s) + kG_2(s)}{G_3(s) + kG_4(s)}.$$

(a) Use Eq. (3.16) to show that [1]

$$S_k^T = \frac{k(G_2G_3 - G_1G_4)}{(G_3 + kG_4)(G_1 + kG_2)}.$$

(b) Determine the sensitivity of the system shown in Fig. P3.13, using the equation verified in part (a).

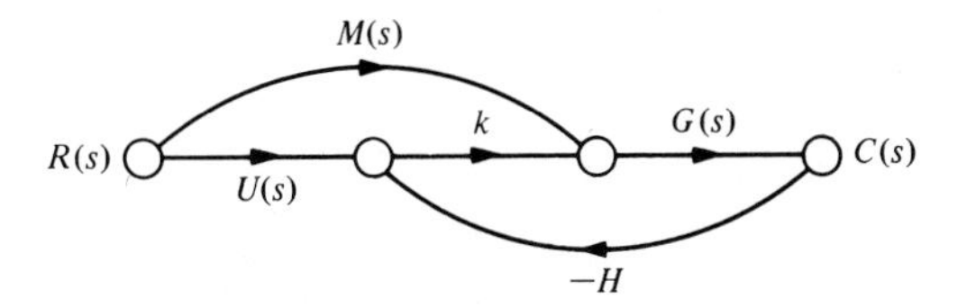

Figure P3.13.

P3.14. A proposed hypersonic plane, as shown in Fig. P3.14(a) on the next page would climb to 100,000 feet and fly 3800 miles per hour, and cross the Pacific in two hours. Control of speed of the aircraft could be represented by the model of Fig. P3.14(b). Find the sensitivity of the closed-loop transfer function, $T(s)$, to a small change in the parameter, a.

P3.15. The steering control of a modern ship may be represented by the system shown in Fig. P3.15 [5, 6]. Find the steady state effect of a constant wind force represented by $D(s) = 1/s$ for $K = 5$ and $K = 30$. (a) Assume that the rudder input $R(s)$ is zero, without any disturbance, and has not been adjusted. (b) Show that the rudder can then be used to bring the ship deviation back to zero.

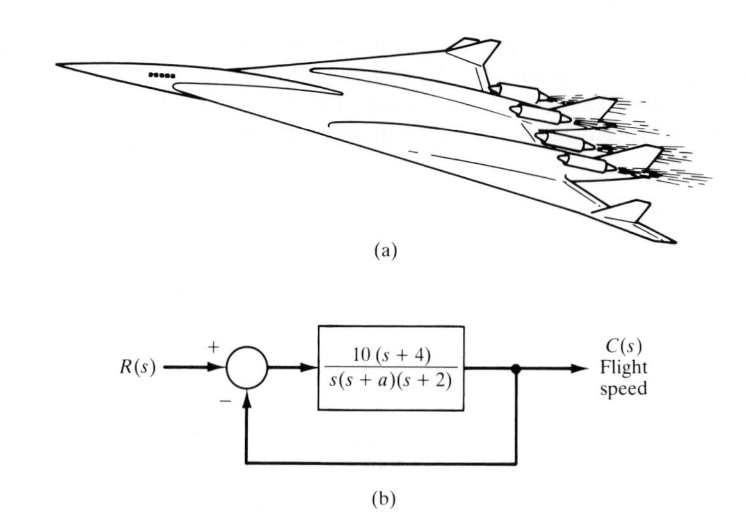

Figure P3.14. Hypersonic airplane speed control.

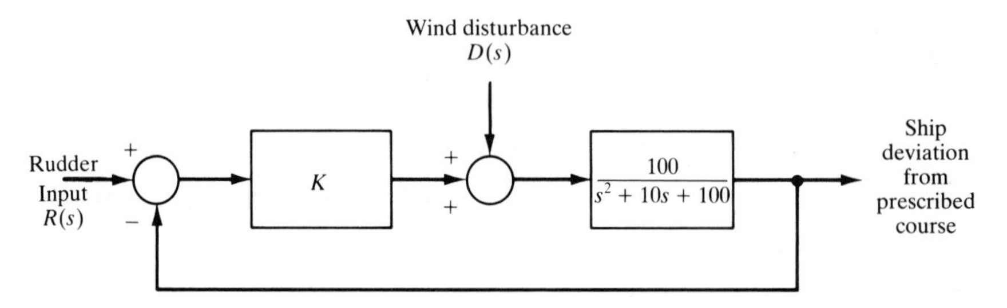

Figure P3.15. Ship steering control.

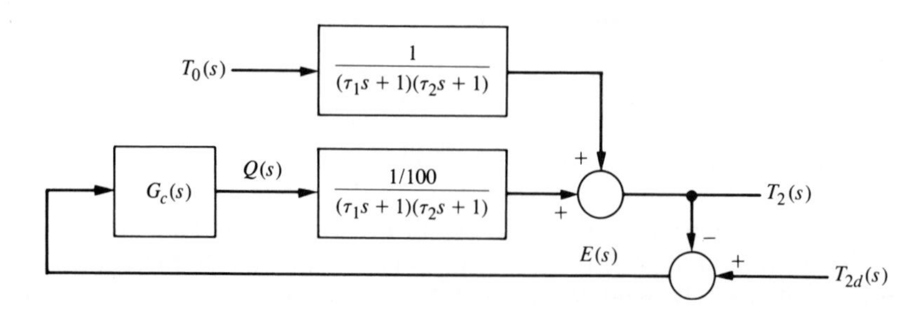

Figure P3.16. Two tank temperature control.

P3.16. A two-tank system containing a heated liquid has the model shown in Fig. P3.16, where T_0 is the temperature of the fluid flowing into the first tank and T_2 is the temperature of the liquid flowing out of the second tank. The system of two tanks has a heater in tank 1 with a controllable heat input Q [23]. The time constants are $\tau_1 = 10s$ and $\tau_2 = 50s$. (a)

Determine $T_2(s)$ in terms of $T_0(s)$ and $T_{2d}(s)$. (b) if $T_{2d}(s)$, the desired output temperature, is changed instantaneously from $T_{2d}(s) = A/s$ to $T_{2d}(s) = 2A/s$ where $T_0(s) = A/s$, determine the transient response of $T_2(t)$ when $G_c(s) = K = 500$. (c) Find the steady-state error, e_{ss}, for the system of part (b), where $E(s) = T_{2d}(s) - T_2(s)$.

P3.17. A robot gripper, shown in part (a) of Fig. P3.17, is to be controlled so that it closes to an angle θ by using a dc motor control system, as shown in part (b). The model of the

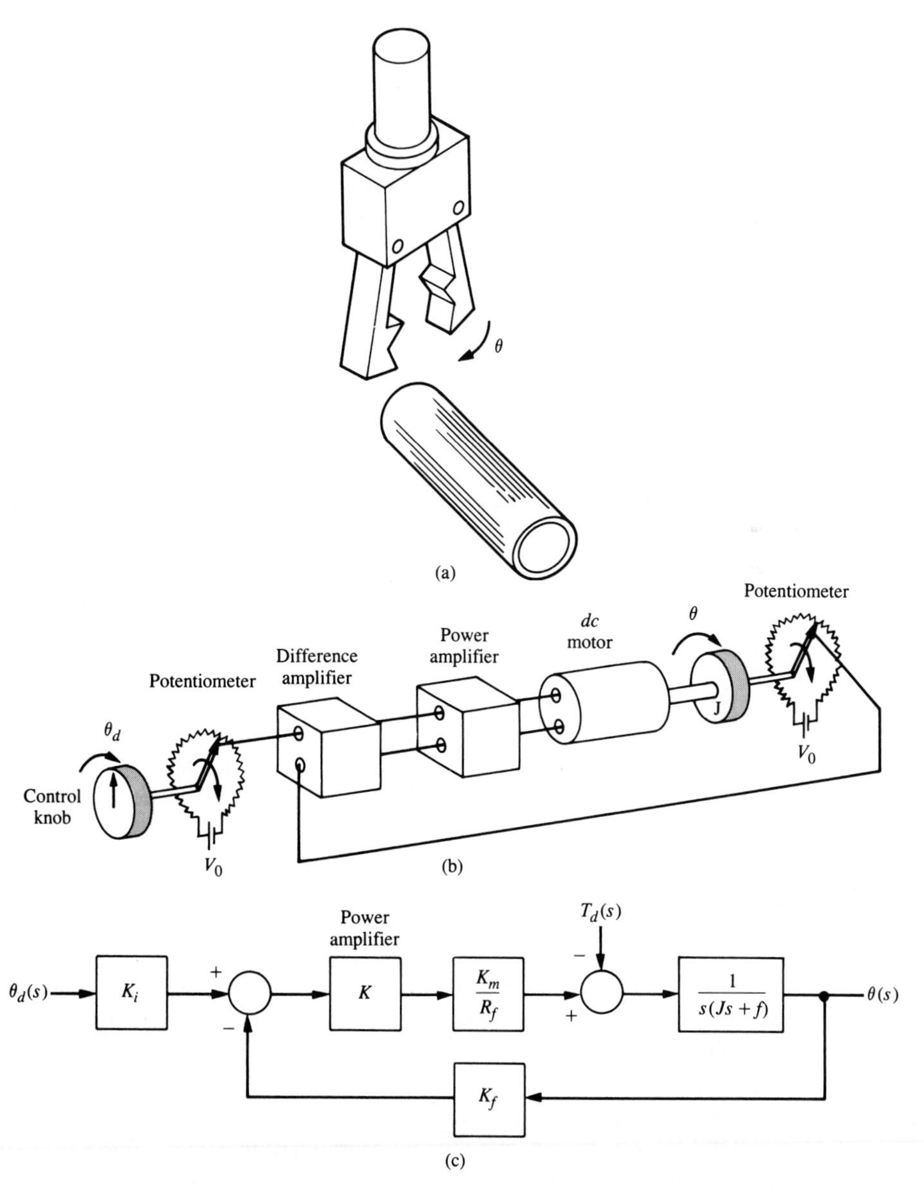

Figure P3.17. Robot gripper control.

control system is shown in part (c), where $K_m = 30$, $R_f = 1\Omega$, $K_f = K_i = 1$, $J = 0.1$, and $f = 1$. (a) Determine the response, $\theta(t)$, of the system to a step change in $\theta_d(t)$ when $K = 20$. (b) Assuming $\theta_d(t) = 0$, find the effect of a load disturbance $T_d(s) = A/s$. (c) Determine the steady-state error, e_{ss}, when the input is $r(t) = t$, $t > 0$. (Assume that $T_d = 0$.)

Design Problems ✦

DP3.1. A closed-loop speed-control system is subjected to a disturbance due to a load, as shown in Fig. DP3.1. The desired speed is $\omega_d(t) = 100$ rad/second and the load disturbance is a unit step input $D(s) = 1/s$. Assume that the speed has attained the no-load speed of 100 rad/s and is in steady state. (a) Determine the steady-state effect of the load disturbance and (b) plot $c(t)$ for the step disturbance for selected values of gain so that $10 \leq K \leq 25$. Determine a suitable value for the gain K.

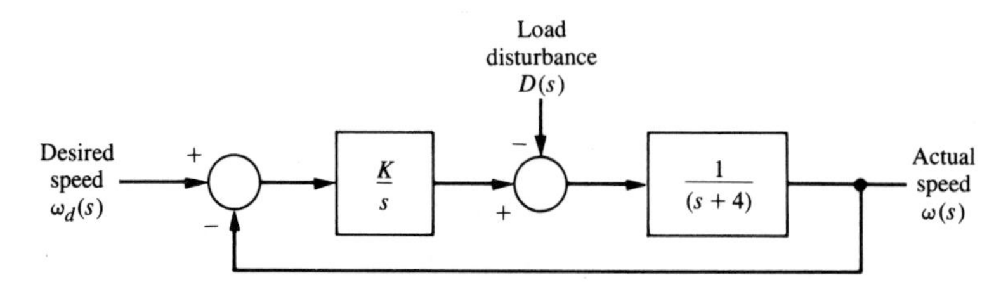

Figure DP3.1. Speed control system.

DP3.2. The control of the roll angle of an airplane is achieved by using the torque developed by the ailerons, as shown in Fig. DP3.2(a) and (b), on the next page. A linear model of the roll control system for a small experimental aircraft is shown in Fig. DP3.2(c), where $q(t)$ is the flow of fluid into a hydraulic cylinder and

$$G(s) = \frac{1}{s^2 + 3s + 9}.$$

The goal is to maintain a small roll angle θ due to disturbances. Select an appropriate gain, KK_1, that will reduce the effect of the disturbance while attaining a desirable transient response to a step disturbance, with $\theta_d(t) = 0$.

DP3.3. Lasers have been used in eye surgery for more than 25 years. They can cut tissue or aid in coagulation [24]. The laser allows the ophthalmologist to apply heat to a location in the eye in a controlled manner. Many procedures use the retina as a laser target. The

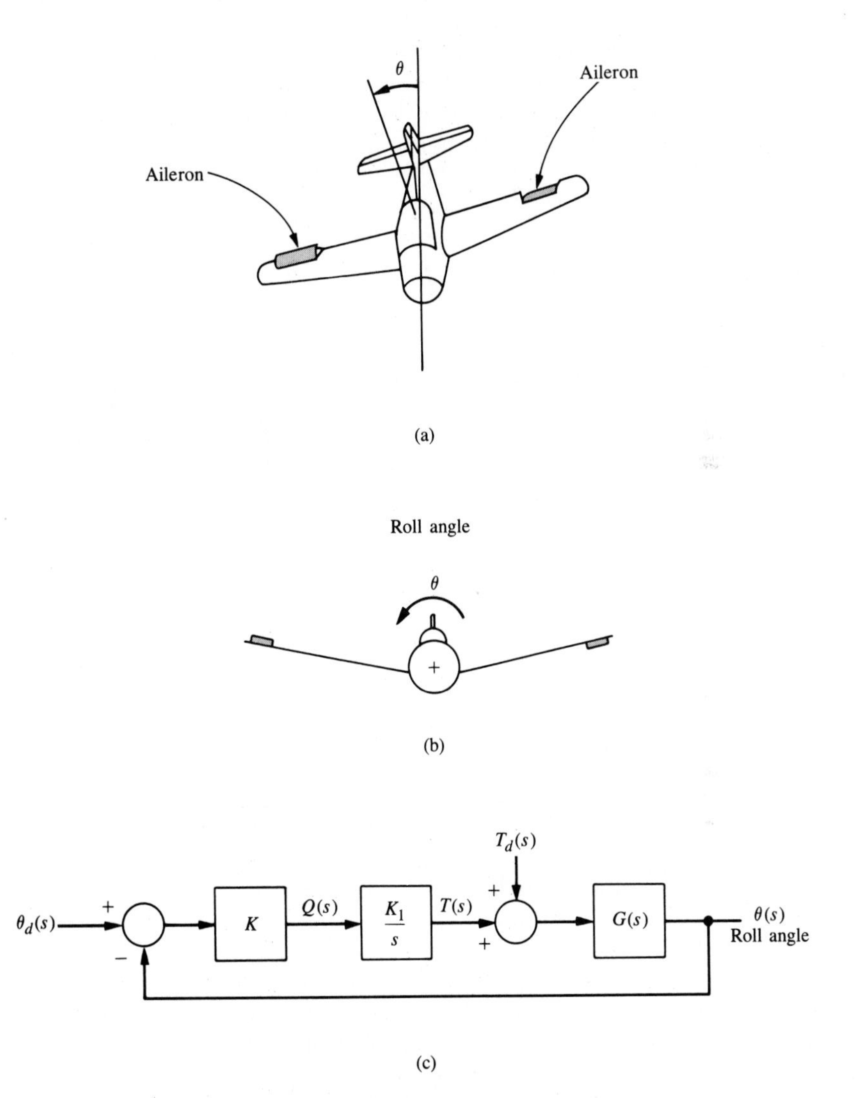

Figure DP3.2. Control of the roll angle of an airplane.

retina is the thin sensory tissue that rests on the inner surface of the back of the eye and is the actual transducer of the eye, converting light energy into electrical pulses. On occasion, this layer will detach from the wall, resulting in the death of the detached area from lack of blood and leading to partial if not total blindness in that eye. A laser can be used to "weld" the retina into its proper place on the inner wall.

Automated control of position enables the ophthalmologist to indicate to the controller where lesions should be inserted. The controller then monitors the retina and controls

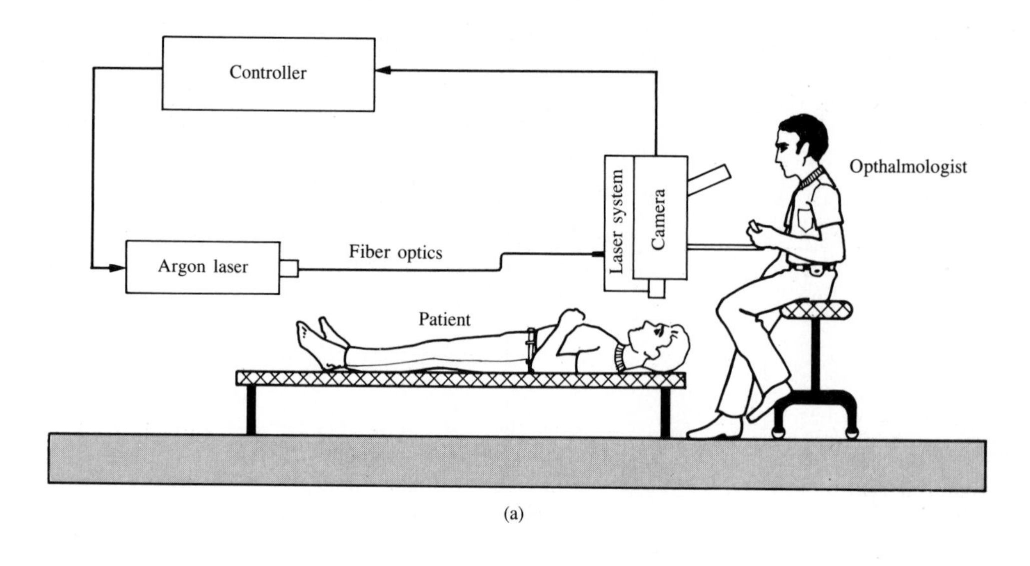

(a)

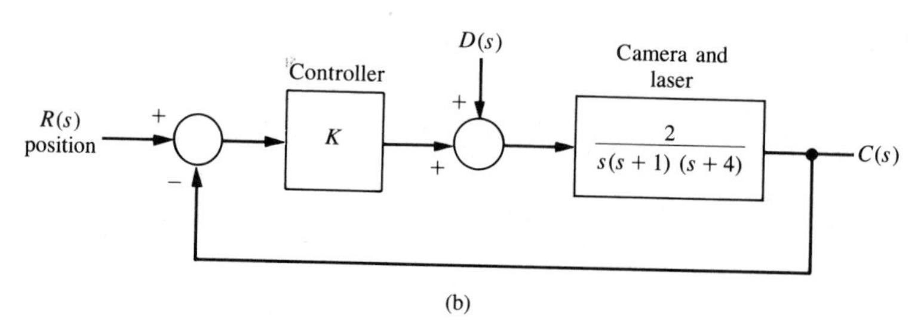

(b)

Figure DP3.3. Laser eye surgery system.

the laser's position such that each lesion is placed at the proper location. A wide-angle video-camera system is required to monitor the movement of the retina, as shown in Fig. DP3.3(a). If the eye moves during the irradiation, the laser must be either redirected or turned off. The position-control system is shown in Fig. DP3.3(b). Select an appropriate gain for the controller so that the transient response to a step change in $r(t)$ is satisfactory and the effect of the disturbance due to noise in the system is minimized. Also ensure that the steady-state error for a step input command is zero.

Terms and Concepts

Disturbance signal An unwanted input signal that affects the system's output signal.

Error signal The difference between the desired output, $R(s)$, and the actual output $C(s)$. Therefore $E(s) = R(s) - C(s)$.

Steady-state error The error when the time period is large and the transient response has decayed, leaving the continuous response.

System sensitivity The proportional change of the transfer function of a system to a proportional change in the system parameter.

Transient response The response of a system as a function of time.

References

1. R. C. Dorf, *Introduction to Electric Circuits,* John Wiley & Sons, New York, 1989.
2. J. W. Nilsson, *Electric Circuits,* 3rd ed., Addison-Wesley, Reading, Mass., 1990.
3. *Motomatic Speed Control,* Electro-Craft Corp., Hopkins, Minn., 1990.
4. D. L. Schilling, *Electronic Circuits,* 3rd ed., McGraw-Hill, New York, 1989.
5. J. Van Amerongen and H. R. Lemke, "Adaptive Control Aspects of a Rudder Roll Stabilization System," *Proceedings of the 1987 IFAC Congress on Automatic Control;* pp. 215–30.
6. C. L. Nachtigal, *Instrumentation and Control,* John Wiley & Sons, New York, 1990.
7. T. Yoshikawa, *Foundations of Robotics,* M.I.T. Press, Cambridge, Mass., 1990.
8. M. Swartz, "Robert Radocy Picks Up the Pieces," *Esquire,* December 1984; pp. 94–97.
9. S. N. Kolpashnikov and I. B. Chelpanov, "Industrial Robot Grasping Devices," *Advanced Manufacturing Engineering,* October 1989; pp. 277–86.
10. B. W. Niebel, *Modern Manufacturing Process Engineering,* McGraw-Hill, New York, 1989.
11. K. Betts, "Process Control Takes Command," *Mechanical Engineering,* July 1990; pp. 64–68.
12. J. Williams, "Take Advantage of Thermal Effects to Solve Circuit Design Problems," *Electronic Design News,* June 28, 1984; pp. 239–48.
13. A. Isidori, *Nonlinear Control Systems,* Springer-Verlag, New York, 1990.
14. R. C. Dorf, *Encyclopedia of Robotics,* John Wiley & Sons, New York, 1988.
15. D. J. Bak, "Dancer Arm Feedback Regulates Tension Control," *Design News,* April 6, 1987; pp. 132–33.
16. B. Frohring, "Waveform Recorder Design for Dynamic Performance," *HP Journal,* February 1988; pp. 39–47.
17. S. C. Jacobsen, "Control Strategies for Tendon-Driven Manipulators," *IEEE Control Systems,* February 1990; pp. 23–28.
18. S. Franco, *Design with Operational Amplifiers and Analog Integrated Circuits,* McGraw-Hill, New York, 1988.
19. "The Smart Projector Demystified," *Science Digest,* May 1985; p. 76.
20. R. Kurzweil, *The Age of Intelligent Machines,* M.I.T. Press, Cambridge, Mass., 1990.
21. C. W. DeSilva, *Control Sensors and Actuators,* Prentice Hall, Englewood Cliffs, N.J. 1990.

22. Y. Dote, *Servo Motor and Motion Control,* Prentice Hall, Englewood Cliffs, N.J., 1990.
23. W. Luyben, *Process Modeling and Control for Chemical Engineers,* 2nd ed., McGraw-Hill, New York, 1990.
24. M. S. Markow, "An Automated Laser System for Eye Surgery," *IEEE Engineering in Medicine and Biology,* December 1989; pp. 24–29.
25. W. R. Perkins, "Sensitivity Function Methods in Control System Education," Proceed. of IFAC Advances in Control Education, June 1991; pp. 14–22.

CHAPTER 4

The Performance of Feedback Control Systems

Preview

The ability to adjust the transient and steady-state response of a control system is a beneficial outcome of structuring a feedback system. We wish to adjust one or more parameters in order to provide a desirable response. Thus we must define the desired response in terms of specifications for the system.

We will use selected input signals to test the response of a control system. This response will be characterized by a selected set of response measures such as the overshoot of a response to a step input. We will then analyze the performance of a system in terms of the *s*-plane location of the poles and zeros of the transfer function of the system.

We will see that one of the most important measures of performance is the steady-state error. The concept of a performance index which adequately represents the system's performance by a single number (or index) will be considered. As is clear, in this chapter, we will strive to delineate a set of quantitative performance measures that adequately represent the performance of the control system. This approach will enable us to adjust the system and thus achieve a performance that we can describe as the best we can achieve.

4.1 Introduction

The ability to adjust the transient and steady-state performance is a distinct advantage of feedback control systems. In order to analyze and design control systems, we must define and measure the performance of a system. Then, based on the desired performance of a control system, the parameters of the system may be adjusted in order to provide the desired response. Because control systems are inherently dynamic systems, the performance is usually specified in terms of both the time response for a specific input signal and the resulting steady-state error.

The design *specifications* for control systems normally include several time-response indices for a specified input command as well as a desired steady-state accuracy. However, often in the course of any design, the specifications are revised in order to effect a compromise. Therefore, specifications are seldom a rigid set of requirements, but rather a first attempt at listing a desired performance. The effective compromise and adjustment of specifications can be graphically illustrated by examining Fig. 4.1. Clearly, the parameter p may minimize the performance measure M_2 by selecting p as a very small value. However, this results in large measure M_1, an undesirable situation. Obviously, if the performance measures are equally important, the crossover point at p_{min} provides the best compromise. This type of compromise is normally encountered in control-

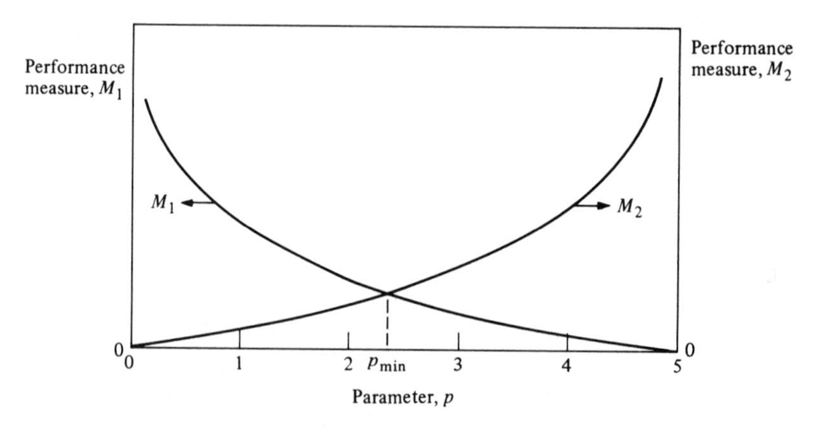

Figure 4.1. Two performance measures vs. parameter p.

system design. It is clear that if the original specifications called for both M_1 and M_2 to be zero, the specifications could not be simultaneously met and would have to be altered to allow for the compromise resulting with $p_{\min}$.

The specifications stated in terms of the measures of performance indicate to the designer the quality of the system. In other words, the performance measures are an answer to the question: How well does the system perform the task it was designed for?

4.2 Time-Domain Performance Specifications

The time-domain performance specifications are important indices because control systems are inherently time-domain systems. That is, the system transient or time performance is the response of prime interest for control systems. It is necessary to determine initially if the system is stable by utilizing the techniques of ensuing chapters. If the system is stable, then the response to a specific input signal will provide several measures of the performance. However, because the actual input signal of the system is usually unknown, a standard *test input signal* is normally chosen. This approach is quite useful because there is a reasonable correlation between the response of a system to a standard test input and the system's ability to perform under normal operating conditions. Furthermore, using a standard input allows the designer to compare several competing designs. Also, many control systems experience input signals very similar to the standard test signals.

The standard test input signals commonly used are (1) the step input, (2) the ramp input, and (3) the parabolic input. These inputs are shown in Fig. 4.2. The equations representing these test signals are given in Table 4.1, where the Laplace transform can be obtained by using Table 2.5. The ramp signal is the integral of the step input, and the parabola is simply the integral of the ramp input. A *unit impulse* function is also useful for test signal purposes. The unit impulse is based on a rectangular function $f_\epsilon(t)$ such that

$$f_\epsilon(t) = \begin{cases} 1/\epsilon, & 0 \leq t \leq \epsilon, \\ 0, & t > \epsilon, \end{cases}$$

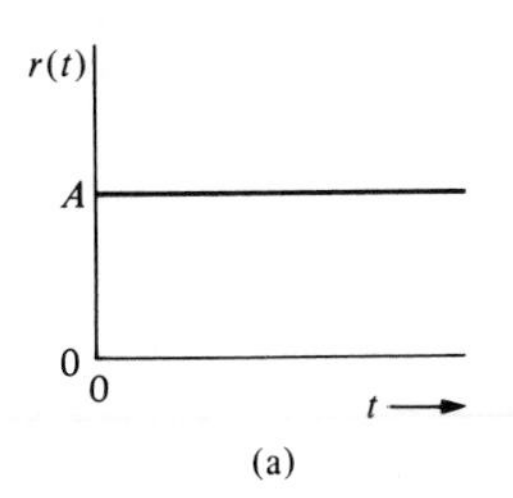

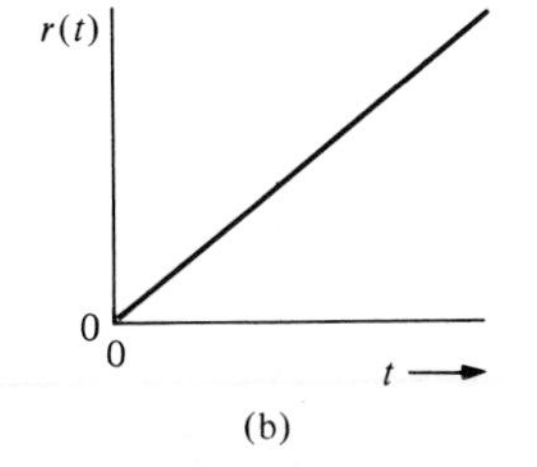

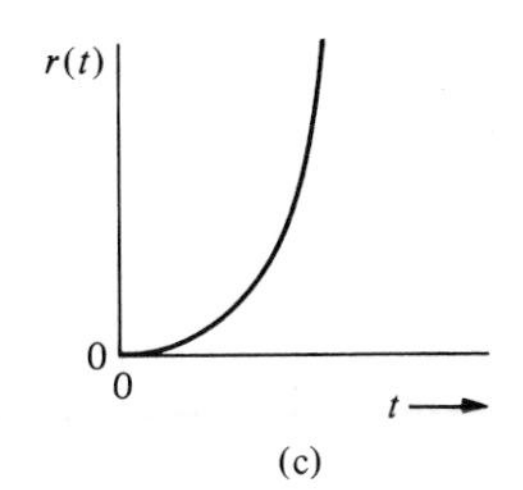

Figure 4.2. Test input signals.

Table 4.1. Test Signal Inputs

Test Signal	$r(t)$	$R(s)$
Step	$r(t) = A,\ t > 0$ $= 0,\ t < 0$	$R(s) = A/s$
Ramp	$r(t) = At,\ t > 0$ $= 0,\ t < 0$	$R(s) = A/s^2$
Parabolic	$r(t) = At^2,\ t > 0$ $= 0,\ t < 0$	$R(s) = 2A/s^3$

where $\epsilon > 0$. As ϵ approaches zero, the function $f_\epsilon(t)$ approaches the impulse function $\delta(t)$, which has the following properties:

$$\int_0^\infty \delta(t)dt = 1,$$
$$\int_0^\infty \delta(t - a)g(t) = g(a). \tag{4.1}$$

The impulse input is useful when one considers the convolution integral for an output $c(t)$ in terms of an input $r(t)$, which is written as

$$\begin{aligned} c(t) &= \int_0^t g(t - \tau)r(\tau)d\tau \\ &= \mathcal{L}^{-1}\{G(s)R(s)\}. \end{aligned} \tag{4.2}$$

This relationship is shown in block diagram form in Fig. 4.3. Clearly, if the input is an impulse function of unit amplitude, we have

$$c(t) = \int_0^t g(t - \tau)\delta(\tau)d\tau. \tag{4.3}$$

The integral has a value only at $\tau = 0$, and therefore

$$c(t) = g(t),$$

the impulse response of the system $G(s)$. The impulse response test signal can often be used for a dynamic system by subjecting the system to a large amplitude, narrow width pulse of area A.

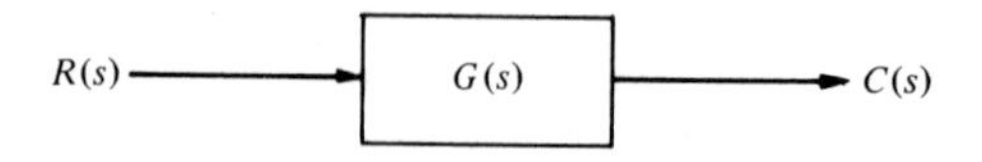

Figure 4.3. Open-loop control system.

The standard test signals are of the general form

$$r(t) = t^n, \tag{4.4}$$

and the Laplace transform is

$$R(s) = \frac{n!}{s^{n+1}}. \tag{4.5}$$

Clearly, one may relate the response to one test signal to the response of another test signal of the form of Eq. (4.4). The step input signal is the easiest to generate and evaluate and is usually chosen for performance tests.

Initially, let us consider a single-loop second-order system and determine its response to a unit step input. A closed-loop feedback control system is shown in Fig. 4.4. The closed-loop output is

$$\begin{aligned} C(s) &= \frac{G(s)}{1 + G(s)} R(s) \\ &= \frac{K}{s^2 + ps + K} R(s). \end{aligned} \tag{4.6}$$

Utilizing the generalized notation of Section 2.4, we may rewrite Eq. (4.6) as

$$C(s) = \frac{\omega_n^2}{s^2 + 2\zeta\omega_n s + \omega_n^2} R(s). \tag{4.7}$$

With a unit step input, we obtain

$$C(s) = \frac{\omega_n^2}{s(s^2 + 2\zeta\omega_n s + \omega_n^2)}, \tag{4.8}$$

for which the transient output, as obtained from the Laplace transform table in Appendix A, is

$$c(t) = 1 - \frac{1}{\beta} e^{-\zeta\omega_n t} \sin(\omega_n \beta t + \theta), \tag{4.9}$$

where $\beta = \sqrt{1 - \zeta^2}$ and $\theta = \tan^{-1}\beta/\zeta$. The transient response of this second-order system for various values of the damping ratio ζ is shown in Fig. 4.5. As ζ decreases, the closed-loop roots approach the imaginary axis and the response becomes increasingly oscillatory. The response as a function of ζ and time is also shown in Fig. 4.5(b) for a step input.

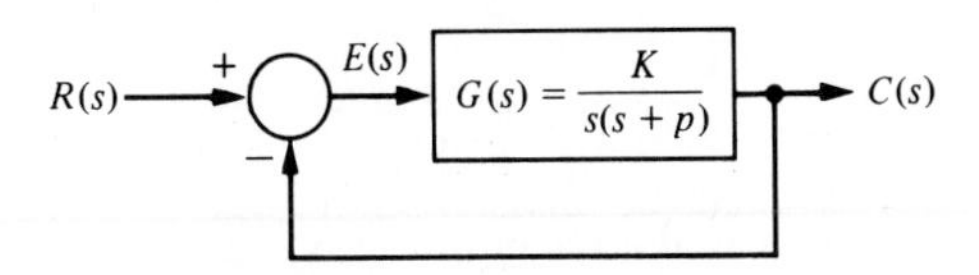

Figure 4.4. Closed-loop control system.

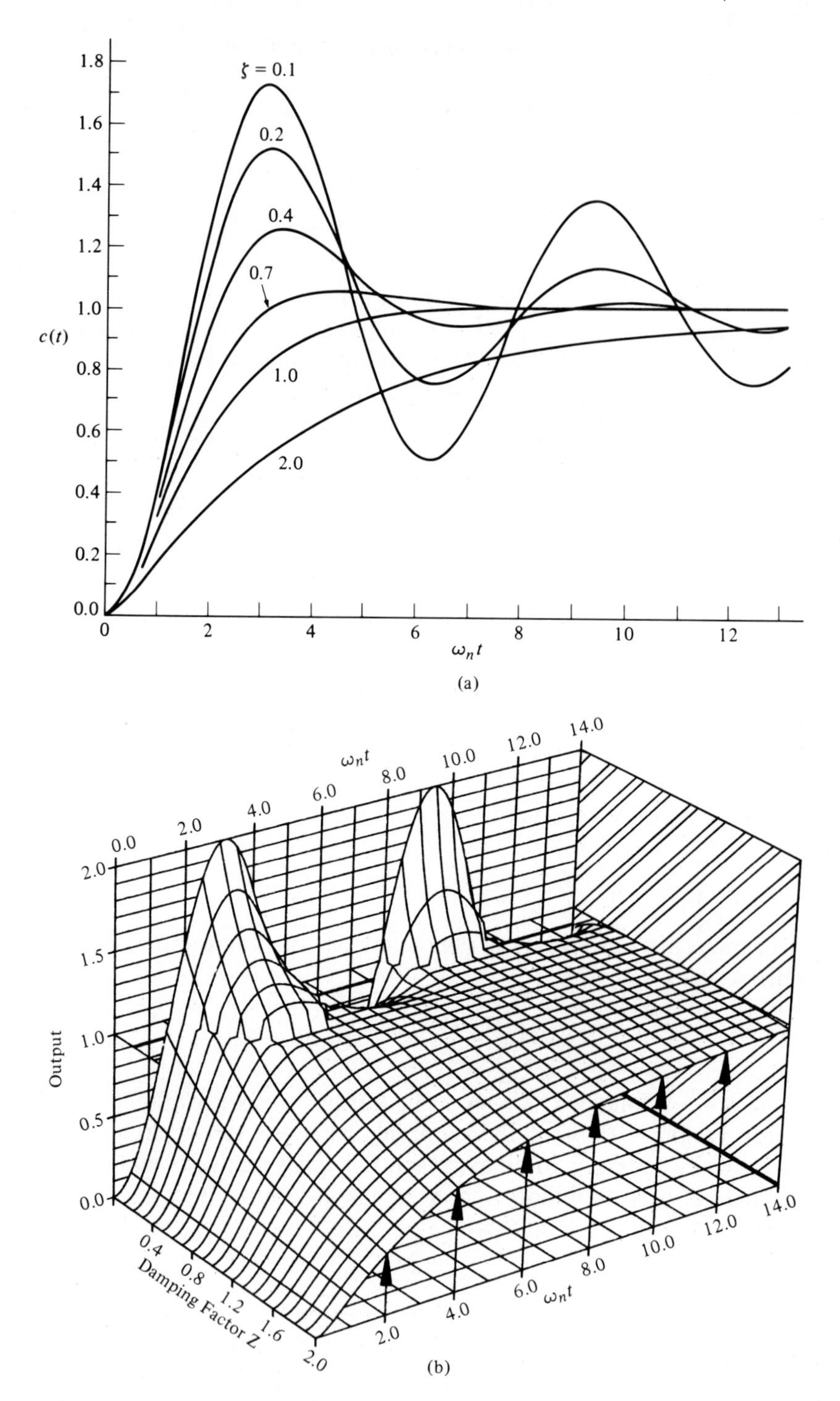

Figure 4.5. (a) Transient response of a second-order system (Eq. 4.9) for a step input. (b) The transient response of a second–order-system (Eq. 4.9) for a step input as a function of $\zeta = Z$ and $\omega_n t$. (Courtesy of Professor R. Jacquot, University of Wyoming.)

The Laplace transform of the unit impulse is $R(s) = 1$, and therefore the output for an impulse is

$$C(s) = \frac{\omega_n^2}{s^2 + 2\zeta\omega_n s + \omega_n^2}, \tag{4.10}$$

which is $T(s) = C(s)/R(s)$, the transfer function of the closed-loop system. The transient response for an impulse function input is then

$$c(t) = \frac{\omega_n}{\beta} e^{-\zeta\omega_n t} \sin \omega_n \beta t, \tag{4.11}$$

which is simply the derivative of the response to a step input. The impulse response of the second-order system is shown in Fig. 4.6 for several values of the damping ratio, ζ. Clearly, one is able to select several alternative performance measures from the transient response of the system for either a step or impulse input.

Standard performance measures are usually defined in terms of the step response of a system as shown in Fig. 4.7. The swiftness of the response is measured by the *rise time* T_r and the *peak time.* For underdamped systems with an overshoot, the 0 to 100% rise time is a useful index. If the system is overdamped, then the peak time is not defined and the 10–90% rise time, T_{r_1}, is normally used. The similarity with which the actual response matches the step input is measured by the percent overshoot and settling time T_s. The *percent overshoot,* P.O., is defined as

$$\text{P.O.} = \frac{M_{p_t} - fv}{fv} \times 100\% \tag{4.12}$$

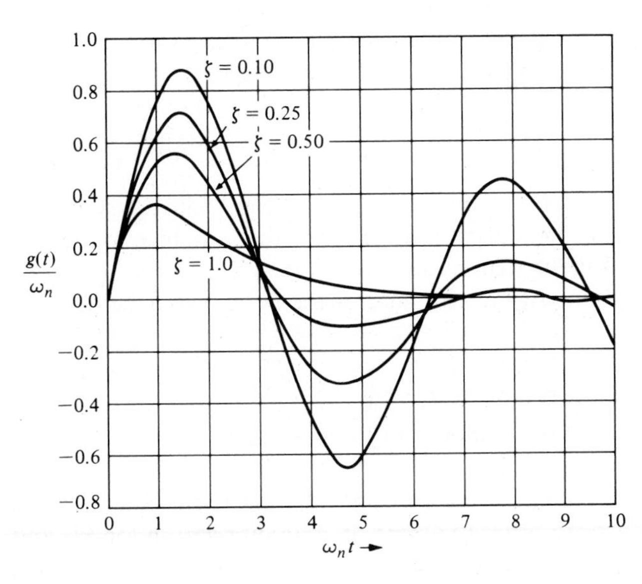

Figure 4.6. Response of a second-order system for an impulse function input.

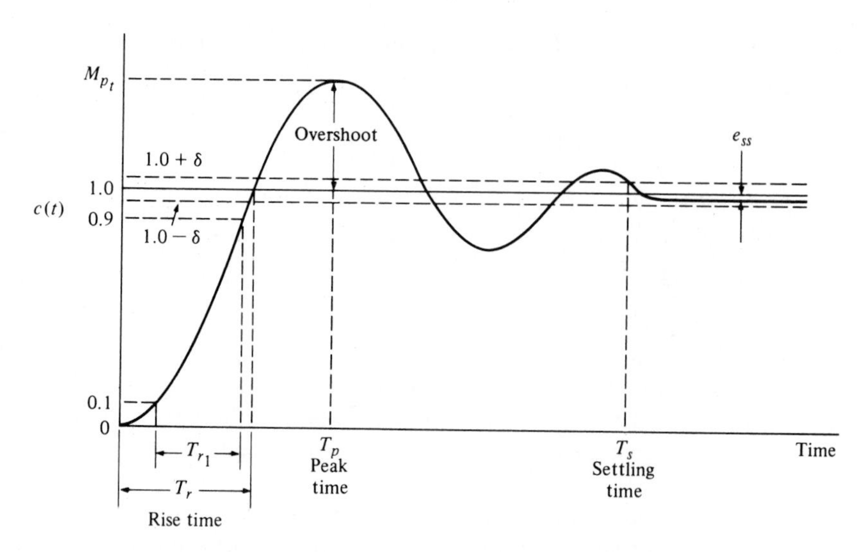

Figure 4.7. Step response of a control system (Eq. 4.9).

for a unit step input, where M_{p_t} is the peak value of the time response and fv is the final value of the response. Normally fv is the magnitude of the input, but many systems have a final value significantly different than the desired input magnitude. For the system with a unit step represented by Eq. (4.8), we have $fv = 1$.

The *settling time,* T_s, is defined as the time required for the system to settle within a certain percentage δ of the input amplitude. This band of $\pm\delta$ is shown in Fig. 4.7. For the second-order system with closed-loop damping constant $\zeta\omega_n$, the response remains within 2% after four time constants, or

$$T_s = 4\tau = \frac{4}{\zeta\omega_n}. \tag{4.13}$$

Therefore we will define the settling time as four time constants of the dominant response. Finally, the steady-state error of the system may be measured on the step response of the system as shown in Fig. 4.7.

Therefore the transient response of the system may be described in terms of two factors:

1. The swiftness of response, T_r and T_p
2. The closeness of the response to the desired M_{p_t} and T_s

As nature would have it, these are contradictory requirements and a compromise must be obtained. In order to obtain an explicit relation for M_{p_t} and T_p as a function of ζ, one can differentiate Eq. (4.9) and set it equal to zero. Alternatively, one may utilize the differentiation property of the Laplace transform, which may be written as

$$\mathcal{L}\left\{\frac{dc(t)}{dt}\right\} = sC(s)$$

when the initial value of $c(t)$ is zero. Therefore we may acquire the derivative of $c(t)$ by multiplying Eq. (4.8) by s and thus obtaining the right side of Eq. (4.10). Taking the inverse transform of the right side of Eq. (4.10) we obtain Eq. (4.11), which is equal to zero when $\omega_n \beta t = \pi$. Therefore we find that the peak time relationship for this second-order system is

$$T_p = \frac{\pi}{\omega_n\sqrt{1-\zeta^2}}, \tag{4.14}$$

and the peak response is

$$M_{p_t} = 1 + e^{-\zeta\pi/\sqrt{1-\zeta^2}}. \tag{4.15}$$

Therefore the percent overshoot is

$$\text{P.O.} = 100e^{-\zeta\pi/\sqrt{1-\zeta^2}}. \tag{4.16}$$

The percent overshoot versus the damping ratio ζ is shown in Fig. 4.8. Also, the normalized peak time, $\omega_n T_p$, is shown versus the damping ratio ζ in Fig. 4.8. The percent overshoot versus the damping ratio is listed in Table 4.2 for selected values of the damping ratio. Again, we are confronted with a necessary compromise between the swiftness of response and the allowable overshoot.

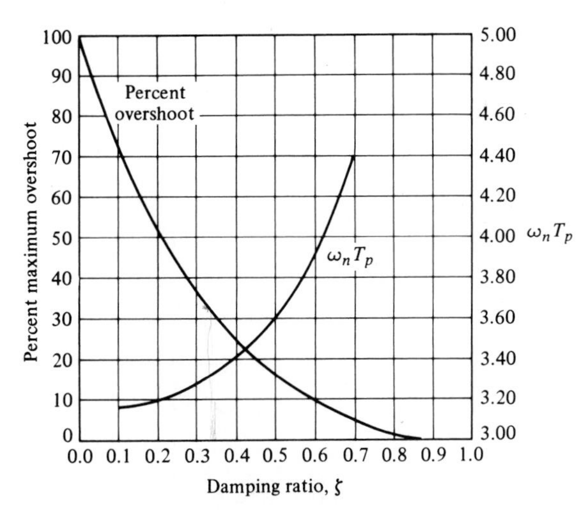

Figure 4.8. Percent overshoot and peak time versus damping ratio ζ for a second-order system (Eq. 4.8).

Table 4.2. Percent Peak Overshoot Versus Damping Ratio for a Second-Order System

Damping ratio	0.9	0.8	0.7	0.6	0.5	0.4	0.3
Percent overshoot	0.2	1.5	4.6	9.5	16.3	25.4	37.2

The curves presented in Fig. 4.8 are exact only for the second-order system of Eq. (4.8). However, they provide a remarkably good source of data, because many systems possess a dominant pair of roots and the step response can be estimated by utilizing Fig. 4.8. This approach, while an approximation, avoids the evaluation of the inverse Laplace transformation in order to determine the percent overshoot and other performance measures. For example, for a third-order system with a closed-loop transfer function

$$T(s) = \frac{1}{(s^2 + 2\zeta s + 1)(\gamma s + 1)}, \qquad (4.17)$$

the s-plane diagram is shown in Fig. 4.9. This third-order system is normalized with $\omega_n = 1$. It was ascertained experimentally that the performance as indicated by the percent overshoot, M_{p_t}, and the settling time, T_s, was represented by the second-order system curves when [4]

$$|1/\gamma| \geq 10|\zeta\omega_n|.$$

In other words, the response of a third-order system can be approximated by the *dominant roots* of the second-order system as long as the real part of the dominant roots is less than 1/10 of the real part of the third root.

Using a computer simulation, when $\zeta = 0.45$ one can determine the response of a system to a unit step input. When $\gamma = 2.25$ we find that the response is overdamped because the real part of the complex poles is -0.45, whereas the real

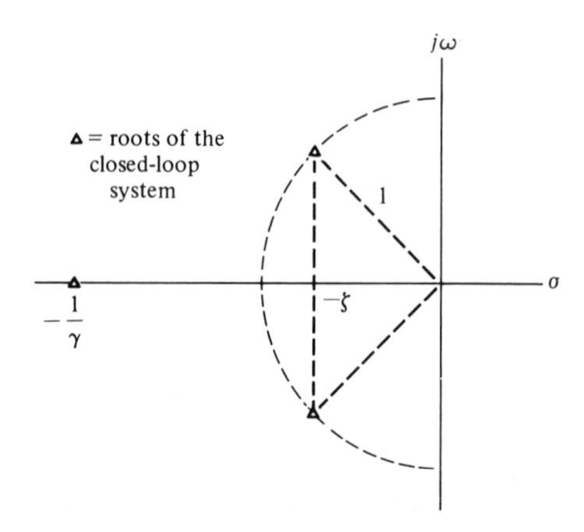

Figure 4.9. An s-plane diagram of a third-order system.

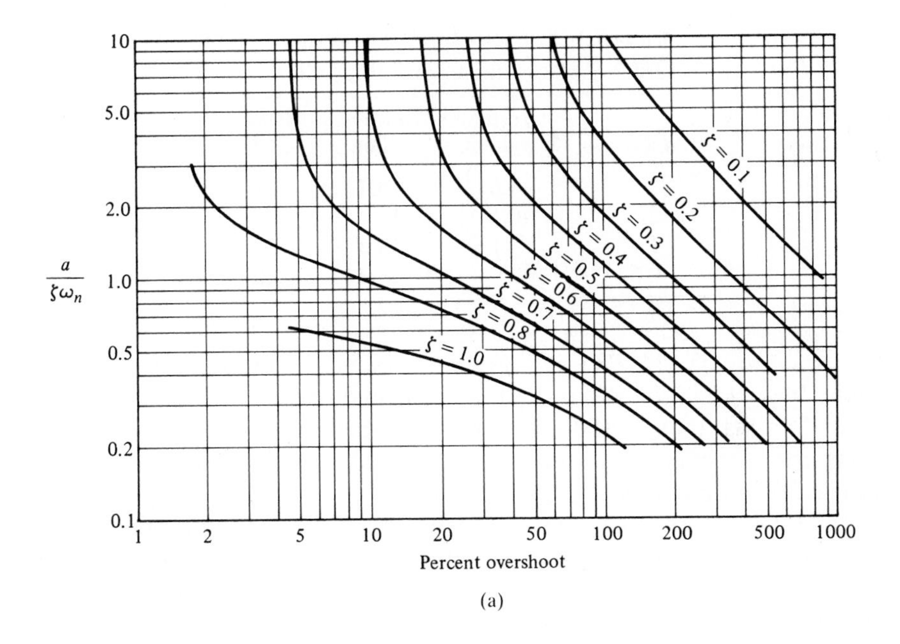

(a)

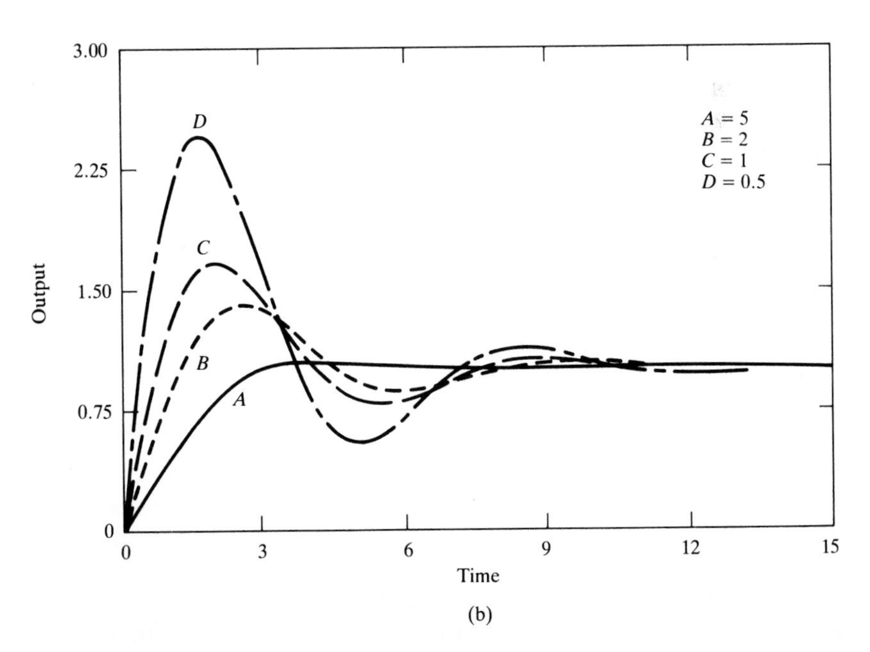

(b)

Figure 4.10. (a) Percent overshoot as a function of ζ and ω_n when a second-order transfer function contains a zero. (From R. N. Clark, *Introduction to Automatic Control Systems,* New York, Wiley, 1962, redrawn with permission.) (b) The response for the second-order transfer function with a zero for four values of the ratio ($a/\zeta\ w_n$): $A = 5$, $B = 2$, $C = 1$ and $D = 0.5$.

Table 4.3. The Response of a Second-Order System with a Zero ($0<\zeta<1$)

$a/\zeta\,\omega_n$	Percent Overshoot	Settling Time	Peak Time
5	4.6	5.7	4.1
2	40.1	10.4	2.6
1	68.1	10.3	2.1
0.5	145.0	13.3	2.5

pole is equal to -0.444. The settling time is found via the simulation to be 12.8 seconds. If $\gamma = 0.90$ or $1/\gamma = 1.11$ is compared to $\zeta\omega_n = 0.45$ of the complex poles we find that the overshoot is 12% and the settling time is 6.4 seconds. If the complex roots were entirely dominant, we would expect the overshoot to be 20% and the settling time to be $4/\zeta\omega_n = 8.9$ seconds.

Also, we must note that the performance measures of Fig. 4.8 are correct only for a transfer function without finite zeros. If the transfer function of a system possesses finite zeros and they are located relatively near the dominant poles, then the zeros will materially affect the transient response of the system [5].

The transient response of a system with one zero and two poles may be affected by the location of the zero [5]. The percent overshoot for a step input as a function of $a/\zeta\omega_n$ is given in Fig. 4.10(a) for the system transfer function

$$T(s) = \frac{(\omega_n^2/a)(s + a)}{s^2 + 2\zeta\omega_n s + \omega_n^2}.$$

The actual transient response for a step input is shown in Fig. 4.10(b) for selected values of $a/\zeta\omega_n$. The actual response for these selected values is summarized in Table 4.3 when $0<\zeta<1$.

The correlation of the time-domain response of a system with the s-plane location of the poles of the closed-loop transfer function is very useful for selecting the specifications of a system. In order to illustrate clearly the utility of the s-plane, let us consider a simple example.

■ Example 4.1 Parameter Selection

A single-loop feedback control system is shown in Fig. 4.11. We desire to select the gain K and the parameter p so that the time-domain specifications will be satisfied. The transient response to a step should be as fast in responding as is reasonable and have an overshoot of less than 5%. Furthermore, the settling time should be less than four seconds. The minimum damping ratio ζ for an overshoot of 4.3% is 0.707. This damping ratio is shown graphically in Fig. 4.12. Because the settling time is

$$T_s = \frac{4}{\zeta\omega_n} \leq 4 \text{ sec}, \tag{4.18}$$

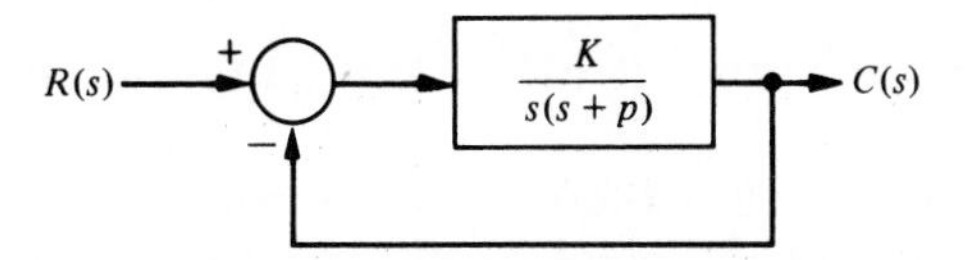

Figure 4.11. Single-loop feedback control system.

we require that the real part of the complex poles of $T(s)$ is

$$\zeta\omega_n \geq 1.$$

This region is also shown in Fig. 4.12. The region that will satisfy both time-domain requirements is shown cross-hatched on the s-plane of Fig. 4.12. If the closed-loop roots are chosen as the limiting point, in order to provide the fastest response, as r_1 and $\hat{r}_1$, then $r_1 = -1 + j1$ and $\hat{r}_1 = -1 - j1$. Therefore $\zeta = 1/\sqrt{2}$ and $\omega_n = 1/\zeta = \sqrt{2}$. The closed-loop transfer function is

$$T(s) = \frac{G(s)}{1 + G(s)} = \frac{K}{s^2 + ps + K}$$
$$= \frac{\omega_n^2}{s^2 + 2\zeta\omega_n s + \omega_n^2}. \tag{4.19}$$

Therefore we require that $K = \omega_n^2 = 2$ and $p = 2\zeta\omega_n = 2$. A full comprehension of the correlation between the closed-loop root location and the system transient response is important to the system analyst and designer. Therefore we shall consider the matter more fully in the following section.

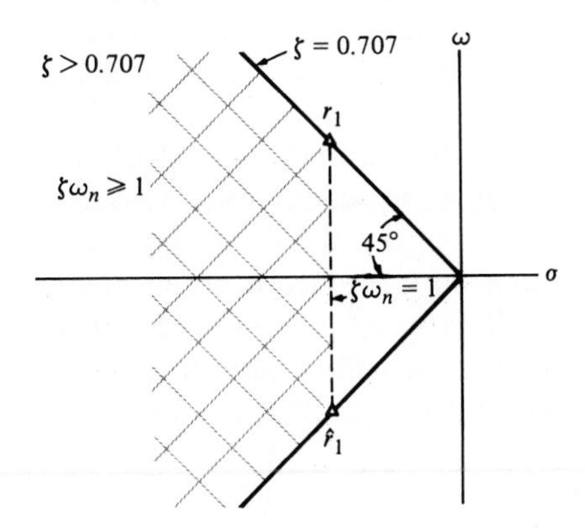

Figure 4.12. Specifications and root locations on the s-plane.

4.3 The *s*-Plane Root Location and the Transient Response

The transient response of a closed-loop feedback control system can be described in terms of the location of the poles of the transfer function. The closed-loop transfer function is written in general as

$$T(s) = \frac{C(s)}{R(s)} = \frac{\Sigma P_i(s)\Delta_i(s)}{\Delta(s)}, \tag{4.20}$$

where $\Delta(s) = 0$ is the characteristic equation of the system. For the single-loop system of Fig. 4.11, the characteristic equation reduces to $1 + G(s) = 0$. It is the poles and zeros of $T(s)$ that determine the transient response. However, for a closed-loop system, the poles of $T(s)$ are the roots of the characteristic $\Delta(s) = 0$ and the poles of $\Sigma P_i(s)\Delta_i(s)$. The output of a system without repeated roots and a unit step input can be formulated as a partial fraction expansion as

$$C(s) = \frac{1}{s} + \sum_{i=1}^{M} \frac{A_i}{s + \sigma_i} + \sum_{k=1}^{N} \frac{B_k}{s^2 + 2\alpha_k s + (\alpha_k^2 + \omega_k^2)}, \tag{4.21}$$

where the A_i and B_k are the residues. The roots of the system must be either $s = -\sigma_i$ or complex conjugate pairs as $s = -\alpha_k \pm j\omega_k$. Then the inverse transform results in the transient response as a sum of terms as follows:

$$c(t) = 1 + \sum_{i=1}^{M} A_i e^{-\sigma_i t} + \sum_{k=1}^{N} B_k \left(\frac{1}{\omega_k}\right) e^{-\alpha_k t} \sin \omega_k t. \tag{4.22}$$

The transient response is composed of the steady-state output, exponential terms, and damped sinusoidal terms. Obviously, in order for the response to be stable—that is, bounded for a step input—one must require that the real part of the roots, σ_i or α_k, be in the left-hand portion of the *s*-plane. The impulse response for various root locations is shown in Fig. 4.13. The information imparted by the location of the roots is graphic, indeed, and usually well worth the effort of determining the location of the roots in the *s*-plane.

It is important for the control system analyst to understand the relationship between the complex-frequency representation of a linear system, through the poles and zeros of its transfer function, and its time-domain response to step and other inputs. Many of the analysis and design calculations in such areas as signal processing and control are done in the complex-frequency plane, where a system model is represented in terms of the poles and zeros of its transfer function $T(s)$. On the other hand, system performance is often analyzed by examining time-domain responses, particularly when dealing with control systems.

The capable system designer will be able to envision the effects on the step and impulse responses of adding, deleting, or moving poles and zeros of $T(s)$ in the *s*-plane. Likewise, the designer should be able to visualize what changes should be made in the poles and zeros of $H(s)$ in order to effect desired changes in the model's step and impulse responses.

An experienced designer is aware of the effects of zero locations on system

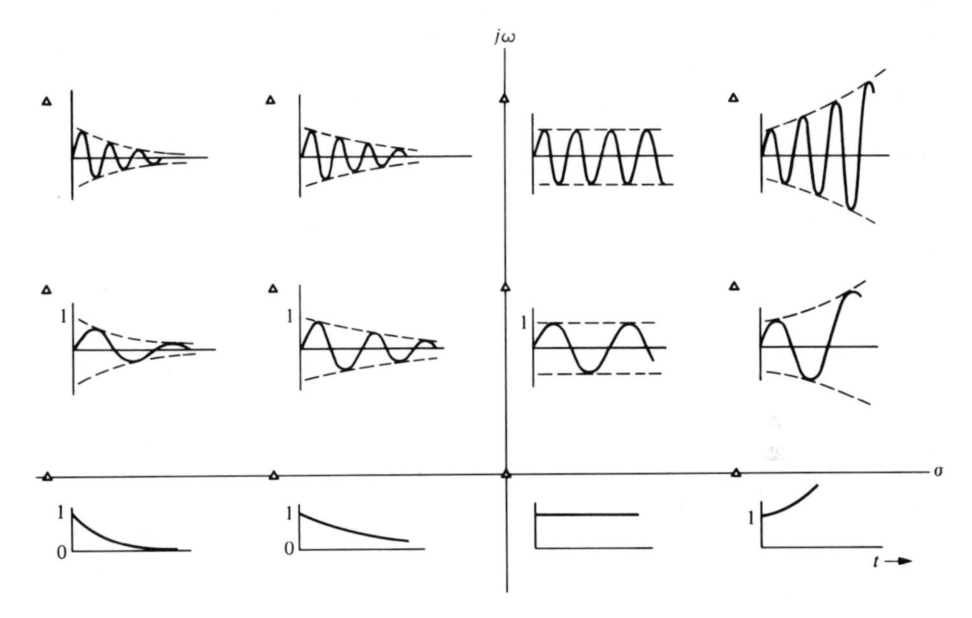

Figure 4.13. Impulse response for various root locations in the s-plane. (The conjugate root is not shown.)

response. The poles of $T(s)$ determine the particular response modes that will be present and the zeros of $T(s)$ establish the relative weightings of the individual mode functions. For example, moving a zero closer to a specific pole will reduce the relative contribution of the mode function corresponding to the pole.

A computer program can be developed that allows a person to specify arbitrary sets of poles and zeros for the transfer function of a linear system. Then the computer will evaluate and plot the system's impulse and step responses individually. It will also display them in reduced form along with the pole-zero plot.

Once the program has been run for a set of poles and zeros, the user can modify the locations of one or more of them. Plots are then presented showing the old and new poles and zeros in the complex plane, and the old and new impulse and step responses.

4.4 The Steady-State Error of Feedback Control Systems

One of the fundamental reasons for using feedback, despite its cost and increased complexity, is the attendant improvement in the reduction of the steady-state error of the system. As was illustrated in Section 3.5, the steady-state error of a stable closed-loop system is usually several orders of magnitude smaller than the error of the open-loop system. The system actuating signal, which is a measure of the system error, is denoted as $E_a(s)$. However, the actual system error is $E(s)$ =

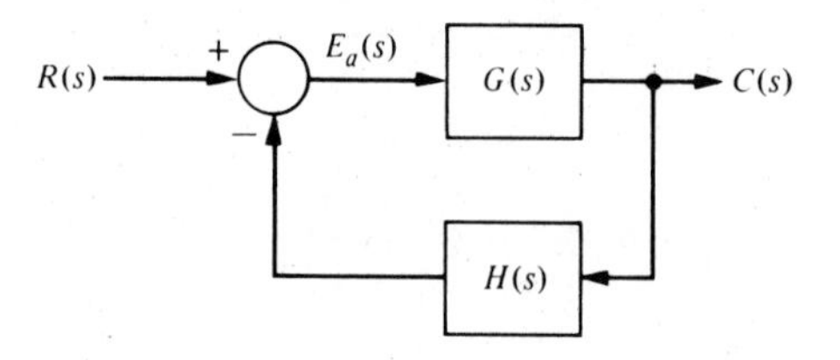

Figure 4.14. Closed-loop control system.

$R(s) - C(s)$. Considering the closed-loop feedback system of Fig. 4.14, we have

$$E(s) = R(s) - \frac{G(s)}{1 + GH(s)} R(s) = \frac{[1 + GH(s) - G(s)]}{1 + GH(s)} R(s). \quad (4.23)$$

The system error is equal to the actuating signal when $H(s) = 1$, which is a common situation. Then

$$E(s) = \frac{1}{1 + G(s)} R(s).$$

The steady-state error, when $H(s) = 1$, is then

$$\lim_{t \to \infty} e(t) = e_{ss} = \lim_{s \to 0} \frac{sR(s)}{1 + G(s)}. \quad (4.24)$$

It is useful to determine the steady-state error of the system for the three standard test inputs for a unity feedback system ($H(s) = 1$).

Step Input. The steady-state error for a step input is therefore

$$e_{ss} = \lim_{s \to 0} \frac{s(A/s)}{1 + G(s)} = \frac{A}{1 + G(0)}. \quad (4.25)$$

Clearly, it is the form of the loop transfer function $GH(s)$ that determines the steady-state error. The loop transfer function is written in general form as

$$G(s) = \frac{K \prod_{i=1}^{M} (s + z_i)}{s^N \prod_{k=1}^{Q} (s + p_k)}, \quad (4.26)$$

where $\prod$ denotes the product of the factors. Therefore the loop transfer function as s approaches zero depends upon the number of integrations N. If N is greater than zero, then $G(0)$ approaches infinity and the steady-state error approaches zero. The number of integrations is often indicated by labeling a system with a *type number* which is simply equal to N.

Therefore, for a type-zero system, $N = 0$, the steady-state error is

$$e_{ss} = \frac{A}{1 + G(0)} = \frac{A}{1 + (K\prod_{i=1}^{M} z_i / \prod_{k=1}^{Q} p_k)}. \tag{4.27}$$

The constant $G(0)$ is denoted by K_p, the position error constant, so that

$$e_{ss} = \frac{A}{1 + K_p}. \tag{4.28}$$

Clearly, the steady-state error for a unit step input with one integration or more, $N \geq 1$, is zero because

$$e_{ss} = \lim_{s \to 0} \frac{A}{1 + (K\prod z_i / s^N \prod p_k)} = \lim_{s \to 0} \frac{As^N}{s^N + (K\prod z_i / \prod p_k)} = 0. \tag{4.29}$$

Ramp Input. The steady-state error for a ramp (velocity) input is

$$e_{ss} = \lim_{s \to 0} \frac{s(A/s^2)}{1 + G(s)} = \lim_{s \to 0} \frac{A}{s + sG(s)} = \lim_{s \to 0} \frac{A}{sG(s)}. \tag{4.30}$$

Again, the steady-state error depends upon the number of integrations N. For a type-zero system, $N = 0$, the steady-state error is infinite. For a type-one system, $N = 1$, the error is

$$e_{ss} = \lim_{s \to 0} \frac{A}{s\{[K\prod(s + z_i)]/[s\prod(s + p_k)]\}} = \frac{A}{(K\prod z_i / \prod p_k)} = \frac{A}{K_v}, \tag{4.31}$$

where K_v is designated the *velocity error constant.* When the transfer function possesses two or more integrations, $N \geq 2$, we obtain a steady-state error of zero.

Acceleration Input. When the system input is $r(t) = At^2/2$, the steady-state error is then

$$e_{ss} = \lim_{s \to 0} \frac{s(A/s^3)}{1 + G(s)} = \lim_{s \to 0} \frac{A}{s^2 G(s)}. \tag{4.32}$$

The steady-state error is infinite for one integration; and for two integrations, $N = 2$, we obtain

$$e_{ss} = \frac{A}{K\prod z_i/\prod p_k} = \frac{A}{K_a}, \tag{4.33}$$

where K_a is designated the acceleration constant. When the number of integrations equals or exceeds three, then the steady-state error of the system is zero.

Control systems are often described in terms of their type number and the error constants, K_p, K_v, and K_a. Definitions for the error constants and the steady-state error for the three inputs are summarized in Table 4.4. The usefulness of the error constants can be illustrated by considering a simple example.

■ Example 4.2 Mobile robot steering control

A severely disabled person could use a mobile robot to serve as an assisting device or servant as shown in Fig. 4.15(a) [31]. The steering control system can be represented by the block diagram shown in Fig. 4.15(b). The steering controller, $G_1(s)$, is

$$G_1(s) = K_1 + K_2/s. \tag{4.34}$$

The steady-state error of the system for a step input when $K_2 = 0$ and $G_1(s) = K_1$ is therefore

$$e_{ss} = \frac{A}{1 + K_p}, \tag{4.35}$$

where $K_p = KK_1$. When K_2 is greater than zero, we have a type-one system,

$$G_1(s) = \frac{K_1 s + K_2}{s},$$

and the steady-state error is zero for a step input.

Table 4.4. Summary of Steady-State Errors

	Input		
Number of Integrations in $G(s)$, type number	**Step, $r(t) = A$, $R(s) = A/s$**	**Ramp, At, A/s^2**	**Parabola, $At^2/2$, A/s^3**
0	$e_{ss} = \frac{A}{1 + K_p}$	Infinite	Infinite
1	$e_{ss} = 0$	$\frac{A}{K_v}$	Infinite
2	$e_{ss} = 0$	0	$\frac{A}{K_a}$

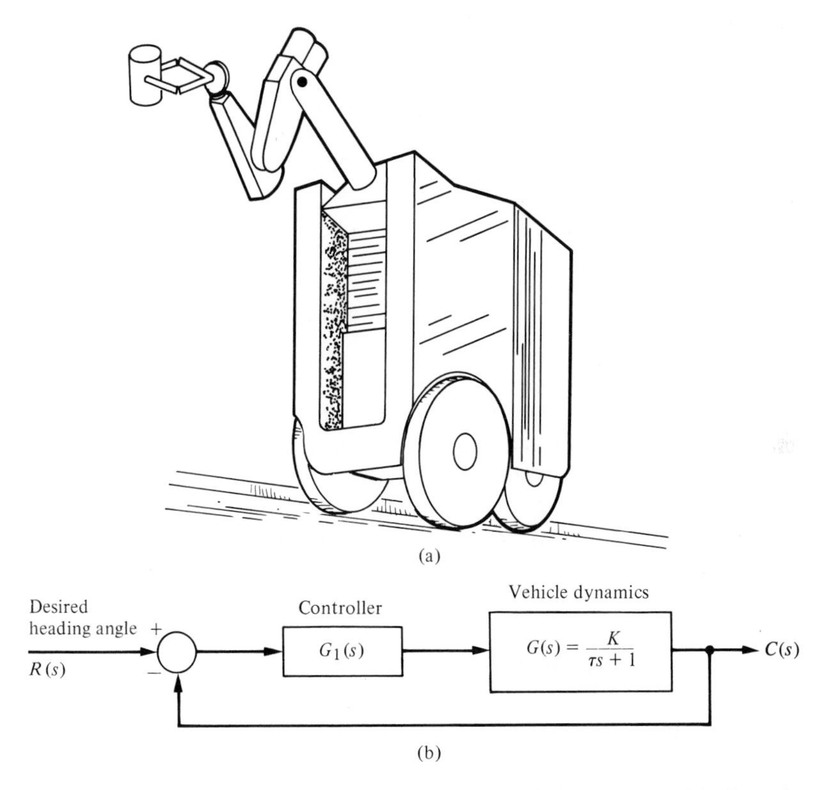

Figure 4.15. (a) A mobile robotic aid for the severely handicapped would allow the disabled person to control the motion of the cart and robot. (b) Block diagram of steering control system.

If the steering command is a ramp input, the steady-state error is then

$$e_{ss} = \frac{A}{K_v}, \tag{4.36}$$

where

$$K_v = \lim_{s \to 0} sG_1(s)G(s) = K_2 K.$$

The transient response of the vehicle to a triangular wave input when $G_1(s) = (K_1 s + K_2)/s$ is shown in Fig. 4.16. The transient response clearly shows the effect of the steady-state error, which may not be objectionable if K_v is sufficiently large.

The error constants, K_p, K_v, and K_a, of a control system describe the ability of a system to reduce or eliminate the steady-state error. They are therefore utilized as numerical measures of the steady-state performance. The designer determines the error constants for a given system and attempts to determine methods of increasing the error constants while maintaining an acceptable transient

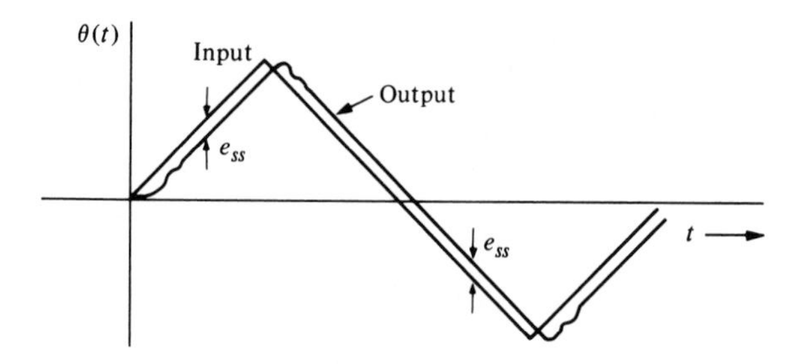

Figure 4.16. Triangular wave response.

response. In the case of the steering control system, it is desirable to increase the gain factor KK_2 in order to increase K_v and reduce the steady-state error. However, an increase in KK_2 results in an attendant decrease in the damping ratio, ζ, of the system and therefore a more oscillatory response to a step input. Again, a compromise would be determined that would provide the largest K_v based on the smallest ζ allowable.

4.5 Performance Indices

An increased amount of emphasis on the mathematical formulation and measurement of control system performance can be found in the recent literature on automatic control. A *performance index* is a quantitative measure of the performance of a system and is chosen so that emphasis is given to the important system specifications. Modern control theory assumes that the systems engineer can specify quantitatively the required system performance. Then a performance index can be calculated or measured and used to evaluate the system's performance. A quantitative measure of the performance of a system is necessary for the operation of modern adaptive control systems, for automatic parameter optimization of a control system, and for the design of optimum systems.

Whether the aim is to improve the design of a system or to design an adaptive control system, a performance index must be chosen and measured. Then the system is considered an *optimum control system* when the system parameters are adjusted so that the index reaches an extremum value, commonly a minimum value. A performance index, in order to be useful, must be a number that is always positive or zero. Then the best system is defined as the system that minimizes this index.

A suitable performance index is the integral of the square of the error, ISE, which is defined as

$$I_1 = \int_0^T e^2(t)\, dt. \tag{4.37}$$

The upper limit T is a finite time chosen somewhat arbitrarily so that the integral approaches a steady-state value. It is usually convenient to choose T as the settling time, T_s. The step response for a specific feedback control system is shown in Fig. 4.17(b); and the error, in Fig. 4.17(c). The error squared is shown in Fig. 4.17(d); and the integral of the error squared, in Fig. 4.17(e). This criterion will discriminate between excessively overdamped systems and excessively underdamped systems. The minimum value of the integral occurs for a compromise value of the damping. The performance index of Eq. (4.37) is easily adapted for practical measurements, because a squaring circuit is readily obtained. Furthermore, the squared error is mathematically convenient for analytical and computational purposes.

Another readily instrumented performance criterion is the integral of the absolute magnitude of the error, IAE, which is written as

$$I_2 = \int_0^T |e(t)|\, dt. \tag{4.38}$$

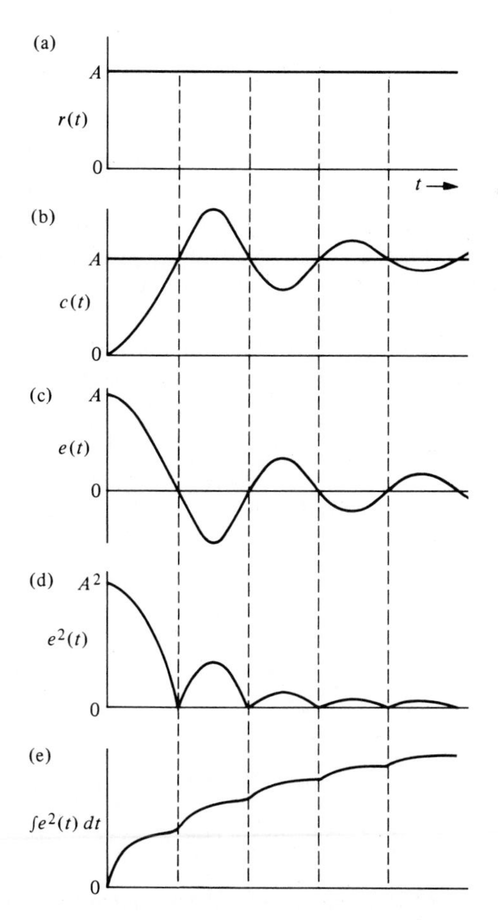

Figure 4.17. The calculation of the integral squared error.

This index is particularly useful for computer simulation studies. In order to reduce the contribution of the large initial error to the value of the performance integral, as well as to place an emphasis on errors occurring later in the response, the following index has been proposed [6]:

$$I_3 = \int_0^T t\,|e(t)|\,dt. \tag{4.39}$$

This performance index is designated the integral of time multiplied by absolute error, ITAE. Another similar index is the integral of time multiplied by the squared error, ITSE, which is

$$I_4 = \int_0^T te^2(t)\,dt. \tag{4.40}$$

The performance index I_3, ITAE, provides the best selectivity of the performance indices; that is, the minimum value of the integral is readily discernible as the system parameters are varied. The general form of the performance integral is

$$I = \int_0^T f(e(t), r(t), c(t), t)\,dt, \tag{4.41}$$

where f is a function of the error, input, output, and time. Clearly, one can obtain numerous indices based on various combinations of the system variables and time. It is worth noting that the minimization of IAE or ISE is often of practical significance. For example, the minimization of a performance index can be directly related to the minimization of fuel consumption for aircraft and space vehicles.

Performance indices are useful for the analysis and design of control systems. Two examples will illustrate the utility of this approach.

■ Example 4.3 Performance criteria

A single-loop feedback control system is shown in Fig. 4.18, where the natural frequency is the normalized value, $\omega_n = 1$. The closed-loop transfer function is then

$$T(s) = \frac{1}{s^2 + 2\zeta s + 1}. \tag{4.42}$$

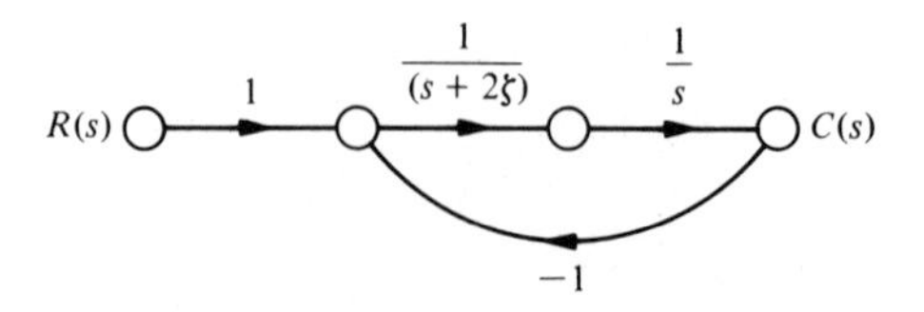

Figure 4.18. Single-loop feedback control system.

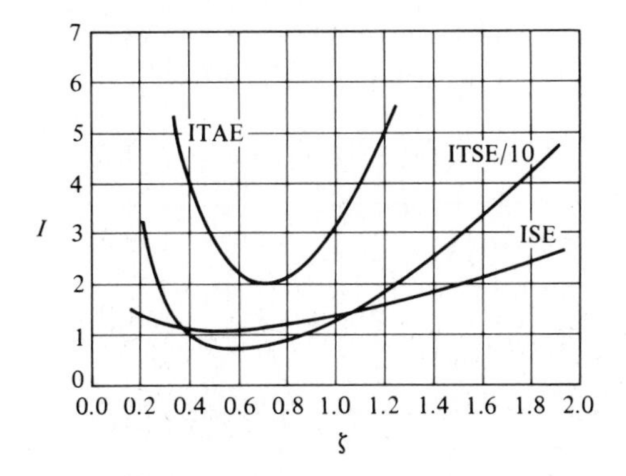

Figure 4.19. Three performance criteria for a second-order system. (Courtesy of Professor R. C. H. Wheeler, U.S. Naval Postgraduate School.)

Three performance indices—ISE, ITSE, and ITAE—calculated for various values of the damping ratio ζ and for a step input are shown in Fig. 4.19. These curves show the selectivity of the ITAE index in comparison with the ISE index. The value of the damping ratio ζ selected on the basis of ITAE is 0.7, which, for a second-order system, results in a swift response to a step with a 5% overshoot.

■ Example 4.4 Space telescope control system

The signal-flow graph of a space telescope pointing control system is shown in Fig. 4.20 [12]. We desire to select the magnitude of the gain K_3 in order to minimize the effect of the disturbance $U(s)$. The disturbance in this case is equivalent to an initial attitude error. The closed-loop transfer function for the disturbance is obtained by using the signal-flow gain formula as

$$\frac{C(s)}{U(s)} = \frac{P_1(s)\,\Delta_1(s)}{\Delta(s)}$$

$$= \frac{1 \cdot (1 + K_1K_3s^{-1})}{1 + K_1K_3s^{-1} + K_1K_2K_p\,s^{-2}} \tag{4.43}$$

$$= \frac{s(s + K_1K_3)}{s^2 + K_1K_3s + K_1K_2K_p}.$$

Typical values for the constants are $K_1 = 0.5$ and $K_1K_2K_p = 2.5$. Then the natural frequency of the vehicle is $f_n = \sqrt{2.5}/2\pi = 0.25$ cycles/sec. For a unit step disturbance, the minimum ISE can be analytically calculated. The attitude $c(t)$ is

$$c(t) = \frac{\sqrt{10}}{\beta}\left[e^{-0.25K_3t}\sin\left(\frac{\beta}{2}t + \psi\right)\right], \tag{4.44}$$

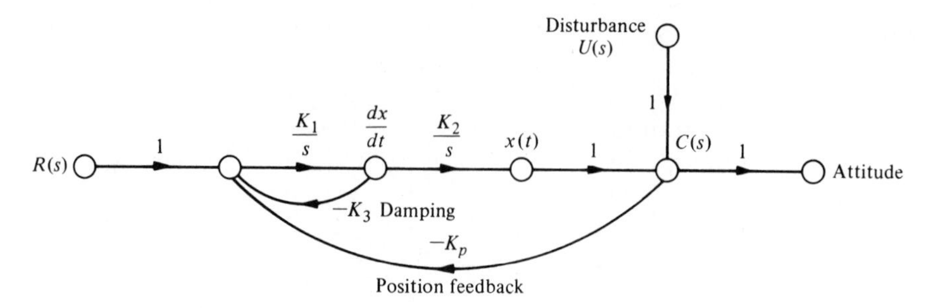

Figure 4.20. A space telescope pointing control system.

where $\beta = K_3\sqrt{(K_3^2/8) - 5}$. Squaring $c(t)$ and integrating the result, we have

$$\begin{aligned} I &= \int_0^\infty \frac{10}{\beta^2} e^{-0.5K_3t} \sin^2\left(\frac{\beta}{2}t + \psi\right) dt \\ &= \int_0^\infty \frac{10}{\beta^2} e^{-0.5K_3t}\left(\frac{1}{2} - \frac{1}{2}\cos(\beta t + 2\psi)\right) dt \qquad (4.45) \\ &= \left(\frac{1}{K_3} + 0.1K_3\right) \end{aligned}$$

Differentiating I and equating the result to zero, we obtain

$$\frac{dI}{dK_3} = -K_3^{-2} + 0.1 = 0. \qquad (4.46)$$

Therefore the minimum ISE is obtained when $K_3 = \sqrt{10} = 3.2$. This value of K_3 corresponds to a damping ratio ζ of 0.50. The values of ISE and IAE for this system are plotted in Fig. 4.21. The minimum for the IAE performance index is obtained when $K_3 = 4.2$ and $\zeta = 0.665$. While the ISE criterion is not as selective as the IAE criterion, it is clear that it is possible to solve analytically for the min-

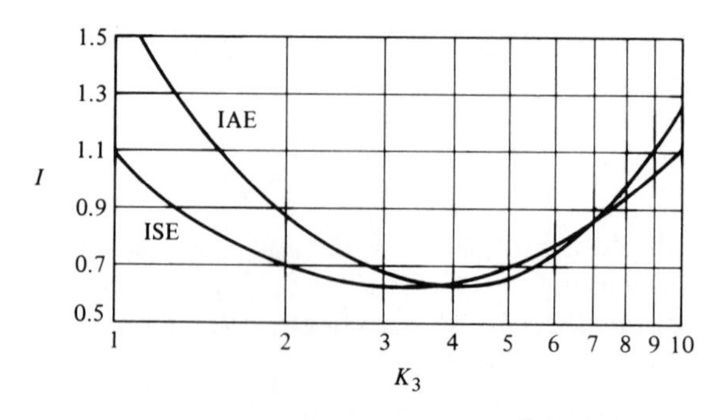

Figure 4.21. The performance indices of the telescope control system versus K_3.

imum value of ISE. The minimum of IAE is obtained by measuring the actual value of IAE for several values of the parameter of interest.

A control system is optimum when the selected performance index is minimized. However, the optimum value of the parameters depends directly upon the definition of optimum, that is, the performance index. Therefore, in the two examples, we found that the optimum setting varied for different performance indices.

The coefficients that will minimize the ITAE performance criterion for a step input have been determined for the general closed-loop transfer function [6]:

$$T(s) = \frac{C(s)}{R(s)} = \frac{b_0}{s^n + b_{n-1}s^{n-1} + \cdots + b_1 s + b_0}. \tag{4.47}$$

This transfer function has a steady-state error equal to zero for a step input. The optimum coefficients for the ITAE criterion are given in Table 4.5. The responses using optimum coefficients for a step input are given in Fig. 4.22 for ISE, IAE, and ITAE. Other standard forms based on different performance indices are available and can be useful in aiding the designer to determine the range of coefficients for a specific problem. A final example will illustrate the utility of the standard forms for ITAE.

■ Example 4.5 Two-camera control

A very accurate and rapidly responding control system is required for a system which allows live actors seemingly to perform inside of complex miniature sets. The two-camera system is shown in Fig. 4.23(a), where one camera is trained on the actor and the other on the miniature set. The challenge is to obtain rapid and accurate coordination of the two cameras by using sensor information from the foreground camera to control the movement of the background camera. The block diagram of the background camera system is shown in Fig. 4.23(b) for one axis of movement of the background camera. The closed-loop transfer function is

$$T(s) = \frac{K_a K_m \omega_0^2}{s^3 + 2\zeta\omega_0 s^2 + \omega_0^2 s + K_a K_m \omega_0^2}. \tag{4.48}$$

Table 4.5. The Optimum Coefficients of $T(s)$ Based on the ITAE Criterion for a Step Input

$s + \omega_n$
$s^2 + 1.4\omega_n s + \omega_n^2$
$s^3 + 1.75\omega_n s^2 + 2.15\omega_n^2 s + \omega_n^3$
$s^4 + 2.1\omega_n s^3 + 3.4\omega_n^2 s^2 + 2.7\omega_n^3 s + \omega_n^4$
$s^5 + 2.8\omega_n s^4 + 5.0\omega_n^2 s^3 + 5.5\omega_n^3 s^2 + 3.4\omega_n^4 s + \omega_n^5$
$s^6 + 3.25\omega_n s^5 + 6.60\omega_n^2 s^4 + 8.60\omega_n^3 s^3 + 7.45\omega_n^4 s^2 + 3.95\omega_n^5 s + \omega_n^6$

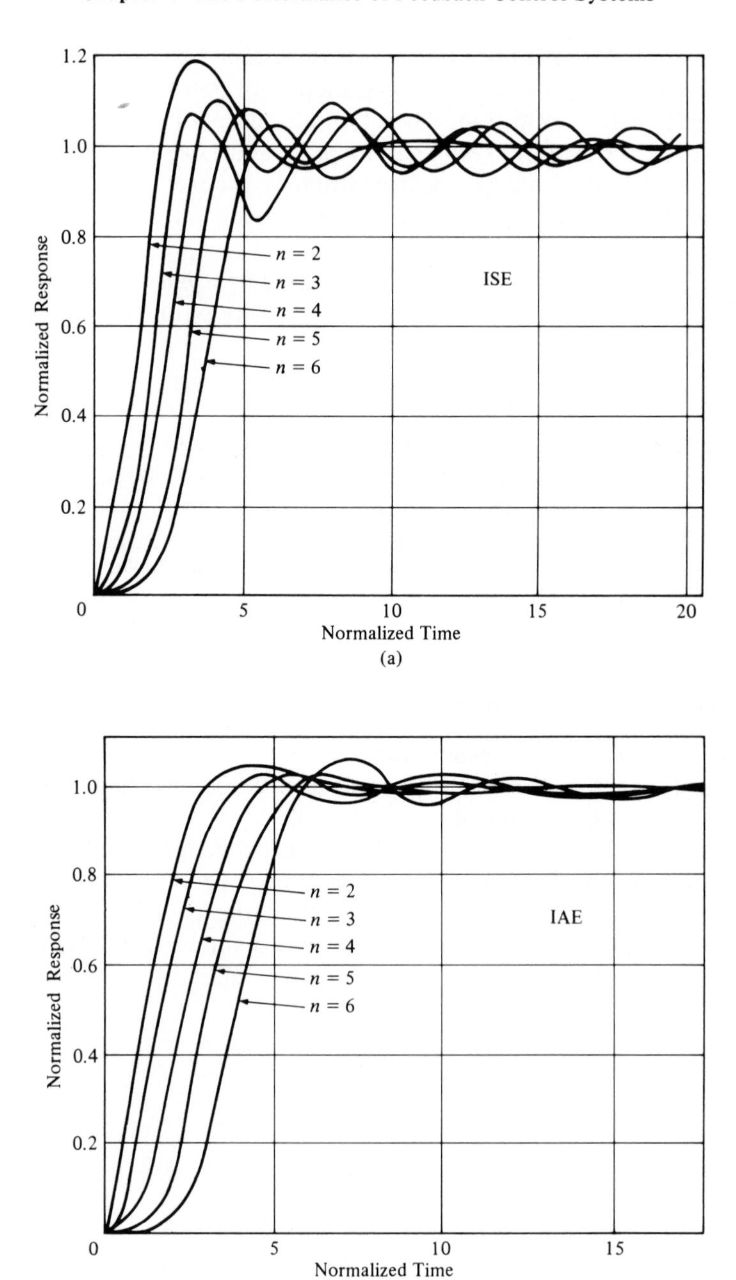

Figure 4.22. Step responses of a normalized transfer function using optimum coefficients for (a) ISE, (b) IAE, and (c) ITAE.

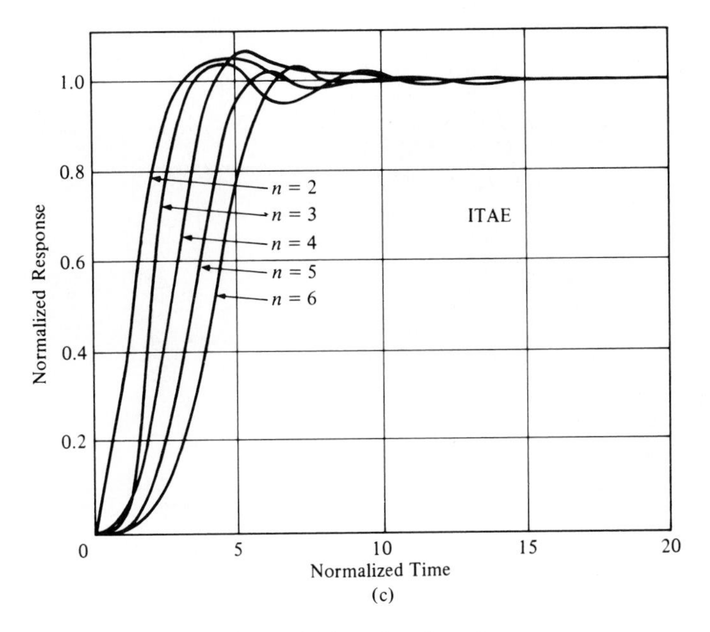

Figure 4.22.—*Continued*

The standard form for a third-order system given in Table 4.5 requires that

$$2\zeta\omega_0 = 1.75\omega_n, \qquad \omega_0^2 = 2.15\omega_n^2, \qquad K_aK_m\omega_0^2 = \omega_n^3.$$

Because a rapid response is required, a large ω_n will be selected so that the settling time will be less than one second. Thus, ω_n will be set equal to 50 rad/sec. Then, for an ITAE system, it is necessary that the parameters of the camera dynamics be

$$\omega_0 = 73 \text{ rad/sec}$$

and

$$\zeta = 0.60.$$

The amplifier and motor gain are required to be

$$K_aK_m = \frac{\omega_n^3}{\omega_0^2} = \frac{\omega_n^3}{2.15\omega_n^2} = \frac{\omega_n}{2.15} = 23.2.$$

Then, the closed-loop transfer function is

$$\begin{aligned} T(s) &= \frac{125{,}000}{s^3 + 87.5s + 5375s + 125{,}000} \\ &= \frac{125{,}000}{(s + 35.5)(s + 26 + j53.4)(s + 26 - j53.4)}. \end{aligned} \tag{4.49}$$

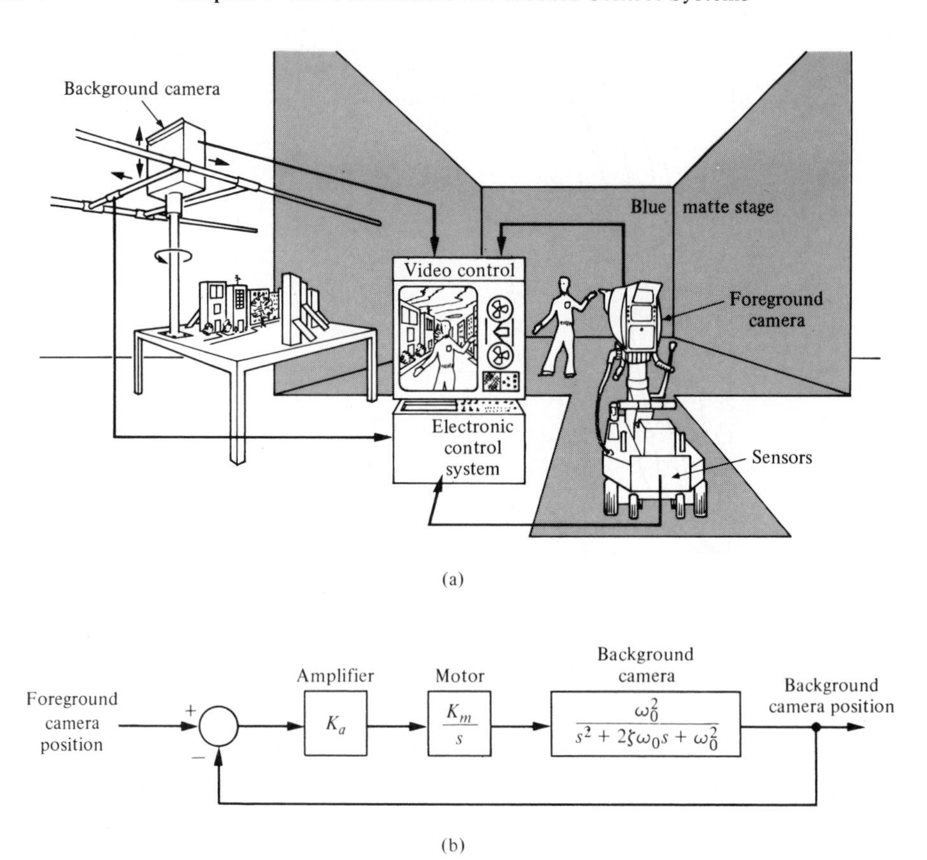

Figure 4.23. The foreground camera, which may be either a film or video camera, is trained on the blue cyclorama stage. The electronic servo control installation permits the slaving, by means of electronic servo devices, of the two cameras. The background camera reaches into the miniature set with a periscope lens and instantaneously reproduces all movements of the foreground camera in the scale of the miniature. The video control installation allows the composite image to be monitored and recorded live. (Part (a) reprinted with permission from Electronic Design 24, 11, May 24, 1976. Copyright © Hayden Publishing Co., Inc., 1976.)

The locations of the closed-loop roots dictated by the ITAE system are shown in Fig. 4.24. The damping ratio of the complex roots is $\zeta = 0.49$, and using Fig. 4.8 one may estimate the overshoot to be approximately 20%. The settling time is approximately $4(1/26) = 0.15$ seconds. This is only an approximation because the complex conjugate roots are not dominant; however, it does indicate the magnitude of the performance measures. The actual response to a step input using a computer simulation with CSDP showed the overshoot to be only 2% and the settling time equal to 0.08 seconds [32]. This illustrates the damping effect of the real root.

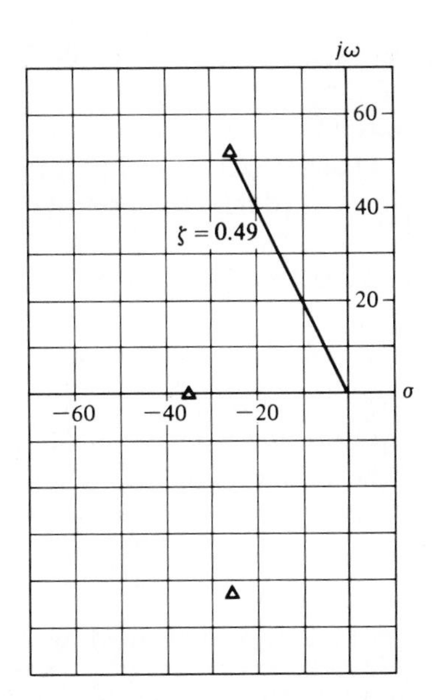

Figure 4.24. The closed-loop roots of a minimum ITAE system.

For a ramp input, the coefficients have been determined that minimize the ITAE criterion for the general closed-loop transfer function [6]:

$$T(s) = \frac{b_1 s + b_0}{s^n + b_{n-1}s^{n-1} + \cdots + b_1 s + b_0}. \tag{4.50}$$

This transfer function has a steady-state error equal to zero for a ramp input. The optimum coefficients for this transfer function are given in Table 4.6. The transfer function, Eq. (4.50), implies that the plant $G(s)$ has two or more pure integrations, as required to provide zero steady-state error.

4.6 The Simplification of Linear Systems

It is quite useful to study complex systems with high-order transfer functions by using lower-order approximate models. Thus, for example, a fourth-order system could be approximated by a second-order system leading to a use of the indices

Table 4.6. The Optimum Coefficients of $T(s)$ Based on the ITAE Criterion for a Ramp Input

$s^2 + 3.2\omega_n s + \omega_n^2$
$s^3 + 1.75\omega_n s^2 + 3.25\omega_n^2 s + \omega_n^3$
$s^4 + 2.41\omega_n s^3 + 4.93\omega_n^2 s^2 + 5.14\omega_n^3 s + \omega_n^4$
$s^5 + 2.19\omega_n s^4 + 6.50\omega_n^2 s^3 + 6.30\omega_n^3 s^2 + 5.24\omega_n^4 s + \omega_n^5$

in Fig. 4.8. There are several methods now available for reducing the order of a systems transfer function [10, 11].

One relatively simple way to delete a certain insignificant pole of a transfer function is to note a pole that has a negative real part that is much larger than the other poles. Thus that pole is expected to insignificantly affect the transient response.

For example, if we have a system plant where

$$G(s) = \frac{K}{s(s+2)(s+30)},$$

we can safely neglect the impact of the pole at $s = 30$. However, we must retain the steady state response of the system and thus we reduce the system to

$$G(s) = \frac{(K/30)}{s(s+2)}.$$

We will let the high-order system be described by the transfer function

$$H(s) = K\frac{a_m s^m + a_{m-1}s^{m-1} + \cdots + a_1 s + 1}{b_n s^n + b_{n-1}s^{n-1} + \cdots + b_1 s + 1}, \tag{4.51}$$

in which the poles are in the left-hand s plane and $m \leq n$. The lower-order approximate transfer function is

$$L(s) = K\frac{c_p s^p + \cdots + c_1 s + 1}{d_g s^g + \cdots + d_1 s + 1}, \tag{4.52}$$

where $p \leq g < n$. Notice that the gain constant K is the same for the original and approximate system in order to ensure the same steady-state response. The method outlined in the following paragraph is based on selecting c_i and d_i in such a way that $L(s)$ has a frequency response (see Chapter 7) very close to that of $H(s)$ [10, 11]. This is equivalent to stating that $H(j\omega)/L(j\omega)$ is required to deviate the least amount from unity for various frequencies. The c and d coefficients are obtained by utilizing the following equation:

$$M^{(k)}(s) = \frac{d^k}{ds^k}M(s) \tag{4.53}$$

and

$$\Delta^{(k)}(s) = \frac{d^k}{ds^k}\Delta(s), \tag{4.54}$$

where $M(s)$ and $\Delta(s)$ are the numerator and denominator polynomials of $H(s)/L(s)$, respectively. We also define

$$M_{2q} = \sum_{k=0}^{2q} \frac{(-1)^{k+q}M^{(k)}(0)M^{(2q-k)}(0)}{k!(2q-k)!}, \qquad q = 0, 1, 2 \ldots \tag{4.55}$$

and a completely identical equation for Δ_{2q}. The solutions for the c and d coefficients are obtained by equating

$$M_{2q} = \Delta_{2q} \tag{4.56}$$

for $q = 1, 2, \ldots$ up to the number required to solve for the unknown coefficients [10].

Let us consider an example in order to clarify the use of these equations.

■ Example 4.6 A simplified model

Consider the third-order system

$$H(s) = \frac{6}{s^3 + 6s^2 + 11s + 6} = \frac{1}{1 + (11/6)s + s^2 + (1/6)s^3}. \tag{4.57}$$

Using the second-order model

$$L(s) = \frac{1}{1 + d_1 s + d_2 s^2}, \tag{4.58}$$

$M(s) = 1 + d_1 s + d_2 s^2$ and $\Delta(s) = 1 + (11/6)s + s^2 + (1/6)s^3$. Then

$$M^0(s) = 1 + d_1 s + d_2 s^2 \tag{4.59}$$

and $M^0(0) = 1$. Similarly,

$$M^1 = \frac{d}{ds}(1 + d_1 s + d_2 s^2) = d_1 + 2\, d_2 s. \tag{4.60}$$

Therefore $M^1(0) = d_1$. Continuing this process, we find that

$$\begin{aligned} M^0(0) &= 1 & \Delta^0(0) &= 1 \\ M^1(0) &= d_1 & \Delta^1(0) &= 11/6 \\ M^2(0) &= 2d_2 & \Delta^2(0) &= 2 \\ M^3(0) &= 0 & \Delta^3(0) &= 1 \end{aligned} \tag{4.61}$$

We now equate $M_{2q} = \Delta_{2q}$ for $q = 1$ and 2. We find that for $q = 1$,

$$\begin{aligned} M_2 &= (-1)\frac{M^0(0)M^2(0)}{2} + \frac{M^1(0)M^1(0)}{1} + (-1)\frac{M^2(0)M^0(0)}{2} \\ &= -d_2 + d_1^2 - d_2 = -2d_2 + d_1^2. \end{aligned} \tag{4.62}$$

Then, because the equation for Δ_2 is identical, we have

$$\begin{aligned} \Delta_2 &= -\frac{\Delta^0(0)\,\Delta^2(0)}{2} + \frac{\Delta^1(0)\,\Delta^1(0)}{1} + (-1)\frac{\Delta^2(0)\,\Delta^0(0)}{2} \\ &= -1 + \frac{121}{36} - 1 = \frac{49}{36}. \end{aligned} \tag{4.63}$$

Therefore, because $M_2 = \Delta_2$, we have

$$-2d_2 + d_1^2 = 49/36. \tag{4.64}$$

Completing the process for $M_4 = \Delta_4$, when $q = 2$, we obtain

$$d_2^2 = 7/18. \tag{4.65}$$

Then the solution for $L(s)$ is $d_1 = 1.615$ and $d_2 = 0.625$. (The other sets of solutions are rejected because they lead to unstable poles.) It is interesting to see that the poles of $H(s)$ are $s = -1, -2, -3$, whereas the poles of $L(s)$ are at $s = -1.029$ and -1.555. The lower-order system transfer function is

$$\begin{aligned} L(s) &= \frac{1}{1 + 1.615s + 0.625s^2} \\ &= \frac{1.60}{s^2 + 2.584s + 1.60}. \end{aligned} \tag{4.66}$$

Because the lower-order model has two poles, we can estimate that we would obtain a slightly overdamped response with a settling time of approximately four seconds.

It is sometimes desirable to retain the dominant poles of the original system, $H(s)$, in the low-order model. This can be accomplished by specifying the denominator of $L(s)$ to be the dominant poles of $H(s)$ and allow the numerator of $L(s)$ to be subject to approximation. A complex system such as the robot system shown in Fig. 4.25 is an example of a high-order system that can be favorably represented by a low-order model.

4.7 Design Example: Hubble Telescope Pointing Control

The Hubble space telescope, the most complex and expensive scientific instrument that has ever been built, is orbiting the earth. Launched to 380 miles above the earth on April 24, 1990, the telescope has pushed technology to new limits. The telescope's 2.4 meter (94.5-inch) mirror has the smoothest surface of any mirror made and its pointing system can center it on a dime 400 miles away [23, 24]. The mirror, as it turns out, has a spherical aberration. However, the Hubble telescope can point accurately. Consider the model of the telescope-pointing system shown in Fig. 4.26.

The goal of the design is to choose K_1 and K so that (1) the percent overshoot of the output to a step command, $r(t)$, is less than or equal to 10%, (2) the steady-state error to a ramp command is minimized, and (3) the effect of a step disturbance is reduced. Since the system has an inner loop, block diagram reduction can be used to obtain the simplified system of Fig. 4.26(b).

Mason's formula can be used to obtain the output due to the two inputs

$$C(s) = T(s)R(s) + [T(s)/K]\, D(s), \tag{4.67}$$

where

$$T(s) = \frac{KG(s)}{1 + KG(s)} = \frac{KG(s)}{1 + L(s)}.$$

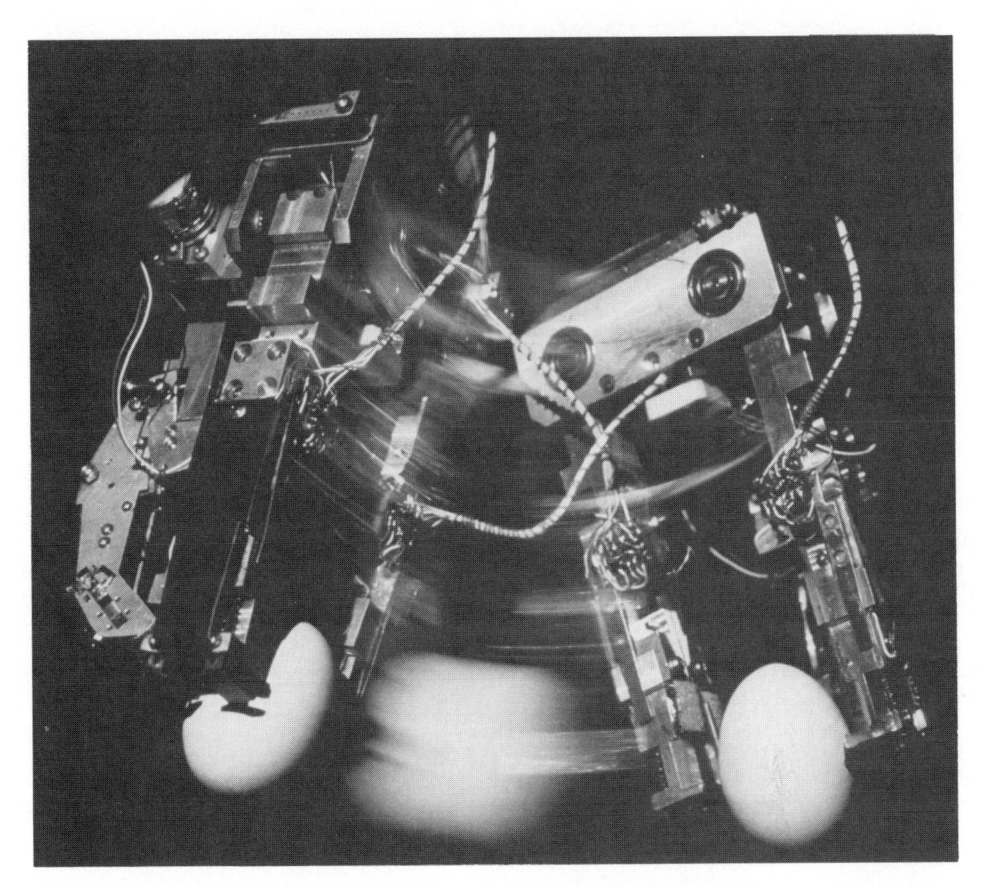

Figure 4.25. A high-performance robot with a delicate touch illustrates the challenge of modern high-performance control systems. (Photo courtesy of Hitachi America Ltd.)

The error $E(s)$ is

$$E(s) = \frac{1}{1 + L(s)} R(s) - \frac{G(s)}{1 + L(s)} D(s). \tag{4.68}$$

First, let us select K and K_1 to meet the percent overshoot requirement for a step input, $R(s) = A/s$. Setting $D(s) = 0$, we have

$$\begin{aligned} C(s) &= \frac{KG(s)}{1 + KG(s)} R(s) \\ &= \frac{K}{s(s + K_1) + K}\left(\frac{A}{s}\right) \\ &= \frac{K}{s^2 + K_1 s + K}\left(\frac{A}{s}\right). \end{aligned} \tag{4.69}$$

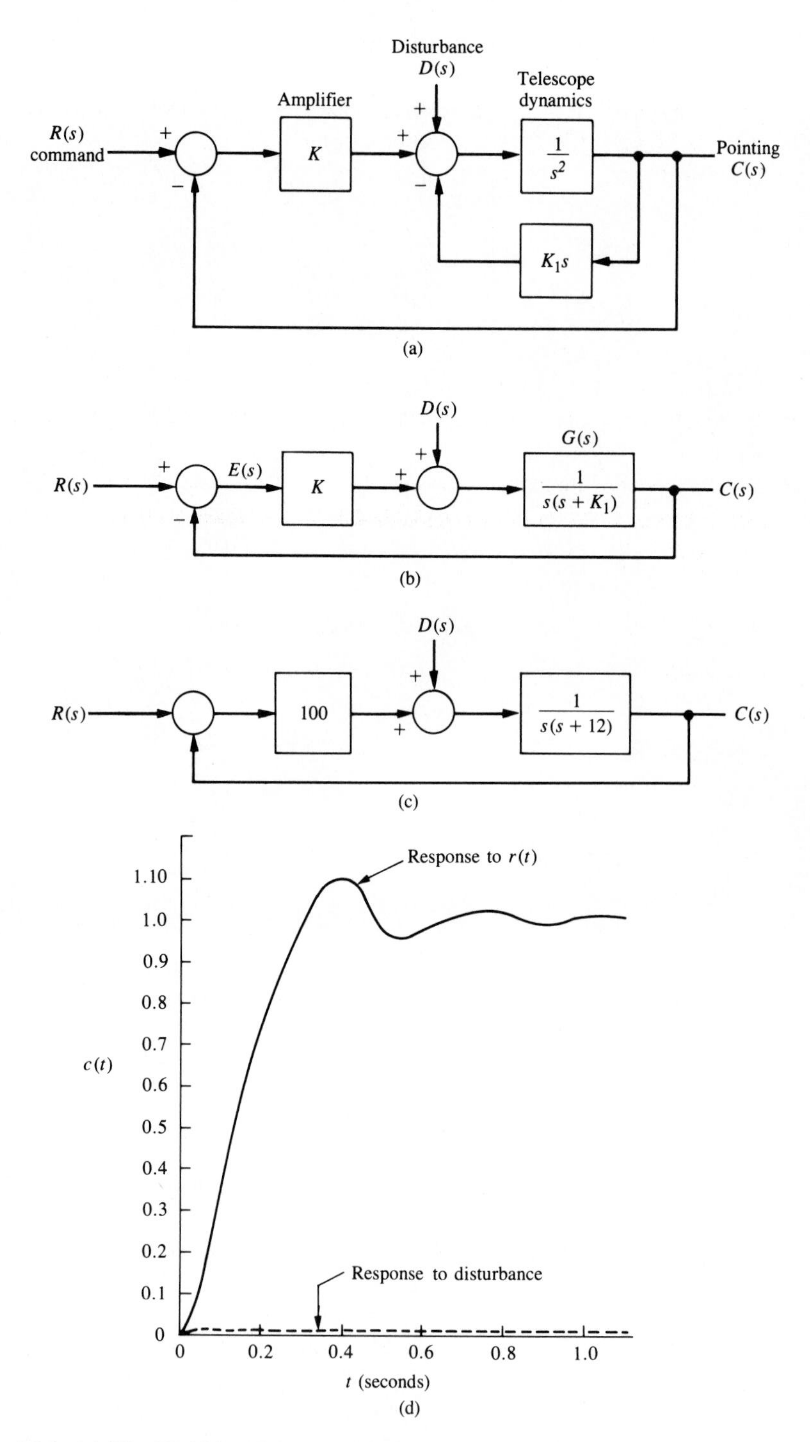

Figure 4.26. (a) The Hubble telescope pointing system, (b) reduced block diagram, (c) system design, and (d) system response to a unit step input command and a unit step disturbance input.

In order to set the overshoot at 10%, we select $\zeta = 0.6$ by examining Fig. 4.8 or using Eq. (4.14) to determine that the overshoot will be 9.5% for $\zeta = 0.6$. We next examine the steady-state error for a ramp, $r(t) = Bt$, $t \geq 0$, using (Eq. 4.30):

$$e_{ss} = \lim_{s\to 0}\left\{\frac{B}{sKG(s)}\right\} = \frac{B}{(K/K_1)}. \tag{4.70}$$

The steady-state error due to a step disturbance is equal to zero. (Can you show this?) The transient response of the error due to the step disturbance input can be reduced by increasing K (see Eq. 4.68). Thus, in summary, we seek a large K, a ζ of the characteristic equation equal to 0.6, and a large value of (K/K_1) in order to obtain a low steady-state error for the ramp input (see Eq. 4.70).

For our design, we need to select K. The characteristic equation of the system is

$$(s^2 + 2\zeta\omega_n s + \omega_n^2) = (s^2 + 2(0.6)\omega_n s + K). \tag{4.71}$$

Therefore $\omega_n = \sqrt{K}$ and the second term of the denominator of Eq. (4.69) requires $K_1 = 2(0.6)\omega_n$. Then, $K_1 = 1.2\sqrt{K}$ or the ratio K/K_1 becomes

$$\frac{K}{K_1} = \frac{K}{1.2\sqrt{K}} = \frac{\sqrt{K}}{1.2}.$$

Selecting $K = 25$, we have $K_1 = 6$ and $K/K_1 = 4.17$. If we select $K = 100$, we have $K_1 = 12$ and $K/K_1 = 8.33$. Realistically, we must limit K so that the system's operation remains linear. Using $K = 100$, we obtain the system shown in Fig. 4.26(c). The responses of the system to a unit step input command and a unit step disturbance input are shown in Fig. 4.26(d). Note how the effect of the disturbance is relatively insignificant.

4.8 Summary

In this chapter we have considered the definition and measurement of the performance of a feedback control system. The concept of a performance measure or index was discussed and the usefulness of standard test signals was outlined. Then several performance measures for a standard step input test signal were delineated. For example, the overshoot, peak time, and settling time of the response of the system under test for a step input signal were considered. The fact that often the specifications on the desired response are contradictory was noted and the concept of a design compromise was proposed. The relationship between the location of the s-plane root of the system transfer function and the system

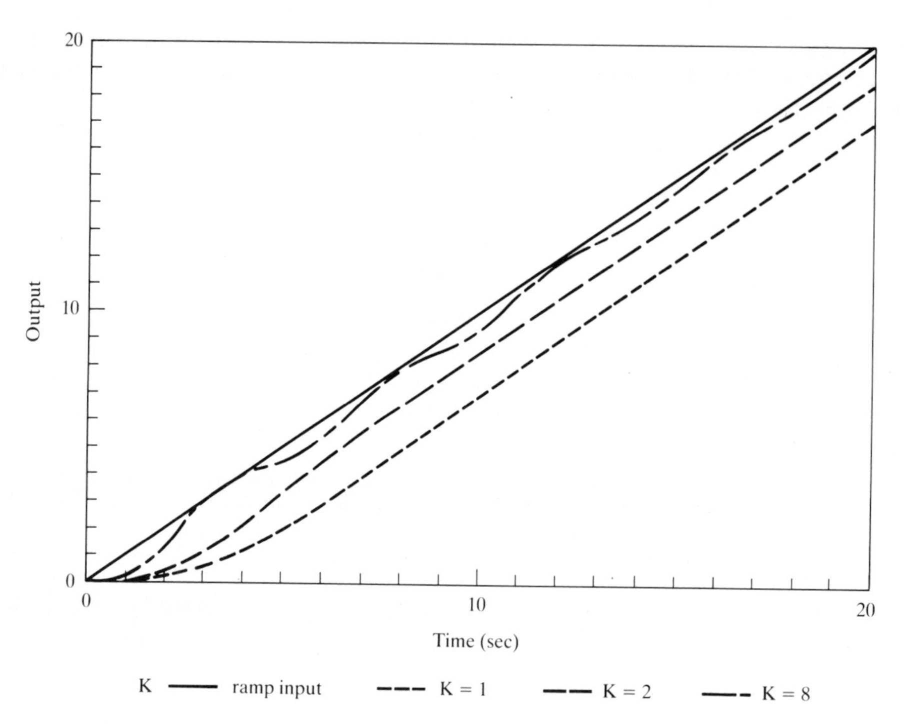

Figure 4.27. The response of a feedback system to a ramp input with $K = 1$, 2, and 8 when $G(s) = K/[s(s+1)(s+3)]$. Note the steady-state error is reduced as K is increased, but the response becomes oscillatory at $K = 8$.

response was discussed. A most important measure of system performance is the steady-state error for specific test input signals. Thus the relationship of the steady-state error of a system in terms of the system parameters was developed by utilizing the final-value theorem. The capability of a feedback control system is demonstrated in Fig. 4.27. Finally, the utility of an integral performance index was outlined and several examples of design which minimized a system's performance index were completed. Thus we have been concerned with the definition and usefulness of quantitative measures of the performance of feedback control systems.

Exercises

E4.1. A motor control system for a computer disk drive must reduce the effect of disturbances and parameter variations, as well as reduce the steady-state error. We desire to have no steady-state error for the head-positioning control system, which is of the form shown in Figure 4.14 where $H(s) = 1$. (a) What type number is required? (How many integrations?) (b) If the input is a ramp signal, then in order to achieve a zero steady-state error, what type number is required?

E4.2. The engine, body, and tires of a racing vehicle affect the acceleration and speed attainable [25]. The speed control of the car is represented by the model shown in Fig. E4.2. (a) Calculate the steady-state error of the car to a step command in speed. (b) Calculate overshoot of the speed to a step command.

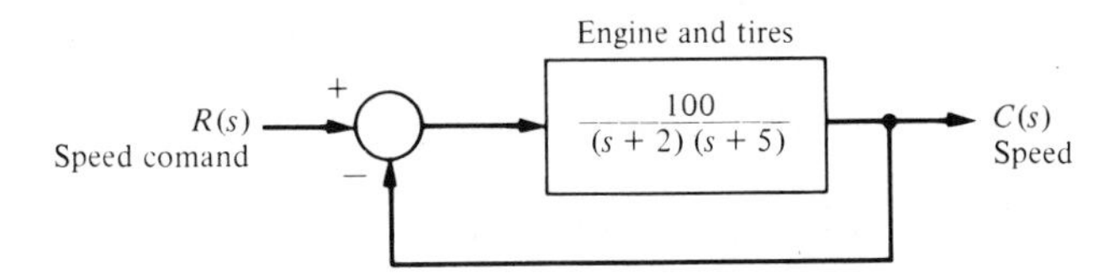

Figure E4.2. Racing car speed control.

E4.3. For years, Amtrak has struggled to attract passengers on its routes in the Midwest, using technology developed decades ago. During the same time, foreign railroads were developing new passenger rail systems that could profitably compete with air travel. Two of these systems, the French TGV and the Japanese Shinkansen, reach speeds of 160 mph [23, 24]. The Transrapid-06, a U.S. experimental magnetic levitation train is shown in Fig. E4.3(a).

(a)

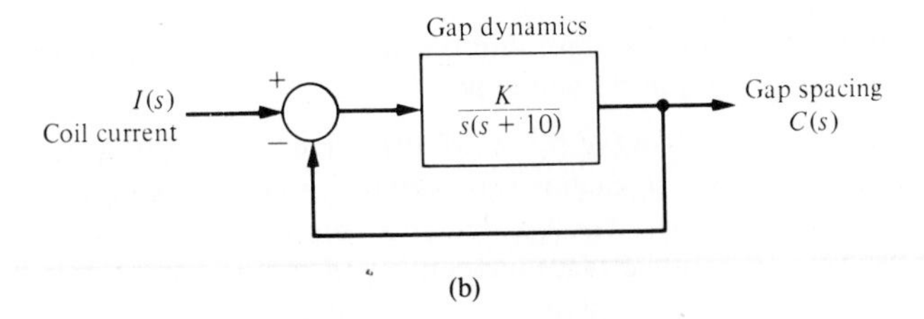

(b)

Figure E4.3. Levitated train control.

The use of magnetic levitation and electromagnetic propulsion to provide contactless vehicle movement makes the Transrapid-06 technology radically different from the existing Metroliner. The underside of the TR-06 carriage (where the wheel trucks would be on a conventional car) wraps around a guideway. Magnets on the bottom of the guideway attract electro magnets on the "wraparound," pulling it up toward the guideway. This suspends the vehicles about one centimeter above the guideway. (See problem 2.27.)

The levitation control is represented by Fig. E4.3(b). (a) Using Table 4.5 for a step input, select K so that the system provides an optimum ITAE response. (b) Using Fig. 4.8, determine the expected overshoot to a step input of $I(s)$.

Answers: K = 50 ; 4.5%

E4.4. A feedback system with negative unity feedback has a plant

$$G(s) = \frac{s + 6}{s(s + 4)}$$

(see Fig. 4.14). (a) Determine the closed-loop transfer function $T(s) = C(s)/R(s)$. (b) Find the time response $c(t)$ for a step input $r(t) = A$ for $t > 0$. (c) Using Fig. 4.10, determine the overshoot of the response. (d) Using the final value theorem, determine the steady-state value of $c(t)$.

E4.5. A low inertia plotter is shown in Fig. E4.5(a), on the next page. This system may be represented by the block diagram shown in Fig. E4.5(b) [19]. (a) Calculate the steady-state error for a ramp input. (b) Select a value of K that will result in zero overshoot to a step input but as rapid response as is attainable.

Plot the poles and zeros of this system and discuss the dominance of the complex poles. What overshoot for a step input do you expect?

E4.6. Effective control of insulin injections can result in better lives for diabetic persons. Automatically controlled insulin injection by means of a pump and a sensor that measures blood sugar can be very effective. A pump and injection system has a feedback control as shown in Fig. E4.6, on the next page. Calculate the suitable gain K so that the overshoot of the step response due to the drug injection is approximately 15%. $R(s)$ is the desired blood-sugar level and $C(s)$ is the actual blood-sugar level. (Hint: use Fig. 4.10.)

Answer: K = 1.67

E4.7. A control system for positioning the head of a floppy disk drive has a closed-loop transfer function

$$T(s) = \frac{0.313(s + 0.8)}{(s + 0.25)(s^2 + 0.3s + 1)}.$$

Plot the poles and zeros of this system and discuss the dominance of the complex poles. What overshoot for a step input do you expect?

E4.8. A prototype for a microwave-powered aircraft was recently tested. The energy to drive the small airplane's electric engine is powered by microwave power beamed up from an earth station. A commercial version will have a 36m wingspan and a 24m long fuselage and it will be used for unmanned reconnaissance and radio relay [21]. If the aircraft flies at 50,000 ft, determine the accuracy of the beam transmission required if it is cruising at 16 mph. Assume the movement is a ramp input and determine the maximum steady-state error allowable.

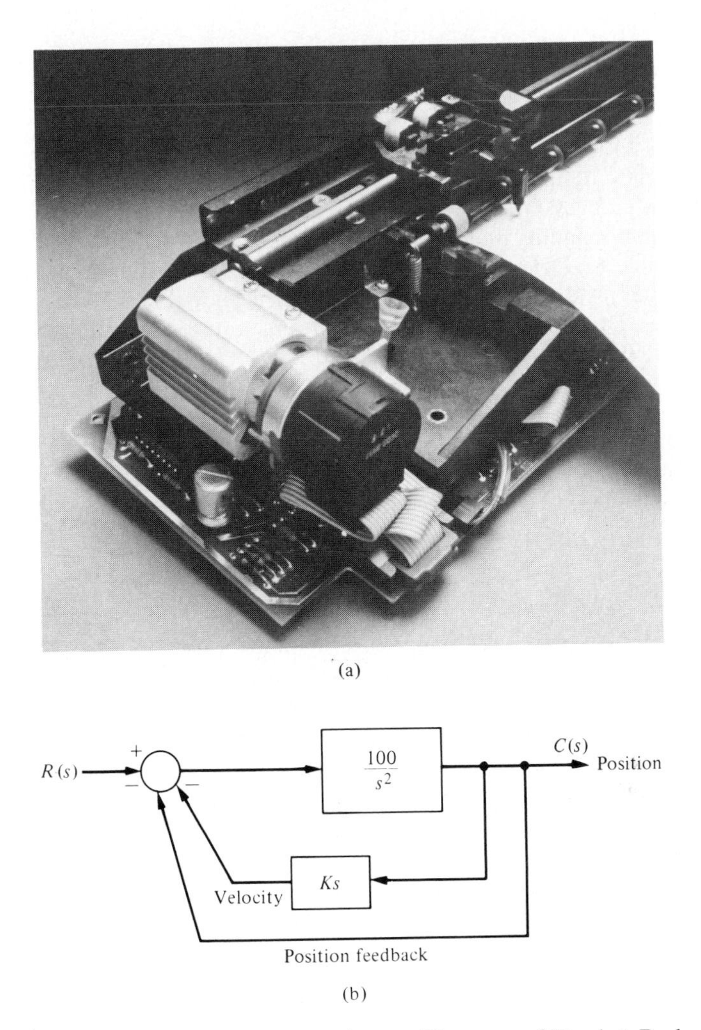

Figure E4.5. (a) The Hewlett-Packard *x-y* plotter. (Courtesy of Hewlett-Packard Co.) (b) Block diagram of plotter.

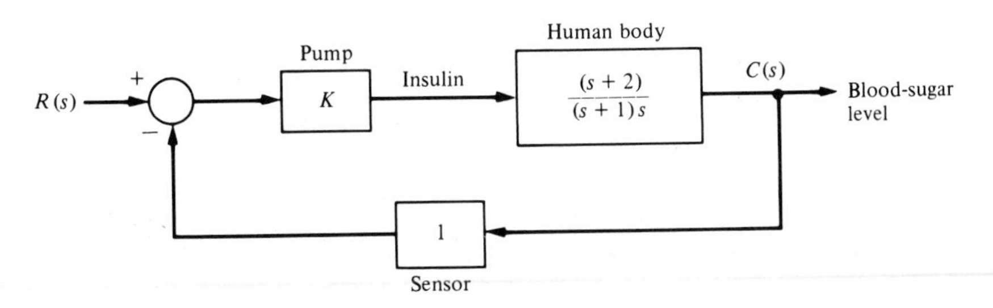

Figure E4.6. Blood-sugar level control.

Problems

P4.1. An important problem for television camera systems is the jumping or wobbling of the picture due to the movement of the camera. This effect occurs when the camera is mounted in a moving truck or airplane. A system called the Dynalens system has been designed which reduces the effect of rapid scanning motion and is shown in Fig. P4.1 [27]. A maximum scanning motion of 25°/sec is expected. Let $K_g = K_t = 1$ and assume that τ_g is negligible. (a) Determine the error of the system $E(s)$. (b) Determine the necessary loop gain, $K_a K_m K_t$, when a 1°/sec steady-state error is allowable. (c) The motor time constant is 0.40 sec. Determine the necessary loop gain so that the settling time of v_b is less than or equal to 0.04 sec.

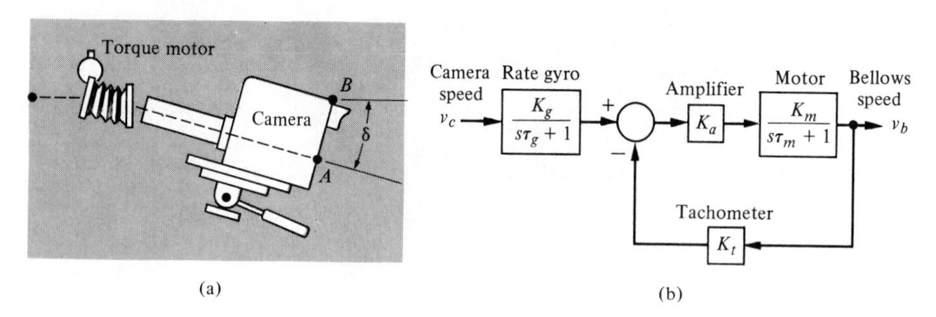

Figure P4.1. Camera wobble control.

P4.2. A laser beam can be used to weld, drill, etch, cut, and mark metals, as shown in Fig. P4.2(a), on the next page [19]. Assume we have a work requirement for an accurate laser to mark a parabolic path with a closed-loop control system as shown in Fig. P4.2(b). Calculate the necessary gain to result in a steady-state error of 1 mm for $r(t) = t^2$, cm/s^2.

P4.3. A specific closed-loop control system is to be designed for an underdamped response to a step input. The specifications for the system are as follows:

$$20\% > \text{percent overshoot} > 10\%$$

$$\text{Settling time} < 0.8 \text{ sec}$$

(a) Identify the desired area for the dominant roots of the system. (b) Determine the smallest value of a third root, r_3, if the complex conjugate roots are to represent the dominant response. (c) The closed-loop system transfer function $T(s)$ is third order and the feedback has a unity gain. Determine the forward transfer function $G(s) = C(s)/E(s)$ when the settling time is 0.8 sec and the percent overshoot is 20%.

P4.4. The open-loop transfer function of a unity negative feedback system is

$$G(s) = \frac{K}{s(s + 2)}.$$

A system response to a step input is specified as follows:

$$\text{peak time } T_p = 1.1 \text{ sec}$$

$$\text{percent overshoot} = 5\%.$$

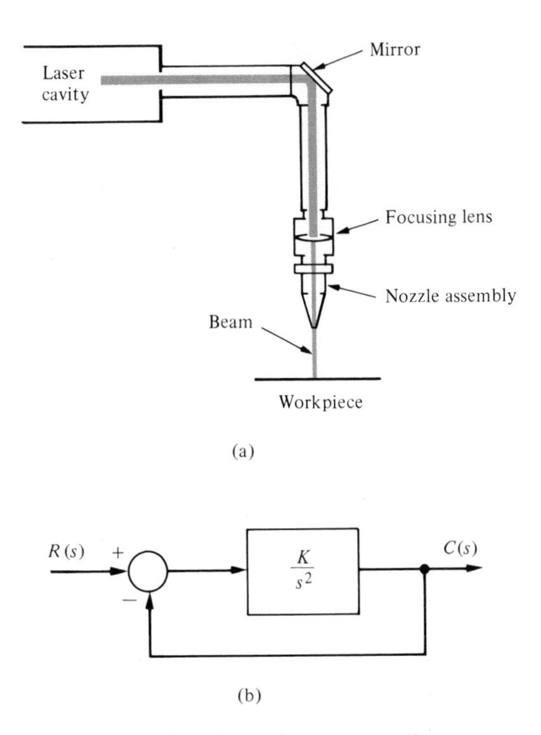

Figure P4.2. Laser beam control.

(a) Determine whether both specifications can be met simultaneously. (b) If the specifications cannot be met simultaneously, determine a compromise value for K so that the peak time and percent overshoot specifications are relaxed the same percentage.

P4.5. A space telescope is to be launched to carry out astronomical experiments [12]. The pointing control system is desired to achieve 0.01 minutes of arc and track solar objects with apparent motion up to 0.21 minutes per second. The system is illustrated in Fig. P4.5 (a), on the next page. The control system is shown in Fig. P4.5(b). Assume that $\tau_1 = 1$ sec and $\tau_2 = 0$ (an approximation). (a) Determine the gain $K = K_1K_2$ required so that the response to a step command is as rapid as reasonable with an overshoot of less than 5%. (b) Determine the steady-state error of the system for a step and a ramp input. (c) Determine the value of K_1K_2 for an ITAE optimal system for (1) a step input and (2) a ramp input.

P4.6. A robot is programmed to have a tool or welding torch follow a prescribed path [15]. Consider a robot tool that is to follow a sawtooth path as shown in Fig. P4.6(a),on the following page. The transfer function of the plant is

$$G(s) = \frac{1000(s + 2)}{s(s + 10)(s + 12)}$$

for the closed-loop system shown in Fig. 4.6(b). Calculate the steady-state error.

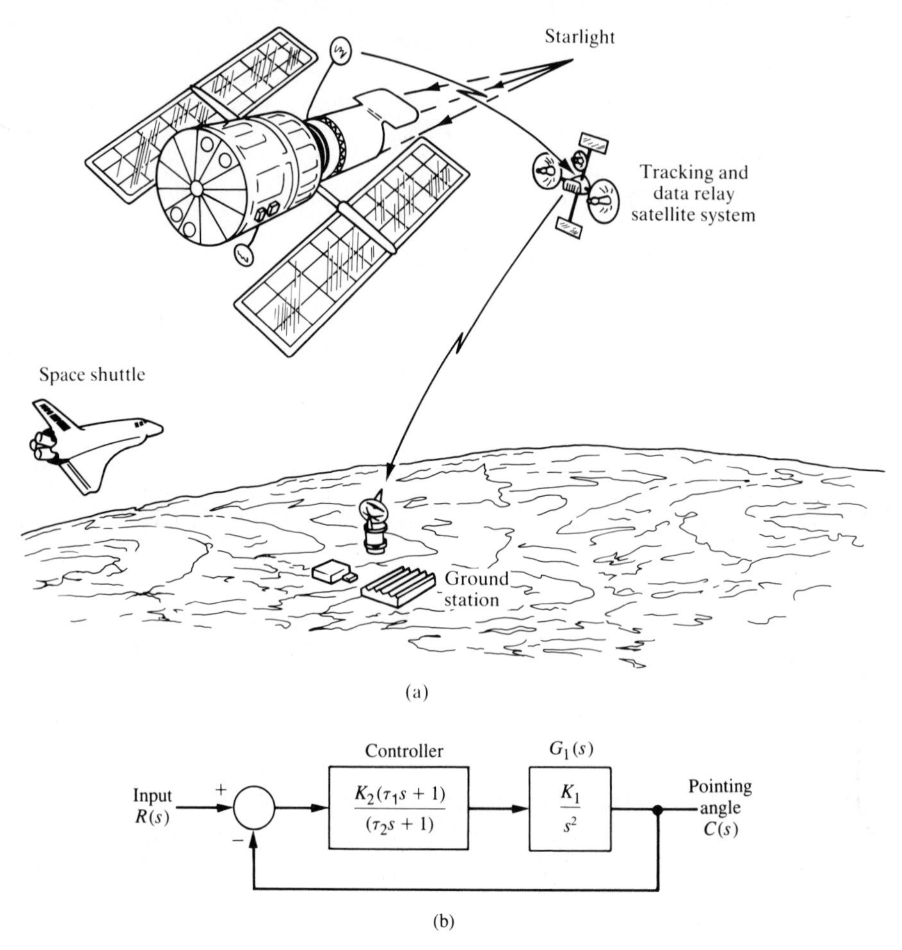

Figure P4.5. (a) The space telescope. (b) The space telescope pointing control system.

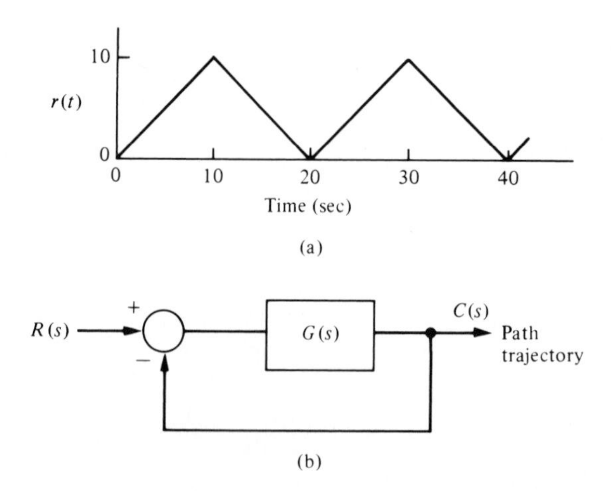

Figure P4.6. Robot path control.

(a)

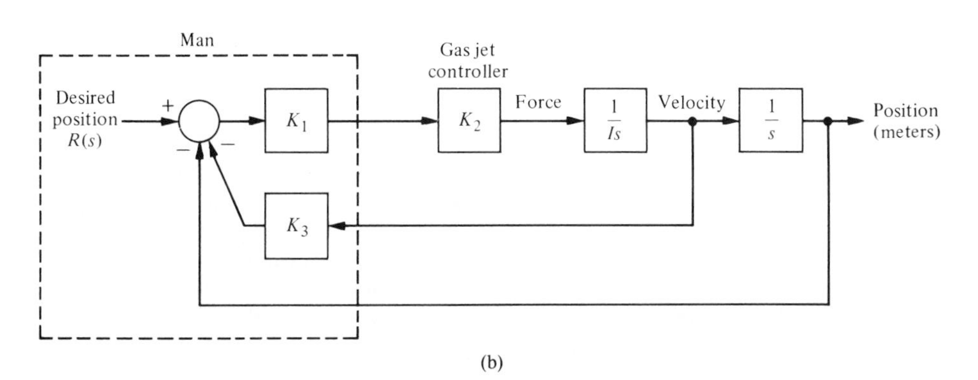

(b)

Figure P4.7. (a) Astronaut Bruce McCandless II is shown a few meters away from the earth-orbiting space shuttle *Challenger*. He used a nitrogen-propelled hand-controlled device called the manned maneuvering unit. (Courtesy of National Aeronautics and Space Administration.) (b) Block diagram of controller.

P4.7. Astronaut Bruce McCandless II took the first untethered walk in space on February 7, 1984, using the gas-jet propulsion device illustrated in Fig. P4.7(a) on the previous page. The controller can be represented by a gain K_2 as shown in Fig. P4.7(b). The inertia of the man and equipment with his arms at his sides is 25 Kg-m^2. (a) Determine the necessary gain K_3 to maintain a steady-state error equal to 1 cm when the input is a ramp $r(t) = t$ (meters). (b) With this gain K_3, determine the necessary gain K_1K_2 in order to restrict the percent overshoot to 10%. (c) Determine analytically the gain K_1K_2 in order to minimize the ISE performance index for a step input.

P4.8. Photovoltaic arrays (solar cells) generate a dc voltage that can be used to drive dc motors or that can be converted to ac power and added to the distribution network. It is desirable to maintain the power out of the array at its maximum available as the solar incidence changes during the day. One such closed-loop system is shown in Fig. P4.8. The transfer function for the process is

$$G(s) = \frac{K}{s+4},$$

where $K = 10$. Find (a) the time constant of the closed-loop system and (b) the settling time of the system when disturbances such as clouds occur.

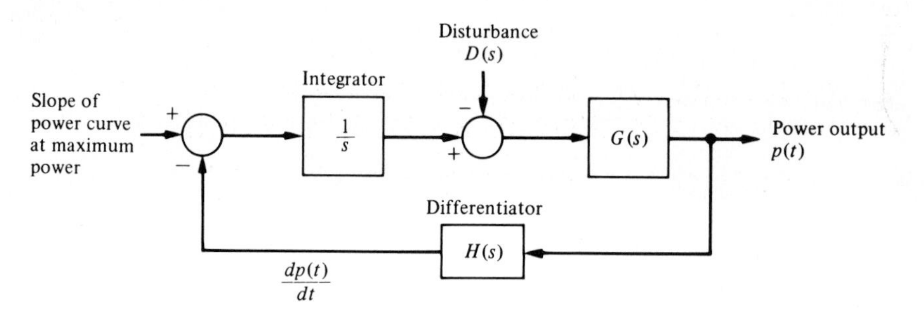

Figure P4.8. Solar cell control.

P4.9. The antenna that receives and transmits signals to the *Telstar* communication satellite is the largest horn antenna ever built. The microwave antenna is 177 ft long, weighs 340 tons, and rolls on a circular track. A photo of the antenna is shown in Fig. P4.9, on the next page. The *Telstar* satellite is 34 inches in diameter and moves about 16,000 mph at an altitude of 2500 miles. The antenna must be positioned accurately to 1/20 of a degree, because the microwave beam is 0.2° wide and highly attenuated by the large distance. If the antenna is following the moving satellite, determine the K_v necessary for the system.

P4.10. A speed-control system of an armature controlled dc-motor uses the back emf voltage of the motor as a feedback signal. (a) Draw the block diagram of this system (see Eq. 2.66). (b) Calculate the steady-state error of this system to a step input command setting the speed to a new level. Assume that $R_a = L_a = J = f = 1$, the motor constant is $k_m = 1$, and $K_b = 1$. (c) Select a feedback gain for the back emf signal to yield a step response with an overshoot of 15%.

Figure P4.9. A model of the antenna for the Telstar System at Andover, Maine. (Photo courtesy of Bell Telephone Laboratories, Inc.)

P4.11. A simple unity feedback control system has a process transfer function

$$\frac{C(s)}{E(s)} = G(s) = \frac{K}{s}.$$

The system input is a step function with an amplitude A. The initial condition of the system at time t_0 is $c(t_0) = Q$, where $c(t)$ is the output of the system. The performance index is defined as

$$I = \int_0^\infty e^2(t)\,dt.$$

(a) Show that $I = (A - Q)^2/2K$. (b) Determine the gain K that will minimize the performance index I. Is this gain a practical value? (c) Select a practical value of gain and determine the resulting value of the performance index.

P4.12. Train travel between cities will increase as trains are developed that travel at high speeds, making the train-travel time from city center to city center equivalent to airline-travel time. The Japanese National Railway has a train called the Bullet Express (Fig. P4.12) that travels between Tokyo and Osaka on the Tokaido line. This train travels the 320 miles in 3 hours and 10 minutes, an average speed of 101 mph [24]. This speed will be increased as new systems are used, such as magnetically levitated systems to float vehicles above an aluminum guideway. In order to maintain a desired speed, a speed control

Figure P4.12. Fifty minutes out of Tokyo, the Bullet Express whizzes past Mt. Fuji. (Photo courtesy of the Japan National Tourist Organization.)

system is proposed which yields a zero steady-state error to a ramp input. A third-order system is sufficient. Determine the optimum system for an ITAE performance criterion and estimate the settling time and overshoot for a step input when $\omega_n = 6$.

P4.13. It is desired to approximate a fourth-order system by a lower-order model. The transfer function of the original system is

$$H(s) = \frac{s^3 + 7s^2 + 24s + 24}{s^4 + 10s^3 + 35s^2 + 50s + 24} = \frac{s^3 + 7s^2 + 24s + 24}{(s+1)(s+2)(s+3)(s+4)}.$$

Show that if we obtain a second-order model by the method of Section 4.6, and we do not specify the poles and the zero of $L(s)$, we have

$$L_1(s) = \frac{0.2917s + 1}{0.399s^2 + 1.375s + 1} = \frac{0.731(s + 3.428)}{(s + 1.043)(s + 2.4)}.$$

P4.14. For the original system of problem 4.13, it is desired to find the lower-order model when the poles of the second-order model are specified as -1 and -2 and the model has one unspecified zero. Show that this low-order model is

$$L(s) = \frac{0.986s + 2}{s^2 + 3s + 2} = \frac{0.986(s + 2.028)}{(s+1)(s+2)}.$$

P4.15. A magnetic amplifier with a low output impedance is shown in Fig. P4.15 in cascade with a low pass filter and a preamplifier. The amplifier has a high input impedance and a gain of one and is used for adding the signals as shown. Select a value for the capacitance C so that the transfer function $V_o(s)/V_{in}(s)$ has a damping ratio of $1/\sqrt{2}$. The time constant of the magnetic amplifier is equal to one second and the gain is $K = 10$. Calculate the settling time of the resulting system.

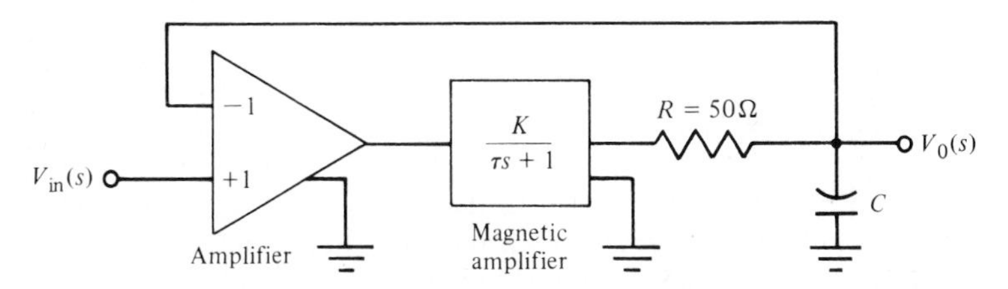

Figure P4.15. Feedback amplifier.

P4.16. Electronic pacemakers for human hearts regulate the speed of the heart pump. A proposed closed-loop system that includes a pacemaker and the measurement of the heart rate is shown in Fig. P4.16. The transfer function of the heart pump and the pacemaker is found to be

$$G(s) = \frac{K}{s(s/12 + 1)}.$$

Design the amplifier gain to yield a tightly controlled system with a settling time to a step disturbance of less than one second. The overshoot to a step in desired heart rate should be less than 10%. (a) Find a suitable range of K. (b) If the nominal value of K is $K = 10$, find the sensitivity of the system to small changes in K. (c) Evaluate the sensitivity at DC (set $s = 0$). (d) Evaluate the magnitude of the sensitivity at the normal heart rate of 60 beats/minute.

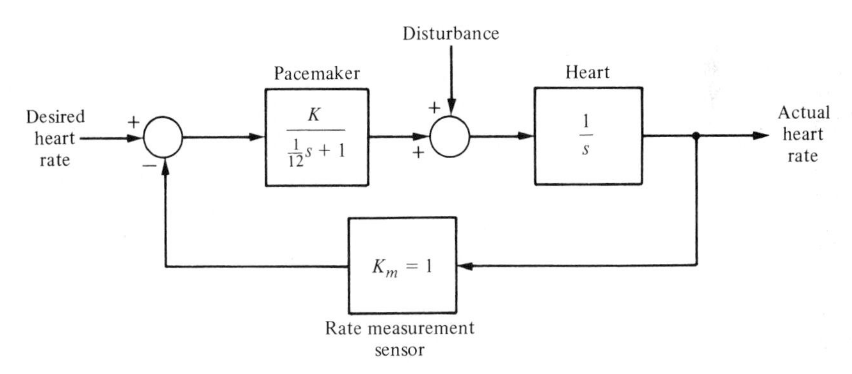

Figure P4.16. Heart pacemaker.

P4.17. Consider the original third-order system given in Example 4.6. Determine a first-order model with one pole unspecified and no zeros that will represent the third-order system.

P4.18. A closed-loop control system with negative unity feedback has a plant with a transfer function

$$G(s) = \frac{8}{s(s^2 + 6s + 12)}.$$

(a) Determine the closed-loop transfer function $T(s)$. (b) Determine a second-order approximation for $T(s)$ using the method of Section 4.6. (c) Using CSDP or a suitable computer program, plot the response of $T(s)$ and the second-order approximation to a unit step input and compare the results.

Design Problems

DP4.1. The roll control autopilot of a jet fighter is shown in Fig. DP4.1. The goal is to select a suitable K so that the response to a unit step command $\phi_d(t) = A, t \geq 0$, will provide a response $\phi(t)$ with a fast response and an overshoot of less than 20%. (a) Determine the closed-loop transfer function $\phi(s)/\phi_d(s)$. (b) Determine the roots of the characteristic equation for $K = 0.7$, 3, and 6. (c) Using the concept of dominant roots, find the expected overshoot and peak time for the approximate second-order system. (d) Plot the actual response and compare with the approximate results of part (c). (e) Select the gain K so that the percentage overshoot is equal to 16%. What is the resulting peak time?

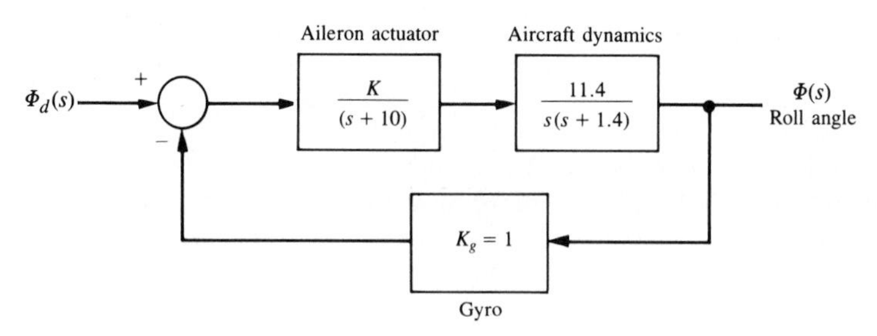

Figure DP4.1. Roll angle control.

DP4.2. The design of the control for a welding arm with a long reach requires the careful selection of the parameters. The system is shown in Fig. DP4.2, where $\zeta = 0.2$ and the gain K and the natural frequency ω_n can be selected. (a) Determine K and ω_n so that the response a unit step input achieves a peak time for the first overshoot (above the desired level of one) is less than or equal to 1 second and the overshoot is less than 5%. (Hint: Try $0.1 < K/\omega_n < 0.3$.) (b) Plot the response of the system designed in part (a) to a step input.

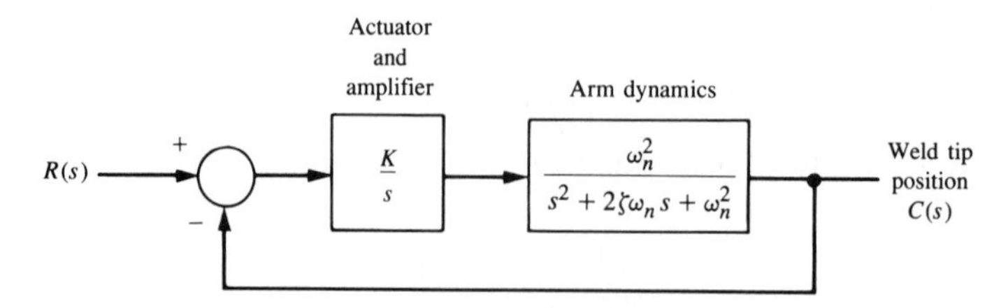

Figure DP4.2. Welding tip position control.

DP4.3. Active suspension systems for modern automobiles provide a quality, firm ride. The design of an active suspension system adjusts the valves of the shock absorber so that the ride fits the conditions. A small electric motor, as shown in Fig. DP4.3, changes the valve settings [33]. Select a design value for K and the parameter q in order to satisfy the ITAE performance for a step command, $R(s)$, and a settling time for the step response of less than or equal to 0.5 second. Upon completion of your design, predict the resulting overshoot for a step input.

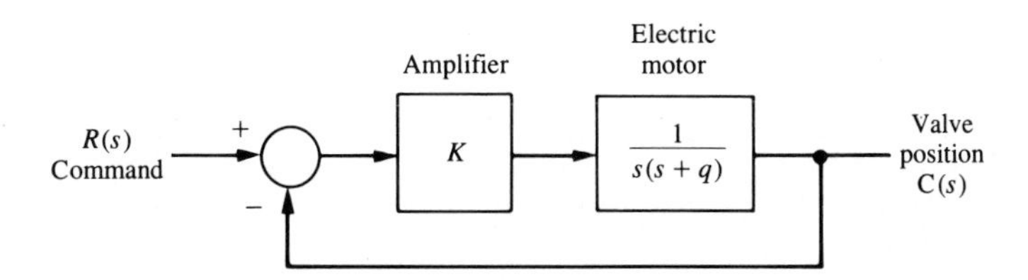

Figure DP4.3. Active suspension system.

Terms and Concepts

Dominant roots The roots of the characteristic equation that cause the dominant transient response of the system.

Optimum control system A system whose parameters are adjusted so that the performance index reaches an extremum value.

Overshoot The amount the system output response proceeds beyond the desired response.

Peak time The time for a system to respond to a step input and rise to a peak response.

Performance index A quantitative measure of the performance of a system.

Rise time The time for a system to respond to a step input and attain a response equal to the magnitude of the input.

Settling time The time required for the system output to settle within a certain percentage of the input amplitude.

Specifications A set of prescribed performance criteria.

Test input signal An input signal used as a standard test of a system's ability to respond adequately.

Type number The number, N, of poles of the transfer function, $G(s)$, at the origin. $G(s)$ is the forward path transfer function.

Velocity error constant, K_v The constant evaluated as $\lim_{s \to 0} [s\,G(s)]$ for a type one system. The steady state error for a ramp input for a type one system is equal to A/K_v.

References

1. J. L. Shearer, *Dynamic Modeling of Engineering Systems,* Macmillan, New York, 1990.
2. R. C. Dorf, *Introduction to Electric Circuits,* John Wiley & Sons, New York, 1989.
3. D. L. Schilling, *Electronic Circuits,* 3rd ed., McGraw-Hill, New York, 1989.
4. P. R. Clement, "A Note on Third-Order Linear Systems," *IRE Transactions on Automatic Control,* June 1960; p. 151.

5. R. N. Clark, *Introduction to Automatic Control Systems,* John Wiley & Sons, New York, 1962; pp. 115–24.
6. D. Graham and R. C. Lathrop, "The Synthesis of Optimum Response: Criteria and Standard Forms, Part 2," *Trans. of the AIEE,* **72,** November 1953; pp. 273–88.
7. S. Franco, *Design with Operational Amplifiers and Analog Integrated Circuits,* McGraw-Hill, New York, 1988.
8. K. J. Astrom and B. Wittenmark, *Adaptive Control,* Addison-Wesley, Reading, Mass., 1989.
9. L. E. Ryan, "Control of an Impact Printer Hammer," *ASME J. of Dynamic Systems,* March 1990; pp. 69–75.
10. T. C. Hsia, "On the Simplification of Linear Systems," *IEEE Transactions on Automatic Control,* June 1972; pp. 372–74.
11. E. J. Davison, "A Method for Simplifying Linear Dynamic Systems," *IEEE Transactions on Automatic Control,* January 1966; pp. 93–101.
12. B. Sridhar, "Design of a Precision Pointing Control System for the Space Infrared Telescope Facility," *IEEE Control Systems,* February 1986; pp. 28–34.
13. T. Beardsley, "Hubble's Legacy," *Scientific American,* June 1990; pp. 18–19.
14. C. Phillips and R. Harbor, *Feedback Control Systems,* Prentice Hall, Englewood Cliffs, N.J., 1991.
15. J. Cloutier, "Assessment of Air-to-Air Missile Guidance and Control Technology," *IEEE Control Systems,* October 1989; pp. 27–34.
16. H. Inooka, "Experimental Studies of Manual Optimization in Control Tasks," *IEEE Control Systems,* August 1990; pp. 20–23.
17. E. J. Stefanides, "Electric Signals Set Air Regulator's Pressures," *Design News,* June 22, 1987; pp. 86–88.
18. D. F. Enns, "Multivariable Flight Control for an Attack Helicopter," *IEEE Control Systems,* April 1987; pp. 86–88.
19. R. Kurzweil, *The Age of Intelligent Machines,* M.I.T. Press, Cambridge, Mass., 1990.
20. E. Kamen, *Signals and Systems,* 2nd ed., Macmillan, New York, 1990.
21. D. McLean, *Automatic Flight Control Systems,* Prentice Hall, Englewood Cliffs, N.J., 1990.
22. E. K. Parsons, "Pointing Control on a Flexible Structure," *IEEE Control Systems,* April 1989; pp. 79–86.
23. H. Kirrmann, "Train Control Systems," *IEEE Micro,* August 1990; pp. 79–80.
24. D. MacKenzie, "French Line Up Europe's High Speed Trains," *New Scientist,* April 28, 1990; p. 41.
25. D. Fuller, "Little Engines That Can," *High Technology,* June 1986; pp. 12–23.
26. D. L. Trumper, "An Electronically Controlled Pressure Regulator," *ASME J. of Dynamic Systems,* March 1989; pp. 75–82.
27. J. De La Cierva, "Rate Servo Keeps TV Picture Clear," *Control Engineering,* May 1965; p. 112.
28. K. K. Chew, "Control of Errors in Disk Drive Systems," *IEEE Control Systems,* January 1990; pp. 16–19.
29. J. J. Moskwa, "Algorithms for Automotive Engine Control," *IEEE Control Systems,* April 1990; pp. 88–92.
30. J. Nilsson, *Electric Circuits,* 3rd ed., Addison-Wesley, Reading, Mass., 1990.
31. R. C. Dorf, *The Encyclopedia of Robotics,* John Wiley & Sons, New York, 1988.
32. R. C. Dorf and R. Jacquot, *The Control System Design Program,* Addison-Wesley, Reading, Mass., 1988.
33. D. Sherman, "Riding on Electrons," *Popular Science,* September 1990; pp. 74–77.

CHAPTER 5

The Stability of Linear Feedback Systems

Preview

The idea of a stable system is familiar to us. We know that an unstable device will exhibit an erratic and destructive response. Thus we seek to ensure that a system is stable and exhibits a bounded transient response.

The stability of a feedback system is related to the location of the roots of the characteristic equation of the system transfer function. Thus we wish to develop a few methods for determining whether a system is stable, and if so, how stable it is.

In this chapter we consider the characteristic equation and examine the determination of the location of its roots. We also will consider a method of the determination of a system's stability that does not require the determination of the roots, but uses only the polynomial coefficients of the characteristic equation.

5.1 The Concept of Stability

The transient response of a feedback control system is of primary interest and must be investigated. A very important characteristic of the transient performance of a system is the *stability* of the system. A *stable system* is defined as a system with a bounded system response. That is, if the system is subjected to a bounded input or disturbance and the response is bounded in magnitude, the system is said to be stable.

The concept of stability can be illustrated by considering a right circular cone placed on a plane horizontal surface. If the cone is resting on its base and is tipped slightly, it returns to its original equilibrium position. This position and response is said to be stable. If the cone rests on its side and is displaced slightly, it rolls with no tendency to leave the position on its side. This position is designated as the neutral stability. On the other hand, if the cone is placed on its tip and released, it falls onto its side. This position is said to be unstable. These three positions are illustrated in Fig. 5.1.

The stability of a dynamic system is defined in a similar manner. The response to a displacement, or initial condition, will result in either a decreasing, neutral, or increasing response. Specifically, it follows from the definition of stability that a linear system is stable if and only if the absolute value of its impulse response, $g(t)$, integrated over an infinite range, is finite. That is, in terms of the convolution integral Eq. (4.1) for a bounded input, one requires that $\int_0^\infty |g(t)|\, dt$ be finite. The location in the s-plane of the poles of a system indicate the resulting transient response. The poles in the left-hand portion of the s-plane result in a decreasing response for disturbance inputs. Similarly, poles on the $j\omega$-axis and in the right-hand plane result in a neutral and an increasing response, respectively, for a disturbance input. This division of the s-plane is shown in Fig. 5.2. Clearly the poles of desirable dynamic systems must lie in the left-hand portion of the s-plane [20].

A common example of the potential destabilizing effect of feedback is that of feedback in audio amplifier and speaker systems used for public address in auditoriums. In this case a loudspeaker produces an audio signal that is an amplified version of the sounds picked up by a microphone. In addition to other audio inputs, the sound coming from the speaker itself may be sensed by the micro-

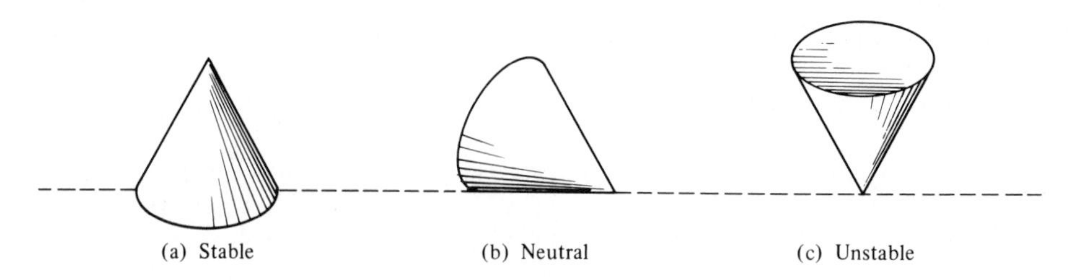

Figure 5.1. The stability of a cone.

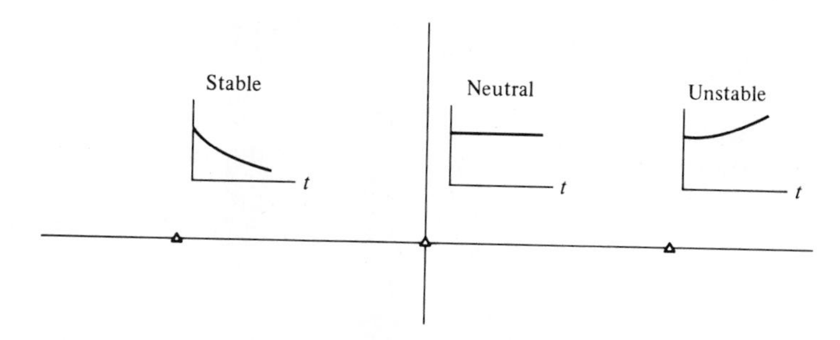

Figure 5.2. Stability in the s-plane.

phone. How strong this particular signal is depends upon the distance between the loud speaker and the microphone. Because of the attenuating properties of air, the larger this distance is, the weaker the signal is that reaches the microphone. In addition, due to the finite propagation speed of sound waves, there is time delay between the signal produced by the loudspeaker and that sensed by the microphone. In this case, the output from the feedback path is added to the external input. This is an example of positive feedback.

As the distance between the loud speaker and the microphone decreases, we find that if the microphone is placed too close to the speaker then the system will be unstable. The result of this instability is an excessive amplification and distortion of audio signals and an oscillatory squeal.

Another example of an unstable system is shown in Fig. 5.3. The first bridge across the Tacoma Narrows at Puget Sound, Washington, was opened to traffic on July 1, 1940. The bridge was found to oscillate whenever the wind blew. After four months, on November 7, 1940, a wind produced an oscillation that grew in amplitude until the bridge broke apart. Figure 5.3(a) shows the condition of beginning oscillation; Fig. 5.3(b) shows the catastrophic failure [11].

In terms of linear systems, we recognize that the stability requirement may be defined in terms of the location of the poles of the closed-loop transfer function. The closed-loop system transfer function is written as

$$T(s) = \frac{p(s)}{q(s)} = \frac{K\prod_{i=1}^{M}(s + z_i)}{s^N\prod_{k=1}^{Q}(s + \sigma_k)\prod_{m=1}^{R}[s^2 + 2\alpha_m s + (\alpha_m^2 + \omega_m^2)]}, \tag{5.1}$$

where $q(s) = \Delta(s)$ is the characteristic equation whose roots are the poles of the closed-loop system. The output response for an impulse function input is then

$$c(t) = \sum_{k=1}^{Q} A_k e^{-\sigma_k t} + \sum_{m=1}^{R} B_m \left(\frac{1}{\omega_m}\right) e^{-\alpha_m t} \sin \omega_m t \tag{5.2}$$

when $N = 0$. Clearly, in order to obtain a bounded response, the poles of the closed-loop system must be in the left-hand portion of the s-plane. Thus, *a necessary and sufficient condition that a feedback system be stable is that all the poles of the system transfer function have negative real parts.*

Figure 5.3(a).

Figure 5.3(b). (Photos courtesy of F. B. Farquharson.)

In order to ascertain the stability of a feedback control system, one could determine the roots of the characteristic equation $q(s)$. However, we are first interested in determining the answer to the question: "Is the system stable?" If we calculate the roots of the characteristic equation in order to answer this question, we have determined much more information than is necessary. Therefore, several methods have been developed which provide the required "yes" or "no" answer to the stability question. The three approaches to the question of stability are: (1) The s-plane approach, (2) the frequency plane ($j\omega$) approach, and (3) the time-domain approach. The real frequency ($j\omega$) approach is outlined in Chapter 8, and the discussion of the time-domain approach is deferred until Chapter 9.

5.2 The Routh-Hurwitz Stability Criterion

The discussion and determination of stability has occupied the interest of many engineers. Maxwell and Vishnegradsky first considered the question of stability of dynamic systems. In the late 1800s, A. Hurwitz and E. J. Routh published independently a method of investigating the stability of a linear system [2, 3]. The Routh-Hurwitz stability method provides an answer to the question of stability by considering the characteristic equation of the system. The characteristic equation in the Laplace variable is written as

$$\Delta(s) = q(s) = a_n s^n + a_{n-1} s^{n-1} + \cdots + a_1 s + a_0 = 0. \tag{5.3}$$

In order to ascertain the stability of the system, it is necessary to determine if any of the roots of $q(s)$ lie in the right half of the s-plane. If Eq. (5.3) is written in factored form, we have

$$a_n(s - r_1)(s - r_2) \cdots (s - r_n) = 0, \tag{5.4}$$

where r_i = ith root of the characteristic equation. Multiplying the factors together, we find that

$$\begin{aligned} q(s) = {} & a_n s^n - a_n(r_1 + r_2 + \cdots + r_n)s^{n-1} \\ & + a_n(r_1 r_2 + r_2 r_3 + r_1 r_3 + \cdots)s^{n-2} \\ & - a_n(r_1 r_2 r_3 + r_1 r_2 r_4 \cdots)s^{n-3} + \cdots \\ & + a_n(-1)^n r_1 r_2 r_3 \cdots r_n = 0. \end{aligned} \tag{5.5}$$

In other words, for an nth-degree equation, we obtain

$$\begin{aligned} q(s) = {} & a_n s^n - a_n(\text{sum of all the roots})s^{n-1} \\ & + a_n(\text{sum of the products of the roots taken 2 at a time})s^{n-2} \\ & - a_n(\text{sum of the products of the roots taken 3 at a time})s^{n-3} \\ & + \cdots + a_n(-1)^n(\text{product of all } n \text{ roots}) = 0. \end{aligned} \tag{5.6}$$

Examining Eq. (5.5), we note that all the coefficients of the polynomial must have the same sign if all the roots are in the left-hand plane. Also, it is necessary for a stable system that all the coefficients be nonzero. However, although these requirements are necessary, they are not sufficient; that is, if they are not satisfied, we immediately know the system is unstable. However, if they are satisfied, we must proceed to ascertain the stability of the system. For example, when the characteristic equation is

$$q(s) = (s + 2)(s^2 - s + 4) = (s^3 + s^2 + 2s + 8), \tag{5.7}$$

the system is unstable and yet the polynomial possesses all positive coefficients.

The *Routh-Hurwitz criterion* is a necessary and sufficient criterion for the stability of linear systems. The method was originally developed in terms of determinants, but we shall utilize the more convenient array formulation.

The Routh-Hurwitz criterion is based on ordering the coefficients of the characteristic equation

$$a_n s^n + a_{n-1}s^{n-1} + a_{n-2}s^{n-2} + \cdots + a_1 s + a_0 = 0 \tag{5.8}$$

into an array or schedule as follows [4]:

$$\begin{array}{c|cccc} s^n & a_n & a_{n-2} & a_{n-4} & \cdots \\ s^{n-1} & a_{n-1} & a_{n-3} & a_{n-5} & \cdots \end{array}$$

Then, further rows of the schedule are completed as follows:

$$\begin{array}{c|ccc} s^n & a_n & a_{n-2} & a_{n-4} \\ s^{n-1} & a_{n-1} & a_{n-3} & a_{n-5} \\ s^{n-2} & b_{n-1} & b_{n-3} & b_{n-5} \\ s^{n-3} & c_{n-1} & c_{n-3} & c_{n-5} \\ \cdot & \cdot & \cdot & \cdot \\ \cdot & \cdot & \cdot & \cdot \\ \cdot & \cdot & \cdot & \cdot \\ s^0 & h_{n-1} & & \end{array}$$

where

$$b_{n-1} = \frac{(a_{n-1})(a_{n-2}) - a_n(a_{n-3})}{a_{n-1}} = \frac{-1}{a_{n-1}} \begin{vmatrix} a_n & a_{n-2} \\ a_{n-1} & a_{n-3} \end{vmatrix},$$

$$b_{n-3} = -\frac{1}{a_{n-1}} \begin{vmatrix} a_n & a_{n-4} \\ a_{n-1} & a_{n-5} \end{vmatrix},$$

and

$$c_{n-1} = \frac{-1}{b_{n-1}} \begin{vmatrix} a_{n-1} & a_{n-3} \\ b_{n-1} & b_{n-3} \end{vmatrix},$$

and so on. The algorithm for calculating the entries in the array can be followed on a determinant basis or by using the form of the equation for b_{n-1}.

The Routh-Hurwitz criterion states that the number of roots of q(s) with positive real parts is equal to the number of changes in sign of the first column of the array. This criterion requires that there be no changes in sign in the first column for a stable system. This requirement is both necessary and sufficient.

There are three distinct cases that must be treated separately, requiring suitable modifications of the array calculation procedure. The three cases are: (1) No element in the first column is zero; (2) there is a zero in the first column, but some other elements of the row containing the zero in the first column are nonzero; and (3) there is a zero in the first column, and the other elements of the row containing the zero are also zero.

In order to clearly illustrate this method, several examples will be presented for each case.

Case 1. *No element in the first column is zero.*

■ Example 5.1 Second-order system

The characteristic equation of a second-order system is

$$q(s) = a_2 s^2 + a_1 s + a_0.$$

The array is written as

$$\begin{array}{c|cc} s^2 & a_2 & a_0 \\ s & a_1 & 0 \\ s^0 & b_1 & 0 \end{array}$$

where

$$b_1 = \frac{a_1 a_0 - (0) a_2}{a_1} = \frac{-1}{a_1} \begin{vmatrix} a_2 & a_0 \\ a_1 & 0 \end{vmatrix} = a_0.$$

Therefore the requirement for a stable second-order system is simply that all the coefficients be positive.

■ Example 5.2 Third-order system

The characteristic equation of a third-order system is

$$q(s) = a_3 s^3 + a_2 s^2 + a_1 s + a_0.$$

The array is

$$\begin{array}{c|cc} s^3 & a_3 & a_1 \\ s^2 & a_2 & a_0 \\ s^1 & b_1 & 0 \\ s^0 & c_1 & 0 \end{array}$$

where

$$b_1 = \frac{a_2 a_1 - a_0 a_3}{a_2} \quad \text{and} \quad c_1 = \frac{b_1 a_0}{b_1} = a_0.$$

For the third-order system to be stable, it is necessary and sufficient that the coefficients be positive and $a_2 a_1 \geq a_0 a_3$. The condition when $a_2 a_1 = a_0 a_3$ results in a borderline stability case, and one pair of roots lies on the imaginary axis in the s-plane. This borderline case is recognized as Case 3 because there is a zero in the first column when $a_2 a_1 = a_0 a_3$, and it will be discussed under Case 3.

As a final example of characteristic equations that result in no zero elements in the first row, let us consider a polynomial

$$q(s) = (s - 1 + j\sqrt{7})(s - 1 - j\sqrt{7})(s + 3) = s^3 + s^2 + 2s + 24. \quad (5.9)$$

The polynomial satisfies all the necessary conditions because all the coefficients exist and are positive. Therefore, utilizing the Routh-Hurwitz array, we have

$$\begin{array}{c|cc} s^3 & 1 & 2 \\ s^2 & 1 & 24 \\ s^1 & -22 & 0 \\ s^0 & 24 & 0 \end{array}$$

Because two changes in sign appear in the first column, we find that two roots of $q(s)$ lie in the right-hand plane, and our prior knowledge is confirmed.

Case 2. Zeros in the first column while some other elements of the row containing a zero in the first column are nonzero.

If only one element in the array is zero, it may be replaced with a small positive number ϵ which is allowed to approach zero after completing the array. For example, consider the following characteristic equation:

$$q(s) = s^5 + 2s^4 + 2s^3 + 4s^2 + 11s + 10. \quad (5.10)$$

The Routh-Hurwitz array is then

$$\begin{array}{c|ccc} s^5 & 1 & 2 & 11 \\ s^4 & 2 & 4 & 10 \\ s^3 & \epsilon & 6 & 0 \\ s^2 & c_1 & 10 & 0 \\ s^1 & d_1 & 0 & 0 \\ s^0 & 10 & 0 & 0 \end{array}$$

where

$$c_1 = \frac{4\epsilon - 12}{\epsilon} = \frac{-12}{\epsilon} \quad \text{and} \quad d_1 = \frac{6c_1 - 10\epsilon}{c_1} \to 6.$$

There are two sign changes due to the large negative number in the first column, $c_1 = -12/\epsilon$. Therefore the system is unstable and two roots lie in the right half of the plane.

■ Example 5.3 Unstable system

As a final example of the type of Case 2, consider the characteristic equation

$$q(s) = s^4 + s^3 + s^2 + s + K, \quad (5.11)$$

where it is desired to determine the gain K that results in borderline stability. The Routh-Hurwitz array is then

$$\begin{array}{c|ccc} s^4 & 1 & 1 & K \\ s^3 & 1 & 1 & 0 \\ s^2 & \epsilon & K & 0 \\ s^1 & c_1 & 0 & 0 \\ s^0 & K & 0 & 0 \end{array}$$

where

$$c_1 = \frac{\epsilon - K}{\epsilon} \rightarrow \frac{-K}{\epsilon}.$$

Therefore, for any value of K greater than zero, the system is unstable. Also, because the last term in the first column is equal to K, a negative value of K will result in an unstable system. Therefore the system is unstable for all values of gain K.

Case 3. Zeros in the first column, and the other elements of the row containing the zero are also zero.

Case 3 occurs when all the elements in one row are zero or when the row consists of a single element which is zero. This condition occurs when the polynomial contains singularities that are symmetrically located about the origin of the s-plane. Therefore Case 3 occurs when factors such as $(s + \sigma)(s - \sigma)$ or $(s + j\omega)(s - j\omega)$ occur. This problem is circumvented by utilizing the *auxiliary equation,* which immediately precedes the zero entry in the Routh array. The order of the auxiliary equation is always even and indicates the number of symmetrical root pairs.

In order to illustrate this approach, let us consider a third-order system with a characteristic equation:

$$q(s) = s^3 + 2s^2 + 4s + K, \tag{5.12}$$

where K is an adjustable loop gain. The Routh array is then

$$\begin{array}{c|cc} s^3 & 1 & 4 \\ s^2 & 2 & K \\ s^1 & \dfrac{8 - K}{2} & 0 \\ s^0 & K & 0. \end{array}$$

Therefore, for a stable system, we require that

$$0 \leq K \leq 8.$$

When $K = 8$, we have two roots on the $j\omega$-axis and a borderline stability case. Note that we obtain a row of zeros (Case 3) when $K = 8$. The auxiliary equation, $U(s)$, is the equation of the row preceding the row of zeros. The equation of the row preceding the row of zeros is, in this case, obtained from the s^2-row. We recall that this row contains the coefficients of the even powers of s and therefore in this case, we have

$$U(s) = 2s^2 + Ks^0 = 2s^2 + 8 = 2(s^2 + 4) = 2(s + j2)(s - j2). \quad (5.13)$$

In order to show that the auxiliary equation, $U(s)$, is indeed a factor of the characteristic equation, we divide $q(s)$ by $U(s)$ to obtain

$$\begin{array}{r|l} & \tfrac{1}{2}s + 1 \\ \hline 2s^2 + 8 & s^3 + 2s^2 + 4s + 8 \\ & s^3 \qquad\quad + 4s \\ \hline & \qquad 2s^2 \qquad + 8 \\ & \qquad 2s^2 \qquad + 8 \end{array}$$

Therefore, when $K = 8$, the factors of the characteristic equation are

$$q(s) = (s + 2)(s + j2)(s - j2). \quad (5.14)$$

Strictly, the borderline case response is an unacceptable oscillation.

■ Example 5.4 Robot control

Let us consider the control of a robot arm as shown in Fig. 5.4. It is predicted that there will be about 100,000 robots in service throughout the world by 1996 [6]. The robot shown in Fig. 5.4 is a six-legged micro robot system using highly flexible legs with high gain controllers that may become unstable and oscillate. Under this condition, we have the characteristic polynomial

$$q(s) = s^5 + s^4 + 4s^3 + 24s^2 + 3s + 63. \quad (5.15)$$

The Routh-Hurwitz array is

$$\begin{array}{c|ccc} s^5 & 1 & 4 & 3 \\ s^4 & 1 & 24 & 63 \\ s^3 & -20 & -60 & 0 \\ s^2 & 21 & 63 & 0 \\ s^1 & 0 & 0 & 0. \end{array}$$

Therefore the auxiliary equation is

$$U(s) = 21s^2 + 63 = 21(s^2 + 3) = 21(s + j\sqrt{3})(s - j\sqrt{3}), \quad (5.16)$$

which indicates that two roots are on the imaginary axis. In order to examine the remaining roots, we divide by the auxiliary equation to obtain

$$\frac{q(s)}{s^2 + 3} = s^3 + s^2 + s + 21.$$

Figure 5.4. A completely integrated, six-legged, micro robot system. The legged design provides maximum dexterity. Legs also provide a unique sensory system for environmental interaction. It is equipped with a sensor network that includes 150 sensors of 12 different types. The legs are instrumented so that it can determine the lay of the terrain, the surface texture, hardness, and even color. The gyro-stabilized camera and rangefinder can be used for gathering data beyond the robot's immediate reach. This high performance system is able to walk quickly, climb over obstacles, and perform dynamic motions. Courtesy of IS Robotics Corporation.

Establishing a Routh-Hurwitz array for this equation, we have

$$\begin{array}{c|cc} s^3 & 1 & 1 \\ s^2 & 1 & 21 \\ s^1 & -20 & 0 \\ s^0 & 21 & 0\,. \end{array}$$

The two changes in sign in the first column indicate the presence of two roots in the right-hand plane, and the system is unstable. The roots in the right-hand plane are $s = +1 \pm j\sqrt{6}$.

■ Example 5.5 Disk-drive control

Large disk-storage devices are used with today's computers [17]. The data head is moved to different positions on the spinning disk and rapid, accurate response is required. A block diagram of a disk-storage data-head positioning system is

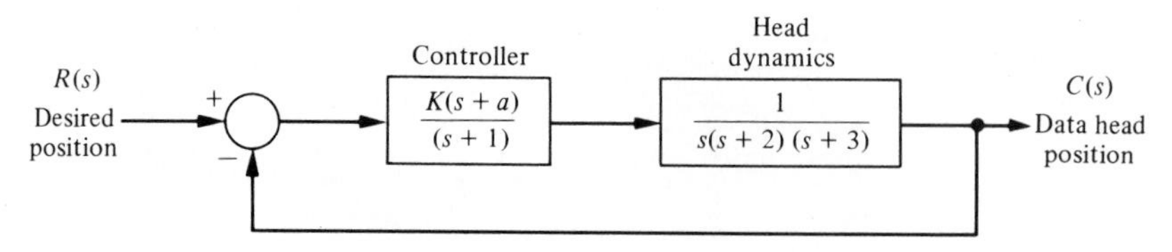

Figure 5.5. Head position control.

shown in Fig. 5.5. It is desired to determine the range of K and a for which the system is stable. The characteristic equation is

$$1 + G(s) = 1 + \frac{K(s+a)}{s(s+1)(s+2)(s+3)} = 0.$$

Therefore $q(s) = s^4 + 6s^3 + 11s^2 + (K+6)s + Ka = 0$. Establishing the Routh array, we have

$$\begin{array}{c|ccc} s^4 & 1 & 11 & Ka \\ s^3 & 6 & (K+6) & \\ s^2 & b_3 & Ka & \\ s^1 & c_3 & & \\ s^0 & Ka & & \end{array}$$

where

$$b_3 = \frac{60 - K}{6}$$

and

$$c_3 = \frac{b_3(K+6) - 6Ka}{b_3}.$$

The coefficient c_3 sets the acceptable range of K and a, while b_3 requires that K be less than 60. Setting $c_3 = 0$ we obtain

$$(K - 60)(K + 6) + 36Ka = 0.$$

The required relationship between K and a is then

$$a \leq \frac{(60 - K)(K + 6)}{36K}$$

when a is positive. Therefore, if $K = 40$, we require $a \leq 0.639$.

5.3 The Relative Stability of Feedback Control Systems

The verification of stability of the Routh-Hurwitz criterion provides only a partial answer to the question of stability. The Routh-Hurwitz criterion ascertains the absolute stability of a system by determining if any of the roots of the character-

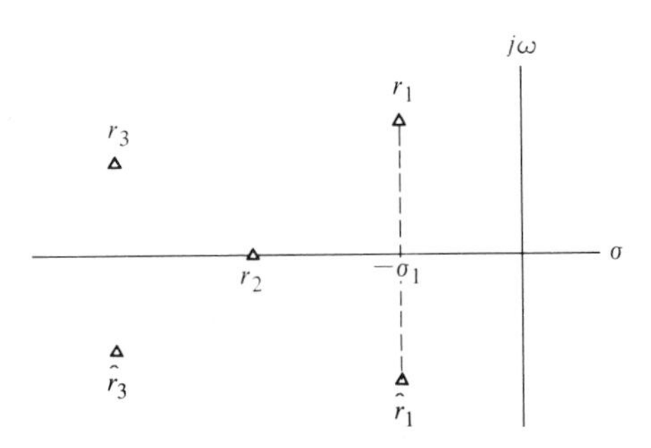

Figure 5.6. Root locations in the s-plane.

istic equation lie in the right half of the s-plane. However, if the system satisfies the Routh-Hurwitz criterion and is absolutely stable, it is desirable to determine the *relative stability;* that is, it is necessary to investigate the relative damping of each root of the characteristic equation. The relative stability of a system may be defined as the property that is measured by the relative settling times of each root or pair of roots. Therefore relative stability is represented by the real part of each root. Thus root r_2 is relatively more stable than the roots r_1, $\hat{r}_1$ as shown in Fig. 5.6. The relative stability of a system can also be defined in terms of the relative damping coefficients ζ of each complex root pair and therefore in terms of the speed of response and overshoot instead of settling time.

Hence the investigation of the relative stability of each root is clearly necessary because, as we found in Chapter 4, the location of the closed-loop poles in the s-plane determines the performance of the system. Thus it is imperative that we reexamine the characteristic equation $q(s)$ and consider several methods for the determination of relative stability.

Because the relative stability of a system is dictated by the location of the roots of the characteristic equation, a first approach using an s-plane formulation is to extend the Routh-Hurwitz criterion to ascertain relative stability. This can be simply accomplished by utilizing a change of variable, which shifts the s-plane axis in order to utilize the Routh-Hurwitz criterion. Examining Fig. 5.6, we notice that a shift of the vertical axis in the s-plane to $-\sigma_1$ will result in the roots r_1, $\hat{r}_1$ appearing on the shifted axis. The correct magnitude to shift the vertical axis must be obtained on a trial-and-error basis. Then, without solving the fifth-order polynomial $q(s)$, one may determine the real part of the dominant roots r_1, $\hat{r}_1$.

■ Example 5.6 Axis shift

Consider the simple third-order characteristic equation

$$q(s) = s^3 + 4s^2 + 6s + 4. \tag{5.17}$$

Initially, one might shift the axis other than one unit and obtain a Routh-Hurwitz array without a zero occurring in the first column. However, upon setting the shifted variable s_n equal to $s + 1$, we obtain

$$(s_n - 1)^3 + 4(s_n - 1)^2 + 6(s_n - 1) + 4 = s_n^3 + s_n^2 + s_n + 1. \tag{5.18}$$

Then the Routh array is established as

$$\begin{array}{c|cc} s_n^3 & 1 & 1 \\ s_n^2 & 1 & 1 \\ s_n^1 & 0 & 0 \\ s_n^0 & 1 & 0. \end{array}$$

Clearly, there are roots on the shifted imaginary axis and the roots can be obtained from the auxiliary equation, which is

$$\begin{aligned} U(s_n) &= s_n^2 + 1 = (s_n + j)(s_n - j) \\ &= (s + 1 + j)(s + 1 - j). \end{aligned} \tag{5.19}$$

The shifting of the s-plane axis in order to ascertain the relative stability of a system is a very useful approach, particularly for higher-order systems with several pairs of closed-loop complex conjugate roots.

5.4 The Determination of Root Locations in the s-Plane

The relative stability of a feedback control system is directly related to the location of the closed-loop roots of the characteristic equation in the s-plane. Therefore it is often necessary and easiest to simply determine the values of the roots of the characteristic equation. This approach has become particularly attractive today due to the availability of digital computer programs for determining the roots of polynomials. However, this approach may even be the most logical when using manual calculations if the order of the system is relatively low. For a third-order system or lower, it is usually simpler to utilize manual calculation methods.

The determination of the roots of a polynomial can be obtained by utilizing *synthetic division* which is based on the remainder theorem; that is, upon dividing the polynomial by a factor, the remainder is zero when the factor is a root of the polynomial. Synthetic division is commonly used to carry out the division process. The relations for the roots of a polynomial as obtained in Eq. (5.6) are utilized to aid in the choice of a first estimate of a root.

■ Example 5.7 Synthetic division

Let us determine the roots of the polynomial

$$q(s) = s^3 + 4s^2 + 6s + 4. \tag{5.20}$$

Establishing a table of synthetic division, we have

$$\begin{array}{cccc|l} 1 & 4 & 6 & 4 & -1 = \text{trial root} \\ & -1 & -3 & -3 & \\ \hline 1 & 3 & 3 & 1 = \text{remainder} & \end{array}$$

for a trial root of $s = -1$. In this table, we multiply by the trial root and successively add in each column. With a remainder of one, we might try $s = -2$, which results in the form

$$\begin{array}{cccc|l} 1 & 4 & 6 & 4 & \underline{-2} \\ & -2 & -4 & -4 & \\ \hline 1 & 2 & 2 & 0 & \end{array}$$

Because the remainder is zero, one root is equal to -2 and the remaining roots may be obtained from the remaining polynomial $(s^2 + 2s + 2)$ by using the quadratic root formula.

The search for a root of the polynomial can be aided considerably by utilizing the rate of change of the polynomial at the estimated root in order to obtain a new estimate. The *Newton-Raphson method* is a rapid method utilizing synthetic division to obtain the value of

$$\left.\frac{dq(s)}{ds}\right|_{s=s_1},$$

where s_1 is a first estimate of the root. The Newton-Raphson method is an iteration approach utilized in many digital computer root-solving programs. A new estimate s_{n+1} of the root is based on the last estimate as [18, 19]

$$s_{n+1} = s_n - \frac{q(s_n)}{q'(s_n)}, \tag{5.21}$$

where

$$q'(s_n) = \left.\frac{dq(s)}{ds}\right|_{s=s_n}.$$

The synthetic division process may be utilized to obtain $q(s_n)$ and $q'(s_n)$. The synthetic division process for a trial root may be written for the polynomial

$$q(s) = a_m s^m + a_{m-1}s^{m-1} + \cdots + a_1 s + a_0$$

as

$$\begin{array}{ccccc|l} a_m & a_{m-1} & \cdots & a_1 & a_0 & \underline{s_n} \\ & b_m s_n & \cdots & & b_1 s_n & \\ \hline b_m & b_{m-1} & \cdots & b_1 & b_0 & \end{array}$$

where $b_0 = q(s_n)$, the remainder of the division process. When s_n is a root of $q(s)$, the remainder is equal to zero and the remaining polynomial

$$b_m s^{m-1} + b_{m-1}s^{m-2} + \cdots + b_1$$

may itself be factored. The derivative evaluated at the nth estimate of the root, $q'(s_n)$, may also be obtained by repeating the synthetic division process on the b_m,

$b_{m-1}, \ldots, b_1$ coefficients. The value of $q'(s_n)$ is the remainder of this repeated synthetic division process. This process converges as the square of the absolute error. The Newton-Raphson method, using synthetic division, is readily illustrated, as can be seen by repeating Example 5.7.

■ Example 5.8 Newton-Raphson method

For the polynomial $q(s) = s^3 + 4s^2 + 6s + 4$, we establish a table of synthetic division for a first estimate as follows:

$$\begin{array}{cccc|l} 1 & 4 & 6 & 4 & \underline{-1} \\ & -1 & -3 & -3 & \\ \hline 1 & 3 & 3 & 1 = q(s_1) & \\ & -1 & -2 & & \\ \hline 1 & 2 & 1 = q'(s_1) & & \end{array}$$

The derivative of $q(s)$ evaluated at s_1 is determined by continuing the synthetic division as shown. Then the second estimate becomes

$$s_2 = s_1 - \frac{q(s_1)}{q'(s_1)} = -1 - \left(\frac{1}{1}\right) = -2.$$

As we found in Example 5.7, s_2 is, in fact, a root of the polynomial and results in a zero remainder.

■ Example 5.9 Third-order system

Let us consider the polynomial

$$q(s) = s^3 + 3.5s^2 + 6.5s + 10. \tag{5.22}$$

From Eq. (5.6), we note that the sum of all the roots is equal to -3.5 and that the product of all the roots is -10. Therefore, as a first estimate, we try $s_1 = -1$ and obtain the following table:

$$\begin{array}{cccc|l} 1 & 3.5 & 6.5 & 10 & \underline{-1} \\ & -1 & -2.5 & -4 & \\ \hline 1 & 2.5 & 4 & 6 = q(s_1) & \\ & -1 & -1.5 & & \\ \hline 1 & 1.5 & 2.5 = q'(s_1) & & \end{array}$$

Therefore a second estimate is

$$s_2 = -1 - \left(\frac{6}{2.5}\right) = -3.40.$$

Now let us use a second estimate that is convenient for these manual calculations. Therefore, on the basis of the calculation of $s_2 = -3.40$, we will choose a second estimate $s_2 = -3.00$. Establishing a table for $s_2 = -3.00$ and completing the synthetic division, we find that

$$s_3 = -s_2 - \frac{q(s_2)}{q'(s_2)} = -3.00 - \frac{(-5)}{12.5} = -2.60.$$

Finally, completing a table for $s_3 = -2.50$, we find that the remainder is zero and the polynomial factors are $q(s) = (s + 2.5)(s^2 + s + 4)$.

The availability of a digital computer or a programmable calculator enables one to readily determine the roots of a polynomial by using root determination programs, which usually use the Newton-Raphson algorithm, Eq. (5.21). This, as is the case for time-shared computers with remote consoles, is a particularly useful approach when a computer is readily available for immediate access. The availability of a console connected in a time-shared manner to a large digital computer or a personal computer is particularly advantageous for a control engineer, because the ability to perform on-line calculations aids in the iterative analysis and design process.

5.5 Design Example: Tracked Vehicle Turning Control

The design of a turning control for a tracked vehicle involves the selection of two parameters [22]. In Fig. 5.7 the system shown in part (a) has the model shown in part (b). The two tracks are operated at different speeds in order to turn the vehi-

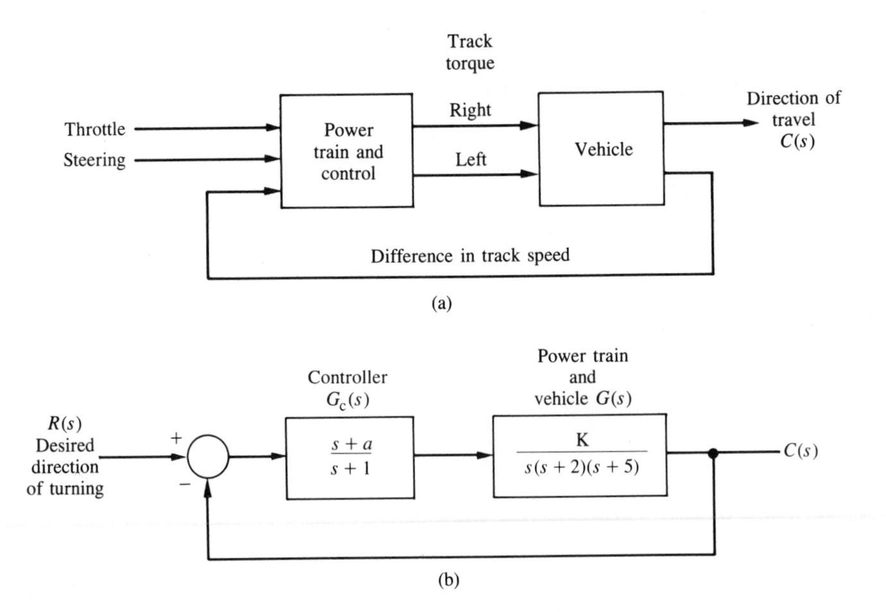

Figure 5.7. Turning control for a two-track vehicle.

cle. Select K and a so that the system is stable and the steady-state error for a ramp command is less than or equal to 24% of the magnitude of the command.

The characteristic equation of the feedback system is

$$1 + G_cG(s) = 0$$

or

$$1 + \frac{K(s + a)}{s(s + 1)(s + 2)(s + 5)} = 0. \qquad (5.23)$$

Therefore we have

$$s(s + 1)(s + 2)(s + 5) + K(s + a) = 0$$

or

$$s^4 + 8s^3 + 17s^2 + (K + 10)s + Ka = 0. \qquad (5.24)$$

In order to determine the stable region for K and a, we establish the Routh array as

$$\begin{array}{c|ccc} s^4 & 1 & 17 & Ka \\ s^3 & 8 & K+10 & 0 \\ s^2 & b_3 & Ka & \\ s^1 & c_3 & & \\ s^0 & Ka & & \end{array}$$

where

$$b_3 = \frac{126 - K}{8} \quad \text{and} \quad c_3 = \frac{b_3(K + 10) - 8Ka}{b_3}.$$

In order for the elements of the first column to be positive, we require that Ka, b_3, and c_3 be positive. Therefore we require

$$\begin{aligned} K &< 126 \\ Ka &> 0 \qquad (5.25) \\ (K + 10)(126 - K) - 64\,Ka &> 0. \end{aligned}$$

The region of stability for $K > 0$ is shown in Fig. 5.8. The steady-state error to a ramp input $r(t) = At$, $t > 0$ is

$$e_{ss} = A/K_v,$$

where

$$K_v = \lim_{s \to 0} sG_cG = Ka/10.$$

Therefore we have

$$e_{ss} = \frac{10A}{Ka}. \qquad (5.26)$$

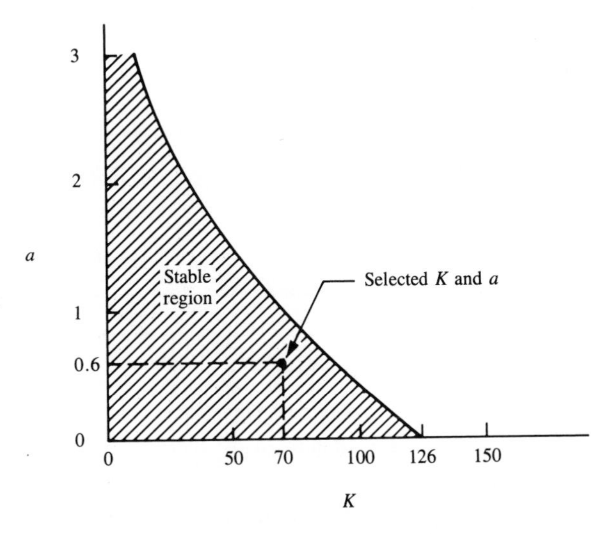

Figure 5.8. The stable region.

When e_{ss} is equal to 23.8% of A, we require that $Ka = 42$. This can be satisfied by the selected point in the stable region when $K = 70$ and $a = 0.6$, as shown in Fig. 5.8. Of course, another acceptable design would be attained when $K = 50$ and $a = 0.84$. We can calculate a series of possible combinations of K and a that can satisfy $Ka = 42$ that lie within the stable region, and all will be acceptable design solutions. However, not all selected values of K and a will lie within the stable region. Note that K cannot exceed 126.

5.6 Summary

In this chapter we have considered the concept of the stability of a feedback control system. A definition of a stable system in terms of a bounded system response was outlined and related to the location of the poles of the system transfer function in the s-plane.

The Routh-Hurwitz stability criterion was introduced and several examples were considered. The relative stability of a feedback control system was also considered in terms of the location of the poles and zeros of the system transfer function in the s-plane. Finally, the determination of the roots of the characteristic equation was considered and the Newton-Raphson method was illustrated.

Exercises

E5.1. A system has a characteristic equation $s^3 + 3Ks^2 + (2 + K)s + 4 = 0$. Determine the range of K for a stable system.

Answer: $K > 0.53$

E5.2. A system has a characteristic equation $s^3 + 9s^2 + 26s + 24 = 0$. (a) Using the Routh criterion, show that the system is stable. (b) Using the Newton-Raphson method, find the three roots.

E5.3. Find the roots of the characteristic equation $s^4 + 9.5s^3 + 30.5s^2 + 37s + 12 = 0$.

E5.4. A control system has the structure shown in Fig. E5.4. Determine the gain at which the system will become unstable.

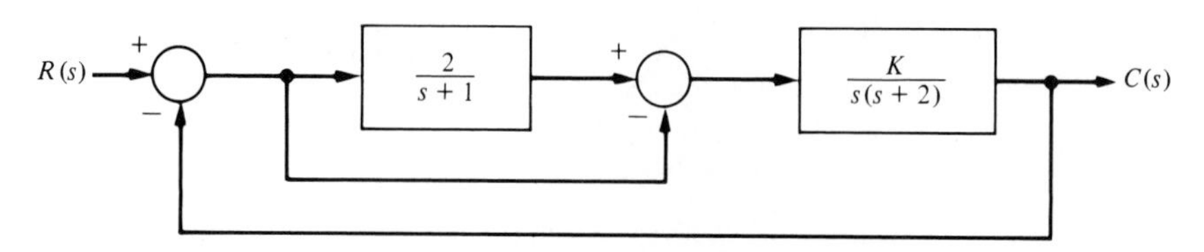

Figure E5.4. Feedforward system.

E5.5. A feedback system has a loop transfer function

$$GH(s) = \frac{K}{(s + 1)(s + 3)(s + 6)},$$

where $K = 10$. Find the roots of this system's characteristic equation.

E5.6. For the feedback system of Exercise E5.5, find the value of K when two roots lie on the imaginary axis. Determine the value of the three roots.

Answer: $s = -10, \pm j5.2$

E5.7. A negative feedback system has a loop transfer function

$$GH(s) = \frac{K(s + 2)}{s(s - 2)}.$$

(a) Find the value of the gain when the ζ of the closed-loop roots is equal to 0.707. (b) Find the value of the gain when the closed-loop system has two roots on the imaginary axis.

E5.8. Designers have developed small, fast, vertical-takeoff fighter aircraft that are invisible to radar (Stealth aircraft). The aircraft concept shown in Fig. E5.8(a), on the next page uses quickly turning jet nozzles to steer the airplane [19]. The control system for the heading or direction control is shown in Fig. E5.8(b). Determine the maximum gain of the system for stable operation.

E5.9. A system has a characteristic equation

$$s^3 + 3s^2 + (K + 1)s + 4 = 0$$

Find the range of K for a stable system.

Answer: $K > 1/3$

E5.10. We all use our eyes and ears to achieve balance. Our orientation system allows us to sit or stand in a desired position even while in motion. This orientation system is primarily run by the information received in the inner ear, where the semicircular canals

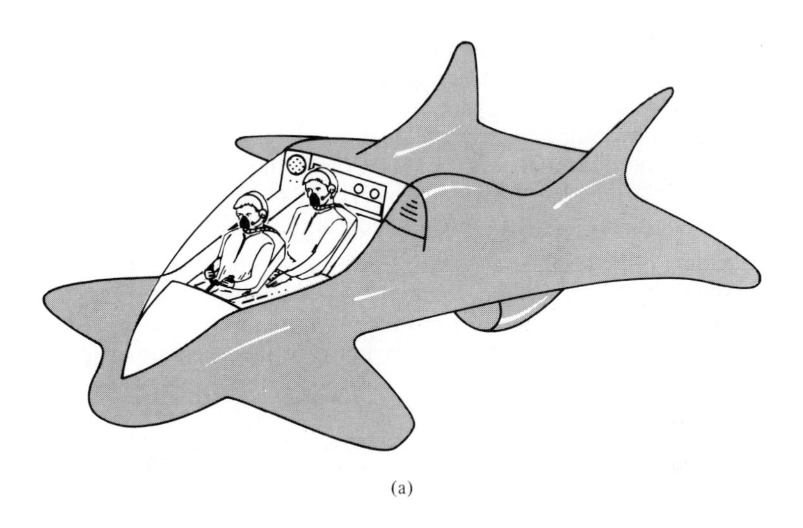

(a)

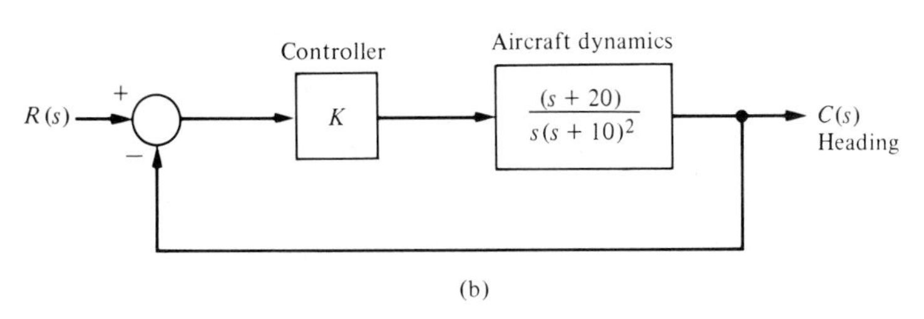

(b)

Figure E5.8. Aircraft heading control.

sense angular acceleration and the otoliths measure linear acceleration. But these acceleration measurements need to be supplemented by visual signals. Try the following experiment: (a) Stand with one foot in front of another and with your hands resting on your hips and your elbows bowed outward. (b) Close your eyes. Did you find that you experienced a low-frequency oscillation that grew until you lost balance? Is this orientation position stable with and without the use of your eyes?

Problems

P5.1. Utilizing the Routh-Hurwitz criterion, determine the stability of the following polynomials:

(a) $s^2 + 4s + 1$ (b) $s^3 + 4s^2 + 5s + 6$

(c) $s^3 + 3s^3 - 6s + 10$ (d) $s^4 + s^3 + 2s^2 + 4s + 8$

(e) $s^4 + s^3 + 3s^2 + 2s + K$ (f) $s^5 + s^4 + 2s^3 + s + 3$

(g) $s^5 + s^4 + 2s^3 + s^2 + s + K$

For all cases, determine the number of roots, if any, in the right-hand plane. Also, when it is adjustable, determine the range of K that results in a stable system.

P5.2. An antenna control system was analyzed in Problem 3.5 and it was determined that in order to reduce the effect of wind disturbances, the gain of the magnetic amplifier k_a should be as large as possible. (a) Determine the limiting value of gain for maintaining a stable system. (b) It is desired to have a system settling time equal to two seconds. Using a shifted axis and the Routh-Hurwitz criterion, determine the value of gain that satisfies this requirement. Assume that the complex roots of the closed-loop system dominate the transient response. (Is this a valid approximation in this case?)

P5.3. Arc welding is one of the most important areas of application for industrial robots [6]. In most manufacturing welding situations, uncertainties in dimensions of the part, geometry of the joint, and the welding process itself require the use of sensors for maintaining weld quality. Several systems use a vision system to measure the geometry of the puddle of melted metal as shown in Fig. P5.3. This system uses a constant rate of feeding the wire to be melted. (a) Calculate the maximum value for K for the system that will result in an oscillatory response. (b) For ½ of the maximum value of K found in part (a), determine the roots of the characteristic equation. (c) Estimate the overshoot of the system of part (b) when it is subjected to a step input.

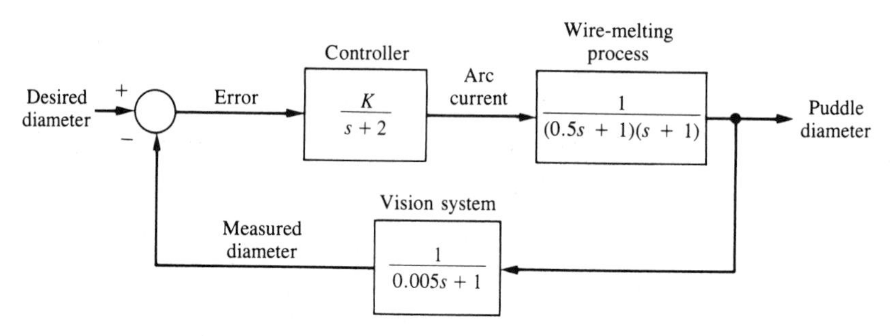

Figure P5.3. Welder control.

P5.4. A feedback control system is shown in Fig. P5.4. The process transfer function is

$$G(s) = \frac{K(s+40)}{s(s+10)},$$

and the feedback transfer function is $H(s) = 1/(s+20)$. (a) Determine the limiting value of gain K for a stable system. (b) For the gain that results in borderline stability, determine the magnitude of the imaginary roots. (c) Reduce the gain to ½ the magnitude of the borderline value and determine the relative stability of the system (1) by shifting the axis and using the Routh-Hurwitz criterion and (2) by determining the root locations. Show the roots are between -1 and -2.

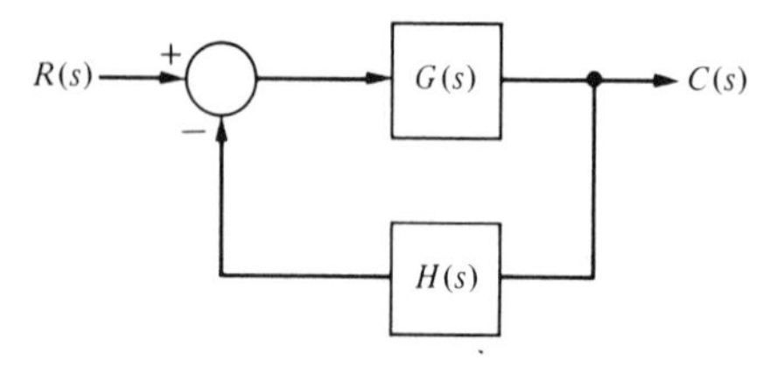

Figure P5.4. Feedback system.

P5.5. Determine the relative stability of the systems with the following characteristic equations (a) by shifting the axis in the s-plane and using the Routh-Hurwitz criterion and (b) by determining the location of the roots in the s-plane:

1. $s^3 + 3s^2 + 4s + 2 = 0$
2. $s^4 + 9s^3 + 30s^2 + 42s + 20 = 0$
3. $s^3 + 19s^2 + 110s + 200 = 0$

P5.6. A unity-feedback control system is shown in Fig. P5.6. Determine the relative stability of the system with the following transfer functions by locating the roots in the s-plane:

(a) $G(s) = \dfrac{65 + 33s}{s^2(s + 9)}$

(b) $G(s) = \dfrac{24}{s(s^3 + 10s^2 + 35s + 50)}$

(c) $G(s) = \dfrac{3(s + 4)(s + 8)}{s(s + 5)^2}$

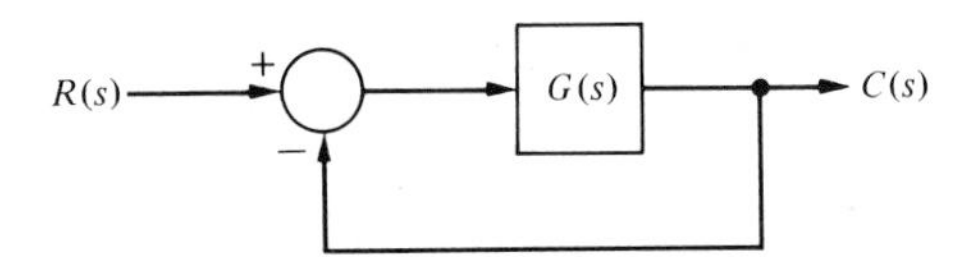

Figure P5.6. Unity feedback system.

P5.7. The linear model of a phase detector (phase-lock loop) can be represented by Fig. P5.7 [7]. The phase-lock systems are designed to maintain zero difference in phase between the input carrier signal and a local voltage-controlled oscillator. Phase-lock loops find application in color television, missile tracking, and space telemetry. The filter for a particular application is chosen as

$$F(s) = \frac{10\,(s + 10)}{(s + 1)(s + 100)}.$$

It is desired to minimize the steady-state error of the system for a ramp change in the phase information signal. (a) Determine the limiting value of the gain $K_a K = K_v$ in order to maintain a stable system. (b) It is decided that a steady-state error equal to 1° is acceptable for a ramp signal of 100 rad/sec. For that value of gain K_v, determine the location of the roots of the system.

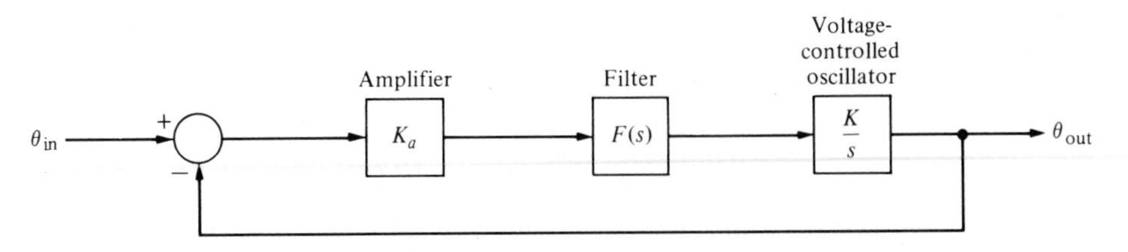

Figure P5.7. Phase-lock loop system.

P5.8. A very interesting and useful velocity control system has been designed for a wheelchair control system [13]. It is desirable to enable people paralyzed from the neck down to drive themselves about in motorized wheelchairs. A proposed system utilizing velocity sensors mounted in a headgear is shown in Fig. P5.8. The headgear sensor provides an output proportional to the magnitude of the head movement. There is a sensor mounted at 90° intervals so that forward, left, right, or reverse can be commanded. Typical values for the time constants are $\tau_1 = 0.5$ sec, $\tau_3 = 1$ sec, and $\tau_4 = \frac{1}{4}$ sec. (a) Determine the limiting gain $K = K_1K_2K_3$ for a stable system. (b) When the gain K is set equal to $\frac{1}{3}$ of the limiting value, determine if the settling time of the system is less than 4 sec. (c) Determine the value of gain that results in a system with a settling time of 4 sec. Also, obtain the value of the roots of the characteristic equation when the settling time is equal to 4 sec.

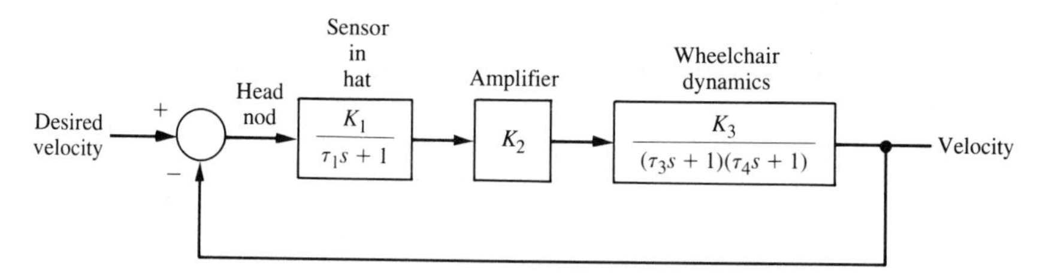

Figure P5.8. Wheelchair control system.

P5.9. A cassette tape storage device has been designed for mass-storage [13]. It is necessary to control accurately the velocity of the tape. The speed control of the tape drive is represented by the system shown in Fig. P5.9. (a) Determine the limiting gain for a stable system. (b) Determine a suitable gain so that the overshoot to a step command is approximately 5%.

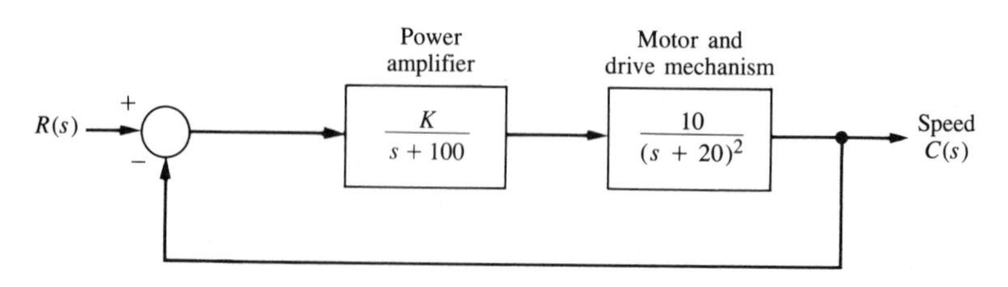

Figure P5.9. Tape drive control.

P5.10. Robots can be used in manufacturing and assembly operations where accurate, fast, and versatile manipulation is required [9]. The open-loop transfer function of a direct-drive arm may be approximated by

$$GH(s) = \frac{K(s + 2)}{s(s + 5)(s^2 + 2s + 5)}.$$

(a) Determine the value of gain K when the system oscillates. (b) Calculate the roots of the closed-loop system for the K determined in part (a).

P5.11. A feedback control system has a characteristic equation:

$$s^3 + (4 + K)s^2 + 6s + 16 + 8K = 0.$$

The parameter K must be positive. What is the maximum value K can assume before the system becomes unstable? When K is equal to the maximum value, the system oscillates. Determine the frequency of oscillation.

P5.12. A feedback control system has a characteristic equation:

$$s^6 + 2s^5 + 5s^4 + 8s^3 + 8s^2 + 8s + 4 + 0.$$

Determine if the system is stable and determine the values of the roots.

P5.13. The stability of a motorcycle and rider is an important area for study because many motorcycle designs result in vehicles that are difficult to control [10]. The handling characteristics of a motorcycle must include a model of the rider as well as one of the vehicle. The dynamics of one motorcycle and rider can be represented by an open-loop transfer function (Fig. P5.4)

$$GH(s) = \frac{K(s^2 + 30s + 1125)}{s(s + 20)(s^2 + 10s + 125)(s^2 + 60s + 3400)}.$$

(a) As an approximation, calculate the acceptable range of K for a stable system when the numerator polynomial (zeros) and the denominator polynomial $(s^2 + 60s + 3400)$ are neglected. (b) Calculate the actual range of acceptable K accounting for all zeros and poles.

P5.14. A system has a transfer function

$$T(s) = \frac{1}{s^3 + 1.3s^2 + 2.0s + 1}.$$

(a) Determine if the system is stable. (b) Determine the roots of the characteristic equation. (c) Plot the response of the system to a unit step input.

Design Problems

DP5.1. The control of the spark ignition of an automotive engine requires constant performance over a wide range of parameters [21]. The control system is shown in Fig. DP5.1, on the next page with a controller gain K to be selected. The parameter p is, equal to 2 for many autos but can equal zero for high performance autos. Select a gain K that will result in a stable system for both values of p.

DP5.2. An automatically guided vehicle on Mars is represented by the system in Fig. DP5.2. The system has a steerable wheel in both the front and back of the vehicle and the design requires the selection of $H(s)$ where $H(s) = Ks + 1$. Determine (a) the value of K required for stability, (b) the value of K when one root of the characteristic equation is equal to $s = -\frac{1}{2}$, and (c) the value of the two remaining roots for the gain selected in part (b). (d) Find the response of the system to a step command for the gain selected in part (b).

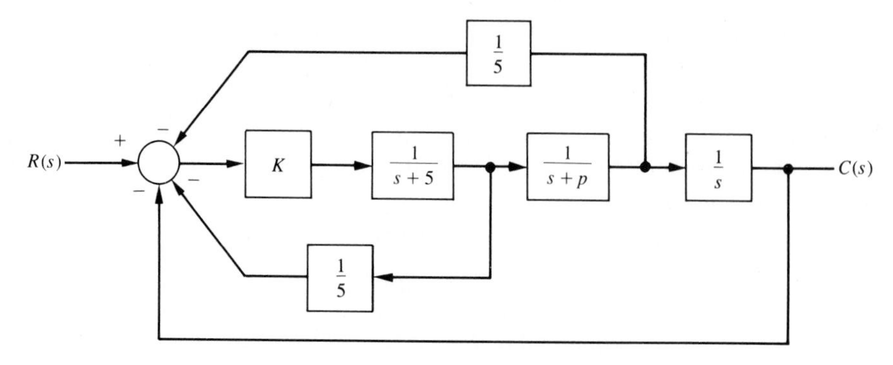

Figure DP5.1. Automobile engine control.

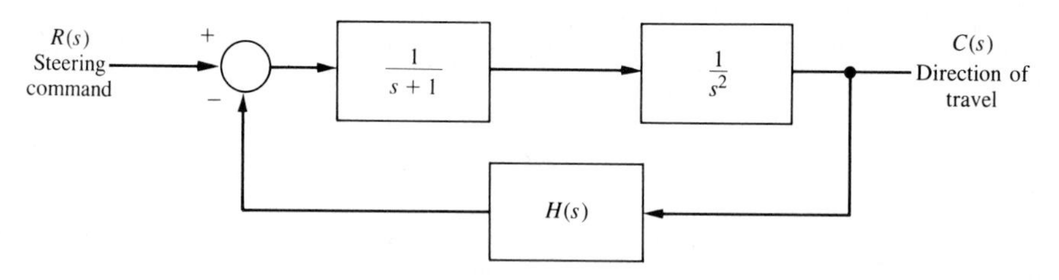

Figure DP5.2. Mars guided vehicle control.

DP5.3. A unity negative feedback system with

$$G(s) = \frac{K(s + 2)}{s(1 + \tau s)(1 + 2s)}$$

has two parameters to be selected. (a) Determine and plot the regions of stability for this system. (b) Select τ and K so that the ready state error to a ramp input is less than or equal to 25% of the input magnitude. (c) Determine the percent overshoot for a step input for the design selected in part (b).

DP5.4. The attitude control system of a space shuttle rocket is shown in Fig. DP5.4 [4]. (a) Determine the range of gain K and parameter m so that the system is stable and plot the region of stability. (b) Select the gain and parameter values so that the steady-state error to a ramp input is less than or equal to 10% of the input magnitude. (c) Determine the percent overshoot for a step input for the design selected in part (b).

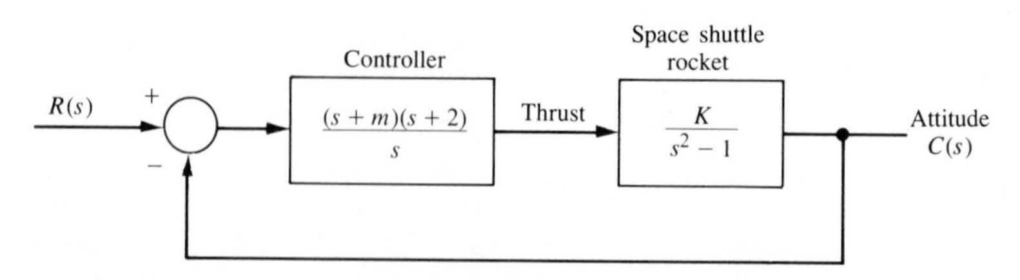

Figure DP5.4. Shuttle attitude control.

Terms and Concepts

Auxiliary equation The equation that immediately precedes the zero entry in the Routh array.

Newton-Raphson method An iterative approach to solving for roots of a polynomial equation.

Relative stability The property that is measured by the relative settling times of each root or pair of roots of the characteristic equation.

Routh-Hurwitz criterion A criterion for determining the stability of a system by examining the characteristic equation of the transfer function. The criterion states that the number of roots of the characteristic equation with positive real parts is equal to the number of changes of sign of the coefficients in the first column of the Routh array.

Stability A performance measure of a system. A system is stable if all the poles of the transfer function have negative real parts.

Stable system A dynamic system with a bounded system response to a bounded input.

Synthetic division A method of determining the roots of the characteristic equation based on the remainder theorem of mathematics.

References

1. R. C. Dorf, *Introduction to Electric Circuits,* John Wiley & Sons, New York, 1989.
2. A. Hurwitz, "On the Conditions Under Which an Equation Has Only Roots with Negative Real Parts," *Mathematische Annalen,* **46,** 1895; pp. 273–84. Also in *Selected Papers on Mathematical Trends in Control Theory,* Dover, New York, 1964; pp. 70–82.
3. E. J. Routh, *Dynamics of a System of Rigid Bodies,* Macmillan, New York, 1892.
4. R. DeMeis, "Shuttling to the Space Station," *Aerospace America,* March 1990; pp. 44–47.
5. J. V. Ringwood, "Shape Control in Sendzimir Mills Using Roll Actuators," *IEEE Transactions on Automatic Control,* April 1990; pp. 453–59.
6. R. C. Dorf, *Encyclopedia of Robotics,* John Wiley & Sons, New York, 1988.
7. J. R. Pierce, *Signals: The Science of Telecommunications,* W. H. Freeman, San Francisco, 1990.
8. H. Tamura, *Large-Scale Systems Control,* Marcel Dekker, New York, 1990.
9. S. Derby, "Mechatronics for Robots," *Mechanical Engineering,* July 1990; pp. 40–42.
10. F. Raven, *Automatic Control Engineering,* 2nd ed., McGraw-Hill, New York, 1990.
11. F. B. Farquharson, "Aerodynamic Stability of Suspension Bridges, with Special Reference to the Tacoma Narrows Bridge," *Bulletin 116, Part I,* The Engineering Experiment Station, University of Washington, 1950.
12. R. C. Dorf, *Introduction to Computers and Computer Science,* 3rd ed., Boyd and Fraser, San Francisco, 1982.
13. P. C. Krause, *Electromechanical Motion Devices,* McGraw-Hill, New York, 1989.

14. R. C. Rosenberg and D. C. Karnopp, *Introduction to Physical System Design,* McGraw-Hill, New York, 1987.
15. T. Yoshikawa, *Foundations of Robotics,* M.I.T. Press, Cambridge, Mass., 1990.
16. B. E. Jeppsen, "A New Family of Dot Matrix Line Printers," *HP Journal,* June 1985; pp. 4–9.
17. R. E. Ziemer, *Signals and Systems,* 2nd ed., Macmillan, New York, 1989.
18. R. C. Dorf and R. Jacquot, *Control System Design Program,* Addison-Wesley, Reading, Mass., 1988.
19. R. A. Hammond, "Fly by Wire Control Systems," *Simulation,* October 1989; pp. 159–67.
20. P. P. Vardyanathan and S. K. Mitra, "A Unified Structural Interpretation of Some Well Known Stability Test Procedures for Linear Systems," *Proceedings of the IEEE,* November 1989; pp. 478–79.
21. P. G. Scotson, "Self-Tuning Optimization of Spark Ignition Automotive Engines," *IEEE Control Systems,* April 1990; pp. 94–99.
22. G. G. Wang, "Design of Turning Control for a Tracked Vehicle," *IEEE Control Systems,* April 1990; pp. 122–25.
23. A. Dane, "Hubble: Heartbreaks and Hope," *Popular Mechanics,* October 1990; pp. 130–31.
24. T. Beardsley, "Hubble's Legacy," *Scientific American,* June 1990; pp. 18–19.
25. C. Phillips and R. Harbor, *Feedback Control Systems,* Prentice Hall, Englewood Cliffs, N.J., 1991.

CHAPTER 6

The Root Locus Method

Preview

Since the performance of a feedback system can be adjusted by changing one or more parameters, we described that performance in terms of the s-plane location of the roots of the characteristic equation in the preceeding chapters. Thus it is very useful to determine how the roots of the characteristic equations move around the s-plane as we change a parameter.

The locus of roots in the s-plane can be determined by a graphical method. Once you understand this graphical method, then we will proceed to demonstrate the utility of computer generation of the locus of roots as a parameter is varied.

Since we can determine how the roots migrate as one parameter varies, it is possible to show they will vary as two parameters vary. This provides us with the opportunity to design a system with two adjustable parameters so as to achieve a very desirable performance.

Using the concept of the root locus as a parameter varies, we also will be able to define a measure of the sensitivity of a specified root to a small incremental change in the parameter.

6.1 Introduction

The relative stability and the transient performance of a closed-loop control system are directly related to the location of the closed-loop roots of the characteristic equation in the s-plane. Also, it is frequently necessary to adjust one or more system parameters in order to obtain suitable root locations. Therefore, it is worthwhile to determine how the roots of the characteristic equation of a given system migrate about the s-plane as the parameters are varied; that is, it is useful to determine the *locus* of roots in the s-plane as a parameter is varied. The *root locus method* was introduced by Evans in 1948 and has been developed and utilized extensively in control engineering practice [1, 2, 3]. The root locus technique is a graphical method for drawing the locus of roots in the s-plane as a parameter is varied. In fact, the root locus method provides the engineer with a measure of the sensitivity of the roots of the system to a variation in the parameter being considered. The root locus technique may be used to great advantage in conjunction with the Routh-Hurwitz criterion and the Newton-Raphson method.

The root locus method provides graphical information, and therefore an approximate sketch can be used to obtain qualitative information concerning the stability and performance of the system. Furthermore, the locus of roots of the characteristic equation of a multiloop system may be investigated as readily as for a single-loop system. If the root locations are not satisfactory, the necessary parameter adjustments can often be readily ascertained from the root locus.

6.2 The Root Locus Concept

The dynamic performance of a closed-loop control system is described by the closed-loop transfer function

$$T(s) = \frac{C(s)}{R(s)} = \frac{p(s)}{q(s)}, \tag{6.1}$$

where $p(s)$ and $q(s)$ are polynomials in s. The roots of the characteristic equation $q(s)$ determine the modes of response of the system. For a closed-loop system, we found in Section 2.7 that by using Mason's signal-flow gain formula, we had

$$\Delta(s) = 1 - \sum_{n=1}^{N} L_n + \sum_{m,q}^{M,N} L_m L_q - \sum L_r L_s L_t + \ldots, \qquad (6.2)$$

where L_q equals the value of the qth self-loop transmittance. Clearly, we have a characteristic equation, which may be written as

$$q(s) = \Delta(s) = 1 + F(s). \qquad (6.3)$$

In order to find the roots of the characteristic equation we set Eq. (6.3) equal to zero and obtain

$$1 + F(s) = 0. \qquad (6.4)$$

Of course, Eq. (6.4) may be rewritten as

$$F(s) = -1, \qquad (6.5)$$

and the roots of the characteristic equation must also satisfy this relation. In the case of the simple single-loop system, as shown in Fig. 6.1, we have the characteristic equation

$$1 + GH(s) = 0, \qquad (6.6)$$

where $F(s) = G(s)H(s)$. The characteristic roots of the system must satisfy Eq. (6.5), where the roots lie in the s-plane. Because s is a complex variable, Eq. (6.5) may be rewritten in polar form as

$$|F(s)| \angle F(s) = -1, \qquad (6.7)$$

and therefore it is necessary that

$$|F(s)| = 1$$

and

$$\angle F(s) = 180° \pm k360°, \qquad (6.8)$$

where $k = 0, \pm 1, \pm 2, \pm 3, \ldots$. The graphical computation required for Eq. (6.8) is readily accomplished by using a protractor for estimating angles.

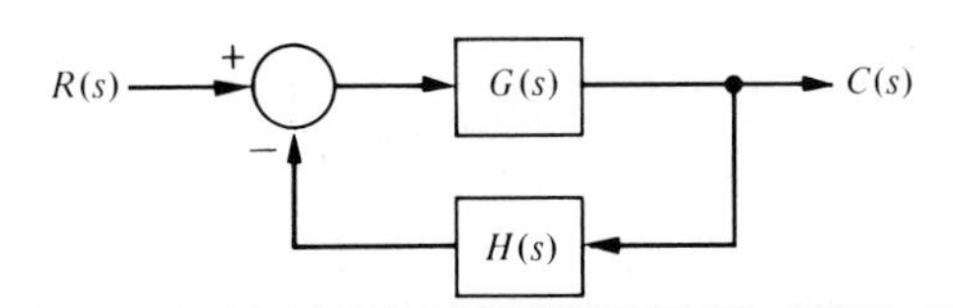

Figure 6.1. Closed-loop control system.

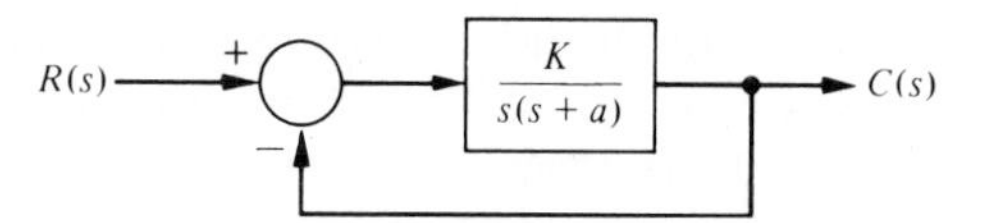

Figure 6.2. Unity feedback control system.

The simple second-order system considered in the previous chapters is shown in Fig. 6.2. The characteristic equation representing this system is

$$\Delta(s) = 1 + G(s) = 1 + \frac{K}{s(s+a)} = 0$$

or, alternatively,

$$q(s) = s^2 + as + K = s^2 + 2\zeta\omega_n s + \omega_n^2 = 0. \tag{6.9}$$

The locus of the roots as the gain K is varied is found by requiring that

$$|G(s)| = \left|\frac{K}{s(s+a)}\right| = 1 \tag{6.10}$$

and

$$\underline{/G(s)} = \pm 180°, \pm 540°, \ldots. \tag{6.11}$$

The gain K may be varied from zero to an infinitely large positive value. For a second-order system, the roots are

$$s_1, s_2 = -\zeta\omega_n \pm \omega_n\sqrt{\zeta^2 - 1}, \tag{6.12}$$

and for $\zeta < 1$, we know that $\theta = \cos^{-1}\zeta$. Graphically, for two open-loop poles as shown in Fig. 6.3, the locus of roots is a vertical line for $\zeta \leq 1$ in order to satisfy

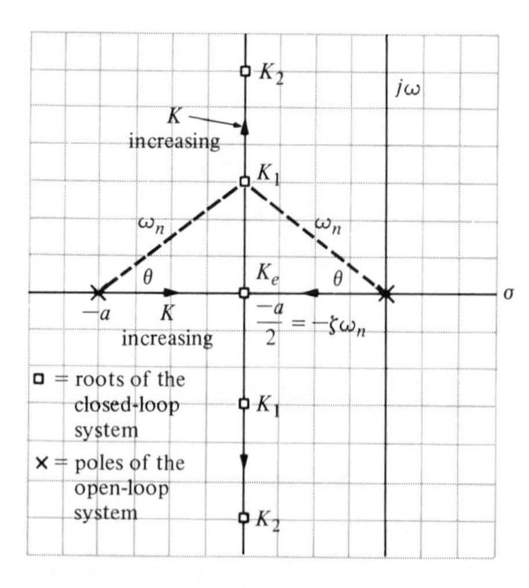

Figure 6.3. Root locus for a second-order system when $K_e < K_1 < K_2$.

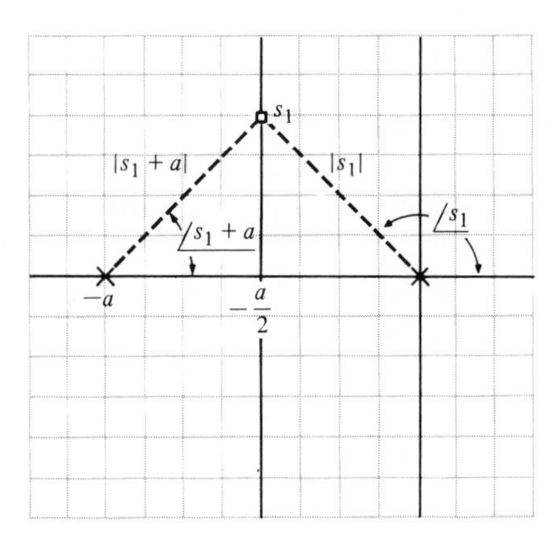

Figure 6.4. Evaluation of the angle and gain at s_1.

the angle requirement, Eq. (6.11). For example, as shown in Fig. 6.4, at a root s_1, the angles are

$$\left.\angle \frac{K}{s(s+a)}\right|_{s=s_1} = -\angle s_1 - \angle(s_1 + a) \tag{6.13}$$
$$= -[(180^\circ - \theta) + \theta] = -180^\circ.$$

This angle requirement is satisfied at any point on the vertical line which is a perpendicular bisector of the line 0 to $-a$. Furthermore, the gain K at the particular point s_1 is found by using Eq. (6.10) as

$$\left|\frac{K}{s(s+a)}\right|_{s=s_1} = \frac{K}{|s_1|\,|s_1+a|} = 1 \tag{6.14}$$

and thus

$$K = |s_1|\,|s_1 + a|, \tag{6.15}$$

where $|s_1|$ is the magnitude of the vector from the origin to s_1, and $|s_1 + a|$ is the magnitude of the vector from $-a$ to s_1.

In general, the function $F(s)$ may be written as

$$F(s) = \frac{K(s+z_1)(s+z_2)(s+z_3)\cdots(s+z_m)}{(s+p_1)(s+p_2)(s+p_3)\cdots(s+p_n)}. \tag{6.16}$$

Then the magnitude and angle requirement for the root locus are

$$|F(s)| = \frac{K|s+z_1|\,|s+z_2|\cdots}{|s+p_1|\,|s+p_2|\cdots} = 1 \tag{6.17}$$

and

$$\angle F(s) = \angle s+z_1 + \angle s+z_2 + \cdots - (\angle s+p_1 + \angle s+p_2 + \cdots). \tag{6.18}$$
$$= 180^\circ \pm k360^\circ$$

The magnitude requirement, Eq. (6.17), enables one to determine the value of K for a given root location s_1. A test point in the s-plane, s_1, is verified as a root location when Eq. (6.18) is satisfied. The angles are all measured in a counterclockwise direction from a horizontal line.

In order to further illustrate the root locus procedure, let us reconsider the second-order system of Fig. 6.2. The effect of varying the parameter, a, can be effectively portrayed by rewriting the characteristic equation for the root locus form with a as the muliplying factor in the numerator. Then the characteristic equation is

$$1 + F(s) = 1 + \frac{K}{s(s + a)} = 0$$

or, alternatively,

$$s^2 + as + K = 0.$$

Dividing by the factor $(s^2 + K)$, we obtain

$$1 + \frac{as}{s^2 + K} = 0. \tag{6.19}$$

Then the magnitude criterion is satisfied when

$$\frac{a|s_1|}{|s_1^2 + K|} = 1 \tag{6.20}$$

at the root s_1. The angle criterion is

$$\underline{/s_1} - (\underline{/s_1 + j\sqrt{K}} + \underline{/s_1 - j\sqrt{K}}) = \pm 180°, \pm 540°, \ldots.$$

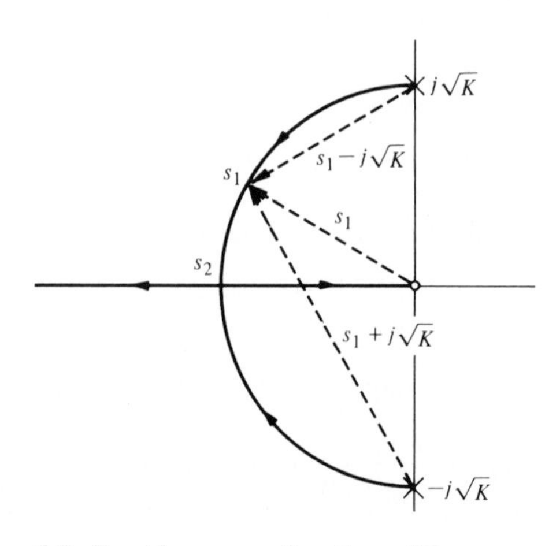

Figure 6.5. Root locus as a function of the parameter a.

In order to construct the root locus, we find the points in the s-plane that satisfy the angle criterion. The points in the s-plane that satisfy the angle criterion are located on a trial-and-error basis by searching in an orderly manner for a point with a total angle of $\pm 180°$, $\pm 540°$, or in general,

$$\frac{\pm(2q + 1)180°}{n_p - n_z}.$$

The algebraic sum of the angles from the poles and zeros is measured with a protractor. Using a protractor we find the locus of roots as shown in Fig. 6.5. Specifically, at the root s_1, the magnitude of the parameter, a, is found from Eq. (6.20) as

$$a = \frac{|s_1 - j\sqrt{K}|\,|s_1 + j\sqrt{K}|}{|s_1|}. \tag{6.21}$$

The roots of the system merge on the real axis at the point s_2 and provide a critically damped response to a step input. The parameter, a, has a magnitude at the critically damped roots $s_2 = \sigma_2$ equal to

$$\begin{aligned} a &= \frac{|\sigma_2 - j\sqrt{K}|\,|\sigma_2 + j\sqrt{K}|}{\sigma_2} \\ &= \frac{1}{\sigma_2}(\sigma_2^2 + K), \end{aligned} \tag{6.22}$$

where σ_2 is evaluated from the s-plane vector lengths as $\sigma_2 = \sqrt{K}$. As a increases beyond the critical value, the roots are both real and distinct; one root is larger than σ_2 and one is smaller.

In general, an orderly process for locating the locus of roots as a parameter varies is desirable. In the following section, we shall develop such an orderly approach to obtaining a root locus diagram.

6.3 The Root Locus Procedure

The characteristic equation of a system provides a valuable insight concerning the response of the system when the roots of the equation are determined. In order to locate the roots of the characteristic equation in a graphical manner on the s-plane, we shall develop an orderly procedure that facilitates the rapid sketching of the locus. First, we write the characteristic equation as

$$1 + F(s) = 0 \tag{6.23}$$

and rearrange the equation, if necessary, so that the parameter of interest, k, appears as the multiplying factor in the form,

$$1 + kP(s) = 0. \tag{6.24}$$

Second, we factor $P(s)$, if necessary, and write the polynomial in the form of poles and zeros as follows:

$$1 + k\frac{\prod_{i=1}^{M}(s + z_i)}{\prod_{j=1}^{n}(s + p_j)} = 0. \tag{6.25}$$

Then we locate the poles and zeros on the s-plane with appropriate markings. Now, we are usually interested in determining the locus of roots as k varies as

$$0 \leq k \leq \infty.$$

Rewriting Eq. (6.25), we have

$$\prod_{j=1}^{n}(s + p_j) + k\prod_{i=1}^{M}(s + z_i) = 0. \tag{6.26}$$

Therefore, when $k = 0$, the roots of the characteristic equation are simply the poles of $P(s)$. Furthermore, when k approaches infinity, the roots of the characteristic equation are simply the zeros of $P(s)$. Therefore we note that *the locus of the roots of the characteristic equation* $1 + kP(s) = 0$ *begins at the poles of* $P(s)$ *and ends at the zeros of* $P(s)$ *as* k *increases from* 0 *to infinity.* For most functions, $P(s)$, that we will encounter, several of the zeros of $P(s)$ lie at infinity in the s-plane.

The root locus on the real axis always lies in a section of the real axis to the left of an odd number of poles and zeros. This fact is clearly ascertained by examining the angle criterion of Eq. (6.18). These useful steps in plotting a root locus will be illustrated by a suitable example.

■ Example 6.1 Second-order system

A single-loop feedback control system possesses the following characteristic equation:

$$1 + GH(s) = 1 + \frac{K(\frac{1}{2}s + 1)}{s(\frac{1}{4}s + 1)} = 0. \tag{6.27}$$

First, the transfer function $GH(s)$ is rewritten in terms of poles and zeros as follows:

$$1 + \frac{2K(s + 2)}{s(s + 4)} = 0, \tag{6.28}$$

and the multiplicative gain parameter is $k = 2K$. In order to determine the locus of roots for the gain $0 \leq K \leq \infty$, we locate the poles and zeros on the real axis as shown in Fig. 6.6(a). Clearly, the angle criterion is satisfied on the real axis between the points 0 and -2, because the angle from pole p_1 is 180° and the angle from the zero and pole p_2 is zero degrees. The locus begins at the pole and ends at the zeros, and therefore the locus of roots appears as shown in Fig. 6.6(b),

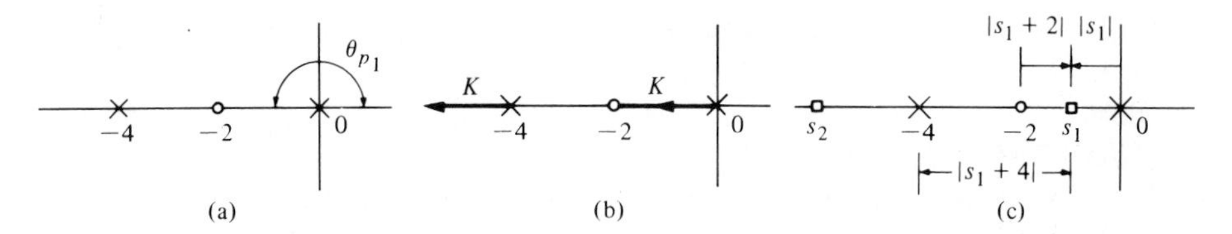

Figure 6.6. Root locus of a second-order system with a zero.

where the direction of the locus at K is increasing ($K \uparrow$) is shown by an arrow. We note that because the system has two real poles and one real zero, the second locus segment ends at a zero at negative infinity. In order to evaluate the gain K at a specific root location on the locus, we utilize the magnitude criterion, Eq. (6.17). For example, the gain K at the root $s = s_1 = -1$ is found from (6.17) as

$$\frac{(2K)|s_1 + 2|}{|s_1|\ |s_1 + 4|} = 1$$

or

$$K = \frac{|-1|\ |-1 + 4|}{2|-1 + 2|} = \frac{3}{2}. \tag{6.29}$$

This magnitude can also be evaluated graphically as is shown in Fig. 6.6(c). Finally, for the gain of $K = 3/2$, one other root exists, located on the locus to the left of the pole at -4. The location of the second root is found graphically to be located at $s = -6$, as shown in Fig. 6.6(c).

Because the loci begin at the poles and end at the zeros, the *number of separate loci* is equal to the number of poles if the number of poles is greater than the number of zeros. In the unusual case when the number of zeros is greater than the number of poles, the number of separate loci would be the number of zeros. Therefore, as we found in Fig. 6.6, the number of separate loci is equal to two because there are two poles and one zero.

The root loci must be symmetrical with respect to the horizontal real axis because the complex roots must appear as pairs of complex conjugate roots.

When the number of finite zeros of $P(s)$, n_z, is less than the number of poles, n_p, by the number $N = n_p - n_z$, then N sections of loci must end at zeros at infinity. These sections of loci proceed to the zeros at infinity along *asymptotes* as k approaches infinity. These linear *asymptotes are centered* at a point on the real axis given by

$$\sigma_A = \frac{\sum \text{poles of } P(s) - \sum \text{zeros of } P(s)}{n_p - n_z} = \frac{\sum_{j=1}^{n} p_j - \sum_{i=1}^{M} z_i}{n_p - n_z}. \tag{6.30}$$

The angle of the asymptotes with respect to the real axis is

$$\phi_A = \frac{(2q + 1)}{n_p - n_z} 180°, \qquad q = 0, 1, 2, \ldots, (n_p - n_z - 1), \tag{6.31}$$

where q is an integer index [3]. The usefulness of this rule is obvious for sketching the approximate form of a root locus. Equation (6.31) can be readily derived by considering a point on a root locus segment at a remote distance from the finite poles and zeros in the s-plane. The net phase angle at this remote point is 180°, because it is a point on a root locus segment. The finite poles and zeros of $P(s)$ are a great distance from the remote point, and so the angle from each pole and zero, ϕ, is essentially equal and therefore the net angle is simply $\phi(n_p - n_z)$, where n_p and n_z are the number of finite poles and zeros, respectively. Thus we have

$$\phi(n_p - n_z) = 180°,$$

or, alternatively,

$$\phi = \frac{180°}{n_p - n_z}.$$

Accounting for all possible root locus segments at remote locations in the s-plane, we obtain Eq. (6.31).

The center of the linear asymptotes, often called the *asymptote centroid*, is determined by considering the characteristic equation $1 + GH(s) = 0$ (Eq. 6.25). For large values of s, only the higher-order terms need be considered so that the characteristic equation reduces to

$$1 + \frac{ks^M}{s^n} = 0.$$

However, this relation, which is an approximation, indicates that the centroid of $(n - M)$ asymptotes is at the origin, $s = 0$. A better approximation is obtained if we consider a characteristic equation of the form

$$1 + \frac{k}{(s - \sigma_A)^{n-M}} = 0$$

with a centroid at σ_A.

The centroid is determined by considering the first two terms of Eq. (6.25), which may be found from the relation

$$1 + \frac{k\prod_{i=1}^{M}(s + z_i)}{\prod_{j=1}^{n}(s + p_j)} = 1 + k\frac{(s^M + b_{M-1}s^{M-1} + \cdots + b_0)}{(s^n + a_{n-1}s^{n-1} + \cdots + a_0)}.$$

From Chapter 5, especially Eq. (5.5), we note that

$$b_{M-1} = \sum_{i=1}^{M} z_i \quad \text{and} \quad a_{n-1} = \sum_{j=1}^{n} p_j.$$

Considering only the first two terms of this expansion we have

$$1 + \frac{k}{s^{n-M} + (a_{n-1} - b_{M-1})s^{n-M-1}} = 0.$$

The first two terms of

$$1 + \frac{k}{(s - \sigma_A)^{n-M}} = 0$$

are

$$1 + \frac{k}{s^{n-M} + (n - M)\,\sigma_A\, s^{n-M-1}} = 0.$$

Equating the term for s^{n-M-1}, we obtain

$$(a_{n-1} - b_{M-1}) = (n - M)\,\sigma_A,$$

which is equivalent to Eq. (6.30).

For example, reexamine the system shown in Fig. 6.2 and discussed in Section 6.2. The characteristic equation is written as

$$1 + \frac{K}{s(s + a)} = 0.$$

Because $n_p - n_z = 2$, we expect two loci to end at zeros at infinity. The asymptotes of the loci are located at a center

$$\sigma_A = \frac{-a}{2},$$

and at angles of

$$\phi_A = 90^\circ, \qquad q = 0,$$

and

$$\phi_A = 270^\circ, \qquad q = 1.$$

Therefore the root locus is readily sketched and the locus as shown in Fig. 6.3 is obtained. An example will further illustrate the process of utilizing the asymptotes.

■ Example 6.2 Fourth-order system

A feedback control system has a characteristic equation as follows:

$$1 + F(s) = 1 + \frac{K(s + 1)}{s(s + 2)(s + 4)^2}. \tag{6.32}$$

We wish to sketch the root locus in order to determine the effect of the gain K. The poles and zeros are located in the s-plane as shown in Fig. 6.7(a). The root loci on the real axis must be located to the left of an odd number of poles and

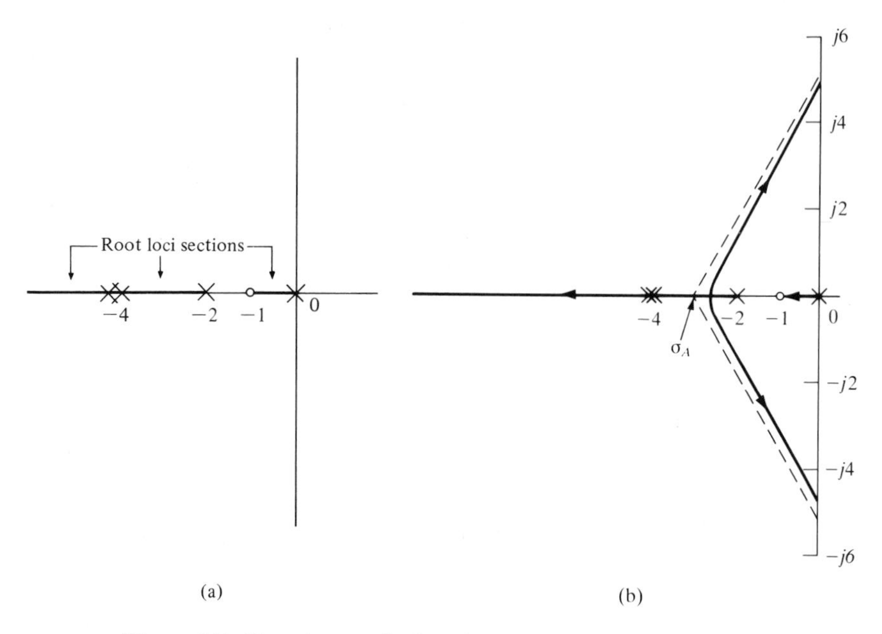

Figure 6.7. Root locus of a fourth-order system with a zero.

zeros and are therefore located as shown in Fig. 6.7(a) as heavy lines. The intersection of the asymptotes is

$$\sigma_A = \frac{(-2) + 2(-4) - (-1)}{4 - 1} = \frac{-9}{3} = -3. \tag{6.33}$$

The angles of the asymptotes are

$$\phi_A = +60^\circ, \qquad q = 0,$$

$$\phi_A = 180^\circ, \qquad q = 1,$$

$$\phi_A = 300^\circ, \qquad q = 2,$$

where there are three asymptotes since $n_p - n_z = 3$. Also, we note that the root loci must begin at the poles, and therefore two loci must leave the double pole at $s = -4$. Then, with the asymptotes drawn in Fig. 6.7(b), we may sketch the form of the root locus as shown in Fig. 6.7(b). The actual shape of the locus in the area near σ_A would be graphically evaluated, if necessary. The actual point at which *the root locus crosses the imaginary axis* is readily evaluated by utilizing the *Routh-Hurwitz criterion.*

The root locus in the previous example left the real axis at a *breakaway point.* The locus breakaway from the real axis occurs where the net change in angle caused by a small vertical displacement is zero. The locus leaves the real axis where there are a multiplicity of roots, typically two roots. The breakaway point

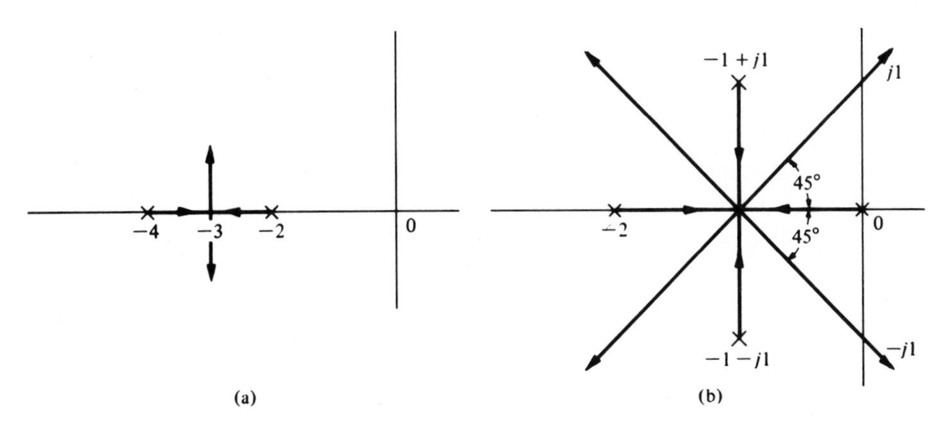

Figure 6.8. Illustration of the breakaway point.

for a simple second-order system is shown in Fig. 6.8(a) and, for a special case of a fourth-order system, in Fig. 6.8(b). In general, due to the phase criterion, *the tangents to the loci at the breakaway point are equally spaced over 360°.* Therefore, in Fig. 6.8(a), we find that the two loci at the breakaway point are spaced 180° apart, whereas in Fig. 6.8(b), the four loci are spaced 90° apart.

The breakaway point on the real axis can be evaluated graphically or analytically. The most straightforward method of evaluating the breakaway point involves the rearranging of the characteristic equation in order to isolate the multiplying factor K. Then the characteristic equation is written as

$$p(s) = K. \tag{6.34}$$

For example, consider a unity feedback closed-loop system with an open-loop transfer function

$$G(s) = \frac{K}{(s + 2)(s + 4)},$$

which has a characteristic equation as follows:

$$1 + G(s) = 1 + \frac{K}{(s + 2)(s + 4)} = 0. \tag{6.35}$$

Alternatively, the equation may be written as

$$K = p(s) = -(s + 2)(s + 4). \tag{6.36}$$

The root loci for this system are shown in Fig. 6.8(a). We expect the breakaway point to be near $s = \sigma = -3$ and plot $p(s)|_{s=\sigma}$ near that point as shown in Fig. 6.9. In this case, $p(s)$ equals zero at the poles $s = -2$ and $s = -4$. The plot of $p(s)$ versus σ is symmetrical and the maximum point occurs at $s = \sigma = -3$, the breakaway point. Analytically, the very same result may be obtained by determining the maximum of $K = p(s)$. In order to find the maximum analytically,

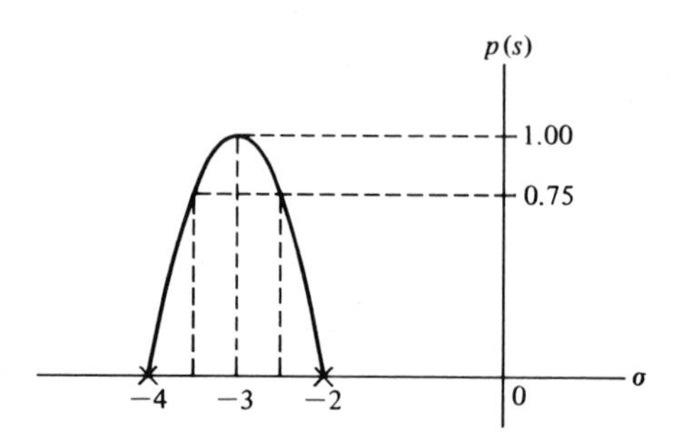

Figure 6.9. A graphical evaluation of the breakaway point.

we differentiate, set the differentiated polynomial equal to zero, and determine the roots of the polynomial. Therefore we may evaluate

$$\frac{dK}{ds} = \frac{dp(s)}{ds} = 0 \tag{6.37}$$

in order to find the breakaway point. Clearly, Eq. (6.37) is an analytical expression of the graphical procedure outlined in Fig. 6.9 and will result in an equation of only one order less than the total number of poles and zeros ($n_p + n_z - 1$). In almost all cases, we will prefer to use the graphical method of locating the breakaway point when it is necessary to do so.

The proof of Eq. (6.37) is obtained from a consideration of the characteristic equation

$$1 + F(s) = 1 + \frac{KY(s)}{X(s)} = 0,$$

which may be written as

$$1 + F(s) = X(s) + KY(s) = 0. \tag{6.38}$$

For a small increment in K, we have

$$X(s) + (K + \Delta K)Y(s) = 1 + \frac{\Delta KY(s)}{X(s) + KY(s)} = 0. \tag{6.39}$$

Because the denominator is the original characteristic equation, at a breakaway point a multiplicity of roots exists and

$$\frac{Y(s)}{X(s) + KY(s)} = \frac{C_i}{(s - s_i)^n} = \frac{C_i}{(\Delta s_i)^n}. \tag{6.40}$$

Then we may write Eq. (6.39) as

$$1 + \frac{\Delta KC_i}{(\Delta s_i)^n} = 0 \tag{6.41}$$

or, alternatively,

$$\frac{\Delta K}{\Delta s} = \frac{-(\Delta s)^{n-1}}{C_i}. \tag{6.42}$$

Therefore, as we let Δs approach zero, we obtain

$$\frac{dK}{ds} = 0 \tag{6.43}$$

at the breakaway points.

Now, reconsidering the specific case where

$$G(s) = \frac{K}{(s+2)(s+4)},$$

we obtain $p(s)$ as

$$K = p(s) = -(s+2)(s+4) = -(s^2 + 6s + 8). \tag{6.44}$$

Then, differentiating, we have

$$\frac{dK}{ds} = -(2s+6) = 0 \tag{6.45}$$

or the breakaway point occurs at $s = -3$. A more complicated example will illustrate the approach and exemplify the advantage of the graphical technique.

■ Example 6.3 Third-order system

A feedback control system is shown in Fig. 6.10. The characteristic equation is

$$1 + G(s)H(s) = 1 + \frac{K(s+1)}{s(s+2)(s+3)} = 0. \tag{6.46}$$

The number of poles, n_p, minus the number of zeros, n_z, is equal to two, and so we have two asymptotes at $\pm 90°$ with a center at $\sigma_A = -2$. The asymptotes and the sections of loci on the real axis are shown in Fig. 6.11(a). A breakaway point

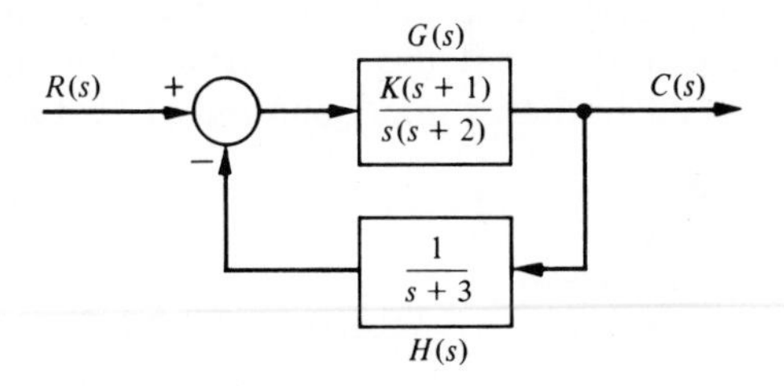

Figure 6.10. Closed-loop system.

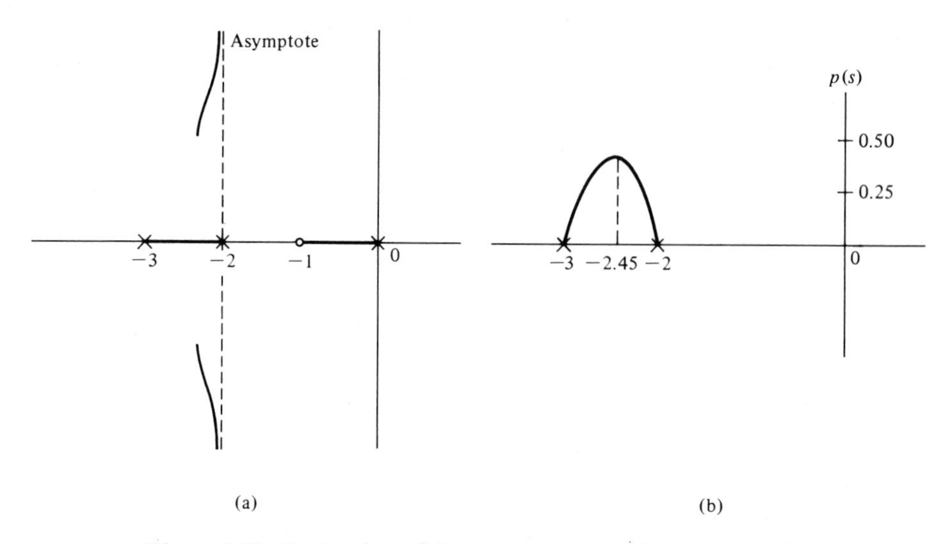

Figure 6.11. Evaluation of the asymptotes and breakaway point.

occurs between $s = -2$ and $s = -3$. In order to evaluate the breakaway point, we rewrite the characteristic equation so that K is separated as follows:

$$s(s+2)(s+3) + K(s+1) = 0$$

or

$$p(s) = \frac{-s(s+2)(s+3)}{(s+1)} = K. \tag{6.47}$$

Then, evaluating $p(s)$ at various values of s between $s = -2$ and $s = -3$, we obtain the results of Table 6.1 as shown in Fig. 6.11(b). Alternatively, we differentiate Eq. (6.47) and set equal to zero to obtain

$$\frac{d}{ds}\left(\frac{-s(s+2)(s+3)}{(s+1)}\right) = \frac{(s^3 + 5s^2 + 6s) - (s+1)(3s^2 + 10s + 6)}{(s+1)^2} = 0$$

$$= 2s^3 + 8s^2 + 10s + 6 = 0. \tag{6.48}$$

Now, in order to locate the maximum of $p(s)$, we locate the roots of Eq. (6.48) by synthetic division or by the Newton-Raphson method to obtain $s = -2.45$. It is evident from this one example that the evaluation of $p(s)$ near the expected breakaway point will result in the simplest method of evaluating the breakaway

Table 6.1.

$p(s)$	0	+0.412	+0.420	+0.417	+0.390	0
s	−2.00	−2.40	−2.45	−2.50	−2.60	−3.0

point. As the order of the characteristic equation increases, the usefulness of the graphical (or tabular) evaluation of the breakaway point will increase in contrast to the analytical approach.

The *angle of departure of the locus from a pole* and the *angle of arrival at the locus at a zero* can be determined from the phase angle criterion. The angle of locus departure from a pole is the difference between the net angle due to all other poles and zeros and the criterion angle of $\pm 180°(2q + 1)$, and similarly for the locus angle of arrival at zero. The angle of departure (or arrival) is particularly of interest for complex poles (and zeros) because the information is helpful in completing the root locus. For example, consider the third-order open-loop transfer function

$$F(s) = G(s)H(s) = \frac{K}{(s + p_3)(s^2 + 2\zeta\omega_n s + \omega_n^2)}. \tag{6.49}$$

The pole locations and the vector angles at one complex pole p_1 are shown in Fig. 6.12(a). The angles at a test point s_1, an infinitesimal distance from p_1, must meet the angle criterion. Therefore, since $\theta_2 = 90°$, we have

$$\theta_1 + \theta_2 + \theta_3 = \theta_1 + 90° + \theta_3 = +180°,$$

or the angle of departure at pole p_1 is

$$\theta_1 = 90° - \theta_3$$

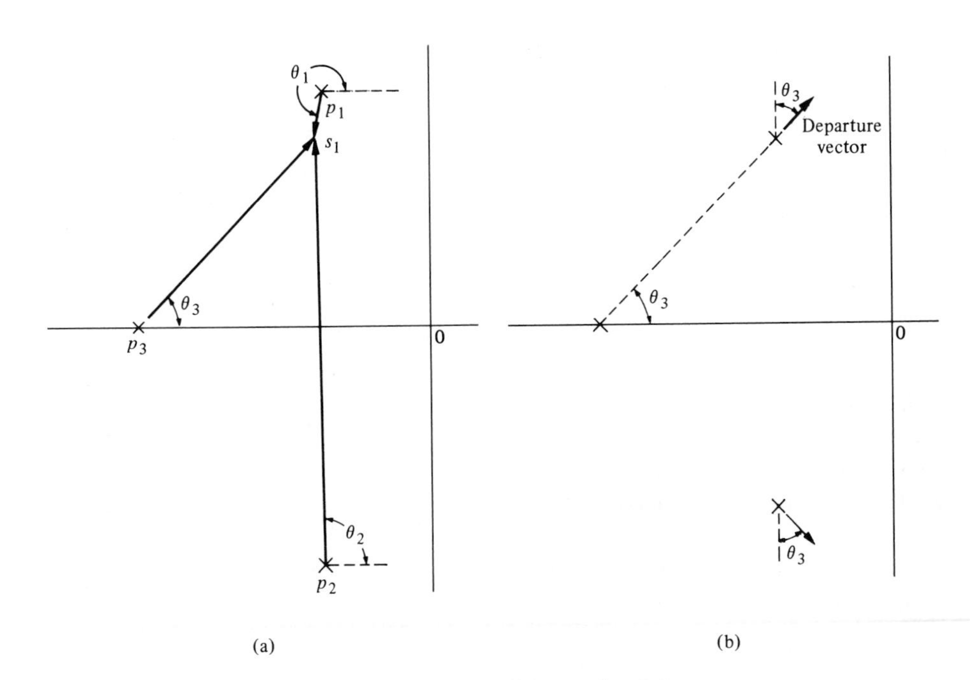

Figure 6.12. Illustration of the angle of departure.

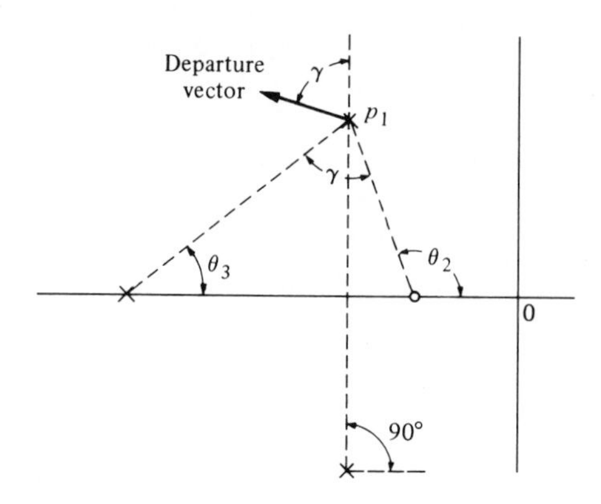

Figure 6.13. Evaluation of the angle of departure.

as shown in Fig. 6.12(b). The departure at pole p_2 is the negative of that at p_1 because p_1 and p_2 are complex conjugates. Another example of a departure angle is shown in Fig. 6.13. In this case, the departure angle is found from

$$\theta_2 - (\theta_1 + \theta_3 + 90°) = 180°.$$

Since $(\theta_2 - \theta_3) = \gamma$, we find that the departure angle is $\theta_1 = 90° + \gamma$.

It is worthwhile at this point to summarize the steps utilized in the root locus method and then illustrate their use in a complete example. The steps utilized in evaluating the locus of roots of a characteristic equation are as follows:

1. Write the characteristic equation in pole-zero form so that the parameter of interest k appears as $1 + kF(s) = 0$.
2. Locate the open-loop poles and zeros of $F(s)$ in the s-plane.
3. Locate the segments of the real axis that are root loci.
4. Determine the number of separate loci.
5. Locate the angles of the asymptotes and the center of the asymptotes.
6. Determine the breakaway point on the real axis (if any).
7. By utilizing the Routh-Hurwitz criterion, determine the point at which the locus crosses the imaginary axis (if it does so).
8. Estimate the angle of locus departure from complex poles and the angle of locus arrival at complex zeros.

■ Example 6.4 Fourth-order system

1. We desire to plot the root locus for the characteristic equation of a system when

$$1 + \frac{K}{s(s+4)(s+4+j4)(s+4-j4)} = 0 \tag{6.50}$$

as K varies from zero to infinity.

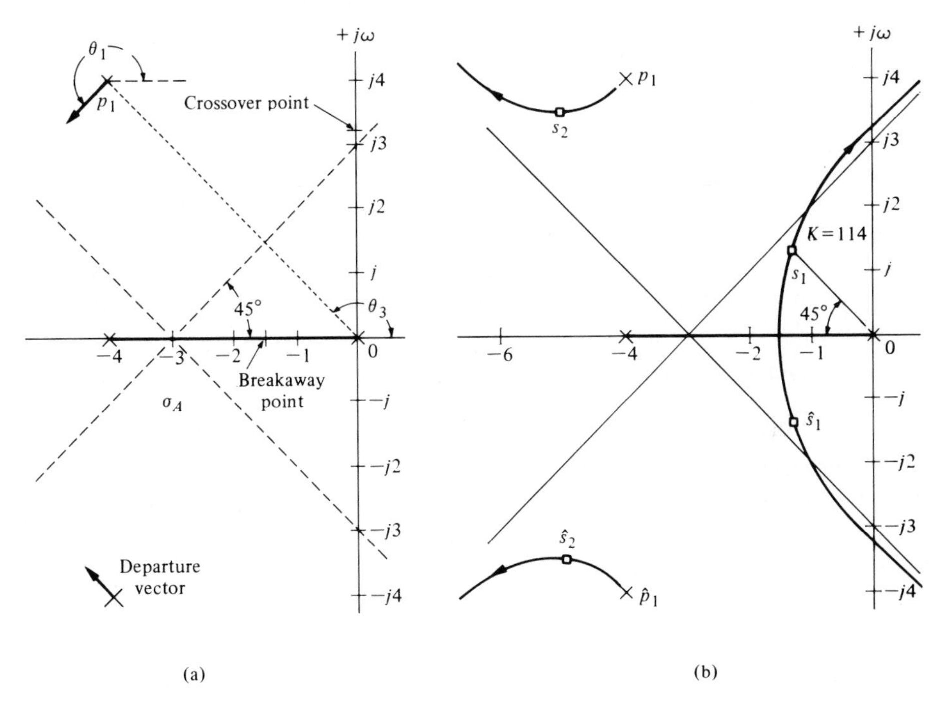

Figure 6.14. The root locus for Example 6.4.

2. The poles are located on the s-plane as shown in Fig. 6.14(a).
3. A segment of the root locus exists on the real axis between $s = 0$ and $s = -4$.
4. Because the number of poles n_p is equal to four, we have four separate loci.
5. The angles of the asymptotes are

$$\phi_A = \frac{(2q+1)}{4} 180^\circ, \qquad q = 0, 1, 2, 3,$$

$$\phi_A = +45^\circ, 135^\circ, 225^\circ, 315^\circ.$$

The center of the asymptotes is

$$\sigma_A = \frac{-4 - 4 - 4}{4} = -3.$$

Then the asymptotes are drawn as shown in Fig. 6.14(a).

6. The breakaway point is estimated by evaluating

$$K = p(s) = -s(s+4)(s+4+j4)(s+4-j4)$$

between $s = -4$ and $s = 0$. We expect the breakaway point to lie between $s = -3$ and $s = -1$ and, therefore, we search for a maximum value of $p(s)$ in that region. The resulting values of $p(s)$ for several values of s are given in

Table 6.2.

$p(s)$	0	51	68.5	80	85	75	0
s	-4.0	-3.0	-2.5	-2.0	-1.5	-1.0	0

Table 6.2. The maximum of $p(s)$ is found to lie at approximately $s = -1.5$ as indicated in the table. A more accurate estimate of the breakaway point is normally not necessary or worthwhile. The breakaway point is then indicated on Fig. 6.14(a).

7. The characteristic equation is rewritten as

$$s(s + 4)(s^2 + 8s + 32) + K = s^4 + 12s^3 + 64s^2 + 128s + K = 0. \quad (6.51)$$

Therefore, the Routh-Hurwitz array is

$$\begin{array}{c|ccc} s^4 & 1 & 64 & K \\ s^3 & 12 & 128 & \\ s^2 & b_1 & K & \\ s & c_1 & & \\ s^0 & K & & \end{array}$$

where

$$b_1 = \frac{12(64) - 128}{12} = 53.33 \quad \text{and} \quad c_1 = \frac{53.33(128) - 12K}{53.33}.$$

Hence the limiting value of gain for stability is $K = 570$ and the roots of the auxiliary equation are

$$53.33s^2 + 570 = 53.33(s^2 + 10.6) = 53.33(s + j3.25)(s - j3.25). \quad (6.52)$$

The points where the locus crosses the imaginary axis are shown in Fig. 6.14(a).

8. The angle of departure at the complex pole p_1 can be estimated by utilizing the angle criterion as follows:

$$\theta_1 + 90° + 90° + \theta_3 = 180°,$$

where θ_3 is the angle subtended by the vector form pole p_3. The angles from the pole at $s = -4$ and $s = -4 - j4$ are each equal to 90°. Since $\theta_3 = 135°$, we find that

$$\theta_1 = -135° = +225°$$

as shown in Fig. 6.14(a).

Utilizing all the information obtained from the eight steps of the root locus method, the complete root locus is plotted by using a protractor to locate points that satisfy the angle criterion. The root locus for this system is shown in Fig.

6.14(b). When complex roots near the origin have a damping ratio of $\zeta = 0.707$, the gain K can be determined graphically as shown in Fig. 6.14(b). The vector lengths to the root location s_1 from the open-loop poles are evaluated and result in a gain at s_1 of

$$K = |s_1|\,|s_1 + 4|\,|s_1 - p_1|\,|s_1 - \hat{p}_1| \tag{6.53}$$
$$= (1.9)(3.0)(3.8)(6.0) = 130.$$

The remaining pair of complex roots occurs at s_2 and $\hat{s}_2$ when $K = 130$. The effect of the complex roots at s_2 and $\hat{s}_2$ on the transient response will be negligible compared to the roots s_1 and $\hat{s}_1$. This fact can be ascertained by considering the damping of the response due to each pair of roots. The damping due to s_1 and $\hat{s}_1$ is

$$e^{-\zeta_1 \omega_n t} = e^{-\sigma_1 t},$$

and the damping factor due to s_2 and $\hat{s}_2$ is

$$e^{-\zeta_2 \omega_{n2} t} = e^{-\sigma_2 t},$$

where σ_2 is approximately five times as large as σ_1. Therefore, the transient response term due to s_2 will decay much more rapidly than the transient response term due to s_1. Thus the response to a unit step input may be written as

$$c(t) = 1 + c_1 e^{-\sigma_1 t} \sin(\omega_1 t + \theta_1) + c_2 e^{-\sigma_2 t} \sin(\omega_2 t + \theta_2) \tag{6.54}$$
$$\cong 1 + c_1 e^{-\sigma_1 t} \sin(\omega_1 t + \theta_1).$$

The complex conjugate roots near the origin of the s-plane relative to the other roots of the closed-loop system are labeled the *dominant roots* of the system because they represent or dominate the transient response. The relative dominance of the roots is determined by the ratio of the real parts of the complex roots and will result in reasonable dominance for ratios exceeding five.

Of course, the dominance of the second term of Eq. (6.54) also depends upon the relative magnitudes of the coefficients c_1 and c_2. These coefficients, which are the residues evaluated at the complex roots, in turn depend upon the location of the zeros in the s-plane. Therefore, the concept of dominant roots is useful for estimating the response of a system but must be used with caution and with a comprehension of the underlying assumptions.

A computer program written in BASIC for the calculation of the root locus is available [4]. The Control System Design Program (CSDP) provides the user with a table of roots for selected values of the parameter K [4] (See Appendix F).

For example, refer back to Example 6.2. After several tries, you should be able to determine that when $K = 1.925$, the breakaway occurs from the real axis with two real roots at $s = -2.6$. In the same way you will find that when $K = 200$, the two roots on the imaginary axis are at $s = \pm j4.82$.

Similarly, for Example 6.4, using CSDP, one can determine that breakaway from the real axis occurs at $K = 84$ when $s = -1.57$. In the same way when $K = 600$ we find that two roots are on the imaginary axis at $s = \pm j3.33$.

6.4 An Example of a Control System Analysis and Design Utilizing the Root Locus Method

The analysis and design of a control system can be accomplished by utilizing the Laplace transform, a signal-flow diagram, the s-plane, and the root locus method. It will be worthwhile at this point in the development to examine a control system and select suitable parameter values based on the root locus method.

An automatic self-balancing scale in which the weighing operation is controlled by the physical balance function through an electrical feedback loop is shown in Fig. 6.15 [5]. The balance is shown in the equilibrium condition, and x is the travel of the counterweight W_c from an unloaded equilibrium condition. The weight to be measured, W, is applied 5 cm from the pivot, and the length of the beam to the viscous damper, l_i is 20 cm. It is desired to accomplish the following items:

1. Select the parameters and the specifications of the feedback system.
2. Obtain a model and signal-flow diagram representing the system.
3. Select the gain K based on a root locus diagram.
4. Determine the dominant mode of response.

The inertia of the beam will be chosen to be equal to 0.05 kg-m^2. We must select a battery voltage that is large enough to provide a reasonable position sensor gain, so let us choose E_{bb} = 24 volts. We will utilize a lead screw of 20 turns/cm and a potentiometer for x equal to 6 cm in length. Accurate balances are required, and

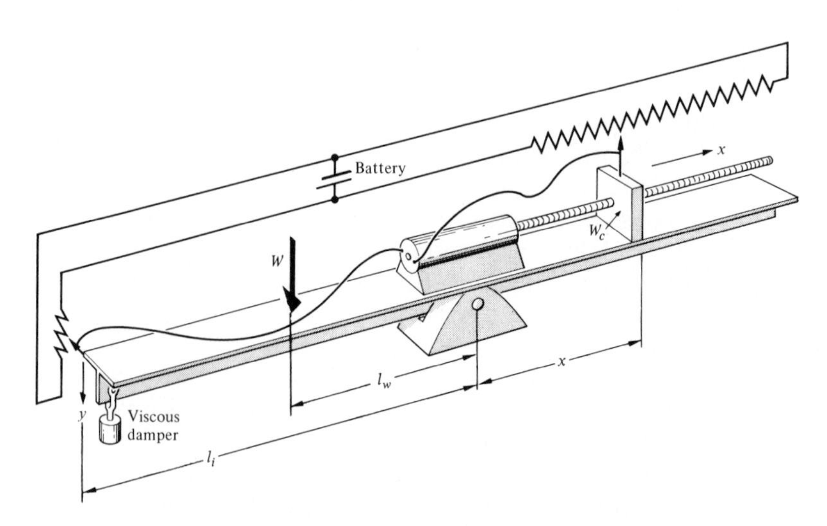

Figure 6.15. An automatic self-balancing scale. (From J. H. Goldberg, *Automatic Controls,* Allyn and Bacon, Boston, 1964, with permission.)

Table 6.3.

$W_c = 2$ N	Lead screw gain $K_s = \dfrac{1}{4000\pi}$ m/rad
$I = 0.05$ kg-m^2	
$l_w = 5$ cm	Input potentiometer gain $K_i = 4800$ v/m
$l_i = 20$ cm	
$f = 10\sqrt{3}$ kg/m/sec	Feedback potentiometer gain $K_f = 400$ v/m

therefore an input potentiometer for y will be chosen to be 0.5 cm in length. A reasonable viscous damper will be chosen with a damping constant $f = 10\sqrt{3}$ kg/m/sec. Finally, a counterweight W_c is chosen so that the expected range of weights W can be balanced. Therefore, in summary, the *parameters* of the system are selected as listed in Table 6.3.

Specifications. A rapid and accurate response resulting in a small steady-state weight measurement error is desired. Therefore we will require the system be at least a type-one system so that a zero measurement error is obtained. An underdamped response to a step change in the measured weight, W, is satisfactory, and therefore a dominant response with $\zeta = 0.5$ will be specified. The settling time of the balance following the introduction of a weight to be measured should be less than 2 sec in order to provide a rapid weight-measuring device. The specifications are summarized in Table 6.4.

The derivation of a model of the electromechanical system may be accomplished by obtaining the equations of motion of the balance. For small deviations from balance, the deviation angle θ is

$$\theta \cong \frac{y}{l_i}. \tag{6.55}$$

The motion of the beam about the pivot is represented by the torque equation:

$$I\frac{d^2\theta}{dt^2} = \Sigma \text{ torques}.$$

Therefore, in terms of the deviation angle, the motion is represented by

$$I\frac{d^2\theta}{dt^2} = l_w W - xW_c - l_i^2 f\frac{d\theta}{dt}. \tag{6.56}$$

Table 6.4. Specifications

Steady-state error	$K_p = \infty$
Underdamped response	$\zeta = 0.5$
Settling time	Less than 2 sec

The input voltage to the motor is

$$v_m(t) = K_i y - K_f x. \tag{6.57}$$

The transfer function of the motor is

$$\frac{\theta_m(s)}{V_m(s)} = \frac{K_m}{s(\tau s + 1)}, \tag{6.58}$$

where τ will be considered to be negligible with respect to the time constants of the overall system, and θ_m is the output shaft rotation. A signal-flow graph representing Eqs. (6.56) through (6.58) is shown in Fig. 6.16. Examining the forward path from W to $X(s)$ we find that the system is a type-one system due to the integration preceding $Y(s)$. Therefore the steady-state error of the system is zero.

The closed-loop transfer function of the system is obtained by utilizing Mason's flow-graph formula and is found to be

$$\frac{X(s)}{W(s)} = \frac{(l_w l_i K_i K_m K_s/Is^3)}{1 + (l_i^2 f/Is) + (K_m K_s K_f/s) + (l_i K_i K_m K_s W_c/Is^3) + (l_i^2 f K_m K_s K_f/Is^2)}, \tag{6.59}$$

where the numerator is the path factor from W to X, the second term in the denominator is the loop L_1, the third term is the loop factor L_2, the fourth term is the loop L_3, and the fifth term is the two nontouching loops $L_1 L_2$. Therefore the closed-loop transfer function is

$$\frac{X(s)}{W(s)} = \frac{l_w l_i K_i K_m K_s}{s(Is + l_i^2 f)(s + K_m K_s K_f) + W_c K_m K_s K_i l_i}. \tag{6.60}$$

The steady-state gain of the system is then

$$\lim_{t\to\infty}\left(\frac{x(t)}{|W|}\right) = \lim_{s\to 0} s\left(\frac{X(s)}{W(s)}\right) = \frac{l_w}{W_c} = 2.5 \text{ cm/kg} \tag{6.61}$$

when $W(s) = |W|/s$. In order to obtain the root locus as a function of the motor constant K_m, we substitute the selected parameters into the characteristic equa-

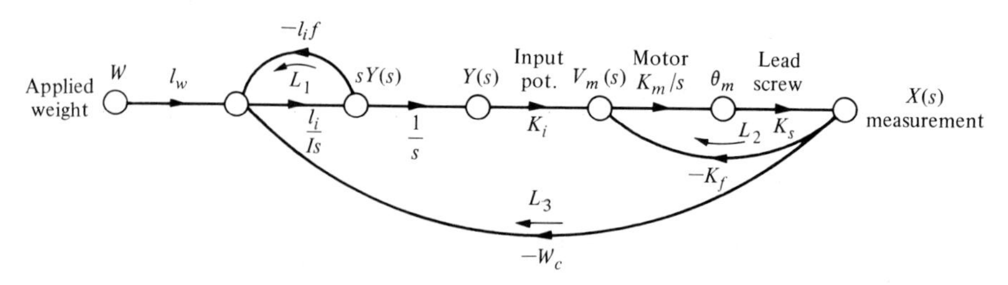

Figure 6.16. Signal-flow graph model of the automatic self-balancing scale.

tion, which is the denominator of Eq. (6.60). Therefore we obtain the following characteristic equation:

$$s(s + 8\sqrt{3})\left(s + \frac{K_m}{10\pi}\right) + \frac{96K_m}{10\pi} = 0. \tag{6.62}$$

Rewriting the characteristic equation in root locus form, we first isolate K_m as follows:

$$s^2(s + 8\sqrt{3}) + s(s + 8\sqrt{3})\frac{K_m}{10\pi} + \frac{96K_m}{10\pi} = 0. \tag{6.63}$$

Then, rewriting Eq. (6.63) in root locus form, we have

$$\begin{aligned} 1 + KP(s) &= 1 + \frac{(K_m/10\pi)[s(s + 8\sqrt{3}) + 96]}{s^2(s + 8\sqrt{3})} = 0 \\ &= 1 + \frac{(K_m/10\pi)(s + 6.93 + j6.93)(s + 6.93 - j6.93)}{s^2(s + 8\sqrt{3})}. \end{aligned} \tag{6.64}$$

The root locus as K_m varies is shown in Fig. 6.17. The dominant roots can be placed at $\zeta = 0.5$ when $K = 25.3 = K_m/10\pi$. In order to achieve this gain,

$$K_m = 795 \frac{\text{rad/sec}}{\text{volt}} = 7600 \frac{\text{rpm}}{\text{volt}}, \tag{6.65}$$

an amplifier would be required to provide a portion of the required gain. The real part of the dominant roots is greater than four and therefore the settling time, $4/\sigma$, is less than 1 sec, and the settling time requirement is satisfied. The third root of the characteristic equation is a real root at $s = -30.2$, and the underdamped

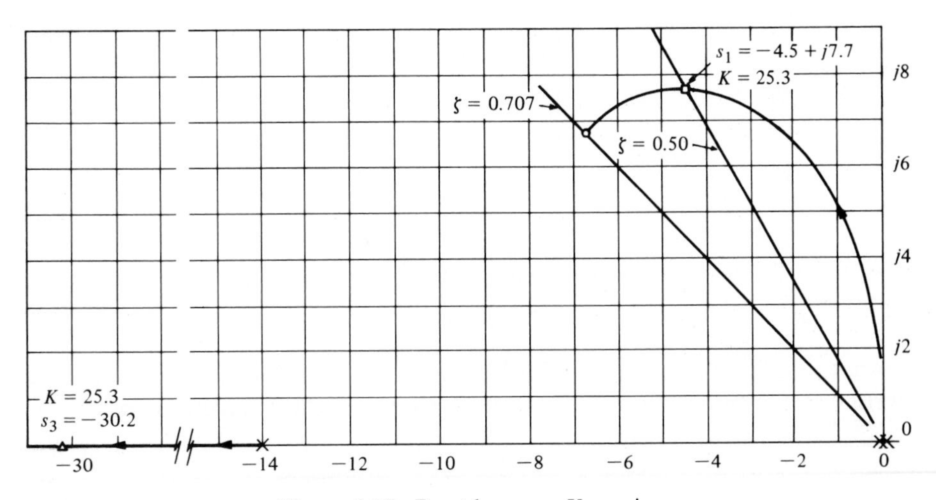

Figure 6.17. Root locus as K_m varies.

roots clearly dominate the response. Therefore the system has been analyzed by the root locus method and a suitable design for the parameter K_m has been achieved. The efficiency of the s-plane and root locus methods is clearly demonstrated by this example.

6.5 Parameter Design by the Root Locus Method

The original development of the root locus method was concerned with the determination of the locus of roots of the characteristic equation as the system gain, K, is varied from zero to infinity. However, as we have seen, the effect of other system parameters may be readily investigated by using the root locus method. Fundamentally, the root locus method is concerned with a characteristic equation (Eq. 6.23), which may be written as

$$1 + F(s) = 0. \tag{6.66}$$

Then the standard root locus method we have studied may be applied. The question arises: How do we investigate the effect of two parameters, α and β? It appears that the root locus method is a single-parameter method; however, fortunately it can be readily extended to the investigation of two or more parameters. This method of *parameter design* uses the root locus approach to select the values of the parameters.

The characteristic equation of a dynamic system may be written as

$$a_n s^n + a_{n-1}s^{n-1} + \cdots + a_1 s + a_0 = 0. \tag{6.67}$$

Clearly, the effect of the coefficient a_1 may be ascertained from the root locus equation

$$1 + \frac{a_1 s}{a_n s^n + a_{n-1}s^{n-1} + \cdots + a_2 s^2 + a_0} = 0. \tag{6.68}$$

If the parameter of interest, α, does not appear solely as a coefficient, the parameter is isolated as

$$a_n s^n + a_{n-1}s^{n-1} + \cdots + (a_{n-q} - \alpha)s^{n-q} + \alpha s^{n-q} + \cdots + a_1 s + a_0 = 0. \tag{6.69}$$

Then, for example, a third-order equation of interest might be

$$s^3 + (3 + \alpha)s^2 + 3s + 6 = 0. \tag{6.70}$$

In order to ascertain the effect of the parameter α, we isolate the parameter and rewrite the equation in root locus form as shown in the following steps:

$$s^3 + 3s^2 + \alpha s^2 + 3s + 6 = 0, \tag{6.71}$$

$$1 + \frac{\alpha s^2}{s^3 + 3s^2 + 3s + 6} = 0. \tag{6.72}$$

Then, in order to determine the effect of two parameters, we must repeat the root locus approach twice. Thus, for a characteristic equation with two variable parameters, α and β, we have

$$a_n s^n + a_{n-1}s^{n-1} + \cdots + (a_{n-q} - \alpha)s^{n-q} + \alpha s^{n-q} + \cdots$$
$$+ (a_{n-r} - \beta)s^{n-r} + \beta s^{n-r} + \cdots + a_1 s + a_0 = 0. \quad (6.73)$$

The two variable parameters have been isolated and the effect of α will be determined, followed by the determination of the effect of β. For example, for a certain third-order characteristic equation with α and β as parameters, we obtain

$$s^3 + s^2 + \beta s + \alpha = 0. \quad (6.74)$$

In this particular case, the parameters appear as the coefficients of the characteristic equation. The effect of varying β from zero to infinity is determined from the root locus equation

$$1 + \frac{\beta s}{s^3 + s^2 + \alpha} = 0. \quad (6.75)$$

One notes that the denominator of Eq. (6.75) is the characteristic equation of the system with $\beta = 0$. Therefore one first evaluates the effect of varying α from zero to infinity by utilizing the equation

$$s^3 + s^2 + \alpha = 0,$$

rewritten as

$$1 + \frac{\alpha}{s^2(s + 1)} = 0, \quad (6.76)$$

where β has been set equal to zero in Eq. (6.74). Then, upon evaluating the effect of α, a value of α is selected and used with Eq. (6.75) to evaluate the effect of β. This two-step method of evaluating the effect of α and then β may be carried out as a two-root locus procedure. First we obtain a locus of roots as α varies, and we select a suitable value of α; the results are satisfactory root locations. Then we obtain the root locus for β by noting that the poles of Eq. (6.75) are the roots evaluated by the root locus of Eq. (6.76). A limitation of this approach is that one will not always be able to obtain a characteristic equation that is linear in the parameter under consideration, for example, α.

In order to illustrate this approach effectively, let us obtain the root locus for α and then β for Eq. (6.74). A sketch of the root locus as α varies for Eq. (6.76) is shown in Fig. 6.18(a), where the roots for two values of gain α are shown. If the gain α is selected as α_1, then the resultant roots of Eq. (6.76) become the poles of Eq. (6.75). The root locus of Eq. (6.75) as β varies is shown in Fig. 6.18(b), and a suitable β can be selected on the basis of the desired root locations.

Using the root locus method, we will further illustrate this parameter design approach by a specific design example.

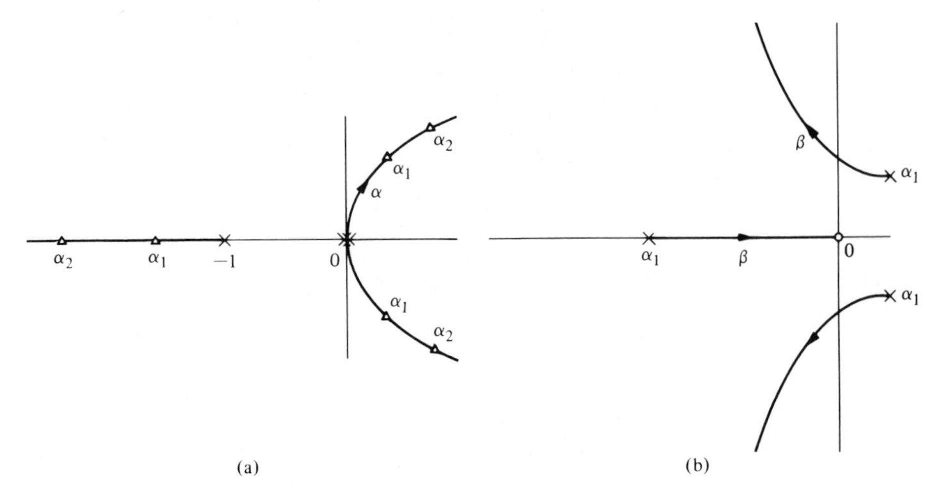

Figure 6.18. Root loci as a function of α and β.

■ Example 6.5 Disk drive control

A feedback control system is to be designed to satisfy the following specifications:

1. Steady-state error for a ramp input $\leq$ 35% of input slope
2. Damping ratio of dominant roots ≥ 0.707
3. Settling time of the system ≤ 3 sec

The structure of the feedback control system is shown in Fig. 6.19, where the amplifier gain K_1 and the derivative feedback gain K_2 are to be selected. The steady-state error specification can be written as follows:

$$e_{ss} = \lim_{t\to\infty} e(t) = \lim_{s\to 0} sE(s) = \lim_{s\to 0} \frac{s(|R|/s^2)}{1 + G_2(s)}, \tag{6.77}$$

where $G_2(s) = G(s)/(1 + G(s)H_1(s))$. Therefore the steady-state error requirement is

$$\frac{e_{ss}}{|R|} = \frac{2 + K_1K_2}{K_1} \leq 0.35. \tag{6.78}$$

Thus we will select a small value of K_2 to achieve a low value of steady-state error. The damping ratio specification requires that the roots of the closed-loop system be below the line at 45° in the left-hand s-plane. The settling time specification can be rewritten in terms of the real part of the dominant roots as

$$T_s = \frac{4}{\sigma} \leq 3 \text{ sec}. \tag{6.79}$$

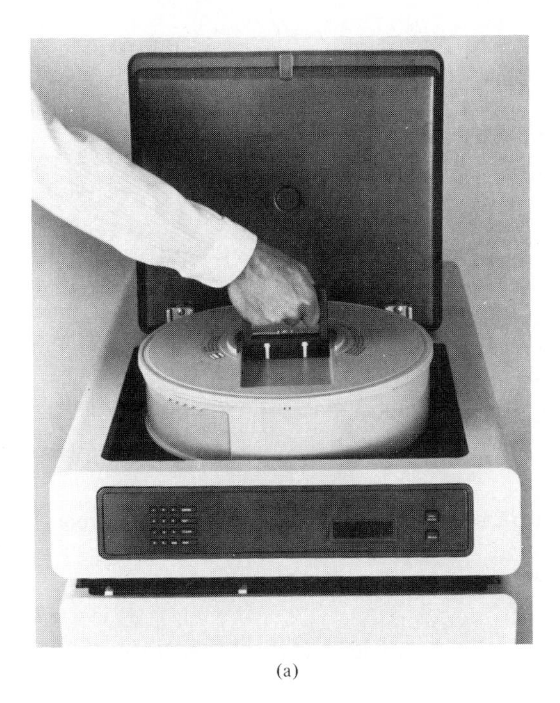

(a)

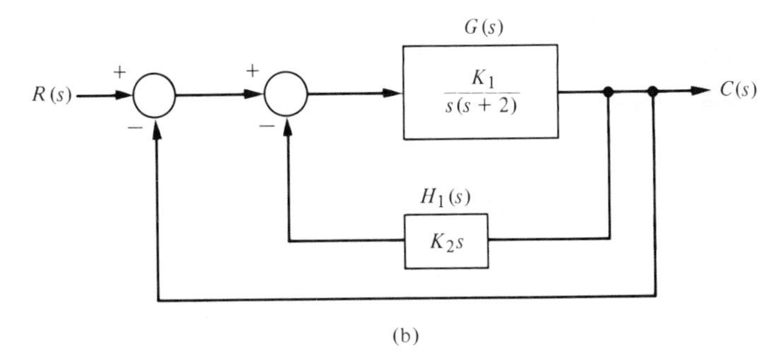

(b)

Figure 6.19. (a) The HP 9795 disk drive and data pack designed to locate stored data in minimum access time [13]. (Courtesy of Hewlett-Packard Co.) (b) Block diagram. © Copyright 1984, Hewlett-Packard Company. Reproduced with permission.

Therefore it is necessary that $\sigma \geq 4/3$ and this area in the left-hand s-plane is indicated along with the ζ-requirement in Fig. 6.20. In order to satisfy the specifications, all the roots must lie within the shaded area of the left-hand plane.

The parameters to be selected are $\alpha = K_1$ and $\beta = K_2K_1$. The characteristic equation is

$$1 + GH(s) = s^2 + 2s + \beta s + \alpha = 0. \tag{6.80}$$

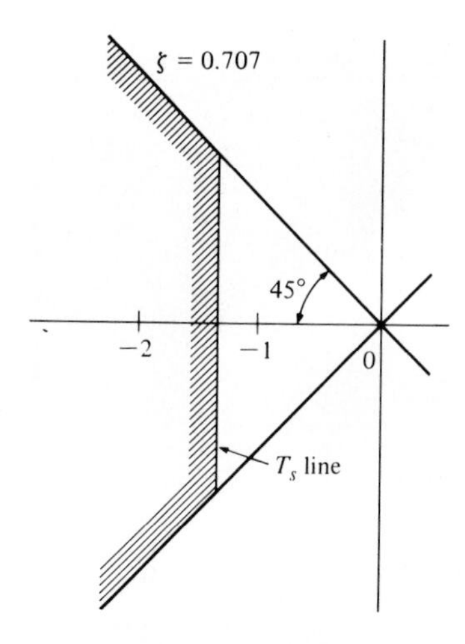

Figure 6.20. A region in the s-plane for desired root location.

The locus of roots as $\alpha = K_1$ varies is determined from the following equation:

$$1 + \frac{\alpha}{s(s+2)} = 0. \tag{6.81}$$

(See Fig. 6.21a.) For a gain of $K_1 = \alpha = 20$, the roots are indicated on the locus. Then the effect of varying $\beta = 20K_2$ is determined from the locus equation

$$1 + \frac{\beta s}{s^2 + 2s + \alpha} = 0, \tag{6.82}$$

where the poles of this root locus are the roots of the locus of Fig. 6.21(a). The root locus for Eq. (6.82) is shown in Fig. 6.21(b) and roots with $\zeta = 0.707$ are obtained when $\beta = 4.3 = 20K_2$ or when $K_2 = 0.215$. The real part of these roots is $\sigma = 3.15$, and therefore the settling time is equal to 1.27 sec, which is considerably less than the specification of 3 sec.

The root locus method may be extended to more than two parameters by extending the number of steps in the method outlined in this section. Furthermore, a family of root loci can be generated for two parameters in order to determine the total effect of varying two parameters. For example, let us determine the effect of varying α and β of the following characteristic equation:

$$s^3 + 3s^2 + 2s + \beta s + \alpha = 0. \tag{6.83}$$

The root locus equation as a function of α is

$$1 + \frac{\alpha}{s(s+1)(s+2)} = 0. \tag{6.84}$$

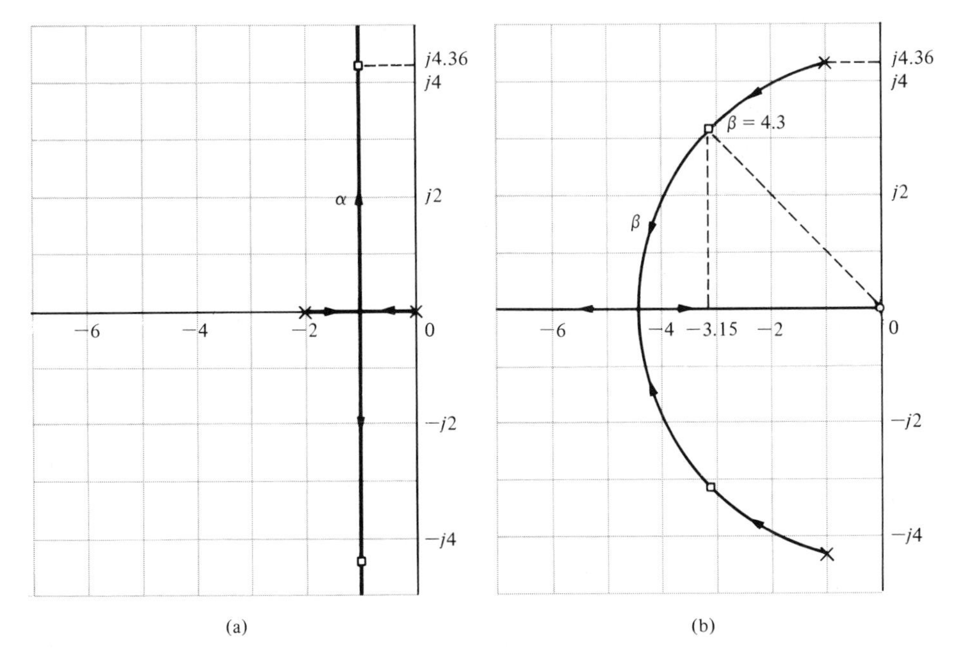

Figure 6.21. Root loci as a function of α and β.

The root locus as a function of β is

$$1 + \frac{\beta s}{s^3 + 3s^2 + 2s + \alpha} = 0. \tag{6.85}$$

The root locus for Eq. (6.84) as a function of α is shown in Fig. 6.22 (unbroken lines). The roots of this locus, indicated by a cross, become the poles for the locus of Eq. (6.85). Then the locus of Eq. (6.85) is continued on Fig. 6.22 (dotted lines), where the locus for β is shown for several selected values of α. This family of loci, often called root contours, illustrates the effect of α and β on the roots of the characteristic equation of a system [3].

6.6 Sensitivity and the Root Locus

One of the prime reasons for the utilization of negative feedback in control systems is to reduce the effect of parameter variations. The effect of parameter variations, as we found in Section 3.2, can be described by a measure of the *sensitivity* of the system performance to specific parameter changes. In Section 3.2, we defined the *logarithmic sensitivity* originally suggested by Bode as

$$S_k^T = \frac{d \ln T}{d \ln k} = \frac{\partial T/T}{\partial k/k}, \tag{6.86}$$

where the system transfer function is $T(s)$ and the parameter of interest is k.

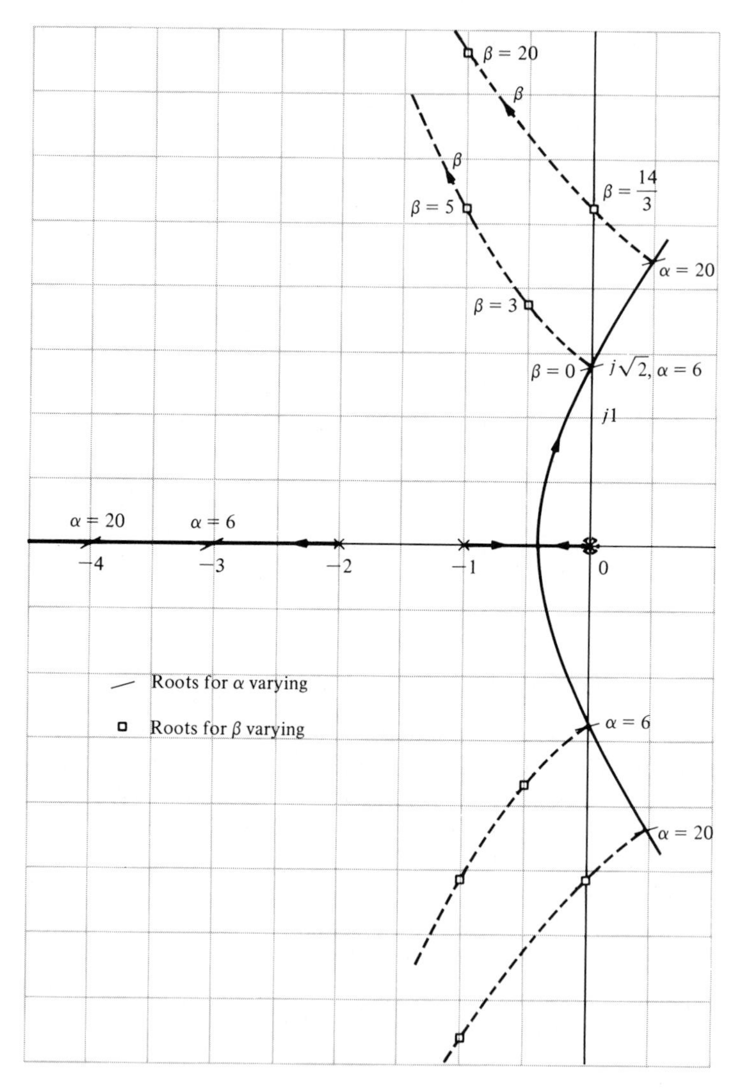

Figure 6.22. Two-parameter root locus.

In recent years, with the increased utilization of the pole–zero (s-plane) approach, it has become useful to define a measure of sensitivity in terms of the positions of the roots of the characteristic equation [6,7]. Because the roots of the characteristic equation represent the dominant modes of transient response, the effect of parameter variations on the position of the roots is an important and useful measure of the sensitivity. The *root sensitivity* of a system $T(s)$ can be defined as

$$S_k^{r_i} = \frac{\partial r_i}{\partial \ln k} = \frac{\partial r_i}{\partial k/k}, \tag{6.87}$$

where r_i equals the ith root of the system so that

$$T(s) = \frac{K_1 \prod_{j=1}^{m} (s + Z_j)}{\prod_{i=1}^{n} (s + r_i)}, \tag{6.88}$$

and k is the parameter. The root sensitivity relates the changes in the location of the root in the s-plane to the change in the parameter. The root sensitivity is related to the Bode logarithmic sensitivity by the relation

$$S_k^T = \frac{\partial \ln K_1}{\partial \ln k} - \sum_{i=1}^{n} \frac{\partial r_i}{\partial \ln k} \cdot \frac{1}{(s + r_i)} \tag{6.89}$$

when the zeros of $T(s)$ are independent of the parameter k so that

$$\frac{\partial Z_j}{\partial \ln k} = 0.$$

This logarithmic sensitivity can be readily obtained by determining the derivative of $T(s)$, Eq. (6.88), with respect to k. For the particular case when the gain of the system is independent of the parameter k, we have

$$S_k^T = -\sum_{i=1}^{n} S_k^{r_i} \cdot \frac{1}{(s + r_i)}, \tag{6.90}$$

and the two sensitivity measures are directly related.

The evaluation of the root sensitivity for a control system can be readily accomplished by utilizing the root locus methods of the preceding section. The root sensitivity $S_k^{r_i}$ may be evaluated at root r_i by examining the root contours for the parameter k. We can change k by a small, finite amount Δk and evaluate the new, modified root $(r_i + \Delta r_i)$ at $k + \Delta k$. Then, using Eq. (6.87) we have

$$S_k^{r_i} = \frac{\Delta r_i}{\Delta k / k} \tag{6.91}$$

An example will illustrate the process of evaluating the root sensitivity.

■ Example 6.6 Root sensitivity of a control system

The characteristic equation of the feedback control system shown in Fig. 6.23 is

$$1 + \frac{K}{s(s + \beta)} = 0$$

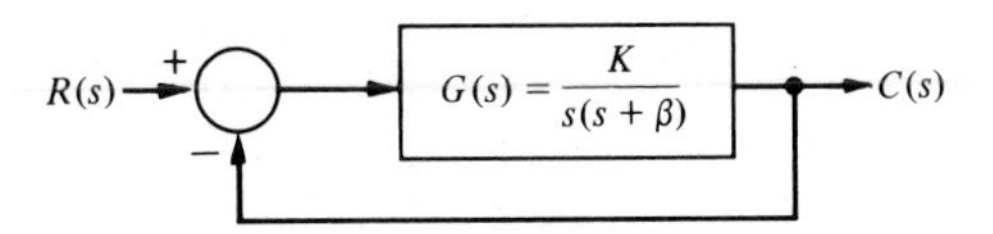

Figure 6.23. A feedback control system.

or, alternatively,

$$s^2 + \beta s + K = 0. \tag{6.92}$$

The gain K will be considered to be the parameter α. Then the effect of a change in each parameter can be determined by utilizing the relations

$$\alpha = \alpha_0 \pm \Delta\alpha$$

$$\beta = \beta_0 \pm \Delta\beta,$$

where α_0 and β_0 are the nominal or desired values for the parameters α and β, respectively. We shall consider the case when the nominal pole value is $\beta_0 = 1$ and the desired gain is $\alpha_0 = K = 0.5$. Then the root locus as a function of $\alpha = K$ can be obtained by utilizing the root locus equation

$$1 + \frac{K}{s(s + \beta_0)} = 1 + \frac{K}{s(s + 1)} = 0 \tag{6.93}$$

as shown in Fig. 6.24. The nominal value of gain $K = \alpha_0 = 0.5$ results in two complex roots, $r_1 = -0.5 + j0.5$ and $r_2 = \hat{r}_1$, as shown in Fig. 6.24. In order to evaluate the effect of unavoidable changes in the gain, the characteristic equation with $\alpha = \alpha_0 \pm \Delta\alpha$ becomes

$$s^2 + s + \alpha_0 \pm \Delta\alpha = s^2 + s + 0.5 \pm \Delta\alpha$$

or

$$1 + \frac{\pm\Delta\alpha}{s^2 + s + 0.5} = 1 + \frac{\pm\Delta\alpha}{(s + r_1)(s + \hat{r}_1)} = 0. \tag{6.94}$$

Therefore the effect of changes in the gain can be evaluated from the root locus of Fig. 6.24. For a 20% change in α, we have $\pm\Delta\alpha = \pm 0.1$. The root locations

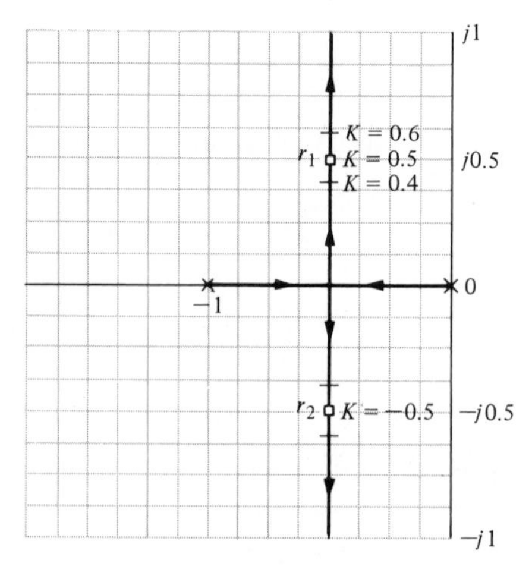

Figure 6.24. The root locus for K.

for a gain $\alpha = 0.4$ and $\alpha = 0.6$ are readily determined by root locus methods, and the root locations for $\pm\Delta\alpha = \pm 0.1$ are shown on Fig. 6.24. When $\alpha = K = 0.6$, the root in the second quadrant of the s-plane is

$$r_1 + \Delta r_1 = -0.5 + j0.59,$$

and the change in the root is $\Delta r_1 = +j0.09$. When $\alpha = K = 0.4$, the root in the second quadrant is

$$r_1 + \Delta r_1 = -0.5 + j0.387,$$

and the change in the root is $\Delta r = -j0.11$. Thus the root sensitivity for r_1 is

$$S^{r_i}_{+\Delta K} = S^{r_i}_{K+} = \frac{\Delta r_1}{\Delta K/K} = \frac{+j0.09}{+0.2} = j0.45 = 0.45\angle{+90^\circ} \tag{6.95}$$

for positive changes of gain. For negative increments of gain, the sensitivity is

$$S^{r_1}_{-\Delta K} = S^{r_i}_{K-} = \frac{\Delta r_1}{\Delta K/K} = \frac{-j0.11}{+0.2} = -j0.55 = 0.55\angle{-90^\circ}.$$

Of course, for infinitesimally small changes in the parameter ∂K, the sensitivity will be equal for negative or positive increments in K. The angle of the root sensitivity indicates the direction the root would move as the parameter varies. The angle of movement for $+\Delta\alpha$ is always 180° minus the angle of movement for $-\Delta\alpha$ at the point $\alpha = \alpha_0$.

The pole β varies due to environmental changes, and it may be represented by $\beta = \beta_0 + \Delta\beta$, where $\beta_0 = 1$. Then the effect of variation of the poles is represented by the characteristic equation

$$s^2 + s + \Delta\beta s + K = 0,$$

or, in root locus form, we have

$$1 + \frac{\Delta\beta s}{s^2 + s + K} = 0. \tag{6.96}$$

Again, the denominator of the second term is the unchanged characteristic equation when $\Delta\beta = 0$. The root locus for the unchanged system ($\Delta\beta = 0$) is shown in Fig. 6.24 as a function of K. For a design specification requiring $\zeta = 0.707$, the complex roots lie at

$$r_1 = -0.5 + j0.5 \quad \text{and} \quad r_2 = \hat{r}_1 = -0.5 - j0.5.$$

Then, because the roots are complex conjugates, the root sensitivity for r_1 is the conjugate of the root sensitivity for $\hat{r}_1 = r_2$. Using the parameter root locus techniques discussed in the preceding section, we obtain the root locus for $\Delta\beta$ as shown in Fig. 6.25. We are normally interested in the effect of a variation for the parameter so that $\beta = \beta_0 \pm \Delta\beta$, for which the locus as $\Delta\beta$ decreases is obtained from the root locus equation

$$1 + \frac{-(\Delta\beta)s}{s^2 + s + K} = 0. \tag{6.97}$$

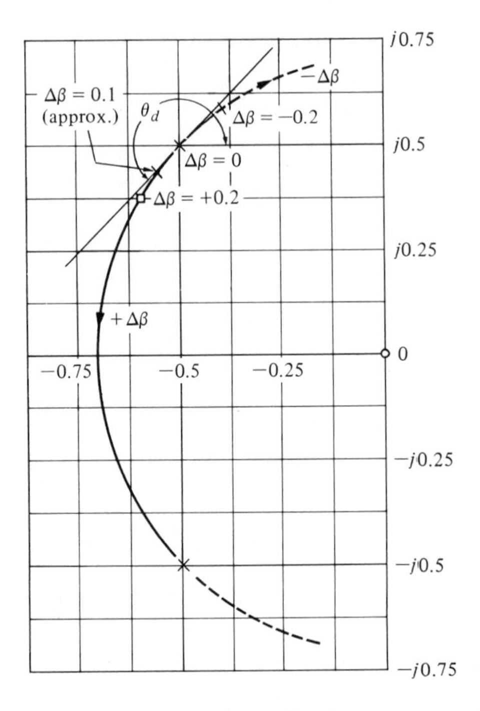

Figure 6.25. The root locus for the parameter β.

Examining Eq. (6.97), we note that the equation is of the form

$$1 - kP(s) = 0.$$

Comparing this equation with Eq. (6.24) (Section 6.3), we find that the sign preceding the gain k is negative in this case. In a manner similar to the development of the root locus method in Section 6.3, we require that the root locus satisfy the equations

$$|kP(s)| = 1, \underline{/P(s)} = 0^\circ \pm q360^\circ, \tag{6.98}$$

where $q = 0, 1, 2, \ldots$. The locus of roots follows a zero-degree locus (Eq. 6.98) in contrast with the 180° locus considered previously. However, the root locus rules of Section 6.3 may be altered to account for the zero-degree phase angle requirement, and then the root locus may be obtained as in the preceding sections. Therefore, in order to obtain the effect of reducing β, one determines the zero-degree locus in contrast to the 180° locus as shown by a dotted locus in Fig. 6.25. Therefore, to find the effect of a 20% change of the parameter β, we evaluate the new roots for $\pm\Delta\beta = \pm 0.20$ as shown in Fig. 6.25. The root sensitivity is readily evaluated graphically and, for a positive change in β, is

$$S_{\beta+}^{r_1} = \frac{\Delta r_1}{(\Delta\beta/\beta)} = \frac{0.16\underline{/-131^\circ}}{0.20} = 0.80\underline{/-131^\circ}. \tag{6.99}$$

The root sensitivity for a negative change in β is

$$S_{\beta-}^{r_1} = \frac{\Delta r_1}{(\Delta\beta/\beta)} = \frac{0.125\angle 38^\circ}{0.20} = 0.625\angle +38^\circ . \tag{6.100}$$

As the percentage change $(\Delta\beta/\beta)$ decreases, the sensitivity measures, $S_{\beta+}^{r_1}$ and $S_{\beta-}^{r_1}$, will approach equality in magnitude and a difference in angle of 180°. Thus for small changes when $\Delta\beta/\beta \leq 0.10$, the sensitivity measures are related as

$$|S_{\beta+}^{r_1}| = |S_{\beta-}^{r_1}| \tag{6.101}$$

and

$$\angle S_{\beta+}^{r} = 180^\circ + \angle S_{\beta-}^{r} . \tag{6.102}$$

Often the desired root sensitivity measure is for small changes in the parameter. When the relative change in the parameter is such that $\Delta\beta/\beta = 0.10$, a root locus approximation is satisfactory. The root locus for Eq. (6.98) when $\Delta\beta$ is varying leaves the pole at $\Delta\beta = 0$ at an angle of departure θ_d. Since θ_d is readily evaluated, one can estimate the increment in the root change by approximating the root locus with the line at θ_d. This approximation is shown in Fig. 6.25 and is accurate for only relatively small changes in $\Delta\beta$. However, the use of this approximation allows the analyst to avoid drawing the complete root locus diagram. Therefore, for Fig. 6.25, the root sensitivity may be evaluated for $\Delta\beta/\beta = 0.10$ along the departure line, and one obtains

$$S_{\beta+}^{r_1} = \frac{0.74\angle -135^\circ}{0.10} = 0.74\angle -135^\circ . \tag{6.103}$$

The root sensitivity measure for a parameter variation is useful for comparing the sensitivity for various design parameters and at different root locations. Comparing Eq. (6.103) for β with Eq. (6.94) for α, we find that the sensitivity for β is greater in magnitude by approximately 50% and the angle for $S_{\beta-}^{r_1}$ indicates that the approach of the root toward the $j\omega$-axis is more sensitive for changes in β. Therefore, the tolerance requirements for β would be more stringent than for α. This information provides the designer with a comparative measure of the required tolerances for each parameter.

■ Example 6.7 Root sensitivity to a parameter

A unity feedback control system has a forward transfer function

$$G(s) = \frac{20.7(s+3)}{s(s+2)(s+\beta)}, \tag{6.104}$$

where $\beta = \beta_0 + \Delta\beta$ and $\beta_0 = 8$. The characteristic equation as a function of $\Delta\beta$ is

$$s(s+2)(s+8+\Delta\beta) + 20.7(s+3) = 0$$

or

$$s(s + 2)(s + 8) + \Delta\beta s(s + 2) + 20.7(s + 3) = 0. \tag{6.105}$$

When $\Delta\beta = 0$, the roots may be determined by the root locus method or the Newton-Raphson method, and thus we evaluate the roots as

$$r_1 = -2.36 + j2.48, \qquad r_2 = \hat{r}_1, \qquad r_3 = -5.27.$$

The root locus for $\Delta\beta$ is determined by using the root locus equation

$$1 + \frac{\Delta\beta s(s + 2)}{(s + r_1)(s + \hat{r}_1)(s + r_3)} = 0. \tag{6.106}$$

The poles and zeros of Eq. (6.106) are shown in Fig. 6.26. The angle of departure at r_1 is evaluated from the angles as follows:

$$\begin{aligned} 180^\circ &= -(\theta_d + 90^\circ + \theta_{p_3}) + (\theta_{z_1} + \theta_{z_2}) \\ &= -(\theta_d + 90^\circ + 40^\circ) + (133^\circ + 98^\circ). \end{aligned} \tag{6.107}$$

Therefore $\theta_d = -80^\circ$ and the locus is approximated near r_1 by the line at an angle of θ_d. For a change of $\Delta r_1 = 0.2\angle -80^\circ$ along the departure line, the $+\Delta\beta$ is evaluated by determining the vector lengths from the poles and zeros. Then we have

$$+\Delta\beta = \frac{4.8(3.75)(0.2)}{(3.25)(2.3)} = 0.48. \tag{6.108}$$

Therefore the sensitivity at r_1 is

$$S_\beta^{r_1} = \frac{\Delta r_1}{\Delta\beta/\beta} = \frac{0.2\angle -80^\circ}{0.48/8} = 3.34\angle -80^\circ, \tag{6.109}$$

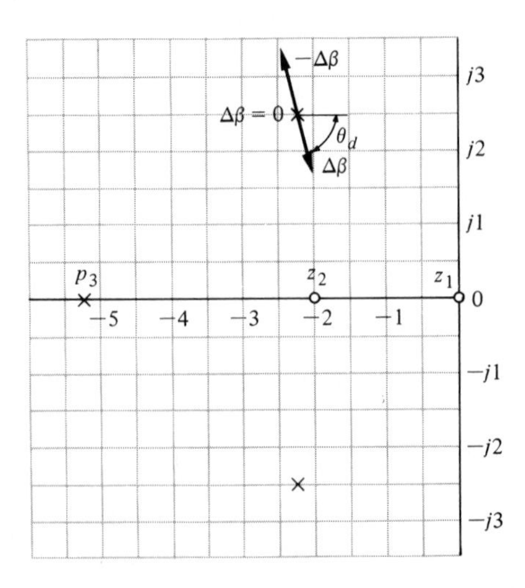

Figure 6.26. Pole and zero diagram for the parameter β.

which indicates that the root is quite sensitive to this 6% change in the parameter β. For comparison, it is worthwhile to determine the sensitivity of the root, r_1, to a change in the zero, $s = -3$. Then the characteristic equation is

$$s(s+2)(s+8) + 20.7(s+3+\Delta\gamma) = 0$$

or

$$1 + \frac{20.7\,\Delta\gamma}{(s+r_1)(s+\hat{r}_1)(s+r_3)} = 0. \tag{6.110}$$

The pole–zero diagram for Eq. (6.110) is shown in Fig. 6.27. The angle of departure at root r_1 is $180° = -(\theta_d + 90° + 40°)$ or

$$\theta_d = +50°. \tag{6.111}$$

For a change of $r_1 = 0.2\,\underline{/+50°}$, the $\Delta\gamma$ is positive, and obtaining the vector lengths, we find

$$|\Delta\gamma| = \frac{5.22(4.18)(0.2)}{20.7} = 0.21. \tag{6.112}$$

Therefore, the sensitivity at r_1 for $+\Delta\gamma$ is

$$S_{\gamma}^{r_1} = \frac{\Delta r_1}{\Delta\gamma/\gamma} = \frac{0.2\underline{/+50°}}{0.21/3} = 2.84\underline{/+50°}\,. \tag{6.113}$$

Thus we find that the magnitude of the root sensitivity for the pole β and the zero γ is approximately equal. However, the sensitivity of the system to the pole can be considered to be less than the sensitivity to the zero because the angle of the sensitivity $S_{\gamma}^{r_1}$ is equal to $+50°$ and the direction of the root change is toward the $j\omega$-axis.

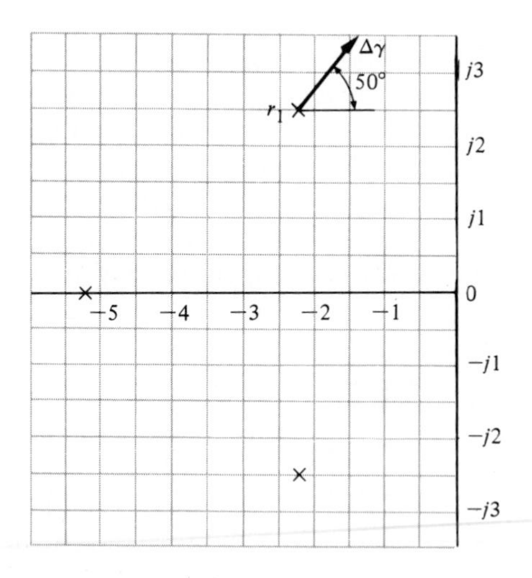

Figure 6.27. Pole–zero diagram for the parameter γ.

Evaluating the root sensitivity in the manner of the preceding paragraphs, we find that for the pole $s = -\delta_0 = -2$ the sensitivity is

$$S_{\delta-}^{r_1} = 2.1\underline{/+27°}. \tag{6.114}$$

Thus, for the parameter δ, the magnitude of the sensitivity is less than for the other parameters, but the direction of the change of the root is more important than for β and γ.

In order to utilize the root sensitivity measure for the analysis and design of control systems, a series of calculations must be performed for various selections of possible root configurations and the zeros and poles of the open-loop transfer function. Therefore the use of the root sensitivity measure as a design technique is somewhat limited by the relatively large number of calculations required and by the lack of an obvious direction for adjusting the parameters in order to provide a minimized or reduced sensitivity. However, the root sensitivity measure can be utilized as an analysis measure, which permits the designer to compare the sensitivity for several system designs based on a suitable method of design. The root sensitivity measure is a useful index of sensitivity of a system to parameter variations expressed in the s-plane. The weakness of the sensitivity measure is that it relies on the ability of the root locations to represent the performance of the system. As we have seen in the preceding chapters, the root locations represent the performance quite adequately for many systems, but due consideration must be given to the location of the zeros of the closed-loop transfer function and the dominancy of the pertinent roots. The root sensitivity measure is a suitable measure of system performance sensitivity and can be used reliably for system analysis and design.

6.7 Design Example: Laser Manipulator Control System

Lasers can be used to drill the hip socket for appropriate insertion of an artificial hip joint. The use of lasers for surgery requires high accuracy for position and velocity response. Let us consider the system shown in Fig. 6.28, which uses a dc motor manipulator for the laser. The amplifier gain K must be adjusted so that the steady-state error for a ramp input, $r(t) = At$ (where $A = 1$ mm/s), is less than or equal to 0.1 mm while a stable response is maintained.

In order to obtain the steady-state error required and a good response, we

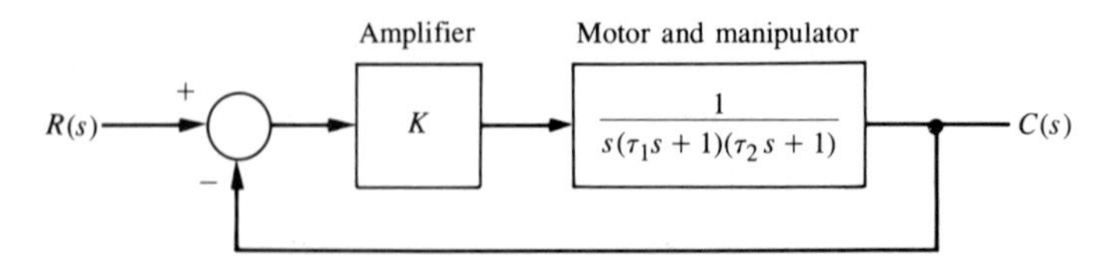

Figure 6.28. Laser manipulator control system.

select a motor with a field time constant $\tau_1 = 0.1s$ and a motor plus load time constant of $\tau_2 = 0.2s$. We then have

$$T(s) = \frac{KG(s)}{1 + KG(s)}$$
$$= \frac{K}{s(\tau_1 s + 1)(\tau_2 s + 1) + K} \tag{6.115}$$
$$= \frac{K}{0.02s^3 + 0.3s^2 + s + K}$$
$$= \frac{50K}{s^3 + 15s^2 + 50s + 50K}.$$

The steady state error for a ramp, $R(s) = A/s^2$, from Eq. (4.31) is

$$e_{ss} = \frac{A}{K_v} = \frac{A}{K}.$$

Since we desire $e_{ss} = 0.1$ mm (or less) and A $= 1$ mm, we require $K = 10$ (or greater).

In order to ensure a stable system, we obtain the characteristic equation from Eq. (6.115) as

$$s^3 + 15s^2 + 50s + 50K = 0.$$

Establishing the Routh-Hurwitz array, we have

$$\begin{array}{c|cc} s^3 & 1 & 50 \\ s^2 & 15 & 50K \\ s^1 & b_1 & 0 \\ s^0 & 50\,K & \end{array}$$

where

$$b_1 = \frac{750 - 50K}{15}.$$

Therefore the system is stable for

$$0 \leq K \leq 15.$$

Then using $K = 10$, where the system is stable, we examine the root locus for $K > 0$. Since there are three loci and the centroid $\sigma = -5$, we obtain the root locus shown in Fig. 6.29. The breakaway point is $s = -2.11$ and the roots at $K = 10$ are $r_2 = -13.98$, $r_1 = -0.51 + j5.96$, and $\hat{r}_1$. The ζ of the complex roots is 0.085 and $\zeta\omega_n = 0.51$. Thus, assuming that the complex roots are dominant, we expect (using Eq. 4.16) an overshoot of 76% and a settling time of

$$T_s = \frac{4}{\zeta\omega_n}$$
$$= \frac{4}{0.51} = 7.8s.$$

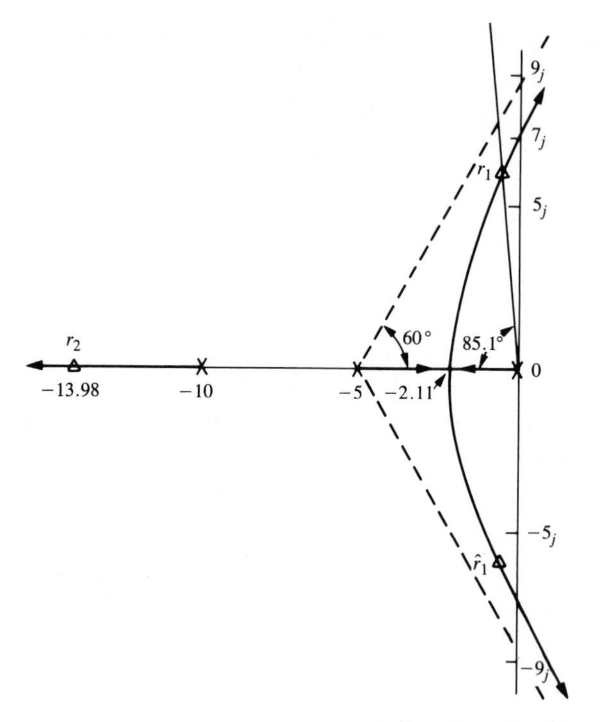

Figure 6.29. Root locus for a laser control system.

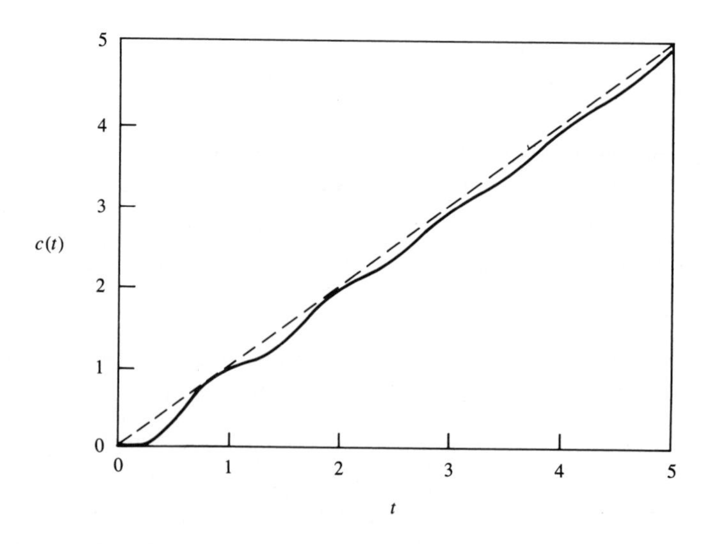

Figure 6.30. The response to a ramp input for a laser control system.

Plotting the actual system response, we find that the overshoot is 72% and the settling time is 7.9 s. Thus the complex roots are essentially dominant. The system response to a step input is highly oscillatory and cannot be tolerated for laser surgery. The command signal must be limited to a low velocity ramp signal. The response to a ramp signal is shown in Fig. 6.30.

6.8 Summary

The relative stability and the transient response performance of a closed-loop control system is directly related to the location of the closed-loop roots of the characteristic equation. Therefore we have investigated the movement of the characteristic roots on the s-plane as the system parameters are varied by utilizing the root locus method. The root locus method, a graphical technique, can be used to obtain an approximate sketch in order to analyze the initial design of a system and determine suitable alterations of the system structure and the parameter values. A computer is commonly used to calculate several accurate roots at important points on the locus. A summary of 15 typical root locus diagrams is shown in Table 8.7.

Furthermore, we extended the root locus method for the design of several parameters for a closed-loop control system. Then the sensitivity of the characteristic roots was investigated for undesired parameter variations by defining a root sensitivity measure. It is clear that the root locus method is a powerful and useful approach for the analysis and design of modern control systems and will continue to be one of the most important procedures of control engineering.

Exercises

E6.1. Let us consider a device that consists of a ball rolling on the inside rim of a hoop [20]. This model is similar to the problem of liquid fuel sloshing in a rocket. The hoop is free to rotate about its horizontal principal axis as shown in Fig. E6.1. The angular position of the hoop may be controlled via the torque T applied to the hoop from a torque motor

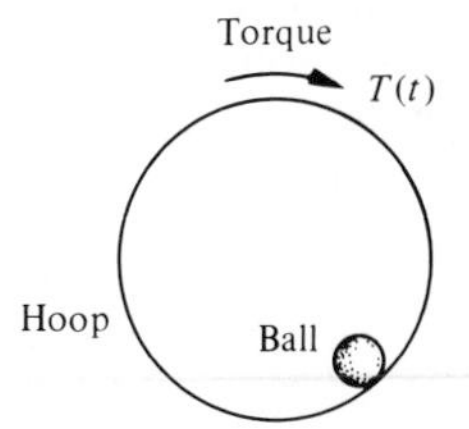

Figure E6.1. Hoop rotated by motor.

attached to the hoop drive shaft. If negative feedback is used, the system characteristic equation is

$$1 + \frac{Ks(s + 4)}{s^2 + 2s + 2} = 0.$$

(a) Sketch the root locus. (b) Find the gain when the roots are both equal. (c) Find these two equal roots. (d) Find the settling time of the system when the roots are equal.

E6.2. A tape recorder has a speed control system so that $H(s) = 1$ with negative feedback and

$$G(s) = \frac{K}{s(s + 2)(s^2 + 4s + 5)}.$$

(a) Draw a root locus for K and show that the dominant roots are $s = -0.35 \pm j0.80$ when $K = 6.5$. (b) For the dominant roots of part (a), calculate the settling time and overshoot for a step input.

E6.3. A control system for an ac induction motor has negative unity feedback and a process [14]

$$G(s) = \frac{K(s^2 + 4s + 8)}{s^2(s + 4)}.$$

It is desired that the dominant roots have a ζ equal to 0.5. Using the root locus, show that $K = 7.35$ is required and the dominant roots are $s = -1.3 \pm j2.2$.

E6.4. Consider a unity feedback system with

$$G(s) = \frac{K(s + 1)}{s^2 + 4s + 5}.$$

(a) Find the angle of departure of the root locus from the complex poles. (b) Find the entry point for the root locus as it enters the real axis.

Answers: $\pm 225°$; -2.4

E6.5. Consider a feedback system with a loop transfer function

$$GH(s) = \frac{K}{(s + 1)(s + 3)(s + 6)}.$$

(a) Find the breakaway point or the real axis. (b) Find the asymptote centroid. (c) Find the value of K at the breakaway point.

E6.6. The United States is planning to have an operating space station in orbit by the mid-1990s. One version of the space station is shown in Fig. E6.6, on the next page. It is critical to keep this station in the proper orientation toward the sun and the earth for generating power and communications. The orientation controller may be represented by a unity feedback system with an actuator and controller,

$$G(s) = \frac{K(s + 30)}{s(s^2 + 20s + 100)}.$$

Sketch the root locus of the system as K increases. Find the value of K that results in an oscillatory response.

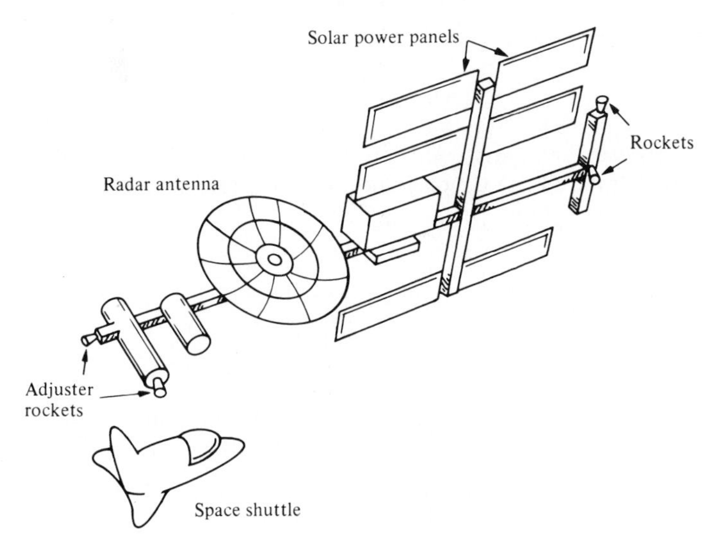

Figure E6.6. Space station.

E6.7. The elevator in a modern office building travels at a top speed of 25 feet per second and is still able to stop within one eighth of an inch of the floor outside. The transfer function of the unity feedback elevator drive is

$$G(s) = \frac{K(s + 10)}{s(s + 1)(s + 20)(s + 50)}.$$

Determine the gain K when the complex roots have a ζ equal to 0.8.

E6.8. Draw the root locus for a unity feedback system (Fig. 6.1) with

$$G(s) = \frac{K(s + 1)}{s^2(s + 9)}.$$

(a) Find the gain when all three roots are real and equal. (b) Find the roots when all the roots are equal as in part (a).

Answers: $K = 27$; $s = -3$

E6.9. The world's largest telescope, completed in 1990, is located in Hawaii [9]. The primary mirror has a diameter of 10 m and consists of a mosaic of 36 hexagonal segments with the orientation of each segment actively controlled. This unity feedback system for the mirror segments (Fig. 6.1) has

$$G(s) = \frac{K}{s(s^2 + s + 1)}.$$

(a) Find the asymptotes and draw them in the s-plane. (b) Find the angle of departure from the complex poles. (c) Determine the gain when two roots lie on the imaginary axis. (d) Sketch the root locus.

E6.10. A unity feedback system (Fig. 6.1) has

$$G(s) = \frac{K(s+2)}{s(s+1)}.$$

(a) Find the breakaway and entry points on the real axis. (b) Find the gain and the roots when the real part of the complex roots is located at -2. (c) Sketch the locus.

Answers: (a) -0.59, -3.41; (b) $K = 3$, $s = -2 \pm j\sqrt{2}$

E6.11. A robot force control system with unity feedback (Fig. 6.1) has a plant [8]

$$G(s) = \frac{K(s+2.5)}{(s^2+2s+2)(s^2+4s+5)}.$$

(a) Using the Control System Design Program or equivalent [4], find the gain K that results in dominant roots with a damping ratio of 0.707. Sketch the locus. (b) Find the actual percent overshoot and peak time for the gain K of part (a).

E6.12. A unity feedback system (Fig. 6.1) has a plant

$$G(s) = \frac{K(s+1)}{s(s^2+4s+8)}.$$

Using the Control System Design Program [4] or equivalent, find (a) The root locus for $K > 0$. (b) The roots when $K = 10$ and 20. (c) The 0–100% rise time, percent overshoot, and settling time of the system for a unit step input when $K = 10$ and 20.

E6.13. A unity feedback system has a process

$$G(s) = \frac{4(s+z)}{s(s+1)(s+3)}.$$

(a) Draw the root locus as z varies from 0 to 100. (b) Using the locus, estimate the percent overshoot and settling time of the system at $z = 0.6$, 2, and 4 for a step input. (c) Use the Control System Design Program to determine the actual overshoot and settling time at $z = 0.6$, 2, and 4.

E6.14. A unity feedback system has the process

$$G(s) = \frac{K(s+10)}{s(s+5)}.$$

(a) Determine the breakaway and entry points of the root locus and sketch the root locus for $K > 0$. (b) Determine the gain K when the two characteristic roots have a ζ of $1/\sqrt{2}$. (c) Calculate the roots.

E6.15. (a) Plot the root locus for

$$GH(s) = \frac{K(s+1)(s+2)}{s^3}.$$

(b) Calculate the range of K for which the system is stable. (c) Predict the steady-state error of the system for a ramp input.

Answers: (b) $K > 2/3$, (c) $e_{ss} = 0$

E6.16. A negative unity feedback system has a plant transfer function

$$G(s) = \frac{Ke^{-sT}}{s + 1},$$

where $T = 0.1$ second. Show that an approximation for the time delay is

$$e^{-sT} \cong \frac{\left(\frac{2}{T} - s\right)}{\left(\frac{2}{T} + s\right)}.$$

Using

$$e^{-0.1s} = \frac{20 - s}{20 + s},$$

obtain the root locus for the system for $K > 0$. Determine the range of K for which the system is stable.

E6.17. A control system as shown in Fig. E6.17 has a plant

$$G(s) = \frac{1}{s(s - 1)}.$$

(a) When $G_c(s) = K$, show that the system is always unstable by sketching the root locus. (b) When

$$G_c(s) = \frac{K(s + 2)}{(s + 10)},$$

sketch the root locus and determine the range of K for which the system is stable. Determine the value of K and the complex roots when two roots lie on the $j\omega$ axis.

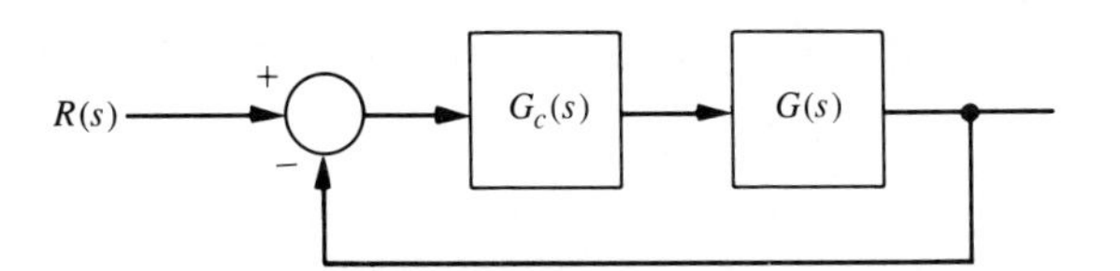

Figure E6.17. Feedback system.

E6.18. A closed-loop negative feedback system is used to control the yaw of the A-6 Intruder attack jet, which was widely used in the Persian Gulf war (Fig. E6.18). When $H(s) = 1$ and

$$G(s) = \frac{K}{s(s + 3)(s^2 + 2s + 2)},$$

determine (a) the root locus breakaway point and (b) the value of the roots on the $j\omega$-axis and the gain required for those roots. Sketch the root locus.

Answer: breakaway: $s = -2.29$
$j\omega$-axis: $s = \pm j1.09$, $K = 8$

Figure E6.18. The A-6 Intruder, a U.S. Navy attack jet.

Problems

P6.1. Draw the root locus for the following open-loop transfer functions of the system shown in Fig. P6.1 when $0<K<\infty$:

(a) $GH(s) = \dfrac{K}{s(s+1)^2}$ (b) $GH(s) = \dfrac{K}{(s^2+s+2)(s+1)}$

(c) $GH(s) = \dfrac{K(s+1)}{s(s+2)(s+3)}$ (d) $GH(s) = \dfrac{K(s^2+4s+8)}{s^2(s+4)}$

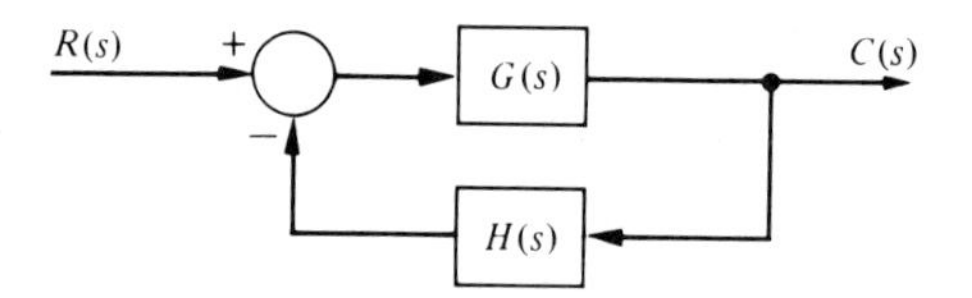

Figure P6.1.

P6.2. The linear model of a phase detector was presented in Problem 5.7. Draw the root locus as a function of the gain $K_v = K_aK$. Determine the value of K_v attained if the complex roots have a damping ratio equal to 0.60 [10].

P6.3. A unity feedback system (Fig. 6.1) has

Find (a) the breakaway point in the real axis and the gain K for this point, (b) the gain and the roots when two roots lie on the imaginary axis, and (c) the roots when $K = 2.5$. (d) Sketch the root locus.

P6.4. The analysis of a large antenna was presented in Problem 3.5. Plot the root locus of the system as $0 < k_a < \infty$. Determine the maximum allowable gain of the amplifier for a stable system.

P6.5. Automatic control of helicopters is necessary because, unlike fixed-wing aircraft, which possess a fair degree of inherent stability, the helicopter is quite unstable. A helicopter control system that utilizes an automatic control loop plus a pilot stick control is shown in Fig. P6.5 [15]. When the pilot is not using the control stick, the switch may be considered to be open. The dynamics of the helicopter are represented by the transfer function

$$G_2(s) = \frac{25(s + 0.03)}{(s + 0.4)(s^2 - 0.36s + 0.16)}.$$

(a) With the pilot control loop open (hands-off control), plot the root locus for the automatic stabilization loop. Determine the gain K_2 that results in a damping for the complex roots equal to $\zeta = 0.707$. (b) For the gain K_2 obtained in part (a), determine the steady-state error due to a wind gust $T_d(s) = 1/s$. (c) With the pilot loop added, draw the root locus as K_1 varies from zero to ∞ when K_2 is set at the value calculated in (a). (d) Recalculate the steady-state error of part (b) when K_1 is equal to a suitable value based on the root locus.

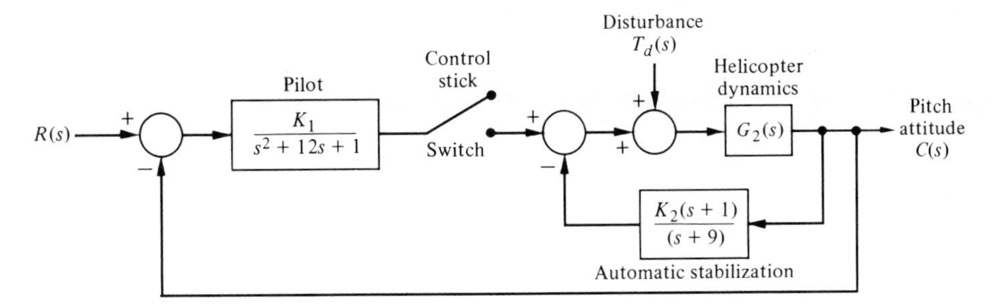

Figure P6.5. Helicopter control.

P6.6. An attitude control system for a satellite vehicle within the earth's atmosphere is shown in Fig. P6.6. The transfer functions of the system are

$$G(s) = \frac{K(s + 0.20)}{(s + 0.90)(s - 0.60)(s - 0.10)},$$

$$G_c(s) = \frac{(s + 1.20 + j1.4)(s + 1.20 - j1.4)}{(s + 4.0)}.$$

(a) Draw the root locus of the system as K varies from 0 to ∞. (b) Determine the gain K which results in a system with a 2.5% settling time less than 12 sec and a damping ratio for the complex roots greater than 0.40.

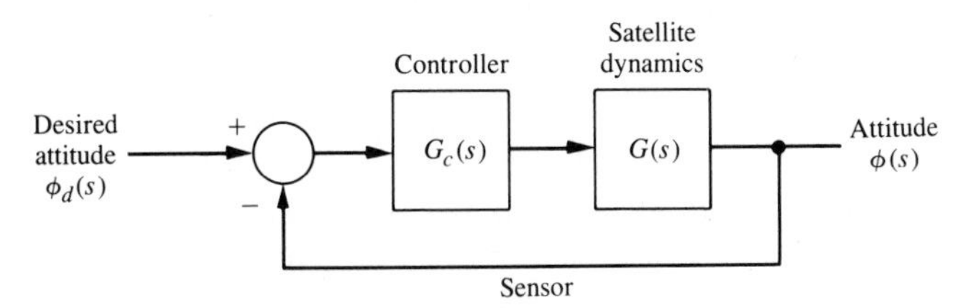

Figure P6.6. Satellite attitude control.

P6.7. The speed control system for an isolated power system is shown in Fig. P6.7. The valve controls the steam flow input to the turbine in order to account for load changes, $\Delta L(s)$, within the power distribution network. The equilibrium speed desired results in a generator frequency equal to 60 cps. The effective rotary inertia, J, is equal to 4000 and the friction constant, f, is equal to 0.75. The steady-state speed regulation factor, R, is represented by the equation $R \cong (\omega_0 - \omega_r)/\Delta L$, where ω_r equals the speed at rated load and ω_0 equals the speed at no load. Clearly, it is desired to obtain a very small R, usually less than 0.10. (a) Using root locus techniques, determine the regulation, R, attainable when the damping ratio of the roots of the system must be greater than 0.60. (b) Verify that the steady-state speed deviation for a load torque change, $\Delta L(s) = \Delta L/s$, is, in fact, approximately equal to $R\Delta L$ when $R \leq 0.1$.

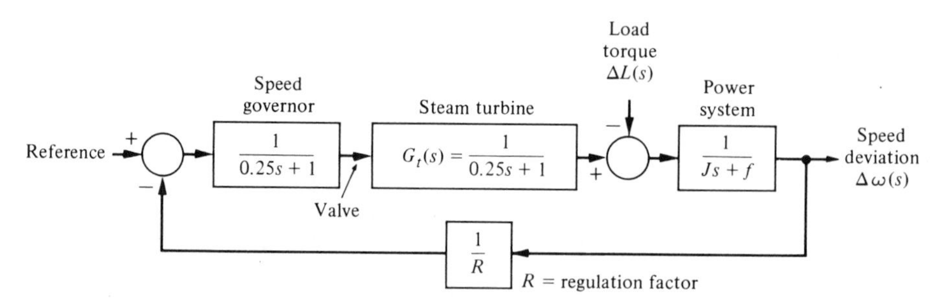

Figure P6.7. Power system control.

P6.8. Reconsider the power control system of the preceding problem when the steam turbine is replaced by a hydroturbine. For hydroturbines, the large inertia of the water used as a source of energy causes a considerably larger time constant. The transfer function of a hydroturbine may be approximated by

$$G_t(s) = \frac{-\tau s + 1}{(\tau/2)s + 1},$$

where $\tau = 1$ sec. With the rest of the system remaining as given in Problem 6.7, repeat parts (a) and (b) of Problem 6.7.

P6.9. The achievement of safe, efficient control of the spacing of automatically controlled guided vehicles is an important part of future use of the vehicles in a manufacturing plant [6]. It is important that the system eliminate the effects of disturbances such as oil on the floor as well as maintain accurate spacing between vehicles on a guideway. The system can be represented by the block diagram of Fig. P6.9, on the next page. The vehicle dynamics can be represented by

$$G(s) = \frac{(s + 0.1)(s^2 + 2s + 289)}{s(s - 0.4)(s + 0.8)(s^2 + 1.45s + 361)}.$$

(a) Neglect the pole of the feedback sensor and draw the root locus of the system. (b) Determine all the roots when the loop gain $K = K_1 K_2 K_4/250$ is equal to 4000.

P6.10. Unlike the present day *Concorde*, a turn-of-the-century supersonic passenger jet would have the range to cross the Pacific in a single hop and the efficiency to make it economic. This new aircraft shown in Fig. P6.10(a) will require the use of temperature resistant, lightweight materials and advanced computer control systems.

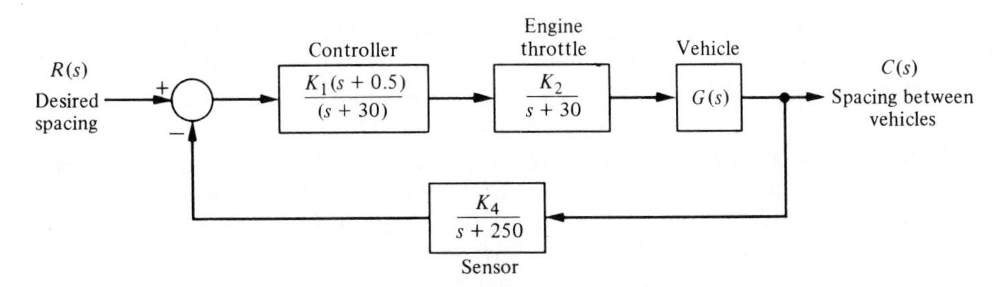

Figure P6.9 Guided vehicle control.

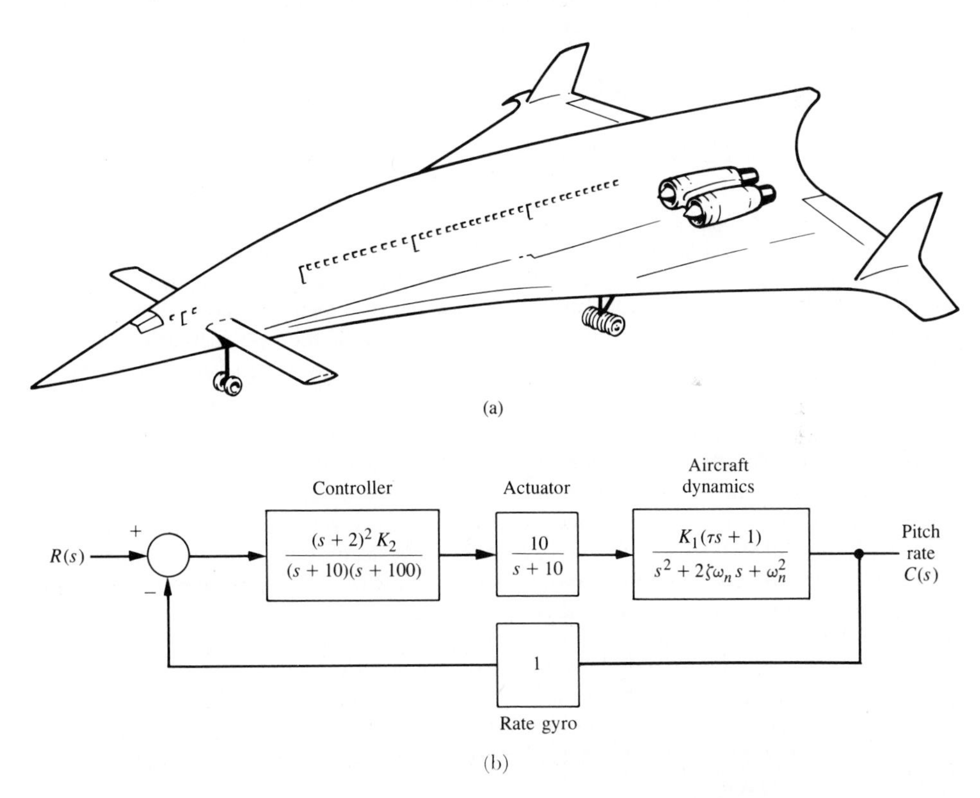

Figure P6.10. (a) A supersonic jet aircraft of the future. (b) Control system.

The plane would carry 300 passengers at three times the speed of sound for up to 7500 miles. The flight control system requires good quality handling and comfortable flying conditions. An automatic flight control system can be designed for SST vehicles. The desired characteristics of the dominant roots of the control system shown in Fig. P6.10(b) have a $\zeta = 0.707$. The characteristics of the aircraft are $\omega_n = 2.5$, $\zeta = 0.30$, and $\tau = 0.1$. The gain factor K_1, however, will vary over the range 0.02 at medium-weight cruise conditions to 0.20 at light-weight descent conditions. (a) Draw the root locus as a function of the loop gain

$K_1 K_2$. (b) Determine the necessary gain K_2 in order to yield roots with $\zeta = 0.707$ when the aircraft is in the medium-cruise condition. (c) With the gain K_2 as found in (b), determine the ζ of the roots when the gain K_1 results from the condition of light-descent.

P6.11. A computer system for a military application requires a high performance magnetic tape transport system [13]. However, the environmental conditions imposed on a military system result in a severe test of control engineering design. A direct-drive dc-motor system for the magnetic tape reel system is shown in Fig. P6.11, on the next page where r equals the reel radius and J equals the reel and rotor inertia. A complete reversal of the tape reel direction is required in 6 msec, and the tape reel must follow a step command in 3 msec or less. The tape is normally operating at a speed of 100 in/sec. The motor and components selected for this system possess the following characteristics:

$$K_b = 0.40 \qquad K_T/LJ = 2.0$$
$$K_p = 1 \qquad r = 0.2$$
$$\tau_1 = \tau_a = 1 \text{ msec} \qquad K_1 = 2.0$$
$$K_2 \text{ is adjustable.}$$

The inertia of the reel and motor rotor is 2.5×10^{-3} when the reel is empty, and 5.0×10^{-3} when the reel is full. A series of photocells is used for an error sensing device. The time constant of the motor is $L/R = 0.5$ msec. (a) Draw the root locus for the system when $K_2 = 10$ and $J = 5.0 \times 10^{-3}$, and $0 < K_a < \infty$. (b) Determine the gain K_a that results in a well-damped system so that the ζ of all the roots is greater than or equal to 0.60. (c) With the K_a determined from part (b), draw a root locus for $0 < K_2 < \infty$.

P6.12. A precision speed control system (Fig. P6.12) on the next page is required for a platform used in gyroscope and inertial system testing where a variety of closely controlled speeds is necessary. A direct-drive dc torque motor system was utilized in order to provide (1) a speed range of 0.01°/sec to 600°/sec, and (2) 0.1% steady-state error maximum for a step input. The direct-drive dc torque motor avoids the use of a gear train with its attendant backlash and friction. Also the direct-drive motor has a high torque capability, high efficiency, and low motor time constants. The motor gain constant is nominally $K_m = 1.8$ but is subject to variations up to 50%. The amplifier gain K_a is normally greater than 10 and subject to a variation of 10%. (a) Determine the minimum loop gain necessary to satisfy the steady-state error requirement. (b) Determine the limiting value of gain for stability. (c) Draw the root locus as K_a varies from 0 to ∞. (d) Determine the roots when $K_a = 50$ and estimate the response to a step input.

P6.13 A unity feedback system (Fig. 6.1) has

$$G(s) = \frac{K}{s(s+2)(s^2+4s+5)}.$$

(a) Find the breakaway points on the real axis and the gain for this point. (b) Find the gain to provide two complex roots nearest the $j\omega$ axis with a damping ratio of 0.707. (c) Are the two roots of part (b) dominant? (d) Determine the settling time of the system when the gain of part (b) is used.

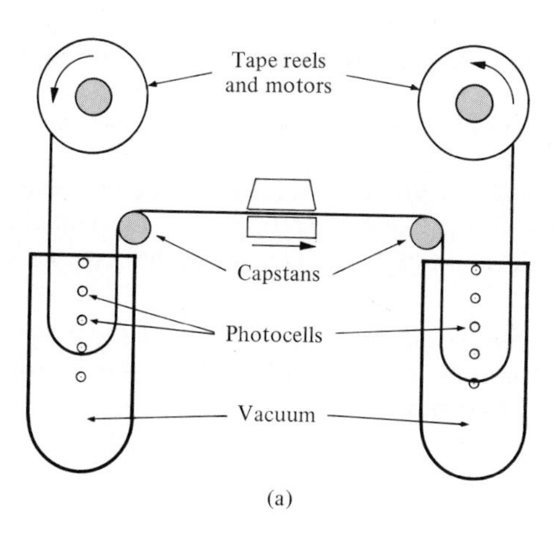

(a)

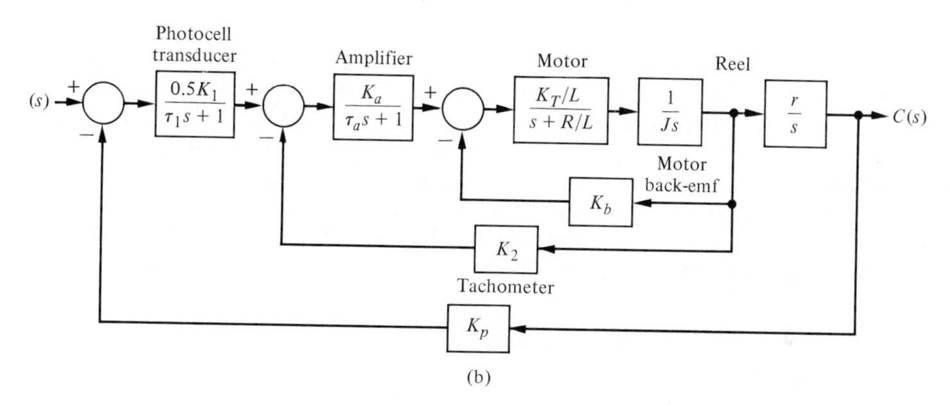

(b)

Figure P6.11. Tape control system.

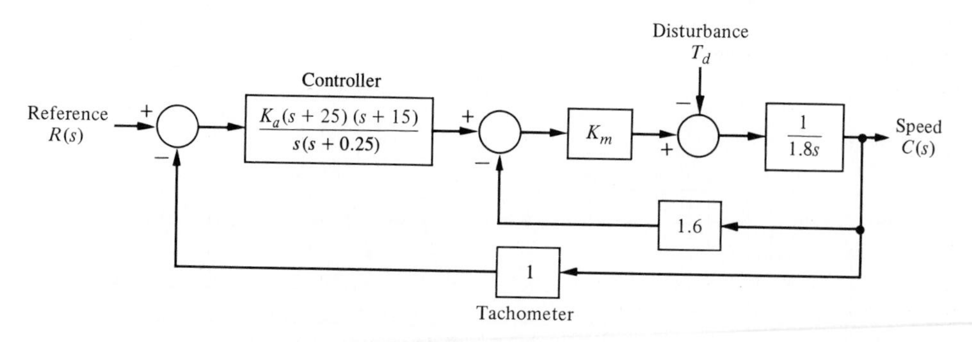

Figure P6.12. Speed control.

P6.14. The open-loop transfer function of a single-loop negative feedback system is

$$GH(s) = \frac{K(s + 2)(s + 3)}{s^2(s + 1)(s + 24)(s + 30)}.$$

This system is called a *conditionally stable* system because the system is stable for only a range of the gain K as follows: $k_1 < K < k_2$. Using the Routh-Hurwitz criteria and the root locus method, determine the range of the gain for which the system is stable. Sketch the root locus for $0 < K < \infty$.

P6.15. Let us again consider the stability and ride of a rider and high performance motorcycle as outlined in Problem 5.13. The dynamics of the motorcycle and rider can be represented by the open-loop transfer function

$$GH(s) = \frac{K(s^2 + 30s + 625)}{s(s + 20)(s^2 + 20s + 200)(s^2 + 60s + 3400)}.$$

Draw the root locus for the system; determine the ζ of the dominant roots when $K = 3 \times 10^4$.

P6.16. Control systems for maintaining constant tension on strip steel in a hot strip finishing mill are called "loopers." A typical system is shown in Fig. P6.16. The looper is an arm 2 to 3 ft long with a roller on the end and is raised and pressed against the strip by a motor. The typical speed of the strip passing the looper is 2000 ft/min. A voltage proportional to the looper position is compared with a reference voltage and integrated where it is assumed that a change in looper position is proportional to a change in the steel strip tension. The time constant of the filter, τ, is negligible relative to the other time constants in the system. (a) Draw the root locus of the control system for $0 < K_a < \infty$. (b) Determine the gain K_a that results in a system whose roots have a damping ratio of $\zeta = 0.707$ or greater. (c) Determine the effect of τ as τ increases from a negligible quantity.

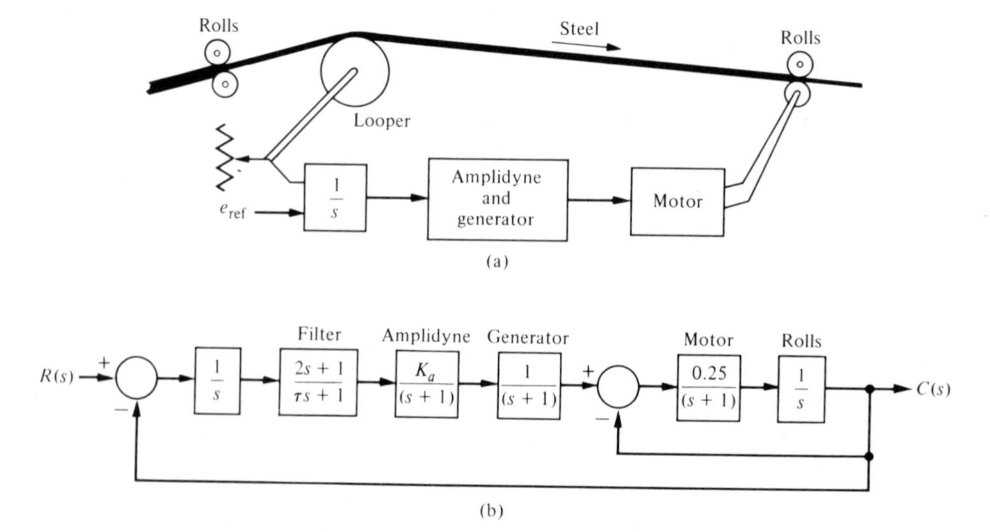

Figure P6.16. Steel mill control system.

P6.17. Reconsider the vibration absorber discussed in Problems 2.2 and 2.10 as a design problem. Using the root locus method, determine the effect of the parameters M_2 and k_{12}. Determine the specific values of the parameters M_2 and k_{12} so that the mass M_1 does not vibrate when $F(t) = a \sin \omega_0 t$. Assume that $M_1 = 1$, $k_1 = 1$, and $f_1 = 1$. Also assume that $k_{12} < 1$ and the term k_{12}^2 may be neglected.

P6.18. A feedback control system is shown in Fig. P6.18. The filter $G_c(s)$ is often called a compensator, and the design problem is that of selecting the parameters α and β. Using the root locus method, determine the effect of varying the parameters. Select a suitable filter so that the settling time is less than 4 sec and the damping ratio of the dominant roots is greater than 0.60.

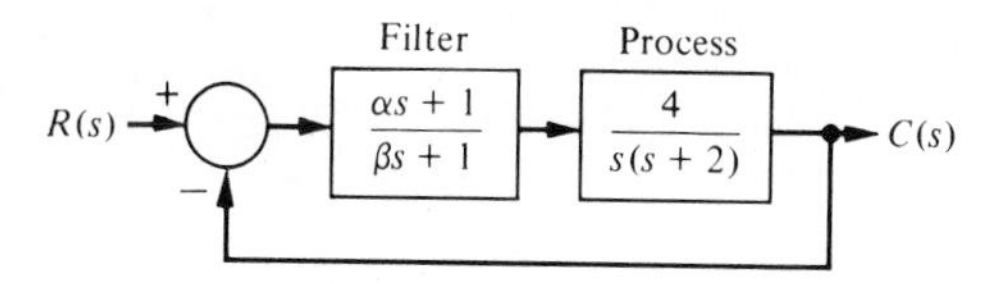

Figure P6.18. Filter design.

P6.19. In recent years, many automatic control systems for guided vehicles in factories have been utilized. One such system uses a guidance cable embedded in the floor to guide the vehicle along the desired lane [6]. An error detector is composed of two coils mounted on the front of the cart which senses a magnetic field produced by the current in the guidance cable. An example of a guided vehicle in a factory is shown in Fig. P6.19(a), on the next page. We have

$$G(s) = \frac{K_a(s^2 + 3.6s + 81)}{s(s + 1)(s + 5)}$$

when K_a equals the amplifier gain. (a) Draw a root locus and determine a suitable gain K_a so that the damping ratio of the complex roots is 0.707. (b) Determine the root sensitivity of the system for the complex root r_1 as a function of (1) K_a and (2) the pole of $G(s)$, $s = -1$.

P6.20. Determine the root sensitivity for the dominant roots of the design for Problem 6.18 for the gain $K = 4\alpha/\beta$ and the pole $s = -2$.

P6.21. Determine the root sensitivity of the dominant roots of the power system of Problem 6.7. Evaluate the sensitivity for variations of (a) the poles at $s = -4$, and (b) the feedback gain, $1/R$.

P6.22. Determine the root sensitivity of the dominant roots of Problem 6.1(a) when K is set so that the damping ratio of the unperturbed roots is 0.707. Evaluate and compare the sensitivity as a function of the poles and zeros of $GH(s)$.

P6.23. Repeat Problem 6.22 for the open-loop transfer function $GH(s)$ of Problem 6.1(c).

P6.24. For systems of relatively high degree, the form of the root locus can often assume an unexpected pattern. The root loci of four different feedback systems of third order or higher are shown in Fig. P6.24, on page 291. The open-loop poles and zeros of $KF(s)$ are shown and the form of the root loci as K varies from zero to infinity is presented. Verify the diagrams of Fig. P6.24 by constructing the root loci.

(a)

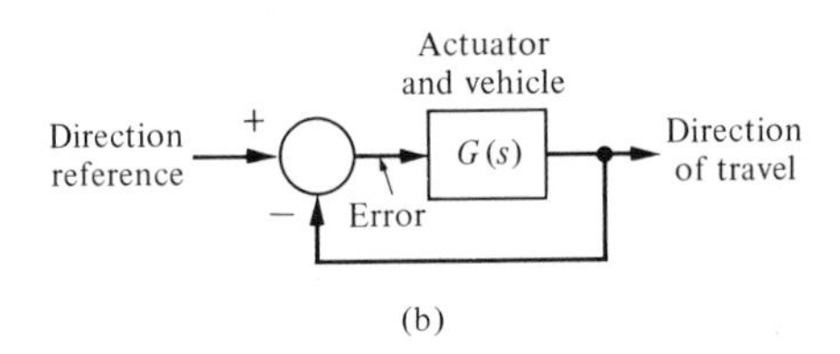

(b)

Figure P6.19. (a) An automatically guided vehicle. (Photo courtesy of Control Engineering Corp.) (b) Block diagram.

P6.25. Solid-state integrated electronic circuits are comprised of distributed R and C elements. Therefore feedback electronic circuits in integrated circuit form must be investigated by obtaining the transfer function of the distributed RC networks. It has been shown that the slope of the attenuation curve of a distributed RC network is n 3db/octave, where n is the order of the RC filter [17]. This attenuation is in contrast with the normal n 6db/octave for the lumped parameter circuits. (The concept of the slope of an attenuation curve is considered in Chapter 7. If the reader is unfamiliar with this concept, this problem may be reexamined following the study of Chapter 7.) An interesting case arises when the distributed RC network occurs in a series-to-shunt feedback path of a transistor amplifier. Then the loop transfer function may be written as

$$GH(s) = \frac{K(s-1)(s+3)^{1/2}}{(s+1)(s+2)^{1/2}}.$$

(a) Using the root locus method, determine the locus of roots as K varies from zero to infinity. (b) Calculate the gain at borderline stability and the frequency of oscillation for this gain.

P6.26. A single-loop negative feedback system has a loop transfer function

$$GH(s) = \frac{K(s+1)^2}{s(s^2+1)(s+4)}.$$

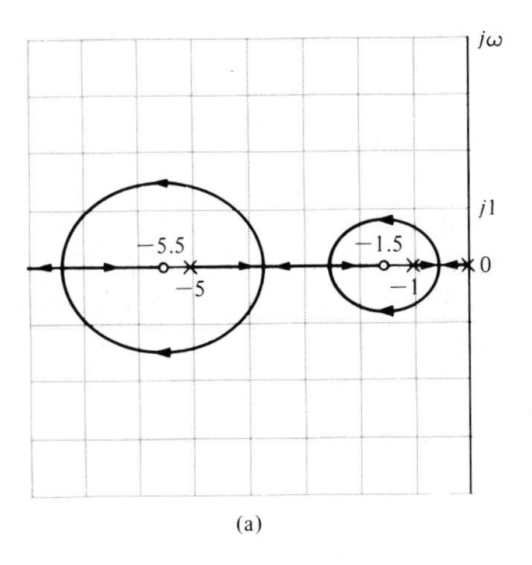

(a)

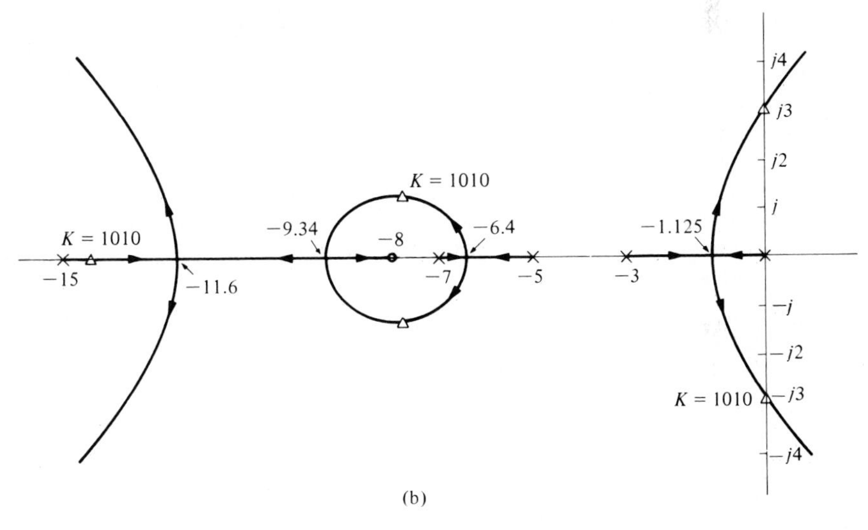

(b)

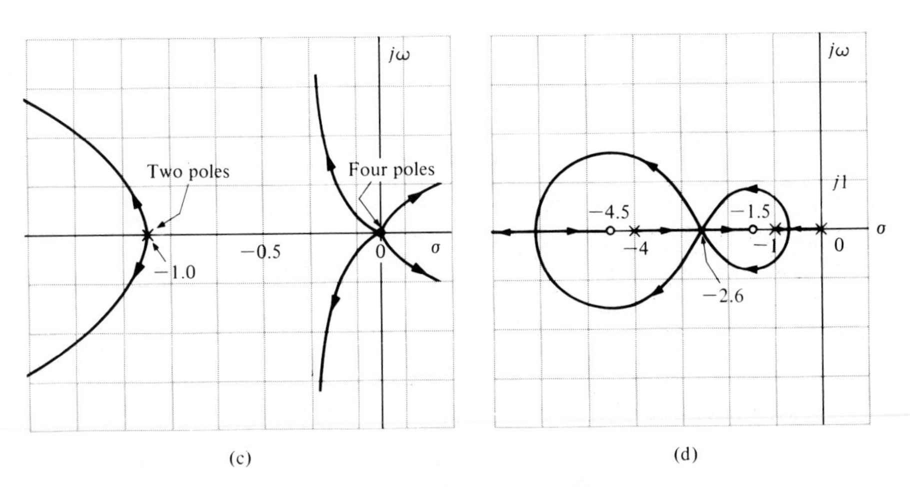

(c) (d)

Figure P6.24. Root loci of four systems.

(a) Sketch the root locus for $0 \leq K \leq \infty$ to indicate the significant features of the locus. (b) Determine the range of the gain K for which the system is stable. (c) For what value of K in the range $K \geq 0$ do purely imaginary roots exist? What are the values of these roots? (d) Would the use of the dominant roots approximation for an estimate of settling time be justified in this case for a large magnitude of gain ($K > 10$)?

P6.27. A unity negative feedback system has a transfer function

$$G(s) = \frac{K(s^2 + 0.105625)}{s(s^2 + 1)} = \frac{K(s + j0.325)(s - j0.325)}{s(s^2 + 1)}.$$

Calculate the root locus as a function of K. Carefully calculate where the segments of the locus enter and leave the real axis.

P6.28. To meet current U.S. emissions standards for automobiles, hydrocarbon (HC) and carbon monoxide (CO) emissions are usually controlled by a catalytic converter in the automobile exhaust. Federal standards for nitrogen oxides (NO_X) emissions are met mainly by exhaust-gas recirculation (EGR) techniques. However, as NO_X emissions standards were tightened from the current limit of 2.0 grams per mile to 1.0 grams per mile, these techniques alone were no longer sufficient.

Although many schemes are under investigation for meeting the emissions standards for all three emissions, one of the most promising employs a three-way catalyst—for HC, CO, and NO_x emissions—in conjunction with a closed-loop engine-control system. The approach is to use a closed-loop engine control as shown in Fig. P6.28 [21]. The exhaust gas sensor gives an indication of a rich or lean exhaust and this is compared to a reference. The difference signal is processed by the controller and the output of the controller modulates the vacuum level in the carburetor to achieve the best air-fuel ratio for proper operation of the catalytic converter. The open-loop transfer function is represented by

$$GH(s) = \frac{K(s + 1)(s + 6)}{s(s + 4)(s + 3)}.$$

Calculate the root locus as a function of K. Carefully calculate where the segments of the locus enter and leave the real axis. Determine the roots when $K = 2$.

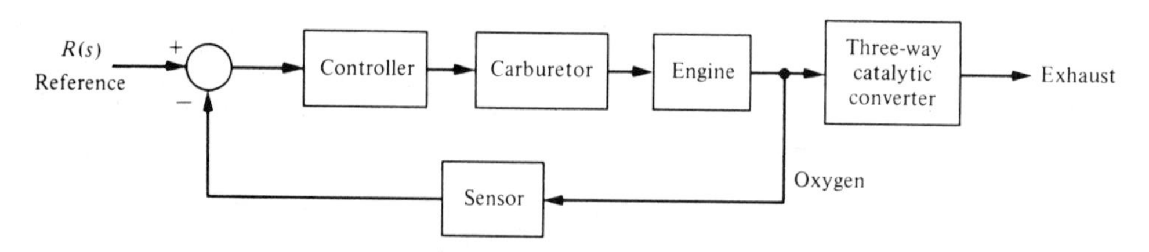

Figure P6.28. Auto engine control.

P6.29. A unity feedback control system has a transfer function

$$G(s) = \frac{K(s^2 + 4s + 8)}{s^2(s + 4)}.$$

It is desired that the dominant roots have a damping ratio equal to 0.5. Find the gain K when this condition is satisfied. Show that at this gain the imaginary roots are $s = -1.3 \pm j2.2$.

P6.30. An *RLC* network is shown in Fig. P6.30. The nominal values (normalized) of the network elements are $L = C = 1$ and $R = 2.5$. Show that the root sensitivity of the two roots of the input impedance $Z(s)$ to a change in R is different by a factor of 4.

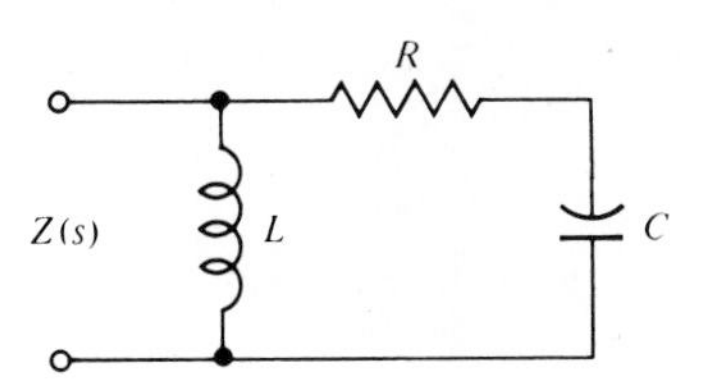

Figure P6.30. RLC network.

P6.31. The development of high speed aircraft and missiles requires information about aerodynamic parameters prevailing at very high speeds. Wind tunnels are used to test these parameters. These wind tunnels are constructed by compressing air to very high pressures and releasing it through a valve to create a wind. Since the air pressure drops as the air escapes it is necessary to open the valve wider to maintain a constant wind speed. Thus a control system is needed to adjust the valve to maintain a constant wind speed. The open-loop transfer function for a unity feedback system is

$$GH(s) = \frac{K(s+4)}{s(s+0.16)(s+7.3+9.7831j)(s+7.3-9.7831j)}$$

Draw the root locus and show the location of the roots for $K = 326$ and $K = 1350$.

P6.32. A mobile robot suitable for nighttime guard duty is shown in Fig. P6.32, on the next page. This guard never sleeps and can tirelessly patrol large warehouses and outdoor yards. The steering control system for the mobile robot has a unity feedback with

$$G(s) = \frac{K(s+1)(s+5)}{s(s+1.5)(s+2)}.$$

(a) Find K for all breakaway and entry points on the real axis. (b) Find K when the damping ratio of the complex roots is 0.707. (c) Find the minimum value of the damping ratio for the complex roots and the associated gain K. (d) Find the overshoot and settling time for a unit step input for the gain, K, determined in (c) and (d).

P6.33. The Bell-Boeing V-22 Osprey Tiltrotor is both an airplane and a helicopter. Its advantage is the ability to rotate its engines to 90° from a vertical position, as shown in Fig. P6.33(a), for takeoffs and landings and then to switch the engines to a horizontal position for cruising as an airplane [19]. The altitude control system in the helicopter mode is shown in Fig. P6.33(b). (a) Determine the root locus as K varies and determine the range of K for a stable system. (b) For $K = 280$ find the actual $c(t)$ for a unit step input $r(t)$ and the percentage overshoot and settling time. (c) When $K = 280$ and $r(t) = 0$, find $c(t)$ for a unit step disturbance, $D(s) = 1/s$. (d) Add a prefilter between $R(s)$ and the summing node so that

$$G_p(s) = \frac{1}{s^2 + 1.5s + 0.5}$$

and repeat (b).

Figure P6.32. Two mobile sentry robots. (Courtesy of Denning Mobile Robots Inc.)

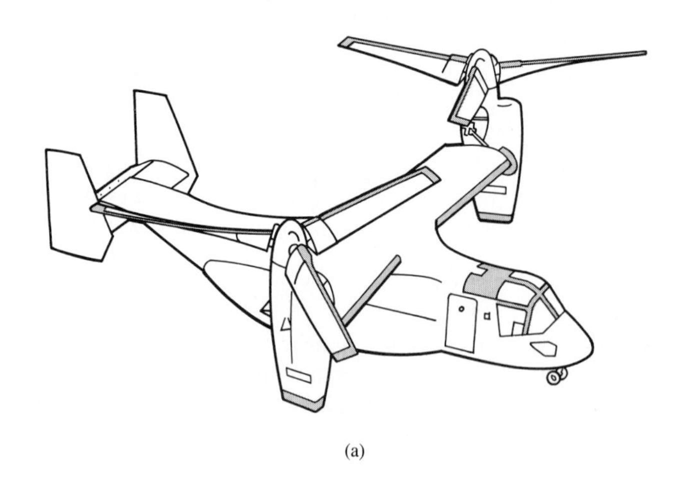

(a)

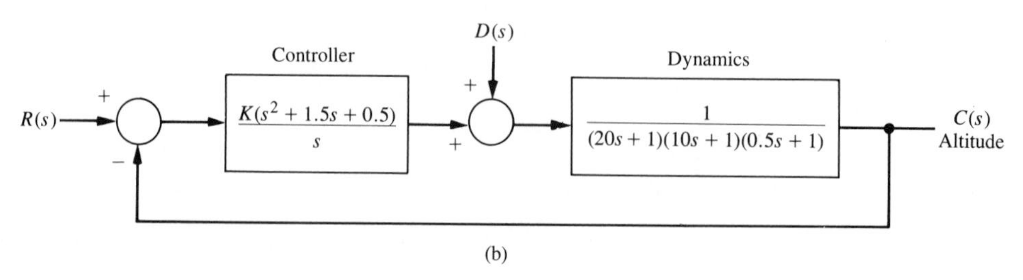

(b)

Figure P6.33. Osprey tiltrotor aircraft control.

P6.34. The fuel control for an automobile uses a diesel pump that is subject to parameter variations [23]. A unity negative feedback of the form shown in Fig. 6.1 has a plant

$$G(s) = \frac{K(s + 1.5)}{(s + 1)(s + 2)(s + 4)(s + 10)}.$$

(a) Sketch the root locus as K varies from 0 to 2000. (b) Find the roots for K equal to 400, 500, and 600. (c) Predict how the percent overshoot to a step will vary for the gain K, assuming dominant roots. (d) Find the actual time response for a step input for all three gains and compare the actual overshoot with the predicted overshoot.

P6.35. A powerful electrohydraulic forklift can be used to lift pallets weighing several tons on top of 35 ft scaffolds at a construction site. The negative unity feedback system has a plant transfer function

$$G(s) = \frac{K(s + 1)^2}{s(s^2 + 1)}.$$

(a) Sketch the root locus for $K > 0$. (b) Find the gain K when two complex roots have a ζ of 0.707 and calculate all three roots. (c) Find the entry point of the root locus at the real axis. (d) Estimate the expected overshoot to a step input and compare it with the actual overshoot determined from CSDP or another computer program.

P6.36. A microrobot with a high-performance manipulator has been designed for testing very small particles such as simple living cells [25]. The single loop unity negative feedback system has a plant transfer function

$$G(s) = \frac{K(s + 1)(s + 2)(s + 3)}{s^3(s - 1)}.$$

(a) Sketch the root locus for $K > 0$. (b) Find the gain and roots when the characteristic equation has two imaginary roots. (c) Determine the characteristic roots when $K = 20$ and $K = 100$. (d) For $K = 20$, estimate the percent overshoot to a step input and compare the estimate to the actual overshoot determined from CSDP or another computer program.

P6.37. Identify the parameters K, a, and b of the system shown in Fig. P6.37. The system is subject to a unit step input and the output response has an overshoot but ultimately attains the final value of 1. When the closed-loop system is subjected to a ramp input, the output response follows the ramp input with a finite steady-state error. When the gain is doubled to $2K$, the output response to an impulse input is a pure sinusoid with a period 0.314 sec. Determine K, a, and b.

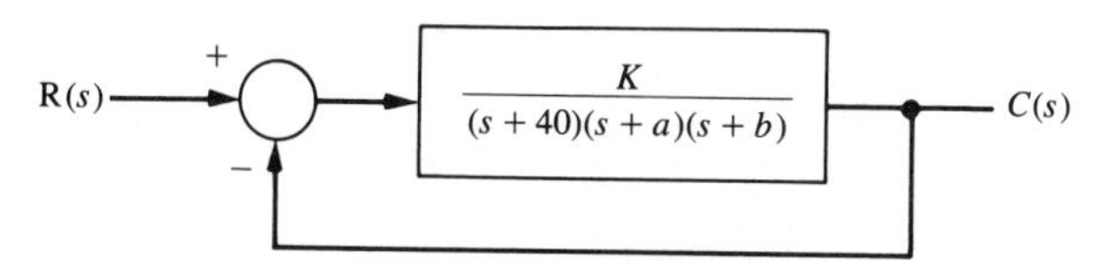

Figure P6.37. Feedback system.

P6.38. A unity feedback system, as shown in Fig. 6.1, has

$$G(s) = \frac{K(s + 1)}{s(s - 3)}.$$

This system is open-loop unstable. (a) Determine the range of K so that the system is stable. (b) Plot the root locus. (c) Determine the roots for $K = 10$. (d) For $K = 10$, predict the percent overshoot for a step input using Fig. 4.10. (e) Determine the actual overshoot by plotting the response.

Design Problems ✦

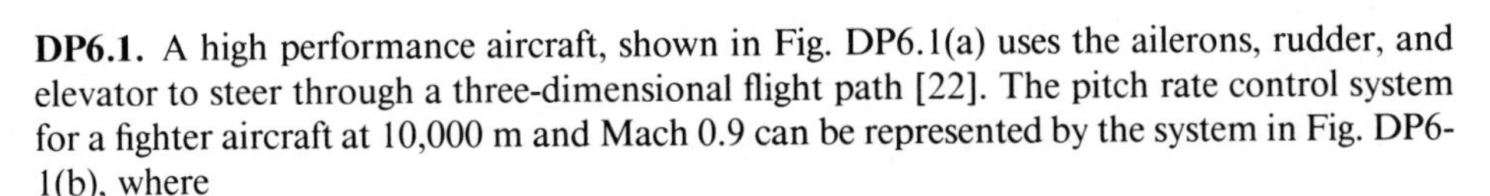

DP6.1. A high performance aircraft, shown in Fig. DP6.1(a) uses the ailerons, rudder, and elevator to steer through a three-dimensional flight path [22]. The pitch rate control system for a fighter aircraft at 10,000 m and Mach 0.9 can be represented by the system in Fig. DP6-1(b), where

$$G(s) = \frac{-18(s + 0.015)(s + 0.45)}{(s^2 + 1.2s + 12)(s^2 + 0.01s + 0.0025)}.$$

(a) Plot the root locus when the controller is a gain, so that $G_c(s) = K$, and determine K when zeta for the roots with $\omega_n > 2$ is larger than 0.15 (seek a maximum ζ). (b) Plot the response, $q(t)$, for a step input $r(t)$. (c) A designer suggests an anticipatory controller so that $G_c(s) = K_1 + K_2 s = K(s + 2)$. Plot the root locus for this system as K varies and determine K so that the ζ of all the closed-loop roots is $0.8 < \zeta < 0.6$. (d) Plot the response, $q(t)$, for a step input $r(t)$.

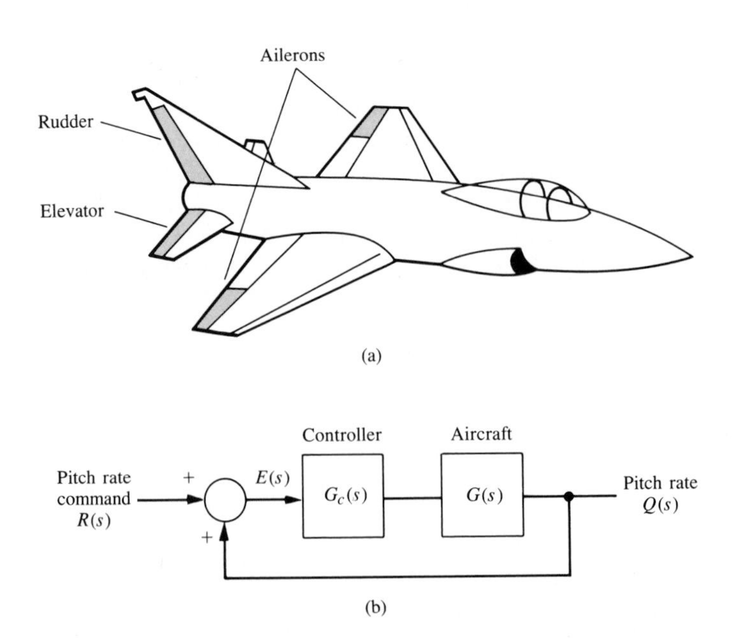

Figure DP6.1. Pitch rate control.

DP6.2. A large helicopter uses two tandem rotors rotating in opposite directions, as shown in Fig. DP6.2(a). The controller adjusts the tilt angle of the main rotor and thus the forward motion as shown in Fig. DP6.2(b). The helicopter dynamics are represented by

$$G(s) = \frac{10}{s^2 + 4.5s + 9},$$

and the controller is selected as

$$G_c(s) = K_1 + \frac{K_2}{s} = \frac{K(s+1)}{s}.$$

(a) Plot the root locus of the system and determine K when ζ of the complex roots is equal to 0.6. (b) Plot the response of the system to a step input $r(t)$ and find the settling time and overshoot for the system of part (a). What is the steady-state error for a step input? (c) Repeat parts (a) and (b) when the ζ of the complex roots is 0.41. Compare the results with those obtained in parts (a) and (b).

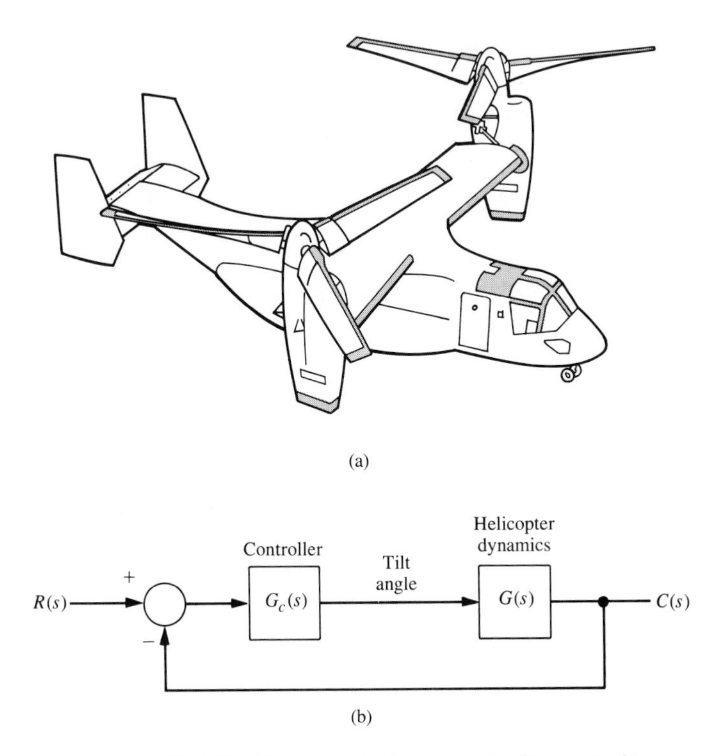

Figure DP6.2. Two-rotor helicopter velocity control.

DP6.3. Rover, the vehicle illustrated in Fig. DP6.3(a), has been designed for maneuvering at 0.25 mph over Martian terrain. Because Mars is 189 million miles from Earth and it would take up to 40 minutes each way to communicate with Earth [24], Rover must act independently and reliably. Resembling a cross between a small flatbed truck and a jacked-up jeep, Rover will be constructed of three articulated sections, each with its own two independent axle-bearing one-meter conical wheels. A pair of sampling arms—one for chipping and drilling, the other for manipulating fine objects—jut from its front end like pincers. The control of the arms can be represented by the system shown in Fig. DP6.3(b). (a) Plot the root locus for K and identify the roots for $K = 4.1$ and 41. (b) Determine the gain K that results in an overshoot to a step of approximately 1%. (c) Determine the gain that minimizes the settling time while maintaining an overshoot of less than 1%.

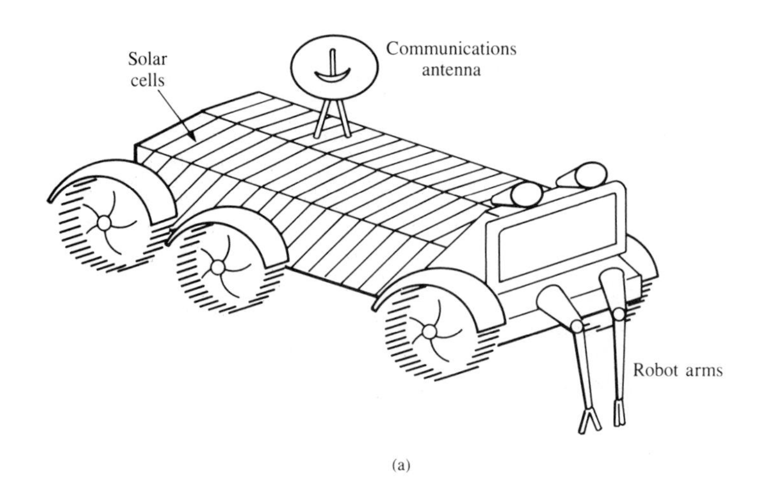

(a)

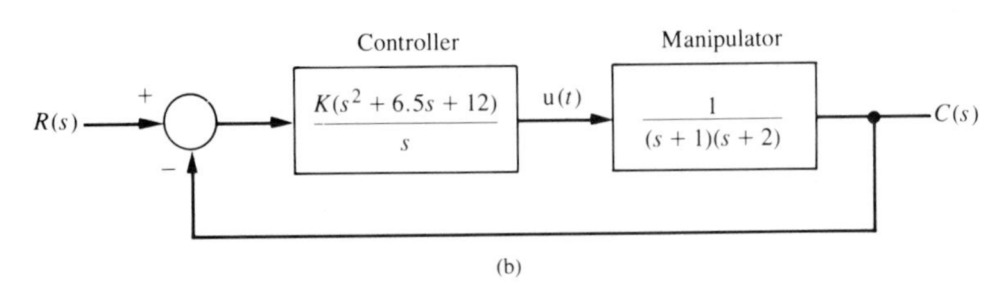

(b)

Figure DP6.3. Mars vehicle robot control.

DP6.4. A welding torch is remotely controlled to achieve high accuracy while operating in changing and hazardous environments [22]. A model of the welding arm position control is shown in Fig. DP6.4, with the disturbance representing the environmental changes. (a) With $D(s) = 0$, select K_1 and K to provide high quality performance of the position control system. Select a set of performance criteria and examine the results of your design. (b) For the system in part (a), let $R(s) = 0$ and determine the effect of a unit step $D(s) = 1/s$ by obtaining $c(t)$.

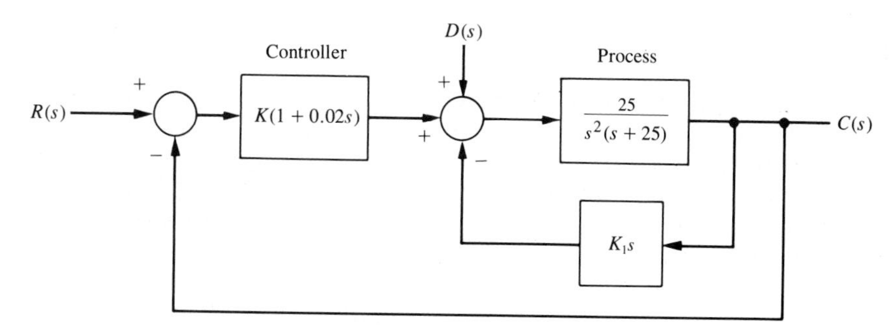

Figure DP6.4. Remotely controlled welder.

Terms and Concepts

Angle of departure The angle at which a locus leaves a complex pole in the s-plane.

Asymptote The path the root locus follows as the parameter becomes very large and approaches infinity. The number of asymptotes is equal to the number of poles minus the number of zeros.

Asymptote centroid The center of the linear asymptotes, σ_A.

Breakaway point The point on the real axis where the locus departs from the real axis of the s-plane.

Locus A path or trajectory that is traced out as a parameter is changed.

Number of separate loci Equal to the number of poles of the transfer function, assuming that the number of poles is greater than the number of zeros of the transfer function.

Parameter design A method of selecting one or two parameters using the root locus method.

Root locus The locus or path of the roots traced out on the s-plane as a parameter is changed.

Root locus method The method for determining the locus of roots of the characteristic equation $1 + KP(s) = 0$ as K varies from 0 to infinity.

Root locus segments on the real axis The root locus lying in a section of the real axis to the left of an odd number of poles and zeros.

Root sensitivity The sensitivity of the roots as a parameter changes from its normal value. The root sensitivity is the incremental change in the root divided by the proportional change of the parameter.

References

1. W. R. Evans, "Graphical Analysis of Control Systems," *Trans. of the AIEE,* **67,** 1948; pp. 547–51. Also in G. J. Thaler, ed., *Automatic Control,* Dowden, Hutchinson, and Ross, Stroudsburg, Pa., 1974; pp. 417–21.
2. W. R. Evans, "Control System Synthesis by Root Locus Method," *Trans. of the AIEE,* **69,** 1950; pp. 1–4. Also in G. J. Thaler, ed., *Automatic Control,* Dowden, Hutchinson, and Ross, Stroudsburg, Pa., 1974; pp. 423–25.
3. W. R. Evans, *Control System Dynamics,* McGraw-Hill, New York, 1954.
4. R. C. Dorf and R. Jacquot, *Control Systems Design Program,* Addison-Wesley, Reading, Mass., 1988.
5. J. G. Goldberg, *Automatic Controls,* Allyn and Bacon, Boston, 1964.
6. R. C. Dorf, *The Encyclopedia of Robotics,* John Wiley & Sons, New York, 1988.
7. H. Ur, "Root Locus Properties and Sensitivity Relations in Control Systems," *I.R.E. Trans. on Automatic Control,* January 1960; pp. 57–65.
8. S. P. Patarinski, "Mechanics and Coordinated Control of Twin Articulated Robot Arms," *Mechatronics,* **1,** 1991; pp. 59–71.
9. J. N. Aubrun et al., "Dynamic Analysis of the Actively Controlled Segmented Mirror of the Keck Ten-Meter Telescope," *IEEE Control Systems,* December 1987; pp. 3–9.
10. C. L. Phillips and R. Harbor, *Feedback Control Systems,* 2nd ed., Prentice Hall, Englewood Cliffs, N.J., 1991.
11. W. L. Brogan, *Modern Control Theory,* Prentice Hall, Englewood Cliffs, N.J., 1991.
12. J. Grossman, "Goalpost Brown's Amazing Flying Camera," *Inc.,* January 1985; pp. 80–88.
13. K. K. Chew, "Control of Errors in Disk Drive Systems," *IEEE Control Systems,* January 1990; pp. 16–19.
14. J. Slotline, *Applied Nonlinear Control,* Prentice Hall, Englewood Cliffs, N.J., 1991.
15. R. C. Nelson, *Flight Stability and Control,* McGraw-Hill, New York, 1989.
16. R. C. Dorf, *Introduction to Electric Circuits,* John Wiley & Sons, New York, 1989.
17. S. Franco, *Design with Operational Amplifiers,* McGraw-Hill, New York, 1988.
18. B. W. Dickenson, *Systems Analysis and Design,* Prentice Hall, Englewood Cliffs, New Jersey, 1991.
19. D. McLean, *Automatic Flight Control Systems,* Prentice Hall, Englewood Cliffs, N.J., 1990.
20. P. E. Wellstead, "The Ball and Hoop System," *Automatica,* **19,** 4, 1983; pp. 401–406.
21. K. J. Astrom and B. Wittenmark, *Adaptive Control,* Addison-Wesley, Reading, Mass., 1988.
22. K. Andersen, "Artificial Neural Networks Applied to Arc Welding Process Control," *IEEE Trans. on Industry Applications,* October 1990; pp. 824–830.
23. H. Kuraoka, "Application of Design to Automotive Fuel Control," *IEEE Control Systems,* April 1990; pp. 102–105.
24. H. E. McCurdy, *The Space Station Decision,* Johns Hopkins University Press, Baltimore, 1990.
25. I. W. Hunter, Manipulation and Dynamic Mechanical Testing of Microscopic Objects Using a Tele-Micro-Robot System," *IEEE Control Systems,* February 1990; pp. 3–8.

CHAPTER 7

Frequency Response Methods

Preview

We have examined the use of test input signals such as a step and a ramp signal. In this chapter, we will use a steady-state sinusoidal input signal and consider the response of the system as the frequency of the sinusoid is varied. Thus we will look at the response of the system to a changing frequency, ω.

We will examine the response of $G(s)$ when $s = j\omega$ and develop several forms of plotting the complex number for $G(j\omega)$ when ω is varied. These plots provide insight regarding the performance of a system. We are able to develop several performance measures for the frequency response of a system. The measures can be used as system specifications and we can adjust parameters in order to meet the specifications.

We will consider the graphical development of one or more forms for the frequency response plot. We can then proceed to use computer-generated data to readily obtain these plots.

7.1 Introduction

In the preceding chapters the response and performance of a system have been described in terms of the complex frequency variable s and the location of the poles and zeros on the s-plane. A very practical and important alternative approach to the analysis and design of a system is the *frequency response* method. *The frequency response of a system is defined as the steady-state response of the system to a sinusoidal input signal.* The sinusoid is a unique input signal, and the resulting output signal for a linear system, as well as signals throughout the system, is sinusoidal in the steady state; it differs from the input waveform only in amplitude and phase angle.

One advantage of the frequency response method is the ready availability of sinusoid test signals for various ranges of frequencies and amplitudes. Thus the experimental determination of the frequency response of a system is easily accomplished and is the most reliable and uncomplicated method for the experimental analysis of a system. Often, as we shall find in Section 7.4, the unknown transfer function of a system can be deduced from the experimentally determined frequency response of a system [1,2]. Furthermore, the design of a system in the frequency domain provides the designer with control of the bandwidth of a system and some measure of the response of the system to undesired noise and disturbances.

A second advantage of the frequency response method is that the transfer function describing the sinusoidal steady-state behavior of a system can be obtained by replacing s with $j\omega$ in the system transfer function $T(s)$. The transfer function representing the sinusoidal steady-state behavior of a system is then a function of the complex variable $j\omega$ and is itself a complex function $T(j\omega)$ which possesses a magnitude and phase angle. The magnitude and phase angle of $T(j\omega)$ are readily represented by graphical plots that provide a significant insight for the analysis and design of control systems.

The basic disadvantage of the frequency response method for analysis and design is the indirect link between the frequency and the time domain. Direct correlations between the frequency response and the corresponding transient response characteristics are somewhat tenuous, and in practice the frequency response characteristic is adjusted by using various design criteria which will normally result in a satisfactory transient response.

The Laplace transform pair was given in Section 2.4 and is written as:

$$F(s) = \mathcal{L}\{f(t)\} = \int_0^\infty f(t)e^{-st}\,dt \tag{7.1}$$

and

$$f(t) = \mathcal{L}^{-1}\{F(s)\} = \frac{1}{2\pi j}\int_{\sigma-j\infty}^{\sigma+j\infty} F(s)e^{st}\,ds, \tag{7.2}$$

where the complex variable $s = \sigma + j\omega$. Similarly, the *Fourier transform* pair is written as

$$F(j\omega) = \mathcal{F}\{f(t)\} = \int_{-\infty}^{\infty} f(t)e^{-j\omega t}\,dt \tag{7.3}$$

and

$$f(t) = \mathcal{F}^{-1}\{F(j\omega)\} = \frac{1}{2\pi}\int_{-\infty}^{\infty} F(j\omega)e^{j\omega t}\,d\omega. \tag{7.4}$$

The Fourier transform exists for $f(t)$ when

$$\int_{-\infty}^{\infty} |f(t)|\,dt < \infty.$$

The Fourier and Laplace transforms are closely related, as we can see by examining Eqs. (7.1) and (7.3). When the function $f(t)$ is defined only for $t \geq 0$, as is often the case, the lower limits on the integrals are the same. Then, we note that the two equations differ only in the complex variable. Thus, if the Laplace transform of a function $f_1(t)$ is known to be $F_1(s)$, we can obtain the Fourier transform of this same time function $F_1(j\omega)$ by setting $s = j\omega$ in $F_1(s)$.

Again, we might ask, because the Fourier and Laplace transforms are so closely related, why not always use the Laplace transform? Why use the Fourier transform at all? The Laplace transform permits us to investigate the s-plane location of the poles and zeros of a transfer $T(s)$ as in Chapter 6. However, the frequency response method allows us to consider the transfer function $T(j\omega)$ and concern ourselves with the amplitude and phase characteristics of the system. This ability to investigate and represent the character of a system by amplitude and phase equations and curves is an advantage for the analysis and design of control systems.

If we consider the frequency response of the closed-loop system, we might have an input $r(t)$ that has a Fourier transform, in the frequency domain, as follows:

$$R(j\omega) = \int_{-\infty}^{\infty} r(t)e^{-j\omega t}\,dt. \tag{7.5}$$

Then the output frequency response of a single-loop control system can be obtained by substituting $s = j\omega$ in the closed-loop system relationship, $C(s) = T(s)R(s)$, so that we have

$$C(j\omega) = T(j\omega)R(j\omega) = \frac{G(j\omega)}{1 + G(j\omega)H(j\omega)}R(j\omega). \tag{7.6}$$

Utilizing the inverse Fourier transform, the output transient response would be

$$c(t) = \mathcal{F}^{-1}\{C(j\omega)\} = \frac{1}{2\pi}\int_{-\infty}^{\infty} C(j\omega)e^{j\omega t}\,d\omega. \tag{7.7}$$

However, it is usually quite difficult to evaluate this inverse transform integral for any but the simplest systems, and a graphical integration may be used. Alternatively, as we will note in succeeding sections, several measures of the transient response can be related to the frequency characteristics and utilized for design purposes.

7.2 Frequency Response Plots

The transfer function of a system $G(s)$ can be described in the frequency domain by the relation†

$$G(j\omega) = G(s)|_{s=j\omega} = R(\omega) + jX(\omega), \tag{7.8}$$

where

$$R(\omega) = Re[G(j\omega)]$$

and

$$X(\omega) = \text{Im}[G(j\omega)].$$

Alternatively, the transfer function can be represented by a magnitude $|G(j\omega)|$ and a phase $\phi(j\omega)$ as

$$G(j\omega) = |G(j\omega)|e^{j\phi(j\omega)} = |G(\omega)|\underline{/\phi(\omega)}, \tag{7.9}$$

where

$$\phi(\omega) = \tan^{-1} X(\omega)/R(\omega)$$

and

$$|G(\omega)|^2 = [R(\omega)]^2 + [X(\omega)]^2.$$

The graphical representation of the frequency response of the system $G(j\omega)$ can utilize either Eq. (7.8) or Eq. (7.9). The *polar plot* representation of the frequency response is obtained by using Eq. (7.8). The coordinates of the polar plot are the real and imaginary parts of $G(j\omega)$ as shown in Fig. 7.1. An example of a polar plot will illustrate this approach.

---

†See Appendix E for a review of complex numbers.

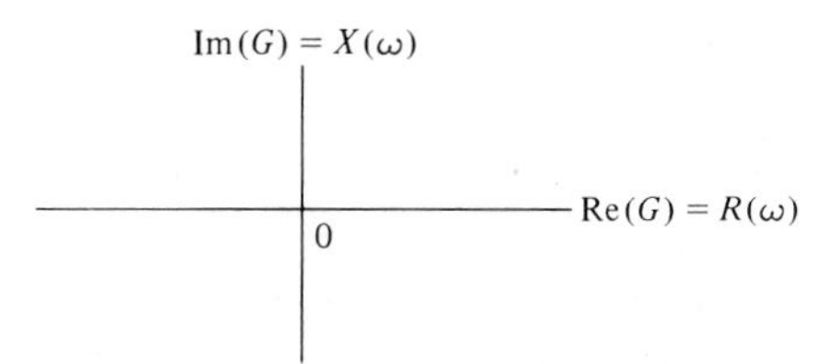

Figure 7.1. The polar plane.

■ Example 7.1 Frequency response of an RC filter

A simple RC filter is shown in Fig. 7.2. The transfer function of this filter is

$$G(s) = \frac{V_2(s)}{V_1(s)} = \frac{1}{RCs + 1}, \tag{7.10}$$

and the sinusoidal steady-state transfer function is

$$G(j\omega) = \frac{1}{j\omega(RC) + 1} = \frac{1}{j(\omega/\omega_1) + 1}, \tag{7.11}$$

where

$$\omega_1 = 1/RC.$$

Then the polar plot is obtained from the relation

$$\begin{aligned} G(j\omega) &= R(\omega) + jX(\omega) \\ &= \frac{1 - j(\omega/\omega_1)}{(\omega/\omega_1)^2 + 1} \\ &= \frac{1}{1 + (\omega/\omega_1)^2} - \frac{j(\omega/\omega_1)}{1 + (\omega/\omega_1)^2}. \end{aligned} \tag{7.12}$$

The locus of the real and imaginary parts is shown in Fig. 7.3 and is easily shown to be a circle with the center at (½, 0). When $\omega = \omega_1$, the real and imaginary parts are equal, and the angle $\phi(\omega) = 45°$. The polar plot can also be readily obtained from Eq. (7.9) as

$$G(j\omega) = |G(\omega)| \underline{/\phi(\omega)}, \tag{7.13}$$

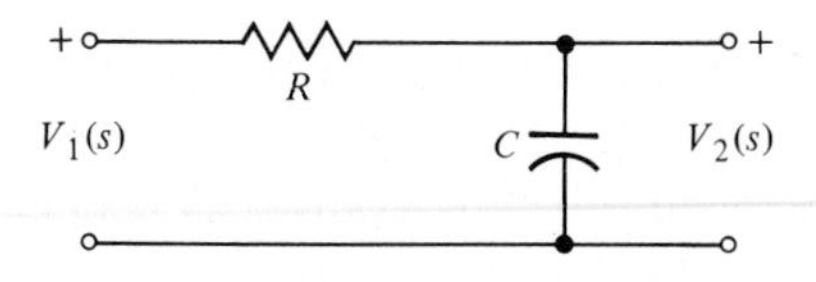

Figure 7.2. An RC filter.

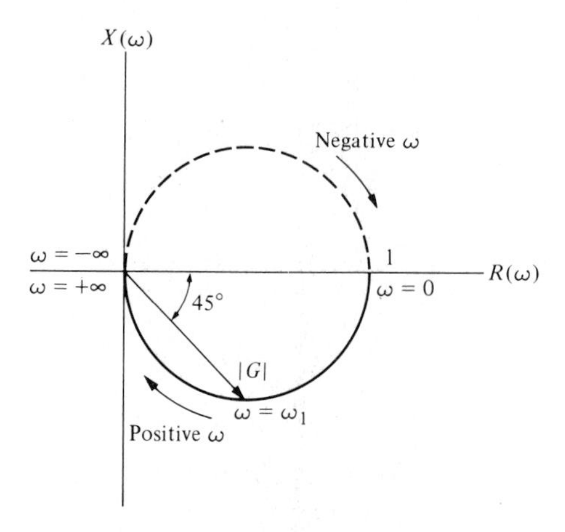

Figure 7.3. Polar plot for *RC* filter.

where

$$|G(\omega)| = \frac{1}{[1 + (\omega/\omega_1)^2]^{1/2}} \quad \text{and} \quad \phi(\omega) = -\tan^{-1}(\omega/\omega_1).$$

Clearly, when $\omega = \omega_1$, magnitude is $|G(\omega_1)| = 1/\sqrt{2}$ and phase $\phi(\omega_1) = -45°$. Also, when ω approaches $+\infty$, we have $|G(\omega)| \to 0$ and $\phi(\omega) = -90°$. Similarly, when $\omega = 0$, we have $|G(\omega)| = 1$ and $\phi(\omega) = 0$.

■ Example 7.2 Polar plot of a transfer function

The polar plot of a transfer function will be useful for investigating system stability and will be utilized in Chapter 8. Therefore it is worthwhile to complete another example at this point. Consider a transfer function

$$|G(s)|_{s=j\omega} = G(j\omega) = \frac{K}{j\omega(j\omega\tau + 1)} = \frac{K}{j\omega - \omega^2\tau}. \tag{7.14}$$

Then the magnitude and phase angle are written as

$$|G(\omega)| = \frac{K}{(\omega^2 + \omega^4\tau^2)^{1/2}} \tag{7.15}$$

and

$$\phi(\omega) = -\tan^{-1}\left(\frac{1}{-\omega\tau}\right).$$

The phase angle and the magnitude are readily calculated at the frequencies $\omega = 0$, $\omega = 1/\tau$, and $\omega = +\infty$. The values of $|G(\omega)|$ and $\phi(\omega)$ are given in Table 7.1, and the polar plot of $G(j\omega)$ is shown in Fig. 7.4.

Table 7.1.

ω	0	$\frac{1}{2}\tau$	$1/\tau$	∞
$\lvert G(\omega)\rvert$	∞	$4K\tau/\sqrt{5}$	$K\tau/\sqrt{2}$	0
$\phi(\omega)$	$-90°$	$-117°$	$-135°$	$-180°$

There are several possibilities for coordinates of a graph portraying the frequency response of a system. As we have seen, we may choose to utilize a polar plot to represent the frequency response (Eq. 7.8) of a system. However, the limitations of polar plots are readily apparent. The addition of poles or zeros to an existing system requires the recalculation of the frequency response as outlined in Examples 7.1 and 7.2. (See Table 7.1.) Furthermore, the calculation of the frequency response in this manner is tedious and does not indicate the effect of the individual poles or zeros.

Therefore the introduction of *logarithmic plots,* often called *Bode plots,* simplifies the determination of the graphical portrayal of the frequency response. The logarithmic plots are called Bode plots in honor of H. W. Bode, who used them extensively in his studies of feedback amplifiers [4,5]. The transfer function in the frequency domain is

$$G(j\omega) = |G(\omega)|e^{j\phi(\omega)}. \tag{7.16}$$

The natural logarithm of Eq. (7.16) is

$$\ln G(j\omega) = \ln |G(\omega)| + j\phi(\omega), \tag{7.17}$$

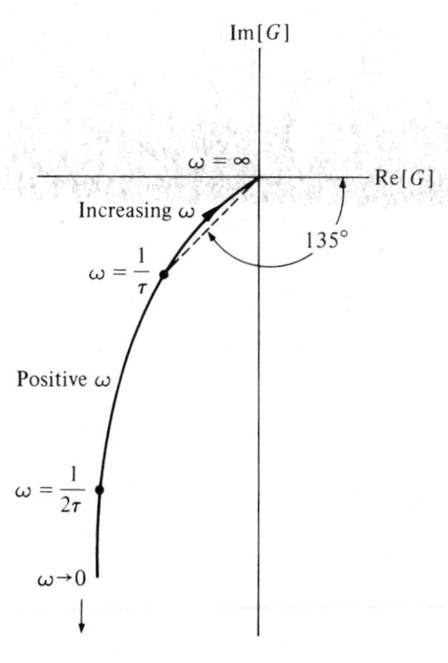

Figure 7.4. Polar plot for $G(j\omega) = K/j\omega(j\omega\tau + 1)$.

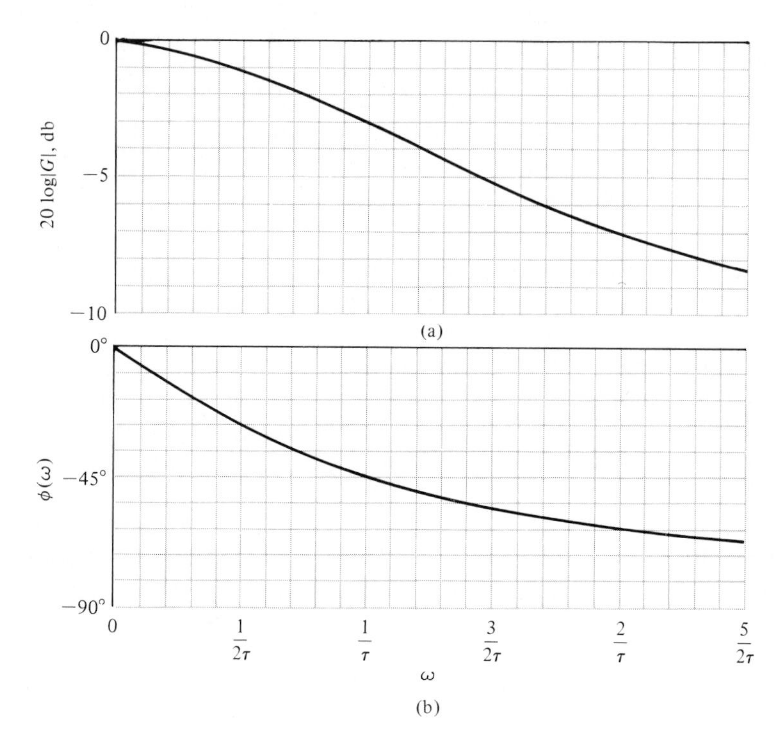

Figure 7.5. Bode diagram for $G(j\omega) = 1/(j\omega\tau + 1)$.

where $\ln |G|$ is the magnitude in nepers. The logarithm of the magnitude is normally expressed in terms of the logarithm to the base 10, so that we use

$$\text{Logarithmic gain} = 20 \log_{10} |G(\omega)|,$$

where the units are decibels (db). A decibel conversion table is given in Appendix D. The logarithmic gain in db and the angle $\phi(\omega)$ can be plotted versus the frequency ω by utilizing several different arrangements. For a Bode diagram, the plot of logarithmic gain in db versus ω is normally plotted on one set of axes, and the phase $\phi(\omega)$ versus ω on another set of axes, as shown in Fig. 7.5. For example, the Bode diagram of the transfer function of Example 7.1 can be readily obtained, as we will find in the following example.

■ Example 7.3 Bode diagram of an RC filter

The transfer function of Example 7.1 is

$$G(j\omega) = \frac{1}{j\omega(RC) + 1} = \frac{1}{j\omega\tau + 1}, \tag{7.18}$$

where

$$\tau = RC,$$

the time constant of the network. The logarithmic gain is

$$20 \log |G| = 20 \log \left(\frac{1}{1 + (\omega\tau)^2} \right)^{1/2} = -10 \log (1 + (\omega\tau)^2). \tag{7.19}$$

For small frequencies, that is $\omega \ll 1/\tau$, the logarithmic gain is

$$20 \log |G| = -10 \log (1) = 0 \text{ db}, \qquad \omega \ll 1/\tau. \tag{7.20}$$

For large frequencies, that is $\omega \gg 1/\tau$, the logarithmic gain is

$$20 \log |G| = -20 \log \omega\tau \qquad \omega \gg 1/\tau, \tag{7.21}$$

and at $\omega = 1/\tau$, we have

$$20 \log |G| = -10 \log 2 = -3.01 \text{ db}.$$

The magnitude plot for this network is shown in Fig. 7.5(a). The phase angle of this network is

$$\phi(j\omega) = -\tan^{-1} \omega\tau. \tag{7.22}$$

The phase plot is shown in Fig. 7.5(b). The frequency $\omega = 1/\tau$ is often called the *break frequency* or *corner frequency.*

Examining the Bode diagram of Fig. 7.5, we find that a linear scale of frequency is not the most convenient or judicious choice and we should consider the use of a logarithmic scale of frequency. The convenience of a logarithmic scale of frequency can be seen by considering Eq. (7.21) for large frequencies $\omega \gg 1/\tau$, as follows:

$$20 \log |G| = -20 \log \omega\tau = -20 \log \tau - 20 \log \omega. \tag{7.23}$$

Then, on a set of axes where the horizontal axis is $\log \omega$, the asymptotic curve for $\omega \gg 1/\tau$ is a straight line, as shown in Fig. 7.6. The slope of the straight line can be ascertained from Eq. (7.21). An interval of two frequencies with a ratio equal

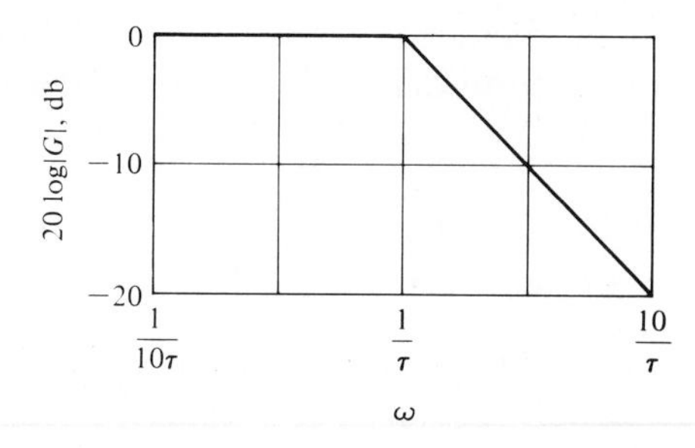

Figure 7.6. Asymptotic curve for $(j\omega\tau + 1)^{-1}$.

to 10 is called a decade, so that the range of frequencies from ω_1 to ω_2, where $\omega_2 = 10\omega_1$, is called a decade. Then, the difference between the logarithmic gains, for $\omega \gg 1/\tau$, over a decade of frequency is

$$\begin{aligned} 20 \log |G(\omega_1)| - 20 \log |G(\omega_2)| &= -20 \log \omega_1\tau - (-20 \log \omega_2\tau) \\ &= -20 \log \frac{\omega_1\tau}{\omega_2\tau} \\ &= -20 \log (1/10) = +20 \text{ db}. \end{aligned} \tag{7.24}$$

That is, the slope of the asymptotic line for this first-order transfer function is -20 db/decade, and the slope is shown for this transfer function in Fig. 7.6. Instead of using a horizontal axis of $\log \omega$ and linear rectangular coordinates, it is simpler to use semilog paper with a linear rectangular coordinate for db and a logarithmic coordinate for ω. Alternatively, we could use a logarithmic coordinate for the magnitude as well as for frequency and avoid the necessity of calculating the logarithm of the magnitude.

The frequency interval $\omega_2 = 2\omega_1$ is often used and is called an octave of frequencies. The difference between the logarithmic gains for $\omega \gg 1/\tau$, for an octave, is

$$\begin{aligned} 20 \log |G(\omega_1)| - 20 \log |G(\omega_2)| &= -20 \log \frac{\omega_1\tau}{\omega_2\tau} \\ &= -20 \log (1/2) = 6.02 \text{ db}. \end{aligned} \tag{7.25}$$

Therefore the slope of the asymptotic line is -6 db/octave or -20 db/decade.

The primary advantage of the logarithmic plot is the conversion of multiplicative factors such as $(j\omega\tau + 1)$ into additive factors $20 \log (j\omega\tau + 1)$ by virtue of the definition of logarithmic gain. This can be readily ascertained by considering a generalized transfer function as

$$G(j\omega) = \frac{K_b \prod_{i=1}^{Q} (1 + j\omega\tau_i)}{(j\omega)^N \prod_{m=1}^{M} (1 + j\omega\tau_m) \prod_{k=1}^{R} [(1 + (2\zeta_k/\omega_{n_k})j\omega + (j\omega/\omega_{n_k})^2)]}. \tag{7.26}$$

This transfer function includes Q zeros, N poles at the origin, M poles on the real axis, and R pairs of complex conjugate poles. Clearly, obtaining the polar plot of such a function would be a formidable task indeed. However, the logarithmic magnitude of $G(j\omega)$ is

$$\begin{aligned} 20 \log |G(\omega)| = 20 \log K_b &+ 20 \sum_{i=1}^{Q} \log |1 + j\omega\tau_i| \\ &- 20 \log |(j\omega)^N| - 20 \sum_{m=1}^{M} \log |1 + j\omega\tau_m| \\ &- 20 \sum_{k=1}^{R} \log \left| 1 + \left(\frac{2\zeta_k}{\omega_{n_k}}\right) j\omega + \left(\frac{j\omega}{\omega_{n_k}}\right)^2 \right|, \end{aligned} \tag{7.27}$$

and the Bode diagram can be obtained by adding the plot due to each individual factor. Furthermore, the separate phase angle plot is obtained as

$$\phi(\omega) = +\sum_{i=1}^{Q} \tan^{-1} \omega\tau_i - N(90^\circ) - \sum_{m=1}^{M} \tan^{-1} \omega\tau_m - \sum_{k=1}^{R} \tan^{-1}\left(\frac{2\zeta_k\omega_{n_k}\omega}{\omega_{n_k}^2 - \omega^2}\right), \tag{7.28}$$

which is simply the summation of the phase angles due to each individual factor of the transfer function.

Therefore the four different kinds of factors that may occur in a transfer function are as follows:

1. Constant gain K_b
2. Poles (or zeros) at the origin $(j\omega)$
3. Poles or zeros on the real axis $(j\omega\tau + 1)$
4. Complex conjugate poles (or zeros) $[1 + (2\zeta/\omega_n)j\omega + (j\omega/\omega_n)^2]$

We can determine the logarithmic magnitude plot and phase angle for these four factors and then utilize them to obtain a Bode diagram for any general form of a transfer function. Typically, the curves for each factor are obtained and then added together graphically to obtain the curves for the complete transfer function. Furthermore, this procedure can be simplified by using the asymptotic approximations to these curves and obtaining the actual curves only at specific important frequencies.

Constant Gain K_b. The logarithmic gain is

$$20 \log K_b = \text{constant in db},$$

and the phase angle is zero. The gain curve is simply a horizontal line on the Bode diagram.

Poles (or Zeros) at the Origin $(j\omega)$. A pole at the origin has a logarithmic magnitude

$$20 \log \left|\frac{1}{j\omega}\right| = -20 \log \omega \text{ db} \tag{7.29}$$

and a phase angle $\phi(\omega) = -90^\circ$. The slope of the magnitude curve is -20 db/decade for a pole. Similarly for a multiple pole at the origin, we have

$$20 \log \left|\frac{1}{(j\omega)^N}\right| = -20\, N \log \omega, \tag{7.30}$$

and the phase is $\phi(\omega) = -90°N$. In this case the slope due to the multiple pole is $-20N$ db/decade. For a zero at the origin, we have a logarithmic magnitude

$$20 \log |j\omega| = +20 \log \omega, \tag{7.31}$$

where the slope is $+20$ db/decade and the phase angle is $+90°$. The Bode diagram of the magnitude and phase angle of $(j\omega)^{\pm N}$ is shown in Fig. 7.7 for $N = 1$ and $N = 2$.

Poles or Zeros on the Real Axis. The pole factor $(1 + j\omega\tau)^{-1}$ has been considered previously and we found that

$$20 \log \left| \frac{1}{1 + j\omega\tau} \right| = -10 \log (1 + \omega^2\tau^2). \tag{7.32}$$

The asymptotic curve for $\omega \ll 1/\tau$ is $20 \log 1 = 0$ db, and the asymptotic curve for $\omega \gg 1/\tau$ is $-20 \log \omega\tau$ which has a slope of -20 db/decade. The intersection of the two asymptotes occurs when

$$20 \log 1 = 0 \text{ db} = -20 \log \omega\tau$$

or when $\omega = 1/\tau$, the *break frequency*. The actual logarithmic gain when $\omega = 1/\tau$ is -3 db for this factor. The phase angle is $\phi(\omega) = -\tan^{-1} \omega\tau$ for the denominator factor. The Bode diagram of a pole factor $(1 + j\omega\tau)^{-1}$ is shown in Fig. 7.8.

The Bode diagram of a zero factor $(1 + j\omega\tau)$ is obtained in the same manner as that of the pole. However, the slope is positive at $+20$ db/decade, and the phase angle is $\phi(\omega) = +\tan^{-1} \omega\tau$.

A linear approximation to the phase angle curve can be obtained as shown in Fig. 7.8. This linear approximation, which passes through the correct phase at the break frequency, is within 6° of the actual phase curve for all frequencies. This approximation will provide a useful means for readily determining the form of the phase angle curves of a transfer function $G(s)$. However, often the accurate phase angle curves are required and the actual phase curve for the first-order factor must be drawn. Therefore it is often worthwhile to prepare a cardboard (or plastic) template which can be utilized repeatedly to draw the phase curves for

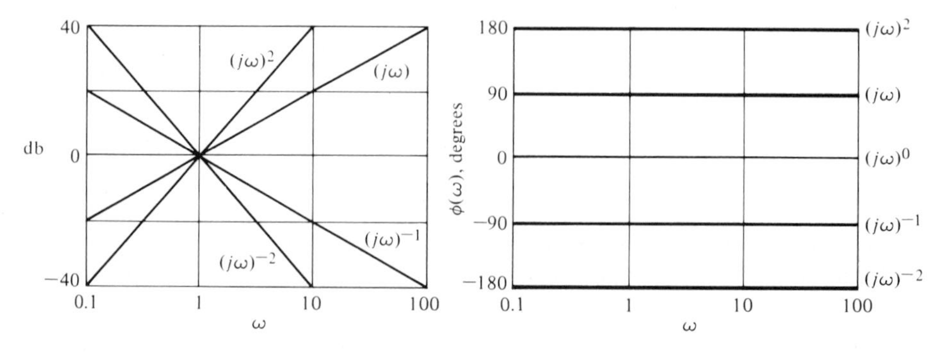

Figure 7.7. Bode diagram for $(j\omega)^{\pm N}$.

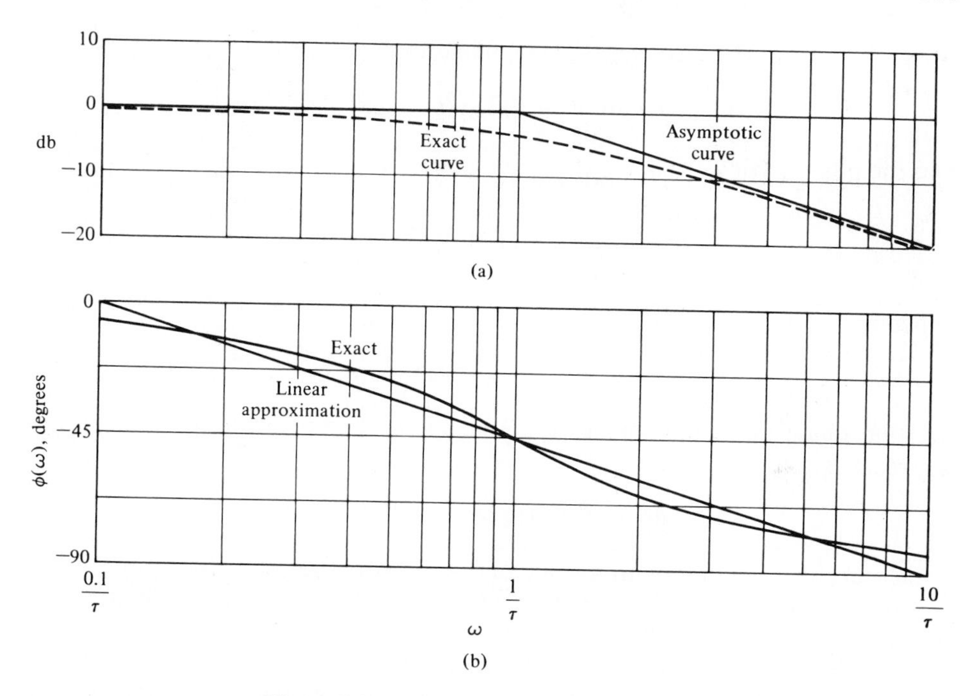

Figure 7.8. Bode diagram for $(1 + j\omega\tau)^{-1}$.

the individual factors. The exact values of the frequency response for the pole $(1 + j\omega\tau)^{-1}$ as well as the values obtained by using the approximation for comparison are given in Table 7.2.

Complex Conjugate Poles or Zeros $[1 + (2\zeta/\omega_n)j\omega + (j\omega/\omega_n)^2]$. The quadratic factor for a pair of complex conjugate poles can be written in normalized form as

$$[1 + j2\zeta u - u^2]^{-1}, \tag{7.33}$$

Table 7.2.

$\omega\tau$	0.10	0.50	0.76	1	1.31	2	5	10
$20 \log \lvert(1 + j\omega\tau)^{-1}\rvert$, db	−0.04	−1.0	−2.0	−3.0	−4.3	−7.0	−14.2	−20.04
Asymptotic approximation, db	0	0	0	0	−2.3	−6.0	−14.0	−20.0
$\phi(\omega)$, degrees	−5.7	−26.6	−37.4	−45.0	−52.7	−63.4	−78.7	−84.3
Linear approximation, degrees	0	−31.5	−39.5	−45.0	−50.3	−58.5	−76.5	−90.0

where $u = \omega/\omega_n$. Then, the logarithmic magnitude is

$$20 \log |G(\omega)| = -10 \log ((1 - u^2)^2 + 4\zeta^2 u^2), \tag{7.34}$$

and the phase angle is

$$\phi(\omega) = -\tan^{-1}\left(\frac{2\zeta u}{1 - u^2}\right). \tag{7.35}$$

When $u \ll 1$, the magnitude is

$$\text{db} = -10 \log 1 = 0 \text{ db},$$

and the phase angle approaches 0°. When $u \gg 1$, the logarithmic magnitude approaches

$$\text{db} = -10 \log u^4 = -40 \log u,$$

which results in a curve with a slope of -40 db/decade. The phase angle, when $u \gg 1$, approaches $-180°$. The magnitude asymptotes meet at the 0-db line when $u = \omega/\omega_n = 1$. However, the difference between the actual magnitude curve and the asymptotic approximation is a function of the damping ratio and must be accounted for when $\zeta < 0.707$. The Bode diagram of a quadratic factor due to a pair of complex conjugate poles is shown in Fig. 7.9. The maximum value of the frequency response, M_{p_ω}, occurs at the *resonant frequency* ω_r. When the damping ratio approaches zero, then ω_r approaches ω_n, the natural frequency. The resonant frequency is determined by taking the derivative of the magnitude of Eq. (7.33) with respect to the normalized frequency, u, and setting it equal to zero. The resonant frequency is represented by the relation

$$\omega_r = \omega_n\sqrt{1 - 2\zeta^2}, \qquad \zeta < 0.707, \tag{7.36}$$

and the maximum value of the magnitude $|G(\omega)|$ is

$$M_{p_\omega} = |G(\omega_r)| = (2\zeta\sqrt{1 - \zeta^2})^{-1}, \qquad \zeta < 0.707, \tag{7.37}$$

for a pair of complex poles. The maximum value of the frequency response M_{p_ω}, and the resonant frequency ω_r are shown as a function of the damping ratio ζ for a pair of complex poles in Fig. 7.10. Assuming the dominance of a pair of complex conjugate closed-loop poles, we find that these curves are useful for estimating the damping ratio of a system from an experimentally determined frequency response.

The frequency response curves can be evaluated graphically on the s-plane by determining the vector lengths and angles at various frequencies ω along the $(s = +j\omega)$-axis. For example, considering the second-order factor with complex conjugate poles, we have

$$G(s) = \frac{1}{(s/\omega_n)^2 + 2\zeta s/\omega_n + 1} = \frac{\omega_n^2}{s^2 + 2\zeta\omega_n s + \omega_n^2}. \tag{7.38}$$

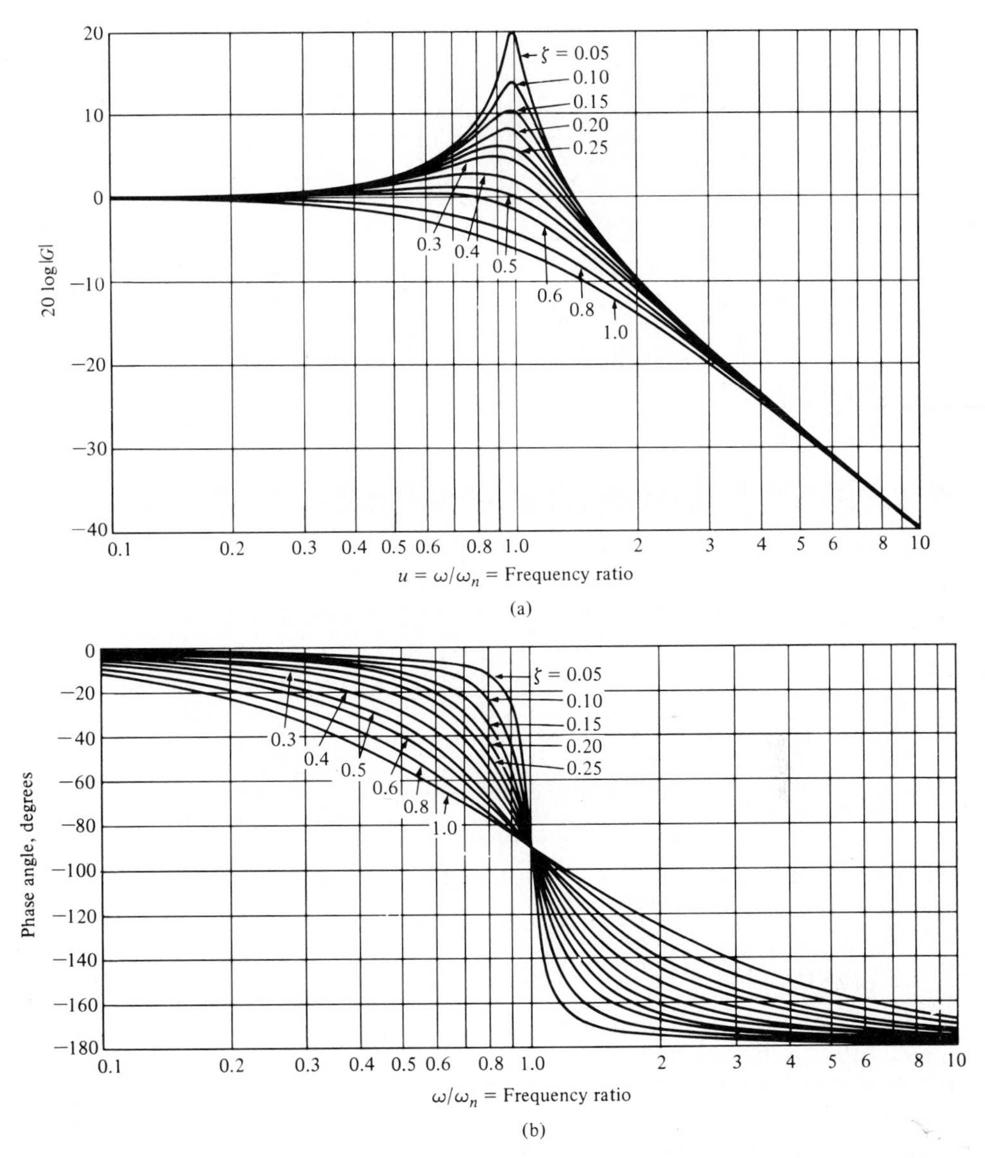

Figure 7.9. Bode diagram for $G(j\omega) = [1 + (2\zeta/\omega_n)j\omega + (j\omega/\omega_n)^2]^{-1}$.

The poles lie on a circle of radius ω_n and are shown for a particular ζ in Fig. 7.11(a). The transfer function evaluated for real frequency $s = j\omega$ is written as

$$G(j\omega) = \left.\frac{\omega_n^2}{(s - s_1)(s - s_1^*)}\right|_{s=j\omega} = \frac{\omega_n^2}{(j\omega - s_1)(j\omega - s_1^*)}, \tag{7.39}$$

where s_1 and s_1^* are the complex conjugate poles. The vectors $(j\omega - s_1)$ and $(j\omega - s_1^*)$ are the vectors from the poles to the frequency $j\omega$, as shown in Fig.

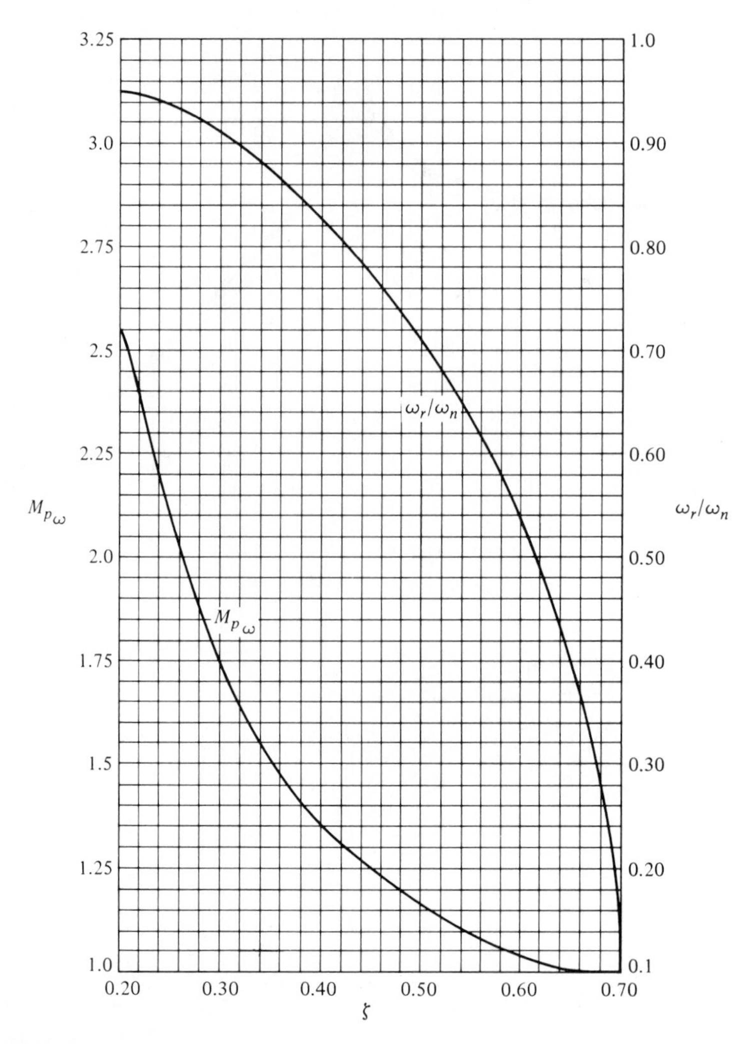

Figure 7.10. The maximum of the frequency response, M_{p_ω}, and the resonant frequency, ω_r, versus ζ for a pair of complex conjugate poles.

7.11(a). Then the magnitude and phase may be evaluated for various specific frequencies. The magnitude is

$$|G(\omega)| = \frac{\omega_n^2}{|j\omega - s_1|\,|j\omega - s_1^*|}, \tag{7.40}$$

and the phase is

$$\phi(\omega) = -\underline{/(j\omega - s_1)} - \underline{/(j\omega - s_1^*)}.$$

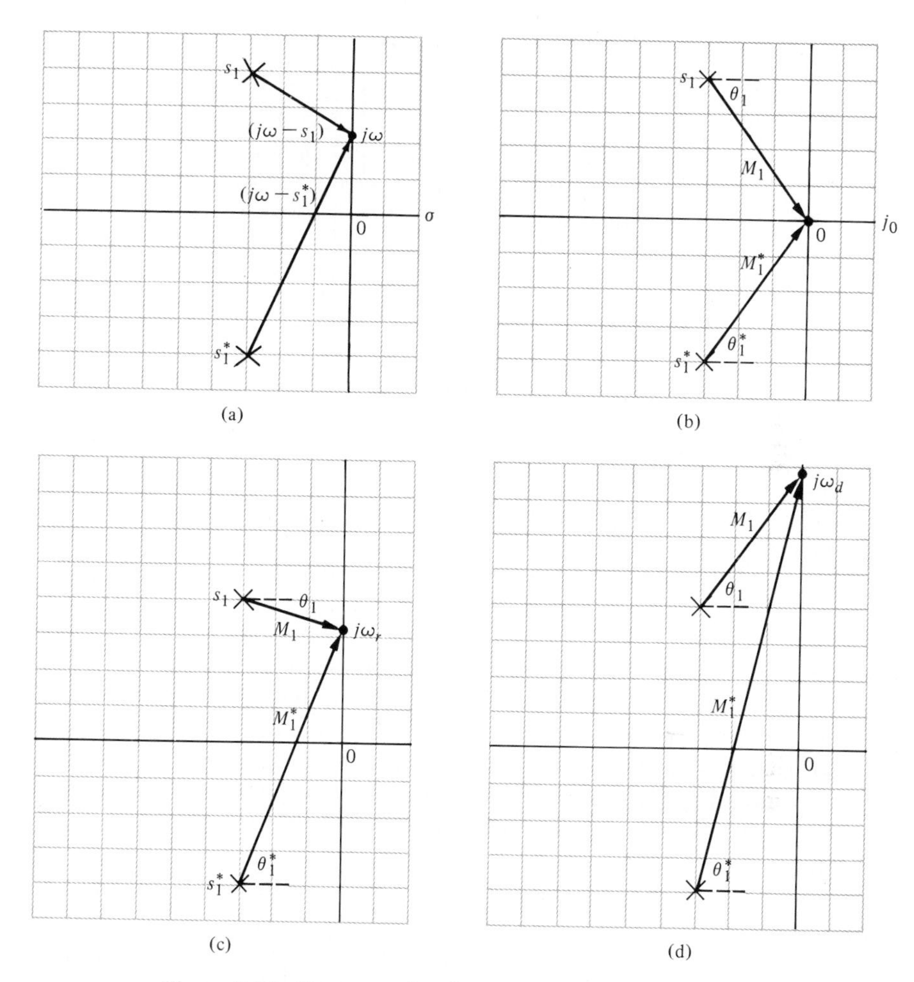

Figure 7.11. Vector evaluation of the frequency response.

The magnitude and phase may be evaluated for three specific frequencies:

$$\omega = 0, \qquad \omega = \omega_r, \qquad \omega = \omega_d,$$

as shown in Fig. 7.11 in parts (b), (c), and (d), respectively. The magnitude and phase corresponding to these frequencies are shown in Fig. 7.12.

■ Example 7.4 Bode diagram of twin-T network

As an example of the determination of the frequency response using the pole-zero diagram and the vectors to $j\omega$, consider the twin-T network shown in Fig. 7.13 [13]. The transfer function of this network is

$$G(s) = \frac{E_{\text{out}}(s)}{E_{\text{in}}(s)} = \frac{(s\tau)^2 + 1}{(s\tau)^2 + 4s\tau + 1}, \tag{7.41}$$

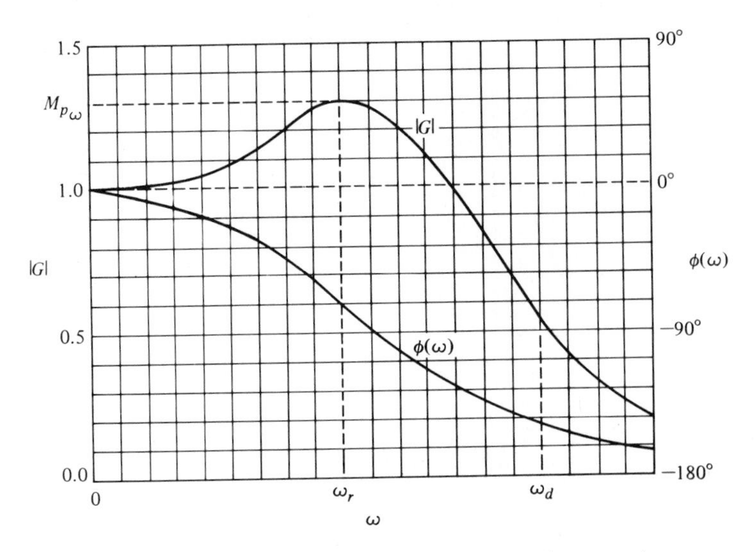

Figure 7.12. Bode diagram for complex conjugate poles.

where $\tau = RC$. The zeros are at $\pm j1$ and the poles are at $-2 \pm \sqrt{3}$ in the $s\tau$-plane as shown in Fig. 7.14(a). Clearly, at $\omega = 0$, we have $|G| = 1$ and $\phi(\omega) = 0°$. At $\omega = 1/\tau$, $|G| = 0$ and the phase angle of the vector from the zero at $s\tau = j1$ passes through a transition of 180°. When ω approaches ∞, $|G| = 1$ and $\phi(\omega) = 0$ again. Evaluating several intermediate frequencies, we can readily obtain the frequency response, as shown in Fig. 7.14(b).

In the previous examples the poles and zeros of $G(s)$ have been restricted to the left-hand plane. However, a system may have zeros located in the right-hand s-plane and may still be stable. Transfer functions with zeros in the right-hand s-plane are classified as *nonminimum phase-shift* transfer functions. If the zeros of a transfer function are all reflected about the $j\omega$-axis, there is no change in the magnitude of the transfer function, and the only difference is in the phase-shift characteristics. If the phase characteristics of the two system functions are compared, it can be readily shown that the net phase shift over the frequency range from zero to infinity is less for the system with all its zeros in the left-hand s-

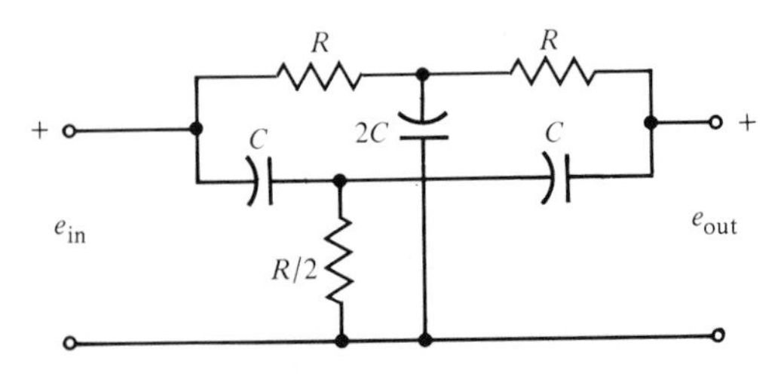

Figure 7.13. Twin-T network.

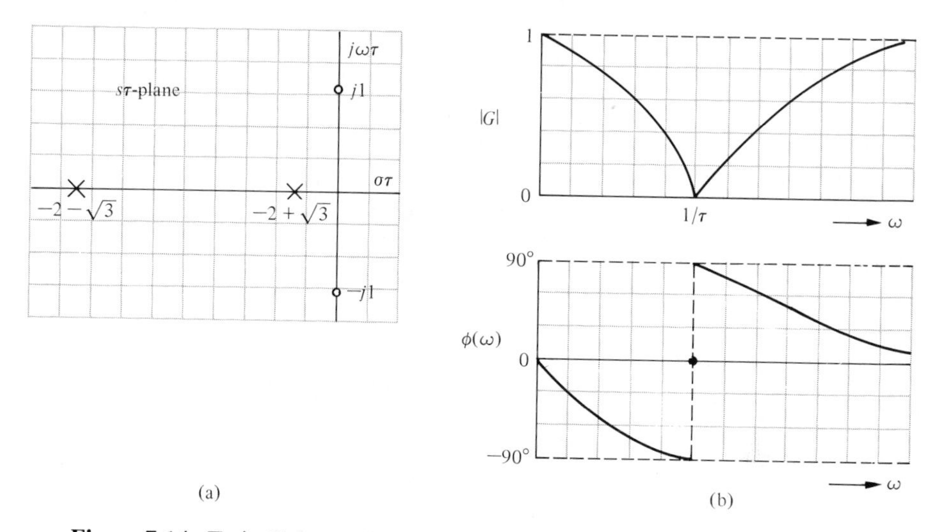

Figure 7.14. Twin-T network. (a) Pole–zero pattern. (b) Frequency response.

plane. Thus the transfer function $G_1(s)$, with all its zeros in the left-hand s-plane, is called a *minimum phase* transfer function. The transfer function $G_2(s)$, with $|G_2(j\omega)| = |G_1(j\omega)|$ and all the zeros of $G_1(s)$ reflected about the $j\omega$-axis into the right-hand s-plane, is called a *nonminimum phase* transfer function. Reflection of any zero or pair of zeros into the right-half plane results in a nonminimum phase transfer function.

The two pole-zero patterns shown in Fig. 7.15(a) and (b) have the same amplitude characteristics as can be deduced from the vector lengths. However, the phase characteristics are different for Fig. 7.15(a) and (b). The minimum phase characteristic of Fig. 7.15(a) and the nonminimum phase characteristic of Fig. 7.15(b) are shown in Fig. 7.16. Clearly, the phase shift of

$$G_1(s) = \frac{s+z}{s+p}$$

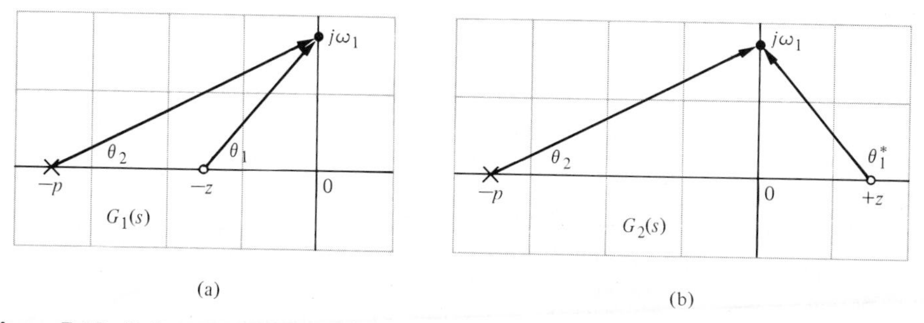

Figure 7.15. Pole-zero patterns giving the same amplitude response and different phase characteristics.

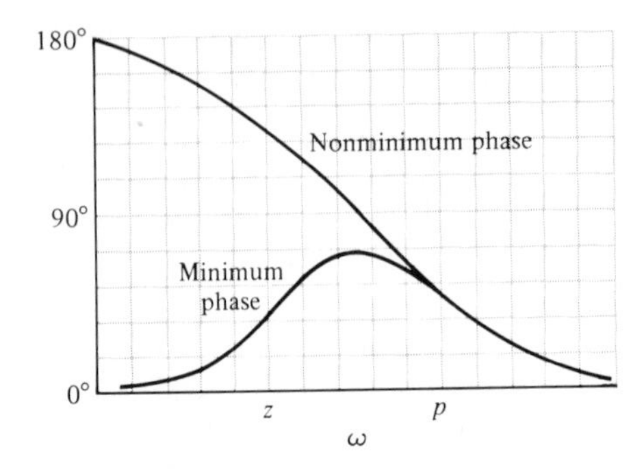

Figure 7.16. The phase characteristics for the minimum phase and nonminimum phase transfer function.

ranges over less than 80°, whereas the phase shift of

$$G_2(s) = \frac{s - z}{s + p}$$

ranges over 180°. The meaning of the term minimum phase is illustrated by Fig. 7.16. The range of phase shift of a minimum phase transfer function is the least

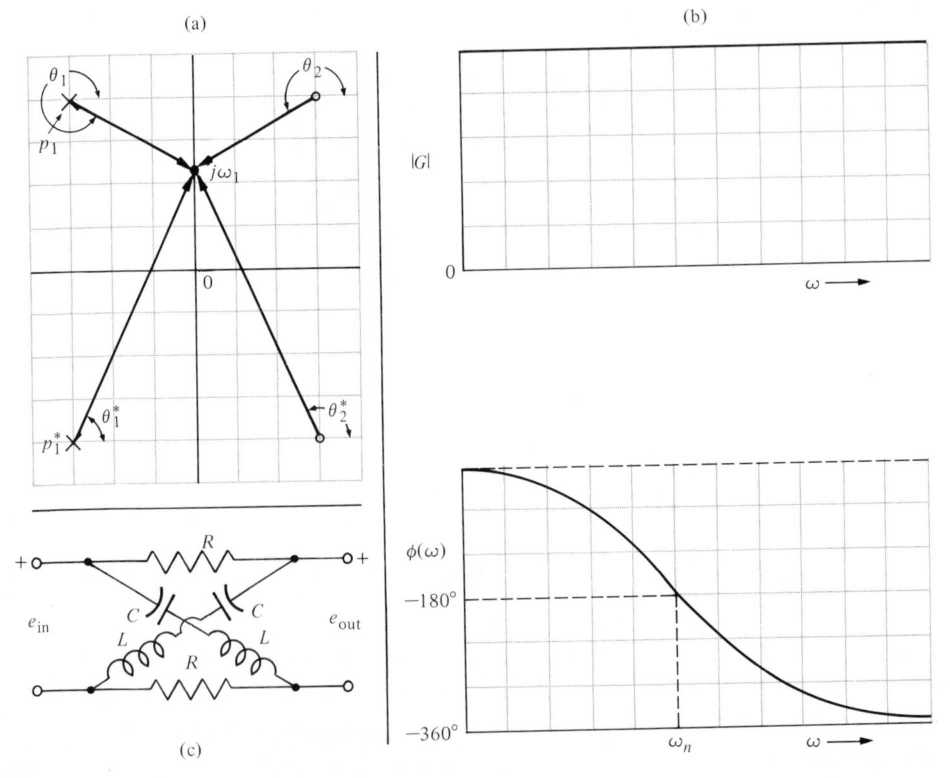

Figure 7.17. The all-pass network pole-zero pattern and frequency response.

possible or minimum corresponding to a given amplitude curve, whereas the range of the nonminimum phase curve is greater than the minimum possible for the given amplitude curve.

A particularly interesting nonminimum phase network is the *all-pass* network, which can be realized with a symmetrical lattice network [8]. A symmetrical pattern of poles and zeros is obtained as shown in Fig. 7.17(a). Again, the magnitude $|G|$ remains constant; in this case, it is equal to unity. However, the angle varies from 0° to −360°. Because $\theta_2 = 180° - \theta_1$ and $\theta_2^* = 180° - \theta_1^*$, the phase is given by $\phi(\omega) = -2(\theta_1 + \theta_1^*)$. The magnitude and phase characteristic of the all-pass network is shown in Fig. 7.17(b). A nonminimum phase lattice network is shown in Fig. 7.17(c).

7.3 An Example of Drawing the Bode Diagram

The Bode diagram of a transfer function $G(s)$, which contains several zeros and poles, is obtained by adding the plot due to each individual pole and zero. The simplicity of this method will be illustrated by considering a transfer function that possesses all the factors considered in the preceding section. The transfer function of interest is

$$G(j\omega) = \frac{5(1 + j0.1\omega)}{j\omega(1 + j0.5\omega)(1 + j0.6(\omega/50) + (j\omega/50)^2)}. \tag{7.42}$$

The factors, in order of their occurrence as frequency increases, are as follows:

1. A constant gain $K = 5$
2. A pole at the origin
3. A pole at $\omega = 2$
4. A zero at $\omega = 10$
5. A pair of complex poles at $\omega = \omega_n = 50$

First, we plot the magnitude characteristic for each individual pole and zero factor and the constant gain.

1. The constant gain is 20 log 5 = 14 db, as shown in Fig. 7.18.
2. The magnitude of the pole at the origin extends from zero frequency to infinite frequencies and has a slope of −20 db/decade intersecting the 0-db line at $\omega = 1$, as shown in Fig. 7.18.
3. The asymptotic approximation of the magnitude of the pole at $\omega = 2$ has a slope of −20 db/decade beyond the break frequency at $\omega = 2$. The asymptotic magnitude below the break frequency is 0 db, as shown in Fig. 7.18.
4. The asymptotic magnitude for the zero at $\omega = +10$ has a slope of +20 db/decade beyond the break frequency at $\omega = 10$, as shown in Fig. 7.18.

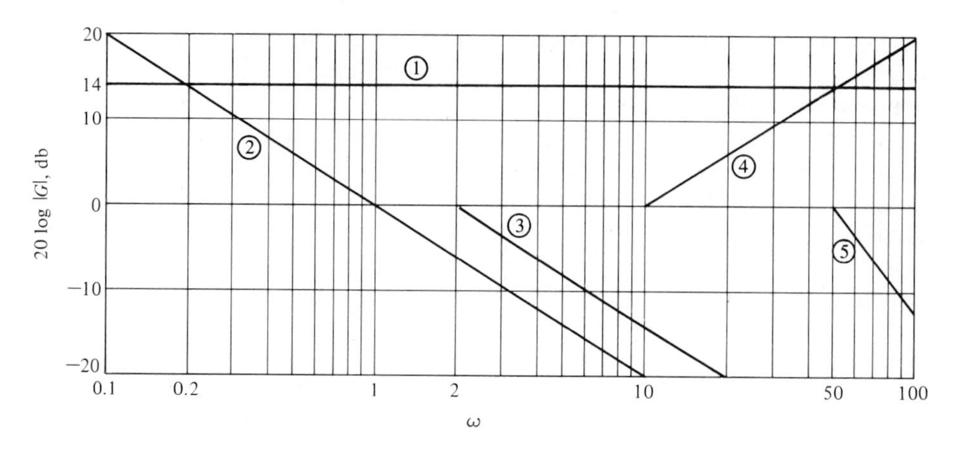

Figure 7.18. Magnitude asymptotes of poles and zeros used in the example.

5. The asymptotic approximation for the pair of complex poles at $\omega = \omega_n = 50$ has a slope of -40 db/decade due to the quadratic forms. The break frequency is $\omega = \omega_n = 50$, as shown in Fig. 7.18. This approximation must be corrected to the actual magnitude because the damping ratio is $\zeta = 0.3$ and the magnitude differs appreciably from the approximation, as shown in Fig. 7.19.

Therefore the total asymptotic magnitude can be plotted by adding the asymptotes due to each factor, as shown by the solid line in Fig. 7.19. Examining the asymptotic curve of Fig. 7.19, one notes that the curve can be obtained

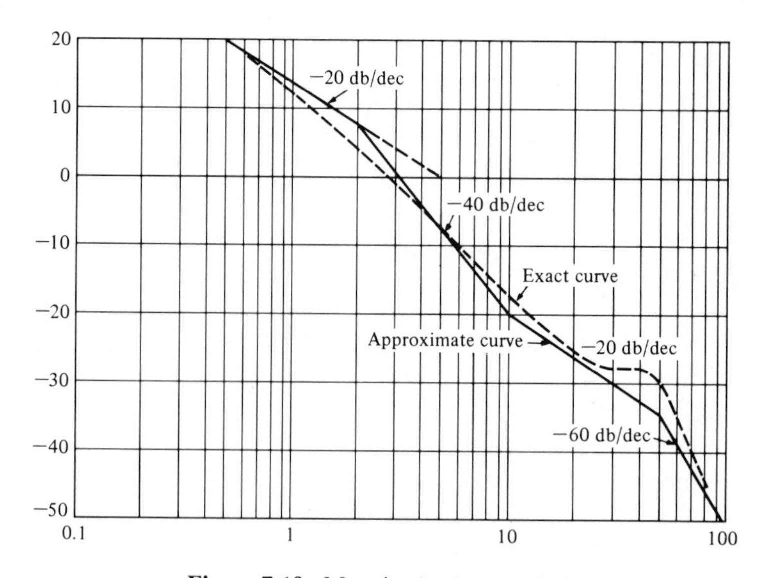

Figure 7.19. Magnitude characteristic.

directly by plotting each asymptote in order as frequency increases. Thus the slope is -20 db/decade due to $(j\omega)^{-1}$ intersecting 14 db at $\omega = 1$. Then at $\omega = 2$, the slope becomes -40 db/decade due to the pole at $\omega = 2$. The slope changes to -20 db/decade due to the zero at $\omega = 10$. Finally, the slope becomes -60 db/decade at $\omega = 50$ due to the pair of complex poles at $\omega_n = 50$.

The exact magnitude curve is then obtained by utilizing Table 7.2, which provides the difference between the actual and asymptotic curves for a single pole or zero. The exact magnitude curve for the pair of complex poles is obtained by utilizing Fig. 7.9(a) for the quadratic factor. The exact magnitude curve for $G(j\omega)$ is shown by a dashed line in Fig. 7.19.

The phase characteristic can be obtained by adding the phase due to each individual factor. Usually, the linear approximation of the phase characteristic for a single pole or zero is suitable for the initial analysis or design attempt. Thus the individual phase characteristics for the poles and zeros are shown in Fig. 7.20.

1. The phase of the constant gain is, of course, 0°.
2. The phase of the pole at the origin is a constant $-90°$.
3. The linear approximation of the phase characteristic for the pole at $\omega = 2$ is shown in Fig. 7.20, where the phase shift is $-45°$ at $\omega = 2$.
4. The linear approximation of the phase characteristic for the zero at $\omega = 10$ is also shown in Fig. 7.20, where the phase shift is $+45°$ at $\omega = 10$.
5. The actual phase characteristic for the pair of complex poles is obtained from Fig. 7.9 and is shown in Fig. 7.20.

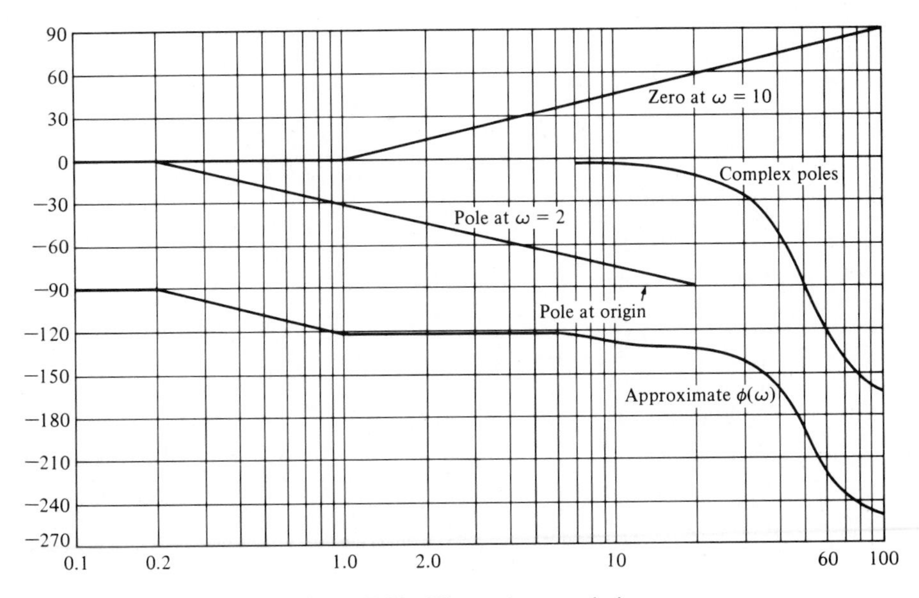

Figure 7.20. Phase characteristic.

Therefore the total phase characteristic, $\phi(\omega)$, is obtained by adding the phase due to each factor as shown in Fig. 7.20. While this curve is an approximation, its usefulness merits consideration as a first attempt to determine the phase characteristic. Thus, for example, a frequency of interest, as we shall note in the following section, is that frequency for which $\phi(\omega) = -180°$. The approximate curve indicates that a phase shift of $-180°$ occurs at $\omega = 46$. The actual phase shift at $\omega = 46$ can be readily calculated as

$$\phi(\omega) = -90° - \tan^{-1} \omega\tau_1 + \tan^{-1} \omega\tau_2 - \tan^{-1} \frac{2\zeta u}{1 - u^2}, \tag{7.43}$$

where

$$\tau_1 = 0.5, \qquad \tau_2 = 0.1, \qquad u = \omega/\omega_n = \omega/50.$$

Then we find that

$$\begin{aligned} \phi(46) &= -90° - \tan^{-1} 23 + \tan^{-1} 4.6 - \tan^{-1} 3.55 \\ &= -175°, \end{aligned} \tag{7.44}$$

and the approximate curve has an error of 5° at $\omega = 46$. However, once the approximate frequency of interest is ascertained from the approximate phase curve, the accurate phase shift for the neighboring frequencies is readily determined by using the exact phase shift relation (Eq. 7.43). This approach is usually preferable to the calculation of the exact phase shift for all frequencies over several decades. In summary, one may obtain approximate curves for the magnitude and phase shift of a transfer function $G(j\omega)$ in order to determine the important frequency ranges. Then, within the relatively small important frequency ranges, the exact magnitude and phase shift can be readily evaluated by using the exact equations, such as Eq. (7.43).

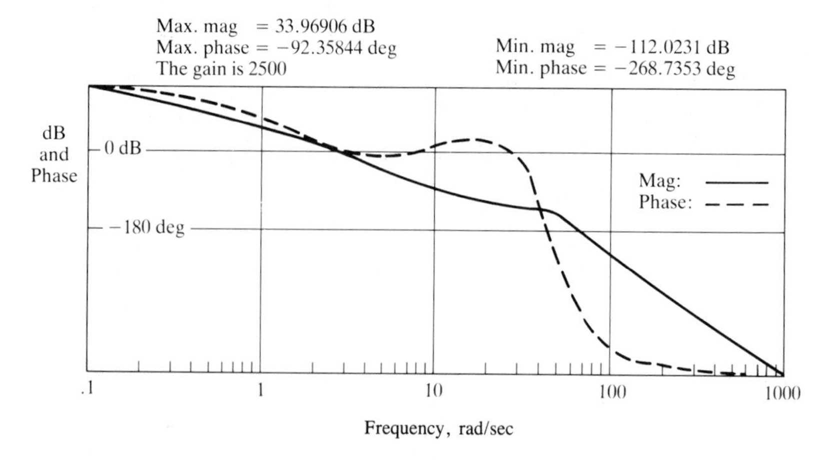

Figure 7.21. The Bode plot of the $G(j\omega)$ of Eq. 7.42 obtained from the Control System Design Program.

The frequency response of $G(j\omega)$ can be calculated and plotted using a computer program. The Control System Design Program (CSDP) will readily generate the frequency response, providing 30 points of actual data over a two-decade range as well as providing the Bode plot [see Appendix F].

The Bode plot for the example in Section 7.3 (Eq. 7.42) can be readily obtained using CSDP, as shown in Fig. 7.21. The plot is generated for four decades and the 0 db line is indicated as well as the -180 degree line. The data above the plot indicates that the magnitude is 34 db and the phase is -92.36 degrees at $\omega = 0.1$. Similarly, the data indicates that the magnitude is -43 db and the phase is -243 degrees at $\omega = 100$. Using the tabular data provided by CSDP, one finds that the magnitude is 0 db at $\omega = 3.0$ and the phase is -180 degrees at $\omega = 50$.

7.4 Frequency Response Measurements

A sine wave can be used to measure the open-loop frequency response of a control system. In practice a plot of amplitude versus frequency and phase versus frequency will be obtained [8,16]. From these two plots the open-loop transfer function $GH(j\omega)$ can be deduced. Similarly, the closed-loop frequency response of a control system, $T(j\omega)$, may be obtained and the actual transfer function deduced.

A device called a wave analyzer can be used to measure the amplitude and phase variations as the frequency of the input sine wave is altered. Also, a device called a transfer function analyzer can be used to measure the open-loop and closed-loop transfer functions [8].

The HP Dynamic Signal Analyzer shown in Fig. 7.22 is an example of a fre-

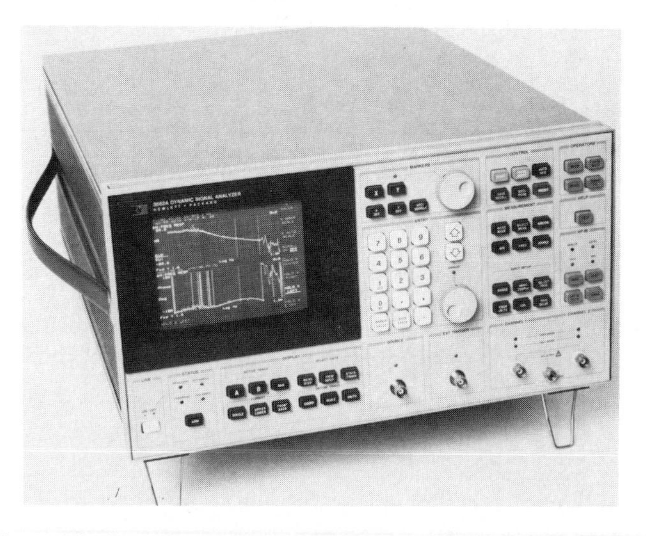

Figure 7.22. The HP 3562A Dynamic Signal Analyzer performs frequency response measurements from dc to 100 kHz. Built-in analysis and modeling capabilities can derive poles and zeros from measured frequency responses or construct phase and magnitude responses from user-supplied models. (Courtesy of Hewlett-Packard Co.)

quency response measurement tool. This device can also synthesize the frequency response of a model of a system allowing a comparison with an actual response.

As an example of determining the transfer function from the Bode plot, let us consider the plot shown in Fig. 7.23. The system is a stable circuit consisting of resistors and capacitors. Because the phase and magnitude decline as ω increases between 10 and 1000, and because the phase is $-45°$ and the gain -3 db at 370 rad/sec, we can deduce that one factor is a pole near $\omega = 370$. Beyond 370 rad/sec, the magnitude drops sharply at -40 db/decade, indicating that another pole exists. However, the phase drops to $-66°$ at $\omega = 1250$ and then starts to rise again, eventually approaching 0° at large values of ω. Also, because the magnitude returns to 0 db as ω exceeds 50,000, we determine that there are

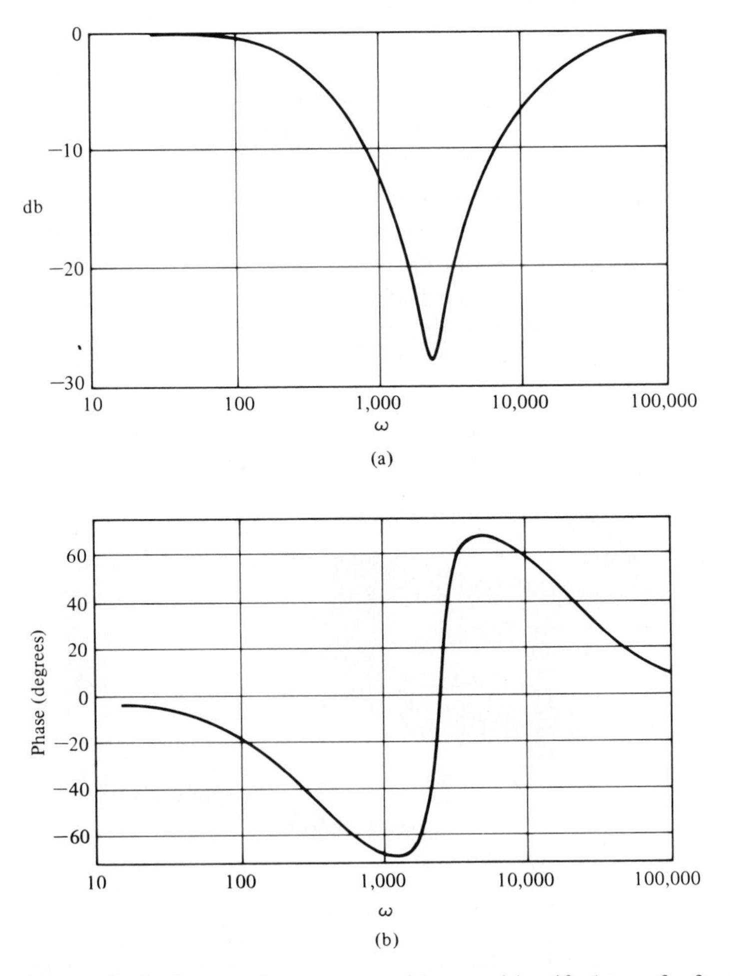

Figure 7.23. A Bode diagram for a system with an unidentified transfer function.

two zeros as well as two poles. We deduce that the numerator is a quadratic factor with a small ζ yielding the sharp phase change. Therefore the transfer function is

$$T(s) = \frac{(s/\omega_n)^2 + (2\zeta/\omega_n)s + 1}{(\tau_1 s + 1)(\tau_2 s + 1)},$$

where we know $\tau_1 = 1/370$. Reviewing Fig. 7.9, we note that the phase passes through $+90°$ for a quadratic numerator at $\omega = \omega_n$. Because τ_1 yields $-90°$ by $\omega = 1000$, we deduce that $\omega_n = 2500$. Drawing the asymptotic curve for the pole $1/\tau_1$ and the numerator, we estimate $\zeta = 0.15$. Finally, the pole $p = 1/\tau_2$ yields a $45°$ phase shift from the asymptotic approximation so that $p = 20{,}000$. Therefore

$$T(s) = \frac{(s/2500)^2 + (0.3/2500)s + 1}{(s/370 + 1)(s/20{,}000 + 1)}.$$

This frequency response is actually obtained from a bridged-T network.

7.5 Performance Specifications in the Frequency Domain

We must continually ask the question: How does the frequency response of a system relate to the expected transient response of the system? In other words, given a set of time-domain (transient performance) specifications, how do we specify the frequency response? For a simple second-order system we have already answered this question by considering the performance in terms of overshoot, settling time, and other performance criteria such as Integral Squared Error. For the second-order system shown in Fig. 7.24, the closed-loop transfer function is

$$T(s) = \frac{\omega_n^2}{s^2 + 2\zeta\omega_n s + \omega_n^2}. \tag{7.45}$$

The frequency response of this feedback system will appear as shown in Fig. 7.25. Because this is a second-order system, the damping ratio of the system is related to the maximum magnitude M_{p_ω}. Furthermore, the resonant frequency ω_r and the -3-db *bandwidth* can be related to the speed of the transient response. Thus, as the bandwidth ω_B increases, the rise time of the step response of the system will decrease. Furthermore, the overshoot to a step input can be related to M_{p_ω} through the damping ratio ζ. The curves of Fig. 7.10 relate the resonance magnitude and

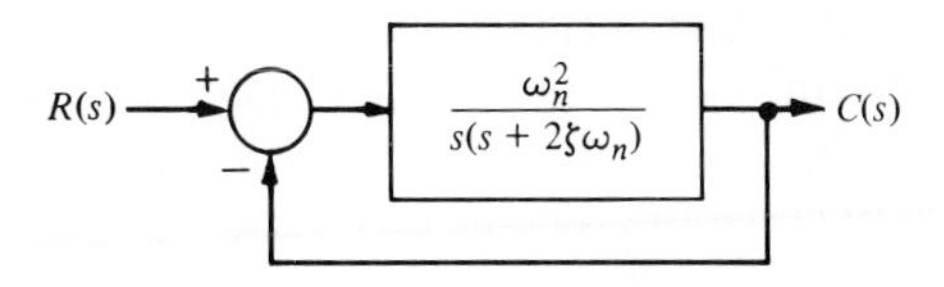

Figure 7.24. A second-order closed-loop system.

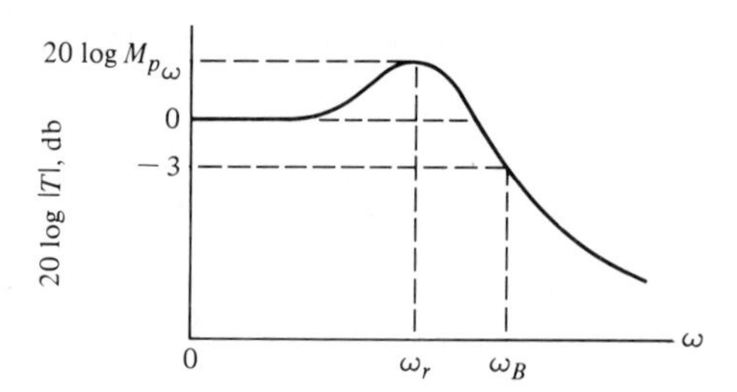

Figure 7.25. Magnitude characteristic of the second-order system.

frequency to the damping ratio of the second-order system. Then the step response overshoot may be estimated from Fig. 4.8 or may be calculated by utilizing Eq. (4.15). Thus we find as the resonant peak M_{p_ω} increases in magnitude, the overshoot to a step input increases. In general, the magnitude M_{p_ω} indicates the relative stability of a system.

The bandwidth of a system ω_B as indicated on the frequency response can be approximately related to the natural frequency of the system. The response of the second-order system to a unit step input is

$$c(t) = 1 + e^{-\zeta\omega_n t} \cos(\omega_1 t + \theta). \tag{7.46}$$

The greater the magnitude of ω_n when ζ is constant, the more rapid the response approaches the desired steady-state value. Thus desirable frequency-domain specifications are as follows:

1. Relatively small resonant magnitudes; $M_{p_\omega} < 1.5$, for example.
2. Relatively large bandwidths so that the system time constant $\tau = 1/\zeta\omega_n$ is sufficiently small.

The usefulness of these frequency-response specifications and their relation to the actual transient performance depend upon the approximation of the system by a second-order pair of complex poles. This approximation was discussed in Section 6.3, and the second-order poles of $T(s)$ are called the *dominant roots.* Clearly, if the frequency response is dominated by a pair of complex poles the relationships between the frequency response and the time response discussed in this section will be valid. Fortunately, a large proportion of control systems satisfy this dominant second-order approximation in practice.

The steady-state error specification can also be related to the frequency response of a closed-loop system. As we found in Section 4.4, the steady-state error for a specific test input signal can be related to the gain and number of integrations (poles at the origin) of the open-loop transfer function. Therefore, for the system shown in Fig. 7.24, the steady-state error for a ramp input is specified in terms of K_v, the velocity constant. The steady-state error for the system is

$$\lim_{t\to\infty} e(t) = R/K_v, \tag{7.47}$$

where R = magnitude of the ramp input. The velocity constant for the closed-loop system of Fig. 7.24 is

$$K_v = \lim_{s\to 0} sG(s) = \lim_{s\to 0} s\left(\frac{\omega_n^2}{s(s + 2\zeta\omega_n)}\right) = \frac{\omega_n}{2\zeta}. \tag{7.48}$$

In Bode diagram form (in terms of time constants), the transfer function $G(s)$ is written as

$$G(s) = \frac{(\omega_n/2\zeta)}{s(s/(2\zeta\omega_n) + 1)} = \frac{K_v}{s(\tau s + 1)}, \tag{7.49}$$

and the gain constant is K_v for this type-one system. For example, reexamining the example of Section 7.3, we had a type-one system with an open-loop transfer function

$$G(j\omega) = \frac{5(1 + j\omega\tau_2)}{j\omega(1 + j\omega\tau_1)(1 + j0.6u - u^2)}, \tag{7.50}$$

where $u = \omega/\omega_n$. Therefore in this case we have $K_v = 5$. In general, if the open-loop transfer function of a feedback system is written as

$$G(j\omega) = \frac{K\prod_{i=1}^{M}(1 + j\omega\tau_i)}{(j\omega)^N\prod_{k=1}^{Q}(1 + j\omega\tau_k)}, \tag{7.51}$$

then the system is type N and the gain K is the gain constant for the steady-state error. Thus for a type-zero system that has an open-loop transfer function

$$G(j\omega) = \frac{K}{(1 + j\omega\tau_1)(1 + j\omega\tau_2)}. \tag{7.52}$$

$K = K_p$ (the position error constant) and appears as the low frequency gain on the Bode diagram.

Furthermore, the gain constant $K = K_v$ for the type-one system appears as the gain of the low frequency section of the magnitude characteristic. Considering only the pole and gain of the type-one system of Eq. (7.50), we have

$$G(j\omega) = \left(\frac{5}{j\omega}\right) = \left(\frac{K_v}{j\omega}\right), \quad \omega < 1/\tau_1, \tag{7.53}$$

and the K_v is equal to the frequency when this portion of the magnitude characteristic intersects the 0-db line. For example, the low frequency intersection of $(K_v/j\omega)$ in Fig. 7.19 is equal to $\omega = 5$, as we expect.

Therefore the frequency response characteristics represent the performance of a system quite adequately, and with some experience they are quite useful for the analysis and design of feedback control systems.

7.6 Log Magnitude and Phase Diagrams

There are several alternative methods of presenting the frequency response of a function $GH(j\omega)$. We have seen that suitable graphical presentations of the frequency response are (1) the polar plot and (2) the Bode diagram. An alternative approach to graphically portraying the frequency response is to plot the logarithmic magnitude in db versus the phase angle for a range of frequencies. Because this information is equivalent to that portrayed by the Bode diagram, it is normally easier to obtain the Bode diagram and transfer the information to the coordinates of the log magnitude versus phase diagram. Alternatively, one can construct templates for first- and second-order factors and work directly on the log-magnitude–phase diagram. The gain and phase of cascaded transfer functions can then be added vectorially directly on the diagram.

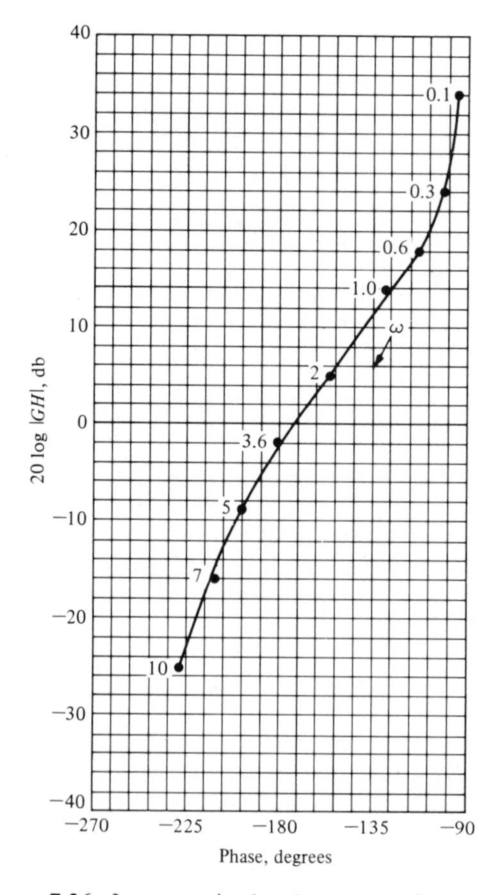

Figure 7.26. Log-magnitude–phase curve for $GH_1(j\omega)$.

An illustration will best portray the use of the log-magnitude–phase diagram. The log-magnitude–phase diagram for a transfer function

$$GH_1(j\omega) = \frac{5}{j\omega(0.5j\omega + 1)((j\omega/6) + 1)} \tag{7.54}$$

is shown in Fig. 7.26. The numbers indicated along the curve are for values of frequency, ω.

The log-magnitude–phase curve for the transfer function

$$GH_2(j\omega) = \frac{5(0.1j\omega + 1)}{j\omega(0.5j\omega + 1)(1 + j0.6(\omega/50) + (j\omega/50)^2)} \tag{7.55}$$

considered in Section 7.3 is shown in Fig. 7.27. This curve is obtained most readily by utilizing the Bode diagrams of Figs. 7.19 and 7.20 to transfer the frequency response information to the log magnitude and phase coordinates. The

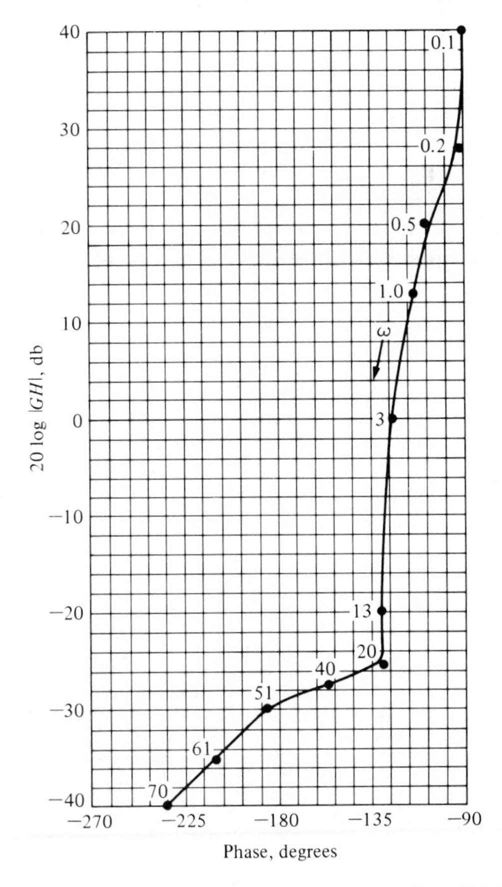

Figure 7.27. Log-magnitude–phase curve for $GH_2(j\omega)$.

shape of the locus of the frequency response on a log-magnitude–phase diagram is particularly important as the phase approaches 180° and the magnitude approaches 0 db. Clearly, the locus of Eq. (7.54) and Fig. 7.26 differs substantially from the locus of Eq. (7.55) and Fig. 7.27. Therefore, as the correlation between the shape of the locus and the transient response of a system is established, we will obtain another useful portrayal of the frequency response of a system. In the following chapter, we will establish a stability criterion in the frequency domain for which it will be useful to utilize the logarithmic-magnitude–phase diagram to investigate the relative stability of closed-loop feedback control systems.

7.7 Design Example: Engraving Machine Control System

The engraving machine shown in Fig. 7.28(a) uses two drive motors and associated lead screws to position the engraving scribe in the x direction. A separate motor is used for both the y and z axis as shown. The block diagram model for

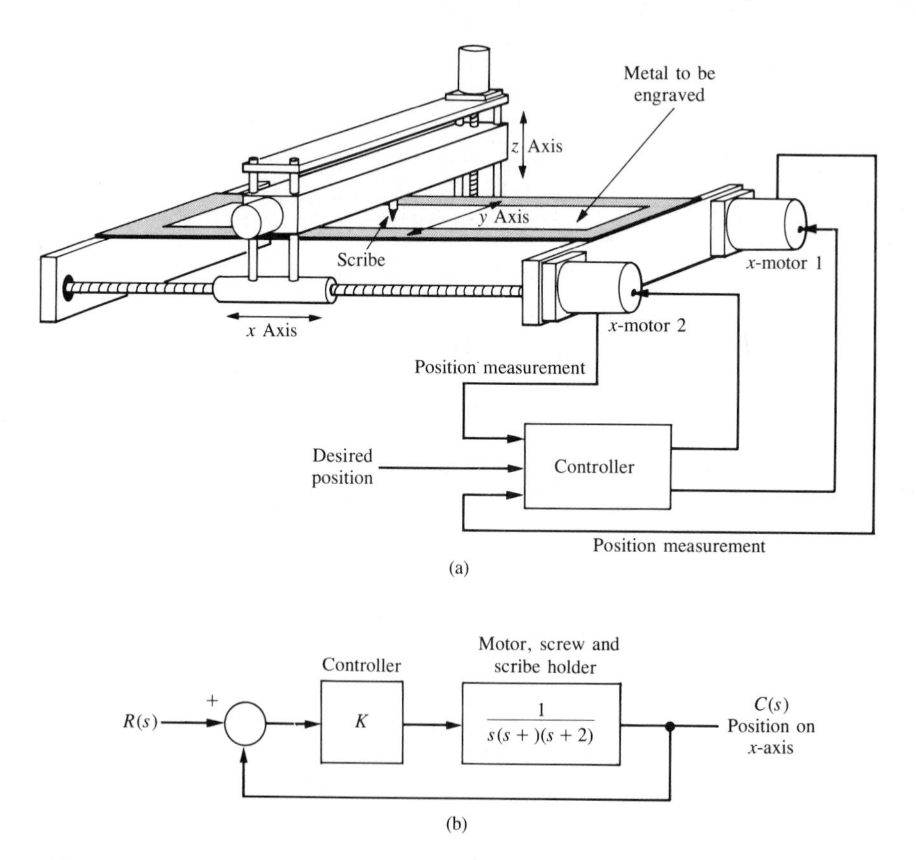

Figure 7.28. (a) Engraving machine control system. (b) Block diagram model.

Table 7.3. Frequency Response for $G(j\omega)$

ω	0.2	0.4	0.8	1.0	1.4	1.8
$20 \log \lvert G \rvert$	14	7	-1	-4	-9	-13
ϕ	$-107°$	$-123°$	$-150.5°$	$-162°$	$-179.5°$	$-193°$

the x-axis position control system is shown in Fig. 7.28(b). The goal is to select an appropriate gain K so that the time response to step commands is acceptable by utilizing frequency response methods.

In order to represent the frequency response of the system, we will first obtain the open-loop Bode diagram and the closed-loop Bode diagram. Then we will use the closed-loop Bode diagram to predict the time response of the system and check the predicted results with the actual results.

In order to plot the frequency response, we arbitrarily select $K = 2$ and proceed with obtaining the Bode diagram. If the resulting system is not acceptable, we will later adjust the gain.

The frequency response of $G(j\omega)$ is partially listed in Table 7.3 and it is plotted in Fig. 7.29. We need the frequency response of the closed-loop transfer function

$$T(s) = \frac{2}{s^3 + 3s^2 + 2s + 2}. \tag{7.56}$$

Therefore we let $s = j\omega$, obtaining

$$T(j\omega) = \frac{2}{(2 - 3\omega^2) + j\omega(2 - \omega^2)}. \tag{7.57}$$

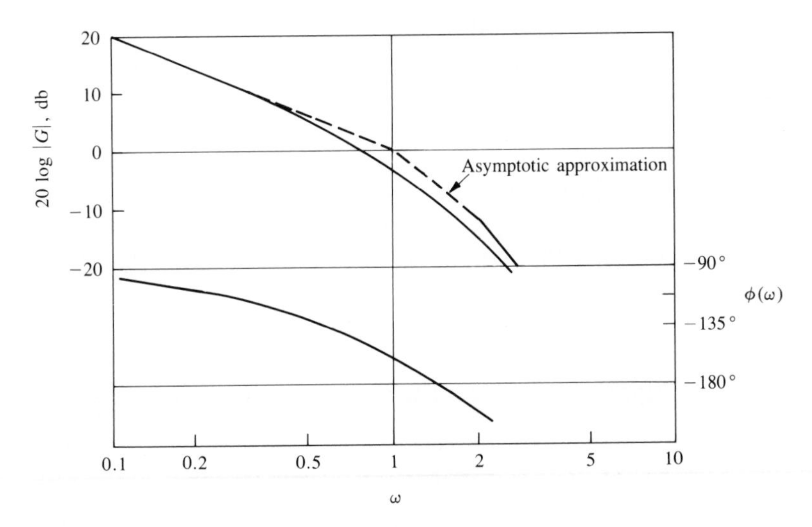

Figure 7.29. Bode diagram for $G(j\omega)$.

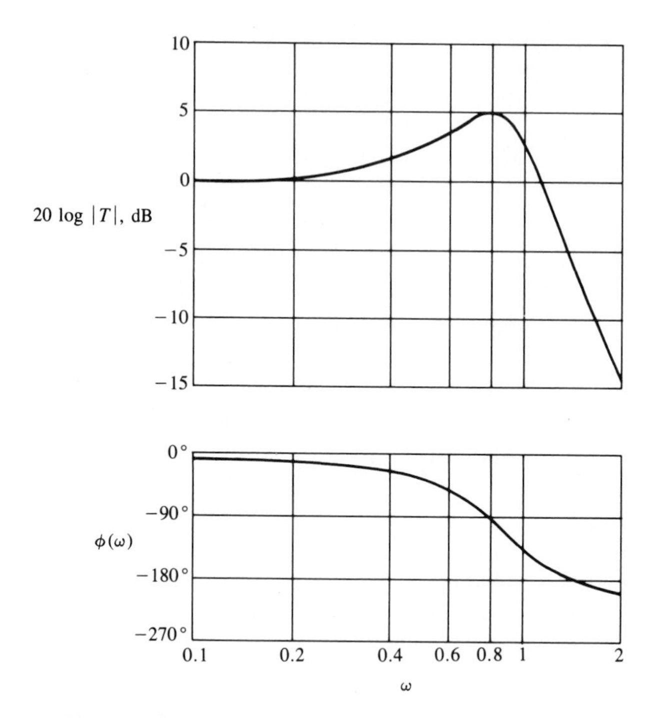

Figure 7.30. Bode diagram for closed-loop system.

The Bode diagram of the closed-loop system is shown in Fig. 7.30, where $20 \log |T| = 5$ db at $\omega_r = 0.8$. Therefore

$$20 \log M_{P_\omega} = 5$$

or

$$M_{P_\omega} = 1.78.$$

If we assume that the system has dominant second-order roots, we can approximate the system with a second-order frequency response of the form shown in Fig. 7.9. Since $M_{P_\omega} = 1.78$, we use Fig. 7.10 to estimate ζ to be 0.29. Using this ζ and $\omega_r = 0.8$, we can use Fig. 7.10 to estimate $\omega_r/\omega_n = 0.91$. Therefore

$$\omega_n = \frac{0.8}{0.91} = 0.88.$$

Since we are now approximating $T(s)$ as a second-order system, we have

$$\begin{aligned} T(s) &\cong \frac{\omega_n^2}{s^2 + 2\zeta\omega_n s + \omega_n^2} \\ &= \frac{0.774}{s^2 + 0.51s + 0.774}. \end{aligned} \tag{7.58}$$

We use Fig. 4.8 to predict the overshoot to a step input as 37% for $\zeta = 0.29$. The settling time is estimated as

$$T_s = \frac{4}{\zeta\omega_n} = \frac{4}{(0.29)0.88} = 15.7 \text{ seconds.}$$

The actual overshoot for a step input is 34% and the actual settling time is 17 seconds. Clearly, the second-order approximation is reasonable in this case and can be used to determine suitable parameters on a system. If we require a system with lower overshoot, we would reduce K to 1 and repeat the procedure.

7.8 Summary

In this chapter we have considered the representation of a feedback control system by its frequency response characteristics. The frequency response of a system was defined as the steady-state response of the system to a sinusoidal input signal. Several alternative forms of frequency response plots were considered. The polar plot of the frequency response of a system $G(j\omega)$ was considered. Also, logarithmic plots, often called Bode plots, were considered and the value of the logarithmic measure was illustrated. The ease of obtaining a Bode plot for the various factors of $G(j\omega)$ was noted, and an example was considered in detail. The asymptotic approximation for drawing the Bode diagram simplifies the computation considerably. Several performance specifications in the frequency domain were discussed; among them were the maximum magnitude M_{p_ω} and the resonant frequency ω_r. The relationship between the Bode diagram plot and the system error constants (K_p and K_v) was noted. Finally, the log magnitude versus phase diagram was considered for graphically representing the frequency response of a system.

Exercises

E7.1. With increased track densities for computer disk drives, it is necessary carefully to design the head positioning control [8]. The transfer function is

$$G(s) = \frac{K}{(s+1)^2}.$$

Plot the polar plot for this system when $K = 4$. Calculate the phase and magnitude at $\omega = 0.5$, 1, 2, and so on.

E7.2. The tendon-operated robotic hand shown in Fig. 1.15 uses a pneumatic actuator [1]. The actuator can be represented by

$$G(s) = \frac{2572}{s^2 + 386s + 15{,}434} = \frac{2572}{(s+45.3)(s+341)}.$$

Plot the frequency response of $G(j\omega)$. Show that the magnitude of $G(j\omega)$ is -15.6 db at $\omega = 10$ and -30 db at $\omega = 200$. Also show that the phase is $-150°$ at $\omega = 700$.

E7.3. A robot arm has a joint control open-loop transfer function

$$G(s) = \frac{300(s + 100)}{s(s + 10)(s + 40)}.$$

Prove that the frequency equals 28.3 rad/sec when the phase angle of $(j\omega)$ is $-180°$. Find the magnitude of $G(j\omega)$ at that frequency.

E7.4. The frequency response for a process of the form

$$G(s) = \frac{Ks}{(s + a)(s^2 + 20s + 100)}$$

is shown in Fig. E7.4. Determine K and a by examining the frequency response curves.

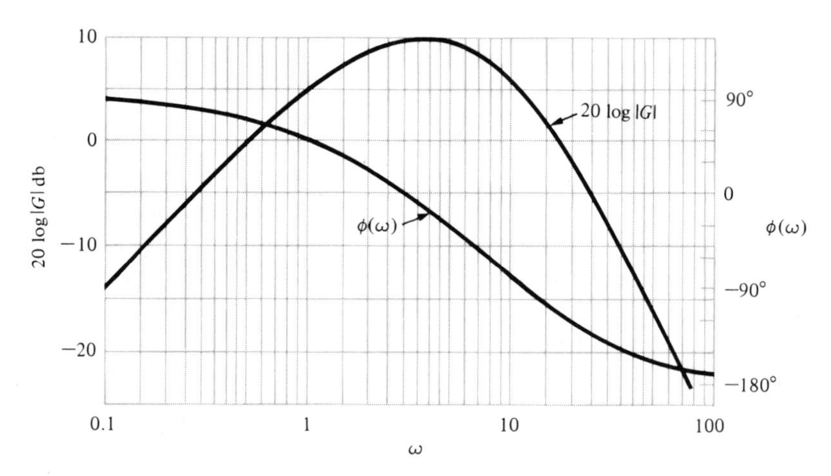

Figure E7.4. Bode diagram.

E7.5. The magnitude plot of a transfer function

$$G(s) = \frac{K(1 + 0.5s)(1 + as)}{s(1 + s/8)(1 + bs)(1 + s/36)}$$

is shown in Fig. E7.5, on the next page. Determine K, a, and b from the plot.

Answers: $K = 8$, $a = 1/4$, $b = 1/24$

E7.6. Several studies have proposed an extravehicular robot that could move about a NASA space station and perform physical tasks at various worksites. One such robot is shown in Fig. E7.6. The arm is controlled by a unity feedback control with

$$G(s) = \frac{K}{s(s/10 + 1)(s/100 + 1)}.$$

Draw the Bode diagram for $K = 100$ and determine the frequency when $20 \log |G|$ is zero db.

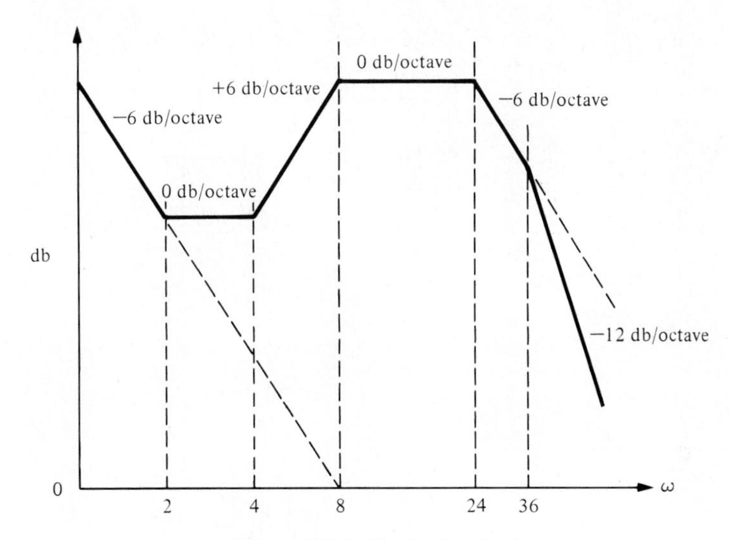

Figure E7.5. Bode diagram.

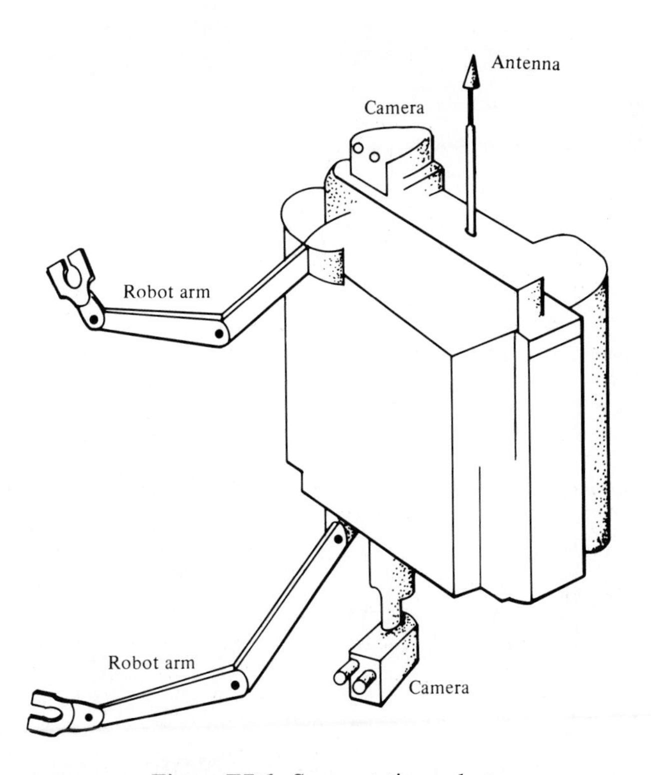

Figure E7.6. Space station robot.

(a)

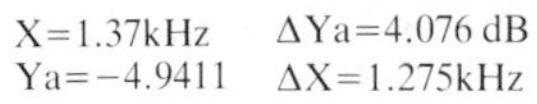

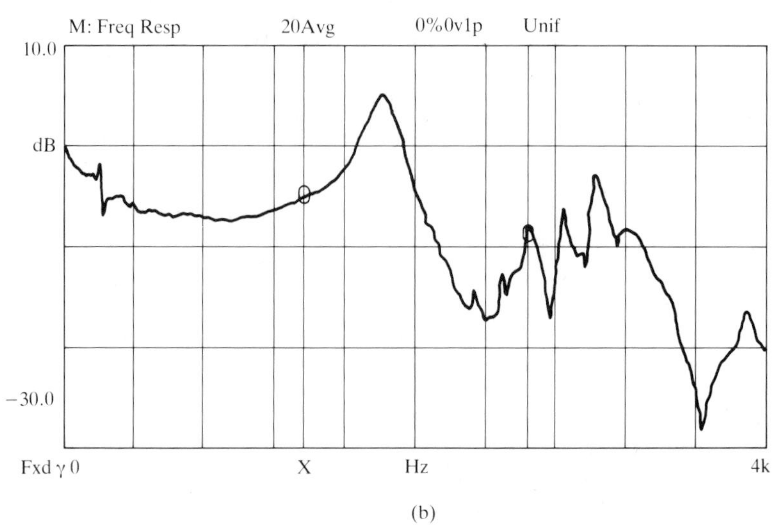

(b)

Figure E7.8. (a) Head positioner (b) frequency response.

E7.7. Consider a system with a closed-loop transfer function

$$T(s) = \frac{C(s)}{R(s)} = \frac{4}{(s^2 + s + 1)(s^2 + 0.4s + 4)}.$$

This system will have no steady-state error for a step input. (a) Using the Control System Design Program (CSDP) or equivalent, plot the frequency response, noting the two peaks in the magnitude response. (b) Predict the time response to a step input, noting that the system has four poles and it cannot be represented as a dominant second-order system. (c) Using CSDP, plot the step response.

E7.8. The dynamic analyzer shown in Fig. 7.22 can be used to display the frequency response of a selected $G(j\omega)$ model. Figure E7.8(a) shows a head positioning mechanism for a disk drive. This device uses a linear motor to position the head. Figure E7.8(b) shows the actual frequency response of the head positioning mechanism. Estimate the poles and zeros of the device. Note $X = 1.37$ kHz at the first cursor and $\Delta X = 1.257$ kHz to the second cursor.

Problems

P7.1. Sketch the polar plot of the frequency response for the following transfer functions:

(a) $GH(s) = \dfrac{1}{(1 + 0.5s)(1 + 2s)}$ (b) $GH(s) = \dfrac{(1 + 0.5s)}{s^2}$

(c) $GH(s) = \dfrac{(s + 3)}{(s^2 + 4s + 16)}$ (d) $GH(s) = \dfrac{30(s + 8)}{s(s + 2)(s + 4)}$

P7.2. Draw the Bode diagram representation of the frequency response for the transfer functions given in problem P7.1.

P7.3. A rejection network that can be utilized instead of the twin-T network of Example 7.4 is the bridged-T network shown in Fig. P7.3. The transfer function of this network is

$$G(s) = \frac{s^2 + \omega_n^2}{s^2 + 2(\omega_n s/Q) + \omega_n^2}$$

(can you show this?), where $\omega_n^2 = 2/LC$ and $Q = \omega_n L/R_1$ and R_2 is adjusted so that $R_2 = (\omega_n L)^2/4R_1$ [13]. (a) Determine the pole-zero pattern and, utilizing the vector approach, evaluate the approximate frequency response. (b) Compare the frequency response of the twin-T and bridged-T networks when $Q = 10$.

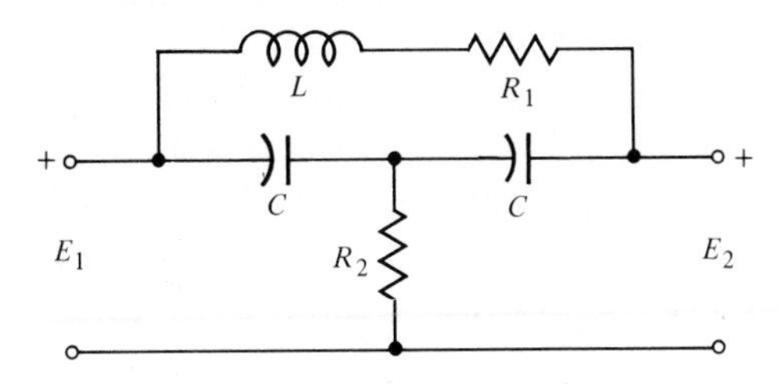

Figure P7.3. Bridged-T network.

P7.4. A control system for controlling the pressure in a closed chamber is shown in Fig. P7.4. The transfer function for the measuring element is

$$H(s) = \frac{450}{s^2 + 90s + 900},$$

and the transfer function for the valve is

$$G_1(s) = \frac{1}{(0.1s + 1)(1/15s + 1)}.$$

The controller transfer function is

$$G_c(s) = (10 + 2s).$$

Obtain the frequency response characteristics for the loop transfer function

$$G_c(s)G_1(s)H(s) \cdot [1/s].$$

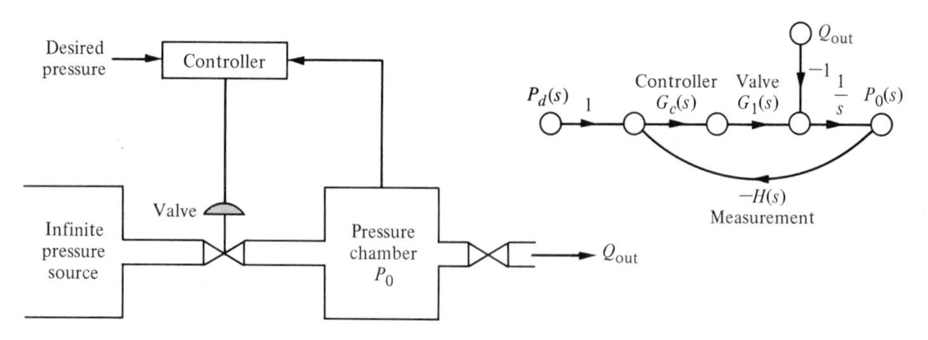

Figure P7.4. (a) Pressure controller (b) flowgraph.

P7.5. The robot industry in the United States is growing at a rate of 30% a year [1]. A typical industrial robot has six axes or degrees of freedom. A position control system for a force-sensing joint has a transfer function

$$G(s) = \frac{K}{(1 + s/2)(1 + s)(1 + s/10)(1 + s/30)},$$

where $H(s) = 1$ and $K = 10$. Plot the Bode diagram of this system.

P7.6. The asymptotic logarithmic magnitude curves for two transfer functions are given in Fig. P7.6, on the next page. Sketch the corresponding asymptotic phase shift curves for each system. Determine the transfer function for each system. Assume that the systems have minimum phase transfer functions.

P7.7. A feedback control system is shown in Fig. P7.7. The specification for the closed-loop system requires that the overshoot to a step input is less than 16%. (a) Determine the corresponding specification in the frequency domain M_{p_ω} for the closed-loop transfer function

$$\frac{C(j\omega)}{R(j\omega)} = T(j\omega).$$

(b) Determine the resonant frequency, ω_r. (c) Determine the bandwidth of the closed-loop system.

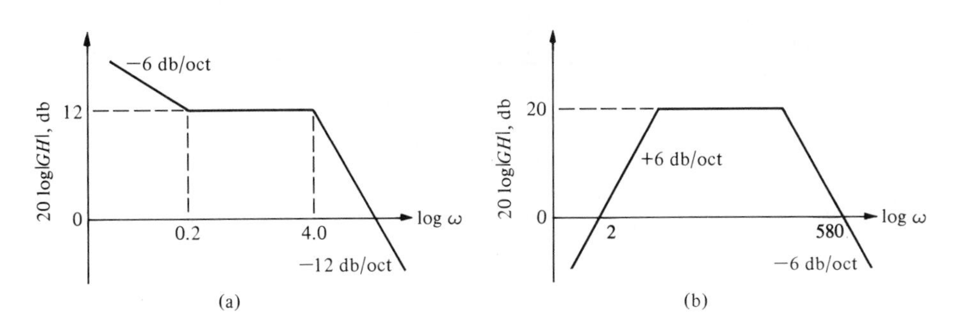

Figure P7.6.

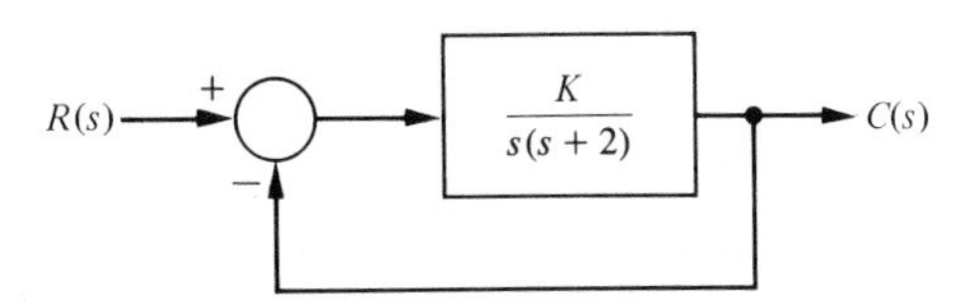

Figure P7.7.

P7.8. Driverless vehicles can be used in warehouses, airports, and many other applications. These vehicles follow a wire imbedded in the floor and adjust the steerable front wheels in order to maintain proper direction, as shown in Fig. P7.8(a), on the next page [1]. The sensing coils, mounted on the front wheel assembly, detect an error in the direction of travel and adjust the steering. The overall control system is shown in Fig. P7.8(b). The open-loop transfer function is

$$GH(s) = \frac{K}{s(s + \pi)^2} = \frac{K_v}{s(s/\pi + 1)^2}.$$

It is desired to have the bandwidth of the closed-loop system exceed 2π rad/sec. (a) Set $K_v = 2\pi$ and plot the Bode diagram. (b) Using the Bode diagram obtain the logarithmic magnitude versus phase angle curve.

P7.9. Draw the logarithmic magnitude versus phase angle curves for the transfer functions *a* and *b* of Problem P7.1.

P7.10. A linear actuator is utilized in the system shown in Fig. P7.10 to position a mass M. The actual position of the mass is measured by a slide wire resistor and thus $H(s) = 1.0$. The amplifier gain is to be selected so that the steady-state error of the system is less than 1% of the magnitude of the position reference $R(s)$. The actuator has a field coil with a resistance $R_f = 0.1$ ohm and $L_f = 0.2$ henries. The mass of the load is 0.1 kg and the friction is 0.2 n-sec/m. The spring constant is equal to 0.4 n/m. (a) Determine the gain K necessary to maintain a steady-state error for a step input less than 1%. That is, K_p must be greater than 99. (b) Draw the Bode diagram of the loop transfer function $GH(s)$. (c) Draw the logarithmic magnitude versus phase angle curve for $GH(j\omega)$. (d) Draw the Bode diagram for the closed-loop transfer function $Y(j\omega)/R(j\omega)$. Determine M_{p_ω}, ω_r, and the bandwidth.

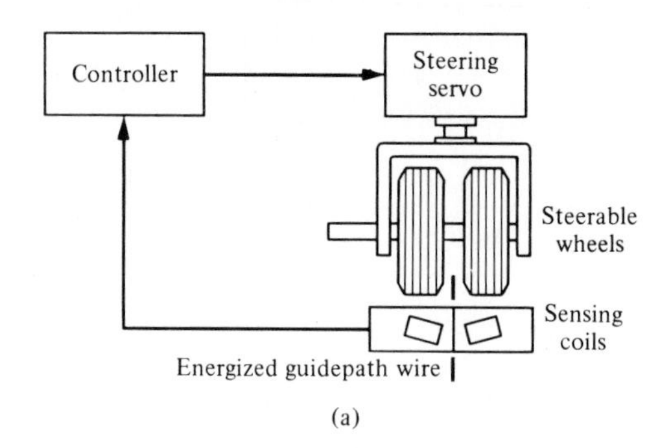

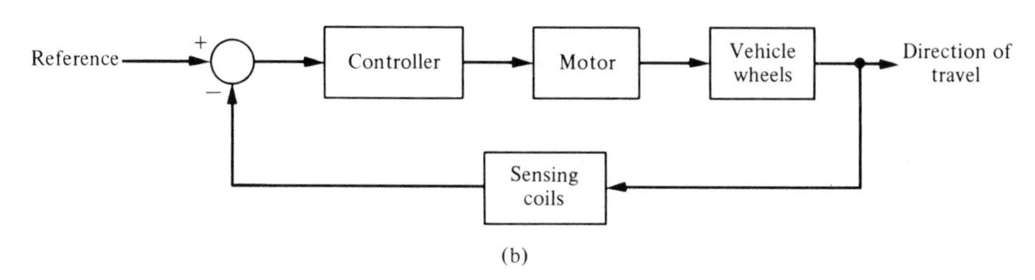

Figure P7.8. Steerable wheel control.

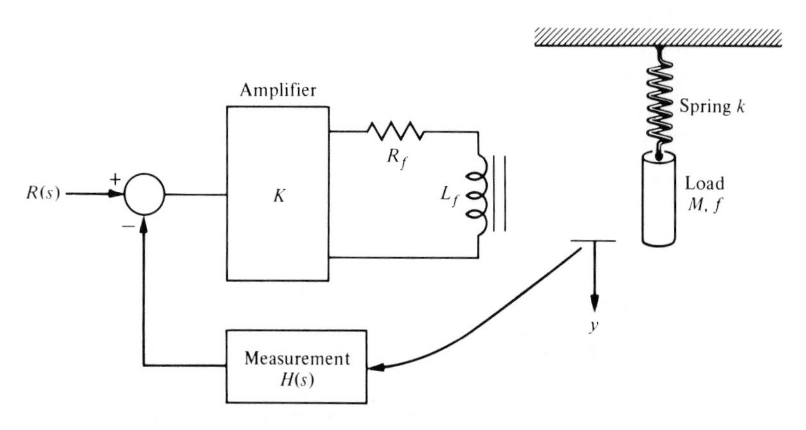

Figure P7.10. Linear actuator control.

P7.11. The block diagram of a feedback control system is shown in Fig. P7.11(a). The transfer functions of the blocks are represented by the frequency response curves shown in Fig. P7.11(b). (a) When G_3 is disconnected from the system, determine the damping ratio ζ of the system. (b) Connect G_3 and determine the damping ratio ζ. Assume that the systems have minimum phase transfer functions.

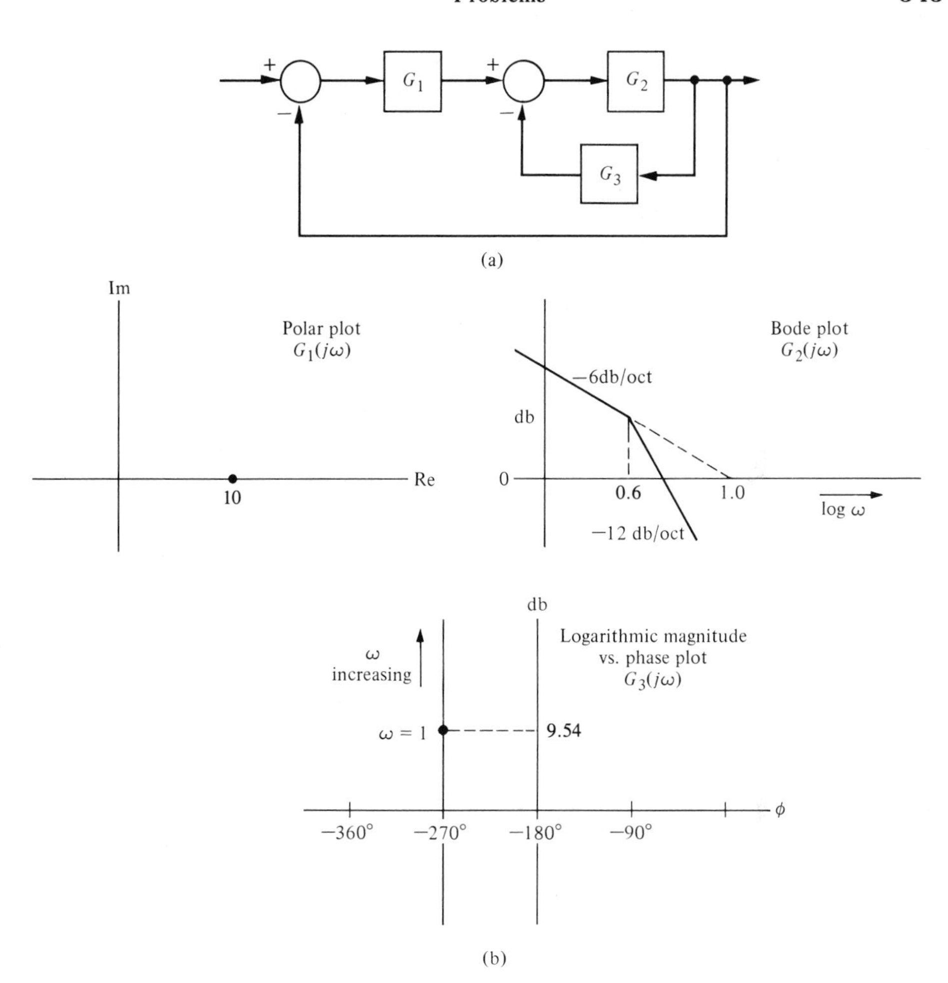

Figure P7.11. Feedback system.

P7.12. A position control system may be constructed by using an ac motor and ac components as shown in Fig. P7.12. The syncro and control transformer may be considered to be a transformer with a rotating winding. The syncro position detector rotor turns with the load through an angle θ_0. The syncro motor is energized with an ac reference voltage, for example, 115 volts, 60 cps. The input signal or command is $R(s) = \theta_{\text{in}}(s)$ and is applied by turning the rotor of the control transformer. The ac two-phase motor operates as a result of the amplified error signal. The advantages of an ac control system are (1) freedom from dc drift effects and (2) the simplicity and accuracy of ac components. In order to measure the open-loop frequency response, one simply disconnects X from Y and X' from Y'. Then one applies a sinusoidal modulation signal generator to the Y–Y' terminals and measures the response at X–X'. [The error $(\theta_0 - \theta_i)$ will be adjusted to zero before applying the ac generator.] The resulting frequency response of the loop, $GH(j\omega)$, is shown in Fig. P7.12(b). Determine the transfer function $GH(j\omega)$. Assume that the system has a minimum phase transfer function.

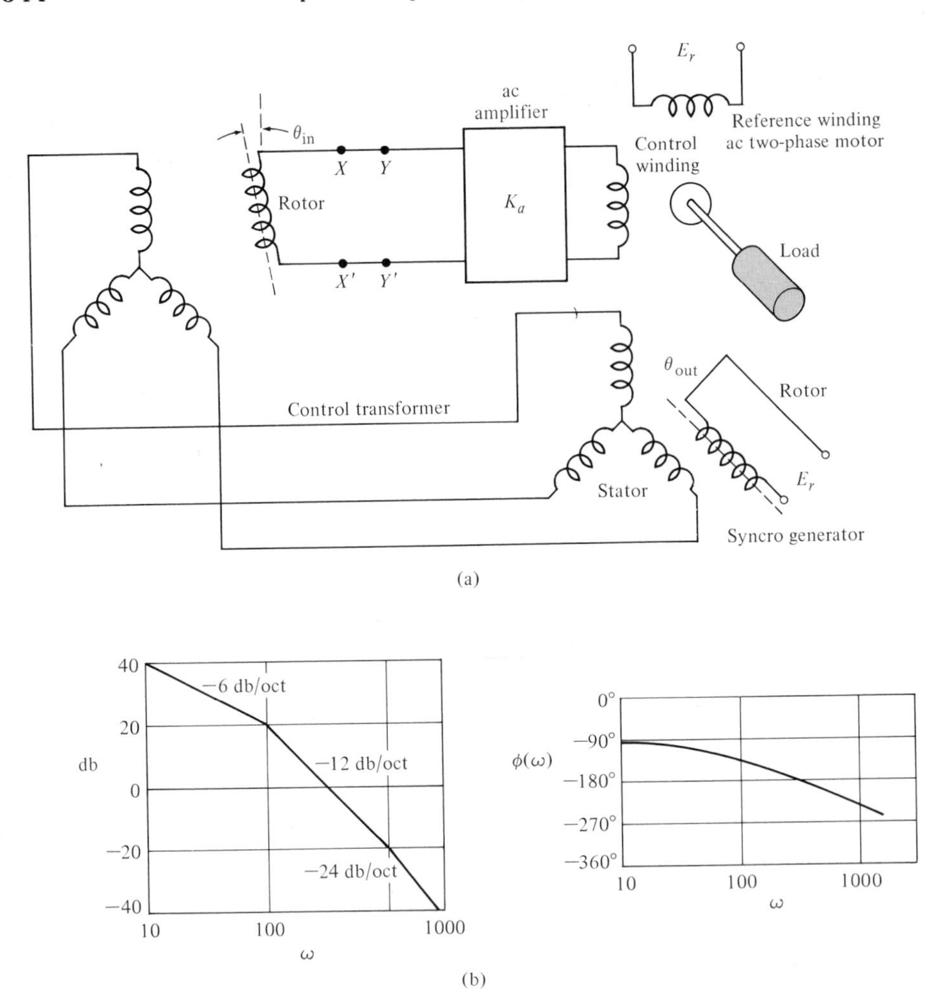

Figure P7.12. (a) AC motor control (b) frequency response.

P7.13. Automatic steering of a ship would be a particularly useful application of feedback control theory. In the case of heavily traveled seas, it is important to maintain the motion of the ship along an accurate track. An automatic system is able to maintain a much smaller error from the desired heading than is a helmsman who recorrects at infrequent intervals. A mathematical model of the steering system has been developed for a ship moving at a constant velocity and for small deviations from the desired track. For a large tanker, the transfer function of the ship is

$$G(s) = \frac{E(s)}{\delta(s)} = \frac{0.164(s + 0.2)(-s + 0.32)}{s^2(s + 0.25)(s - 0.009)},$$

where $E(s)$ is the Laplace transform of the deviation of the ship from the desired heading and $\delta(s)$ is the Laplace transform of the angle of deflection of the steering rudder.

Verify that the frequency response of the ship, $E(j\omega)/\delta(j\omega)$, is that shown in Fig. P7.13.

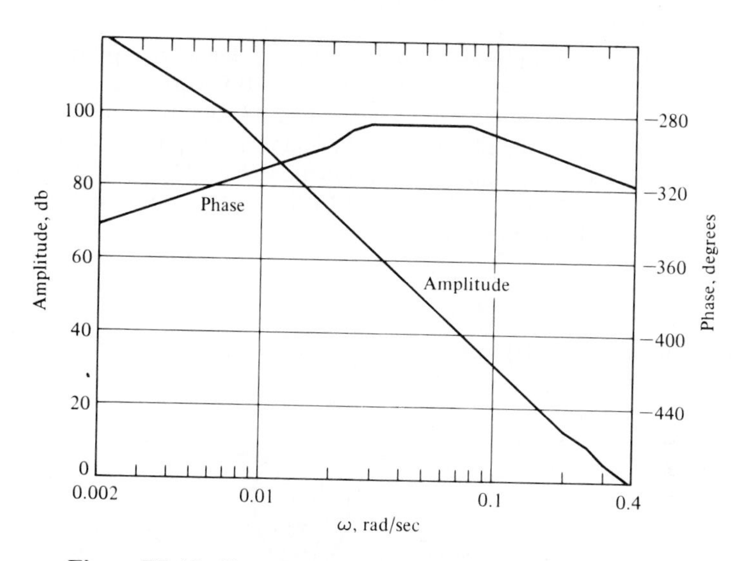

Figure P7.13. Frequency response of ship control system.

P7.14. In order to determine the transfer function of a plant $G(s)$, the frequency response may be measured using a sinusoidal input. One system yields the following data. Determine the transfer function $G(s)$.

ω, rad/sec	0.1	1	2	4	5	6.3	8	10	12.5	20	31
$\lvert G(j\omega)\rvert$	50	5.02	2.57	1.36	1.17	1.03	0.97	0.97	0.74	0.13	0.026
Phase, degrees	−90	−92.4	−96.2	−100	−104	−110	−120	−143	−169	−245	−258

P7.15. A bandpass amplifier may be represented by the circuit model shown in Fig. P7.15 [13]. When $R_1 = R_2 = 1\ \text{k}\Omega$, $C_1 = 100$ pf, $C_2 = 1\ \mu$f, and $K = 100$, show that

$$G(s) = \frac{10^9 s}{(s + 1000)(s + 10^7)}.$$

(a) Sketch the Bode diagram of $G(j\omega)$. (b) Find the midband gain (in db). (c) Find the high and low frequency −3 db points.

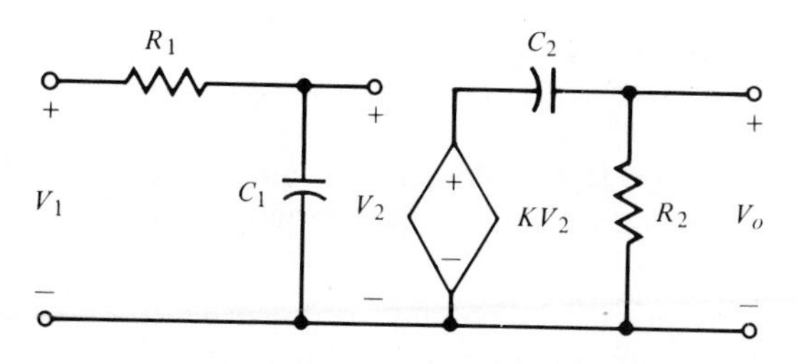

Figure P7.15. Bandpass amplifier.

P7.16. On April 13, 1985, a communications satellite was launched, but the ignited rocket failed to drive the satellite to its desired orbit. In August 1985, the space shuttle Discovery salvaged the satellite using a complicated scheme. Fig. P7.16 illustrates how a crew member, with his feet strapped to the platform on the end of the shuttle's robot arm, used his arms to stop the satellite's spin and ignite the engine switch. The control system has the form shown in Fig. 3.14, where $G_1 = K = 8$ and $H(s) = 1$. The control system of the robot arm has a closed-loop transfer function

$$\frac{C(s)}{R(s)} = \frac{8}{s^2 + 4s + 8}.$$

(a) Determine the response $c(t)$ to a unit step disturbance. (b) Determine the bandwidth of the system.

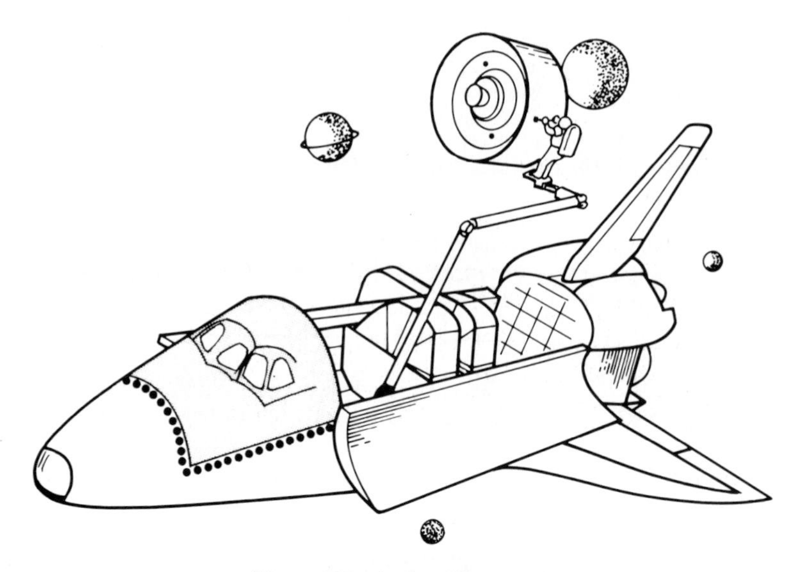

Figure P7.16. Satellite repair.

P7.17. The experimental Oblique Wing Aircraft (OWA) has a wing that pivots as shown in Fig. P7.17, on the next page. The wing is in the normal unskewed position for low speeds and can move to a skewed position for improved supersonic flight [14]. The aircraft control system has $H(s) = 1$ and

$$G(s) = \frac{4\,(0.5s + 1)}{s(2s + 1)\left[\left(\frac{s}{8}\right)^2 + \left(\frac{s}{20}\right) + 1\right]}.$$

(a) Using the Control System Design Program or equivalent, find the Bode diagram. (b) Find the frequency, ω_1, when the magnitude is 0 db and the frequency, ω_2, when the phase is -180 degrees.

P7.18. Remote operation plays an important role in hostile environments, such as those in nuclear or high-temperature environments and in deep space. In spite of the efforts of many researchers, a teleoperation system that is comparable to the human's direct operation has not been developed. Research engineers have been trying to improve teleopera-

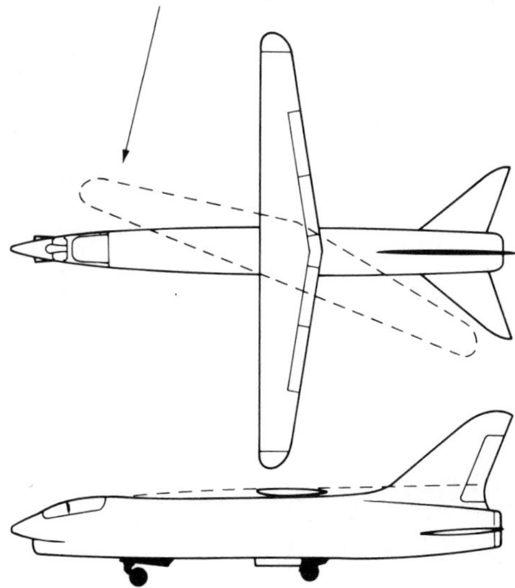

Figure P7.17. The Oblique Wing Aircraft top and side view.

tions by feeding back rich sensory information acquired by the robot to the operator with a sensation of presence. This concept is called tele-existence or telepresence [19].

The tele-existence master-slave system consists of a master system with a visual and auditory sensation of presence, a computer control system, and an anthropomorphic slave robot mechanism with an arm having seven degrees of freedom and a locomotion mechanism. The operator's head movement, right arm movement, right hand movement, and other auxiliary motion are measured by the master system. A specially designed stereo visual and auditory input system mounted on the neck mechanism of the slave robot gathers visual and auditory information of the remote environment. These pieces of information are sent back to the master system and are applied to the specially designed stereo display system to evoke the sensation of presence of the operator. A diagram of the locomation system and robot is shown in Fig. P7.18, on the next page. The locomotion control system has the loop transfer

$$GH(s) = \frac{12(s + 0.5)}{s^2 + 13s + 30}.$$

Obtain the Bode diagram for $GH(j\omega)$ and determine the frequency when $20 \log |GH|$ is very close to 0 db.

P.7.19. Low altitude wind shear is a major cause of air carrier accidents in the United States. Most of these accidents have been caused by either microbursts (small scale, low altitude, intense thunderstorm downdrafts that impact the surface and cause strong divergent outflows of wind) or by the gust front at the leading edge of expanding thunderstorm outflows. A microburst encounter is a serious problem for either *landing* or *departing* aircraft since the aircraft is at low altitudes and is traveling at just over 25% above its stall speed [20].

The design of the control of an aircraft encountering wind shear after takeoff may be treated as a problem of stabilizing the climb rate about a desired value of the climb rate. The resulting controller is a feedback one utilizing only climb rate information.

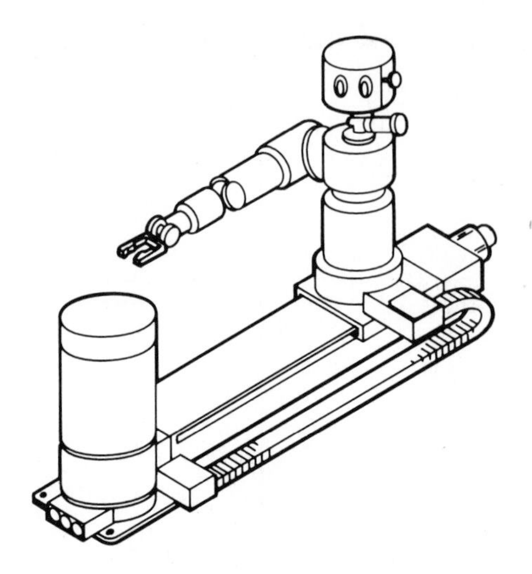

Figure P7.18. A tele-existence robot.

The standard negative unity feedback system of Fig. 7.24 has a loop transfer function

$$G(s) = \frac{-200s^2}{s^3 + 14s^2 + 44s + 40}.$$

Note the negative gain in $G(s)$. This system represents the control system for climb rate. Draw the Bode diagram and determine gain (in db) when the phase is $-180°$.

P7.20. Space robotics is an emerging field. For the successful development of space projects, robotics and automation will be a key technology. Autonomous and dexterous space robots can reduce the workload of astronauts and increase operational efficiency in many missions. Figure P7.20 shows a concept called a free-flying robot [1,19]. A major characteristic of space robots, which clearly distinguishes them from robots operated on earth, is the lack of a fixed base. Any motion of the manipulator arm will induce reaction forces and moments in the base, which disturb its position and attitude.

The control of one of the joints of the robot can be represented by the loop transfer function

$$GH(s) = \frac{600(s + 6)}{s^2 + 10s + 400}.$$

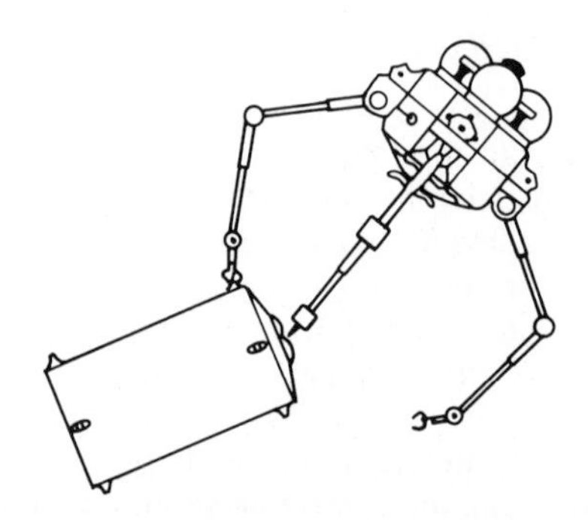

Figure P7.20. A space robot with three arms, shown capturing a satellite.

(a) Plot the Bode diagram of $GH(j\omega)$. (b) Determine the maximum value of $20 \log |GH|$, the frequency at which it occurs, and the phase at that frequency.

P7.21. A dc motor controller used extensively in automobiles is shown in Fig. P7.21(a). The measured plot of $\Theta(s)/I(s)$ is shown in Fig. P7.21(b). Determine the transfer function of $\Theta(s)/I(s)$.

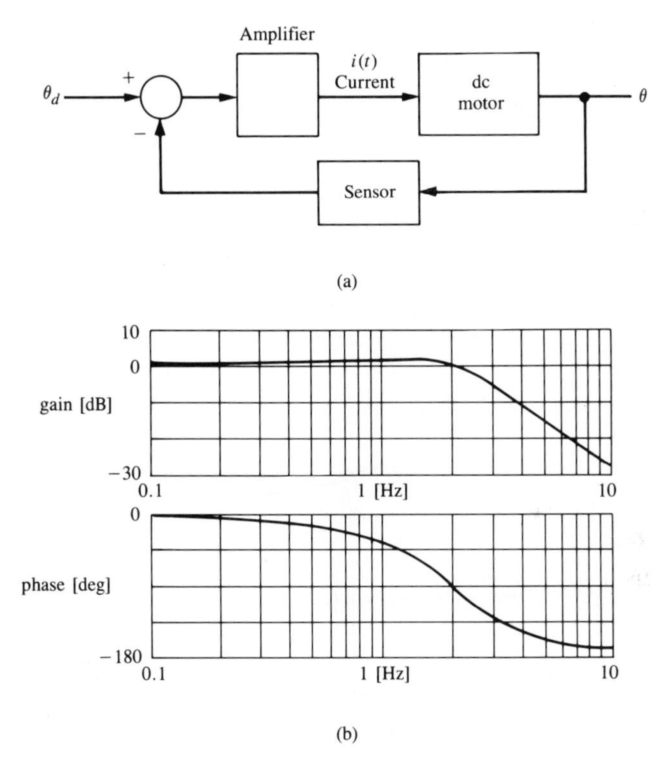

Figure P7.21. Motor controller.

Design Problems 

DP7.1. The behavior of a human steering an automobile remains interesting [6,17]. The design and development of systems for four-wheel steering, active suspensions, active, independent braking, and "drive-by-wire" steering provide the engineer with considerably more freedom in altering vehicle handling qualities than existed in the past.

The vehicle and the driver are represented by the model in Fig. DP7.1, where the driver develops anticipation of the vehicle deviation from the center line. For $K = 1$, plot the Bode diagram of (a) the open-loop transfer function $G_c(s)G(s)$ and (b) the closed-loop transfer function $T(s)$. (c) Repeat parts (a) and (b) when $K = 10$. (d) A driver can select the gain K. Determine the appropriate gain so that $M_{p_\omega} \leq 2$ and the bandwidth is the maximum attainable for the closed-loop system. (e) Determine the steady-state error of the system for a ramp input, $r(t) = t$.

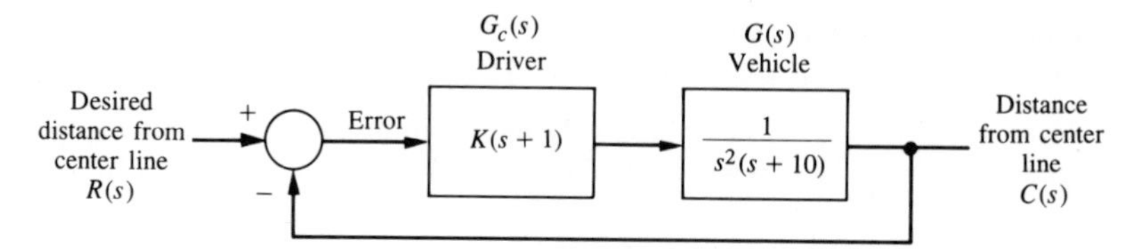

Figure DP7.1.

DP7.2. The unmanned exploration of planets such as Mars requires a high level of autonomy because of the communication delays between robots in space and their Earth-based stations. This impacts all the components of the system: planning, sensing, and mechanism. In particular, such a level of autonomy can be achieved only if each robot has a perception system that can reliably build and maintain models of the environment. The Robotics Institute of Carnegie-Mellon University has proposed a perception system that is designed for application to autonomous planetary exploration. The perception system is a major part of the development of a complete system that includes planning and mechanism design. The target vehicle is the Ambler, a six-legged walking machine being developed at CMU, shown in Fig. DP7.2(a) [18]. The control system of one leg is shown in Fig. DP7.2(b).

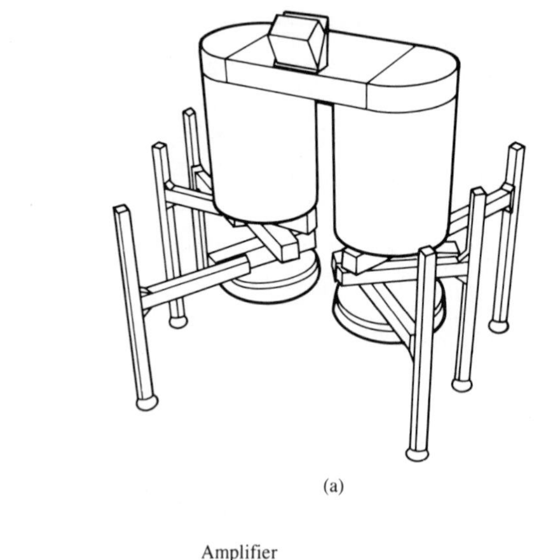

(a)

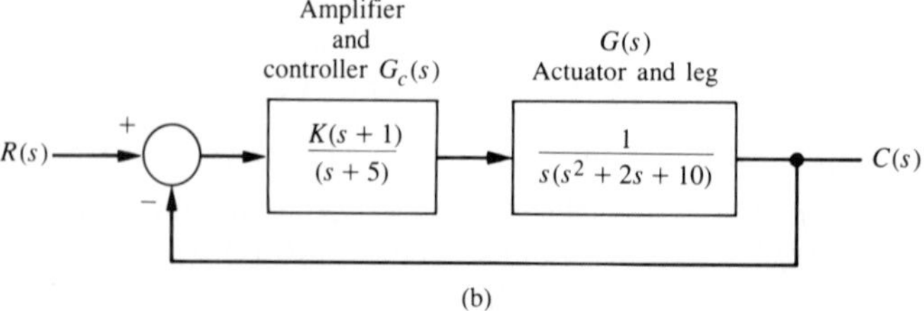

(b)

Figure DP7.2. (a) The six-legged Ambler. (b) Block diagram of the control system for one leg.

(a) Draw the Bode diagram for $G_c(s)G(s)$ for $0.1 \leq \omega \leq 100$ when $K = 20$. Determine (1) the frequency when the phase is $-180°$ and (2) the frequency when $20 \log |GG_c| = 0$ db. (b) Plot the Bode diagram for the closed-loop transfer function $T(s)$ when $K = 20$. (c) Determine M_{p_ω}, ω_r, and ω_B for the closed-loop system when $K = 20$ and $K = 40$. (d) Select the best gain of the two specified in part (c) when it is desired that the overshoot of the system to a step input, $r(t)$, is less than 35% and the settling time is as short as feasible.

Terms and Concepts

Bandwidth The frequency at which the frequency response has declined 3 db from its low-frequency value.

Bode plot The logarithm of the magnitude of the transfer function is plotted versus the logarithm of ω, the frequency. The phase, ϕ, of the transfer function is separately plotted versus the logarithm of the frequency.

Break frequency The frequency at which the asymptotic approximation of the frequency response for a pole (or zero) changes slope.

Corner frequency See Break frequency.

Decibel (db) The units of the logarithmic gain.

Fourier transform The transformation of a function of time, $f(t)$, into the frequency domain.

Frequency response The steady-state response of a system to a sinusoidal input signal.

Logarithmic magnitude The logarithm of the magnitude of the transfer function, $20 \log_{10} |G|$.

Logarithmic plot See Bode plot.

Maximum value of the frequency response A pair of complex poles will result in a maximum value for the frequency response occuring at the resonant frequency.

Minimum phase All the zeros of a transfer function lie in the left-hand side of the s-plane.

Natural frequency The frequency of natural oscillation that would occur for two complex poles if the damping was equal to zero.

Nonminimum phase Transfer functions with zeros in the right-hand s-plane.

Polar plot A plot of the real part of $G(j\omega)$ versus the imaginary part of $G(j\omega)$.

Resonant frequency The frequency, ω_r, at which the maximum value of the frequency response of a complex pair of poles is attained.

Transfer function in the frequency domain The ratio of the output to the input signal where the input is a sinusoid. It is expressed as $G(j\omega)$.

References

1. R. C. Dorf, *Encyclopedia of Robotics,* John Wiley & Sons, New York, 1988.
2. V. Feliu, "Adaptive Control of a Single-Link Flexible Manipulator," *IEEE Control Systems,* February 1990; pp. 29–32.
3. E. K. Parsons, "An Experiment Demonstrating Pointing Control," *IEEE Control Systems,* April 1989; pp. 79–86.
4. H. W. Bode, "Relations Between Attenuation and Phase in Feedback Amplifier Design," *Bell System Tech. J.,* July 1940; pp. 421–54. Also in *Automatic Control: Classical Linear Theory,* G. J. Thaler, ed., Dowden, Hutchinson, and Ross, Stroudsburg, Pa., 1974; pp. 145–78.
5. M. D. Fagen, *A History of Engineering and Science in the Bell System,* Bell Telephone Laboratories, Murray Hill, N.J., 1978; Ch. 3.
6. R. A. Hess, "A Control Theoretic Model of Driver Steering Behavior," *IEEE Control Systems,* August 1990; pp. 3–8.
7. S. Winchester, "Leviathans of the Sky," *Atlantic Monthly,* October 1990; pp. 107–18.
8. L. B. Jackson, *Signals, Systems, and Transforms,* Addison-Wesley, Reading, Mass., 1991.
9. M. Rimer and D. K. Frederick, "Solutions of the Grumman F-14 Benchmark Control Problem," *IEEE Control Systems,* August 1987; pp. 36–40.
10. K. Egilmez, "A Logical Approach to Knowledge-Based Control," *J. of Intelligent Manufacturing,* no. 1, 1990; pp. 59–76.
11. C. Philips and R. Harbor, *Feedback Control Systems,* Prentice Hall, Englewood Cliffs, N.J., 1991.
12. J. F. Manji, "Smart Cars of the Twenty-first Century," *Automation,* October 1990; pp. 18–25.
13. S. Franco, *Design with Operational Amplifiers and Analog Integrated Circuits,* McGraw-Hill, New York, 1988.
14. R. C. Nelson, *Flight Stability and Automatic Control,* McGraw-Hill, New York, 1989.
15. R. C. Dorf and R. G. Jacquot, *Control System Design Program,* Addison-Wesley, Reading, Mass., 1988.
16. R. J. Smith and R. C. Dorf, *Circuits, Devices, and Systems,* 5th ed., John Wiley & Sons, New York, 1991.
17. J. Krueger, "Developments in Automotive Electronics," *Automotive Engineering,* September 1990; pp. 37–44.
18. M. Hebert, "A Perception System for a Planetary Explorer," *Proceed. of the IEEE Conference on Decision and Control,* 1989; pp. 80–89.
19. S. Tachi, "Tele-Existence Master-Slave System for Remote Manipulation," *Proceed. of IEEE Conference on Decision and Control,* December 1990; pp. 85–90.
20. G. Leitman, "Aircraft Control Under Conditions of Windshear," *Proceed. of IEEE Conference on Decision and Control,* December 1990; pp. 747–49.
21. K. Yoshida, "Control of Space Free-Flying Robot," *Proceed. of IEEE Conference on Decision and Control,* December 1990; pp. 97–101.
22. T. Kawabe, "Controller for Servo Positioning System of an Automobile," *Proceed. of IEEE Conference on Decision and Control,* December 1990; pp. 2170–74.

CHAPTER 8

Stability in the Frequency Domain

Preview

As we noted in earlier chapters, it is important to determine whether a system is stable. If it is stable, then the degree of stability is important to determine. We may use the frequency response of a transfer function around a feedback loop $GH(j\omega)$ to provide answers to our inquiry about the system's relative stability.

We will use some concepts developed in the theory of complex variables to obtain a stability criterion in the frequency domain. Then this criterion can be extended to indicate relative stability by indicating how close we come to operating at the edge of instability.

We will then demonstrate how we can examine the frequency response of the closed-loop transfer function, $T(j\omega)$, as well as the loop transfer function $GH(j\omega)$.

Finally, we will use these methods to analyze the response and performance of a system with a pure time delay, without attenuation, located within the feedback loop of a closed-loop control system.

8.1 Introduction

For a control system, it is necessary to determine whether the system is stable. Furthermore, if the system is stable, it is often necessary to investigate the relative stability. In Chapter 5, we discussed the concept of stability and several methods of determining the absolute and relative stability of a system. The Routh-Hurwitz method discussed in Chapter 5 is a useful method for investigating the characteristic equation expressed in terms of the complex variable $s = \sigma + j\omega$. Then in Chapter 6, we investigated the relative stability of a system utilizing the root locus method, which is also in terms of the complex variable s. In this chapter, we are concerned with investigating the stability of a system in the real frequency domain, that is, in terms of the frequency response discussed in Chapter 7.

The frequency response of a system represents the sinusoidal steady-state response of a system and provides sufficient information for the determination of the relative stability of the system. The frequency response of a system can readily be obtained experimentally by exciting the system with sinusoidal input signals; therefore it can be utilized to investigate the relative stability of a system when the system parameter values have not been determined. Furthermore, a frequency-domain stability criterion would be useful for determining suitable approaches to altering a system in order to increase its relative stability.

A frequency domain stability criterion was developed by H. Nyquist in 1932 and remains a fundamental approach to the investigation of the stability of linear control systems [1, 2]. The *Nyquist stability criterion* is based upon a theorem in the theory of the function of a complex variable due to Cauchy. Cauchy's theorem is concerned with *mapping contours* in the complex s-plane, and fortunately the theorem can be understood without a formal proof, which uses complex variable theory.

In order to determine the relative stability of a closed-loop system, we must investigate the characteristic equation of the system:

$$F(s) = 1 + P(s) = 0. \tag{8.1}$$

For a multiloop system, we found in Section 2.7 that, in terms of signal-flow graphs, the characteristic equation is

$$F(s) = \Delta(s) = 1 - \Sigma L_n + \Sigma L_m L_q \cdots,$$

where $\Delta(s)$ is the graph determinant. Therefore we can represent the characteristic equation of single-loop or multiple-loop systems by Eq. (8.1), where $P(s)$ is a rational function of s. In order to ensure stability, we must ascertain that all the zeros of $F(s)$ lie in the left-hand s-plane. In order to investigate this, Nyquist proposed a mapping of the right-hand s-plane into the $F(s)$-plane. Therefore, to utilize and understand Nyquist's criterion, we shall first consider briefly the mapping of contours in the complex plane.

8.2 Mapping Contours in the s-Plane

We are concerned with the mapping of contours in the s-plane by a function $F(s)$. A *contour map* is a contour or trajectory in one plane mapped or translated into another plane by a relation $F(s)$. Since s is a complex variable, $s = \sigma + j\omega$, the function $F(s)$ is itself complex and can be defined as $F(s) = u + jv$ and can be represented on a complex $F(s)$-plane with coordinates u and v. As an example, let us consider a function $F(s) = 2s + 1$ and a contour in the s-plane as shown in Fig. 8.1(a). The mapping of the s-plane unit square contour to the $F(s)$ plane is accomplished through the relation $F(s)$, and so

$$u + jv = F(s) = 2s + 1 = 2(\sigma + j\omega) + 1. \tag{8.2}$$

Therefore, in this case, we have

$$u = 2\sigma + 1 \tag{8.3}$$

and

$$v = 2\omega. \tag{8.4}$$

Thus the contour has been mapped by $F(s)$ into a contour of an identical form, a square, with the center shifted by one unit and the magnitude of a side multiplied by 2. This type of mapping, which retains the angles of the s-plane contour

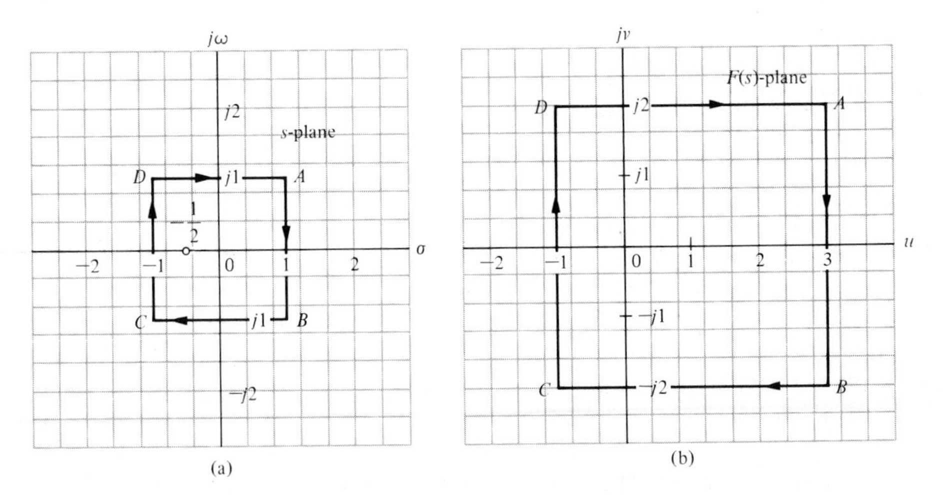

Figure 8.1. Mapping a square contour by $F(s) = 2s + 1 = 2(s + \frac{1}{2})$.

on the $F(s)$-plane, is called a *conformal mapping.* We also note that a closed contour in the s-plane results in a closed contour in the $F(s)$-plane.

The points A, B, C, and D, as shown in the s-plane contour, map into the points A, B, C, and D shown in the $F(s)$-plane. Furthermore, a direction of traversal of the s-plane contour can be indicated by the direction $ABCD$ and the arrows shown on the contour. Then a similar traversal occurs on the $F(s)$-plane contour as we pass $ABCD$ in order, as shown by the arrows. By convention, the area within a contour to the right of the traversal of the contour is considered to be the *area enclosed* by the contour. Therefore we will assume *clockwise traversal* of a contour to be positive and the area enclosed within the contour to be on the right. This convention is opposite to that usually employed in complex variable theory but is equally applicable and is generally used in control system theory. Readers might consider the area on the right as they walk along the contour in a clockwise direction and call this rule "clockwise and eyes right."

Typically, we are concerned with an $F(s)$ that is a rational function of s. Therefore it will be worthwhile to consider another example of a mapping of a contour. Let us again consider the unit square contour for the function

$$F(s) = \frac{s}{s+2}. \tag{8.5}$$

Several values of $F(s)$ as s traverses the square contour are given in Table 8.1, and the resulting contour in the $F(s)$-plane is shown in Fig. 8.2(b). The contour in the $F(s)$-plane encloses the origin of the $F(s)$-plane because the origin lies within the enclosed area of the contour in the $F(s)$-plane.

Cauchy's theorem is concerned with mapping a function $F(s)$, which has a finite number of poles and zeros within the contour so that we may express $F(s)$ as

$$F(s) = \frac{K\prod_{i=1}^{n}(s+s_i)}{\prod_{k=1}^{M}(s+s_k)}, \tag{8.6}$$

where s_i are the zeros of the function $F(s)$ and s_k are the poles of $F(s)$. The function $F(s)$ is the characteristic equation, and so

$$F(s) = 1 + P(s), \tag{8.7}$$

where

$$P(s) = \frac{N(s)}{D(s)}.$$

Table 8.1.

$s = \sigma + j\omega$	Point A $1 + j1$	1	Point B $1 - j1$	$-j1$	Point C $-1 - j1$	-1	Point D $-1 + j1$	$j1$
$F(s) = u + jv$	$\frac{4+2j}{10}$	$\frac{1}{3}$	$\frac{4-2j}{10}$	$\frac{1-2j}{5}$	$-j$	-1	$+j$	$\frac{1+2j}{5}$

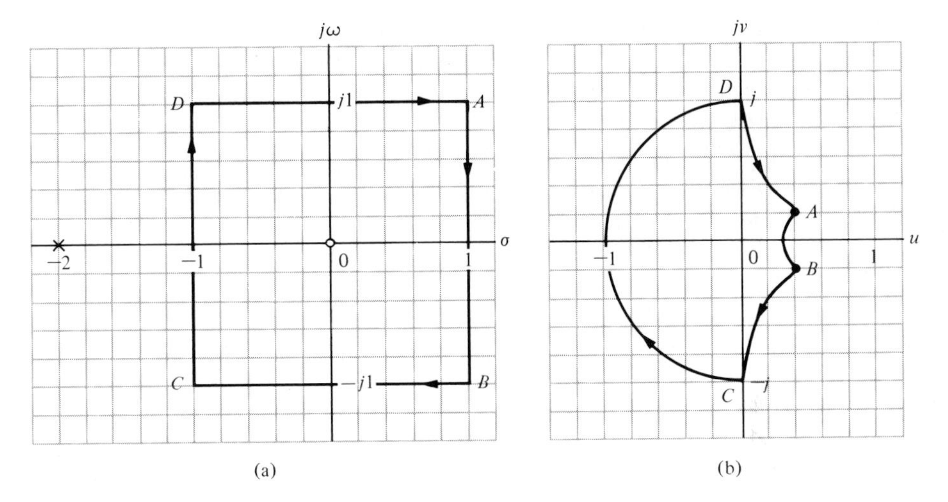

Figure 8.2. Mapping for $F(s) = s/(s + 2)$.

Therefore we have

$$F(s) = 1 + \frac{N(s)}{D(s)} = \frac{D(s) + N(s)}{D(s)} = \frac{K \prod_{i=1}^{n} (s + s_i)}{\prod_{k=1}^{M} (s + s_k)}, \tag{8.8}$$

and the poles of $P(s)$ are the poles of $F(s)$. However, it is the zeros of $F(s)$ that are the characteristic roots of the system and that indicate the response of the system. This is clear if we recall that the output of the system is

$$C(s) = T(s)R(s) = \frac{\Sigma P_k \Delta_k}{\Delta(s)} R(s) = \frac{\Sigma P_k \Delta_k}{F(s)} R(s), \tag{8.9}$$

where P_k and Δ_k are the path factors and cofactors as defined in Section 2.7.

Now, reexamining the example when $F(s) = 2(s + ½)$, we have one zero of $F(s)$ at $s = -½$, as shown in Fig. 8.1. The contour that we chose (that is, the unit square) enclosed and encircled once the zero within the area of the contour. Similarly, for the function $F(s) = s/(s + 2)$, the unit square encircled the zero at the origin but did not encircle the pole at $s = -2$. The encirclement of the poles and zeros of $F(s)$ can be related to the encirclement of the origin in the $F(s)$-plane by a *theorem* of *Cauchy*, commonly known as the *principle of the argument,* which states [3, 7]:

> If a contour Γ_s in the s-plane encircles Z zeros and P poles of $F(s)$ and does not pass through any poles or zeros of $F(s)$ as the traversal is in the clockwise direction along the contour, the corresponding contour Γ_F in the $F(s)$-plane encircles the origin of the $F(s)$-plane $N = Z - P$ times in the clockwise direction.

Thus for the examples shown in Figs. 8.1 and 8.2, the contour in the $F(s)$-plane encircles the origin once, because $N = Z - P = 1$, as we expect. As another example, consider the function $F(s) = s/(s + \frac{1}{2})$. For the unit square contour shown in Fig. 8.3(a), the resulting contour in the $F(s)$ plane is shown in Fig. 8.3(b). In this case, $N = Z - P = 0$ as is the case in Fig. 8.3(b), since the contour Γ_F does not encircle the origin.

Cauchy's theorem can be best comprehended by considering $F(s)$ in terms of the angle due to each pole and zero as the contour Γ_s is traversed in a clockwise direction. Thus let us consider the function

$$F(s) = \frac{(s + z_1)(s + z_2)}{(s + p_1)(s + p_2)}, \tag{8.10}$$

where z_i is a zero of $F(s)$ and p_k is a pole of $F(s)$. Equation (8.10) can be written as

$$\begin{aligned} F(s) &= |F(s)| \underline{/F(s)} \\ &= \frac{|s + z_1|\,|s + z_2|}{|s + p_1|\,|s + p_2|} (\underline{/s + z_1} + \underline{/s + z_2} - \underline{/s + p_1} - \underline{/s + p_2}) \qquad (8.11) \\ &= |F(s)|(\phi_{z_1} + \phi_{z_2} - \phi_{p_1} - \phi_{p_2}). \end{aligned}$$

Now, considering the vectors as shown for a specific contour Γ_s (Fig. 8.4a), we can determine the angles as s traverses the contour. Clearly, the net angle change as s traverses along Γ_s a full rotation of 360° for ϕ_{p_1}, ϕ_{p_2} and ϕ_{z_2} is zero degrees. However, for ϕ_{z_1} as s traverses 360° around Γ_s, the angle ϕ_{z_1} traverses a full 360° clockwise. Thus, as Γ_s is completely traversed, the net angle of $F(s)$ is equal to 360° since only one zero is enclosed. If Z zeros were enclosed within Γ_s, then the net angle would be equal to $\phi_z = 2\pi(Z)$ rad. Following this reasoning, if Z zeros and P poles are encircled as Γ_s is traversed, then $2\pi(Z) - 2\pi(P)$ is the net resul-

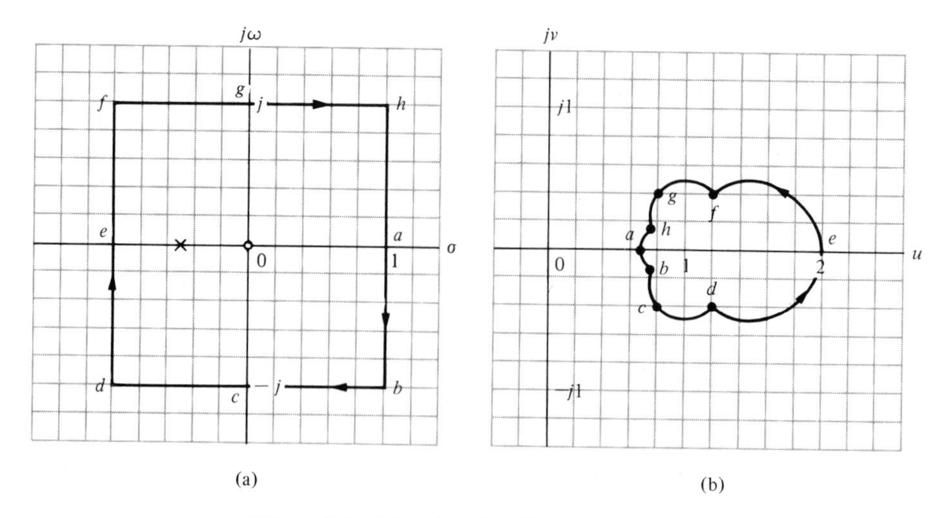

Figure 8.3. Mapping for $F(s) = s/(s + \frac{1}{2})$.

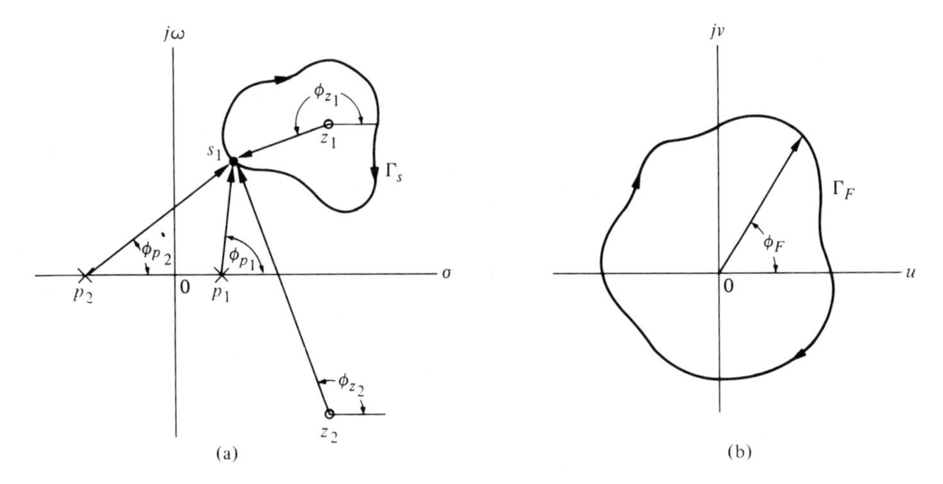

Figure 8.4. Evaluation of the net angle of Γ_F.

tant angle of $F(s)$. Thus the net angle of Γ_F of the contour in the $F(s)$-plane, ϕ_F, is simply

$$\phi_F = \phi_Z - \phi_P$$

or

$$2\pi N = 2\pi Z - 2\pi P, \tag{8.12}$$

and the net number of encirclements of the origin of the $F(s)$-plane is $N = Z - P$. Thus for the contour shown in Fig. 8.4(a), which encircles one zero, the contour Γ_F shown in Fig. 8.4(b) encircles the origin once in the clockwise direction.

As an example of the use of Cauchy's theorem, consider the pole-zero pattern shown in Fig. 8.5(a) with the contour Γ_s to be considered. The contour encloses and encircles three zeros and one pole. Therefore we obtain

$$N = 3 - 1 = +2,$$

and Γ_F completes two clockwise encirclements of the origin in the $F(s)$-plane as shown in Fig. 8.5(b).

For the pole and zero pattern shown and the contour Γ_s as shown in Fig. 8.6(a), one pole is encircled and no zeros are encircled. Therefore we have

$$N = Z - P = -1,$$

and we expect one encirclement of the origin by the contour Γ_F in the $F(s)$-plane. However, since the sign of N is negative, we find that the encirclement moves in the counterclockwise direction as shown in Fig. 8.6(b).

Now that we have developed and illustrated the concept of mapping of contours through a function $F(s)$, we are ready to consider the stability criterion proposed by Nyquist.

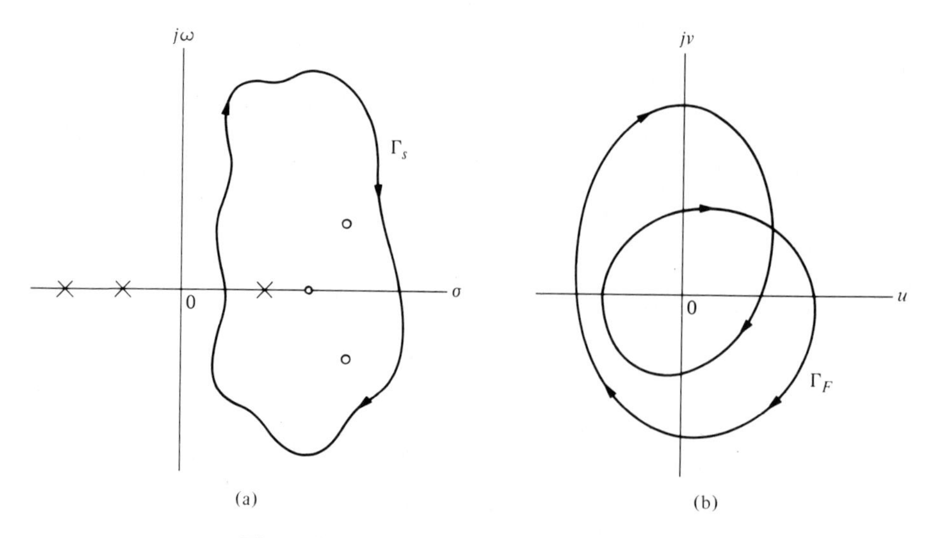

Figure 8.5. Example of Cauchy's theorem.

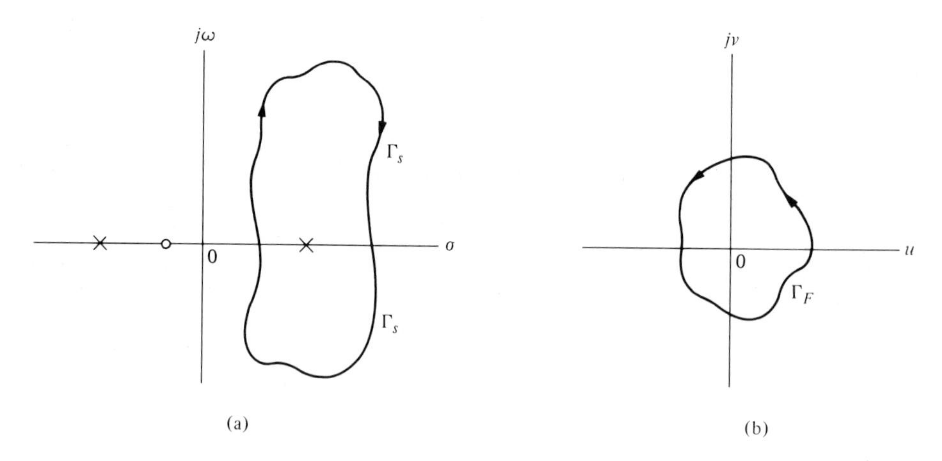

Figure 8.6. Example of Cauchy's theorem.

8.3 The Nyquist Criterion

In order to investigate the stability of a control system, we consider the characteristic equation, which is $F(s) = 0$, so that

$$F(s) = 1 + P(s) = \frac{K\prod_{i=1}^{n}(s + s_i)}{\prod_{k=1}^{M}(s + s_k)} = 0. \tag{8.13}$$

For a system to be stable, all the zeros of $F(s)$ must lie in the left-hand s-plane. Thus we find that the roots of a stable system [the zeros of $F(s)$] must lie to the left of the $j\omega$-axis in the s-plane. Therefore we chose a contour Γ_s in the s-plane which encloses the entire right-hand s-plane, and we determine whether any zeros of $F(s)$ lie within Γ_s by utilizing Cauchy's theorem. That is, we plot Γ_F in the $F(s)$-plane and determine the number of encirclements of the origin N. Then the number of zeros of $F(s)$ within the Γ_s contour [and therefore unstable zeros of $F(s)$] is

$$Z = N + P. \tag{8.14}$$

Thus if $P = 0$, as is usually the case, we find that the number of unstable roots of the system is equal to N, the number of encirclements of the origin of the $F(s)$ plane.

The Nyquist contour that encloses the entire right-hand s-plane is shown in Fig. 8.7. The contour Γ_s passes along the $j\omega$-axis from $-j\infty$ to $+j\infty$, and this part of the contour provides the familiar $F(j\omega)$. The contour is completed by a semicircular path of radius r where r approaches infinity.

Now, the Nyquist criterion is concerned with the mapping of the characteristic equation

$$F(s) = 1 + P(s) \tag{8.15}$$

and the number of encirclements of the origin of the $F(s)$-plane. Alternatively, we may define the function $F'(s)$ so that

$$F'(s) = F(s) - 1 = P(s). \tag{8.16}$$

The change of functions represented by Eq. (8.16) is very convenient because $P(s)$ is typically available in factored form, while $1 + P(s)$ is not. Then the mapping of Γ_s in the s-plane will be through the function $F'(s) = P(s)$ into the $P(s)$-plane. In this case the number of clockwise encirclements of the origin of the $F(s)$-plane becomes the number of clockwise encirclements of the -1 point in the

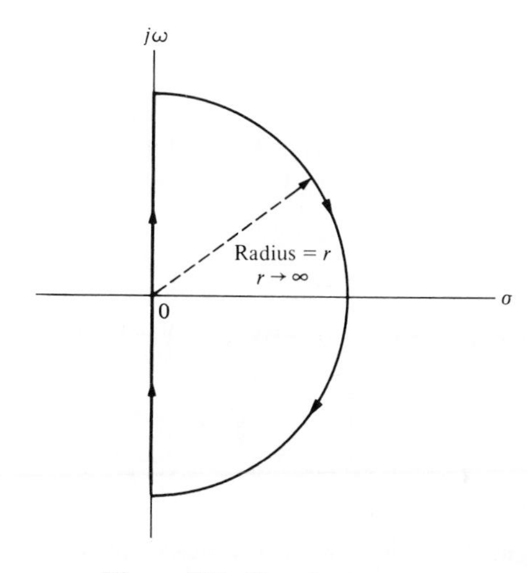

Figure 8.7. Nyquist contour.

$F'(s) = P(s)$ plane because $F'(s) = F(s) - 1$. Therefore the *Nyquist stability criterion* can be stated as follows:

> A feedback system is stable if and only if the contour Γ_p in the $P(s)$-plane does not encircle the $(-1, 0)$ point when the number of poles of $P(s)$ in the right-hand s-plane is zero ($P = 0$).

When the number of poles of $P(s)$ in the right-hand s-plane is other than zero, the Nyquist criterion is:

> A feedback control system is stable if and only if, for the contour Γ_p, the number of counterclockwise encirclements of the $(-1, 0)$ point is equal to the number of poles of $P(s)$ with positive real parts.

The basis for the two statements is the fact that for the $F'(s) = P(s)$ mapping, the number of roots (or zeros) of $1 + P(s)$ in the right-hand s-plane is represented by the expression

$$Z = N + P.$$

Clearly, if the number of poles of $P(s)$ in the right-hand s-plane is zero ($P = 0$), we require for a stable system that $N = 0$ and the contour Γ_p must not encircle the -1 point. Also, if P is other than zero and we require for a stable system that $Z = 0$, then we must have $N = -P$, or P counterclockwise encirclements.

It is best to illustrate the use of the Nyquist criterion by completing several examples.

■ Example 8.1 System with two real poles

A single-loop control system is shown in Fig. 8.8, where

$$GH(s) = \frac{K}{(\tau_1 s + 1)(\tau_2 s + 1)}. \tag{8.17}$$

In this case, $P(s) = GH(s)$, and we utilize a contour $\Gamma_P = \Gamma_{GH}$ in the $GH(s)$ plane. The contour Γ_s in the s-plane is shown in Fig. 8.9(a), and the contour Γ_{GH} is shown in Fig. 8.9(b) for $\tau_1 = 1$, $\tau_2 = 1/10$, and $K = 100$. The magnitude and phase of

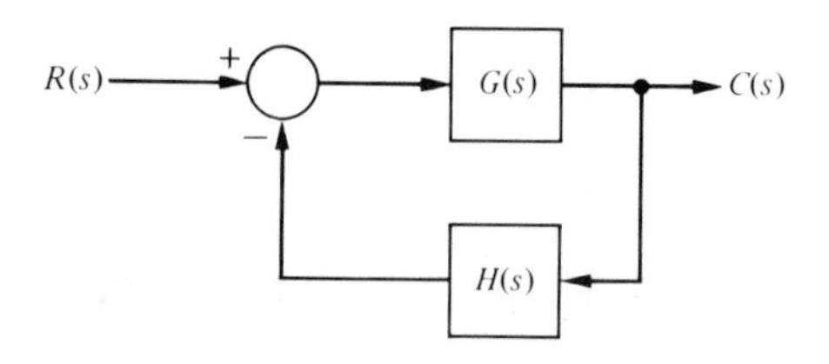

Figure 8.8. Single-loop feedback control system.

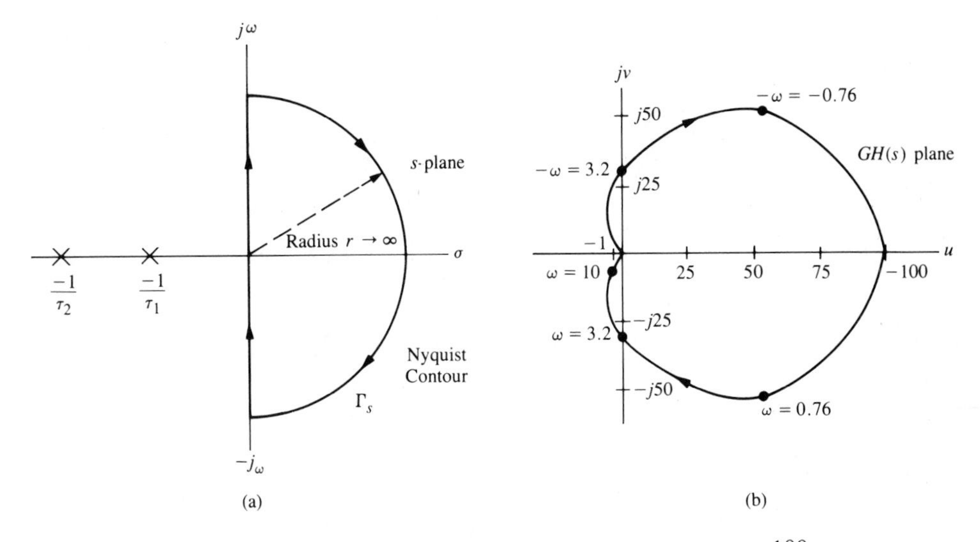

Figure 8.9. Nyquist contour and mapping for $GH(s) = \dfrac{100}{(s+1)(s/10+1)}$

$GH(j\omega)$ for selected values of ω are given in Table 8.2. We use these values to obtain the polar plot of Fig. 8.9(b).

We note that the number of poles of $GH(s)$ in the right-hand s-plane is zero and thus $P = 0$. Therefore, for this system to be stable we require $N = Z = 0$, and the contour must not encircle the -1 point in the $GH(s)$ plane. Examining Fig. 8.9(b) and Eq. (8.17), we find that, irrespective of the value of K, the contour does not encircle the -1 point and the system is always stable for all K greater than zero.

■ Example 8.2 System with a pole at the origin

A single-loop control system is shown in Fig. 8.8, where

$$GH(s) = \frac{K}{s(\tau s + 1)}.$$

In this single-loop case, $P(s) = GH(s)$ and we determine the contour $\Gamma_p = \Gamma_{GH}$ in the $GH(s)$-plane. The contour Γ_s in the s-plane is shown in Fig. 8.10(a), where an

Table 8.2.

ω	0	0.1	0.76	1	2	10	20	100	∞
$\lvert GH(j\omega)\rvert$	100	96	79.6	70.7	50.2	6.8	2.24	0.10	0
$\underline{/GH(j\omega)}$ (degrees)	0	−5.7	−41.5	−50.7	−74.7	−129.3	−150.5	−173.7	−180

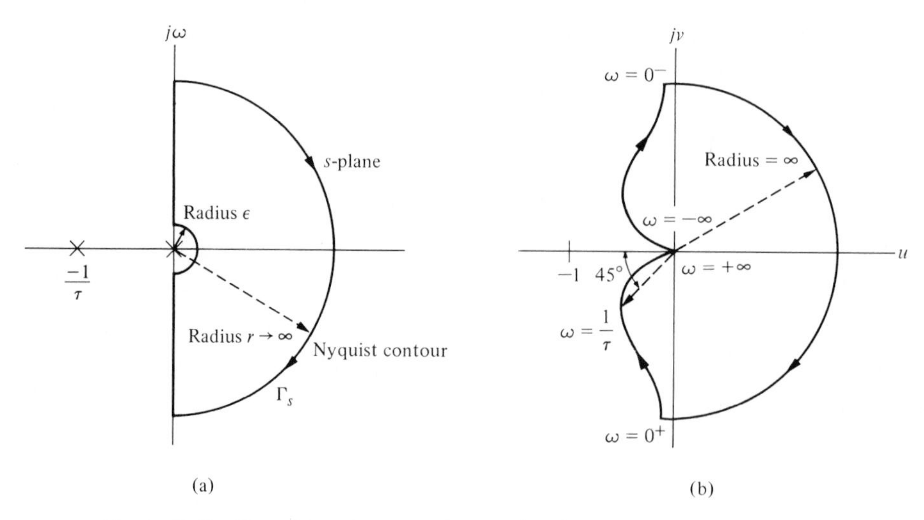

Figure 8.10. Nyquist contour and mapping for $GH(s) = K/s(\tau s + 1)$.

infinitesimal detour around the pole at the origin is effected by a small semicircle of radius ϵ, where $\epsilon \to 0$. This detour is a consequence of the condition of Cauchy's theorem which requires that the contour cannot pass through the pole of the origin. A sketch of the contour Γ_{GH} is shown in Fig. 8.10(b). Clearly, the portion of the contour Γ_{GH} from $\omega = 0^+$ to $\omega = +\infty$ is simply $GH(j\omega)$, the real frequency polar plot. Let us consider each portion of the Nyquist contour Γ_s in detail and determine the corresponding portions of the $GH(s)$-plane contour Γ_{GH}.

(a) The Origin of the s-Plane. The small semicircular detour around the pole at the origin can be represented by setting $s = \epsilon e^{j\phi}$ and allowing ϕ to vary from $-90°$ at $\omega = 0^-$ to $+90°$ at $\omega = 0^+$. Because ϵ approaches zero, the mapping $GH(s)$ is

$$\lim_{\epsilon \to 0} GH(s) = \lim_{\epsilon \to 0} \left(\frac{K}{\epsilon e^{j\phi}} \right) = \lim_{\epsilon \to 0} \left(\frac{K}{\epsilon} \right) e^{-j\phi}. \tag{8.18}$$

Therefore the angle of the contour in the $GH(s)$-plane changes from $90°$ at $\omega = 0^-$ to $-90°$ at $\omega = 0^+$, passing through $0°$ at $\omega = 0$. The radius of the contour in the $GH(s)$-plane for this portion of the contour is infinite, and this portion of the contour is shown in Fig. 8.10(b).

(b) The Portion from $\omega = 0^+$ to $\omega = +\infty$. The portion of the contour Γ_s from $\omega = 0^+$ to $\omega = +\infty$ is mapped by the function $GH(s)$ as the real frequency polar plot because $s = j\omega$ and

$$GH(s)|_{s=j\omega} = GH(j\omega) \tag{8.19}$$

for this part of the contour. This results in the real frequency polar plot shown in Fig. 8.10(b). When ω approaches $+\infty$, we have

$$\lim_{\omega \to +\infty} GH(j\omega) = \lim_{\omega \to +\infty} \frac{K}{+j\omega(j\omega\tau + 1)} \tag{8.20}$$
$$= \lim_{\omega \to \infty} \left| \frac{K}{\tau\omega^2} \right| \underline{/-(\pi/2) - \tan^{-1} \omega\tau}.$$

Therefore the magnitude approaches zero at an angle of $-180°$.

(c) The Portion from $\omega = +\infty$ to $\omega = -\infty$. The portion of Γ_s from $\omega = +\infty$ to $\omega = -\infty$ is mapped into the point zero at the origin of the $GH(s)$-plane by the function $GH(s)$. The mapping is represented by

$$\lim_{r \to \infty} GH(s)|_{s=re^{j\phi}} = \lim_{r \to \infty} \left| \frac{K}{r^2} \right| e^{-2j\phi} \tag{8.21}$$

as ϕ changes from $\phi = +90°$ at $\omega = +\infty$ to $\phi = -90°$ at $\omega = -\infty$. Thus the contour moves from an angle of $-180°$ at $\omega = +\infty$ to an angle of $+180°$ at $\omega = -\infty$. The magnitude of the $GH(s)$ contour when r is infinite is always zero or a constant.

(d) The Portion from $\omega = -\infty$ to $\omega = 0^-$. The portion of the contour Γ_s from $\omega = -\infty$ to $\omega = 0^-$ is mapped by the function $GH(s)$ as

$$GH(s)|_{s=-j\omega} = GH(-j\omega). \tag{8.22}$$

Thus we obtain the complex conjugate of $GH(j\omega)$, and the plot for the portion of the polar plot from $\omega = -\infty$ to $\omega = 0^-$ is symmetrical to the polar plot from $\omega = +\infty$ to $\omega = 0^+$. This symmetrical polar plot is shown on the $GH(s)$-plane in Fig. 8.10(b).

Now, in order to investigate the stability of this second-order system, we first note that the number of poles P within the right-hand s-plane is zero. Therefore, for this system to be stable, we require $N = Z = 0$, and the contour Γ_{GH} must not encircle the -1 point in the GH-plane. Examining Fig. 8.10(b), we find that irrespective of the value of the gain K and the time constant τ, the contour does not encircle the -1 point, and the system is always stable. As in Chapter 6, we are considering positive values of gain K. If negative values of gain are to be considered, one should use $-K$, where $K \geq 0$.

We may draw two general conclusions from this example as follows:

1. The plot of the contour Γ_{GH} for the range $-\infty < \omega < 0^-$ will be the complex conjugate of the plot for the range $0^+ < \omega < +\infty$ and the polar plot of $GH(s)$ will be symmetrical in the $GH(s)$-plane about the u-axis. Therefore *it is sufficient to construct the contour Γ_{GH} for the frequency range $0^+ < \omega < +\infty$ in order to investigate the stability.*

2. The magnitude of $GH(s)$ as $s = re^{j\phi}$ and $r \to \infty$ will normally approach zero or a constant.

■ Example 8.3 System with three poles

Let us again consider the single-loop system shown in Fig. 8.8 when

$$GH(s) = \frac{K}{s(\tau_1 s + 1)(\tau_2 s + 1)}. \tag{8.23}$$

The Nyquist contour Γ_s is shown in Fig. 8.10(a). Again this mapping is symmetrical for $GH(j\omega)$ and $GH(-j\omega)$ so that it is sufficient to investigate the $GH(j\omega)$-locus. The origin of the s-plane maps into a semicircle of infinite radius as in the last example. Also, the semicircle $re^{j\phi}$ in the s-plane maps into the point $GH(s) = 0$ as we expect. Therefore, in order to investigate the stability of the system, it is sufficient to plot the portion of the contour Γ_{GH} which is the real frequency polar plot $GH(j\omega)$ for $0^+ < \omega < 0^+ + \infty$. Therefore, when $s = +j\omega$, we have

$$\begin{aligned} GH(j\omega) &= \frac{K}{j\omega(j\omega\tau_1 + 1)(j\omega\tau_2 + 1)} \\ &= \frac{-K(\tau_1 + \tau_2) - jK(1/\omega)(1 - \omega^2\tau_1\tau_2)}{1 + \omega^2(\tau_1^2 + \tau_2^2) + \omega^4\tau_1^2\tau_2^2} \\ &= \frac{K}{[\omega^4(\tau_1 + \tau_2)^2 + \omega^2(1 - \omega^2\tau_1\tau_2)^2]^{1/2}} \\ &\quad \times \angle -\tan^{-1}\omega\tau_1 - \tan^{-1}\omega\tau_2 - (\pi/2). \end{aligned} \tag{8.24}$$

When $\omega = 0^+$, the magnitude of the locus is infinite at an angle of $-90°$ in the $GH(s)$-plane. When ω approaches $+\infty$, we have

$$\begin{aligned} \lim_{\omega\to\infty} GH(j\omega) &= \lim_{\omega\to\infty} \left|\frac{1}{\omega^3}\right| \angle -(\pi/2) - \tan^{-1}\omega\tau_1 - \tan^{-1}\omega\tau_2 \\ &= \left(\lim_{\omega\to\infty} \left|\frac{1}{\omega^3}\right|\right) \angle -(3\pi/2). \end{aligned} \tag{8.25}$$

Therefore $GH(j\omega)$ approaches a magnitude of zero at an angle of $-270°$. In order for the locus to approach at an angle of $-270°$, the locus must cross the u-axis in the $GH(s)$-plane as shown in Fig. 8.11. Thus it is possible to encircle the -1 point as is shown in Fig. 8.11. The number of encirclements, when the -1 point lies within the locus as shown in Fig. 8.11, is equal to two and the system is unstable with two roots in the right-hand s-plane. The point where the $GH(s)$-locus intersects the real axis can be found by setting the imaginary part of $GH(j\omega) = u + jv$ equal to zero. We then have from Eq. (8.24)

$$v = \frac{-K(1/\omega)(1 - \omega^2\tau_1\tau_2)}{1 + \omega^2(\tau_1^2 + \tau_2^2) + \omega^4\tau_1^2\tau_2^2} = 0. \tag{8.26}$$

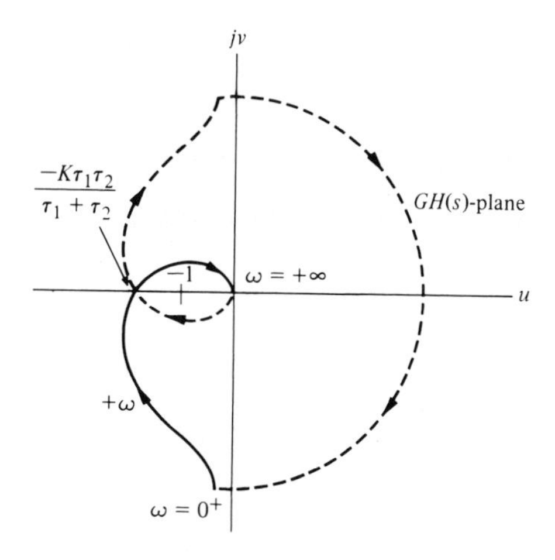

Figure 8.11. Nyquist diagram for $GH(s) = K/s(\tau_1 s + 1)(\tau_2 s + 1)$.

Thus $v = 0$ when $1 - \omega^2\tau_1\tau_2 = 0$ or $\omega = 1/\sqrt{\tau_1\tau_2}$. The magnitude of the real part, u, of $GH(j\omega)$ at this frequency is

$$\begin{aligned} u &= \left.\frac{-K(\tau_1 + \tau_2)}{1 + \omega^2(\tau_1^2 + \tau_2^2) + \omega^4\tau_1^2\tau_2^2}\right|_{\omega^2 = 1/\tau_1\tau_2} \\ &= \frac{-K(\tau_1 + \tau_2)\tau_1\tau_2}{\tau_1\tau_2 + (\tau_1^2 + \tau_2^2) + \tau_1\tau_2} = \frac{-K\tau_1\tau_2}{\tau_1 + \tau_2}. \end{aligned} \tag{8.27}$$

Therefore the system is stable when

$$\frac{-K\tau_1\tau_2}{\tau_1 + \tau_2} \geq -1$$

or

$$K \leq \frac{\tau_1 + \tau_2}{\tau_1\tau_2}. \tag{8.28}$$

■ Example 8.4 System with two poles at the origin

Again let us determine the stability of the single-loop system shown in Fig. 8.8 when

$$GH(s) = \frac{K}{s^2(\tau s + 1)}. \tag{8.29}$$

The real frequency polar plot is obtained when $s = j\omega$, and we have

$$\begin{aligned} GH(j\omega) &= \frac{K}{-\omega^2(j\omega\tau + 1)} \\ &= \frac{K}{[\omega^4 + \tau^2\omega^6]^{1/2}} \underline{/-\pi - \tan^{-1}\omega\tau}. \end{aligned} \tag{8.30}$$

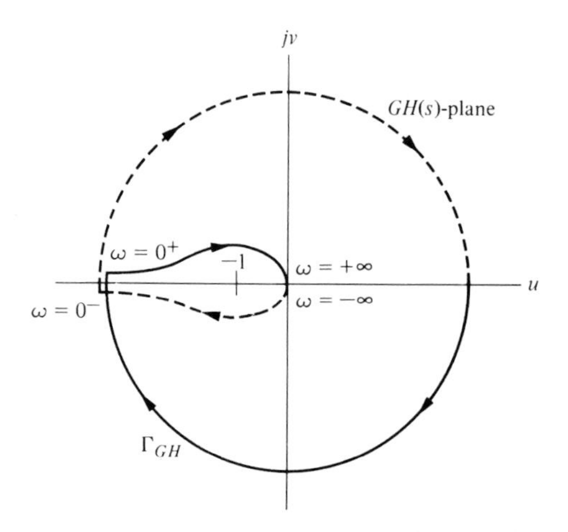

Figure 8.12. Nyquist contour plot for $GH(s) = K/s^2(\tau s + 1)$.

We note that the angle of $GH(j\omega)$ is always $-180°$ or greater, and the locus of $GH(j\omega)$ is above the u-axis for all values of ω. As ω approaches 0^+, we have

$$\lim_{\omega \to 0+} GH(j\omega) = \left(\lim_{\omega \to 0+} \left| \frac{K}{\omega^2} \right| \right) \angle -\pi. \tag{8.31}$$

As ω approaches $+\infty$, we have

$$\lim_{\omega \to +\infty} GH(j\omega) = \left(\lim_{\omega \to +\infty} \frac{K}{\omega^3} \right) \angle -3\pi/2. \tag{8.32}$$

At the small semicircular detour at the origin of the s-plane where $s = \epsilon e^{j\phi}$, we have

$$\lim_{\epsilon \to 0} GH(s) = \lim_{\epsilon \to 0} \frac{K}{\epsilon^2} e^{-2j\phi}, \tag{8.33}$$

where $-\pi/2 \le \phi \le \pi/2$. Thus the contour Γ_{GH} ranges from an angle of $+\pi$ at $\omega = 0^+$ to $-\pi$ at $\omega = 0^+$ and passes through a full circle of 2π rad as ω changes from $\omega = 0^-$ to $\omega = 0^+$. The complete contour plot of Γ_{GH} is shown in Fig. 8.12. Because the contour encircles the -1 point twice, there are two roots of the closed-loop system in the right-hand plane and the system, irrespective of the gain K, is unstable.

■ Example 8.5 System with a pole in the right-hand s-plane

Let us consider the control system shown in Fig. 8.13 and determine the stability of the system. First, let us consider the system without derivative feedback so that $K_2 = 0$. We then have the open-loop transfer function

$$GH(s) = \frac{K_1}{s(s-1)}. \tag{8.34}$$

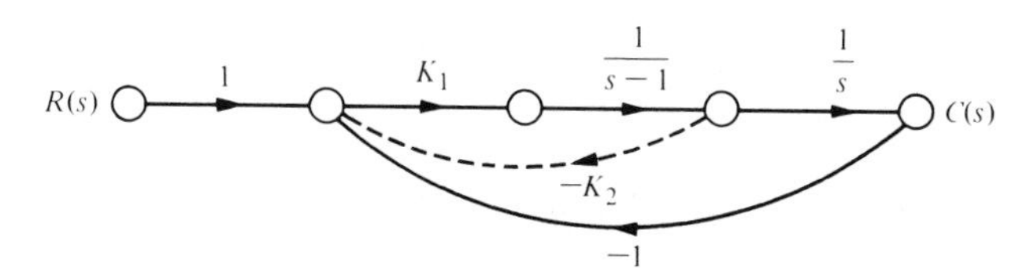

Figure 8.13. Second-order feedback control system.

Thus the open-loop transfer function has one pole in the right-hand s-plane, and therefore $P = 1$. In order for this system to be stable, we require $N = -P = -1$, one counterclockwise encirclement of the -1 point. At the semicircular detour at the origin of the s-plane, we let $s = \epsilon e^{j\phi}$ when $-\pi/2 \leq \phi \leq \pi/2$. Then we have, when $s = \epsilon e^{j\phi}$,

$$\lim_{\epsilon\to 0} GH(s) = \lim_{\epsilon\to 0} \frac{K_1}{-\epsilon e^{j\phi}} = \left(\lim_{\epsilon\to 0} \left|\frac{K_1}{\epsilon}\right|\right) \angle -180^\circ - \phi. \tag{8.35}$$

Therefore this portion of the contour Γ_{GH} is a semicircle of infinite magnitude in the left-hand GH-plane, as shown in Fig. 8.14. When $s = j\omega$, we have

$$\begin{aligned} GH(j\omega) &= \frac{K_1}{j\omega(j\omega - 1)} = \frac{K_1}{(\omega^2 + \omega^4)^{1/2}} \angle (-\pi/2) - \tan^{-1}(-\omega) \\ &= \frac{K_1}{(\omega^2 + \omega^4)^{1/2}} \angle (+\pi/2) + \tan^{-1}\omega. \end{aligned} \tag{8.36}$$

Finally, for the semicircle of radius r as r approaches infinity, we have

$$\lim_{r\to\infty} GH(s)|_{s=re^{j\phi}} = \left(\lim_{r\to\infty} \left|\frac{K_1}{r^2}\right|\right) e^{-2j\phi}, \tag{8.37}$$

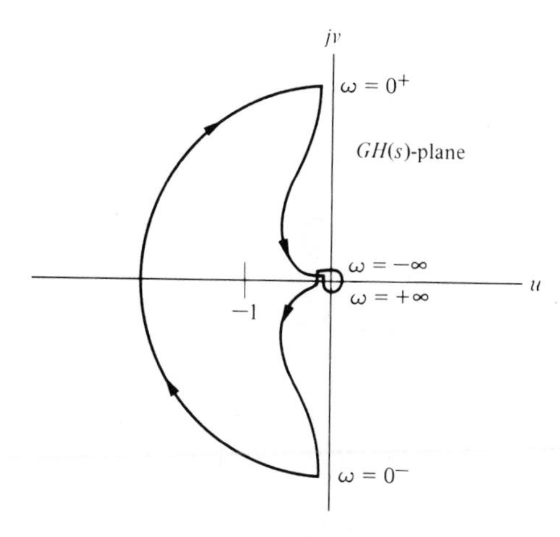

Figure 8.14. Nyquist diagram for $GH(s) = K_1/s(s - 1)$.

Table 8.3.

s	$j0^-$	$j0^+$	$j1$	$+j\infty$	$-j\infty$
$\lvert GH\rvert/K_1$	∞	∞	$1/\sqrt{2}$	0	0
$\underline{/GH}$	$-90°$	$+90°$	$+135°$	$+180°$	$-180°$

where ϕ varies from $\pi/2$ to $-\pi/2$ in a clockwise direction. Therefore the contour Γ_{GH}, at the origin of the GH-plane, varies 2π rad in a counterclockwise direction, as shown in Fig. 8.14. Several important values of the $GH(s)$-locus are given in Table 8.3. The contour Γ_{GH} in the $GH(s)$-plane encircles the -1 point once in the clockwise direction and $N = +1$. Therefore

$$Z = N + P = 2. \tag{8.38}$$

and the system is unstable because two zeros of the characteristic equation, irrespective of the value of the gain K, lie in the right half of the s-plane.

Let us now reconsider the system when the derivative feedback is included in the system shown in Fig. 8.13. Then the open-loop transfer function is

$$GH(s) = \frac{K_1(1 + K_2 s)}{s(s - 1)}. \tag{8.39}$$

The portion of the contour Γ_{GH} when $s = \epsilon e^{j\phi}$ is the same as the system without derivative feedback, as is shown in Fig. 8.15. However, when $s = re^{j\phi}$ as r approaches infinity, we have

$$\lim_{r\to\infty} GH(s)\big|_{s=re^{j\phi}} = \lim_{r\to\infty} \left|\frac{K_1K_2}{r}\right| e^{-j\phi}, \tag{8.40}$$

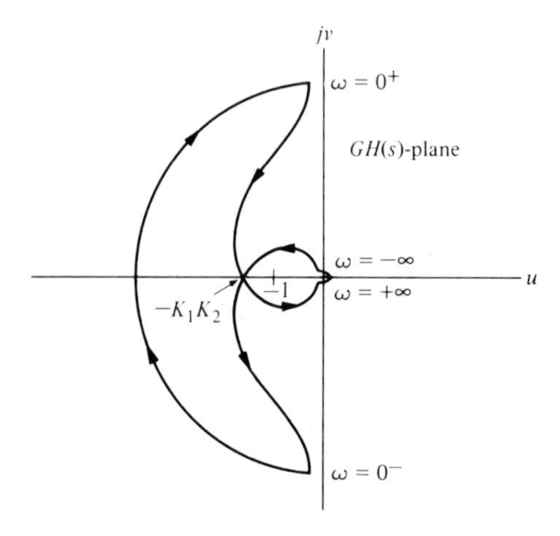

Figure 8.15. Nyquist diagram for $GH(s) = K_1(1 + K_2 s)/s(s - 1)$.

and the Γ_{GH}-contour at the origin of the GH-plane varies π rad in a counterclockwise direction, as shown in Fig. 8.15. The frequency locus $GH(j\omega)$ crosses the u-axis and is determined by considering the real frequency transfer function

$$GH(j\omega) = \frac{K_1(1 + K_2 j\omega)}{-\omega^2 - j\omega} = \frac{-K_1(\omega^2 + \omega^2 K_2) + j(\omega - K_2\omega^3)K_1}{\omega^2 + \omega^4}. \tag{8.41}$$

The $GH(j\omega)$-locus intersects the u-axis at a point where the imaginary part of $GH(j\omega)$ is zero. Therefore

$$\omega - K_2\omega^3 = 0$$

at this point, or $\omega^2 = 1/K_2$. The value of the real part of $GH(j\omega)$ at the intersection is then

$$u\,|_{\omega^2 = 1/K_2} = \left.\frac{-\omega^2 K_1(1 + K_2)}{\omega^2 + \omega^4}\right|_{\omega^2 = 1/K_2} = -K_1 K_2. \tag{8.42}$$

Therefore, when $-K_1K_2 < -1$ or $K_1K_2 > 1$, the contour Γ_{GH} encircles the -1 point once in a counterclockwise direction, and therefore $N = -1$. Then Z, the number of zeros of the system in the right-hand plane, is

$$Z = N + P = -1 + 1 = 0. \tag{8.43}$$

Thus the system is stable when $K_1K_2 > 1$. Often, it may be useful to utilize a computer or calculator program to calculate the Nyquist diagram [5].

8.4 Relative Stability and the Nyquist Criterion

We discussed the relative stability of a system in terms of the s-plane in Section 5.3. For the s-plane, we defined the relative stability of a system as the property measured by the relative settling time of each root or pair of roots. We would like to determine a similar measure of relative stability useful for the frequency-response method. The Nyquist criterion provides us with suitable information concerning the absolute stability and, furthermore, can be utilized to define and ascertain the relative stability of a system.

The Nyquist stability criterion is defined in terms of the $(-1, 0)$ point on the polar plot or the 0 db, 180° point on the Bode diagram or log-magnitude–phase diagram. Clearly, the proximity of the $GH(j\omega)$-locus to this stability point is a measure of the relative stability of a system. The polar plot for $GH(j\omega)$ for several values of K and

$$GH(j\omega) = \frac{K}{j\omega(j\omega\tau_1 + 1)(j\omega\tau_2 + 1)} \tag{8.44}$$

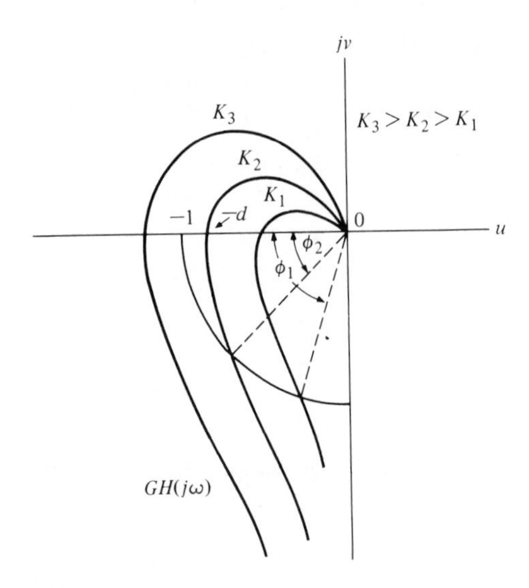

Figure 8.16. Polar plot for $GH(j\omega)$ for three values of gain.

is shown in Fig. 8.16. As K increases, the polar plot approaches the -1 point and eventually encircles the -1 point for a gain $K = K_3$. We determined in Section 8.3 that the locus intersects the u-axis at a point

$$u = \frac{-K\tau_1\tau_2}{\tau_1 + \tau_2}. \tag{8.45}$$

Therefore the system has roots on the $j\omega$-axis when

$$u = -1 \quad \text{or} \quad K = \left(\frac{\tau_1 + \tau_2}{\tau_1\tau_2}\right).$$

As K is decreased below this marginal value, the stability is increased and the margin between the gain $K = (\tau_1 + \tau_2)/\tau_1\tau_2$ and a gain $K = K_2$ is a measure of the relative stability. This measure of relative stability is called the *gain margin* and is defined as *the reciprocal of the gain* $|GH(j\omega)|$ *at the frequency at which the phase angle reaches 180°* (that is, $v = 0$). The gain margin is a measure of the factor by which the system gain would have to be increased for the $GH(j\omega)$ locus to pass through the $u = -1$ point. Thus, for a gain $K = K_2$ in Fig. 8.16, the gain margin is equal to the reciprocal of $GH(j\omega)$ when $v = 0$. Because $\omega = 1/\sqrt{\tau_1\tau_2}$ when the phase shift is 180°, we have a gain margin equal to

$$\frac{1}{|GH(j\omega)|} = \left[\frac{K_2\tau_1\tau_2}{\tau_1 + \tau_2}\right]^{-1} = \frac{1}{d}. \tag{8.46}$$

The gain margin can be defined in terms of a logarithmic (decibel) measure as

$$20 \log\left(\frac{1}{d}\right) = -20 \log d \text{ db}. \tag{8.47}$$

For example, when $\tau_1 = \tau_2 = 1$, the system is stable when $K \leq 2$. Thus when $K = K_2 = 0.5$, the gain margin is equal to

$$\frac{1}{d} = \left[\frac{K_2\tau_1\tau_2}{\tau_1 + \tau_2}\right]^{-1} = 4, \tag{8.48}$$

or, in logarithmic measure,

$$20 \log 4 = 12 \text{ db}. \tag{8.49}$$

Therefore the gain margin indicates that the system gain can be increased by a factor of four (12 db) before the stability boundary is reached.

An alternative measure of relative stability can be defined in terms of the phase angle margin between a specific system and a system that is marginally stable. Several roots of the characteristic equation lie on the $j\omega$-axis when the $GH(j\omega)$-locus intersects the $u = -1$, $v = 0$ point in the GH-plane. Therefore a measure of relative stability, the *phase margin,* is defined as *the phase angle through which the $GH(j\omega)$ locus must be rotated in order that the unity magnitude $|GH(j\omega)| = 1$ point passes through the $(-1, 0)$ point in the $GH(j\omega)$ plane.* This measure of relative stability is called the phase margin and is equal to the additional phase lag required before the system becomes unstable. This information can be determined from the Nyquist diagram shown in Fig. 8.16. For a gain $K = K_2$, an additional phase angle, ϕ_2, may be added to the system before the system becomes unstable. Furthermore, for the gain K_1, the phase margin is equal to ϕ_1, as shown in Fig. 8.16.

The gain and phase margins are easily evaluated from the Bode diagram, and because it is preferable to draw the Bode diagram in contrast to the polar plot, it is worthwhile to illustrate the relative stability measures for the Bode diagram. The critical point for stability is $u = -1$, $v = 0$ in the $GH(j\omega)$ plane which is equivalent to a logarithmic magnitude of 0 db and a phase angle of 180° (or −180°) on the Bode diagram.

The gain margin and phase margin can be readily calculated by utilizing a computer program [6]. A computer program for accomplishing this calculation is given in Table 8.4 in the computer language BASIC [7]. This program can readily be converted to other languages. The symbols used are W = ω; G2 = $|G(s)|^2$; P = phase of $G(s)$; and PM = phase margin. The program is shown for the case where $GH(j\omega)$ is as given in Eq. (8.50). The calculations commence at $\omega = 0.1$ and increase by 2% at each iteration at line 50.

The Bode diagram of

$$GH(j\omega) = \frac{1}{j\omega(j\omega + 1)(0.2j\omega + 1)} \tag{8.50}$$

is shown in Fig. 8.17. The phase angle when the logarithmic magnitude is 0 db is equal to 137°. Thus the phase margin is 180° − 137° = 43°, as shown in Fig. 8.17. The logarithmic magnitude when the phase angle is −180° is −15 db, and therefore the gain margin is equal to 15 db, as shown in Fig. 8.17.

The frequency response of a system can be graphically portrayed on the

Table 8.4. A Computer Program in BASIC Computer Language for Calculating the Gain Margin and Phase Margin for the Third-Order System $GH(j\omega) = 1/j\omega(j\omega + 1)(0.2j\omega + 1)$

```
10 LET W = 0.1
20 GOSUB 100
30 IF G2 < =1 THEN 60
35 IF G2 < =100 THEN 50
40 LET W = 2*W
45 GO TO 20
50 LET W = 1.02*W
55 GO TO 20
60 IF P> =180 THEN 140
65 PRINT "UNITY GAIN", "W =" W, "P=" P
70 LET W=1.02*W
75 GOSUB 100
80 IF P> =180 THEN 90
85 GO TO 70
90 PRINT "W=" W, "GAIN MARGIN =" 4.343*LOG(1/G2)
95 GO TO 200
100 LET P = 57.3*(ATN(W) + ATN(0.2*W) + 1.571)
110 LET X = W*W
120 LET G2 = 1/((1 + X)*(1 + 0.04*X)*X)
130 RETURN
140 PRINT "W =" W, "SYSTEM UNSTABLE"
200 END
```

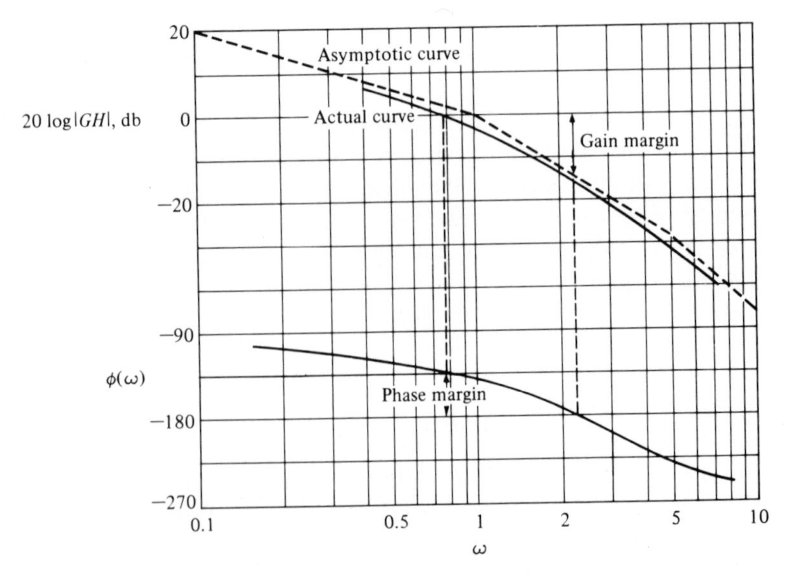

Figure 8.17. Bode diagram for $GH_1(j\omega) = 1/j\omega(j\omega + 1)(0.2j\omega + 1)$.

logarithmic-magnitude–phase-angle diagram. For the log-magnitude–phase diagram, the critical stability point is the 0 db, $-180°$ point, and the gain margin and phase margin can be easily determined and indicated on the diagram. The log-magnitude–phase locus of

$$GH_1(j\omega) = \frac{1}{j\omega(j\omega + 1)(0.2j\omega + 1)} \tag{8.51}$$

is shown in Fig. 8.18. The indicated phase margin is 43° and the gain margin is 15 db. For comparison, the locus for

$$GH_2(j\omega) = \frac{1}{j\omega(j\omega + 1)^2} \tag{8.52}$$

is also shown in Fig. 8.18. The gain margin for GH_2 is equal to 5.7 db, and the phase margin for GH_2 is equal to 20°. Clearly, the feedback system $GH_2(j\omega)$ is relatively less stable than the system $GH_1(j\omega)$. However, the question still remains: How much less stable is the system $GH_2(j\omega)$ in comparison to the system $GH_1(j\omega)$? In the following paragraph we shall answer this question for a

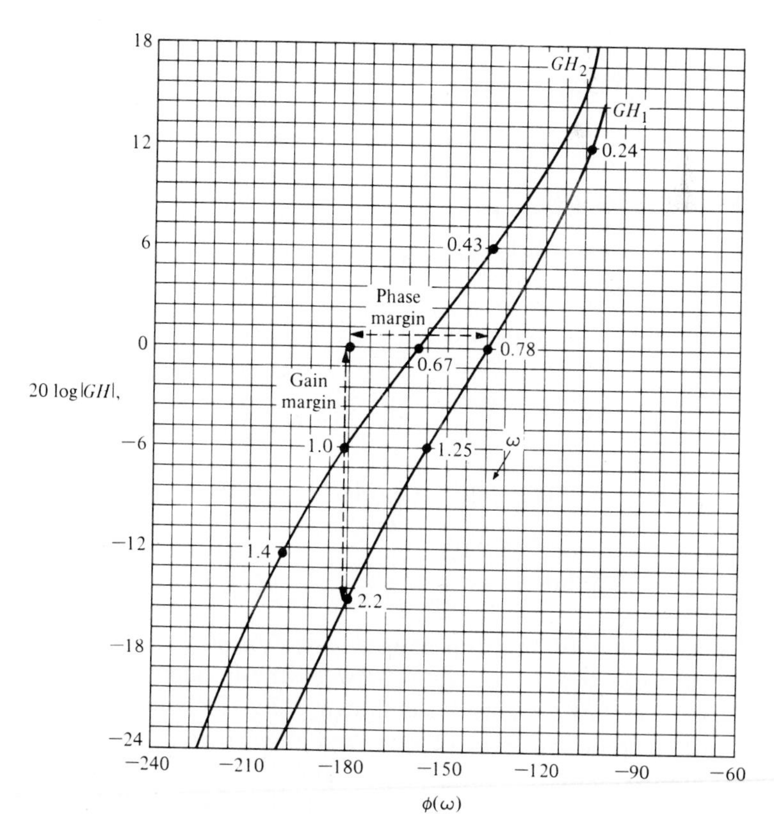

Figure 8.18. Log-magnitude–phase curve for GH_1 and GH_2.

second-order system, and the usefulness of the relation that we develop will depend upon the presence of dominant roots.

Let us now determine the phase margin of a second-order system and relate the phase margin to the damping ratio ζ of an underdamped system. Consider the loop-transfer function

$$GH(j\omega) = \frac{\omega_n^2}{j\omega(j\omega + 2\zeta\omega_n)}. \tag{8.53}$$

The characteristic equation for this second-order system is

$$s^2 + 2\zeta\omega_n s + \omega_n^2 = 0.$$

Therefore the closed-loop roots are

$$s = -\zeta\omega_n \pm j\omega_n\sqrt{1 - \zeta^2}.$$

The magnitude of the frequency response is equal to 1 at a frequency ω_c, and thus

$$\frac{\omega_n^2}{\omega_c(\omega_c^2 + 4\zeta^2\omega_n^2)^{1/2}} = 1. \tag{8.54}$$

Rearranging Eq. (8.54), we obtain

$$(\omega_c^2)^2 + 4\zeta^2\omega_n^2(\omega_c^2) - \omega_n^4 = 0. \tag{8.55}$$

Solving for ω_c, we find that

$$\frac{\omega_c^2}{\omega_n^2} = (4\zeta^4 + 1)^{1/2} - 2\zeta^2. \tag{8.56}$$

The phase margin for this system is

$$\begin{aligned}\phi_{pm} &= 180^\circ - 90^\circ - \tan^{-1}\left(\frac{\omega_c}{2\zeta\omega_n}\right)\\ &= 90^\circ - \tan^{-1}\left(\frac{1}{2\zeta}[(4\zeta^4 + 1)^{1/2} - 2\zeta^2]^{1/2}\right)\\ &= \tan^{-1}\left(2\zeta\left[\frac{1}{(4\zeta^4 + 1)^{1/2} - 2\zeta^2}\right]^{1/2}\right).\end{aligned} \tag{8.57}$$

Equation (8.57) is the relationship between the damping ratio ζ and the phase margin ϕ_{pm} that provides a correlation between the frequency response and the time response. A plot of ζ versus ϕ_{pm} is shown in Fig. 8.19. The actual curve of ζ versus ϕ_{pm} can be approximated by the dashed line shown in Fig. 8.19. The slope of the linear approximation is equal to 0.01, and therefore an approximate linear relationship between the damping ratio and the phase margin is

$$\zeta = 0.01\phi_{pm}, \tag{8.58}$$

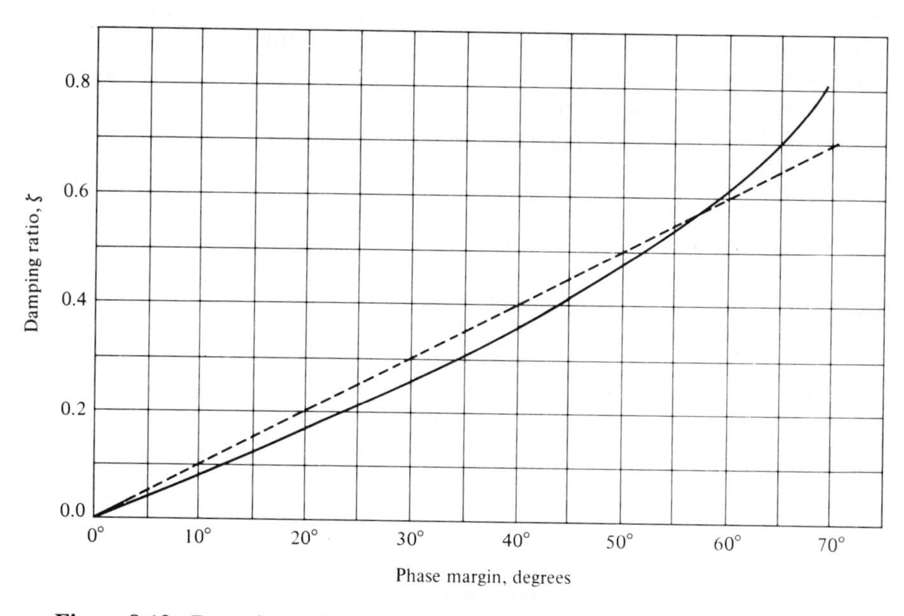

Figure 8.19. Damping ratio versus phase margin for a second-order system.

where the phase margin is measured in degrees. This approximation is reasonably accurate for $\zeta \leq 0.7$, and is a useful index for correlating the frequency response with the transient performance of a system. Equation (8.58) is a suitable approximation for a second-order system and may be used for higher-order systems if one can assume that the transient response of the system is primarily due to a pair of dominant underdamped roots. The approximation of a higher-order system by a dominant second-order system is a useful approximation indeed! Although it must be used with care, control engineers find this approach to be a simple, yet fairly accurate, technique of setting the specifications of a control system.

Therefore, for the system with a loop-transfer function

$$GH(j\omega) = \frac{1}{j\omega(j\omega + 1)(0.2j\omega + 1)}, \tag{8.59}$$

we found that the phase margin was 43°, as shown in Fig. 8.17. Thus the damping ratio is approximately

$$\zeta \simeq 0.01\phi_{pm} = 0.43. \tag{8.60}$$

Then the peak response to a step input for this system is approximately

$$M_{pt} = 1.22 \tag{8.61}$$

as obtained from Fig. 4.8 for $\zeta = 0.43$.

The phase margin of a system is a quite suitable frequency response measure for indicating the expected transient performance of a system. Another useful index of performance in the frequency domain is M_{p_ω}, the maximum magnitude of the closed-loop frequency response, and we shall now consider this practical index.

8.5 The Closed-Loop Frequency Response

The transient performance of a feedback system can be estimated from the closed-loop frequency response. The *closed-loop frequency response* is the frequency response of the closed-loop transfer function $T(j\omega)$. The open- and closed-loop frequency responses for a single-loop system are related as follows:

$$\frac{C(j\omega)}{R(j\omega)} = T(j\omega) = \frac{G(j\omega)}{1 + GH(j\omega)}. \tag{8.62}$$

The Nyquist criterion and the phase margin index are defined for the open-loop transfer function $GH(j\omega)$. However, as we found in Section 7.2, the maximum magnitude of the closed-loop frequency response can be related to the damping ratio of a second-order system of

$$M_{p_\omega} = |T(\omega_r)| = (2\zeta\sqrt{1 - \zeta^2})^{-1}, \qquad \zeta < 0.707. \tag{8.63}$$

This relation is graphically portrayed in Fig. 7.10. Because this relationship between the closed-loop frequency response and the transient response is a useful relationship, we would like to be able to determine M_{p_ω} from the plots completed for the investigation of the Nyquist criterion. That is, it is desirable to be able to obtain the closed-loop frequency response (Eq. 8.62) from the open-loop frequency response. Of course, we could determine the closed-loop roots of $1 + GH(s)$ and plot the closed-loop frequency response. However, once we have invested all the effort necessary to find the closed-loop roots of a characteristic equation, then a closed-loop frequency response is not necessary.

The relation between the closed-loop and open-loop frequency response is easily obtained by considering Eq. (8.62) when $H(j\omega) = 1$. If the system is not in fact a unity feedback system where $H(j\omega) = 1$, we will simply redefine the system output to be equal to the output of $H(j\omega)$. Then Eq. (8.62) becomes

$$T(j\omega) = M(\omega)e^{j\phi(\omega)} = \frac{G(j\omega)}{1 + G(j\omega)}. \tag{8.64}$$

The relationship between $T(j\omega)$ and $G(j\omega)$ is readily obtained in terms of complex variables utilizing the $G(j\omega)$-plane. The coordinates of the $G(j\omega)$-plane are u and v, and we have

$$G(j\omega) = u + jv. \tag{8.65}$$

Therefore the magnitude of the closed-loop response $M(\omega)$ is

$$M = \left|\frac{G(j\omega)}{1 + G(j\omega)}\right| = \left|\frac{u + jv}{1 + u + jv}\right| = \frac{(u^2 + v^2)^{1/2}}{((1 + u)^2 + v^2)^{1/2}}. \tag{8.66}$$

Squaring Eq. (8.66) and rearranging, we obtain

$$(1 - M^2)u^2 + (1 - M^2)v^2 - 2M^2u = M^2. \tag{8.67}$$

Dividing Eq. (8.67) by $(1 - M^2)$ and adding the term $[M^2/(1 - M^2)]^2$ to both sides of Eq. (8.67), we have

$$u^2 + v^2 - \frac{2M^2u}{1 - M^2} + \left(\frac{M^2}{1 - M^2}\right)^2 = \left(\frac{M^2}{1 - M^2}\right) + \left(\frac{M^2}{1 - M^2}\right)^2. \tag{8.68}$$

Rearranging, we obtain

$$\left(u - \frac{M^2}{1 - M^2}\right)^2 + v^2 = \left(\frac{M}{1 - M^2}\right)^2, \tag{8.69}$$

which is the equation of a circle on the u, v-plane with the center at

$$u = \frac{M^2}{1 - M^2}, \qquad v = 0.$$

The radius of the circle is equal to $|M/(1 - M^2)|$. Therefore we can plot several circles of constant magnitude M in the $G(j\omega) = u + jv$ plane. Several constant M circles are shown in Fig. 8.20. The circles to the left of $u = -\frac{1}{2}$ are for $M > 1$, and the circles to the right of $u = -\frac{1}{2}$ are for $M < 1$. When $M = 1$, the circle becomes the straight line $u = -\frac{1}{2}$, which is evident from inspection of Eq. (8.67).

The open-loop frequency response for a system is shown in Fig. 8.21 for two gain values where $K_2 > K_1$. The frequency response curve for the system with gain K_1 is tangent to magnitude circle M_1 at a frequency ω_{r_1}. Similarly, the fre-

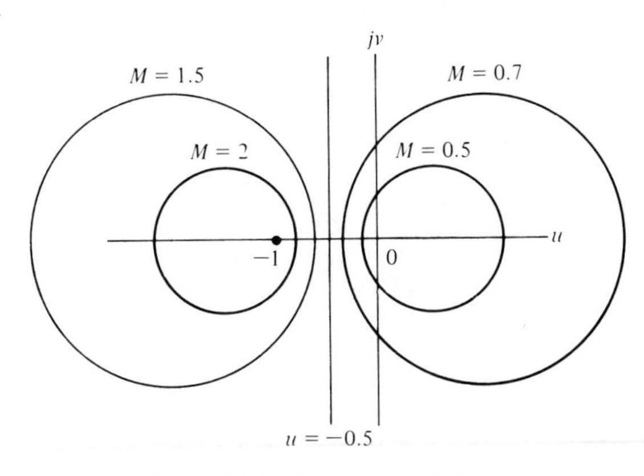

Figure 8.20. Constant M circles.

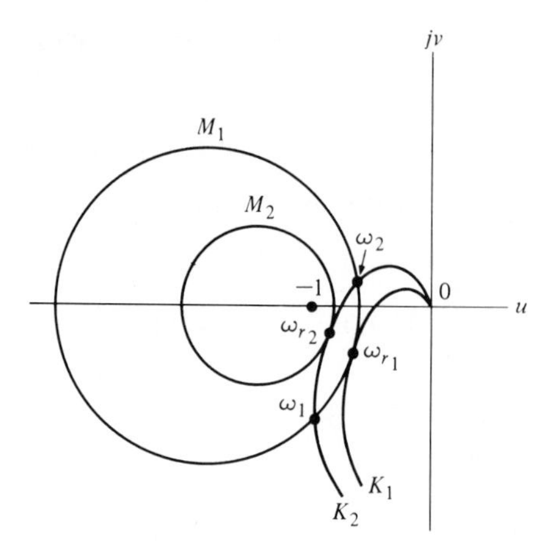

Figure 8.21. Polar plot of $G(j\omega)$ for two values of a gain.

quency response curve for gain K_2 is tangent to magnitude circle M_2 at the frequency ω_{r_2}. Therefore the closed-loop frequency response magnitude curves are estimated as shown in Fig. 8.22. Clearly, we can obtain the closed-loop frequency response of a system from the $(u + jv)$ plane. If the maximum magnitude, M_{p_ω}, is the only information desired, then it is sufficient to read this value directly from the polar plot. The maximum magnitude of the closed-loop frequency response, M_{p_ω}, is the value of the M circle that is tangent to the $G(j\omega)$-locus. The point of tangency occurs at the frequency ω_r, the resonant frequency. The complete closed-loop frequency response of a system can be obtained by reading the magnitude M of the circles that the $G(j\omega)$-locus intersects at several frequencies. Therefore the system with a gain $K = K_2$ has a closed-loop magnitude M_1 at the frequencies ω_1 and ω_2. This magnitude is read from Fig. 8.21 and is shown on the closed-loop frequency response in Fig. 8.22. The *bandwidth* for K_1, is shown as ω_{B_1}.

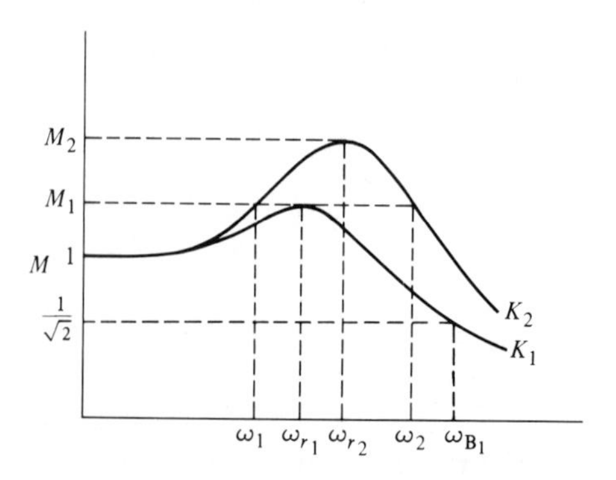

Figure 8.22. Closed-loop frequency response of $T(j\omega) = G(j\omega)/1 + G(j\omega)$.

It may be empirically shown that the crossover frequency on the open-loop Bode diagram, ω_c, is related to the closed-loop system bandwidth, ω_B, by the approximation $\omega_B = 1.6\omega_c$ for ζ in the range 0.2 to 0.8.

In a similar manner, we can obtain circles of constant closed-loop phase angles. Thus, for Eq. (8.64), the angle relation is

$$\begin{aligned}\phi = \underline{/T(j\omega)} &= \underline{/(u + jv)/(1 + u + jv)} \\ &= \tan^{-1}\left(\frac{v}{u}\right) - \tan^{-1}\left(\frac{v}{1 + u}\right).\end{aligned} \tag{8.70}$$

Taking the tangent of both sides and rearranging, we have

$$u^2 + v^2 + u - \frac{v}{N} = 0, \tag{8.71}$$

where $N = \tan\phi =$ constant. Adding the term $\frac{1}{4}[1 + (1/N^2)]$ to both sides of the equation and simplifying, we obtain

$$(u + 0.5)^2 + \left(v - \frac{1}{2N}\right)^2 = \frac{1}{4}\left(1 + \frac{1}{N^2}\right), \tag{8.72}$$

which is the equation of a circle with its center at $u = -0.5$ and $v = +(1/2N)$. The radius of the circle is equal to $\frac{1}{2}[1 + (1/N^2)]^{1/2}$. Therefore the constant phase angle curves can be obtained for various values of N in a manner similar to the M circles.

The constant M and N circles can be used for analysis and design in the polar plane. However, it is much easier to obtain the Bode diagram for a system, and it would be preferable if the constant M and N circles were translated to a logarithmic gain phase. N. B. Nichols transformed the constant M and N circles to the log-magnitude–phase diagram, and the resulting chart is called the *Nichols chart* [7]. The M and N circles appear as contours on the Nichols chart shown in Fig. 8.23. The coordinates of the log-magnitude–phase diagram are the same as those used in Section 7.5. However, superimposed on the log-magnitude–phase plane we find constant M and N lines. The constant M lines are given in decibels and the N lines in degrees. An example will illustrate the use of the Nichols chart to determine the closed-loop frequency response.

■ Example 8.6 Stability using the Nichols chart

Consider a feedback system with a loop transfer function

$$G(j\omega) = \frac{1}{j\omega(j\omega + 1)(0.2j\omega + 1)}. \tag{8.73}$$

The $G(j\omega)$-locus is plotted on the Nichols chart and is shown in Fig. 8.24. The maximum magnitude, M_{p_ω}, is equal to +2.5 db and occurs at a frequency $\omega_r =$ 0.8. The closed-loop phase angle at ω_r is equal to $-72°$. The 3-db closed-loop bandwidth where the closed-loop magnitude is -3 db is equal to $\omega_B = 1.33$, as

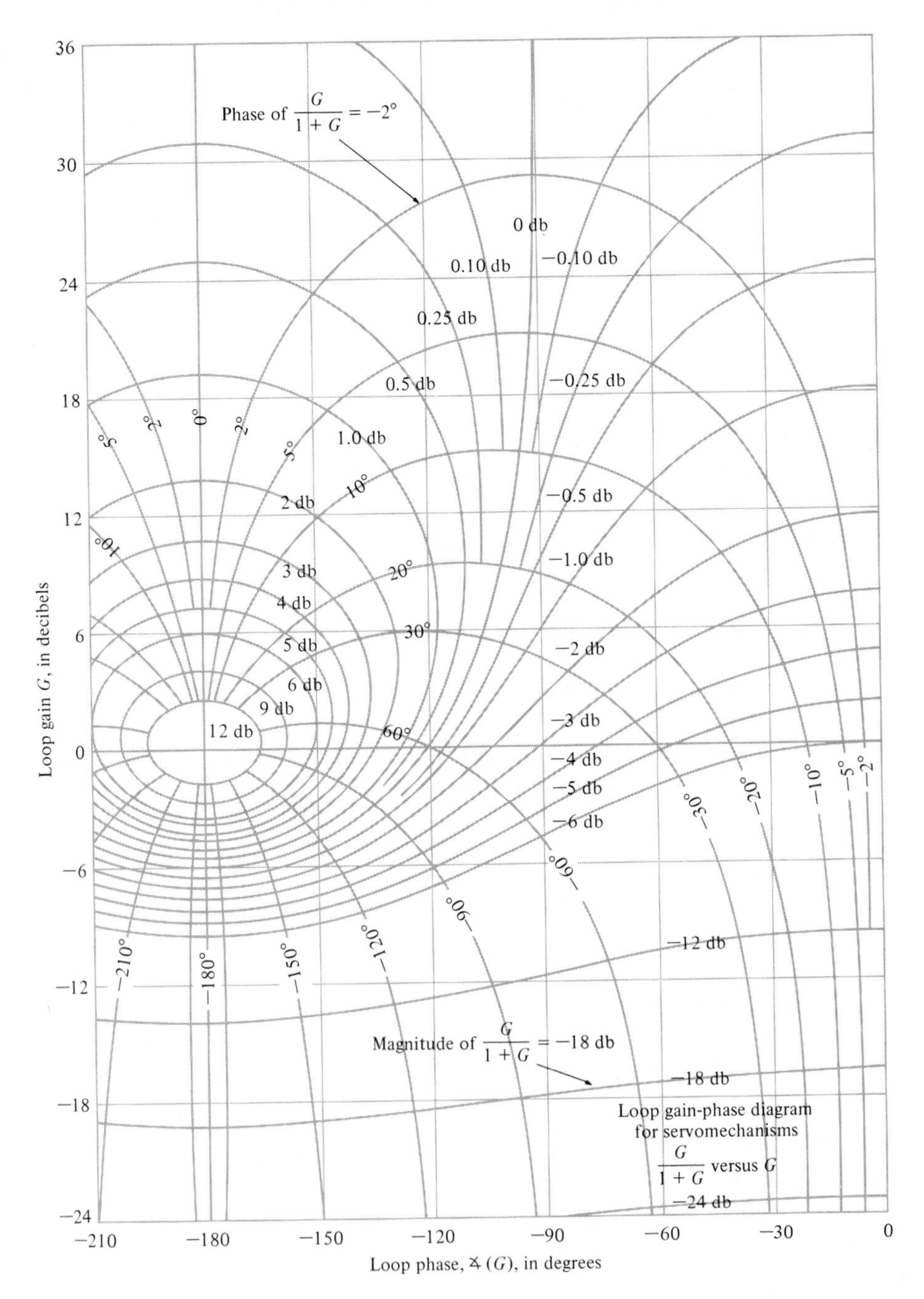

Figure 8.23. Nichols chart.

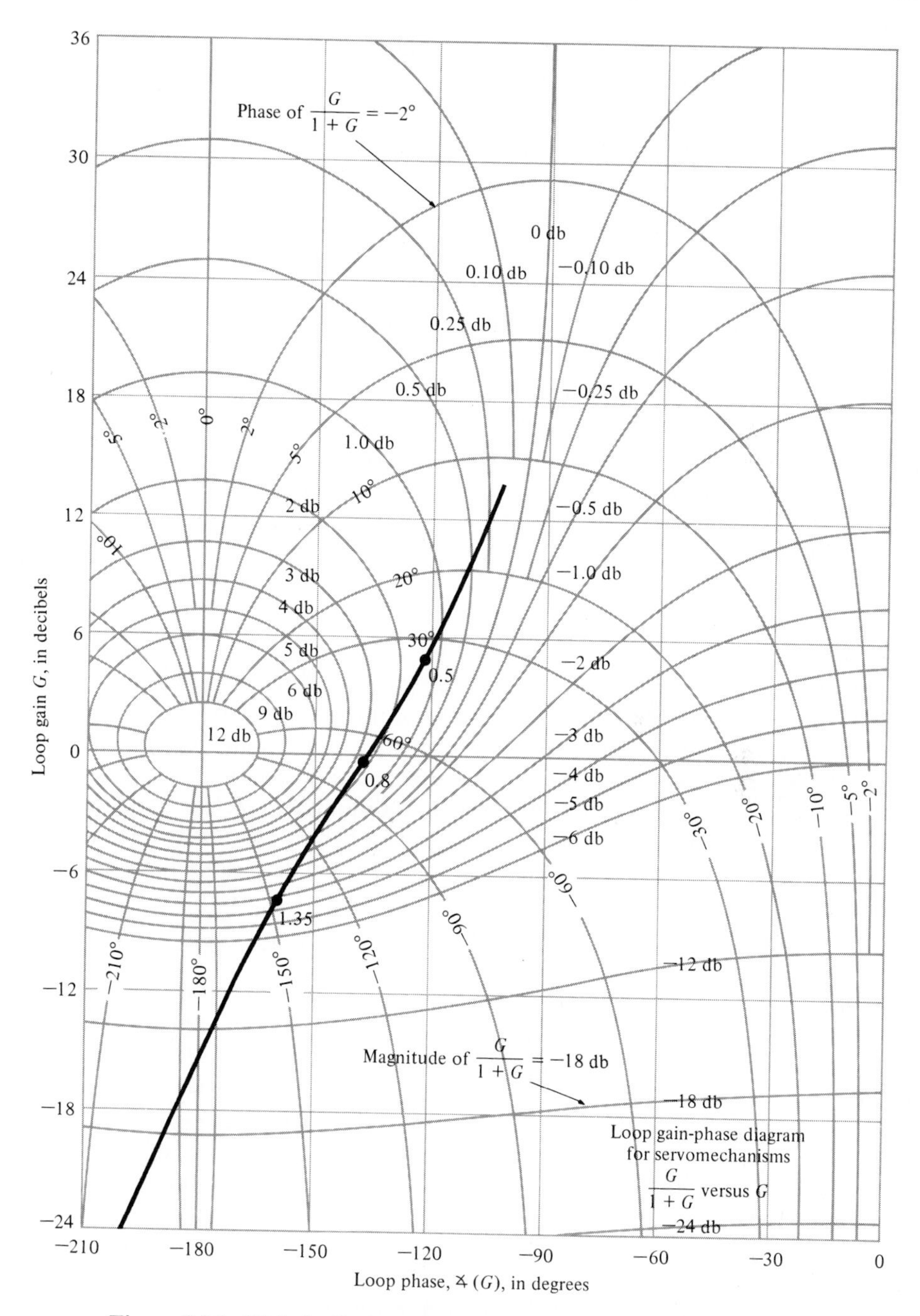

Figure 8.24. Nichols diagram for $G(j\omega) = 1/j\omega(j\omega + 1)(0.2j\omega + 1)$.

shown in Fig. 8.24. The closed-loop phase angle at ω_B is equal to $-142°$. One can also use the Control Systems Design Program to obtain the closed-loop frequency response. Using CSDP it was verified that the bandwidth was 1.33 and the closed-loop phase angle was $-142°$.

■ Example 8.7 Third order system

Let us consider a system with an open-loop transfer function

$$G(j\omega) = \frac{0.64}{j\omega[(j\omega)^2 + j\omega + 1]}, \tag{8.74}$$

where $\zeta = 0.5$ for the complex poles and $H(j\omega) = 1$. The Nichols diagram for this system is shown in Fig. 8.25. The phase margin for this system as it is determined from the Nichols chart is 30°. On the basis of the phase, we estimate the system damping ratio as $\zeta = 0.30$. The maximum magnitude is equal to +9 db occurring at a frequency $\omega_r = 0.88$. Therefore

$$20 \log M_{p_\omega} = 9 \text{ db}$$

or

$$M_{p_\omega} = 2.8.$$

Utilizing Fig. 7.10 to estimate the damping ratio, we find that $\zeta \simeq 0.175$.

We are confronted with two conflicting damping ratios, where one is obtained from a phase margin measure and another from a peak frequency-response measure. In this case, we have discovered an example in which the correlation between the frequency domain and the time domain is unclear and uncertain. This apparent conflict is caused by the nature of the $G(j\omega)$-locus, which slopes rapidly toward the 180° line from the 0-db axis. If we determine the roots of the characteristic equation for $1 + GH(s)$, we obtain

$$q(s) = (s + 0.77)(s^2 + 0.225s + 0.826) = 0. \tag{8.75}$$

The damping ratio of the complex conjugate roots is equal to 0.124, where the complex roots do not dominate the response of the system. Therefore the real root will add some damping to the system and one might estimate the damping ratio as being approximately the value determined from the M_{p_ω} index; that is, $\zeta = 0.175$. A designer must use the frequency-domain to time-domain correlations with caution. However, one is usually safe if the lower value of the damping ratio resulting from the phase margin and the M_{p_ω} relation is utilized for analysis and design purposes.

The Nichols chart can be used for design purposes by altering the $G(j\omega)$-locus in a suitable manner in order to obtain a desirable phase margin and M_{p_ω}. The system gain K is readily adjusted in order to provide a suitable phase margin and M_{p_ω} by inspecting the Nichols chart. For example, let us reconsider the previous example, where

$$G(j\omega) = \frac{K}{j\omega[(j\omega)^2 + j\omega + 1]}. \tag{8.76}$$

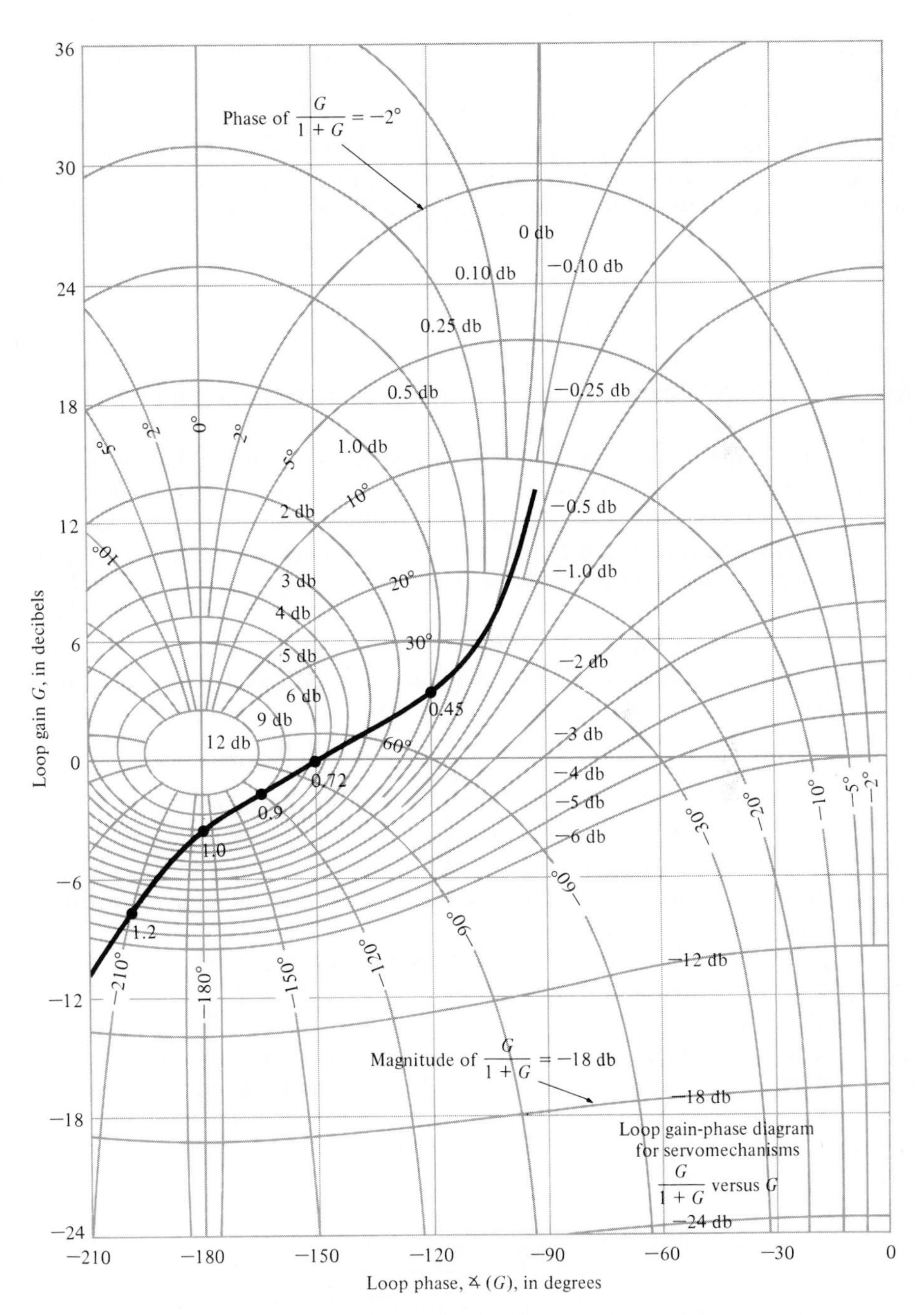

Figure 8.25. Nichols diagram for $G(j\omega) = 0.64/j\omega[(j\omega)^2 + j\omega + 1]$.

The $G(j\omega)$-locus on the Nichols chart for $K = 0.64$ is shown in Fig. 8.25. Let us determine a suitable value for K so that the system damping ratio is greater than 0.30. Examining Fig. 7.10, we find that it is required that M_{p_ω} be less than 1.75 (4.9 db). From Fig. 8.25, we find that the $G(j\omega)$-locus will be tangent to the 4.9 db curve if the $G(j\omega)$-locus is lowered by a factor of 2.2 db. Therefore, K should be reduced by 2.2 db or the factor antilog (2.2/20) = 1.28. Thus the gain K must be less than 0.64/1.28 = 0.50 if the system damping ratio is to be greater than 0.30.

8.6 The Stability of Control Systems with Time Delays

The Nyquist stability criterion has been discussed and illustrated in the previous sections for control systems whose transfer functions are rational polynomials of $j\omega$. There are many control systems that have a time delay within the closed loop of the system which affects the stability of the system. A *time delay* is the time interval between the start of an event at one point in a system and its resulting action at another point in the system. Fortunately, the Nyquist criterion can be utilized to determine the effect of the time delay on the relative stability of the feedback system. A pure time delay, without attenuation, is represented by the transfer function

$$G_d(s) = e^{-sT}, \tag{8.77}$$

where T is the delay time. The Nyquist criterion remains valid for a system with a time delay because the factor e^{-sT} does not introduce any additional poles or zeros within the contour. The factor adds a phase shift to the frequency response without altering the magnitude curve.

This type of time delay occurs in systems that have a movement of a material that requires a finite time to pass from an input or control point to an output or measured point [8, 12].

For example, a steel rolling mill control system is shown in Fig. 8.26. The motor adjusts the separation of the rolls so that the thickness error is minimized.

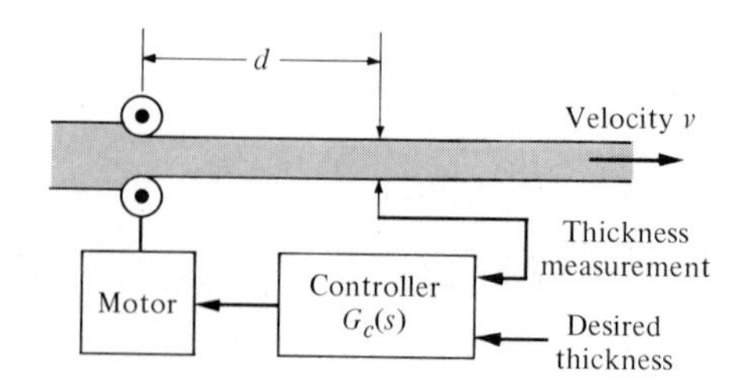

Figure 8.26. Steel rolling mill control system.

If the steel is traveling at a velocity v, then the time delay between the roll adjustment and the measurement is

$$T = \frac{d}{v}. \tag{8.78}$$

Therefore, in order to have a negligible time delay, we must decrease the distance to the measurement and increase the velocity of the flow of steel. Usually, we cannot eliminate the effect of time delay and thus the loop transfer function is

$$G(s)G_c(s)e^{-sT}. \tag{8.79}$$

However, one notes that the frequency response of this system is obtained from the loop-transfer function

$$GH(j\omega) = GG_c(j\omega)e^{-j\omega T}. \tag{8.80}$$

The usual loop-transfer function is plotted on the $GH(j\omega)$-plane and the stability ascertained relative to the -1 point. Alternatively, we can plot the Bode diagram including the delay factor, and investigate the stability relative to the 0 db, $-180°$ point. The delay factor $e^{-j\omega T}$ results in a phase shift

$$\phi(\omega) = -\omega T \tag{8.81}$$

and is readily added to the phase shift resulting from $GG_c(j\omega)$. Note that the angle is in radians in Eq. (8.81). An example will show the simplicity of this approach on the Bode diagram.

■ Example 8.8 Liquid level control system

A level control system is shown in Fig. 8.27(a) and the block diagram in Fig. 8.27(b). The time delay between the valve adjustment and the fluid output is $T = d/v$. Therefore, if the flow rate is 5 m^3/sec, the cross-sectional area of the pipe is 1 m^2, and the distance is equal to 5 m, then we have a time delay $T = 1$ sec. The loop-transfer function is then

$$\begin{aligned} GH(s) &= G_A(s)G(s)G_f(s)e^{-sT} \\ &= \frac{31.5}{(s+1)(30s+1)[(s^2/9)+(s/3)+1]}e^{-sT}. \end{aligned} \tag{8.82}$$

The Bode diagram for this system is shown in Fig. 8.28. The phase angle is shown both for the denominator factors alone and with the additional phase lag due to the time delay. The logarithmic gain curve crosses the 0-db line at $\omega = 0.8$. Therefore the phase margin of the system without the pure time delay would be 40°. However, with the time delay added, we find that the phase margin is equal to $-3°$, and the system is unstable. Therefore the system gain must be reduced in order to provide a reasonable phase margin. In order to provide a phase margin of 30°, the gain would have to be decreased by a factor of 5 db to $K = 31.5/1.78 = 17.7$.

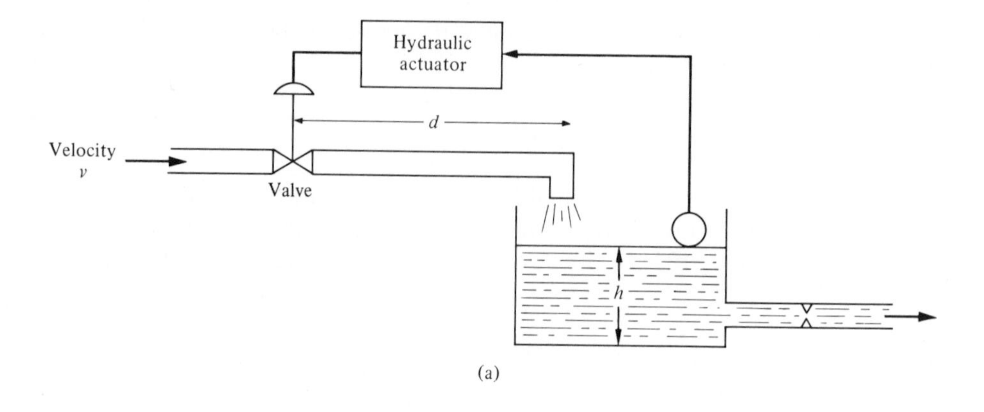

(a)

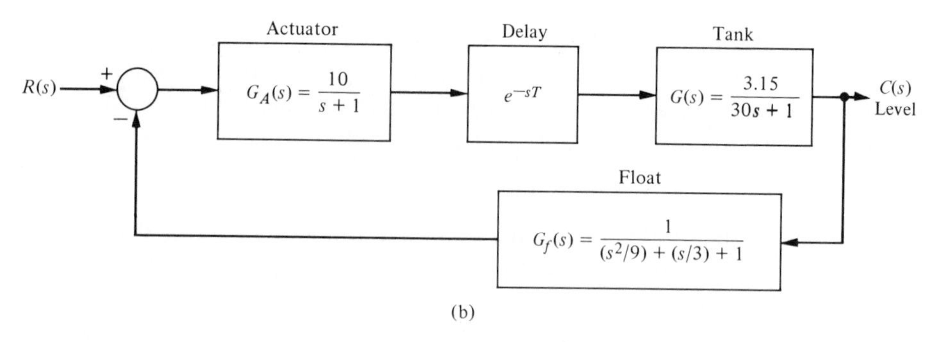

(b)

Figure 8.27. Liquid level control system.

A time delay, e^{-sT}, in a feedback system introduces an additional phase lag and results in a less stable system. Therefore as pure time delays are unavoidable in many systems, it is often necessary to reduce the loop gain in order to obtain a stable response. However, the cost of stability is the resulting increase in the steady-state error of the system as the loop gain is reduced.

8.7 System Bandwidth

The bandwidth of the closed-loop control system is an excellent measurement of the range of fidelity of response of the system. In systems where the low frequency magnitude is 0 db on the Bode diagram, the bandwidth is measured at the -3 db frequency. The speed of response to a step input will be roughly proportional to ω_B and the settling time is inversely proportional to ω_B. Thus we seek a large bandwidth consistent with reasonable system components.

Consider the following two closed-loop system transfer functions:

$$T_1(s) = \frac{1}{s+1}$$

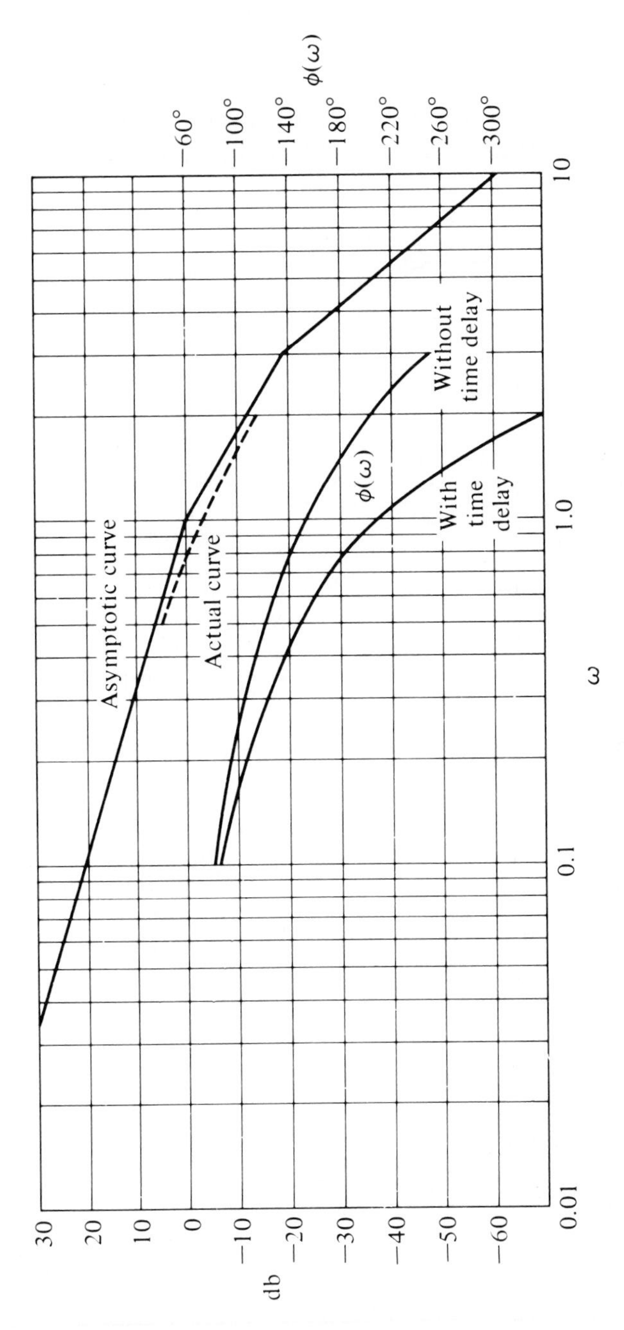

Figure 8.28. Bode diagram for level control system.

and

$$T_2(s) = \frac{1}{5s + 1}. \tag{8.83}$$

The frequency response of the two systems is contrasted in part (a) of Fig. 8.29 and the step response of the systems is shown in part (b). Also, the response to a

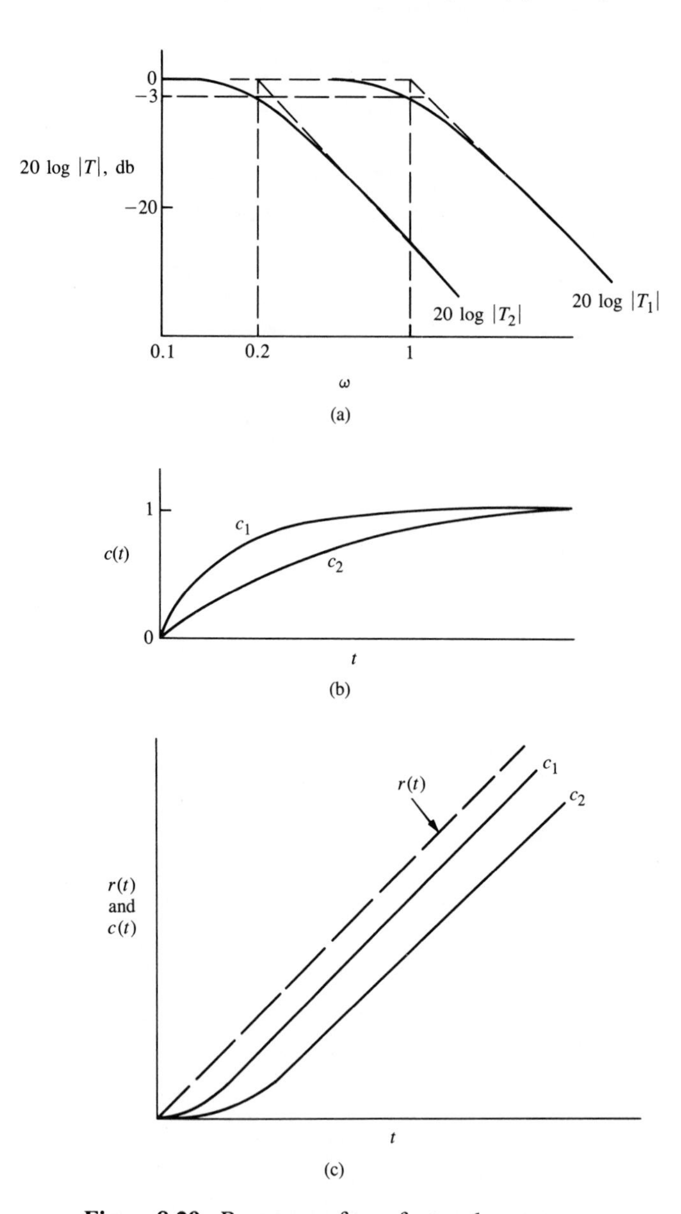

Figure 8.29. Response of two first-order systems.

ramp is shown in part (c) of that figure. Clearly, the system with the larger bandwidth provides the faster step response and higher fidelity ramp response.

Consider two second-order systems with closed-loop transfer functions:

$$T_3(s) = \frac{100}{s^2 + 10s + 100}$$

and

$$T_4(s) = \frac{900}{s^2 + 30s + 900}. \tag{8.84}$$

Both systems have a ζ of 0.5. The frequency response of both closed-loop systems is shown in Fig. 8.30(a). The natural frequency is 10 and 30 for systems three and four, respectively. The bandwidth is 15 and 40 for systems three and four, respec-

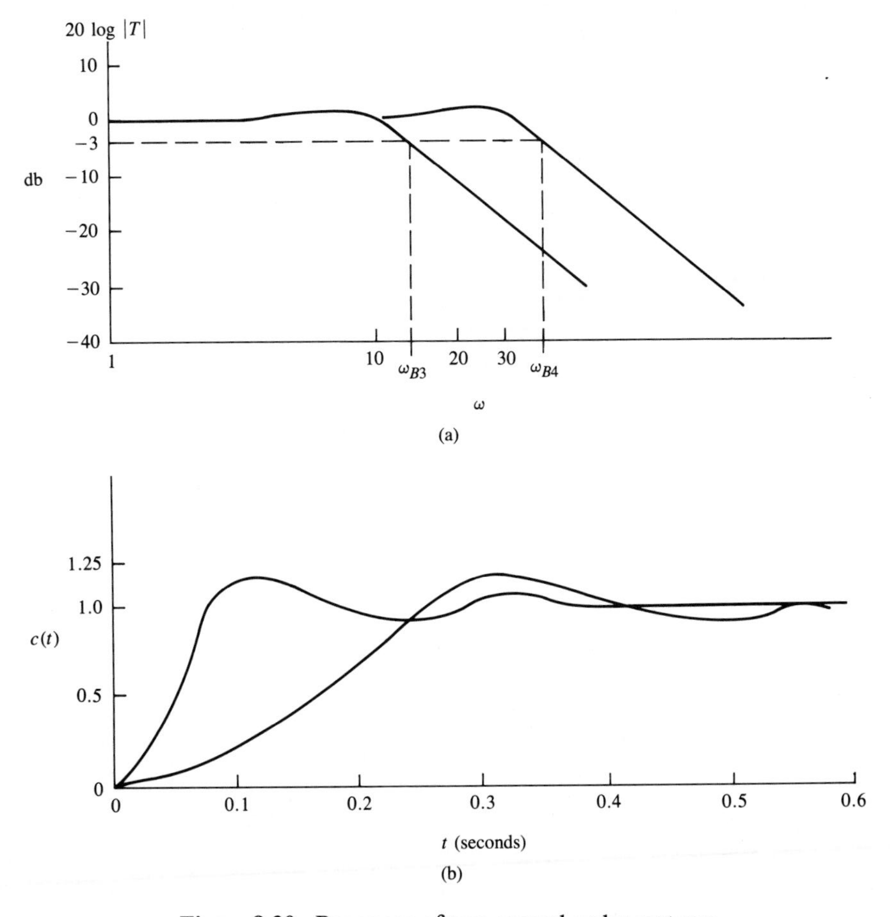

Figure 8.30. Response of two second-order systems.

tively. Both systems have a 15% overshoot, but T_4 has a peak time of 0.1 second compared to 0.3 for T_3, as shown in Fig. 8.30(b). Also, note that the settling time for T_4 is 0.37 second, while it is 0.8 second for T_3. Clearly, the system with a larger bandwidth provides a faster response. In general, we will pursue the design of systems with good stability and large bandwidth.

8.8 Design Example: Remotely Controlled Battlefield Vehicle

The use of remotely controlled vehicles for reconnaissance on the battlefield may be an idea whose time has come. One concept of a roving vehicle is shown in Fig. 8.31(a) and a proposed speed-control system is shown in Fig. 8.31(b). The desired speed $R(s)$ is transmitted by radio to the vehicle and the disturbance $D(s)$ represents hills and rocks. The goal is to achieve good overall control with low steady-state error and low-overshoot response to step commands, $R(s)$.

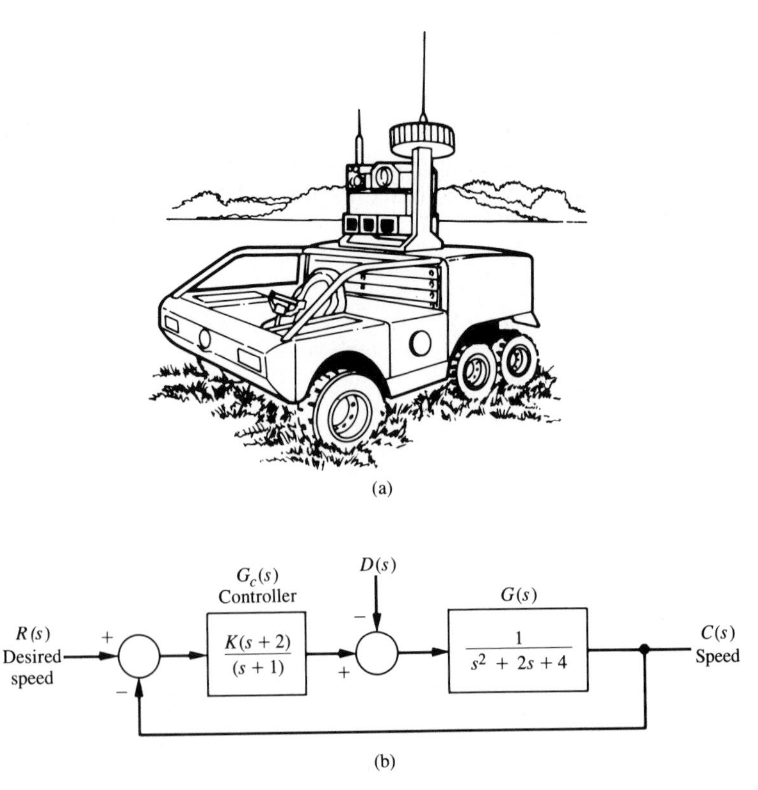

Figure 8.31. (a) Remotely controlled reconnaissance vehicle. (b) Speed-control system.

First, in order to achieve low steady-state error for a unit step command, we calculate e_{ss} as

$$e_{ss} = \lim_{s\to 0} sE(s)$$

$$= \lim_{s\to 0} s\left[\frac{R(s)}{1 + G_cG(s)}\right] \tag{8.85}$$

$$= \frac{1}{1 + G_cG(s)} = \frac{1}{1 + K/2}.$$

If we select $K = 20$, we will obtain a steady-state error of 9% of the magnitude of the input command. Using $K = 20$, we reformulate $G(s)$ for Bode diagram calculations, obtaining

$$G(s) = \frac{10(1 + s/2)}{(1 + s)(1 + s/2 + s^2/4)}. \tag{8.86}$$

The calculations for $0 \leq \omega \leq 6$ provide the data summarized in Table 8.5. The Nichols diagram for $K = 20$ is shown in Fig. 8.32. Examining the Nichols chart, we find that M_{P_ω} is 12 db and the phase margin is 25 degrees. The step response of this system is underdamped and we predict an excessive overshoot of approximately 50%.

In order to reduce the overshoot to a step input, we can reduce the gain to achieve a predicted overshoot. In order to limit the overshoot to 25%, we select a desired ζ of the dominant roots as 0.4 (from Fig. 4.8) and thus require $M_{P_\omega} = 1.35$ (from Fig. 7.10) or $20 \log M_{P_\omega} = 2.6$ db. In order to lower the gain, we will move the frequency response vertically down on the Nichols chart, as shown in Fig. 3.32. At $\omega_1 = 2.8$, we just intersect the 2.6-db closed-loop curve. The reduction (vertical drop) in gain is equal to 13 db or a factor of 4.5. Thus $K = 20/4.5 = 4.44$. For this reduced gain, the steady-state error is

$$e_{ss} = \frac{1}{(1 + 4.4/2)} = 0.31,$$

or we have a 31% steady-state error.

The actual step response when $K = 4.44$, as shown in Fig. 8.33, has an overshoot of 5%. If we use a gain of 10, we have an overshoot of 30% with a steady-state error of 17%. The performance of the system is summarized in Table 8.6. As a suitable compromise, we select $K = 10$ and draw the frequency response on

Table 8.5. Frequency Response Data for Design Example

ω	0	1.2	1.6	2.0	2.8	4	6
db	20	18.4	17.8	16.0	10.5	2.7	−5.2
degrees	0	−65	−86	−108	−142	−161	−170°

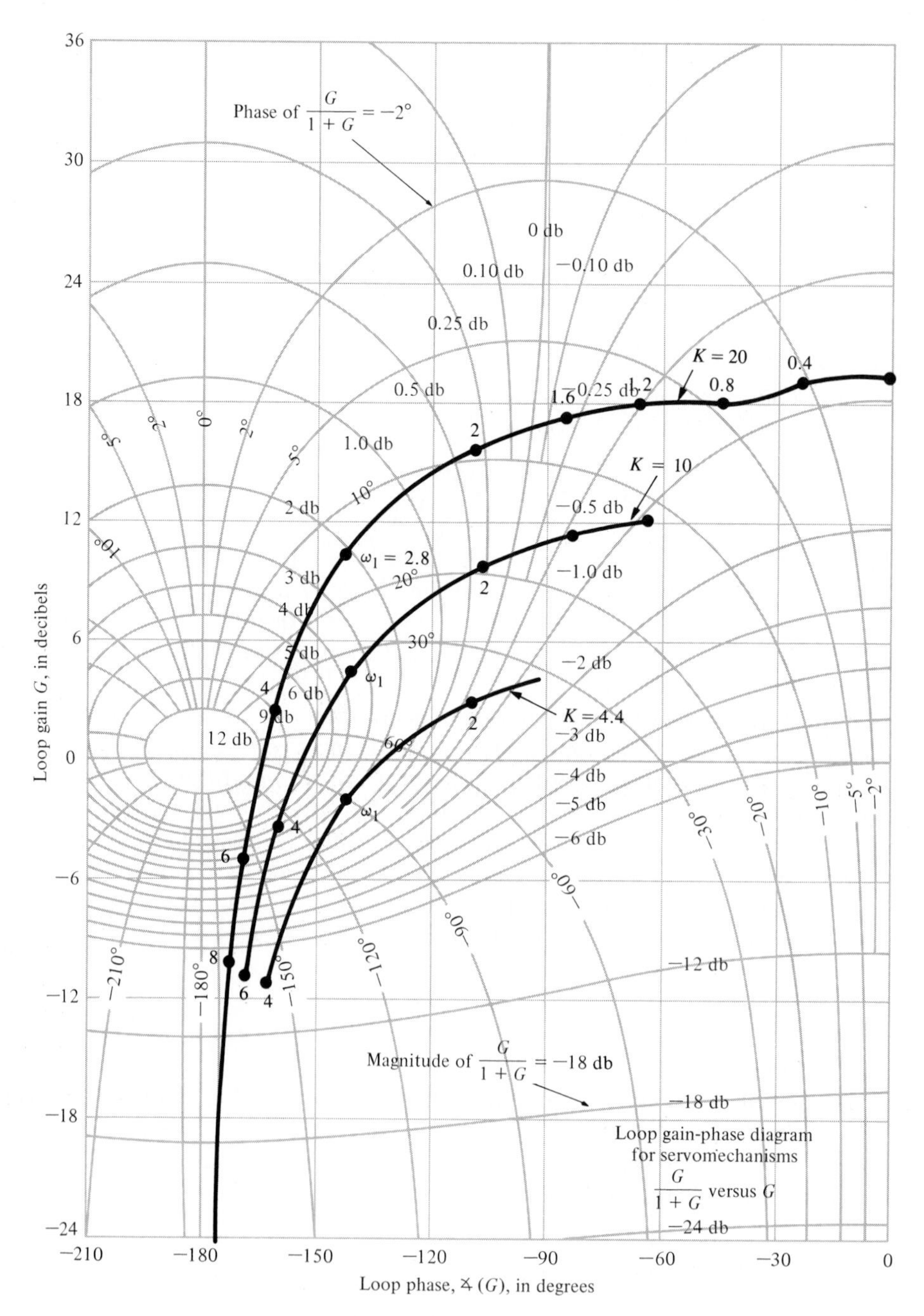

Figure 8.32. Nichol's diagram for the design example when $K = 20$ and for two reduced gains.

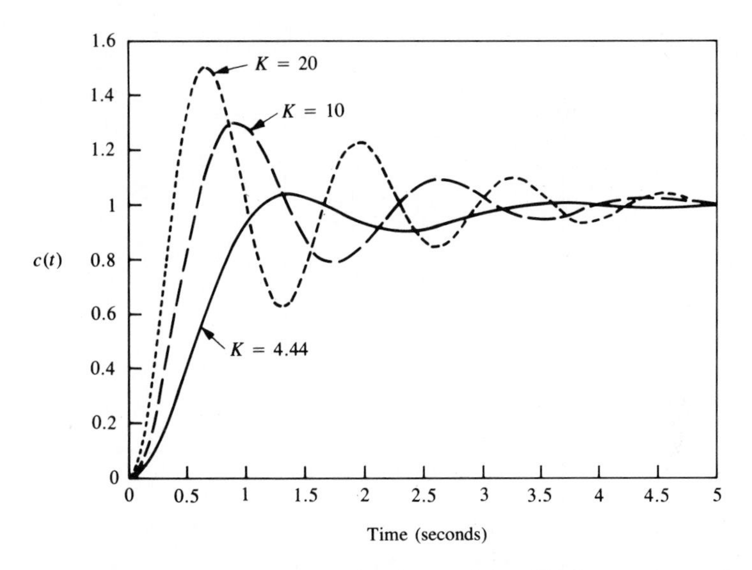

Figure 8.33. The response of the system for three values of K for a unit step input, $r(t)$.

the Nichols chart by moving the response for $K = 20$ down by $20 \log 2 = 6$ db, as shown on Fig. 8.32.

Examining the Nichols chart for $K = 10$, we have $M_{P_\omega} = 7$ db, and the phase margin is 34 degrees. Thus we estimate a ζ for the dominant roots of 0.34, which should result in an overshoot to a step input of 30%. The actual response, as recorded in Table 8.6, is 30%, as expected. The bandwidth of the system is $\omega_B \cong 5$. Thus we predict a settling time

$$T_s = \frac{4}{\zeta\omega_n}$$

$$= \frac{4}{(0.34)(\omega_B/1.4)} = 3.3 \text{ seconds},$$

since $\omega_B \cong 1.4\ \omega_n$ for $\zeta = 0.34$. The actual settling time is approximately five seconds, as shown in Fig. 8.33.

Table 8.6. Actual Response for Selected Gains

K	4.44	10	20
Percent overshoot	5%	30%	50%
Settling time (seconds)	3.5	5	6
Peak time (seconds)	1.4	1.0	0.7
e_{ss}	31%	17%	9%

Table 8.7. Transfer-Function Plots for Typical Transfer Function

	$G(s)$	Polar Plot	Bode Diagram
1.	$\dfrac{K}{s\tau_1 + 1}$	$-\omega$; -1; $\omega = \infty$; $\omega = 0$; $+\omega$	0°; −45°; −90°; ϕ M; 0 db/oct; K_{db}; Phase margin; −180°; 0 db; $\frac{1}{\tau_1}$; log ω; −6 db/oct
2.	$\dfrac{K}{(s\tau_1 + 1)(s\tau_2 + 1)}$	$-\omega$; -1; $\omega = \infty$; $\omega = 0$; $+\omega$	0°; ϕ; 0; ϕ M; Phase margin; −6; −180°; 0 db; $\frac{1}{\tau_1}$; $\frac{1}{\tau_2}$; log ω; −12 db/oct
3.	$\dfrac{K}{(s\tau_1 + 1)(s\tau_2 + 1)(s\tau_3 + 1)}$	$-\omega$; -1; $\omega = \infty$; $\omega = 0$; $+\omega$	0°; ϕ; 0; ϕ M; −6; Gain margin; −12 db/oct; −180°; 0 db; $\frac{1}{\tau_1}$; $\frac{1}{\tau_2}$; $\frac{1}{\tau_3}$; log ω; −270°; Phase margin; −18 db/oct
4.	$\dfrac{K}{s}$	$\omega = 0$; $-\omega$; -1; $\omega = \infty$; $+\omega$; $\omega \to 0$	−90°; ϕ M; Phase margin; −180°; 0 db; log ω; −6 db/oct

Nichols Diagram	Root Locus	Comments
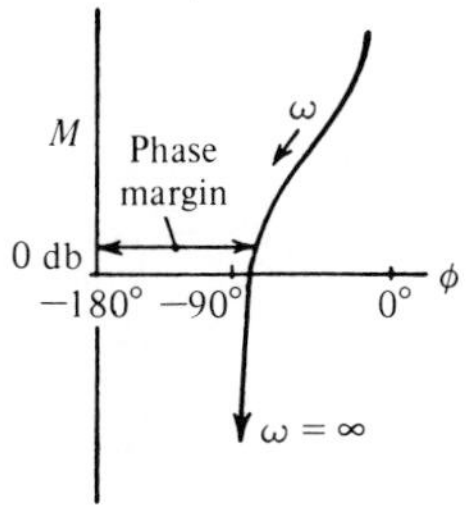	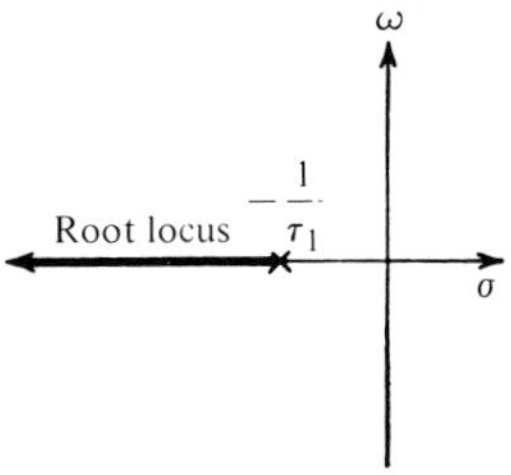	Stable; gain margin = ∞
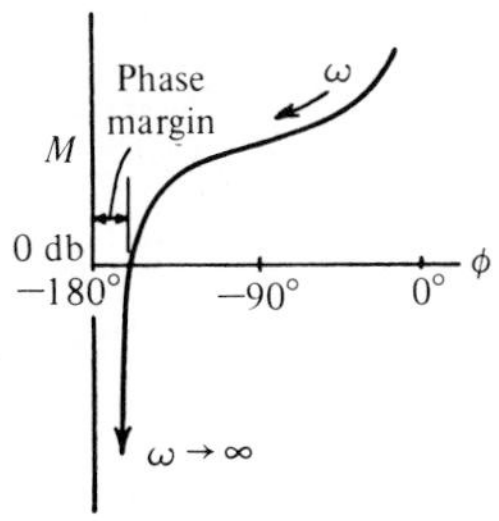	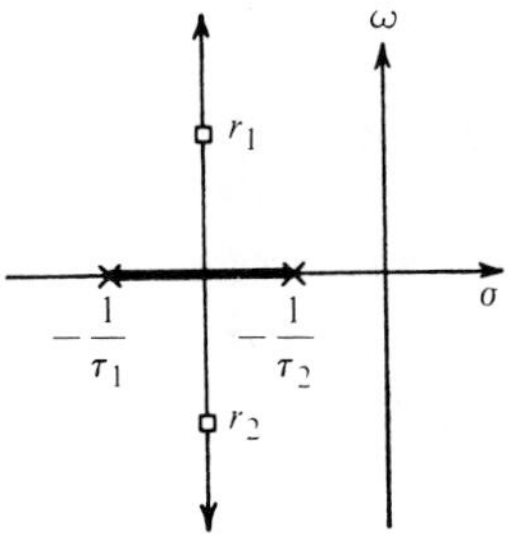	Elementary regulator; stable; gain margin = ∞
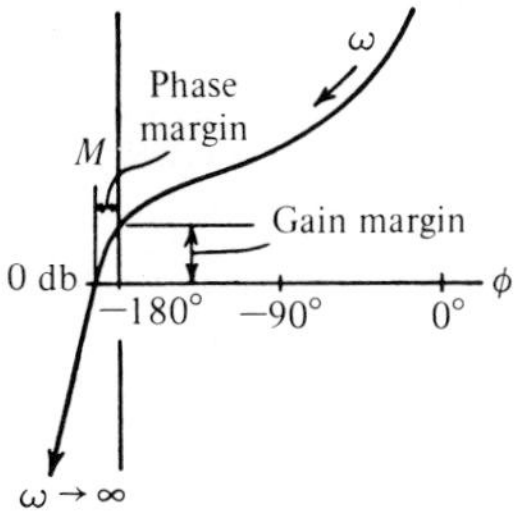	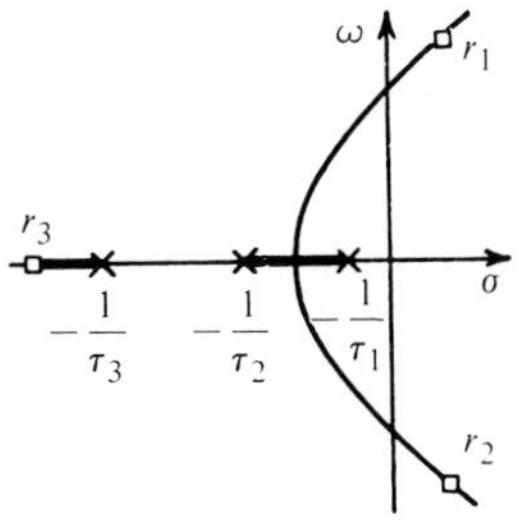	Regulator with additional energy-storage component; unstable, but can be made stable by reducing gain
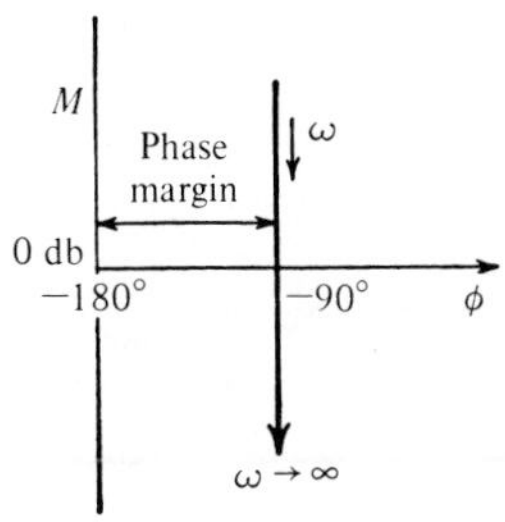	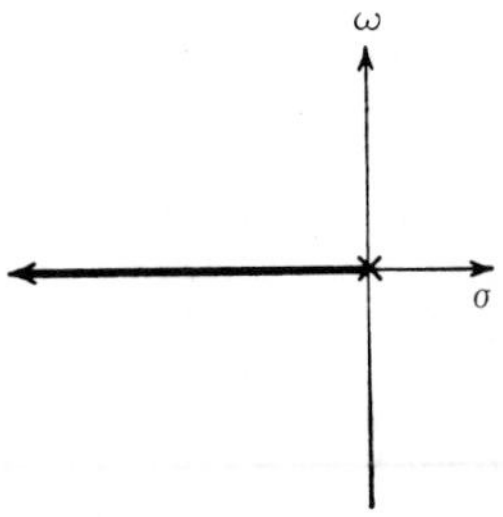	Ideal integrator; stable

Continued

Table 8.7.—*Continued*

	$G(s)$	Polar Plot	Bode Diagram
5.	$\dfrac{K}{s(s\tau_1 + 1)}$	$\omega \to 0$; $-\omega$; -1; $\omega = \infty$; $+\omega$; $\omega \to 0$	-6 db/oct; $-90°$; ϕ M; Phase margin; $-180°$; 0 db; $\frac{1}{\tau_1}$; log ω; -12 db/oct
6.	$\dfrac{K}{s(s\tau_1 + 1)(s\tau_2 + 1)}$	$\omega \to 0$; $-\omega$; -1; $\omega \to \infty$; $+\omega$; $\omega \to 0$	$-90°$; ϕ M; -6; Phase margin; Gain margin; $1/\tau_2$; $-180°$; 0 db; $\frac{1}{\tau_1}$; -12; log ω; $-270°$; -18 db/oct
7.	$\dfrac{K(s\tau_a + 1)}{s(s\tau_1 + 1)(s\tau_2 + 1)}$	$\omega \to 0$; $-\omega$; -1; $\omega = \infty$; $+\omega$; $\omega \to 0$	-6 db/oct; $-90°$; ϕ M; Phase margin; -12; $1/\tau_2$; ϕ; $-180°$; 0 db; $\frac{1}{\tau_1}$; $\frac{1}{\tau_a}$; -6; log ω; -12 db/oct
8.	$\dfrac{K}{s^2}$	$\omega \to -\omega$; -1; $\omega = \infty$; $+\omega$	Gain margin = 0; Phase margin = 0; ϕ M; -12 db/oct; $-180°$; ϕ; 0 db; log ω

Nichols Diagram	Root Locus	Comments
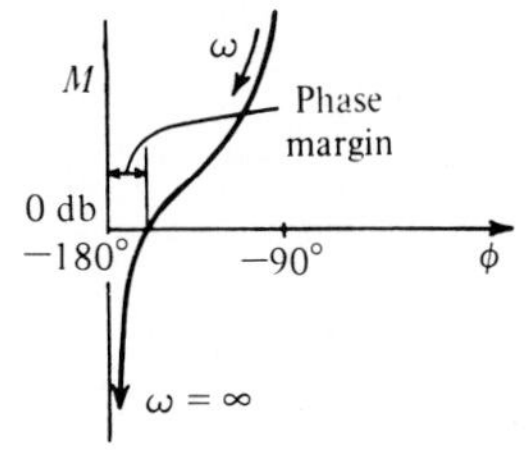	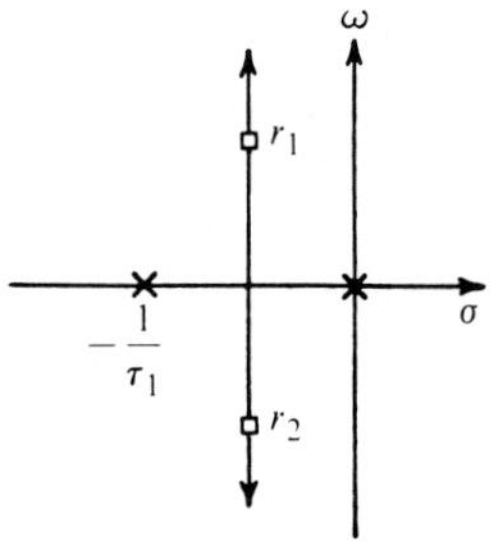	Elementary instrument servo; inherently stable; gain margin = ∞
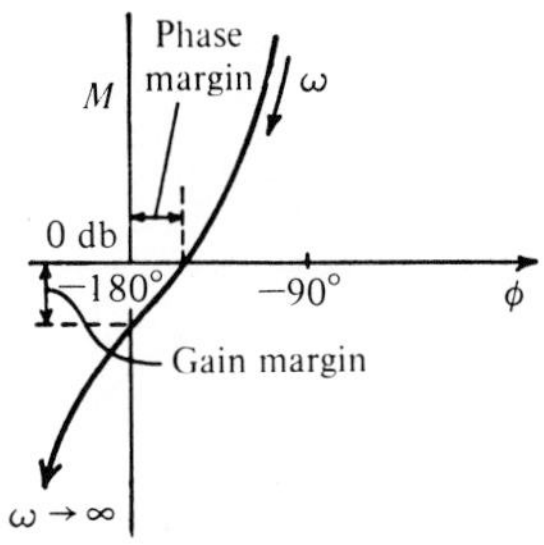	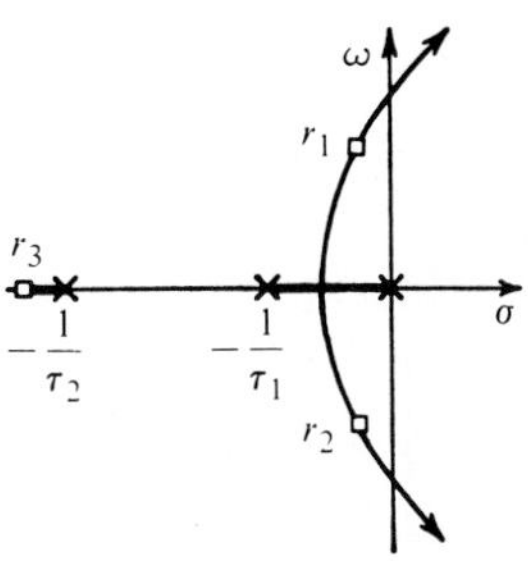	Instrument servo with field-control motor or power servo with elementary Ward–Leonard drive; stable as shown, but may become unstable with increased gain
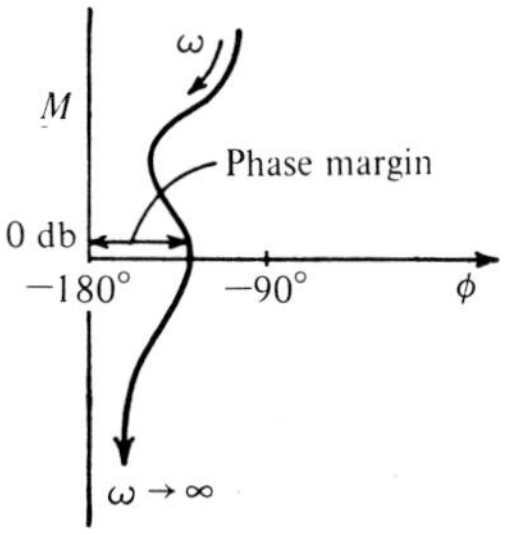	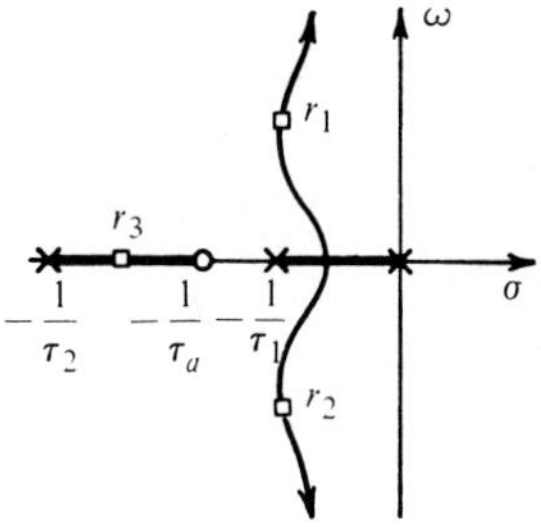	Elementary instrument servo with phase-lead (derivative) compensator; stable
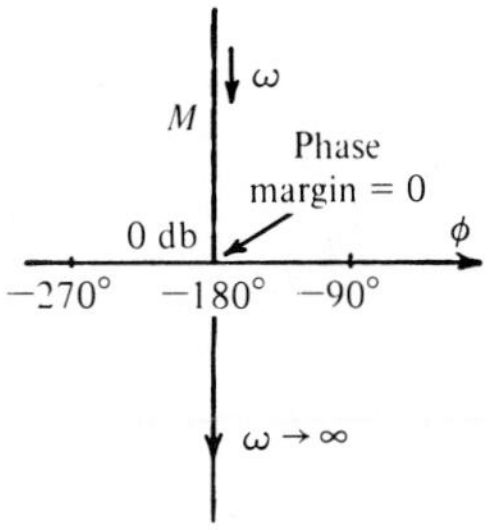	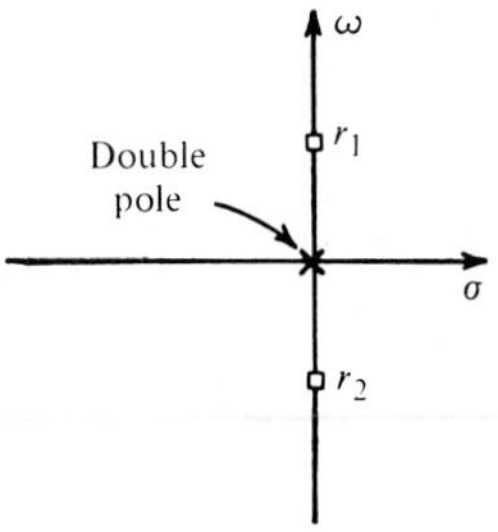	Inherently unstable; must be compensated

Continued

Table 8.7.—*Continued*

$G(s)$	Polar Plot	Bode Diagram
9. $\dfrac{K}{s^2(s\tau_1 + 1)}$	$+\omega$; -1; $\omega = \infty$; $-\omega$	ϕ M; -12 db/oct; Phase margin; $-180°$; 0 db; $\frac{1}{\tau_1}$; log ω; $-270°$; ϕ; -18 db/oct
10. $\dfrac{K(s\tau_a + 1)}{s^2(s\tau_1 + 1)}$ $\tau_a > \tau_1$	$-\omega$; -1; $\omega = \infty$; $+\omega$	ϕ M; -12 db/oct; Phase margin; ϕ; $1/\tau_1$; $-180°$; 0 db; $\frac{1}{\tau_a}$; -6 db/oct; log ω; -12 db/oct
11. $\dfrac{K}{s^3}$	$+\omega$; -1; $\omega = \infty$; $-\omega$	ϕ M; -18 db/oct; $-180°$; 0 db; Phase margin; log ω; $-270°$; ϕ
12. $\dfrac{K(s\tau_a + 1)}{s^3}$	$+\omega$; -1; $\omega = \infty$; $-\omega$	ϕ M; -18 db/oct; Phase margin; log ω; $-180°$; 0 db; $1/\tau_a$; -12 db/oct; $-270°$; ϕ

Nichols Diagram	Root Locus	Comments
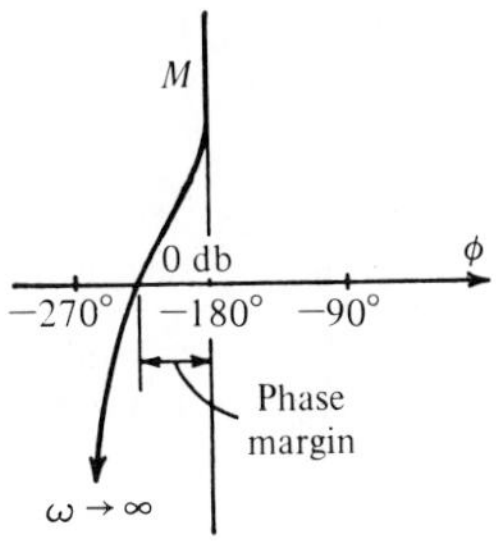	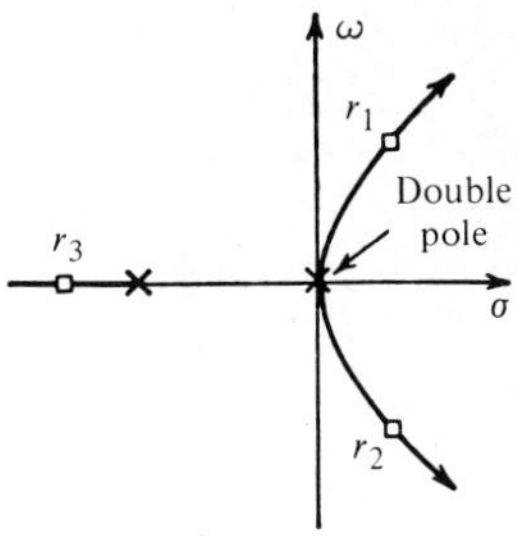	Inherently unstable; must be compensated
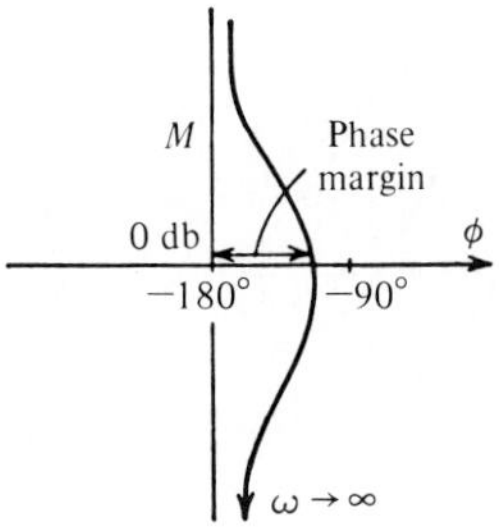	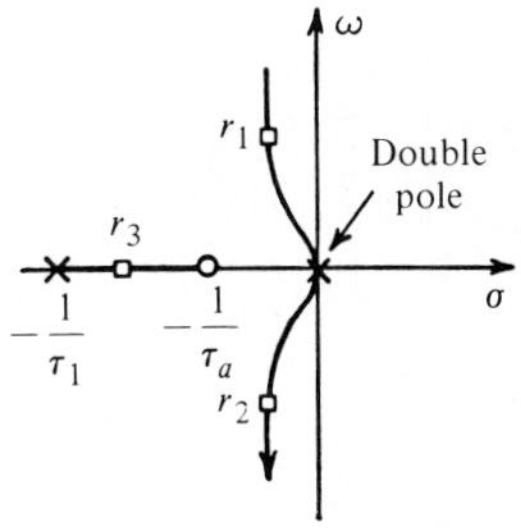	Stable for all gains
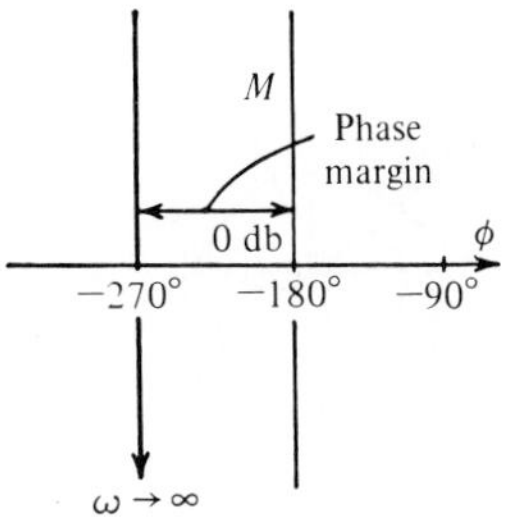	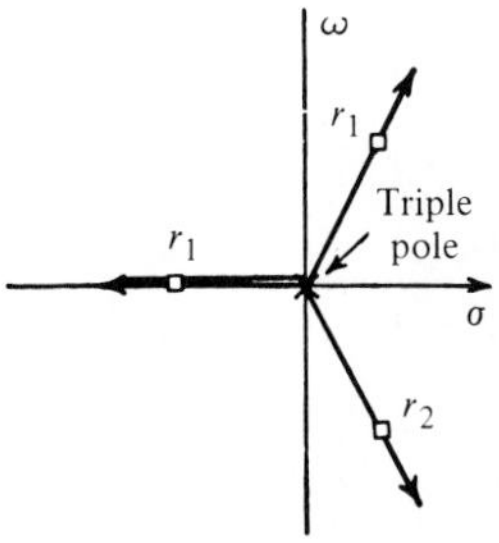	Inherently unstable
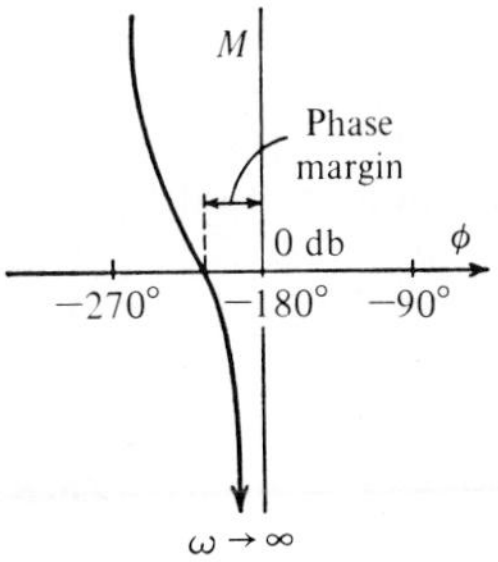	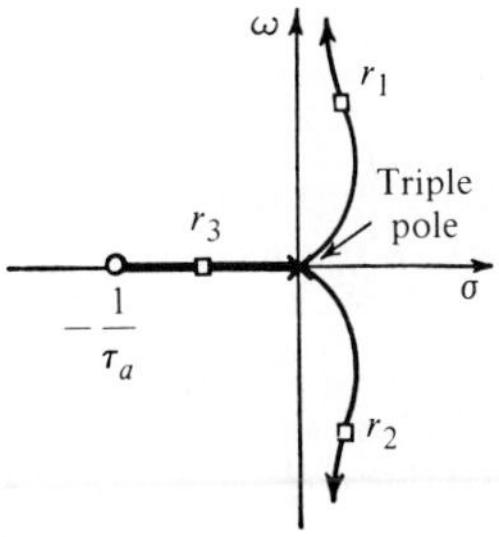	Inherently unstable

Continued

Table 8.7.—*Continued*

$G(s)$	Polar Plot	Bode Diagram
13. $\dfrac{K(s\tau_a + 1)(s\tau_b + 1)}{s^3}$	$+\omega$ -1 $\omega = \infty$ $-\omega$	$-90°$ ϕ M $-180°$ $-270°$ -18 db/oct -12 db/oct ϕ Phase margin 0 db $\frac{1}{\tau_a}$ $\frac{1}{\tau_b}$ log ω Gain margin -6 db/oct
14. $\dfrac{K(s\tau_a + 1)(s\tau_b + 1)}{s(s\tau_1 + 1)(s\tau_2 + 1)(s\tau_3 + 1)(s\tau_4 + 1)}$	$-\omega$ -1 $\omega = \infty$ $+\omega$	$-90°$ ϕ M $-180°$ $-270°$ -6 -12 -18 -12 Gain margin $\frac{1}{\tau_4}$ log ω $\frac{1}{\tau_3}$ $\frac{1}{\tau_b}$ 0 db $\frac{1}{\tau_1}$ $\frac{1}{\tau_2}$ $1/\tau_a$ -6 -12 Phase margin -18
15. $\dfrac{K(s\tau_a + 1)}{s^2(s\tau_1 + 1)(s\tau_2 + 1)}$	$-\omega$ -1 $\omega = \infty$ $+\omega$	ϕ M $-180°$ -12 Phase margin -6 $\frac{1}{\tau_1}$ $\frac{1}{\tau_2}$ 0 db $\frac{1}{\tau_a}$ log ω -12 Gain margin -18

The steady-state effect of a unit step disturbance can be determined by using the final value theorem with $R(s) = 0$, as follows:

$$c(\infty) = \lim_{s \to 0} s \left[\frac{G(s)}{1 + GG_c(s)} \right] \left(\frac{1}{s} \right) \tag{8.87}$$

$$= \frac{1}{4 + 2K}.$$

Nichols Diagram	Root Locus	Comments
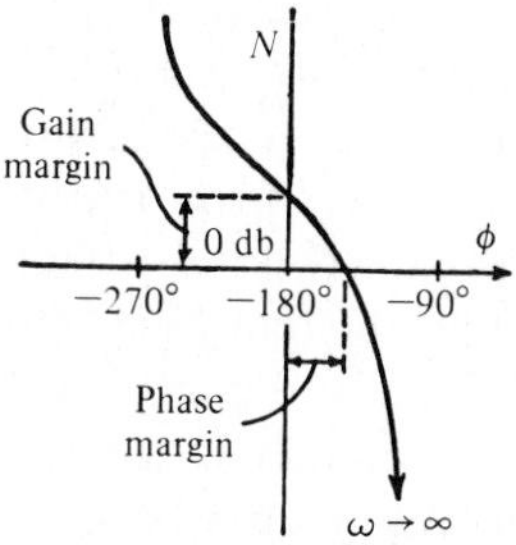	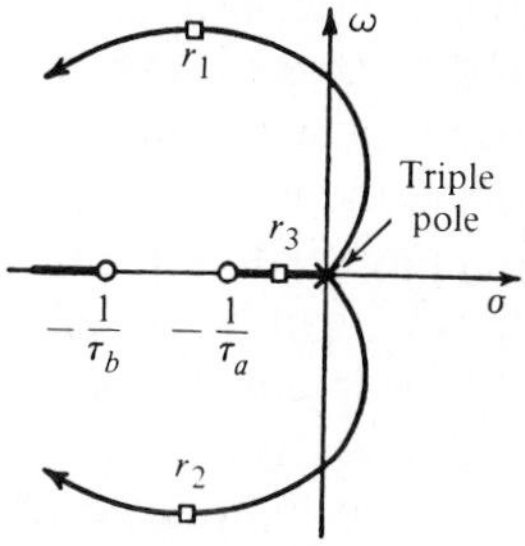	Conditionally stable; becomes unstable if gain is too low
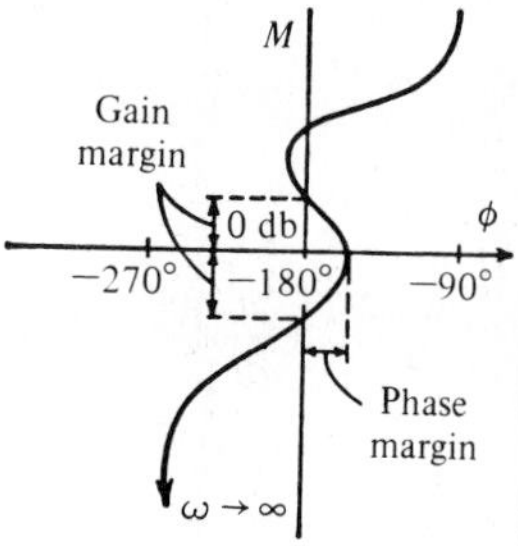	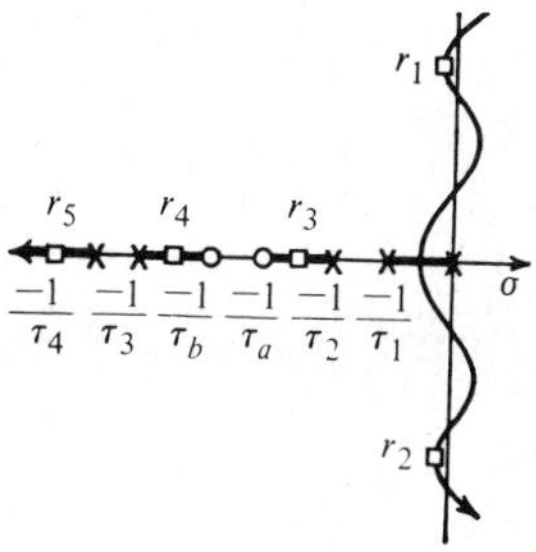	Conditionally stable; stable at low gain, becomes unstable as gain is raised, again becomes stable as gain is further increased, and becomes unstable for very high gains
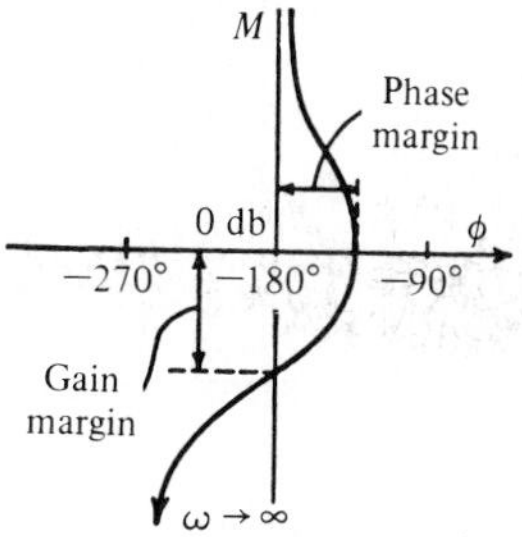	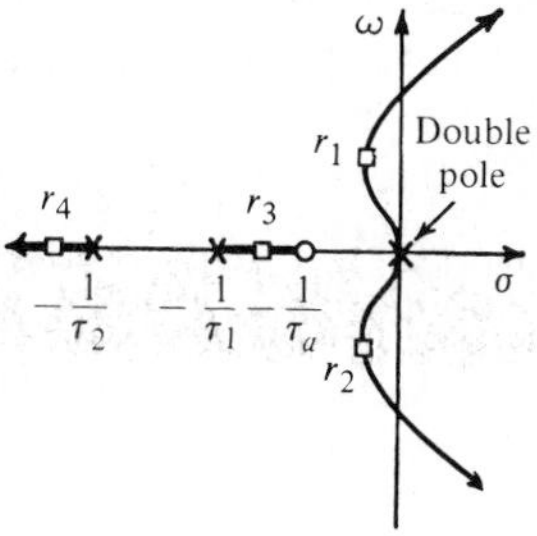	Conditionally stable; becomes unstable at high gain

Thus the unit disturbance is reduced by the factor $(4 + 2K)$. For $K = 10$, we have $c(\infty) = 1/24$ or the steady-state disturbance is reduced to 4% of the disturbance magnitude. Thus we have achieved a reasonable design with $K = 10$.

8.9 Summary

The stability of a feedback control system can be determined in the frequency domain by utilizing Nyquist's criterion. Furthermore, Nyquist's criterion provides us with two relative stability measures: (1) gain margin and (2) phase margin. These relative stability measures can be utilized as indices of the transient performance on the basis of correlations established between the frequency domain and the transient response. The magnitude and phase of the closed-loop system can be determined from the frequency response of the open-loop transfer function by utilizing constant magnitude and phase circles on the polar plot. Alternatively, we can utilize a log-magnitude–phase diagram with closed-loop magnitude and phase curves superimposed (called the Nichols chart) to obtain the closed-loop frequency response. A measure of relative stability, the maximum magnitude of the closed-loop frequency response, M_{p_ω}, is available from the Nichols chart. The frequency measure, M_{p_ω}, can be correlated with the damping ratio of the time response and is a useful index of performance. Finally, a control system with a pure time delay can be investigated in a similar manner to systems without time delay. A summary of the Nyquist criterion, the relative stability measures, and the Nichols diagram are given in Table 8.7 for several transfer functions.

Exercises

E8.1. A system has a transfer function

$$G(s) = \frac{4(1 + s/2)}{s(1 + 2s)(1 + s/20 + s^2/64)}.$$

Plot the Bode diagram for the frequency range of 0.1 to 10. Show that the phase margin is approximately 60° and the gain margin is approximately 10 db.

E8.2. A system has a transfer function

$$G(s) = \frac{K(1 + s/5)}{s(1 + s/2)(1 + s/10)},$$

where $K = 6.14$. Using a computer or calculator program show that the system crossover (0 db) frequency is 3.5 rad/sec and the phase margin is 45°.

E8.3. An integrated circuit is available to serve as a feedback system to regulate the output voltage of a power supply. The Bode diagram of the required loop transfer function $GH(j\omega)$ is shown in Fig. E8.3, on the next page.

Estimate the gain and phase margin of the regulator.

Answer: $GM = 25$ db, $PM = 75°$

E8.4. An integrated CMOS digital circuit can be represented by the Bode diagram shown in Fig. E8.4, on the next page. (a) Find the gain and phase margin of the circuit. (b) Estimate how much you would need to reduce the system gain (db) to obtain a phase margin of 60°.

E8.5. Consider a system with a loop transfer function

$$G(s) = \frac{100}{s(s + 10)},$$

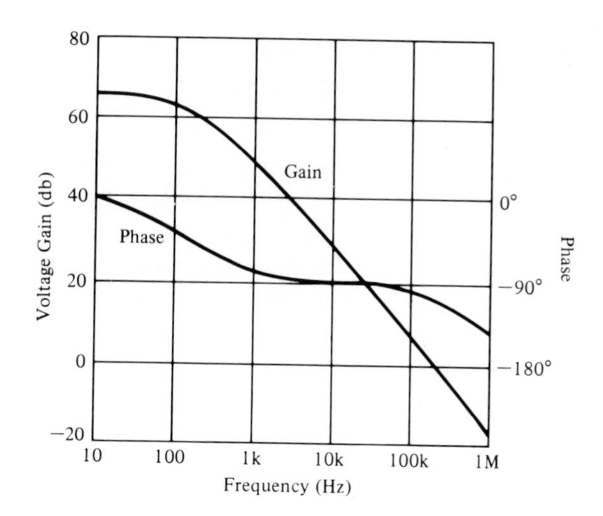

Figure E8.3. Power supply regulator.

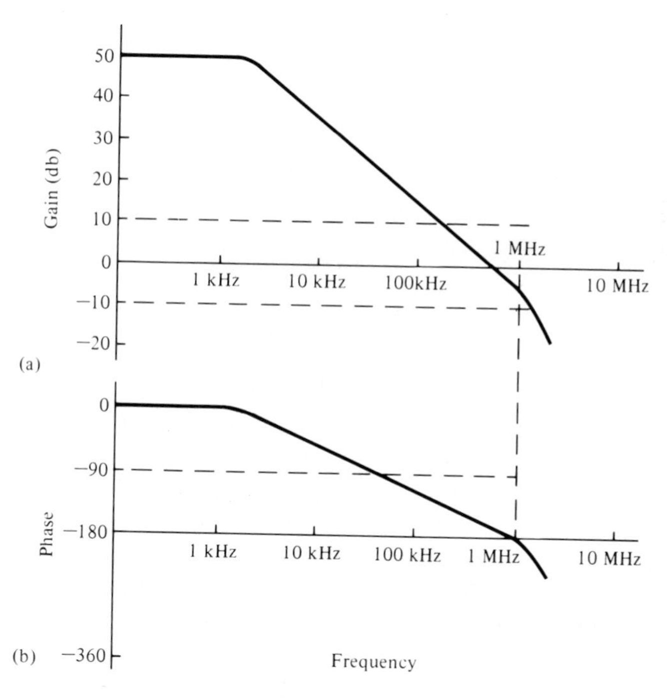

Figure E8.4. CMOS circuit.

where $H(s) = 1$. We wish to obtain a resonant peak $M_{p_\omega} = 3.0$ db for the closed-loop system. The peak occurs between 8 and 9 rad/sec and is only 1.25 db at 8.66 rad/sec. Plot the Nichols chart for the range of frequency from 8 to 15 rad/sec. Show that the system gain needs to be raised by 4.6 db to 171. Determine the resonant frequency for the adjusted system.

E8.6. A system has an open-loop transfer function

$$G(s) = \frac{K(s + 100)}{s(s + 10)(s + 40)}.$$

When $K = 500$, the system is unstable. If we reduce the gain to 50, show that the resonant peak is 3.5 db. Find the phase margin of the system with $K = 50$.

E8.7. A Nichols chart is given in Fig. E8.7, on the next page for a system where $G(j\omega)$ is plotted. Using the table below, find (a) the peak resonance M_{p_ω} in db; (b) the resonant frequency ω_r; (c) the 3 db bandwidth; (d) the phase margin of the system.

	ω_1	ω_2	ω_3	ω_4
rad/sec	1	3	6	10

E8.8. Consider a unity feedback system with

$$G(s) = \frac{K}{s(s + 1)(s + 2)}.$$

(a) For $K = 4$ show that the gain margin is 3.5 db. (b) If we wish to achieve a gain margin equal to 16 db, determine the value of the gain K.

Answer: (b) $K = 0.98$

E8.9. For the system of E8.8 find the phase margin of the system for $K = 3$.

E8.10. Consider the wind tunnel control system of Problem P6.31. Draw the Bode diagram and show that the phase margin is 25° and the gain margin is 10 db. Also, show that the bandwidth of the closed-loop system is 6 rad/sec.

E8.11. Consider a unity feedback system with

$$G(s) = \frac{40(1 + 0.4s)}{s(1 + 2s)(1 + 0.24s + 0.04s^2)}.$$

(a) Using the Control System Design Program or equivalent plot the Bode diagram. (b) Find the gain margin and the phase margin.

E8.12. An actuator for a disk drive uses a shock mount to absorb vibrational energy at approximately 60 Hz [15]. The Bode diagram of $G(s)$ of the control system is shown in Fig. E8.12, on page 408. (a) Find the expected percent overshoot for a step input for the closed-loop system, (b) estimate the bandwidth of the closed loop system, and (c) estimate the settling time of the system.

E8.13. A unity feedback system has a process

$$G(s) = \frac{150}{s(s + 5)}.$$

(a) Find the maximum magnitude of the closed-loop frequency response using the Nichol's chart or the Control System Design Program. (b) Find the bandwidth and the resonant frequency of this system. (c) Use these frequency measures to estimate the overshoot of the system to a step response.

Answers: (a) 7.5 db, (b) $\omega_B = 19$, $\omega_r = 12.6$

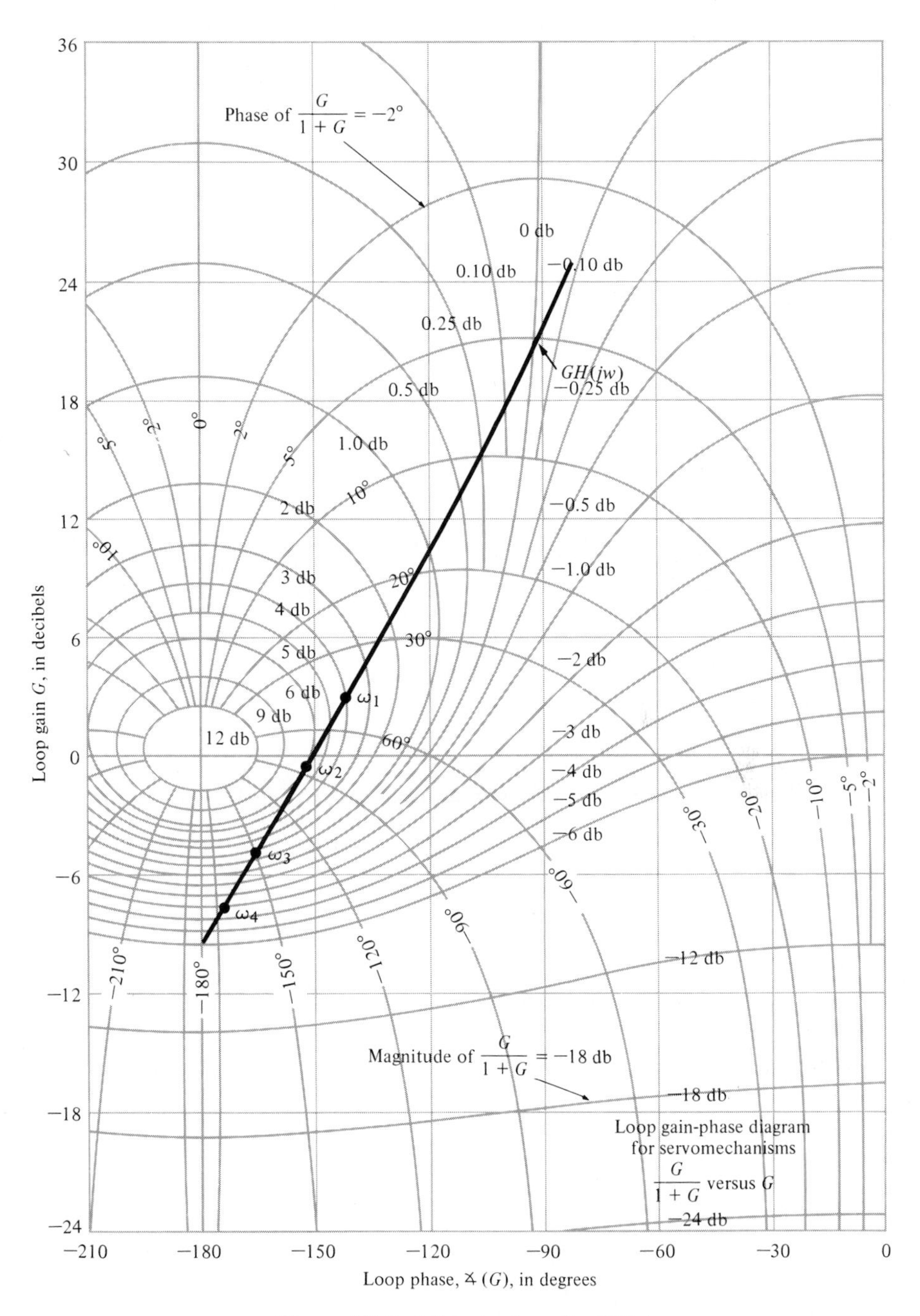

Figure E8.7. Nichols chart for G($j\omega$).

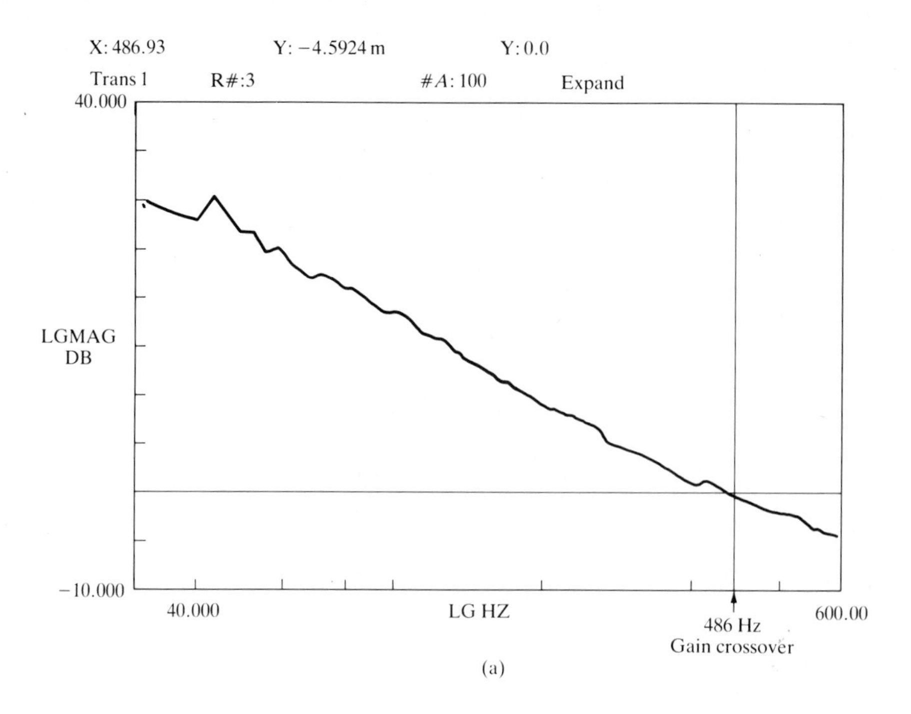

(a)

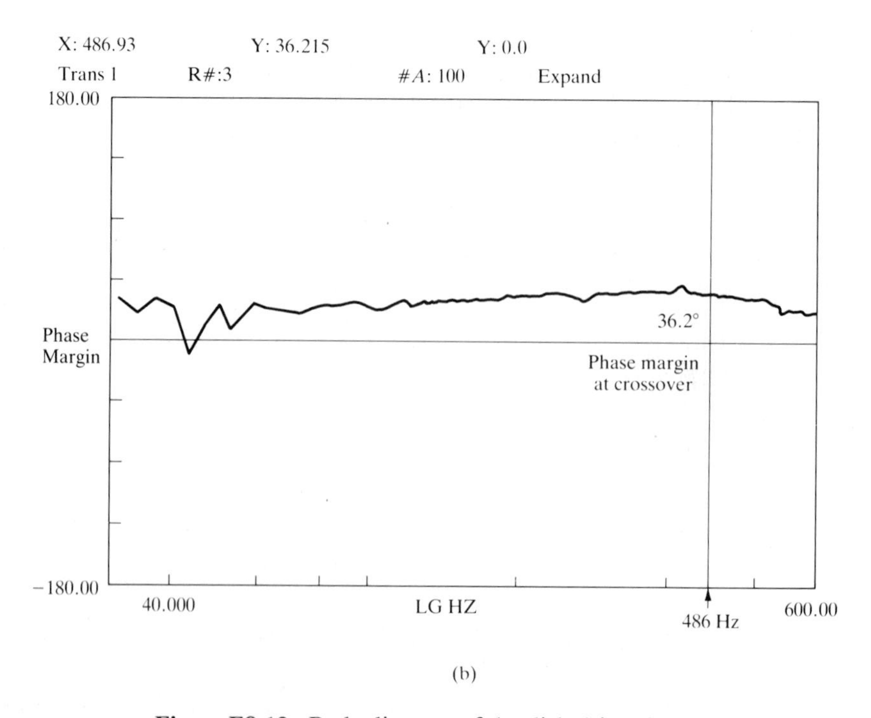

(b)

Figure E8.12. Bode diagram of the disk drive $G(s)$.

E8.14. A unity feedback system has a process

$$G(s) = \frac{K}{(s - 1)}.$$

Determine the range of K for which the system is stable by drawing the polar plot.

E8.15. Consider a unity feedback system with

$$G(s) = \frac{1000}{(s + 100)}$$

Find the bandwidth of the open-loop system and the closed-loop system and compare the results.

Answers: open-loop $\omega_B = 100$, closed-loop $\omega_B = 1100$

E8.16. The pure time delay e^{-sT} may be approximated by a transfer function as

$$e^{-sT} \cong \frac{(1 - Ts/2)}{(1 + Ts/2)}$$

for $0 < \omega < 2/T$. Obtain the Bode diagram for the actual transfer function and the approximation for $T = 2$ for $0 < \omega < 1$.

E8.17. A unity feedback system has a plant

$$G(s) = \frac{K(s + 3)}{s(s + 1)(s + 5)}.$$

(a) Plot the Bode diagram and (b) determine the gain K required in order to obtain a phase margin of 40°. What is the steady-state error for a ramp input for the gain of part (b)?

E8.18. A closed-loop system, as shown in Fig. 8.8, has $H(s) = 1$ and

$$G(s) = \frac{K}{s(\tau_1 s + 1)(\tau_2 s + 1)},$$

where $\tau_1 = 0.02$ and $\tau_2 = 0.2$ seconds. (a) Select a gain K so that the steady-state error for a ramp input is 10% of the magnitude of the ramp function A, where $r(t) = At$, $t \geq 0$. (b) Plot the Bode plot of $G(s)$ and determine the phase and gain margin. (c) Using the Nichols chart, determine the bandwidth ω_B, the resonant peak M_{p_ω}, and the resonant frequency ω_r of the closed-loop system.

Answers: (a) $K = 10$
(b) $P.M. = 32°$, $G.M. = 15$ db
(c) $\omega_B = 10.3$, $M_{p_\omega} = 1.84$, $\omega_r = 6.5$

Problems

P8.1. For the polar plots of Problem P7.1 use the Nyquist criterion to ascertain the stability of the various systems. In each case specify the values of N, P, and Z.

P8.2. Sketch the polar plots of the following loop transfer functions $GH(s)$ and determine whether the system is stable by utilizing the Nyquist criterion.

$$\text{(a) } GH(s) = \frac{K}{s(s^2 + s + 4)},$$

$$\text{(b) } GH(s) = \frac{K(s + 2)}{s^2(s + 4)}.$$

If the system is stable, find the maximum value for K by determining the point where the polar plot crosses the u-axis.

P8.3. The polar plot of a conditionally stable system is shown in Fig. P8.3 for a specific gain K. Determine whether the system is stable and find the number of roots (if any) in the right-hand s-plane. The system has no poles of $GH(s)$ in the right-half plane.

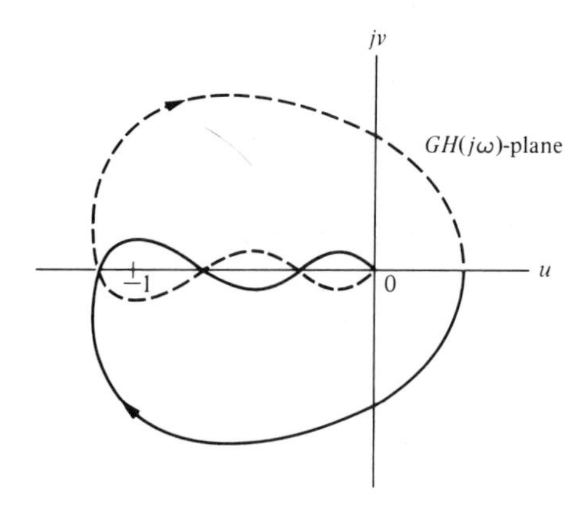

Figure P8.3 Polar plot of conditionally stable system.

P8.4. (a) Find a suitable contour Γ_s in the s-plane that can be used to determine whether all roots of the characteristic equation have damping ratios greater than ζ_1. (b) Find a suitable contour Γ_s in the s-plane that can be used to determine whether all the roots of the characteristic equation have real parts less than $s = -\sigma_1$. (c) Using the contour of part (b) and Cauchy's theorem, determine whether the following characteristic equation has roots with real parts less than $s = -1$:

$$q(s) = s^3 + 8s^2 + 30s + 36.$$

P8.5. A speed control for a gasoline engine is shown in Fig. P8.5. Because of the restriction at the carburetor intake and the capacitance of the reduction manifold, the lag τ_t occurs

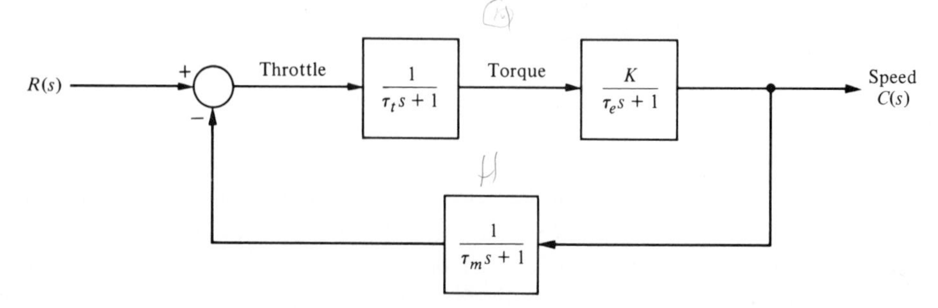

Figure P8.5 Engine speed control.

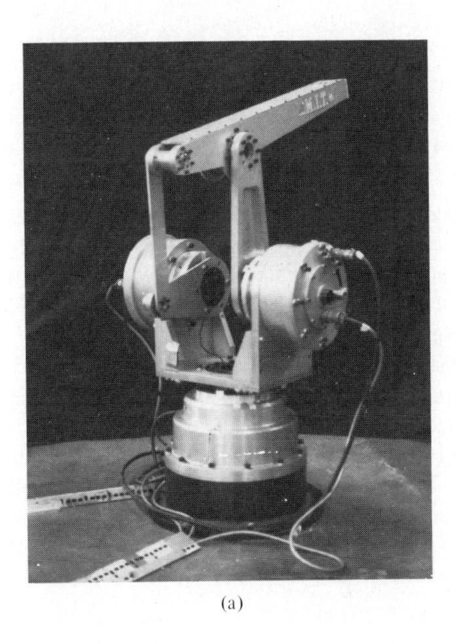

(a)

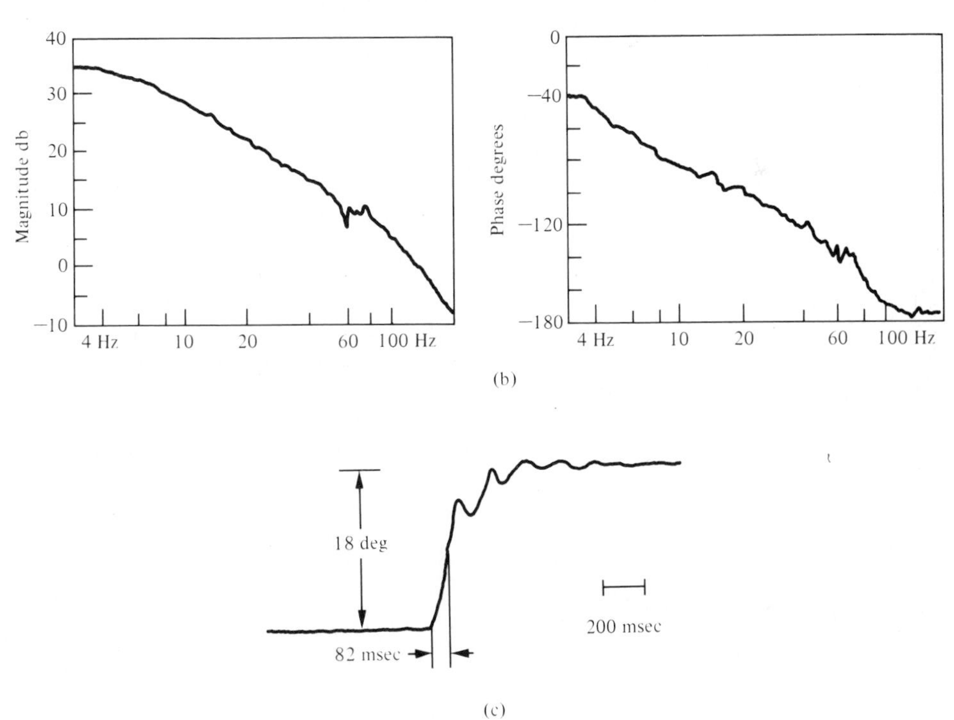

Figure P8.6. (a) The MIT arm. (b) Frequency response; (c) position response. (Photo courtesy of Massachusetts Institute of Technology.)

and is equal to 1 sec. The engine time constant τ_e is equal to $J/f = 2$ sec. The speed measurement time constant is $\tau_m = 0.5$ sec. (a) Determine the necessary gain K if the steady-state speed error is required to be less than 10% of the speed reference setting. (b) With the gain determined from (a), utilize the Nyquist criterion to investigate the stability of the system. (c) Determine the phase and gain margins of the system.

P8.6. A direct-drive arm is an innovative mechanical arm in which no reducers are used between motors and their loads. Because the motor rotors are directly coupled to the loads, the drive systems have no backlash, small friction, and high mechanical stiffness, which are all important features for fast and accurate positioning and dexterous handling using sophisticated torque control.

The goal of the MIT direct-drive arm project is to achieve arm speeds of 10 m/sec [13]. The arm has torques of up to 660 n-m (475 ft-lb). Feedback and a set of position and velocity sensors are used with each motor. The arm is shown in Fig. P8.6(a), with a motor visible on the side facing the reader. The frequency response of one joint of the arm is shown in Fig. P8.6(b). The two poles appear at 3.7 Hz. and 68 Hz. Figure P8.6(c) shows the step response with position and velocity feedback used. The time constant of the closed-loop system is 82 msec. Develop the block diagram of the drive system and prove that 82 msec is a reasonable result.

P8.7. A vertical takeoff (VTOL) aircraft is an inherently unstable vehicle and requires an automatic stabilization system. An attitude stabilization system for the K-16B U.S. Army VTOL aircraft has been designed and is shown in block diagram form in Fig. P8.7 [11]. At 40 knots, the dynamics of the vehicle are approximately represented by the transfer function

$$G(s) = \frac{10}{(s^2 + 0.36)}.$$

The actuator and filter is represented by the transfer function

$$G_1(s) = \frac{K_1(s + 8)}{(s + 2)}.$$

(a) Draw the Bode diagram of the loop transfer function $G_1(s)G(s)H(s)$ when the gain is $K_1 = 2$. (b) Determine the gain and phase margins of this system. (c) Determine the steady-state error for a wind disturbance of $T_d(s) = 1/s$. (d) Determine the maximum amplitude of the resonant peak of the closed-loop frequency response and the frequency of the resonance. (e) Estimate the damping ratio of the system from M_{p_ω} and the phase margin.

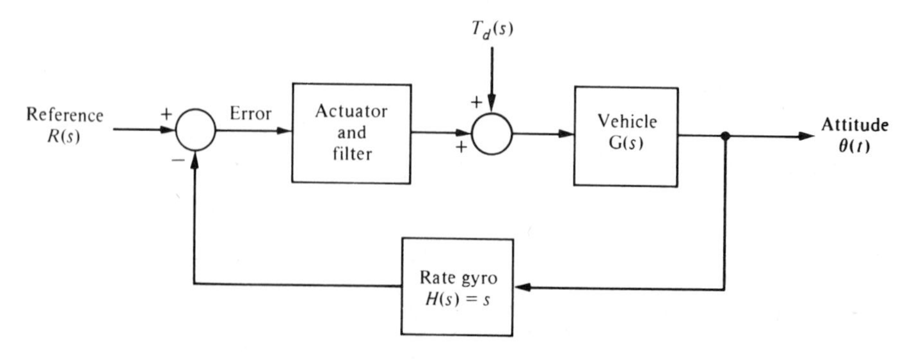

Figure P8.7 VTOL aircraft stabilization system.

P8.8. Electrohydraulic servomechanisms are utilized in control systems requiring a rapid response for a large mass. An electrohydraulic servomechanism can provide an output of 100 kW or greater. A photo of a servo valve and actuator is shown in Fig. P8.8(a). The output sensor yields a measurement of actuator position, which is compared with V_{in}. The

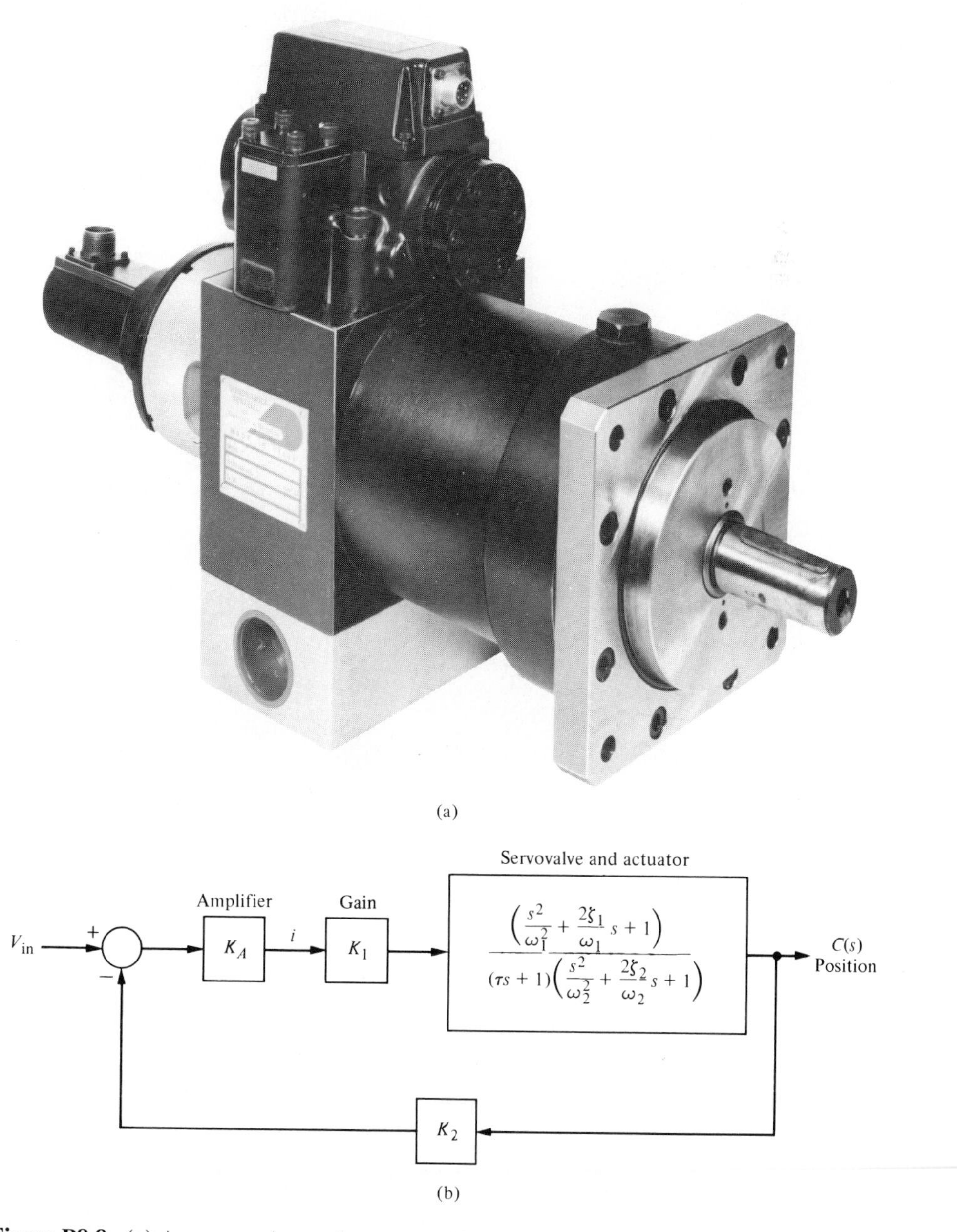

Figure P8.8. (a) A servo valve and actuator. (Courtesy of Moog, Inc., Industrial Division); (b) block diagram.

error is amplified and controls the hydraulic valve position, thus controlling the hydraulic fluid flow to the actuator. The block diagram of a closed-loop electrohydraulic servomechanism using pressure feedback to obtain damping is shown in Fig. P8.8(b) [9, 13]. Typical values for this system are $\tau = 0.02$ sec, and for the hydraulic system are $\omega_2 = 7(2\pi)$ and $\zeta_2 = 0.05$. The structural resonance ω_1 is equal to $10(2\pi)$ and the damping is $\zeta_1 = 0.05$. The loop gain is $K_A K_1 K_2 = 1.0$. (a) Sketch the Bode diagram and determine the phase margin of the system. (b) The damping of the system can be increased by drilling a small hole in the piston so that $\zeta_2 = 0.25$. Sketch the Bode diagram and determine the phase margin of this system.

P8.9. The key to future exploration and use of space is the reusable earth-to-orbit transport system, popularly known as the space shuttle. The shuttle, shown in Fig. P8.9(a), carries large payloads into space and returns them to earth for reuse. The shuttle, roughly the size

(a)

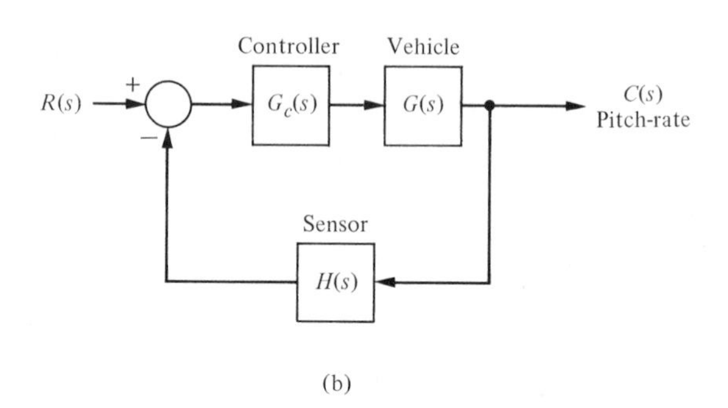

(b)

Figure P8.9. (a) The Earth orbiting space shuttle *Columbia* against the blackness of space on June 22, 1983. The remote manipulator robot is shown with the cargo bay doors open in this top view, taken by a satellite. (b) Pitch rate control system. (Courtesy of NASA.)

of a DC-9 with an empty weight of 75,000 kg, uses elevons at the trailing edge of the wing and a brake on the tail to control the flight. The block diagram of a pitch rate control system is shown in Fig. P8.9(b). The sensor is represented by a gain, $H(s) = 0.5$, and the vehicle by the transfer function

$$G(s) = \frac{0.30(s + 0.05)(s^2 + 1600)}{(s^2 + 0.05s + 16)(s + 70)}.$$

The controller $G_c(s)$ can be a gain or any suitable transfer function. (a) Draw the Bode diagram of the system when $G_c(s) = 2$ and determine the stability margin. (b) Draw the Bode diagram of the system when

$$G_c(s) = K_1 + K_2/s \quad \text{and} \quad K_2/K_1 = 0.5.$$

The gain K_1 should be selected so that the gain margin is 10 db.

P8.10. Machine tools are often automatically controlled by a punched tape reader as shown in Fig. P8.10. These automatic systems are often called numerical machine controls. Considering one axis, the desired position of the machine tool is compared with the actual position and is used to actuate a solenoid coil and the shaft of a hydraulic actuator. The transfer function of the actuator (see Table 2.6) is

$$G_a(s) = \frac{X(s)}{Y(s)} = \frac{K_a}{s(\tau_a s + 1)},$$

where $K_a = 1$ and $\tau_a = 0.4$ sec. The output voltage of the difference amplifier is

$$E_0(s) = K_1(X(s) - X_d(s)),$$

where $x_d(t)$ is the desired position input from the tape reader. The force on the shaft is proportional to the current i so that $F = K_2 i(t)$, where $K_2 = 3.0$. The spring constant K_s is equal to 1.5 and $R = 0.1$ and $L = 0.2$. (a) Determine the gain K_1 that results in a system with a phase margin of 30°. (b) For the gain K_1 of part (a), determine M_{p_ω}, ω_r and the closed-loop system bandwidth. (c) Estimate the percent overshoot of the transient response for a step input, $X_d(s) = 1/s$, and the settling time.

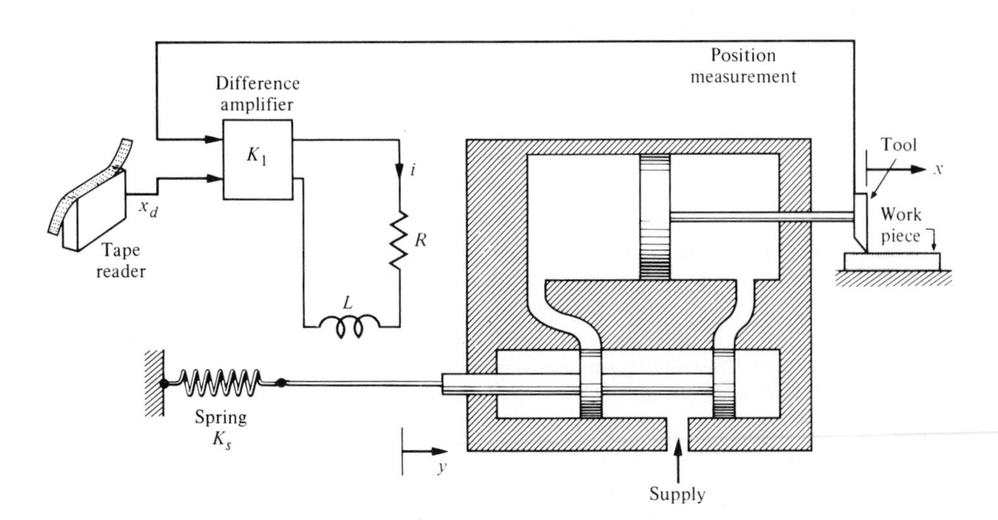

Figure P8.10. Machine tool control.

P8.11. A control system for a chemical concentration control system is shown in Fig. P8.11. The system receives a granular feed of varying composition, and it is desired to maintain a constant composition of the output mixture by adjusting the feed-flow valve. The transfer function of the tank and output valve is

$$G(s) = \frac{5}{5s + 1}$$

and the controller is

$$G_c(s) = K_1 + \frac{K_2}{s}.$$

The transport of the feed along the conveyor requires a transport (or delay) time, $T = 2$ sec. (a) Sketch the Bode diagram when $K_1 = K_2 = 1$, and investigate the stability of the system. (b) Sketch the Bode diagram when $K_1 = 0.1$ and $K_2 = 0.05$, and investigate the stability of the system. (c) When $K_1 = 0$, use the Nyquist criterion to calculate the maximum allowable gain K_2 for the system to remain stable.

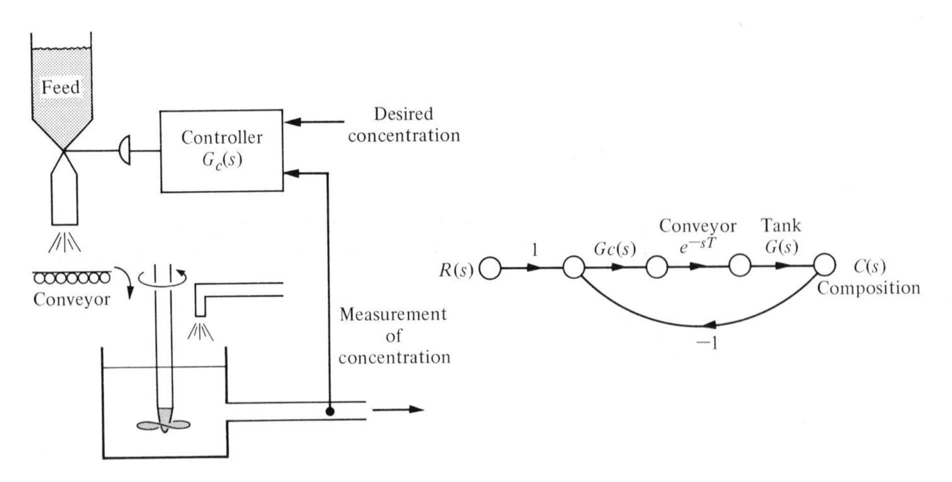

Figure P8.11. Chemical concentration control.

P8.12. A simplified model of the control system for regulating the pupillary aperture in the human eye is shown in Fig. P8.12, on the next page [14]. The gain K represents the pupillary gain and τ is the pupil time constant which is 0.5 sec. The time delay T is equal to 0.1 sec. The pupillary gain is equal to 4.0. (a) Assuming the time delay is negligible, draw the Bode diagram for the system. Determine the phase margin of the system. (b) Include the effect of the time delay by adding the phase shift due to the delay. Determine the phase margin of the system with the time delay included.

P8.13. A controller is used to regulate the temperature of a mold for plastic part fabrication, as shown in Fig. P8.13. The value of the delay time is estimated as 2 sec. (a) Utilizing the Nyquist criterion, determine the stability of the system for $K_a = K = 1$. (b) Determine a suitable value for K_a for a stable system when $K = 1$.

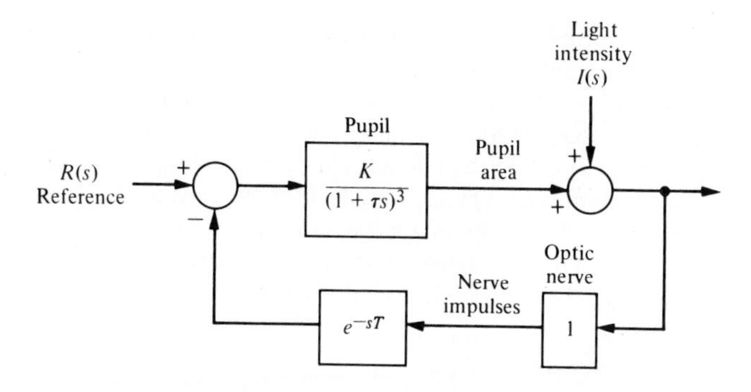

Figure P8.12. Human pupil aperature control.

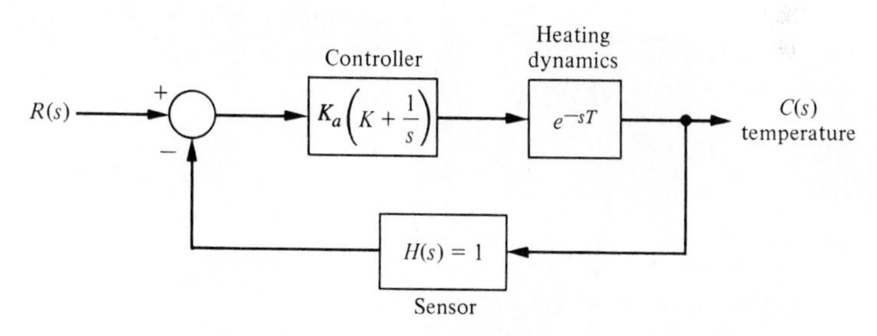

Figure P8.13. Temperature regulator.

P8.14. The closed-loop performance of a high-frequency operational amplifier can be predicted using the Nichols chart. One inverting operational amplifier has an open-loop response as given below.

f(MHz)	GH (dB)	$\angle GH$ (degrees)
1	33	−132
2	26	−120
4	21	−114
6	17	−114
10	12	−120
15	6	−128
17	5	−132
20	3	−137
25	0	−142
30	−2	−148
38	−5	−155
42	−7	−160

Obtain the Nichols chart and show that the $M_{p\omega}$ is $+4.0$ db when $f = 25$ MHz.

P8.15. Electronics and computers are being used to control automobiles. An example of an automobile control system, the steering control for the General Motors Firebird III research automobile, is the control stick shown in Fig. P8.15(a). The control stick, called a Unicontrol, is used for steering and controlling the throttle. A typical driver has a reaction time of $T = 0.2$ sec. (a) Using the Nichols chart, determine the magnitude of the gain K that will result in a system with a peak magnitude of the closed-loop frequency response M_{p_ω} less than or equal to 2 db. (b) Estimate the damping ratio of the system based on (1) M_{p_ω} and (2) the phase margin. Compare the results and explain the difference, if any. (c) Determine the closed-loop 3-db bandwidth of the system.

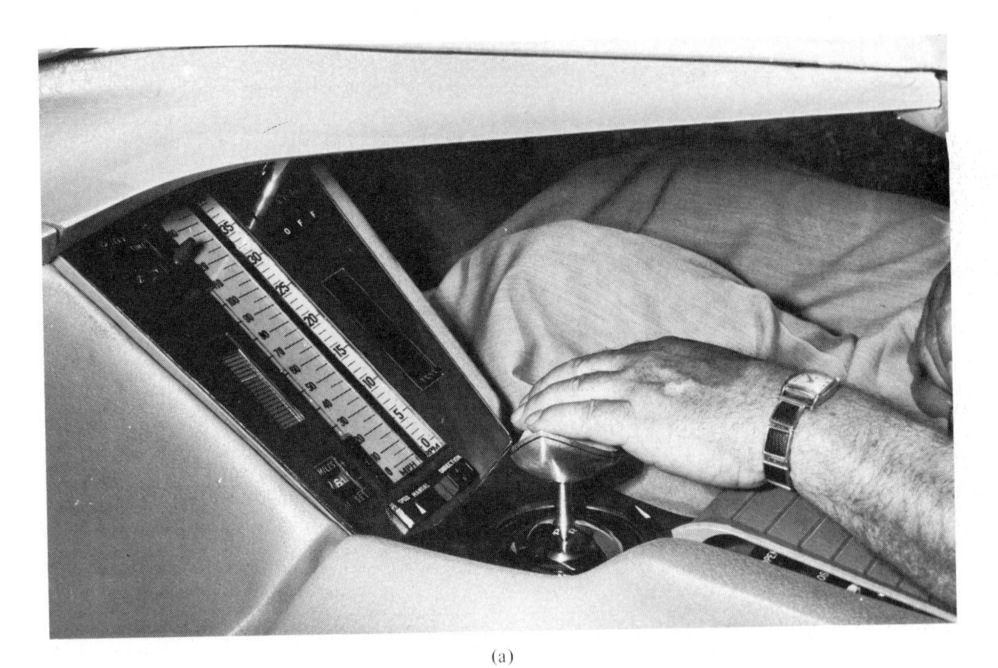

(a)

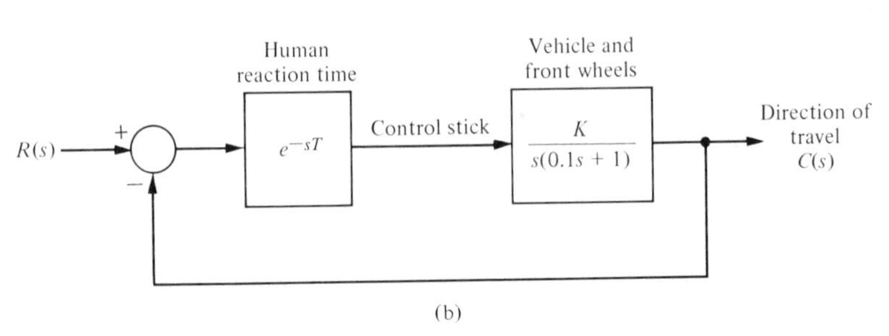

(b)

Figure P8.15. Automobile steering control.

P8.16. Consider the automatic ship steering system discussed in Problem P7.13. The frequency response of the open-loop portion of the ship steering control system is shown in Fig. P7.13. The deviation of the tanker from the straight track is measured by radar and is used to generate the error signal as shown in Fig. P8.16. This error signal is used to control the rudder angle $\delta(s)$. (a) Is this system stable? Discuss what an unstable ship steering system indicates in terms of the transient response of the system. Recall that the system under consideration is a ship attempting to follow a straight track. (b) Is it possible to stabilize this system by lowering the gain of the transfer function $G(s)$? (c) Is it possible to stabilize this system? Can you suggest a suitable feedback compensator? (d) Repeat parts (a), (b), and (c) when switch S is closed.

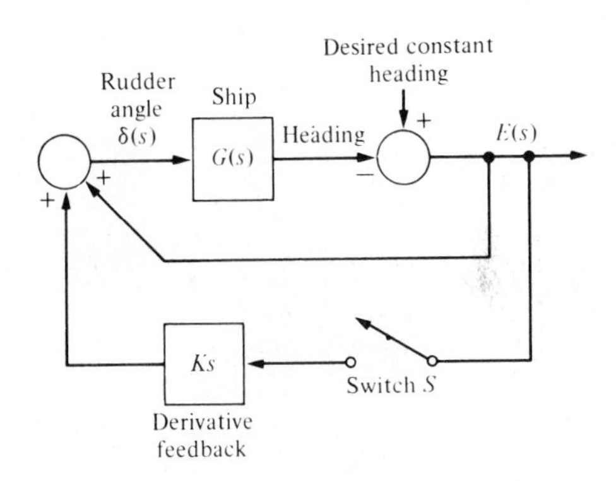

Figure P8.16. Automatic ship steering.

P8.17. An electric carrier that automatically follows a tape track laid out on a factory floor is shown in Fig. P8.17(a), on the next page [13]. Closed-loop feedback systems are used to control the guidance and speed of the vehicle. The cart senses the tape path by means of an array of 16 phototransistors. The block diagram of the steering system is shown in Fig. P8.17(b). Select a gain K so that the phase margin is approximately 30°.

P8.18. The primary objective of many control systems is to maintain the output variable at the desired or reference condition when the system is subjected to a disturbance [10]. A typical chemical reactor control scheme is shown in Fig. P8.18. The disturbance is represented by $U(s)$ and the chemical process by G_3 and G_4. The controller is represented by G_1 and the valve by G_2. The feedback sensor is $H(s)$ and will be assumed to be equal to one. We will assume that G_2, G_3, and G_4 are all of the form

$$G_i(s) = \frac{K_i}{1 + \tau_i s},$$

where $\tau_3 = \tau_4 = 4$ sec and $K_3 = K_4 = 0.1$. The valve constants are $K_2 = 20$ and $\tau_2 = 0.5$ sec. It is desired to maintain a steady-state error less than 5% of the desired reference position. (a) When $G_1(s) = K_1$, find the necessary gain to satisfy the error constant requirement. For this condition, determine the expected overshoot to a step change in the reference signal $r(t)$. (b) If the controller has a proportional term plus an integral term so that $G_1(s) = K_1(1 + 1/s)$, determine a suitable gain to yield a system with an overshoot less than 30% but greater than 5%. For parts (a) and (b) use the approximation of damping ratio as a function of phase margin that yields $\zeta = 0.01\ \phi_{pm}$. For these calculations, assume that $U(s) = 0$. (c) Estimate the settling time of the step response of the system for the controller of parts (a) and (b). (d) The system is expected to be subjected to a step disturbance $U(s) = A/s$. For ease, assume that the desired reference is $r(t) = 0$ when the system has settled. Determine the response of the system of part (b) to the disturbance.

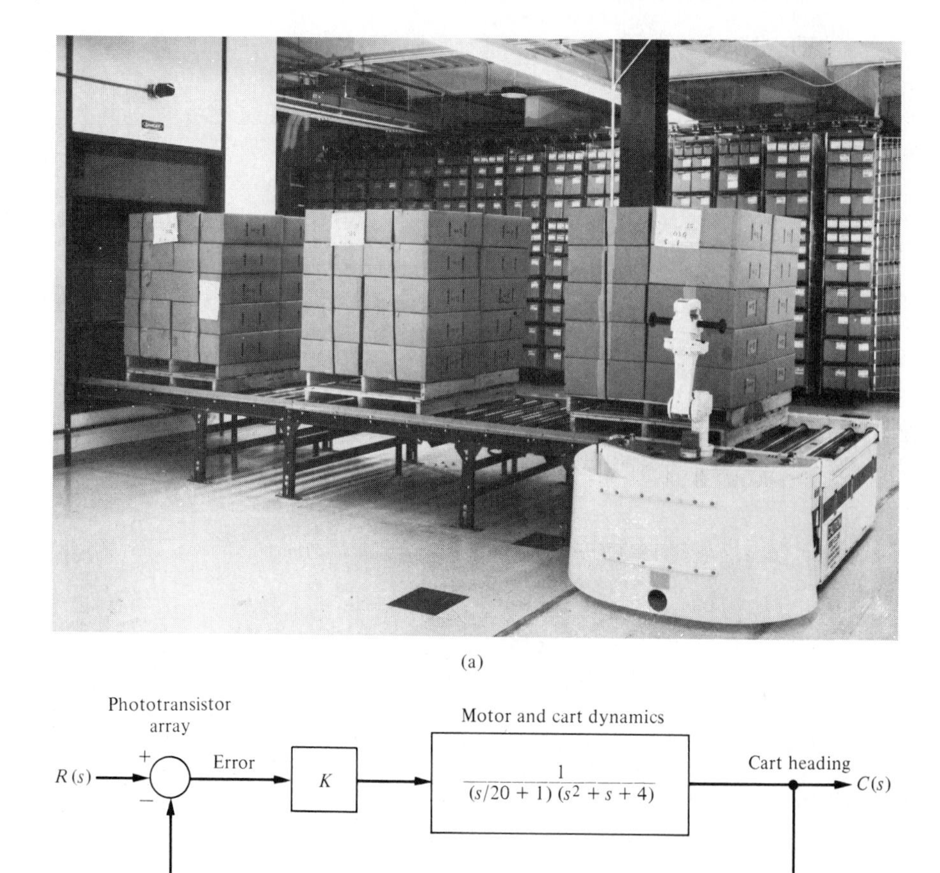

(b)

Figure P8.17. (a) An electric carrier vehicle (photo courtesy of Control Engineering Corporation). (b) Block diagram.

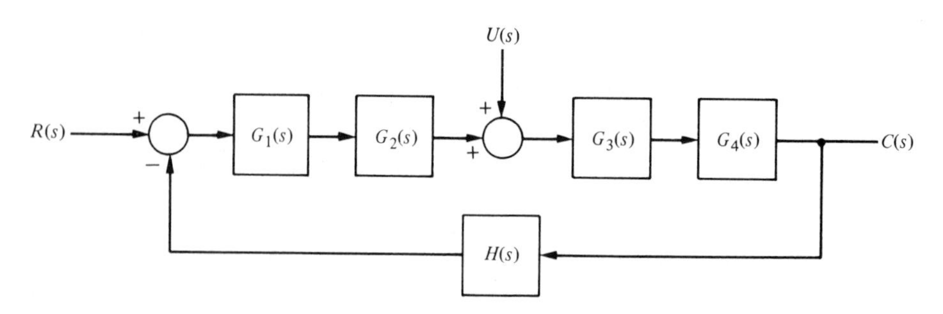

Figure P8.18. Chemical reactor control.

P8.19. A model of a driver of an automobile attempting to steer a course is shown in Fig. P8.19, where $K = 5.3$. (a) Find the frequency response and the gain and phase margin when the reaction time T is zero. (b) Find the phase margin when the reaction time is 0.1 second. (c) Find the reaction time that will cause the system to be borderline stable (phase margin = 0°).

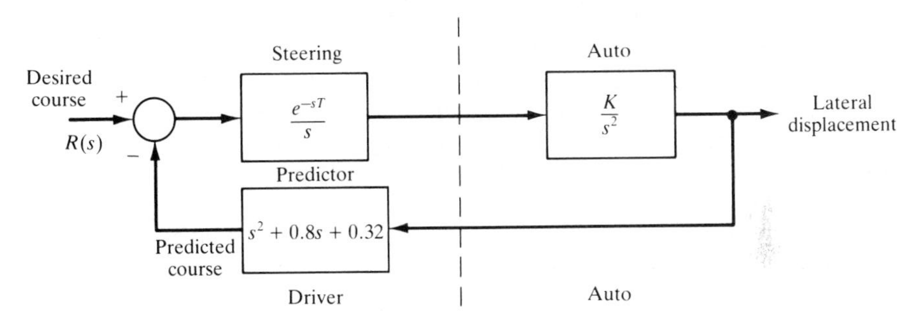

Figure P8.19. Automobile and driver control.

P8.20. In the United States over $4.3 billion are spent annually for solid waste collection and disposal. One system, which uses a remote control pick-up arm for collecting waste bags, is shown in Fig. P8.20 [20]. The open-loop transfer of the remote pick-up arm is

$$GH(s) = \frac{0.3}{s(1 + 3s)(1 + s)}.$$

(a) Plot the Nichols chart and show that the gain margin is approximately 11.5 db. (b) Determine the phase margin and the M_{p_ω} for the closed loop. Also determine the closed-loop bandwidth.

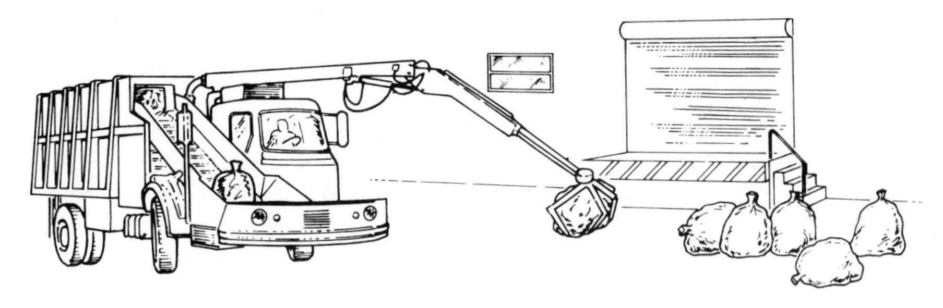

Figure P8.20. Waste collection system.

P8.21. The Bell-Boeing V-22 Osprey Tiltrotor is both an airplane and a helicopter. Its advantage is the ability to rotate its engines to 90° from a vertical position, as shown in Fig. P8.21(a), for takeoffs and landings and then switch the engines to horizontal for cruising as an airplane. The altitude control system in the helicopter mode is shown in Fig. P8.21(b). (a) Obtain the frequency response of the system for $K = 100$. (b) Find the gain margin and the phase margin for this system. (c) Select a suitable gain K so that the phase margin is 40°. (Increase the gain above $K = 100$.) (d) Find the response $c(t)$ of the system for the gain selected in part (c).

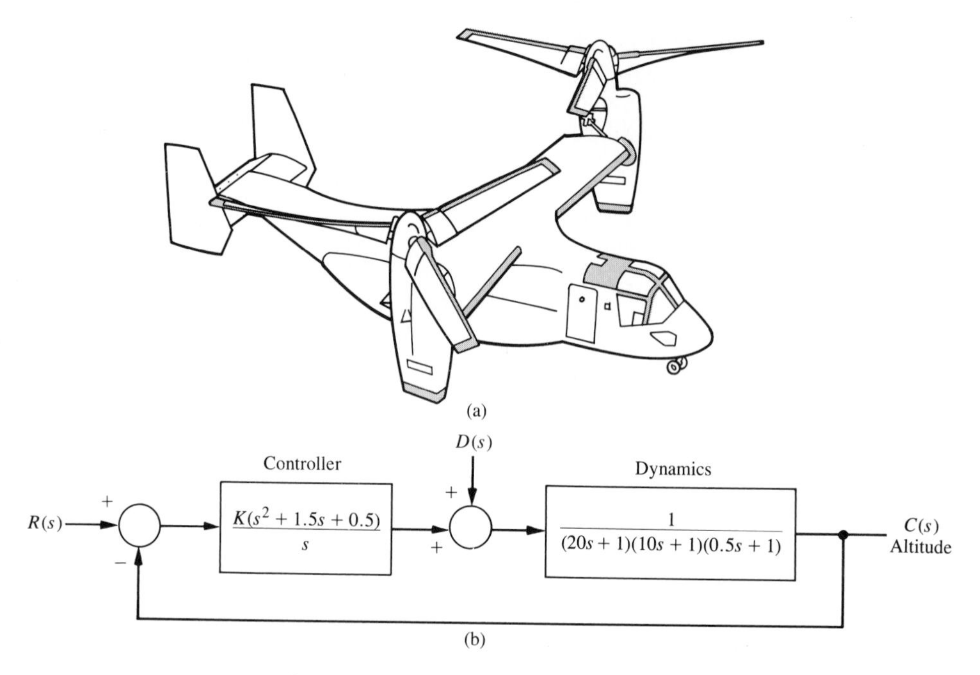

Figure P8.21. Tiltrotor aircraft control.

P8.22. Consider a unity feedback system with

$$G(s) = \frac{K}{s(s+1)(s+2)}.$$

(a) Draw the Bode diagram accurately for $K = 4$. Determine (b) the gain margin, (c) the value of K required to provide a gain margin equal to 16 db, and (d) the value of K to yield a steady-state error of 10% of the magnitude A for the ramp input $r(t) = At,\ t > 0$. Can this gain be utilized and achieve acceptable performance?

P8.23. The Nichols diagram for a $GH(j\omega)$ of a closed-loop system is shown in Fig. P8.23, on the next page. The frequency for each point on the graph is given in the table below.

Point	1	2	3	4	5	6	7	8	9
ω	1	2.0	2.6	3.4	4.2	5.2	6.0	7.0	8.0

Determine (a) the resonant frequency, (b) the bandwidth, (c) the phase margin, and (d) the gain margin. (e) Estimate the overshoot and settling time of the response to a step input.

P8.24. A specialty machine shop is improving the efficiency of its surface grinding process. The existing machine is sound mechanically but manually operated. Automating the machine will free the operator for other tasks, which will increase overall throughput of the machine shop. The grinding machine is shown in Fig. P8.24(a), on page 424 with all three axes automated with motors and feedback systems. The control system for the y-axis is shown in Fig. P8.24(b). In order to achieve a low steady-state error to a ramp command, we choose $K = 10$. Draw the Bode diagram of the open-loop system and obtain the Nichols chart plot. Determine the gain and phase margin of the system and the bandwidth of the closed-loop system. Estimate the ζ of the system and the predicted overshoot and settling time.

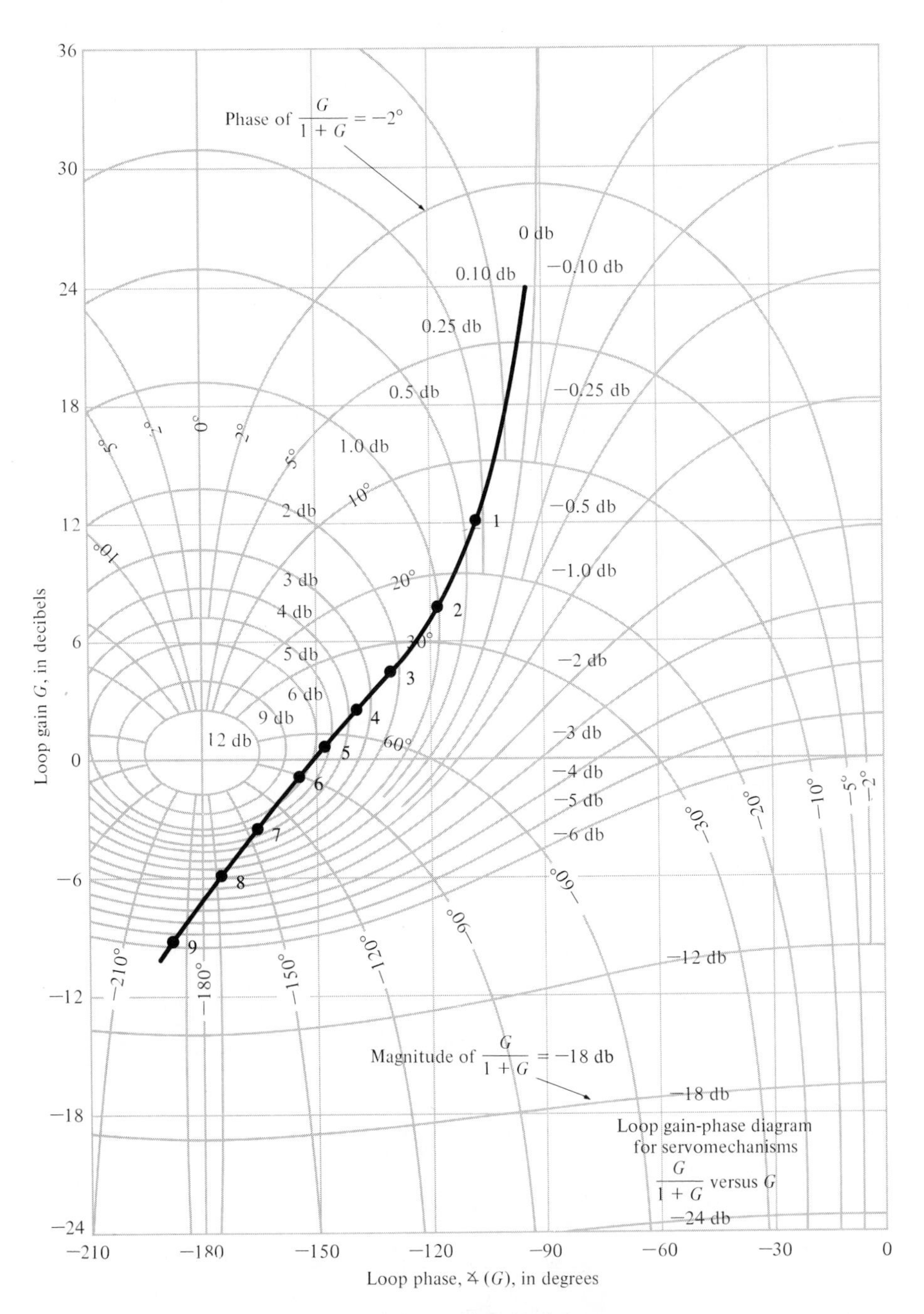

Figure P8.23. Nichols chart.

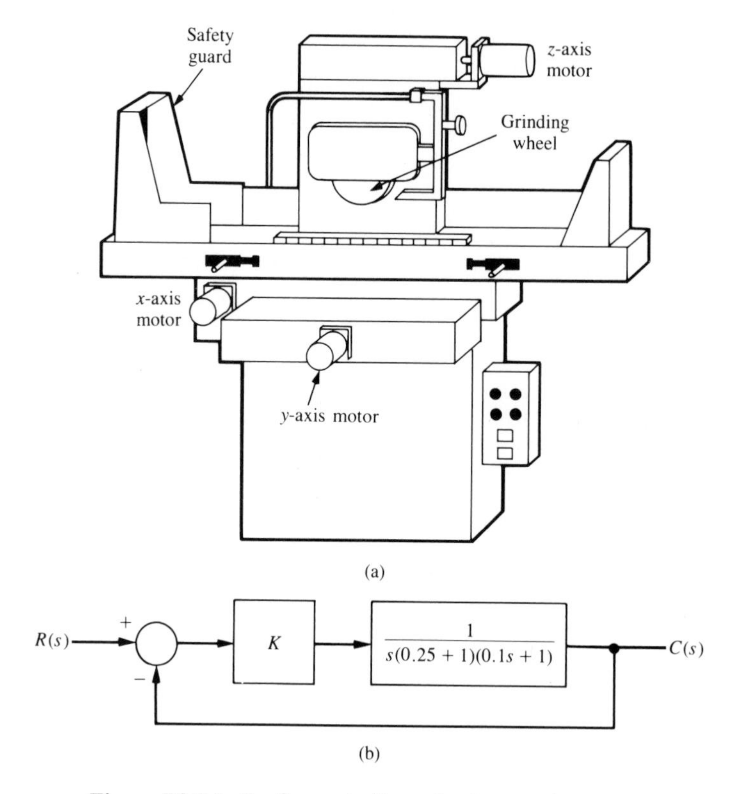

Figure P8.24. Surface grinding wheel control system.

Design Problems 

DP8.1. A mobile robot for toxic waste cleanup is shown in Fig. DP8.1a [24]. The closed-loop speed control is represented by Fig. 8.8 with $H(s) = 1$. The Nichol's chart in Fig. DP8.1(b) shows the plot of $G(j\omega)/k$ versus ω. The value of the frequency at points indicated

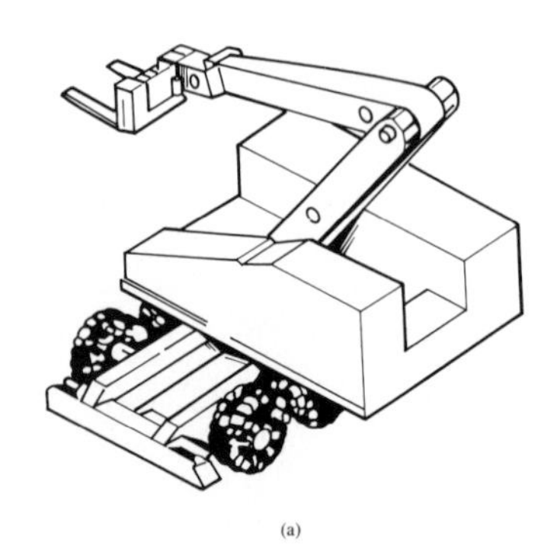

Figure DP8.1(a). Mobile robot for toxic waste cleanup.

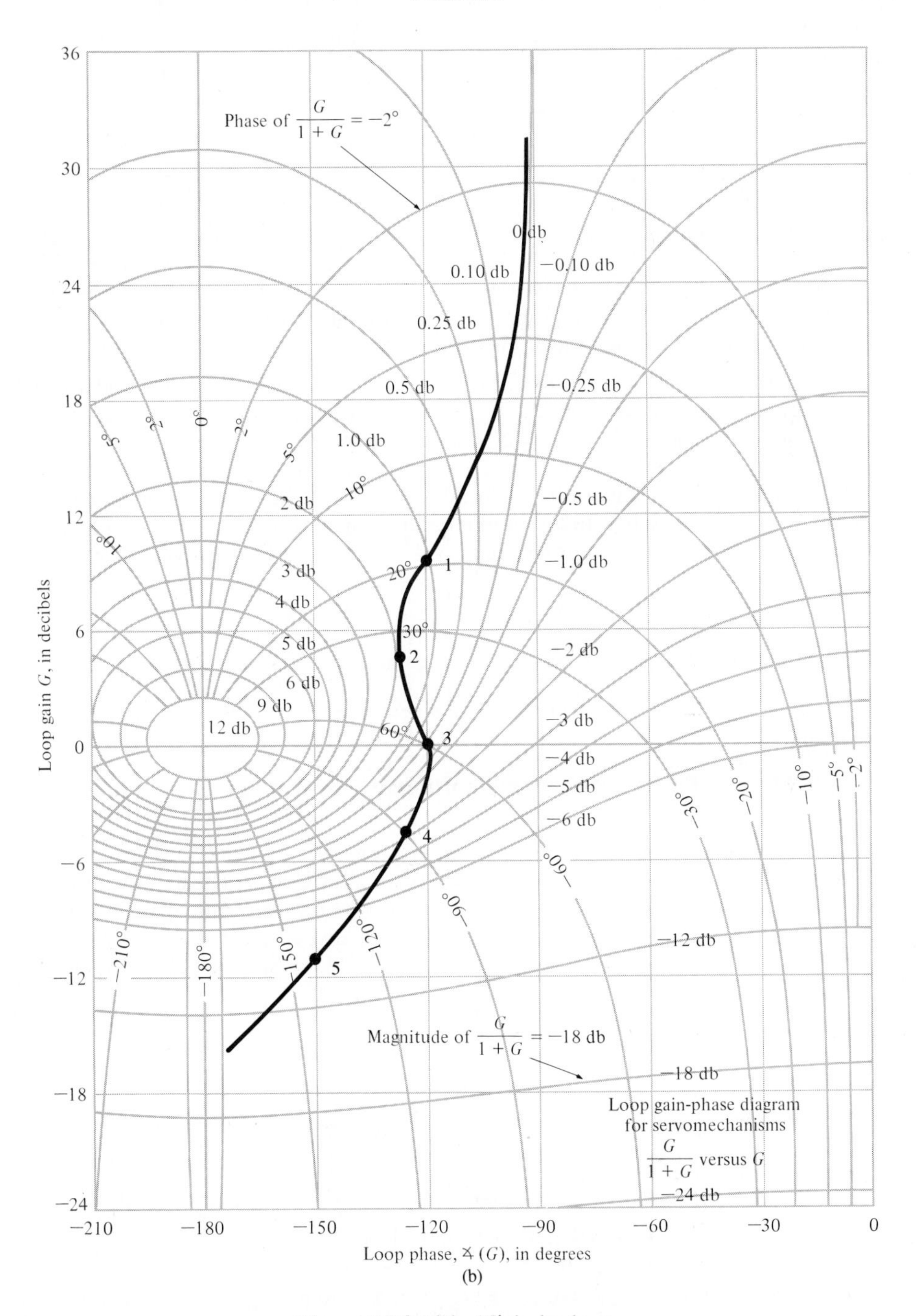

Figure DP8.1(b). Nichols chart.

is recorded in the table below.

Point	1	2	3	4	5
ω	2	5	10	20	50

(a) Determine the gain and phase margin of the closed-loop system when $k = 1$. (b) Determine the resonant peak in db and the resonant frequency for $k = 1$. (c) Determine the system bandwidth and estimate the settling time and percent overshoot of this system for a step input. (d) Determine the appropriate gain k so that the overshoot to a step input is 30%, and estimate the settling time of the system.

DP8.2. Flexible-joint robot arms are constructed of lightweight materials and exhibit lightly damped open-loop dynamics [23]. A feedback control system for a flexible arm is shown in Fig. DP8.2. Select K so that the system has maximum phase margin. Predict the overshoot for a step input based on the phase margin attained and compare it to the actual overshoot for a step input. Determine the bandwidth of the closed-loop system and predict the settling time of the system to a step input and compare it to the actual settling time. Discuss the suitability of this control system.

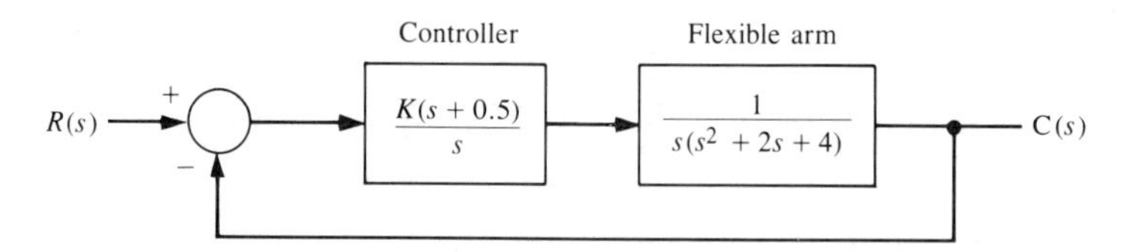

Figure DP8.2. Control of a flexible robot arm.

DP8.3. An automatic drug delivery system is used in the regulation of critical care patients suffering from cardiac failure [18]. The goal is to maintain stable patient status within narrow bounds. Consider the use of a drug delivery system for the regulation of blood pressure by the infusion of a drug. The feedback control system is shown in Fig. DP8.3. Select an appropriate gain K that maintains narrow deviation for blood pressure while achieving a good dynamic response.

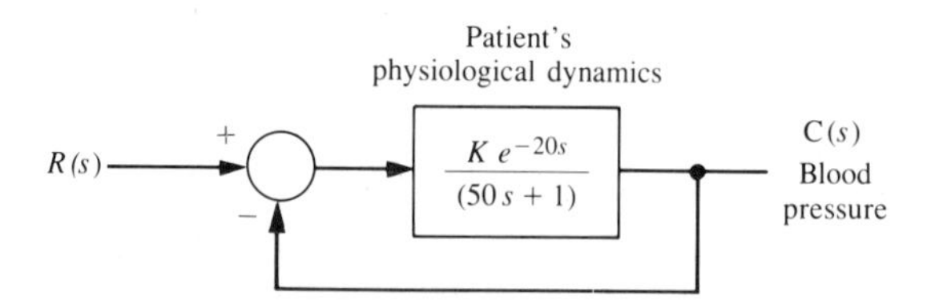

Figure DP8.3. Automatic drug delivery.

DP8.4. A robot tennis player is shown in Fig. DP8.4(a), and a simplified control system for $\Theta_2(t)$ is shown in Fig. DP8.4(b). The goal of the control system is to attain the best step response while attaining a high K_v for the system. Select $K_{v_1} = 0.325$ and $K_{v_2} = 0.45$ and determine the phase margin, gain margin, and closed-loop bandwidth for each case. Estimate the step response for each case and select the best value for K.

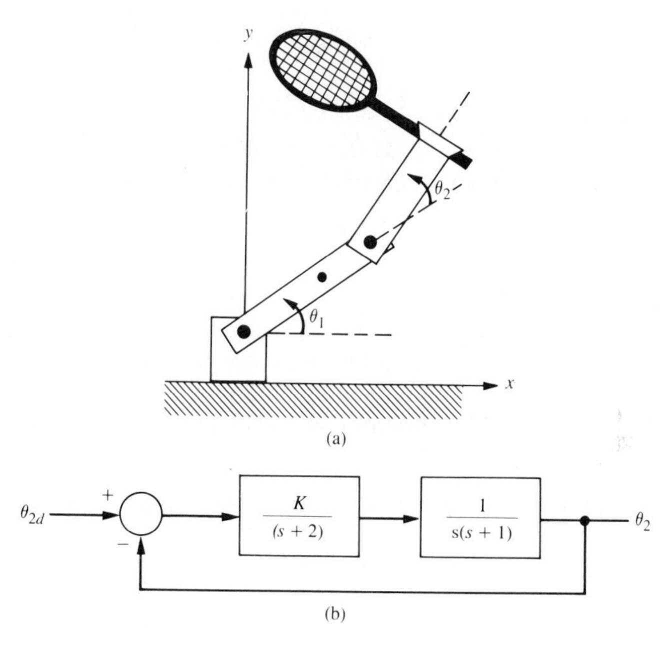

Figure DP8.4. An articulated two-link tennis player robot.

Terms and Concepts

Cauchy's Theorem If a contour encircles Z zero and P poles of $F(s)$ traversing clockwise, the corresponding contour in the $F(s)$-plane encircles the origin of the $F(s)$-plane $N = Z - P$ times clockwise.

Closed-loop frequency response The frequency response of the closed-loop transfer function $T(j\omega)$.

Conformal mapping A contour mapping that retains the angles on the s-plane on the $F(s)$-plane.

Contour map A contour or trajectory in one plane is mapped into another plane by a relation $F(s)$.

Gain margin The reciprocal of the gain $|GH|$ at the frequency at which the phase angle reaches 180°.

Nichols chart A chart displaying the curves for the relationship between the open-loop and closed-loop frequency response.

Nyquist stability criterion A feedback system is stable if, and only if, the contour in the $G(s)$ plane does not encircle the $(-1, 0)$ point when the number of poles of $G(s)$ in the right-hand s-plane is zero. If $G(s)$ has P poles in the right-hand plane, then the number of counterclockwise encirclements of the $(-1, 0)$ point must be equal to P for a stable system.

Phase margin The phase angle through which the $GH(j\omega)$ locus must be rotated in order that the unity magnitude point passes through the $(-1, 0)$ point in the $GH(j\omega)$ plane.

Principle of the argument See Cauchy's Theorem.

Time delay A pure time delay, T, so that events occurring at time t at one point in the system occur at another point in the system at a later time, $t + T$.

References

1. H. Nyquist, "Regeneration Theory," *Bell Systems Tech. J.*, January 1932; pp. 126–47. Also in *Automatic Control: Classical Linear Theory*, G. J. Thaler, ed., Dowden, Hutchinson, and Ross, Stroudsburg, Pa.; 1932; pp. 105–26.
2. M. D. Fagen, *A History of Engineering and Science in the Bell System*, Bell Telephone Laboratories, Inc., Murray Hill, N.J., 1978; Chapter 5.
3. E. Kamen, *Introduction to Signals and Systems*, Macmillan, New York, 1988.
4. W. Shepard and L. N. Hulley, *Power Electronics and Motor Control*, Cambridge University Press, New York, 1987.
5. R. C. Dorf and R. Jacquot, *Control System Design Program*, Addison-Wesley, Reading, Mass., 1988.
6. L. B. Jackson, *Signals, Systems and Transforms*, Addison-Wesley, Reading, Mass., 1991.
7. H. M. James, N. B. Nichols, and R. S. Phillips, *Theory of Servomechanisms*, McGraw-Hill, New York, 1947.
8. J. V. Ringwood, "Shape Control in Sendzimir Mills Using Roll Actuators," *IEEE Trans. on Automatic Control*, April 1990; pp. 453–59.
9. V. D. Hunt, *Mechatronics*, Chapman and Hall, New York, 1989.
10. W. L. Brogan, *Modern Control Theory*, Prentice Hall, Englewood Cliffs, N.J., 1991.
11. R. C. Nelson, *Flight Stability and Control*, McGraw-Hill, New York, 1989.
12. R. C. Dorf, *Robotics and Automated Manufacturing*, Reston Publishing, Reston, Virginia, 1983.
13. R. C. Dorf, *The Encyclopedia of Robotics*, John Wiley & Sons, New York, 1988.
14. A. T. Bahill and L. Stark, "The Trajectories of Saccadic Eye Movements," *Scientific American*, January 1979; pp. 108–17.
15. K. K. Chew, "Control of Errors in Disk Drive Systems," *IEEE Control Systems*, January 1990; pp. 16–19.
16. J. Slottine, *Applied Nonlinear Control*, Prentice Hall, Englewood Cliffs, N.J., 1991.
17. J. F. Engleberger, *Robotics in Service*, M.I.T. Press, Cambridge, Mass., 1989.
18. G. W. Neat, "Expert Adaptive Control for Drug Delivery Systems," *IEEE Control Systems*, June 1989; pp. 20–23.
19. J. C. Maciejowski, *Multivariable Feedback Design*, Addison-Wesley, Reading, Mass., 1990.
20. R. Kurzweil, *The Age of Intelligent Machines*, M.I.T. Press, Cambridge, Mass., 1990.
21. R. P. Brennan, *Levitating Trains*, John Wiley & Sons, New York, 1990.
22. D. M. Schneider, "Control of Processes with Time Delays," *IEEE Trans. on Industry Applications*, April 1988; pp. 186–91.
23. F. Ghorbel, "Adaptive Control of Flexible-Joint Manipulators," *IEEE Control Systems*, December 1989; pp. 9–12.
24. N. A. Hootsmans, "Large Motion Control of Mobile Manipulators," *Proceed. of the 1991 IEEE Conference on Robotics*, April 1991; pp. 2336–41.

✦

CHAPTER 9

Time-Domain Analysis of Control Systems

Preview

In the preceding chapters we used the Laplace transform or sinusoidal steady-state frequency response to describe the performance of a feedback system. These methods are attractive since they provide a practical approach to analysis of a system. However, we recall that it is the response of the system over a time period that is actually our focus of interest. Thus, in this chapter, we turn to a method of analysis in the time domain.

In this chapter, we will reconsider the differential equations describing a control system and select a certain form of differential equations. We will use a set of variables that can be used to establish a set of first-order differential equations.

Using matrix methods we will be able to determine the transient response of a control system and examine the stability of these systems. These time-domain matrix methods lend themselves readily to computer solution. Furthermore, it is feasible to propose new feedback structures based on the utilization of this new set of time-domain variables, called state variables.

9.1 Introduction

In the preceding chapters, we have developed and studied several useful approaches to the analysis and design of feedback systems. The Laplace transform was utilized to transform the differential equations representing the system into an algebraic equation expressed in terms of the complex variable s. Utilizing this algebraic equation, we were able to obtain a transfer function representation of the input–output relationship. Then the root locus and s-plane methods were developed on the basis of the Laplace transform representation. Furthermore, the steady-state representation of the system in terms of the real frequency variable ω was developed, and several useful techniques for analysis were studied. The frequency-domain approach, in terms of the complex variable s or the real frequency variable ω, is extremely useful; it is and will remain one of the primary tools of the control engineer. However, the limitations of the frequency-domain techniques and the recently acquired attractiveness of the time-domain approach require a reconsideration of the time-domain formulation of the equations representing control systems.

The frequency-domain techniques are limited in applicability to linear, time-invariant systems. Furthermore, they are particularly limited in their usefulness for multivariable control systems due to the emphasis on the input-output relationship of transfer functions. By contrast, the time-domain techniques can be readily utilized for nonlinear, time-varying, and multivariable systems. *A time-varying control system is a system for which one or more of the parameters of the system may vary as a function of time.* For example, the mass of a missile varies as a function of time as the fuel is expended during flight. A multivariable system, as discussed in Section 2.6, is a system with several input and output signals. The solution of a time-domain formulation of a control system problem is facilitated by the availability and ease of use of digital and analog computers. Therefore we are interested in reconsidering the time-domain description of dynamic systems as they are represented by the system differential equation. The *time-domain* is the mathematical domain that incorporates the response and description of a system in terms of time, t.

The time-domain representation of control systems is an essential basis for modern control theory and system optimization. In Chapter 10, we shall have an opportunity to design an optimum control system by utilizing time-domain

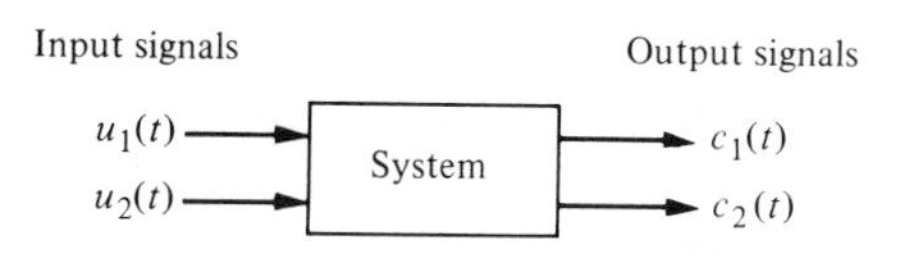

Figure 9.1. System block diagram.

methods. In this chapter, we shall develop the time-domain representation of control systems, investigate the stability of these systems, and illustrate several methods for the solution of the system time response.

9.2 The State Variables of a Dynamic System

The time-domain analysis and design of control systems utilize the concept of the state of a system [1, 2, 3, 5]. *The state of a system is a set of numbers such that the knowledge of these numbers and the input functions will, with the equations describing the dynamics, provide the future state and output of the system.* For a dynamic system, the state of a system is described in terms of a set of *state variables* $[x_1(t), x_2(t), \ldots, x_n(t)]$. The state variables are those variables that determine the future behavior of a system when the present state of the system and the excitation signals are known. Consider the system shown in Fig. 9.1, where $c_1(t)$ and $c_2(t)$ are the output signals and $u_1(t)$ and $u_2(t)$ are the input signals. A set of state variables $(x_1, x_2, \ldots, x_n)$ for the system shown in Fig. 9.1 is a set such that knowledge of the initial values of the state variables $[x_1(t_0), x_2(t_0), \ldots, x_n(t_0)]$ at the initial time t_0, and of the input signals $u_1(t)$ and $u_2(t)$ for $t \geq t_0$, suffices to determine the future values of the outputs and state variables [2].

A simple example of a state variable is the state of an On-Off light switch. The switch can be in either the On or the Off position and thus the state of the switch can assume one of two possible values. Thus if we know the present state (position) of the switch at t_0 and if an input is applied, we are able to determine the future value of the state of the element.

The concept of a set of state variables that represent a dynamic system can be illustrated in terms of the spring-mass-damper system shown in Fig. 9.2. The

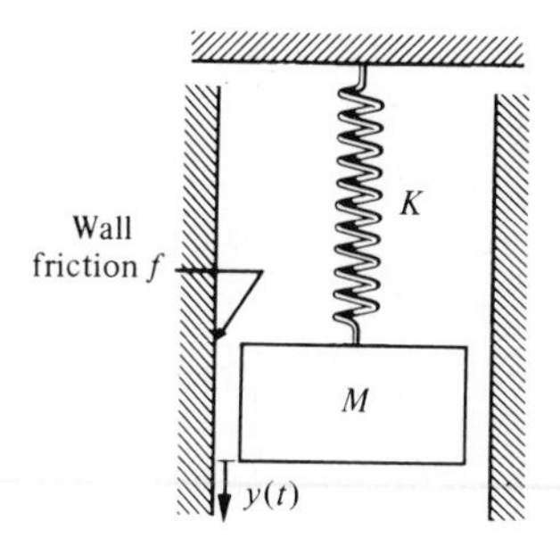

Figure 9.2. A spring-mass-damper system.

number of state variables chosen to represent this system should be as few as possible in order to avoid redundant state variables. A set of state variables sufficient to describe this system is the position and the velocity of the mass. Therefore we will define a set of state variables as (x_1, x_2), where

$$x_1(t) = y(t) \quad \text{and} \quad x_2(t) = \frac{dy(t)}{dt}.$$

The differential equation describes the behavior of the system and is usually written as

$$M\frac{d^2y}{dt^2} + f\frac{dy}{dt} + Ky = u(t). \tag{9.1}$$

In order to write Eq. (9.1) in terms of the state variables, we substitute the definition of the state variables and obtain

$$M\frac{dx_2}{dt} + fx_2 + Kx_1 = u(t). \tag{9.2}$$

Therefore we can write the differential equations that describe the behavior of the spring-mass-damper system as a set of two first-order differential equations as follows:

$$\frac{dx_1}{dt} = x_2. \tag{9.3}$$

$$\frac{dx_2}{dt} = \frac{-f}{M}x_2 - \frac{K}{M}x_1 + \frac{1}{M}u. \tag{9.4}$$

This set of differential equations describes the behavior of the state of the system in terms of the rate of change of each state variable.

As another example of the state variable characterization of a system, let us consider the *RLC* circuit shown in Fig. 9.3. The state of this system can be described in terms of a set of state variables (x_1, x_2), where x_1 is the capacitor voltage $v_c(t)$ and x_2 is equal to the inductor current $i_L(t)$. This choice of state variables is intuitively satisfactory because the stored energy of the network can be described in terms of these variables as

$$\mathcal{E} = (1/2)Li_L^2 + (1/2)Cv_c^2. \tag{9.5}$$

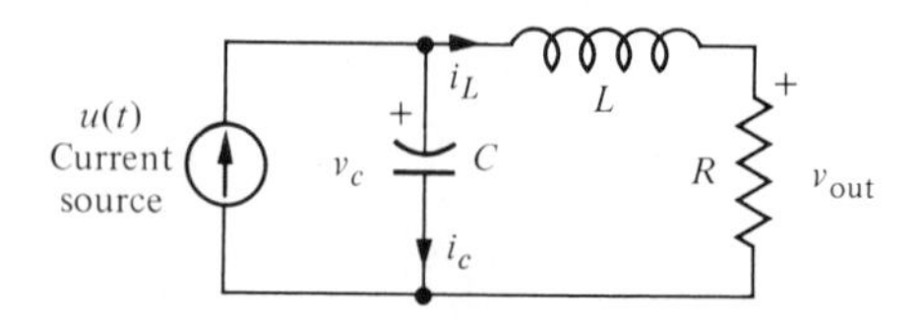

Figure 9.3. An *RLC* circuit.

Therefore, $x_1(t_0)$ and $x_2(t_0)$ represent the total initial energy of the network and thus the state of the system at $t = t_0$. For a passive *RLC* network, the number of state variables required is equal to the number of independent energy-storage elements. Utilizing Kirchhoff's current law at the junction, we obtain a first-order differential equation by describing the rate of change of capacitor voltage as

$$i_c = C\frac{dv_c}{dt} = +u(t) - i_L. \tag{9.6}$$

Kirchhoff's voltage law for the right-hand loop provides the equation describing the rate of change of inductor current as

$$L\frac{di_L}{dt} = -Ri_L + v_c. \tag{9.7}$$

The output of this system is represented by the linear algebraic equation

$$v_{\text{out}} = Ri_L(t).$$

We can rewrite Eqs. (9.6) and (9.7) as a set of two first-order differential equations in terms of the state variables x_1 and x_2 as follows:

$$\frac{dx_1}{dt} = -\frac{1}{C}x_2 + \frac{1}{C}u(t), \tag{9.8}$$

$$\frac{dx_2}{dt} = +\frac{1}{L}x_1 - \frac{R}{L}x_2. \tag{9.9}$$

The output signal is then

$$c_1(t) = v_{\text{out}}(t) = Rx_2. \tag{9.10}$$

Utilizing Eqs. (9.8) and (9.9) and the initial conditions of the network represented by $[x_1(t_0), x_2(t_0)]$, we can determine the system's future behavior and its output.

The state variables that describe a system are not a unique set, and several alternative sets of state variables can be chosen. For example, for a second-order system, such as the mass-spring-damper or *RLC* circuit, the state variables may be any two independent linear combinations of $x_1(t)$ and $x_2(t)$. Therefore, for the *RLC* circuit, we might choose the set of state variables as the two voltages, $v_c(t)$ and $v_L(t)$, where v_L is the voltage drop across the inductor. Then the new state variables, x_1^* and x_2^*, are related to the old state variables, x_1 and x_2, as

$$x_1^* = v_c = x_1, \tag{9.11}$$

$$x_2^* = v_L = v_c - Ri_L = x_1 - Rx_2. \tag{9.12}$$

Equation (9.12) represents the relation between the inductor voltage and the former state variables v_c and i_L. Thus, in an actual system, there are several choices of a set of state variables that specify the energy stored in a system and therefore adequately describe the dynamics of the system. A widely used choice is a set of state variables that can be readily measured.

An alternative approach to developing a model of a device is the use of the bond graph. Bond graphs can be used for electrical, mechanical, hydraulic, and thermal devices or systems as well as for combinations of various types of elements. Bond graphs produce a set of equations in the state-variable form [2].

The state variables of a system characterize the dynamic behavior of a system. The engineer's interest is primarily in physical systems, where the variables are voltages, currents, velocities, positions, pressures, temperatures, and similar physical variables. However, the concept of system state is not limited to the analysis of physical systems and is particularly useful for analyzing biological, social, and economic systems as well as physical systems. For these systems, the concept of state is extended beyond the concept of energy of a physical system to the broader viewpoint of variables that describe the future behavior of the system.

9.3 The State Vector Differential Equation

The state of a system is described by the set of first-order differential equations written in terms of the state variables ($x_1, x_2, \ldots, x_n$). These first-order differential equations can be written in general form as

$$
\begin{aligned}
\dot{x}_1 &= a_{11}x_1 + a_{12}x_2 + \cdots + a_{1n}x_n + b_{11}u_1 + \cdots + b_{1m}u_m, \\
\dot{x}_2 &= a_{21}x_1 + a_{22}x_2 + \cdots + a_{2n}x_n + b_{21}u_1 + \cdots + b_{2m}u_m, \\
&\vdots \\
\dot{x}_n &= a_{n1}x_1 + a_{n2}x_2 + \cdots + a_{nn}x_n + b_{n1}u_1 + \cdots + b_{nm}u_m,
\end{aligned}
\qquad (9.13)
$$

where $\dot{x} = dx/dt$. Thus this set of simultaneous differential equations can be written in matrix form as follows [5, 7]:

$$
\frac{d}{dt}\begin{bmatrix} x_1 \\ x_2 \\ \vdots \\ x_n \end{bmatrix} = \begin{bmatrix} a_{11} & a_{11} \cdots a_{1n} \\ a_{21} & a_{22} \cdots a_{2n} \\ \vdots & \cdots \vdots \\ a_{n1} & a_{n2} \cdots a_{nn} \end{bmatrix} \begin{bmatrix} x_1 \\ x_2 \\ \vdots \\ x_n \end{bmatrix}
+ \begin{bmatrix} b_{11} \\ \vdots \\ b_{n1} \end{bmatrix} \begin{bmatrix} \cdots \\ \\ \cdots \end{bmatrix} \begin{bmatrix} b_{1m} \\ \vdots \\ b_{nm} \end{bmatrix} \begin{bmatrix} u_1 \\ \vdots \\ u_m \end{bmatrix}. \qquad (9.14)
$$

The column matrix consisting of the state variables is called the *state vector* and is written as

$$
\mathbf{x} = \begin{bmatrix} x_1 \\ x_2 \\ \vdots \\ x_n \end{bmatrix}, \qquad (9.15)
$$

where the boldface indicates a matrix. The matrix of input signals is defined as $\mathbf{u}$. Then the system can be represented by the compact notation of the *state vector differential equation* as

$$\dot{\mathbf{x}} = \mathbf{A}\mathbf{x} + \mathbf{B}\mathbf{u}. \tag{9.16}$$

The matrix $\mathbf{A}$ is an $n \times m$ square matrix and $\mathbf{B}$ is an $n \times m$ matrix.* The vector matrix differential equation relates the rate of change of the state of the system to the state of the system and the input signals. In general, the outputs of a linear system can be related to the state variables and the input signals by the vector matrix equation

$$\mathbf{c} = \mathbf{D}\mathbf{x} + \mathbf{H}\mathbf{u}, \tag{9.17}$$

where $\mathbf{c}$ is the set of output signals expressed in column vector form.

The solution of the state vector differential equation (Eq. 9.16) can be obtained in a manner similar to the approach we utilize for solving a first-order differential equation. Consider the first-order differential equation

$$\dot{x} = ax + bu, \tag{9.18}$$

where $x(t)$ and $u(t)$ are scalar functions of time. We expect an exponential solution of the form e^{at}. Taking the Laplace transform of Eq. (9.18), we have

$$sX(s) - x(0) = aX(s) + bU(s),$$

and therefore

$$X(s) = \frac{x(0)}{s-a} + \frac{b}{s-a}U(s). \tag{9.19}$$

The inverse Laplace transform of Eq. (9.19) results in the solution

$$x(t) = e^{at}x(0) + \int_0^t e^{+a(t-\tau)}bu(\tau)\,d\tau. \tag{9.20}$$

We expect the solution of the vector differential equation to be similar to Eq. (9.20) and to be of exponential form. The matrix exponential function is defined as

$$e^{\mathbf{A}t} = \exp(\mathbf{A}t) = \mathbf{I} + \mathbf{A}t + \frac{\mathbf{A}^2t^2}{2!} + \cdots + \frac{\mathbf{A}^kt^k}{k!} + \cdots, \tag{9.21}$$

which converges for all finite t and any $\mathbf{A}$ [7]. Then the solution of the vector differential equation is found to be [1]

$$\mathbf{x}(t) = \exp(\mathbf{A}t)\mathbf{x}(0) + \int_0^t \exp[\mathbf{A}(t-\tau)]\mathbf{B}\mathbf{u}(\tau)\,d\tau. \tag{9.22}$$

---

*Boldfaced lowercase letters denote vector quantities and boldfaced uppercase letters denote matrices. For an introduction to matrices and elementary matrix operations, refer to Appendix C and references [3] and [7].

Equation (9.22) may be obtained by taking the Laplace transform of Eq. (9.16) and rearranging to obtain

$$\mathbf{X}(s) = [s\mathbf{I} - \mathbf{A}]^{-1}\mathbf{x}(0) + [s\mathbf{I} - \mathbf{A}]^{-1}\mathbf{BU}(s), \tag{9.23}$$

where we note that $[s\mathbf{I} - \mathbf{A}]^{-1} = \boldsymbol{\phi}(s)$, which is the Laplace transform of $\boldsymbol{\phi}(t) = \exp(\mathbf{A}t)$. Taking the inverse Laplace transform of Eq. (9.23) and noting that the second term on the right-hand side involves the product $\boldsymbol{\phi}(s)\mathbf{BU}(s)$, we obtain Eq. (9.22). The matrix exponential function describes the unforced response of the system and is called the *fundamental* or *transition matrix* $\boldsymbol{\phi}(t)$. Therefore Eq. (9.22) can be written as

$$\mathbf{x}(t) = \boldsymbol{\phi}(t)\mathbf{x}(0) + \int_0^t \boldsymbol{\phi}(t - \tau)\mathbf{Bu}(\tau)\, d\tau. \tag{9.24}$$

The solution to the unforced system (that is, when $\mathbf{u} = 0$) is simply

$$\begin{bmatrix} x_1(t) \\ x_2(t) \\ \vdots \\ x_n(t) \end{bmatrix} = \begin{bmatrix} \phi_{11}(t) \cdots \phi_{1n}(t) \\ \phi_{21}(t) \cdots \phi_{2n}(t) \\ \vdots \qquad \vdots \\ \phi_{n1}(t) \cdots \phi_{nn}(t) \end{bmatrix} \begin{bmatrix} x_1(0) \\ x_2(0) \\ \vdots \\ x_n(0) \end{bmatrix}. \tag{9.25}$$

We note, therefore, that in order to determine the transition matrix, all initial conditions are set to 0 except for one state variable, and the output of each state variable is evaluated. That is, the term $\phi_{ij}(t)$ is the response of the ith state variable due to an initial condition on the jth state variable when there are zero initial conditions on all the other states. We shall utilize this relationship between the initial conditions and the state variables to evaluate the coefficients of the transition matrix in a later section. However, first we shall develop several suitable signal-flow state models of systems and investigate the stability of the systems by utilizing these flow graphs.

9.4 Signal-Flow Graph State Models

The state of a system describes that system's dynamic behavior where the dynamics of the system are represented by a series of first-order differential equations. Alternatively, the dynamics of the system can be represented by a vector differential equation as in Eq. (9.16). In either case, it is useful to develop a state flow graph model of the system and use this model to relate the state variable concept to the familiar transfer function representation.

As we have learned in previous chapters, a system can be meaningfully described by an input-output relationship, the transfer function $G(s)$. For example, if we are interested in the relation between the output voltage and the input voltage of the network of Fig. 9.3, we can obtain the transfer function

$$G(s) = \frac{V_0(s)}{U(s)}.$$

The transfer function for the *RLC* network is of the form

$$G(s) = \frac{V_0(s)}{U(s)} = \frac{\alpha}{s^2 + \beta s + \gamma}, \tag{9.26}$$

where α, β, and γ are functions of the circuit parameters R, L, and C. The values of α, β, and γ can be determined from the flow graph representing the differential equations that describe the circuit. For the *RLC* circuit (see Eqs. 9.8 and 9.9), we have

$$\dot{x}_1 = -\frac{1}{C}x_2 + \frac{1}{C}u(t), \tag{9.27}$$

$$\dot{x}_2 = \frac{1}{L}x_1 - \frac{R}{L}x_2, \tag{9.28}$$

$$v_{\text{out}} = Rx_2. \tag{9.29}$$

The flow graph representing these simultaneous equations is shown in Fig. 9.4, where $1/s$ indicates an integration. Using Mason's signal-flow gain formula, we obtain the transfer function

$$\frac{V_{\text{out}}(s)}{U(s)} = \frac{+R/LCs^2}{1 + (R/Ls) + (1/LCs^2)} = \frac{+R/LC}{s^2 + (R/L)s + (1/LC)}. \tag{9.30}$$

Unfortunately, many electric circuits, electromechanical systems, and other control systems are not so simple as the *RLC* circuit of Fig. 9.3, and it is often a difficult task to determine a series of first-order differential equations describing the system. Therefore it is often simpler to derive the transfer function of the system by the techniques of Chapter 2 and then derive the state model from the transfer function.

The signal-flow graph state model can be readily derived from the transfer function of a system. However, as we noted in Section 9.3, there is more than one alternative set of state variables, and therefore there is more than one possible form for the signal-flow graph state model. In general, we can represent a transfer function as

$$G(s) = \frac{C(s)}{U(s)} = \frac{s^m + b_{m-1}s^{m-1} + \cdots + b_1 s + b_0}{s^n + a_{n-1}s^{n-1} + \cdots + a_1 s + a_0}, \tag{9.31}$$

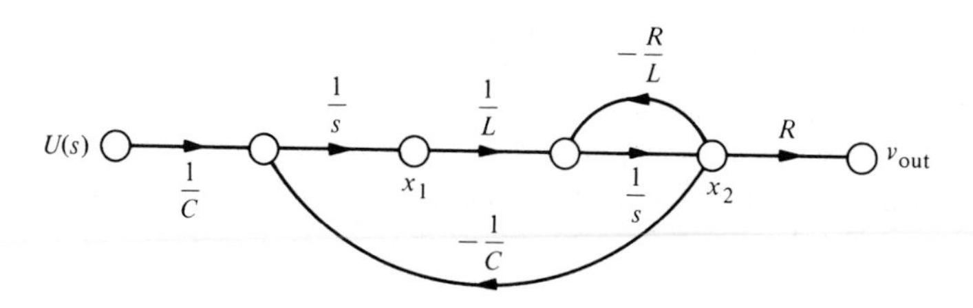

Figure 9.4. Flow graph for the *RLC* network.

where $n \geq m$ and all the a coefficients are real positive numbers. If we multiply the numerator and denominator by s^{-n}, we obtain

$$G(s) = \frac{s^{-(n-m)} + b_{m-1}s^{-(n-m+1)} + \cdots + b_1 s^{-(n-1)} + b_0 s^{-n}}{1 + a_{n-1}s^{-1} + \cdots + a_1 s^{-(n-1)} + a_0 s^{-n}}. \tag{9.32}$$

Our familiarity with Mason's flow graph gain formula causes us to recognize the familiar feedback factors in the denominator and the forward-path factors in the numerator. Mason's flow graph formula was discussed in Section 2.7 and is written as

$$G(s) = \frac{C(s)}{U(s)} = \frac{\Sigma_k P_k \Delta_k}{\Delta}. \tag{9.33}$$

When all the feedback loops are touching and all the forward paths touch the feedback loops, Eq. (9.33) reduces to

$$G(s) = \frac{\Sigma_k P_k}{1 - \Sigma_{q=1}^{N} L_q} = \frac{\text{Sum of the forward-path factors}}{1 - \text{sum of the feedback loop factors}}. \tag{9.34}$$

There are several flow graphs that could represent the transfer function. Two flow-graph configurations are of particular interest, and we will consider these in greater detail.

In order to illustrate the derivation of the signal-flow graph state model, let us initially consider the fourth-order transfer function

$$G(s) = \frac{C(s)}{U(s)} = \frac{b_0}{s^4 + a_3 s^3 + a_2 s^2 + a_1 s + a_0} = \frac{b_0 s^{-4}}{1 + a_3 s^{-1} + a_2 s^{-2} + a_1 s^{-3} + a_0 s^{-4}}. \tag{9.35}$$

First we note that the system is fourth order, and hence we identify four state variables (x_1, x_2, x_3, x_4). Recalling Mason's gain formula, we note that the denominator can be considered to be one minus the sum of the loop gains. Furthermore, the numerator of the transfer function is equal to the forward-path factor of the flow graph. The flow graph must utilize a minimum number of integrators equal to the order of the system. Therefore we use four integrators to represent this system. The necessary flow-graph nodes and the four integrators are shown in Fig. 9.5. Considering the simplest series interconnection of integrators, we can represent the transfer function by the flow graph of Fig. 9.6. Examining Fig. 9.6, we note that all the loops are touching and that the transfer function of this flow

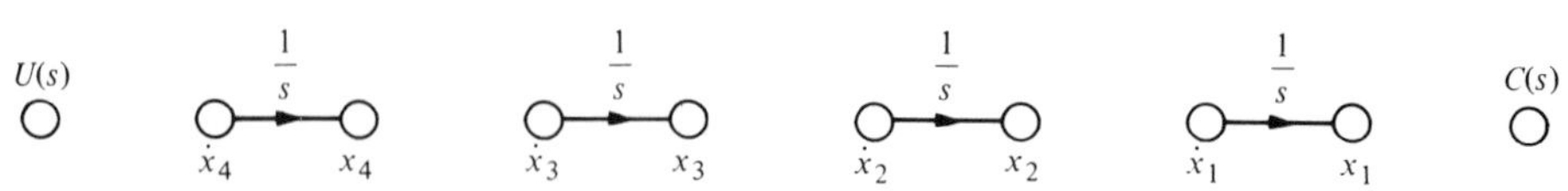

$U(s)$ $\quad \dot{x}_4 \xrightarrow{1/s} x_4 \quad \dot{x}_3 \xrightarrow{1/s} x_3 \quad \dot{x}_2 \xrightarrow{1/s} x_2 \quad \dot{x}_1 \xrightarrow{1/s} x_1 \quad$ $C(s)$

Figure 9.5. Flow-graph nodes and integrators for fourth-order system.

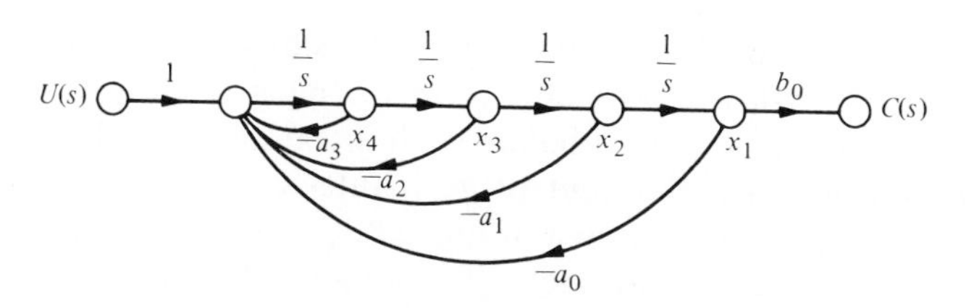

Figure 9.6. Flow-graph state model for $G(s)$ of Eq. (9.35).

graph is indeed Eq. (9.35). This can be readily verified by the reader by noting that the forward-path factor of the flow graph is b_0/s^4 and the denominator is equal to 1 minus the sum of the loop gains.

Now, consider the fourth-order transfer function when the numerator polynomial is a polynomial in s so that we have

$$\begin{aligned} G(s) &= \frac{b_3 s^3 + b_2 s^2 + b_1 s + b_0}{s^4 + a_3 s^3 + a_2 s^2 + a_1 s + a_0} \\ &= \frac{b_3 s^{-1} + b_2 s^{-2} + b_1 s^{-3} + b_0 s^{-4}}{1 + a_3 s^{-1} + a_2 s^{-2} + a_1 s^{-3} + a_0 s^{-4}}. \end{aligned} \tag{9.36}$$

The numerator terms represent forward-path factors in Mason's gain formula. The forward paths will touch all the loops, and a suitable signal-flow graph realization of Eq. (9.36) is shown in Fig. 9.7. The forward-path factors are b_3/s, b_2/s^2, b_1/s^3, and b_0/s^4 as required to provide the numerator of the transfer function. Recall that Mason's flow-graph gain formula indicates that the numerator of the transfer function is simply the sum of the forward-path factors. This general form of a signal-flow graph can represent the general transfer function of Eq. (9.36) by utilizing n feedback loops involving the a_n coefficients and m forward-path factors involving the b_m coefficients.

The state variables are identified in Fig. 9.7 as the output of each energy storage element; that is, the output of each integrator. In order to obtain the set of

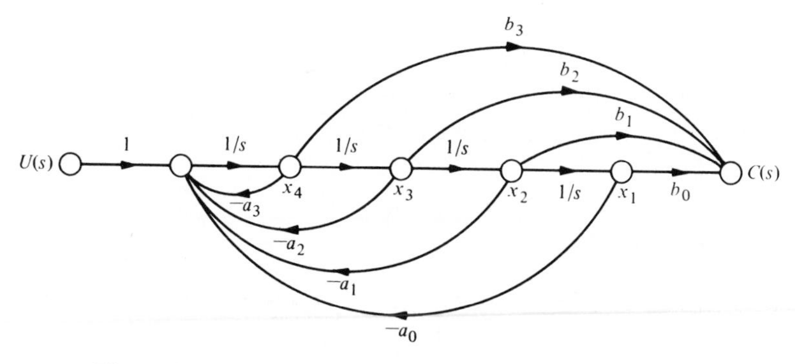

Figure 9.7. Flow-graph state model for $G(s)$ of Eq. (9.36).

first-order differential equations representing the state model of Fig. 9.7, we will introduce a new set of flow graph nodes immediately preceding each integrator of Fig. 9.7 [1]. The nodes are placed before each integrator, and therefore they represent the derivative of the output of each integrator. The signal-flow graph, including the added nodes, is shown in Fig. 9.8. Using the flow graph of Fig. 9.8, we are able to obtain the following set of first-order differential equations describing the state of the model:

$$\begin{aligned}\dot{x}_1 &= x_2,\\ \dot{x}_2 &= x_3,\\ \dot{x}_3 &= x_4,\\ \dot{x}_4 &= -a_0x_1 - a_1x_2 - a_2x_3 - a_3x_4 + u.\end{aligned} \tag{9.37}$$

Furthermore, the output is simply

$$c(t) = b_0x_1 + b_1x_2 + b_2x_3 + b_3x_4. \tag{9.38}$$

Then, in matrix form, we have

$$\dot{\mathbf{x}} = \mathbf{Ax} + \mathbf{b}u \tag{9.39}$$

or

$$\frac{d}{dt}\begin{bmatrix} x_1 \\ x_2 \\ x_3 \\ x_4 \end{bmatrix} = \begin{bmatrix} 0 & 1 & 0 & 0 \\ 0 & 0 & 1 & 0 \\ 0 & 0 & 0 & 1 \\ -a_0 & -a_1 & -a_2 & -a_3 \end{bmatrix}\begin{bmatrix} x_1 \\ x_2 \\ x_3 \\ x_4 \end{bmatrix} + \begin{bmatrix} 0 \\ 0 \\ 0 \\ 1 \end{bmatrix} u(t), \tag{9.40}$$

and the output is

$$c(t) = \mathbf{Dx} = [b_0, b_1, b_2, b_3]\begin{bmatrix} x_1 \\ x_2 \\ x_3 \\ x_4 \end{bmatrix}. \tag{9.41}$$

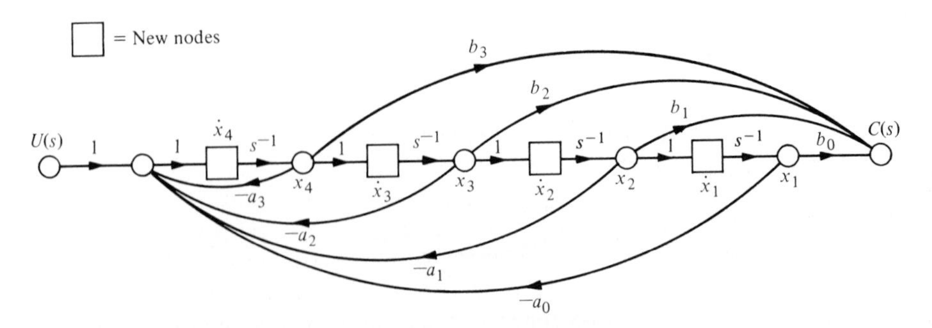

Figure 9.8. Flow graph of Fig. 9.7 with nodes inserted.

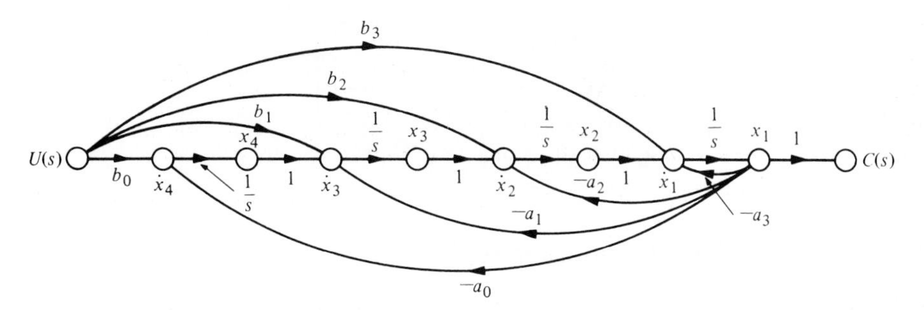

Figure 9.9. Alternative flow-graph state model for Eq. (9.36).

The flow-graph structure of Fig. 9.7 is not a unique representation of Eq. (9.36); another equally useful structure can be obtained. A flow graph that represents Eq. (9.36) equally well is shown in Fig. 9.9. In this case, the forward-path factors are obtained by feeding forward the signal $U(s)$.

Then the output signal $c(t)$ is equal to the first state variable $x_1(t)$. This flow graph structure has the forward-path factors b_0/s^4, b_1/s^3, b_2/s^2, b_3/s, and all the forward paths touch the feedback loops. Therefore the resulting transfer function is indeed equal to Eq. (9.36).

Using the flow graph of Fig. 9.9 to obtain the set of first-order differential equations, we obtain

$$\begin{aligned} \dot{x}_1 &= -a_3x_1 + x_2 + b_3u, \\ \dot{x}_2 &= -a_2x_1 + x_3 + b_2u, \\ \dot{x}_3 &= -a_1x_1 + x_4 + b_1u, \\ \dot{x}_4 &= -a_0x_1 + b_0u. \end{aligned} \tag{9.42}$$

Thus in matrix form we have

$$\frac{d\mathbf{x}}{dt} = \begin{bmatrix} -a_3 & 1 & 0 & 0 \\ -a_2 & 0 & 1 & 0 \\ -a_1 & 0 & 0 & 1 \\ -a_0 & 0 & 0 & 0 \end{bmatrix} \mathbf{x} + \begin{bmatrix} b_3 \\ b_2 \\ b_1 \\ b_0 \end{bmatrix} u(t). \tag{9.43}$$

Although the flow graph of Fig. 9.9 represents the same transfer function as the flow graph of Fig. 9.7, the state variables of each graph are not equal because the structure of each flow graph is different. The signal-flow graphs can be recognized as being equivalent to an analog computer diagram. Furthermore, we recognize that the initial conditions of the system can be represented by the initial conditions of the integrators, $x_1(0), x_2(0), \ldots, x_n(0)$. Let us consider a control system and determine the state vector differential equation by utilizing the two forms of flow-graph state models.

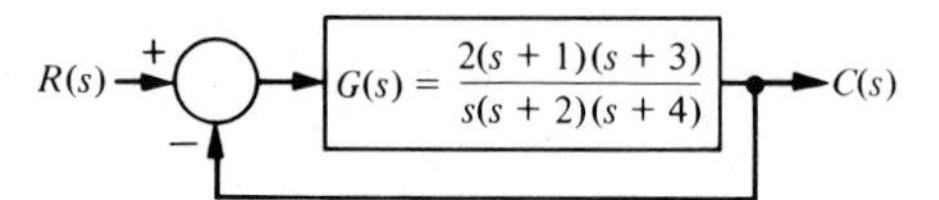

Figure 9.10. Single-loop control system.

■ Example 9.1 Third-order system

A single-loop control system is shown in Fig. 9.10. The closed-loop transfer function of the system is

$$T(s) = \frac{C(s)}{R(s)} = \frac{2s^2 + 8s + 6}{s^3 + 8s^2 + 16s + 6}. \tag{9.44}$$

Multiplying the numerator and denominator by s^{-3}, we have

$$T(s) = \frac{C(s)}{R(s)} = \frac{2s^{-1} + 8s^{-2} + 6s^{-3}}{1 + 8s^{-1} + 16s^{-2} + 6s^{-3}}. \tag{9.45}$$

The signal-flow graph state model using the feedforward of the state variables to provide the output signal is shown in Fig. 9.11. The vector differential equation for this flow graph is

$$\dot{\mathbf{x}} = \begin{bmatrix} 0 & 1 & 0 \\ 0 & 0 & 1 \\ -6 & -16 & -8 \end{bmatrix} \mathbf{x} + \begin{bmatrix} 0 \\ 0 \\ 1 \end{bmatrix} u(t), \tag{9.46}$$

and the output is

$$c(t) = [6, 8, 2] \begin{bmatrix} x_1 \\ x_2 \\ x_3 \end{bmatrix}. \tag{9.47}$$

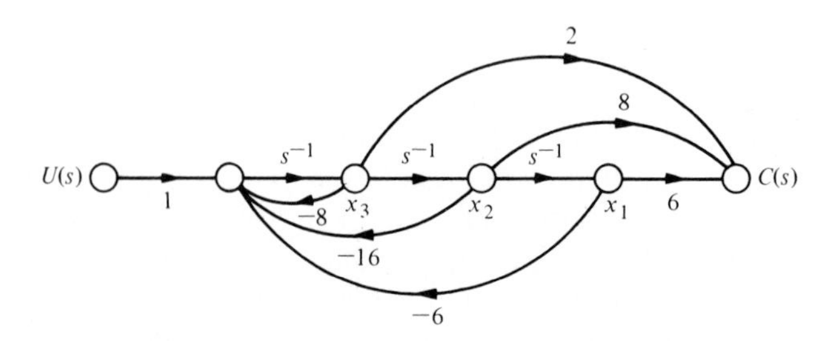

Figure 9.11. Flow-graph state model for $T(s)$.

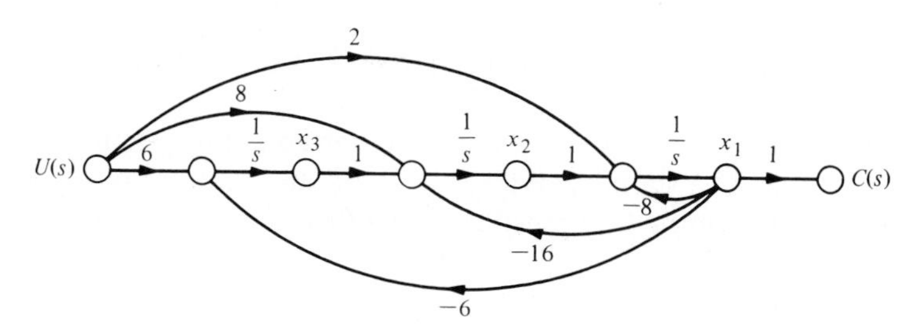

Figure 9.12. Alternative flow-graph state model for $T(s)$.

The flow-graph state model using the feedforward of the input variable is shown in Fig. 9.12. The vector differential equation for this flow graph is

$$\dot{\mathbf{x}} = \begin{bmatrix} -8 & 1 & 0 \\ -16 & 0 & 1 \\ -6 & 0 & 0 \end{bmatrix} \mathbf{x} + \begin{bmatrix} 2 \\ 8 \\ 6 \end{bmatrix} u(t), \tag{9.48}$$

and the output is $c(t) = x_1(t)$.

We note that both signal-flow graph representations of the transfer function $T(s)$ are readily obtained. Furthermore, it was not necessary to factor the numerator or denominator polynomial in order to obtain the state differential equations. Avoiding the factoring of polynomials permits us to avoid the tedious effort involved. Each of the two signal-flow graph state models represents an analog computer simulation of the transfer function. Both models require three integrators because the system is third order. However, it is important to emphasize that the state variables of the state model of Fig. 9.11 are not identical to the state variables of the state model of Fig. 9.12. Of course, one set of state variables is related to the other set of state variables by a suitable linear transformation of variables. A linear matrix transformation is represented by $\mathbf{y} = \mathbf{Bx}$, which transforms the $\mathbf{x}$-vector into the $\mathbf{y}$-vector by means of the $\mathbf{B}$-matrix (see Appendix C, especially Section C-3, for an introduction to matrix algebra). Finally, we note that the transfer function of Eq. (9.31) represents a single output linear constant coefficient system, and thus the transfer function can represent an nth-order differential equation

$$\frac{d^n c}{dt^n} + a_{n-1}\frac{d^{n-1}c}{dt^{n-1}} + \cdots + a_0 c(t) = \frac{d^m u}{dt^m} + b_{m-1}\frac{d^{m-1}u}{dt^{m-1}} + \cdots + b_0 u(t). \tag{9.49}$$

Thus we can obtain the n first-order equations for the nth-order differential equation by utilizing the signal-flow graph state models of this section.

9.5 The Stability of Systems in the Time Domain

The stability of a system modeled by a state variable flow-graph model can be readily ascertained. The stability of a system with an input–output transfer function $T(s)$ can be determined by examining the denominator polynomial of $T(s)$. Therefore if the transfer function is written as

$$T(s) = \frac{p(s)}{q(s)},$$

where $p(s)$ and $q(s)$ are polynomials in s, then the stability of the system is represented by the roots of $q(s)$. The polynomial $q(s)$, when set equal to zero, is called the characteristic equation and is discussed in Section 2.4. The roots of the characteristic equation must lie in the left-hand s-plane for the system to exhibit a stable time response. Therefore in order to ascertain the stability of a system represented by a transfer function, we investigate the characteristic equation and utilize the Routh-Hurwitz criterion. If the system we are investigating is represented by a signal-flow graph state model, we obtain the characteristic equation by evaluating the flow graph determinant. As an illustration of this method, let us investigate the stability of the system of Example 9.1.

■ Example 9.2 Stability of a system

The transfer function $T(s)$ examined in Example 9.1 is

$$T(s) = \frac{2s^2 + 8s + 6}{s^3 + 8s^2 + 16s + 6}. \tag{9.50}$$

Clearly, the characteristic equation for this system is

$$q(s) = s^3 + 8s^2 + 16s + 6. \tag{9.51}$$

Of course, this characteristic equation is also readily obtained from either the flow-graph model shown in Fig. 9.7 or the one shown in Fig. 9.9. Using the Routh-Hurwitz criterion, we find that the system is stable and all the roots of $q(s)$ lie in the left-hand s-plane.

We often determine the flow-graph state model directly from a set of state differential equations. In this case, we can use the flow graph directly to determine the stability of the system by obtaining the characteristic equation from the flow-graph determinant $\Delta(s)$. An illustration of this approach will aid in comprehending this method.

■ Example 9.3 Spread of an epidemic disease

The spread of an epidemic disease can be described by a set of differential equations. The population under study is made up of three groups, x_1, x_2, and x_3, such that the group x_1 is susceptible to the epidemic disease, group x_2 is infected with the disease, and group x_3 has been removed from x_1 and x_2. The removal of x_3 will be due to immunization, death, or isolation from x_1. The feedback system can be represented by the following equations:

$$\frac{dx_1}{dt} = -\alpha x_1 - \beta x_2 + u_1(t), \tag{9.52}$$

$$\frac{dx_2}{dt} = \beta x_1 - \gamma x_2 + u_2(t), \tag{9.53}$$

$$\frac{dx_3}{dt} = \alpha x_1 + \gamma x_2. \tag{9.54}$$

The rate at which new susceptibles are added to the population is equal to $u_1(t)$ and the rate at which new infectives are added to the population is equal to $u_2(t)$. For a closed population, we have $u_1(t) = u_2(t) = 0$. It is interesting to note that these equations could equally well represent the spread of information of a new idea through a populace.

The state variables for this system are x_1, x_2, and x_3. The signal-flow diagram that represents this set of differential equations is shown in Fig. 9.13. The vector differential equation is equal to

$$\frac{d}{dt}\begin{bmatrix} x_1 \\ x_2 \\ x_3 \end{bmatrix} = \begin{bmatrix} -\alpha & -\beta & 0 \\ \beta & -\gamma & 0 \\ \alpha & \gamma & 0 \end{bmatrix} \begin{bmatrix} x_1 \\ x_2 \\ x_3 \end{bmatrix} + \begin{bmatrix} 1 & 0 \\ 0 & 1 \\ 0 & 0 \end{bmatrix} \begin{bmatrix} u_1(t) \\ u_2(t) \end{bmatrix}. \tag{9.55}$$

By examining Eq. (9.55) and the signal-flow graph, we find that the state variable x_3 is dependent on x_1 and x_2 and does not affect the variables x_1 and x_2.

Let us consider a closed population so that $u_1(t) = u_2(t) = 0$. The equilibrium point in the state space for this system is obtained by setting $d\mathbf{x}/dt = \mathbf{0}$. The equilibrium point in the state space is the point to which the system settles in the equilibrium, or rest, condition. Examining Eq. (9.55), we find that the equilibrium

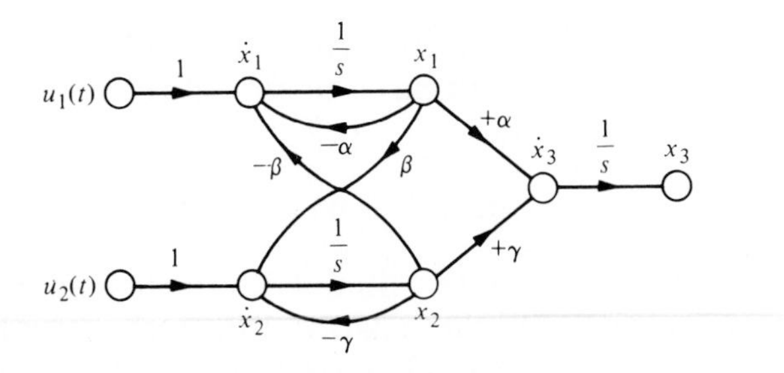

Figure 9.13. State model flow graph for the spread of an epidemic disease.

point for this system is $x_1 = x_2 = 0$. Thus, in order to determine if the system is stable and the epidemic disease is eliminated from the population, we must obtain the characteristic equation of the system. From the signal-flow graph shown in Fig. 9.13, we obtain the flow graph determinant

$$\Delta(s) = 1 - (-\alpha s^{-1} - \gamma s^{-1} - \beta^2 s^{-2}) + (\alpha\gamma s^{-2}), \tag{9.56}$$

where there are three loops, two of which are nontouching. Thus the characteristic equation is

$$q(s) = s^2\,\Delta(s) = s^2 + (\alpha + \gamma)s + (\alpha\gamma + \beta^2) = 0. \tag{9.57}$$

Examining Eq. (9.57), we find that this system is stable when $(\alpha + \gamma) > 0$ and $(\alpha\gamma + \beta^2) > 0$.

A method of obtaining the characteristic equation directly from the vector differential equation is based on the fact that the solution to the unforced system is an exponential function. The vector differential equation without input signals is

$$\dot{\mathbf{x}} = \mathbf{A}\mathbf{x}, \tag{9.58}$$

where $\mathbf{x}$ is the state vector. The solution is of exponential form, and we can define a constant λ such that the solution of the system for one state might be $x_i(t) = k_i e^{\lambda_i t}$. The λ_i are called the characteristic roots of the system, which are simply the roots of the characteristic equation. If we let $\mathbf{x} = \mathbf{c}e^{\lambda t}$ and substitute into Eq. (9.58), we have

$$\lambda\mathbf{c}e^{\lambda t} = \mathbf{A}\mathbf{c}e^{\lambda t} \tag{9.59}$$

or

$$\lambda\mathbf{x} = \mathbf{A}\mathbf{x}. \tag{9.60}$$

Equation (9.60) can be rewritten as

$$(\lambda\mathbf{I} - \mathbf{A})\mathbf{x} = \mathbf{0}, \tag{9.61}$$

where $\mathbf{I}$ equals the identity matrix and $\mathbf{0}$ equals the null matrix. The solution of this set of simultaneous equations has a nontrivial solution if and only if the determinant vanishes; that is, only if

$$\det(\lambda\mathbf{I} - \mathbf{A}) = 0. \tag{9.62}$$

The nth-order equation in λ resulting from the evaluation of this determinant is the characteristic equation, and the stability of the system can be readily ascertained. Let us reconsider Example 9.3 in order to illustrate this approach.

■ Example 9.4 Closed epidemic system

The vector differential equation of the epidemic system is given in Eq. (9.55). The characteristic equation is then

$$\det(\lambda\mathbf{I} - \mathbf{A}) = \det\left\{\begin{bmatrix} \lambda & 0 & 0 \\ 0 & \lambda & 0 \\ 0 & 0 & \lambda \end{bmatrix} - \begin{bmatrix} -\alpha & -\beta & 0 \\ \beta & -\gamma & 0 \\ \alpha & \gamma & 0 \end{bmatrix}\right\}$$

$$= \det\begin{bmatrix} (\lambda + \alpha) & \beta & 0 \\ -\beta & (\lambda + \gamma) & 0 \\ -\alpha & -\gamma & \lambda \end{bmatrix} \tag{9.63}$$

$$= \lambda[(\lambda + \alpha)(\lambda + \gamma) + \beta^2]$$

$$= \lambda[\lambda^2 + (\alpha + \gamma)\lambda + (\alpha\gamma + \beta^2)] = 0.$$

Thus we obtain the characteristic equation of the system, and it is similar to that obtained in Eq. (9.57) by flow-graph methods. The additional root $\lambda = 0$ results from the definition of x_3 as the integral of $(\alpha x_1 + \gamma x_2)$, and x_3 does not affect the other state variables. Thus the root $\lambda = 0$ indicates the integration connected with x_3. The characteristic equation indicates that the system is stable when $(\alpha + \gamma) > 0$ and $(\alpha\gamma + \beta^2) > 0$.

■ Example 9.5 Inverted pendulum stabilty

The problem of balancing a broomstick on the end of one's finger is not unlike the problem of controlling the attitude of a missile during the initial stages of launch. This problem is the classic and intriguing problem of the inverted pendulum mounted on a cart, as shown in Fig. 9.14. The cart must be moved so that mass m is always in an upright position. The state variables must be expressed

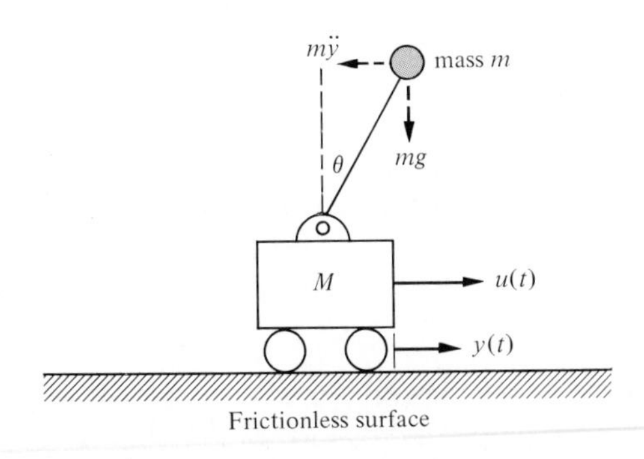

Figure 9.14. A cart and an inverted pendulum.

in terms of the angular rotation $\theta(t)$ and the position of the cart $y(t)$. The differential equations describing the motion of the system can be obtained by writing the sum of the forces in the horizontal direction and the sum of the moments about the pivot point [2, 3, 15]. We will assume that $M \gg m$ and the angle of rotation θ is small so that the equations are linear. The sum of the forces in the horizontal direction is

$$M\ddot{y} + ml\ddot{\theta} - u(t) = 0, \tag{9.64}$$

where $u(t)$ equals the force on the cart and l is the distance from the mass m to the pivot point. The sum of the torques about the pivot point is

$$ml\ddot{y} + ml^2\ddot{\theta} - mlg\theta = 0. \tag{9.65}$$

The state variables for the two second-order equations are chosen as $(x_1, x_2, x_3, x_4) = (y, \dot{y}, \theta, \dot{\theta})$. Then Eqs. (9.64) and (9.65) are written in terms of the state variables as

$$M\dot{x}_2 + ml\dot{x}_4 - u(t) = 0 \tag{9.66}$$

and

$$\dot{x}_2 + l\dot{x}_4 - gx_3 = 0. \tag{9.67}$$

In order to obtain the necessary first-order differential equations, we solve for $l\dot{x}_4$ in Eq. (9.67) and substitute into Eq. (9.66) to obtain

$$M\dot{x}_2 + mgx_3 = u(t), \tag{9.68}$$

since $M \gg m$. Substituting $\dot{x}_2$ from Eq. (9.66) into Eq. (9.67), we have

$$Ml\dot{x}_4 - Mgx_3 + u(t) = 0. \tag{9.69}$$

Therefore the four first-order differential equations can be written as

$$\begin{aligned}
\dot{x}_1 &= x_2, \\
\dot{x}_2 &= \frac{-mg}{M}x_3 + \frac{1}{M}u(t), \\
\dot{x}_3 &= x_4, \\
\dot{x}_4 &= \frac{g}{l}x_3 - \frac{1}{Ml}u(t).
\end{aligned} \tag{9.70}$$

Thus the system matrix is

$$\mathbf{A} = \begin{bmatrix} 0 & 1 & 0 & 0 \\ 0 & 0 & -(mg/M) & 0 \\ 0 & 0 & 0 & 1 \\ 0 & 0 & g/l & 0 \end{bmatrix}. \tag{9.71}$$

The characteristic equation can be obtained from the determinant of $(\lambda\mathbf{I} - \mathbf{A})$ as follows:

$$\det \begin{bmatrix} \lambda & -1 & 0 & 0 \\ 0 & \lambda & m/Mg & 0 \\ 0 & 0 & \lambda & -1 \\ 0 & 0 & -(g/l) & \lambda \end{bmatrix} = \lambda\left[\lambda\left(\lambda^2 - \frac{g}{l}\right)\right] = \lambda^2\left(\lambda^2 - \frac{g}{l}\right) = 0. \quad (9.72)$$

The characteristic equation indicates that there are two roots at $\lambda = 0$, a root at $\lambda = +\sqrt{g/l}$, and a root at $\lambda = -\sqrt{g/l}$. Clearly, the system is unstable, because there is a root in the right-hand plane at $\lambda = +\sqrt{g/l}$.

A control system can be designed so that if $u(t)$ is a function of the state variables, a stable system will result. The design of a stable feedback control system is based on a suitable selection of a feedback system structure. Therefore, considering the control of the cart and the unstable inverted pendulum shown in Fig. 9.14, we must measure and utilize the state variables of the system in order to control the cart. Thus if we desire to measure the state variable $x_3 = \theta$, we could use a potentiometer connected to the shaft of the pendulum hinge. Similarly, we could measure the rate of change of the angle, $x_4 = \dot{\theta}$, by using a tachometer generator. The state variables, x_1 and x_2, which are the position and velocity of the cart, can also be measured by suitable sensors. If the state variables are all measured, then they can be utilized in a feedback controller so that $u = \mathbf{hx}$, where $\mathbf{h}$ is the feedback matrix. The state vector $\mathbf{x}$ represents the state of the system; therefore, knowledge of $\mathbf{x}(t)$ and the equations describing the system dynamics provide sufficient information for control and stabilization of a system. This design approach is called *state-variable feedback* [4, 7]. In order to illustrate the utilization of state-variable feedback, let us reconsider the unstable portion of the inverted pendulum system and design a suitable state-variable feedback control system.

■ Example 9.6 Inverted pendulum control

In order to investigate the unstable portion of the inverted pendulum system, let us consider a reduced system. If we assume that the control signal is an acceleration signal and that the mass of the cart is negligible, we can focus on the unstable dynamics of the pendulum. When $u(t)$ is an acceleration signal, Eq. (9.67) becomes

$$gx_3 - l\dot{x}_4 = \dot{x}_2 = \ddot{y} = u(t). \quad (9.73)$$

For the reduced system, where the control signal is an acceleration signal, the position and velocity of the cart are integral functions of $u(t)$. The portion of the state vector under consideration is $[x_3, x_4] = [\theta, \dot{\theta}]$. Thus the state vector differential equation reduces to

$$\frac{d}{dt}\begin{bmatrix} x_3 \\ x_4 \end{bmatrix} = \begin{bmatrix} 0 & 1 \\ g/l & 0 \end{bmatrix}\begin{bmatrix} x_3 \\ x_4 \end{bmatrix} + \begin{bmatrix} 0 \\ -(1/l) \end{bmatrix} u(t). \quad (9.74)$$

Clearly, the **A** matrix of Eq. (9.74) is simply the lower right-hand portion of the **A** matrix of Eq. (9.71) and the system has the characteristic equation $[\lambda^2 - (g/l)]$ with one root in the right-hand s-plane. In order to stabilize the system, we generate a control signal that is a function of the two state variables x_3 and x_4. Then we have

$$\begin{aligned} u(t) &= \mathbf{hx} \\ &= [h_1, h_2]\begin{bmatrix} x_3 \\ x_4 \end{bmatrix} \\ &= h_1x_3 + h_2x_4. \end{aligned} \tag{9.75}$$

Substituting this control signal relationship into Eq. (9.74), we have

$$\begin{bmatrix} \dot{x}_3 \\ \dot{x}_4 \end{bmatrix} = \begin{bmatrix} 0 & 1 \\ g/l & 0 \end{bmatrix}\begin{bmatrix} x_3 \\ x_4 \end{bmatrix} + \begin{bmatrix} 0 \\ -(1/l)(h_1x_3 + h_2x_4) \end{bmatrix}. \tag{9.76}$$

Combining the two additive terms on the right side of the equation, we find

$$\begin{bmatrix} \dot{x}_3 \\ \dot{x}_4 \end{bmatrix} = \begin{bmatrix} 0 & 1 \\ 1/l(g - h_1) & -(h_2/l) \end{bmatrix}\begin{bmatrix} x_3 \\ x_4 \end{bmatrix}. \tag{9.77}$$

Therefore, obtaining the characteristic equation, we have

$$\begin{aligned} \det\begin{bmatrix} +\lambda & (-1) \\ -(1/l)(g - h_1) & (\lambda + h_2/l) \end{bmatrix} &= \lambda\left(\lambda + \frac{h_2}{l}\right) - \frac{1}{l}(g - h_1) \\ &= \lambda^2 + \left(\frac{h_2}{l}\right)\lambda + \frac{1}{l}(h_1 - g). \end{aligned} \tag{9.78}$$

Thus for the system to be stable, we require that $(h_2/l) > 0$ and $h_1 > g$. Hence we have stabilized an unstable system by measuring the state variables x_3 and x_4 and using the control function $u = h_1x_3 + h_2x_4$ to obtain a stable system.

In this section we have developed a useful method of investigating the stability of a system represented by a state vector differential equation. Furthermore, we have established an approach for the design of a feedback control system by using the state variables as the feedback variables in order to increase the stability of the system. In the next two sections, we will be concerned with developing two methods for obtaining the time response of the state variables.

9.6 The Time Response and the Transition Matrix

It is often desirable to obtain the time response of the state variables of a control system and thus examine the performance of the system. The transient response of a system can be readily obtained by evaluating the solution to the state vector differential equation. We found in Section 9.3 that the solution for the state vector differential equation (Eq. 9.34) was

$$\mathbf{x}(t) = \boldsymbol{\phi}(t)\mathbf{x}(0) + \int_0^t \boldsymbol{\phi}(t - \tau)\mathbf{Bu}(\tau)\, d\tau. \tag{9.79}$$

Clearly, if the initial conditions $\mathbf{x}(0)$, the input $\mathbf{u}(\tau)$, and the transition matrix $\phi(t)$ are known, the time response of $\mathbf{x}(t)$ can be numerically evaluated. Thus the problem focuses on the evaluation of $\phi(t)$, the transition matrix that represents the response of the system. Fortunately, the transition matrix can be readily evaluated by using the signal-flow graph techniques with which we are already familiar.

However, before proceeding to the evaluation of the transition matrix using signal-flow graphs, we should note that several other methods exist for evaluating the transition matrix, such as the evaluation of the exponential series

$$\phi(t) = e^{At} = \sum_{k=0}^{\infty} \frac{\mathbf{A}^k t^k}{k!}$$

in a truncated form [7, 14]. Several efficient methods exist for the evaluation of $\phi(t)$ by means of a computer algorithm.

A series of computer programs to assist in the calculation of the state variable response of systems is often available in local computer centers. Personal computers can be used to obtain the state variable response of a dynamic system [14]. The personal computer affords the user the advantage of interactive use of the computer for design purposes. Control System Design Program uses a truncated series to solve Eq. (9.16) and yields the state variable response, and it is available for a personal computer [14].

In addition, we found in Eq. (9.23) that $\phi(s) = [s\mathbf{I} - \mathbf{A}]^{-1}$. Therefore, if $\phi(s)$ is obtained by completing the matrix inversion, we can obtain $\phi(t)$ by noting that $\phi(t) = \mathcal{L}^{-1}\{\phi(s)\}$. However, the matrix inversion process is unwieldy for higher-order systems.

The usefulness of the signal-flow graph state model for obtaining the transition matrix becomes clear upon consideration of the Laplace transformation version of Eq. (9.79) when the input is zero. Taking the Laplace transformation of Eq. (9.79), when $\mathbf{u}(\tau) = 0$, we have

$$\mathbf{X}(s) = \phi(s)\mathbf{x}(0). \tag{9.80}$$

Therefore we can evaluate the Laplace transform of the transition matrix from the signal-flow graph by determining the relation between a state variable $X_i(s)$ and the state initial conditions $[x_1(0), x_2(0), \ldots, x_n(0)]$. Then the state transition matrix is simply the inverse transform of $\phi(s)$; that is,

$$\phi(t) = \mathcal{L}^{-1}\{\phi(s)\}. \tag{9.81}$$

The relationship between a state variable $X_i(s)$ and the initial conditions $\mathbf{x}(0)$ is obtained by using Mason's gain formula. Thus for a second-order system, we would have

$$\begin{aligned} X_1(s) &= \phi_{11}(s)x_1(0) + \phi_{12}(s)x_2(0), \\ X_2(s) &= \phi_{21}(s)x_1(0) + \phi_{22}(s)x_2(0), \end{aligned} \tag{9.82}$$

and the relation between $X_2(s)$ as an output and $x_1(0)$ as an input can be evaluated by Mason's formula. All the elements of the transition matrix, $\phi_{ij}(s)$, can be obtained by evaluating the individual relationships between $X_i(s)$ and $x_j(0)$ from the state model flow graph. An example will illustrate this approach to the determination of the transition matrix.

■ Example 9.7 Evaluation of the transition matrix

The signal-flow graph state model of the *RLC* network of Fig. 9.3 is shown in Fig. 9.4. This *RLC* network, which was discussed in Sections 9.3 and 9.4, can be represented by the state variables $x_1 = v_c$ and $x_2 = i_L$. The initial conditions, $x_1(0)$ and $x_2(0)$, represent the initial capacitor voltage and inductor current, respectively. The flow graph, including the initial conditions of each state variable, is shown in Fig. 9.15. The initial conditions appear as the initial value of the state variable at the output of each integrator.

In order to obtain $\phi(s)$, we set $U(s) = 0$. When $R = 3$, $L = 1$, and $C = ½$, we obtain the signal-flow graph shown in Fig. 9.16, where the output and input nodes are deleted because they are not involved in the evaluation of $\phi(s)$. Then, using Mason's gain formula, we obtain $X_1(s)$ in terms of $x_1(0)$ as

$$X_1(s) = \frac{1 \cdot \Delta_1(s) \cdot [x_1(0)/s]}{\Delta(s)}, \tag{9.83}$$

where $\Delta(s)$ is the graph determinant and $\Delta_1(s)$ is the path cofactor. The graph determinant is

$$\Delta(s) = 1 + 3s^{-1} + 2s^{-2}.$$

The path cofactor is $\Delta_1 = 1 + 3s^{-1}$ because the path between $x_1(0)$ and $X_1(s)$ does not touch the loop with the factor $-3s^{-1}$. Therefore the first element of the transition matrix is

$$\begin{aligned}\phi_{11}(s) &= \frac{(1 + 3s^{-1})(1/s)}{1 + 3s^{-1} + 2s^{-2}} \\ &= \frac{(s + 3)}{(s^2 + 3s + 2)}.\end{aligned} \tag{9.84}$$

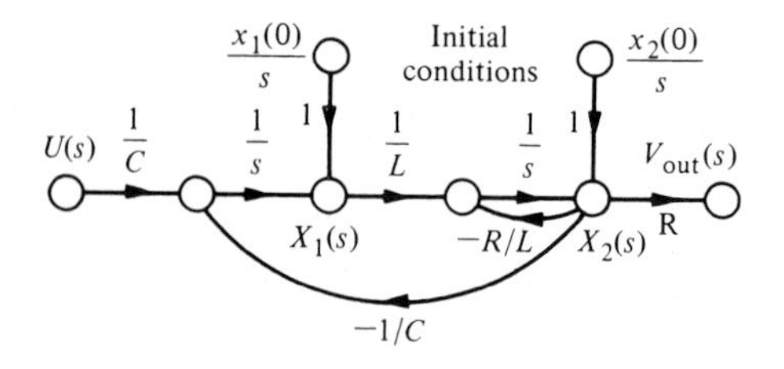

Figure 9.15. Flow graph of the *RLC* network.

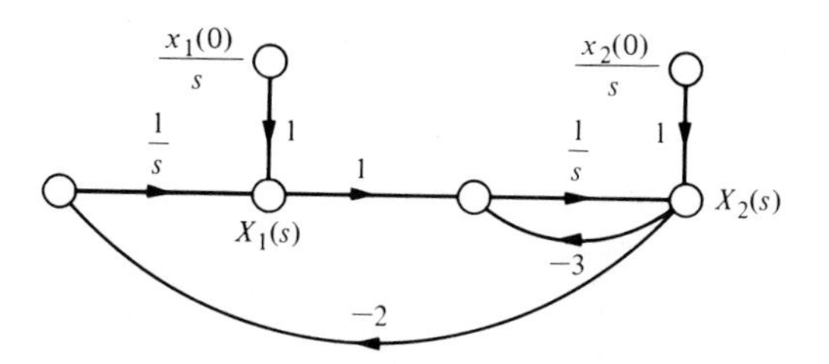

Figure 9.16. Flow graph of the *RLC* network with $U(s) = 0$.

The element $\phi_{12}(s)$ is obtained by evaluating the relationship between $X_1(s)$ and $x_2(0)$ as

$$X_1(s) = \frac{(-2s^{-1})(x_2(0)/s)}{1 + 3s^{-1} + 2s^{-2}}. \tag{9.85}$$

Therefore we obtain

$$\phi_{12}(s) = \frac{-2}{s^2 + 3s + 2}. \tag{9.86}$$

Similarly, for $\phi_{21}(s)$ we have

$$\begin{aligned} \phi_{21}(s) &= \frac{(s^{-1})(1/s)}{1 + 3s^{-1} + 2s^{-2}} \\ &= \frac{1}{s^2 + 3s + 2}. \end{aligned} \tag{9.87}$$

Finally, for $\phi_{22}(s)$ we obtain

$$\begin{aligned} \phi_{22}(s) &= \frac{1(1/s)}{1 + 3s^{-1} + 2s^{-2}} \\ &= \frac{s}{s^2 + 3s + 2}. \end{aligned} \tag{9.88}$$

Therefore, the transition matrix in Laplace transformation forms is

$$\boldsymbol{\phi}(s) = \begin{bmatrix} (s + 3)/(s^2 + 3s + 2) & -2/(s^2 + 3s + 2) \\ 1/(s^2 + 3s + 2) & s/(s^2 + 3s + 2) \end{bmatrix}. \tag{9.89}$$

The factors of the characteristic equation are $(s + 1)$ and $(s + 2)$ so that

$$(s + 1)(s + 2) = s^2 + 3s + 2.$$

Then the transition matrix is

$$\boldsymbol{\phi}(t) = \mathcal{L}^{-1}\{\boldsymbol{\phi}(s)\} = \begin{bmatrix} (2e^{-t} - e^{-2t}) & (-2e^{-t} + 2e^{-2t}) \\ (e^{-t} - e^{-2t}) & (-e^{-t} + 2e^{-2t}) \end{bmatrix}. \tag{9.90}$$

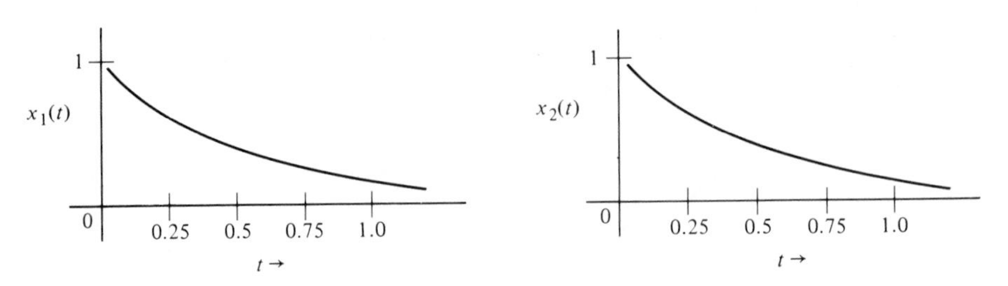

Figure 9.17. Time response of the state variables of the *RLC* network for $x_1(0) = x_2(0) = 1$.

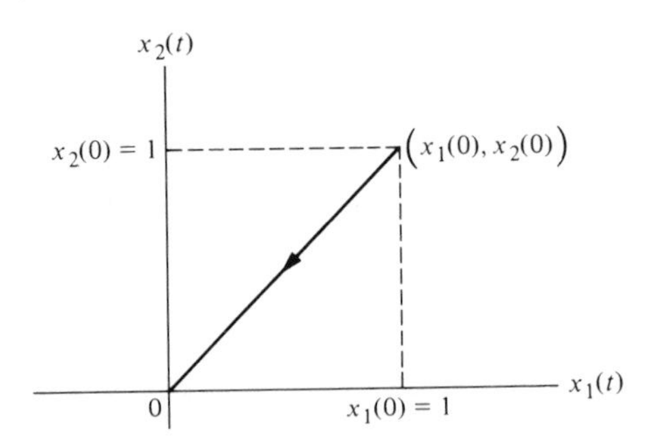

Figure 9.18. Trajectory of the state vector in the (x_1 vs. x_2)-plane.

The evaluation of the time response of the *RLC* network to various initial conditions and input signals can now be evaluated by utilizing Eq. (9.79). For example, when $x_1(0) = x_2(0) = 1$ and $u(t) = 0$, we have

$$\begin{bmatrix} x_1(t) \\ x_2(t) \end{bmatrix} = \phi(t) \begin{bmatrix} 1 \\ 1 \end{bmatrix} = \begin{bmatrix} e^{-2t} \\ e^{-2t} \end{bmatrix}. \tag{9.91}$$

The response of the system for these initial conditions is shown in Fig. 9.17. The trajectory of the state vector $[x_1(t), x_2(t)]$ on the (x_1 vs. x_2)-plane is shown in Fig. 9.18.

The evaluation of the time response is facilitated by the determination of the transition matrix. Although this approach is limited to linear systems, it is a powerful method and utilizes the familiar signal-flow graph to evaluate the transition matrix.

9.7 A Discrete-Time Evaluation of the Time Response

The response of a system represented by a state vector differential equation can be obtained by utilizing a *discrete-time approximation*. The discrete-time approximation is based on the division of the time axis into sufficiently small time increments. Then the values of the state variables are evaluated at the successive time intervals; that is, $t = 0, T, 2T, 3T, \ldots,$ where T is the increment of time $\Delta t = T$. This approach is a familiar method utilized in numerical analysis and digital computer numerical methods. If the time increment, T, is sufficiently small compared with the time constants of the system, the response evaluated by discrete-time methods will be reasonably accurate.

The linear state vector differential equation is written as

$$\dot{\mathbf{x}} = \mathbf{A}\mathbf{x} + \mathbf{B}\mathbf{u}. \tag{9.92}$$

The basic definition of a derivative is

$$\dot{\mathbf{x}}(t) = \lim_{\Delta t \to 0} \frac{\mathbf{x}(t + \Delta t) - \mathbf{x}(t)}{\Delta t}. \tag{9.93}$$

Therefore we can utilize this definition of the derivative and determine the value of $\mathbf{x}(t)$ when t is divided in small intervals $\Delta t = T$. Thus, approximating the derivative as

$$\dot{\mathbf{x}} = \frac{\mathbf{x}(t + T) - \mathbf{x}(t)}{T}, \tag{9.94}$$

we substitute into Eq. (9.92) to obtain

$$\frac{\mathbf{x}(t + T) - \mathbf{x}(t)}{T} = \mathbf{A}\mathbf{x}(t) + \mathbf{B}\mathbf{u}(t). \tag{9.95}$$

Solving for $\mathbf{x}(t + T)$, we have

$$\begin{aligned} \mathbf{x}(t + T) &= T\mathbf{A}\mathbf{x}(t) + \mathbf{x}(t) + T\mathbf{B}\mathbf{u}(t) \\ &= (T\mathbf{A} + \mathbf{I})\mathbf{x}(t) + T\mathbf{B}\mathbf{u}(t), \end{aligned} \tag{9.96}$$

where t is divided into intervals of width T. Therefore the time t is written as $t = kT$, where k is an integer index so that $k = 0, 1, 2, 3, \ldots$. Then Eq. (9.96) is written as

$$\mathbf{x}[(k + 1)T] = (T\mathbf{A} + \mathbf{I})\mathbf{x}(kT) + T\mathbf{B}\mathbf{u}(kT). \tag{9.97}$$

Therefore the value of the state vector at the $(k + 1)$st time instant is evaluated in terms of the values of $\mathbf{x}$ and $\mathbf{u}$ at the kth time instant. Equation (9.97) can be rewritten as

$$\mathbf{x}(k + 1) = \psi(t)\mathbf{x}(k) + T\mathbf{B}\mathbf{u}(k), \tag{9.98}$$

where $\psi(T) = (T\mathbf{A} + \mathbf{I})$ and the symbol T is omitted from the arguments of the variables. Equation (9.98) clearly relates the resulting operation for obtaining $\mathbf{x}(t)$ by evaluating the discrete-time approximation $\mathbf{x}(k + 1)$ in terms of the previous value $\mathbf{x}(k)$. This recurrence operation is a sequential series of calculations and is very suitable for digital computer calculation. In order to illustrate this approximate approach, let us reconsider the evaluation of the response of the RLC network of Fig. 9.3.

■ Example 9.8 Response of RLC network

We shall evaluate the time response of the RLC network without determining the transition matrix by using the discrete-time approximation. Therefore, as in Example 9.7, we will let $R = 3$, $L = 1$, and $C = ½$. Then, as we found in Eqs. (9.27) and (9.28), the state vector differential equation is

$$\dot{\mathbf{x}} = \begin{bmatrix} 0 & -(1/C) \\ 1/L & -(R/L) \end{bmatrix} \mathbf{x} + \begin{bmatrix} +(1/C) \\ 0 \end{bmatrix} u(t)$$
$$= \begin{bmatrix} 0 & -2 \\ 1 & -3 \end{bmatrix} \mathbf{x} + \begin{bmatrix} +2 \\ 0 \end{bmatrix} u(t). \tag{9.99}$$

Now we must choose a sufficiently small time interval T so that the approximation of the derivative (Eq. 9.94) is reasonably accurate. A suitable interval T must be chosen so that the solution to Eq. (9.97) is stable. Usually we choose T to be less than one-half of the smallest time constant of the system. Therefore, since the shortest time constant of this system is 0.5 sec [recalling that the characteristic equation is $[(s + 1)(s + 2)]$, we might choose $T = 0.2$. Alternatively, if a digital computer is used for the calculations and the number of calculations is not important, we would choose $T = 0.05$ in order to obtain greater accuracy. However, we note that as we decrease the increment size, the number of calculations increases proportionally if we wish to evaluate the output from 0 to 10 sec, for example. Using $T = 0.2$ sec, Eq. (9.97) is

$$\mathbf{x}(k + 1) = (0.2\mathbf{A} + \mathbf{I})\mathbf{x}(k) + 0.2\mathbf{B}\mathbf{u}(k). \tag{9.100}$$

Therefore

$$\psi(T) = \begin{bmatrix} 1 & -0.4 \\ 0.2 & 0.4 \end{bmatrix} \tag{9.101}$$

and

$$T\mathbf{B} = \begin{bmatrix} +0.4 \\ 0 \end{bmatrix}. \tag{9.102}$$

Now let us evaluate the response of the system when $x_1(0) = x_2(0) = 1$ and $u(t) = 0$ as in Example 9.7. The response at the first instant, when $t = T$, or $k = 0$, is

$$\mathbf{x}(1) = \begin{bmatrix} 1 & -0.4 \\ 0.2 & 0.4 \end{bmatrix} \mathbf{x}(0) = \begin{bmatrix} 0.6 \\ 0.6 \end{bmatrix}. \tag{9.103}$$

Then the response at the time $t = 2T = 0.4$ sec, or $k = 1$, is

$$\mathbf{x}(2) = \begin{bmatrix} 1 & -0.4 \\ 0.2 & 0.4 \end{bmatrix} \mathbf{x}(1) = \begin{bmatrix} 0.36 \\ 0.36 \end{bmatrix}. \tag{9.104}$$

The value of the response as $k = 2, 3, 4, \ldots$ is then evaluated in a similar manner.

Table 9.1.

Time t	0	0.2	0.4	0.6	0.8
Exact $x_1(t)$	1	0.67	0.448	0.30	0.20
Approximate $x_1(t)$, $T = 0.1$	1	0.64	0.41	0.262	0.168
Approximate $x_1(t)$, $T = 0.2$	1	0.60	0.36	0.216	0.130

Now let us compare the actual response of the system evaluated in the previous section using the transition matrix with the approximate response determined by the discrete-time approximation. We found in Example 9.7 that the exact value of the state variables, when $x_1(0) = x_2(0) = 1$, is $x_1(t) = x_2(t) = e^{-2t}$. Therefore the exact values can be readily calculated and compared with the approximate values of the time response in Table 9.1. The approximate time response values for $T = 0.1$ sec are also given in Table 9.1. The error, when $T = 0.2$, is approximately a constant equal to 0.07, and thus the percentage error compared to the initial value is 7%. When T is equal to 0.1 sec, the percentage error compared to the initial value is approximately 3.5%. If we use $T = 0.05$, the value of the approximation, when time $t = 0.2$ sec, is $x_1(t) = 0.655$, and the error has been reduced to 1.5% of the initial value.

Therefore, if one is using a digital computer to evaluate the transient response by evaluating the discrete-time response equations, a value of T equal to one-tenth of the smallest time constant of the system would be selected. The value of this method merits another illustration of the evaluation of the time response of a system.

■ Example 9.9 Time response of an epidemic

Let us reconsider the state variable representation of the spread of an epidemic disease presented in Example 9.3. The state vector differential equation was given in Eq. (9.55). When the constants are $\alpha = \beta = \gamma = 1$, we have

$$\dot{\mathbf{x}} = \begin{bmatrix} -1 & -1 & 0 \\ 1 & -1 & 0 \\ 1 & 1 & 0 \end{bmatrix} \mathbf{x} + \begin{bmatrix} 1 & 0 \\ 0 & 1 \\ 0 & 0 \end{bmatrix} \mathbf{u}. \tag{9.105}$$

The characteristic equation of this system, as determined in Eq. (9.57), is $s(s^2 + s + 2) = 0$, and thus the system has complex roots. Let us determine the transient response of the spread of disease when the rate of new susceptibles is zero; that is, when $u_1 = 0$. The rate of adding new infectives is represented by $u_2(0) = 1$ and $u_2(k) = 0$ for $k \geq 1$; that is, one new infective is added at the initial time only (this is equivalent to a pulse input). The time constant of the complex roots is $1/\zeta\omega_n = 2$ sec, and therefore we will use $T = 0.2$ sec. (Note that the actual time units might be months and the units of the input in thousands.)

Then the discrete-time equation is

$$\mathbf{x}(k+1) = \begin{bmatrix} 0.8 & -0.2 & 0 \\ 0.2 & 0.8 & 0 \\ 0.2 & 0.2 & 1 \end{bmatrix} \mathbf{x}(k) + \begin{bmatrix} 0 \\ 0.2 \\ 0 \end{bmatrix} u_2(k). \tag{9.106}$$

Therefore the response at the first instant, $t = T$, is obtained when $k = 0$ as

$$\mathbf{x}(1) = \begin{bmatrix} 0 \\ 0.2 \\ 0 \end{bmatrix}, \tag{9.107}$$

when $x_1(0) = x_2(0) = x_3(0) = 0$. Then the input $u_2(k)$ is zero for $k \geq 1$ and the response at $t = 2T$ is

$$\mathbf{x}(2) = \begin{bmatrix} 0.8 & -0.2 & 0 \\ 0.2 & 0.8 & 0 \\ 0.2 & 0.2 & 1 \end{bmatrix} \begin{bmatrix} 0 \\ 0.2 \\ 0 \end{bmatrix} = \begin{bmatrix} -0.04 \\ 0.16 \\ 0.04 \end{bmatrix}. \tag{9.108}$$

The response at $t = 3T$ is then

$$\mathbf{x}(3) = \begin{bmatrix} 0.8 & -0.2 & 0 \\ 0.2 & 0.8 & 0 \\ 0.2 & 0.2 & 1 \end{bmatrix} \begin{bmatrix} -0.04 \\ 0.16 \\ 0.04 \end{bmatrix} = \begin{bmatrix} -0.064 \\ 0.120 \\ 0.064 \end{bmatrix}, \tag{9.109}$$

and the ensuing values can then be readily evaluated. Of course, the actual physical value of x_1 cannot become negative. The negative value of x_1 is obtained as a result of an inadequate model.

The discrete-time approximate method is particularly useful for evaluating the time response of nonlinear systems. The transition matrix approach is limited to linear systems, but the discrete-time approximation is not limited to linear systems and can be readily applied to nonlinear and time-varying systems. The basic state vector differential equation can be written as

$$\dot{\mathbf{x}} = \mathbf{f}(\mathbf{x}, \mathbf{u}, t), \tag{9.110}$$

where **f** is a function, not necessarily linear, of the state vector **x** and the input vector **u**. The column vector **f** is the column matrix of functions of **x** and **u**. If the system is a linear function of the control signals, Eq. (9.110) becomes

$$\dot{\mathbf{x}} = \mathbf{f}(\mathbf{x}, t) + \mathbf{B}\mathbf{u}. \tag{9.111}$$

If the system is not time-varying—that is, if the coefficients of the differential equation are constants—Eq. (9.111) is then

$$\dot{\mathbf{x}} = \mathbf{f}(\mathbf{x}) + \mathbf{B}\mathbf{u}. \tag{9.112}$$

Let us consider Eq. (9.112) for a nonlinear system and determine the discrete-time approximation. Using Eq. (9.94) as the approximation to the derivative we have

$$\frac{\mathbf{x}(t + T) - \mathbf{x}(t)}{T} = \mathbf{f}(\mathbf{x}(t)) + \mathbf{Bu}(t). \tag{9.113}$$

Therefore, solving for $\mathbf{x}(k + 1)$ when $t = kT$, we obtain

$$\mathbf{x}(k + 1) = \mathbf{x}(k) + T[\mathbf{f}(\mathbf{x}(k)) + \mathbf{Bu}(k)]. \tag{9.114}$$

Similarly, the general discrete-time approximation to Eq. (9.110) is

$$\mathbf{x}(k + 1) = \mathbf{x}(k) + T\mathbf{f}(\mathbf{x}(k), \mathbf{u}(k), k). \tag{9.115}$$

Now let us reconsider the previous example when the system is nonlinear.

■ Example 9.10 Improved model of an epidemic

The spread of an epidemic disease is actually best represented by a set of nonlinear equations as

$$\begin{aligned} \dot{x}_1 &= -\alpha x_1 - \beta x_1 x_2 + u_1(t), \\ \dot{x}_2 &= \beta x_1 x_2 - \gamma x_2 + u_2(t), \\ \dot{x}_3 &= \alpha x_1 + \gamma x_2, \end{aligned} \tag{9.116}$$

where the interaction between the groups is represented by the nonlinear term $x_1 x_2$. Now, the transition matrix approach and the characteristic equation are not applicable because the system is nonlinear. As in the previous example, we will let $\alpha = \beta = \gamma = 1$ and $u_1(t) = 0$. Also, $u_2(0) = 1$ and $u_2(k) = 0$ for $k \geq 1$. We will again select the time increment as $T = 0.2$ sec and the initial conditions as $\mathbf{x}^T(0) = [1, 0, 0]$. Then, substituting the $t = kT$ and

$$\dot{x}_i(k) = \frac{x_i(k + 1) - x_i(k)}{T} \tag{9.117}$$

into Eq. (9.116), we obtain

$$\begin{aligned} \frac{x_1(k + 1) - x_1(k)}{T} &= -x_1(k) - x_1(k)x_2(k), \\ \frac{x_2(k + 1) - x_2(k)}{T} &= +x_1(k)x_2(k) - x_2(k) + u_2(k), \\ \frac{x_3(k + 1) - x_3(k)}{T} &= x_1(k) + x_2(k). \end{aligned} \tag{9.118}$$

Solving these equations for $x_i(k+1)$ and recalling that $T = 0.2$, we have

$$\begin{aligned} x_1(k+1) &= 0.8x_1(k) - 0.2x_1(k)x_2(k), \\ x_2(k+1) &= 0.8x_2(k) + 0.2x_1(k)x_2(k) + 0.2u_2(k), \\ x_3(k+1) &= x_3(k) + 0.2x_1(k) + 0.2x_2(k). \end{aligned} \tag{9.119}$$

Then the response of the first instant, $t = T$, is

$$\begin{aligned} x_1(1) &= 0.8x_1(0) = 0.8, \\ x_2(1) &= 0.2u_2(k) = 0.2, \\ x_3(1) &= 0.2x_1(0) = 0.2. \end{aligned} \tag{9.120}$$

Again, using Eq. (9.119) and noting that $u_2(1) = 0$, we have

$$\begin{aligned} x_1(2) &= 0.8x_1(1) - 0.2x_1(1)x_2(1) = 0.608, \\ x_2(2) &= 0.8x_2(1) + 0.2x_1(1)x_2(1) = 0.192, \\ x_3(2) &= x_3(1) + 0.2x_1(1) + 0.2x_2(1) = 0.40. \end{aligned} \tag{9.121}$$

At the third instant, when $t = 3T$, we obtain

$$\begin{aligned} x_1(3) &= 0.463, \\ x_2(3) &= 0.177, \\ x_3(3) &= 0.56. \end{aligned} \tag{9.122}$$

The evaluation of the ensuing values follows in a similar manner. We note that the response of the nonlinear system differs considerably from the response of the linear model considered in the previous example.

Finally, in order to illustrate the utility of the time-domain approach, let us consider a system that is both nonlinear and time-varying.

■ Example 9.11 Seasonal model of an epidemic

Let us again consider the spread of an epidemic disease which is represented by the nonlinear differential equations of Eq. (9.116) when the coefficient β is time-varying. The time variation, $\beta(t)$, might represent the cyclic seasonal variation of interaction between the susceptible group and the infected group; that is, the interaction between $x_1(t)$ and $x_2(t)$ is greatest in the winter when people are together indoors and is least during the summer. Therefore we represent the time variation of β as

$$\begin{aligned} \beta(t) &= \beta_0 + \sin \omega_0 t \\ &= 1 + \sin\left(\frac{\pi}{2}\right) t. \end{aligned} \tag{9.123}$$

Then, the nonlinear time-varying differential equations are

$$\begin{aligned}\dot{x}_1(t) &= -\alpha_1 x_1(t) - \beta(t)x_1(t)x_2(t) + u_1(t),\\ \dot{x}_2(t) &= \beta(t)x_1(t)x_2(t) - \gamma x_2(t) + u_2(t),\\ \dot{x}_3(t) &= \alpha x_1(t) + \gamma x_2(t).\end{aligned} \tag{9.124}$$

As in Example 9.10, we will let $\alpha = \gamma = 1$, $u_1(t) = 0$, $u_2(0) = 1$ and $u_2(k) = 0$ for $k \geq 1$. Also, we will again select the time increment as $T = 0.2$ sec and the initial conditions as $\mathbf{x}^T(0) = [1, 0, 0]$. Then substituting $t = kT$ and

$$\dot{x}_i(k) = \frac{x_i(k+1) - x_i(k)}{T} \tag{9.125}$$

into Eq. (9.124), we obtain

$$\begin{aligned}\frac{x_1(k+1) - x_1(k)}{T} &= -x_1(k) - \beta(k)x_1(k)x_2(k),\\ \frac{x_2(k+1) - x_2(k)}{T} &= \beta(k)x_1(k)x_2(k) - x_2(k) + u_2(k),\\ \frac{x_3(k+1) - x_3(k)}{T} &= x_1(k) + x_2(k).\end{aligned} \tag{9.126}$$

Solving these equations for $x_i(k+1)$ and again recalling that $T = 0.2$, we obtain

$$\begin{aligned}x_1(k+1) &= 0.8x_1(k) - 0.2\beta(k)x_1(k)x_2(k),\\ x_2(k+1) &= 0.8x_2(k) + 0.2\beta(k)x_1(k)x_2(k) + 0.2u_2(k),\\ x_3(k+1) &= x_3(k) + 0.2x_1(k) + 0.2x_2(k).\end{aligned} \tag{9.127}$$

Then the response at the first instant, $t = T$, is

$$\begin{aligned}x_1(1) &= 0.8x_1(0) = 0.8,\\ x_2(1) &= 0.2u_2(0) = 0.2,\\ x_3(1) &= 0.2x_1(0) = 0.2.\end{aligned} \tag{9.128}$$

Noting that

$$\beta(k) = 1 + \sin(\pi/2)kT = 1 + \sin(0.314k), \tag{9.129}$$

we evaluate the response at the second instant as

$$\begin{aligned}x_1(2) &= 0.8x_1(1) - 0.2\beta(1)x_1(1)x_2(1) = 0.598,\\ x_2(2) &= 0.8x_2(1) + 0.2\beta(1)x_1(1)x_2(1) = 0.202,\\ x_3(2) &= x_3(1) + 0.2x_1(1) + 0.2x_2(1) = 0.40.\end{aligned} \tag{9.130}$$

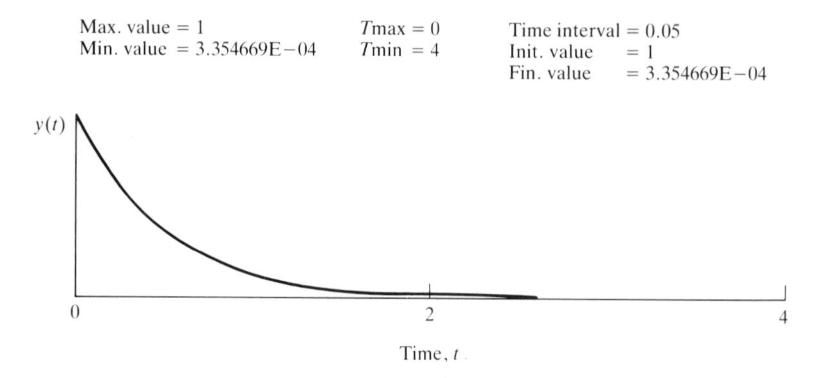

Figure 9.19. The response of x_1 for Example 9.8 using the Control System Design Program.

The evaluation of the ensuing values of the time response follows in a similar manner by using Eq. (6.129) to account for the time-varying parameter.

The evaluation of the time response of the state variables of linear systems is readily accomplished by using either (1) the transition matrix approach or (2) the discrete-time approximation. The transition matrix of linear systems is readily obtained from the signal-flow graph state model. For a nonlinear system, the discrete-time approximation provides a suitable approach, and the discrete-time approximation method is particularly useful if a digital computer is used for numerical calculations.

The Control System Design Program (CSDP) will readily calculate the state response, obtaining

$$x[(n + 1)T] = \phi(T)x(nT) + \mathbf{T}u(nT).$$

For example, the response of x_1 for Example 9.8 is shown in Fig. 9.19 [14].

9.8 Design Example: Automatic Test System

An automatic test and inspection system uses a dc motor to move a set of test probes, as shown in Fig. 9.20. Low throughput and a high degree of error are

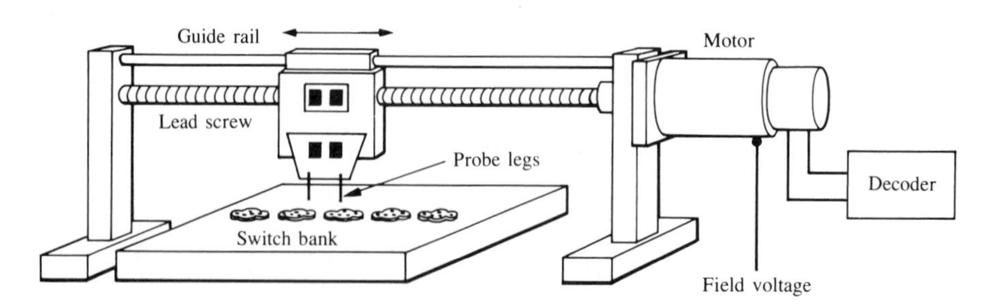

Figure 9.20. Automatic test system.

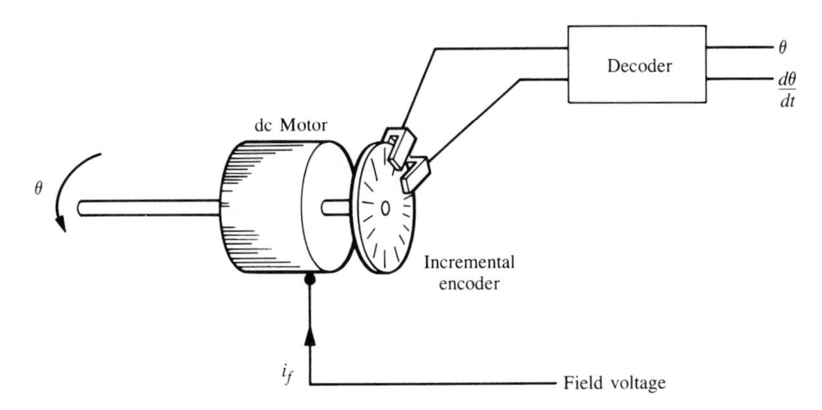

Figure 9.21. A dc motor with mounted encoder wheel.

possible from manually testing various panels of switches, relay, and indicator lights. Automating the test from a controller requires placing a plug across the leads of a part, and testing for continuity, resistance, or functionality. The system uses a dc motor with an encoded disk to measure position and velocity, as shown in Fig. 9.21. The parameters of the system are shown in Fig. 9.22 with K representing the required power amplifier.

We select the state variables as $x_1 = \Theta$, $x_2 = d\Theta/dt$, and $x_3 = i_f$, as shown in Fig. 9.23. State variable feedback is available, and we let

$$u = [-K_1, -K_2, -K_3]\mathbf{x}$$

or

$$u = -K_1 x_1 - K_2 x_2 - K_3 x_3, \tag{9.131}$$

as shown in Fig. 9.24. The goal is to select the gains so that the response to a step command has a settling time of less than two seconds and an overshoot of less than 4.0%.

In order to achieve accurate output position, we let $K_1 = 1$ and determine K, K_2, and K_3. The characteristic equation of the system may be obtained several ways.

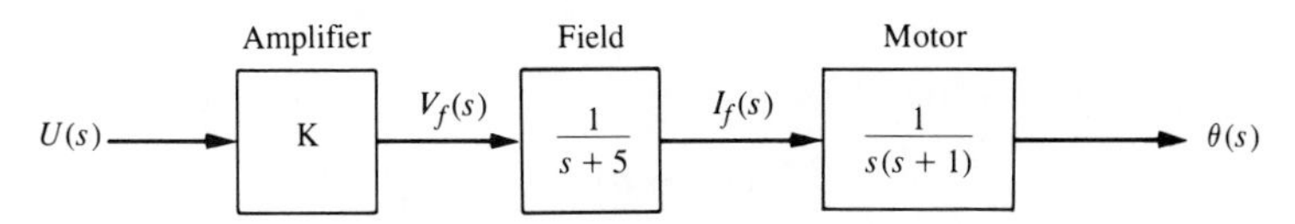

Figure 9.22. Block diagram.

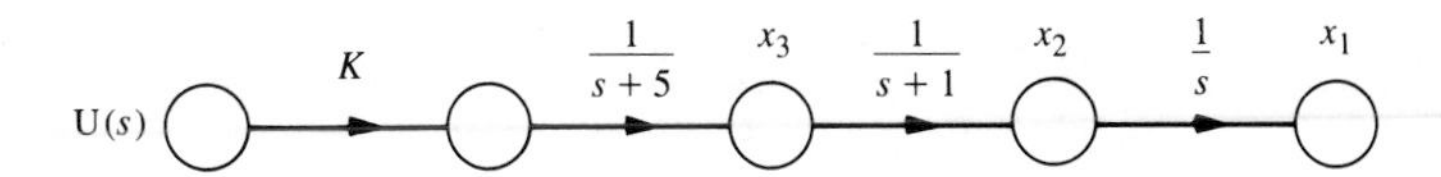

Figure 9.23. Signal flow graph.

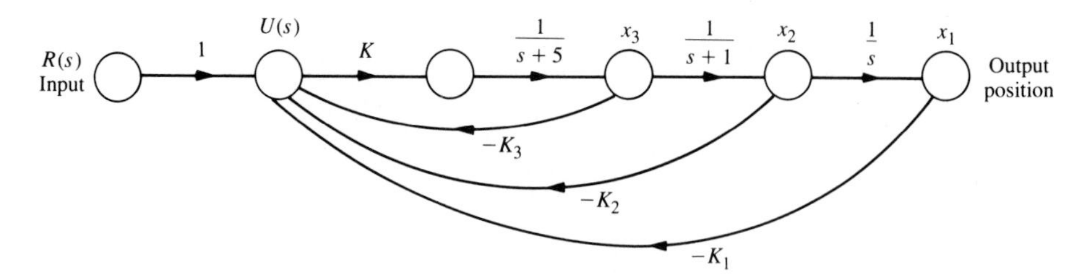

Figure 9.24. Feedback system.

Since $u = \mathbf{hx}$ is already given in Eq. (9.131), we use Fig. 9.23 to obtain

$$\dot{\mathbf{x}} = \mathbf{Ax} + \mathbf{bu}$$

$$= \begin{bmatrix} 0 & 1 & 0 \\ 0 & -1 & 1 \\ 0 & 0 & -5 \end{bmatrix} \mathbf{x} + \begin{bmatrix} 0 \\ 0 \\ K \end{bmatrix} u \tag{9.132}$$

Adding u as defined by Eq. (9.131), we have

$$\dot{\mathbf{x}} = \begin{bmatrix} 0 & 1 & 0 \\ 0 & -K & 1 \\ -K & -KK_2 & -(5 + K_3K) \end{bmatrix} \mathbf{x} \tag{9.133}$$

when $K_1 = 1$. The characteristic equation can also be readily obtained from Fig. 9.24 by letting $G(s)$ be the forward path transfer function and letting $H(s)$ be the equivalent feedback transfer function of Fig. 9.25. Then

$$G(s) = \frac{K}{s(s + 1)(s + 5)}$$

and

$$\begin{aligned} H(s) &= K_1 + sK_2 + K_3(s + 1)s \\ &= K_3\left[s^2 + \frac{K_2 + K_3}{K_3}s + \frac{K_1}{K_2}\right] \end{aligned} \tag{9.134}$$

Thus we can plot a root locus for K_3K as

$$1 + \frac{KK_3(s^2 + as + b)}{s(s + 1)(s + 5)} = 0, \tag{9.135}$$

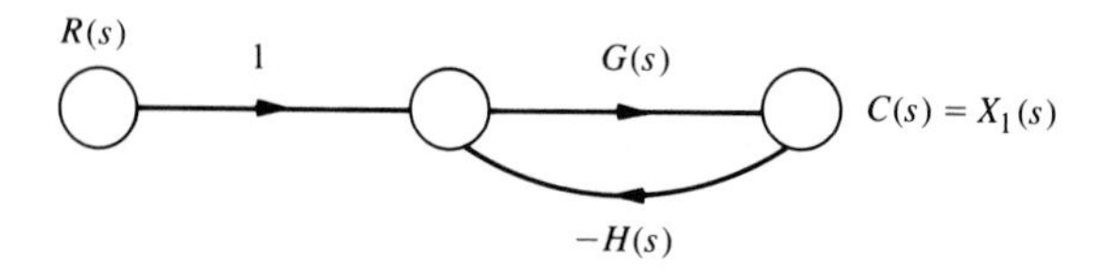

Figure 9.25. Equivalent model of feedback system.

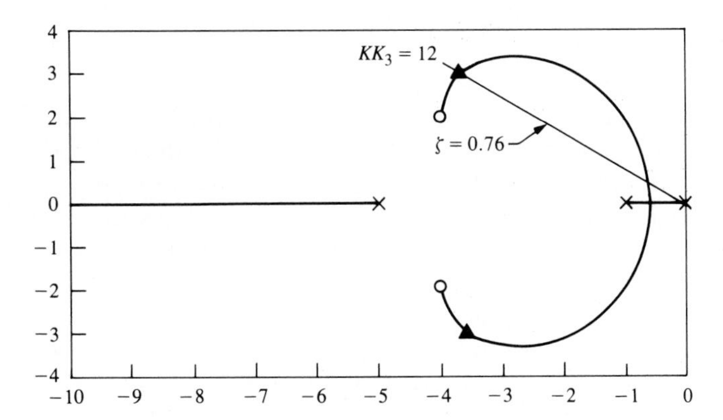

Figure 9.26. Root locus for automatic test system.

where a and b are chosen by selecting K_2 and K_3. Letting $K_1 = 1$ and setting $a = 8$ and $b = 30$, we place the zeros at $s = -4 \pm j2$ in order to pull the locus to the left in the s-plane. Then,

$$\frac{K_2 + K_3}{K_3} = 8 \qquad \text{and} \qquad \frac{1}{K_3} = 20.$$

Therefore $K_1 = 1$, $K_2 = 0.35$, and $K_3 = 0.05$. A plot of the root locus is shown in Fig. 9.26. When $KK_3 = 12$, the roots lie on the $\zeta = 0.76$ line, as shown in Fig. 9.26. Then, since $K_3 = 0.05$, we have $K = 240$. The roots at $K = 240$ are

$$s = -10.62, \qquad s = -3.69 \pm j3.00.$$

The step response of this system is shown in Fig. 9.27. The overshoot is 3% and the settling time is 1.8 seconds. Thus the design is quite acceptable.

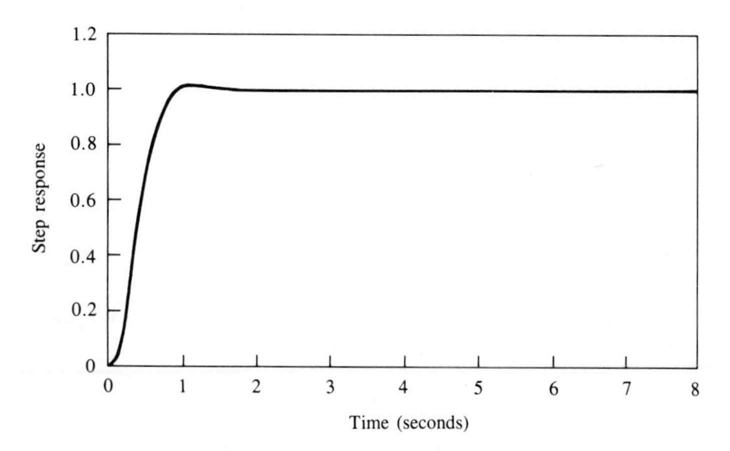

Figure 9.27. Step response.

9.9 Summary

In this chapter we have considered the description and analysis of systems in the time domain. The concept of the state of a system and the definition of the state variables of a system were discussed. The selection of a set of state variables in terms of the variables that describe the stored energy of a system was examined, and the nonuniqueness of a set of state variables was noted. The state vector differential equation and the solution for $\mathbf{x}(t)$ were discussed. Two alternative signal-flow graph model structures were considered for representing the transfer function (or differential equation) of a system. Using Mason's gain formula, we noted the ease of obtaining the flow-graph model. The vector differential equation representing these flow-graph models was also examined. Then the stability of a system represented by a state variable formulation was considered in terms of the vector differential equation. The time response of a linear system and its associated transition matrix was discussed and the utility of Mason's gain formula for obtaining the transition matrix was illustrated. Finally, a discrete-time evaluation of the time response of a nonlinear system and a time-varying system were considered. It was noted that the flow-graph state model is equivalent to an analog computer diagram where the output of each integrator is a state variable. Also, we found that the discrete-time approximation for a time response, as well as the transition matrix formulation for linear systems, is readily applicable for programming and solution by using a digital computer. The time-domain approach is applicable to biological, chemical, sociological, business, and physiological systems as well as to physical systems and thus appears to be an approach of general interest.

Exercises

E9.1. For the circuit shown in Fig. E9.1 identify a set of state variables.

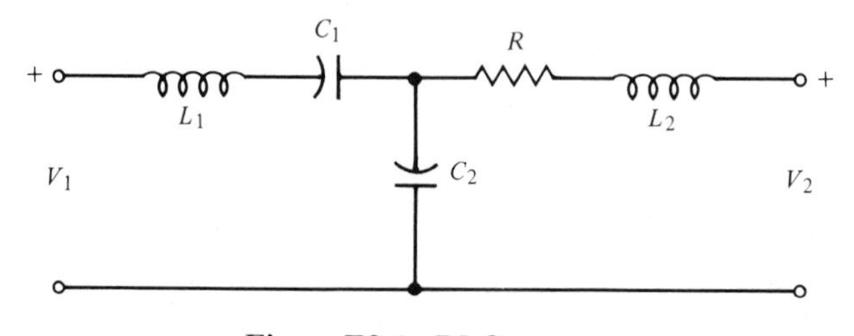

Figure E9.1. RLC motor.

E9.2. A robot-arm drive system for one joint can be represented by the differential equation [1]

$$\frac{dv(t)}{dt} = -k_1 v(t) - k_2 y(t) + k_3 i(t),$$

where $v(t)$ = velocity, $y(t)$ = position, and $i(t)$ is the control-motor current. Put the equations in state-variable form and set up the matrix form for $k_1 = k_2 = 1$.

E9.3. A system can be represented by the state vector differential equation of Eq. (9.16) where

$$\mathbf{A} = \begin{bmatrix} 0 & 1 \\ -1 & -1 \end{bmatrix}.$$

Find the characteristic roots of the system.

Answer: $s = -\frac{1}{2} \pm j\sqrt{3}/2$

E9.4. A system is represented by Eq. (9.16) where

$$\mathbf{A} = \begin{bmatrix} 0 & 1 & 0 \\ 0 & 0 & 1 \\ -1 & -k & -2 \end{bmatrix}.$$

Find the range of k where the system is stable.

E9.5. A system is represented by a flow graph as shown in Fig. E9.5. Write the state equations for this flow graph in the form of Eqs. (9.16) and (9.17).

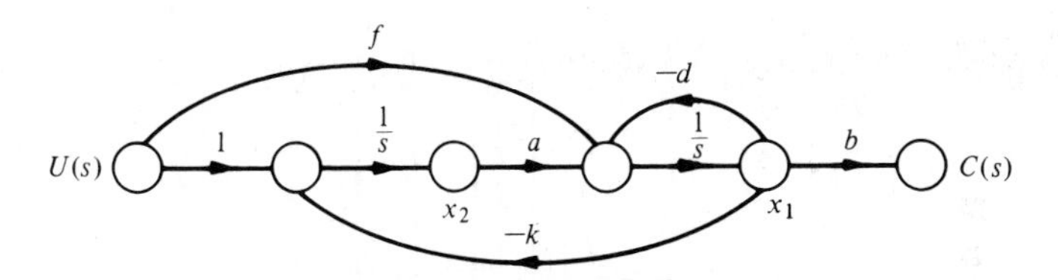

Figure E9.5. System flow graph.

E9.6. A system is represented by Eq. (9.16) where

$$\mathbf{A} = \begin{bmatrix} 0 & 1 \\ 0 & 0 \end{bmatrix}.$$

(a) Find the matrix $\phi(t)$. (b) For the initial conditions $x_1(0) = x_2(0) = 1$, find $\mathbf{x}(t)$.

Answer: (b) $x_1 = (1 + t)$, $x_2 = 1$, $t \geq 0$

E9.7. Consider the spring and mass shown in Fig. 9.2 where $M = 1$, $K = 100$, and $f = 20$. (a) Find the state vector differential equation. (b) Find the roots of the characteristic equation for this system.

E9.8. The manual low-altitude hovering task above a moving landing deck of a small ship is very demanding, in particular, in adverse weather and sea conditions. The hovering condition is shown in Fig. E9.8. The **A** matrix is

$$\mathbf{A} = \begin{bmatrix} 0 & 1 & 0 \\ 0 & 0 & 1 \\ 0 & -4 & -1 \end{bmatrix}.$$

Find the roots of the characteristic equation.

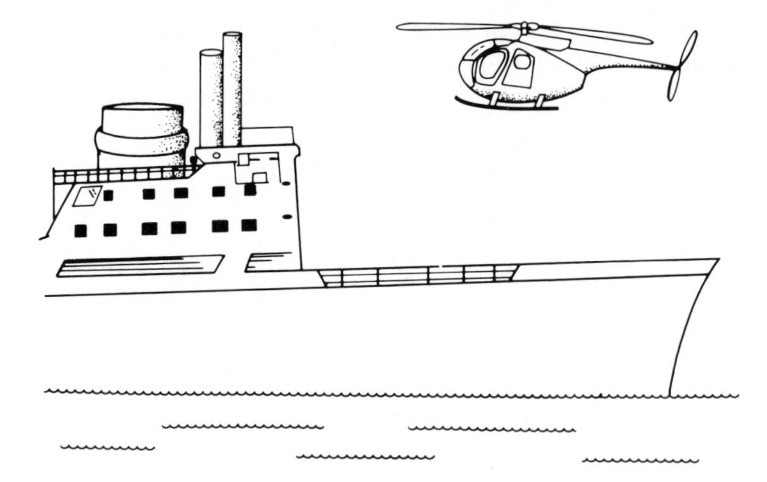

Figure E9.8. Helicopter hovering task.

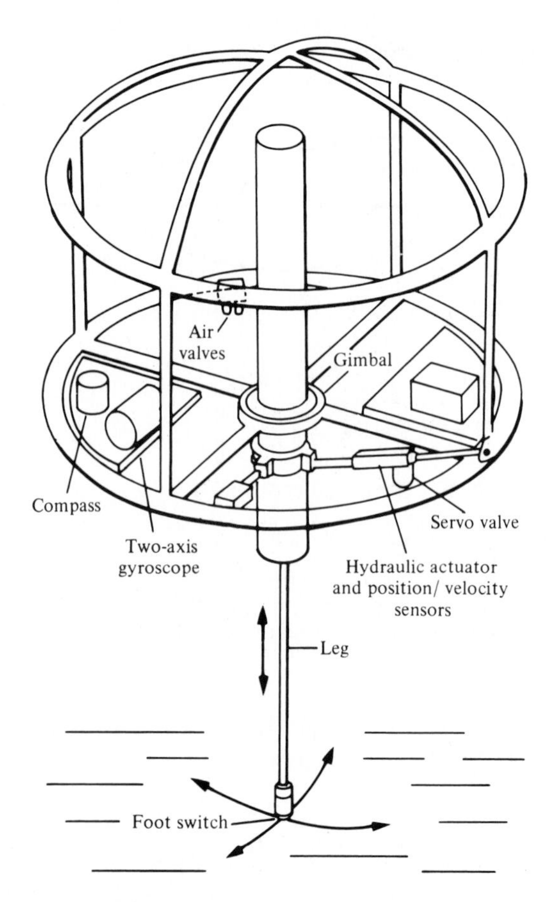

Figure E9.9. Single leg control.

E9.9. The ability to balance actively is a key ingredient in the mobility of a device that hops and runs on one springy leg as shown in Fig. E9.9 [21], on the previous page. The control of the attitude of the device uses a gyroscope as shown and a feedback such that $\mathbf{u} = \mathbf{H}\mathbf{x}$ where

$$\mathbf{H} = \begin{bmatrix} -1 & 0 \\ 0 & -k \end{bmatrix}$$

and

$$\mathbf{A} = \begin{bmatrix} 0 & 1 \\ 1 & 0 \end{bmatrix} \text{ and } \mathbf{B} = \mathbf{I}.$$

Determine a value for k so that the response of each hop is critically damped.

E9.10. A hovering vehicle (flying saucer) control system is represented by two state variables and [17]

$$A = \begin{bmatrix} 0 & 6 \\ -1 & -5 \end{bmatrix}.$$

(a) Find the roots of the characteristic equation. (b) Find the transition matrix $\phi(t)$.

E9.11. A fruit-picker robot is shown in Fig. E.9.11. This system may be represented by the state equations of Fig. 9.11. Using the Control System Design Program obtain x_1 and $c(t)$.

Figure E9.11. Fruit-picker robot.

E9.12. Use a state variable model to describe the circuit of Fig. E9.12. Obtain the response to an input unit step when the initial current is zero and the initial capacitor voltage is zero. Use the Control System Design Program to obtain the plot.

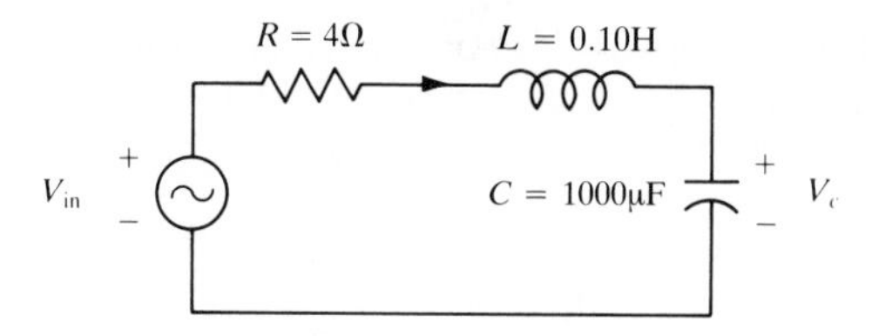

Figure E9.12. RLC series circuit.

E9.13. A system is described by two differential equations as

$$\frac{dy}{dt} + y - 2u + aw = 0$$

$$\frac{dw}{dt} - by + 4u = 0,$$

where w and y are functions of time, and u is an input $u(t)$. (a) Select a set of state variables. (b) Write the matix differential equation and specify the elements of the matrices. (c) Determine the range of values of the parameters a and b for a stable system. (d) Find the characteristic roots of the system in terms of the parameters a and b.

Answers: (c) $ab \geq 0$, (d) $s = -½ \pm \sqrt{1 - 4ab}/2$

E9.14. Develop the state space representation of a radioactive material of mass M to which additional radioactive material is added at the rate $r(t) = Ku(t)$, where K is a constant. Identify the state variables.

Problems

P9.1. An *RLC* circuit is shown in Fig. P9.1. (a) Identify a suitable set of state variables. (b) Obtain the set of first-order differential equations in terms of the state variables. (c) Write the state equations in matrix form. (d) Draw the state variable flow graph.

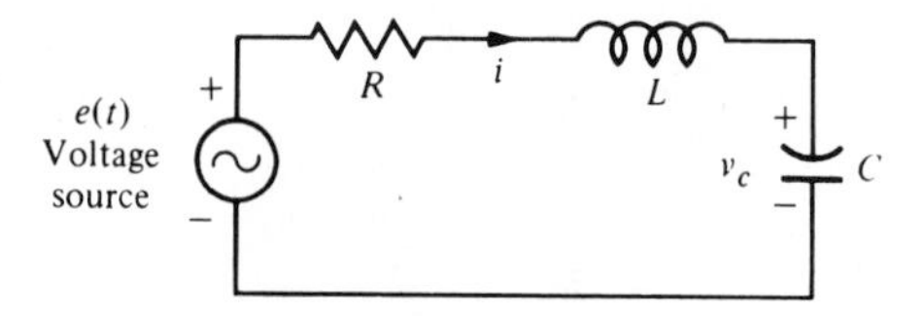

Figure P9.1. RLC circuit.

P9.2. A *balanced* bridge network is shown in Fig. P9.2. (a) Show that the **A** and **B** matrices for this circuit are

$$\mathbf{A} = \begin{bmatrix} -(2/(R_1 + R_2)C & 0 \\ 0 & -2R_1R_2/(R_1 + R_2)L \end{bmatrix},$$

$$\mathbf{B} = 1/(R_1 + R_2)\begin{bmatrix} 1/C & 1/C \\ R_2/L & -R_2/L \end{bmatrix}.$$

(b) Draw the state model flow graph. The state variables are $(x_1, x_2) = (v_c, i_l)$.

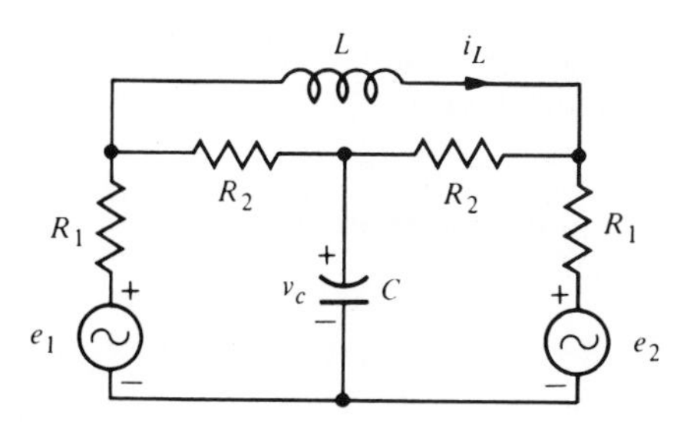

Figure P9.2. Balanced bridge network.

P9.3. An *RLC* network is shown in Fig. P9.3. Define the state variables as $x_1 = i_L$ and $x_2 = v_c$. (a) Obtain the vector differential equation. (b) Draw the state model flow graph.

Partial answer:

$$\mathbf{A} = \begin{bmatrix} 0 & 1/L \\ -(1/C) & -(1/RC) \end{bmatrix}.$$

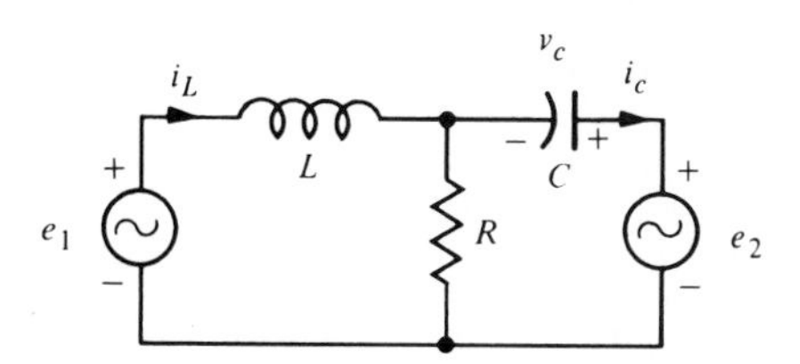

Figure P9.3. RLC circuit.

P9.4. The transfer function of a system is

$$T(s) = \frac{C(s)}{R(s)} = \frac{s^2 + 2s + 3}{s^3 + 2s^2 + 3s + 4}.$$

(a) Draw the flow-graph state model where all the state variables are fed back to the input node and the state variables are fed forward to the ouput signal, as in Fig. 9.7. (b) Determine the vector differential equation for the flow graph of (a). (c) Draw the flow-graph state model where the input signal is fed forward to the state variables, as in Fig. 9.9. (d) Determine the vector differential equation for the flow graph of (c).

P9.5. A closed-loop control system is shown in Fig. P9.5. (a) Determine the closed-loop transfer function $T(s) = C(s)/R(s)$. (b) Draw the state model flow graph for the system using the form of Fig. 9.7, where the state variables are fed back to the input node. (c) Determine the state vector differential equation.

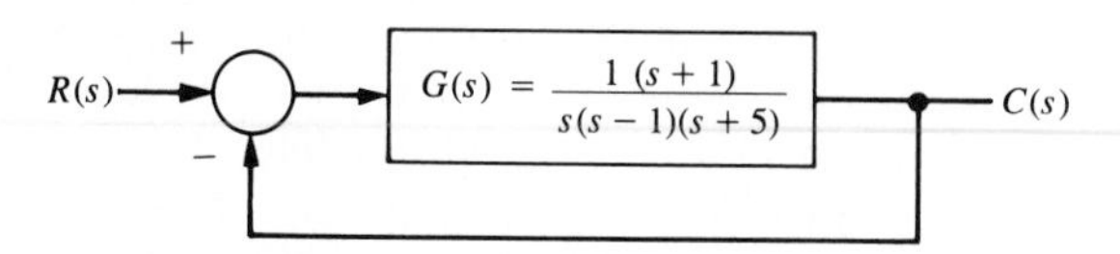

Figure P9.5. Closed-loop control system.

P9.6. Consider the case of the rabbits and foxes in Australia. The number of rabbits is x_1 and if left alone would grow indefinitely (until the food supply was exhausted) so that [7]

$$\dot{x}_1 = kx_1.$$

However, with foxes present on the continent, we have

$$\dot{x}_1 = kx_1 - ax_2,$$

where x_2 is the number of foxes. Now, if the foxes must have rabbits to exist, we have

$$\dot{x}_2 = -hx_2 + bx_1.$$

Determine if this system is stable and thus decays to the condition $x_1(t) = x_2(t) = 0$ at $t = \infty$. What are the requirements on a, b, h, and k for a stable system? What is the result when k is greater than h?

P9.7. An automatic depth-control system for a robot submarine is shown in Fig. P9.7. The depth is measured by a pressure transducer. The gain of the stern plane actuator is $K = 1$ when the velocity is 25 m/sec. The submarine has the approximate transfer function

$$G(s) = \frac{(s + 1)^2}{(s^2 + 1)},$$

and the feedback transducer is $H(s) = 1$. (a) Obtain a flow-graph state model. (b) Determine the vector differential equation for the system. (c) Determine whether the system is stable.

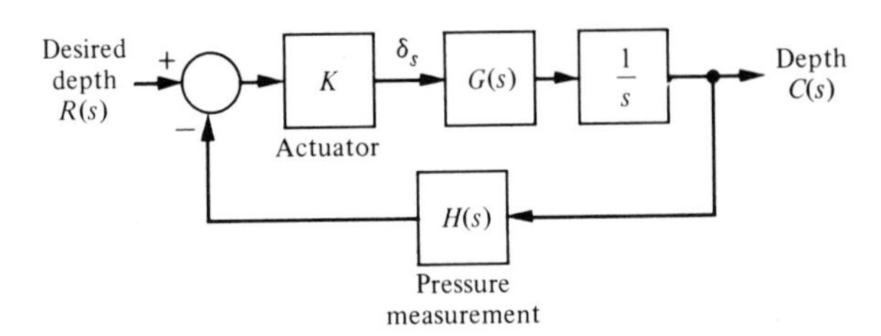

Figure P9.7. Submarine depth control.

P9.8. The soft landing of a lunar module descending on the moon can be modeled as shown in Fig. P9.8, on the next page. Define the state variables as $x_1 = y$, $x_2 = dy/dt$, $x_3 = m$ and the control as $u = dm/dt$. Assume that g is the gravity constant on the moon. Find the state-space equations for this system. Is this a linear model?

P9.9. A speed-control system utilizing fluid flow components can be designed. The system is a pure fluid-control system because the system does not have any moving mechanical parts. The fluid may be a gas or a liquid. A system is desired which maintains the speed within 0.5% of the desired speed by using a tuning fork reference and a valve actuator. Fluid-control systems are insensitive and reliable over a wide range of temperature, electromagnetic and nuclear radiation, acceleration, and vibration. The amplification within the system is achieved by using a fluid jet deflection amplifier. The system can be designed for a 500-kw steam turbine with a speed of 12,000 rpm. The block diagram of the system is shown in Fig. P9.9. The friction of the large inertia turbine is negligible and thus $f = 0$.

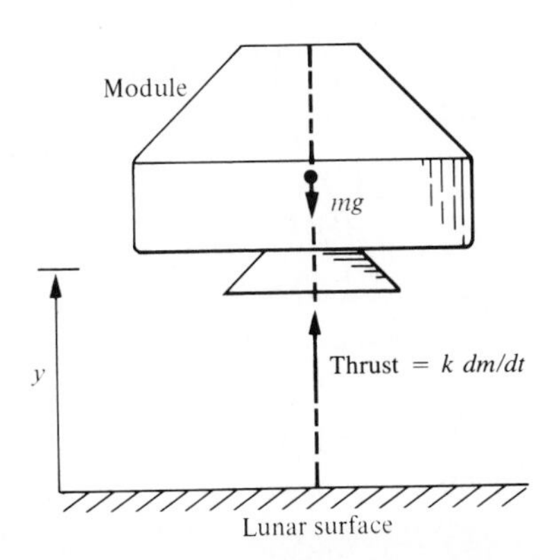

Figure P9.8. Lunar module landing control.

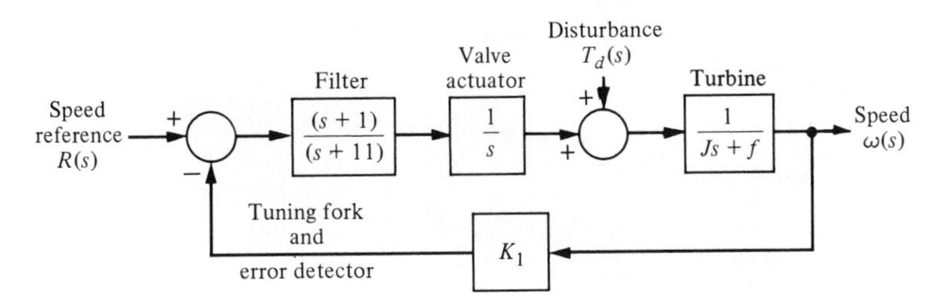

Figure P9.9. Steam turbine control.

The closed-loop gain is $K_1/J = 1$, where $K_1 = J = 10^4$. (a) Determine the closed-loop transfer function

$$T(s) = \frac{\omega(s)}{R(s)},$$

and draw the state model flow graph for the form of Fig. 9.7, where all the state variables are fed back to the input node. (b) Determine the state vector differential equation. (c) Determine whether the system is stable by investigating the characteristic equation obtained from the **A** matrix.

P9.10. Many control systems must operate in two dimensions, for example, the x- and and y-axes. A two-axis control system is shown in Fig. P9.10, on the next page, where a set of state variables is identified. The gain of each axis is K_1 and K_2 respectively. (a) Obtain the state vector differential equation. (b) Find the characteristic equation from the **A** matrix, and determine whether the system is stable for suitable values of K_1 and K_2.

P9.11. Consider the problem of rabbits and foxes in Australia as discussed in Problem P9.6. The values of the constants are $k = 1$, $h = 3$, and $a = b = 2$. Is this system stable? (a) Determine the transition matrix for this system. (b) Using the transition matrix, determine the response of the system when $x_1(0) = x_2(0) = 10$.

P9.12. The potential for an electrically propelled car for general use has been demonstrated [19]. One electric vehicle that has been built using a microprocessor for control of

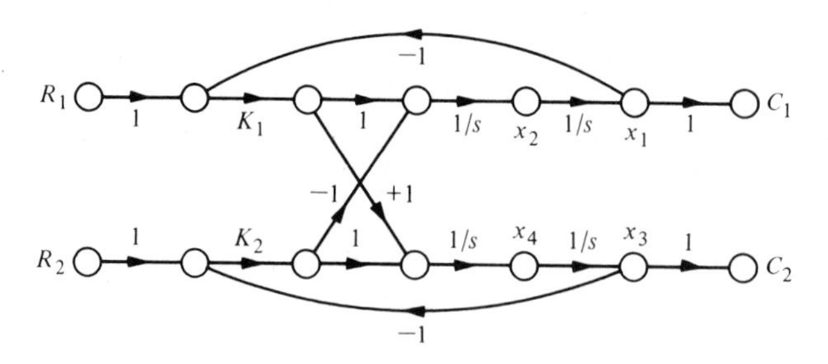

Figure P9.10. Two-axis control system.

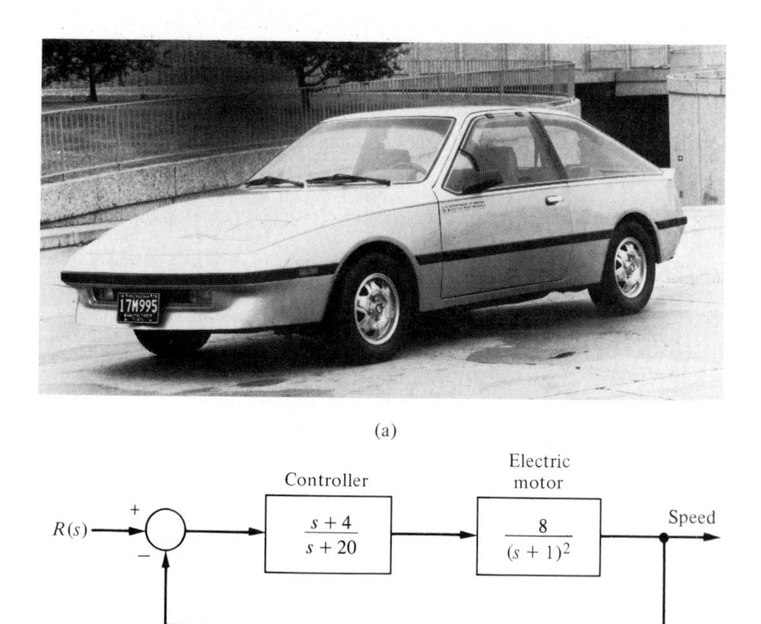

(b)

Figure P9.12. (a) An electric vehicle that has been built using a microprocessor for control of the motor and diagnostics. (b) A block diagram of the motor control system. (Photo courtesy of General Electric.)

the motor and diagnostics is shown in Fig. P9.12(a). This vehicle has a 75-mile range and is able to travel at up to 50 mph on a 5% grade. The block diagram of the motor control system is shown in Fig. P9.12(b). (a) Determine a flow-graph representation of this system. (b) Obtain the transition matrix $\phi(t)$. (c) Is the system stable?

P9.13. Reconsider the *RLC* circuit of Problem P9.1 when $R = 2.5$, $L = \frac{1}{4}$, and $C = \frac{1}{6}$. (a) Determine whether the system is stable by finding the characteristic equation with the aid of the **A** matrix. (b) Determine the transition matrix of the network. (c) When the initial inductor current is 0.1 amp, $v_c(0) = 0$, and $e(t) = 0$, determine the response of the system. (d) Repeat part (c) when the initial conditions are zero and $e(t) = E$, for $t > 0$, where E is a constant.

P9.14. Consider the system discussed in Problem P9.5. (a) Determine whether the system is stable by obtaining the characteristic equation from the **A** matrix. (b) An additional feedback signal is to be added in order to stabilize the system. If $R(s) = -KsC(s)$, the state variable x_2 is fed back so that $r(t) = Kx_2(t)$. Determine the minimum value of K so that the system is stable.

P9.15. Reconsider the rabbits-and-foxes ecology discussed in Problems P9.6 and P9.11 when we take the depletion of food into account. Then we have

$$\dot{x}_1 = kx_1 - ax_2 + ax_3,$$

$$\dot{x}_2 = -hx_2 + bx_1,$$

$$\dot{x}_3 = \beta x_3 - \gamma x_1,$$

where x_3 = the amount of rabbit food per unit area. Again, assume that $k = 1$, $h = 3$, and $a = b = 2$. In this case $\beta = 1$ and $\alpha = 0.1$. Determine a suitable value for γ in order to eliminate the rabbits.

P9.16. A system for dispensing radioactive fluid into capsules is shown in Fig. P9.16(a). The horizontal axis moving the tray of capsules is actuated by a linear motor. The x-axis control is shown in Fig. P9.16(b). Obtain (a) a state-variable representation and (b) the time-response of the system. (c) Determine the characteristic roots of the system.

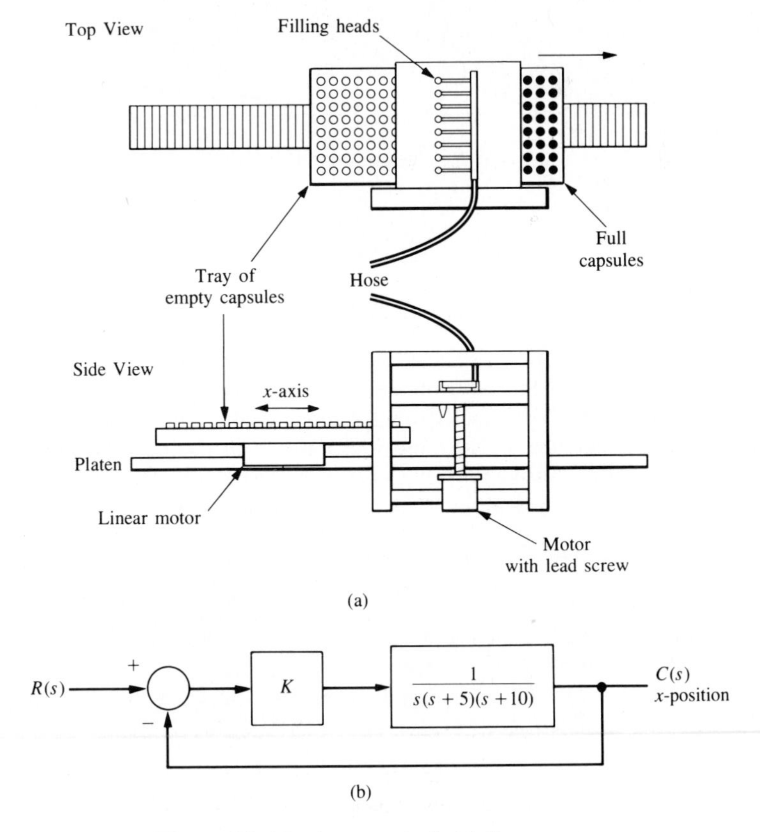

Figure P9.16. Automatic fluid dispenser.

P9.17. The dynamics of a controlled submarine are significantly different from those of an aircraft, missile, or surface ship. This difference results primarily from the moment in the vertical plane due to the buoyancy effect. Therefore it is interesting to consider the control of the depth of a submarine. The equations describing the dynamics of a submarine can be obtained by using Newton's laws and the angles defined in Fig. P9.17. In order to simplify the equations, we will assume that θ is a small angle and the velocity v is constant and equal to 25 ft/sec. The state variables of the submarine, considering only vertical control, are $x_1 = \theta$, $x_2 = d\theta/dt$ and $x_3 = \alpha$, where α is the angle of attack. Thus the state vector differential equation for this system, when the submarine has an Albacore type hull, is

$$\dot{\mathbf{x}} = \begin{bmatrix} 0 & 1 & 0 \\ -0.0071 & -0.111 & 0.12 \\ 0 & 0.07 & -0.3 \end{bmatrix} \mathbf{x} + \begin{bmatrix} 0 \\ -0.095 \\ +0.072 \end{bmatrix} u(t),$$

where $u(t) = \delta_s(t)$, the deflection of the stern plane. (a) Determine whether the system is stable. (b) Using the discrete-time approximation, determine the response of the system to a stern plane step command of 0.285° with the initial conditions equal to zero. Use a time increment of T equal to 2 sec. (c) (optional) Using a time increment of $T = 1$ sec and a digital computer, obtain the transient response for each state for 80 sec. Compare the response calculated for (b) and (c).

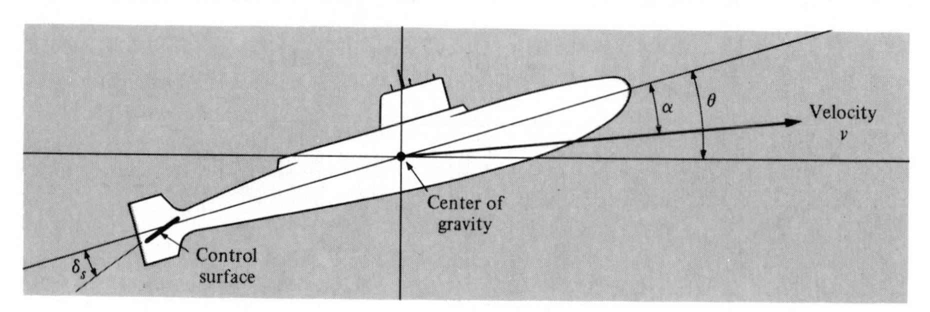

Figure P9.17. Submarine depth control.

P9.18. An interesting mechanical system with a challenging control problem is the ball and beam shown in Fig. P9.18(a), on the next page. It consists of a rigid beam which is free to rotate in the plane of the paper around a center pivot, with a solid ball rolling along a groove in the top of the beam. The control problem is to position the ball at a desired point on the beam using a torque applied to the beam as a control input at the pivot.

A linear model of the system with a measured value of the angle ϕ and its angular velocity $\dfrac{d\phi}{dt} = \omega$ is available. Select a feedback scheme so that the response of the closed-loop system has an overshoot of 4% and the settling time is 1 sec for a step input.

P9.19. Consider the control of the robot shown in Fig. P9.19(a), on page 478. The motor turning at the elbow moves the wrist through the forearm, which has some flexibility as represented by Fig. 9.19(b). The spring has a spring constant k and friction damping constant f. Let the state variables be $x_1 = \phi_1 - \phi_2$ and $x_2 = \omega_2/\omega_0$, where

$$\omega_0^2 = \frac{k(J_1 + J_2)}{J_1 J_2}.$$

Write the state variable equation in matrix form.

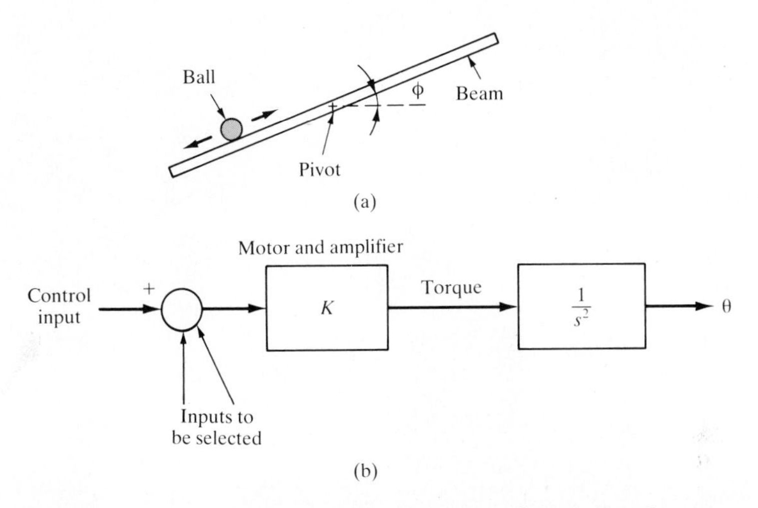

Figure P9.18. (a) Ball and beam. (b) Model of the ball and beam.

P9.20. The derivative of a state variable can be approximated by the equation

$$\dot{x}(t) \simeq \frac{1}{2T}[3x(k+1) - 4x(k) + x(k-1)].$$

This approximation of the derivative utilizes two past values to estimate the derivative, whereas Eq. (9.94) uses one past value of the state variable. Using this approximation for the derivative, repeat the calculations for Example 9.8. Compare the resulting approximation for $x_1(t)$, $T = 0.2$, with the results given in Table 9.1. Is this approximation more accurate?

P9.21. There are several forms that are equivalent signal-flow graph state models. Two equivalent state flow-graph models for a fourth-order equation (Eq. 9.36) are shown in Fig. 9.7 and Fig. 9.9. Another alternative structure for a state flow-graph model is shown in Fig. P9.21, on the next page. In this case the system is second order and the input-output transfer function is

$$G(s) = \frac{C(s)}{U(s)} = \frac{b_1 s + b_0}{s^2 + a_1 s + a_0}.$$

(a) Verify that the flow graph of Fig. P9.21 is in fact a model of $G(s)$. (b) Show that the vector differential equation representing the flow-graph model of Fig. P9.21 is

$$\dot{\mathbf{x}} = \begin{bmatrix} 0 & 1 \\ -a_0 & -a_1 \end{bmatrix} \mathbf{x} + \begin{bmatrix} h_1 \\ h_0 \end{bmatrix} u(t),$$

where $h_1 = b_1$ and $h_0 = b_0 - b_1 a_1$.

P9.22. A nuclear reactor that has been operating in equilibrium at a high thermal-neutron flux level is suddenly shut down. At shutdown, the density of xenon 135(X) and iodine 135(I) are 3×10^{15} and 7×10^{16} atoms per unit volume, respectively. The half lives of I 135 and Xe 135 nucleides are 6.7 and 9.2 hours, respectively. The decay equations are

$$\dot{I} = -\frac{0.693}{6.7} I,$$

$$\dot{X} = -\frac{0.693}{9.2} X - I.$$

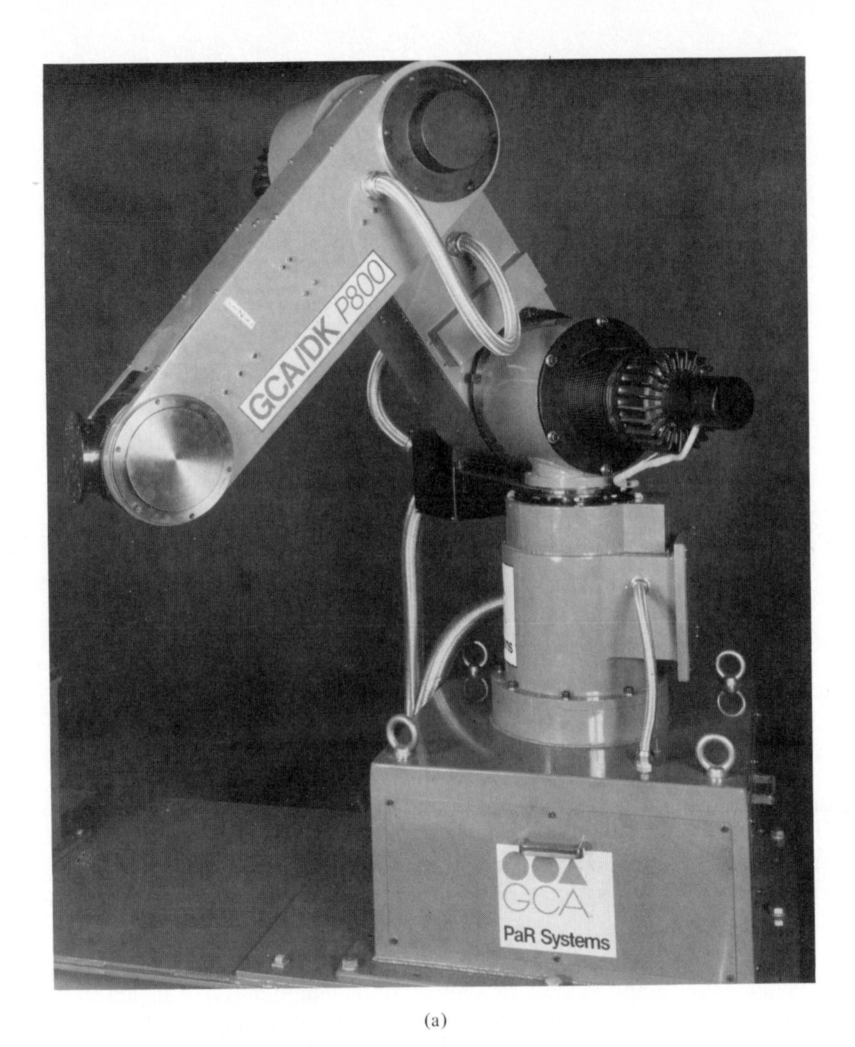

(a)

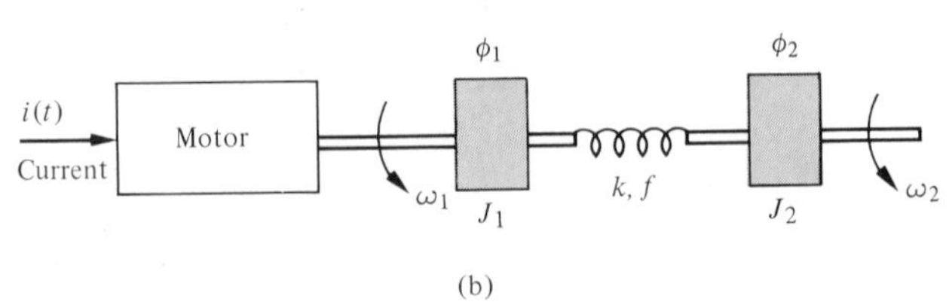

(b)

Figure P9.19. (a) An industrial robot. (Courtesy of GCA Corporation.)

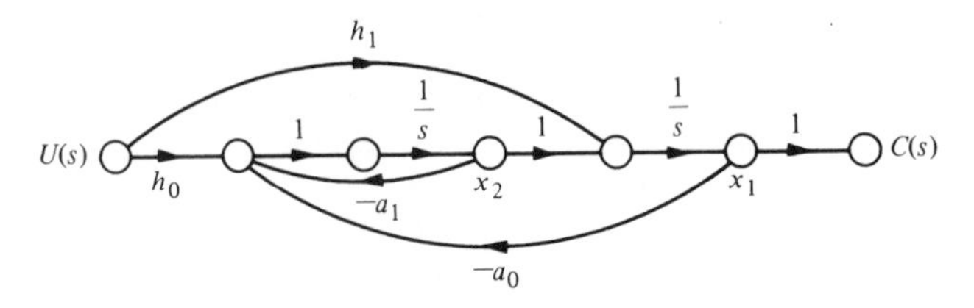

Figure P9.21. Model of second-order system.

Determine the concentrations of I 135 and Xe 135 as functions of time following shutdown by determining (a) the transition matrix and the system response, and (b) a discrete-time evaluation of the time response. Verify that the response of the system is that shown in Fig. P9.22.

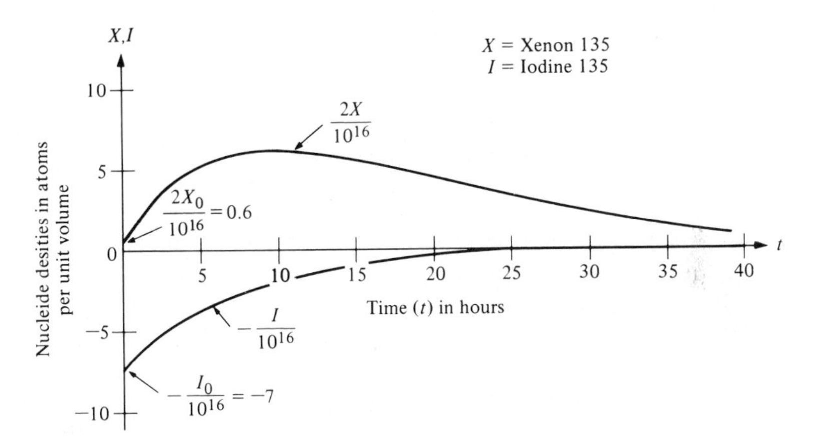

Figure P9.22. Nuclear reactor response.

P9.23. Consider the following mathematical model of the social interaction of humans that is often called group dynamics. The four system variables of interest are (1) the intensity of interaction (or communication) among the members of the group, $x_1(t)$; (2) the amount of friendliness or group identification among group members, $x_2(t)$; (3) the total amount of activity carried on by a member of the group, $x_3(t)$; (4) the amount of activity imposed on the group by its external environment, $u(t)$. A simple model might be represented by the set of equations [7]

$$\dot{x}_1 = a_1 x_2 + a_2 x_3,$$

$$\dot{x}_2 = b(x_1 - \beta x_2),$$

$$\dot{x}_3 = c_1(x_2 - \gamma x_3) + c_2(u - x_3).$$

The first equation indicates that interaction results from friendliness or activity. The second equation indicates that friendliness will increase as the amount of interaction grows larger than friendliness. The third equation relates the effect of all variables on a change of activity. (a) Determine the requirements on the coefficients for a stable system. When the system is stable, the changes in the variables evidently decay to zero, and equilibrium is attained when $u(t) = 0$. Determine the values of the variables at equilibrium when $u(t) = 0$. Is this representative of the social disintegration of the group? Is an external force $u(t)$ required to maintain the group activity? (b) The problem of group morale has also been studied with the aid of this model. A group is said to have positive morale when the activity $x_3(t)$ exceeds that required by an external social force $u(t) = U$, where U is a constant. Determine the necessary relationship for the coefficients a_1, a_2, β, and γ for positive morale. (c) Determine the transient response of a group, such as a college social fraternity, which is highly active and is subjected to a high level of external social forces, $u(t) = U$. Assume that initially $x_1(0) = x_2(0) = x_3(0) = 0$, that $a_1 = a_2 = b = \beta = c_1 = c_2 = 1$, and $\gamma = -2$.

P9.24. Consider the automatic ship steering system discussed in Problems P7.13 and P8.16. The state variable form of the system differential equation is

$$\dot{\mathbf{x}}(t) = \begin{bmatrix} -0.05 & -6 & 0 & 0 \\ -10^{-3} & -0.15 & 0 & 0 \\ 1 & 0 & 0 & 13 \\ 0 & 1 & 0 & 0 \end{bmatrix} \mathbf{x}(t) + \begin{bmatrix} -0.2 \\ 0.03 \\ 0 \\ 0 \end{bmatrix} \delta(t),$$

where $\mathbf{x}^T(t) = [\dot{v}, \omega_s, y, \theta]$. The state variables are $x_1 = \dot{v}$ = the transverse velocity; $x_2 = \omega_s$ = angular rate of ship's coordinate frame relative to response frame; $x_3 = y$ = deviation distance on an axis perpendicular to the track; $x_4 = \theta$ = deviation angle. (a) Determine whether the system is stable. (b) Feedback can be added so that

$$\delta(t) = -k_1x_1 - k_3x_3.$$

Determine whether this system is stable for suitable values of k_1 and k_3.

P9.25. It is desirable to use well-designed controllers to maintain building temperature with solar collector space heating systems. One solar heating system can be described by [10]

$$\frac{dx_1}{dt} = 3x_1 + u_1 + u_2,$$

$$\frac{dx_2}{dt} = 2x_2 + u_2 + d,$$

where x_1 = temperature deviation from desired equilibrium and x_2 = temperature of the storage material (such as a water tank). Also, u_1 and u_2 are the respective flow rates of conventional and solar heat, where the transport medium is forced air. A solar disturbance on the storage temperature (such as overcast skies) is represented by d. Write the matrix equations and solve for the system response from equilibrium when $u_1 = 0$, $u_2 = 1$, and $d = 1$.

P9.26. For the fourth-order system of Problem P4.12, determine the state vector equations. Then for the approximate second-order model determine the state vector equations and compare with the fourth-order equations.

P9.27. Consider a model of the interaction of the OPEC nations and the United States. The OPEC nations want to increase the price of their oil and maintain control over their destiny. The United States wishes to decrease the price of the imported oil and decrease the OPEC nations' control. The two state variables are price, p, and control, c. One model is then

$$p(k+1) = p(k) - u_1(k) + u_2(k),$$

$$c(k+1) = c(k) - u_1(k) + u_2(k),$$

where $u_1(k)$ = action by the United States, and $u_2(k)$ = action by OPEC. The United States selects a control action so that $u_1(k) = 0.5p(k)$, and OPEC selects $u_2(k) = 0.4c(k-1)$. Examine the response of this system for several time periods. What will be the ultimate outcome if this model is a true representation? Assume that $p(0) = c(0) = 10$ and $c(-1) = 10$. Reexamine the situation if $u_2(k) = 0.6c(k)$.

P9.28. A gyroscope with a single degree of freedom is shown in Fig. P9.28. Gyroscopes sense the angular motion of a system and are used in automatic flight control systems. The gimbal moves about the output axis OB. The input is measured around the input axis OA.

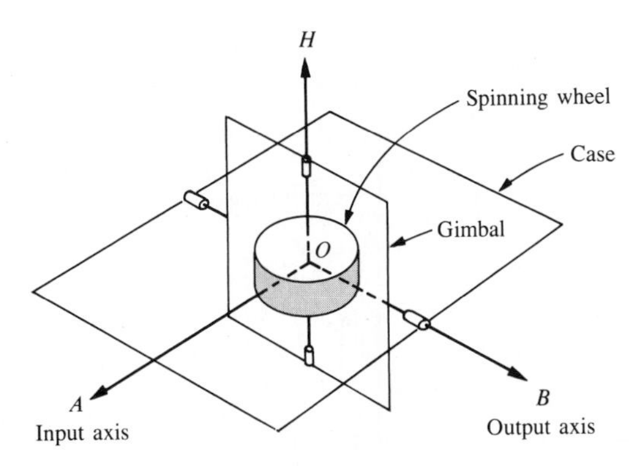

Figure P9.28. Gyroscope.

The equation of motion about the output axis is obtained by equating the rate of change of angular momentum to the sum of torques. Obtain a state-space representation of the gyro system.

P9.29. An *RL* circuit is shown in Fig. P9.29. (a) Select the two stable variables and obtain the vector differential equation where the output is $v_0(t)$. (b) Determine if the state variables are observable when $R_1/L_1 = R_2/L_2$. (c) Find the conditions when the system has two equal roots.

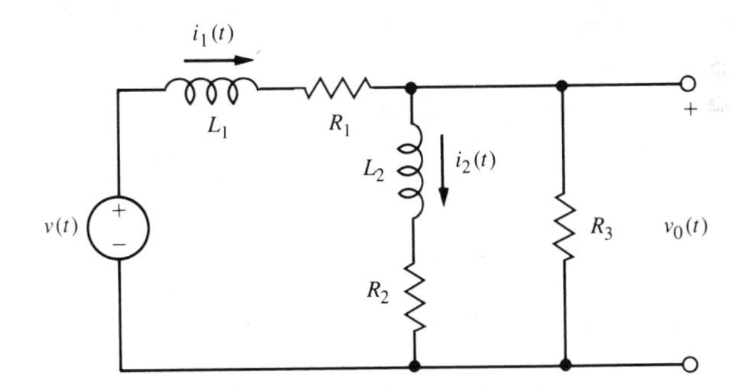

Figure P9.29. RLC circuit.

P9.30. There has been considerable engineering effort directed at finding ways to perform manipulative operations in space—for example, assembling a space station and acquiring target satellites. To perform such tasks, space shuttles carry a remote manipulator system (RMS) in the cargo bay [11]. The RMS has proven its effectiveness on recent shuttle missions, but now a new design approach is being considered—a manipulator with inflatable arm segments. Such a design might reduce manipulator weight by a factor of four while producing a manipulator that, prior to inflation, occupies only one-eighth as much space in the shuttle's cargo bay as the present RMS.

The use of an RMS for constructing a space structure in the shuttle bay is shown in Fig. P9.30(a), and a model of the flexible RMS arm is shown in Fig. P9.30(b), where J is the inertia of the drive motor and L is the distance to the center of gravity of the load component. Derive the state equations for this system.

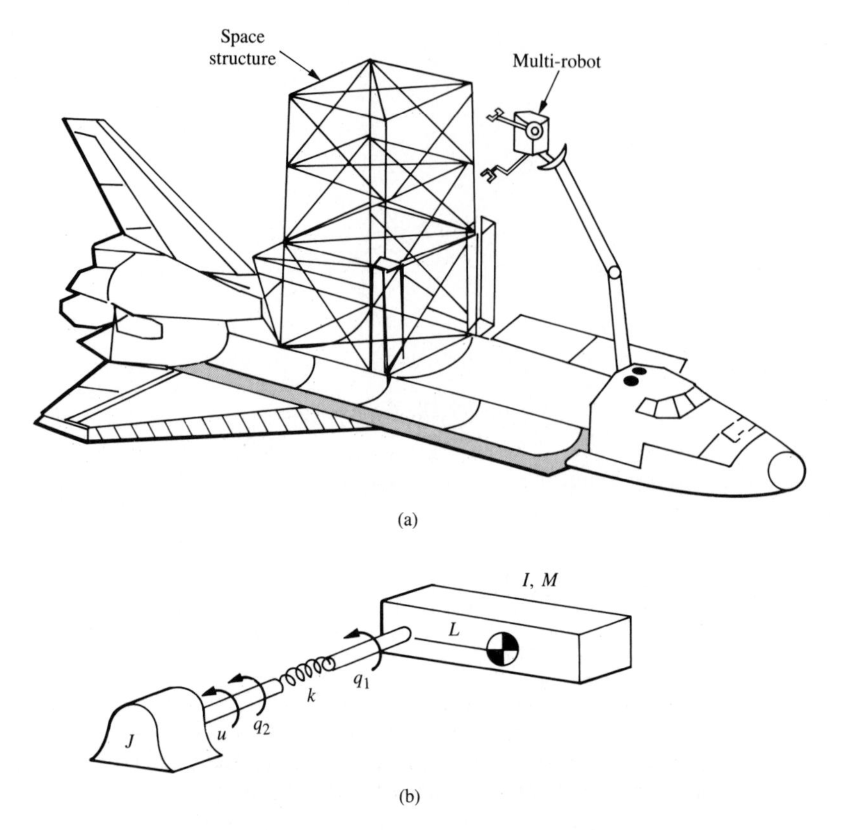

Figure P9.30. Remote manipulator system.

P9.31. A manipulator control system has a plant transfer function of

$$G(s) = \frac{1}{s(s + 0.4)}$$

and negative unity feedback [22]. Represent this system by a state-variable signal-flow graph and a vector differential equation. (a) Plot the response of the closed-loop system to a step input. (b) Use state variable feedback so that the overshoot is 5% and the settling time is 1.35 sec. (c) Plot the response of the state variable feedback system to a step input.

P9.32. Obtain the state equations for the two-input and one-output circuit shown in Fig. P9.32, where the output is i_2.

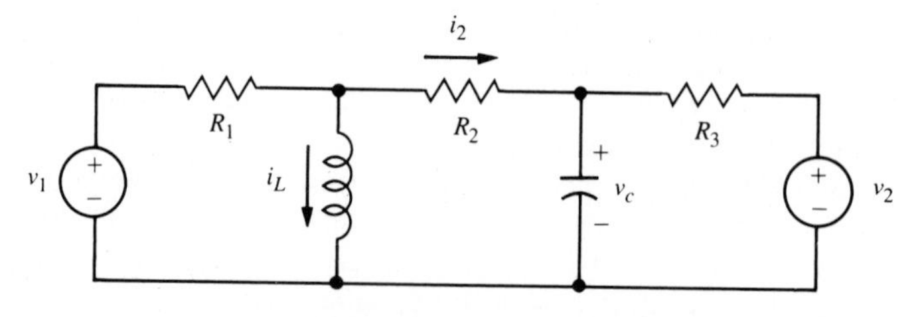

Figure P9.32. Two-input RLC circuit.

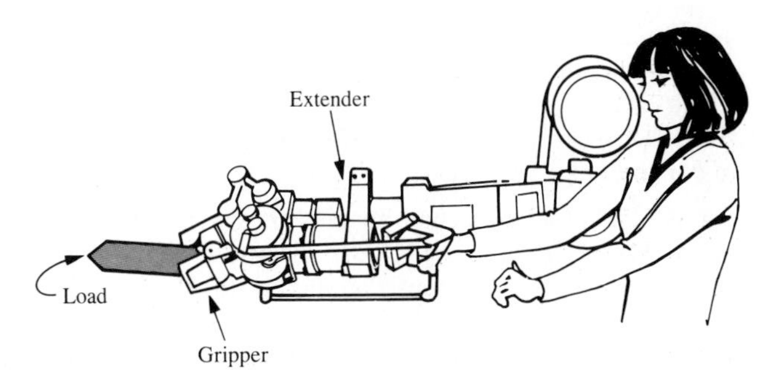

Figure P9.33. Extender for increasing the strength of the human arm in load maneuvering tasks.

P9.33. Extenders are robot manipulators which extend (i.e., increase) the strength of the human arm in load maneuvering tasks (Fig. P9.33) [23]. The system is represented by the transfer function

$$\frac{C(s)}{U(s)} = G(s) = \frac{1}{s^2 + 4s + 3}.$$

Using the form in Fig. 9.6, determine the state variable equations and the state transition matrix.

P9.34. A drug taken orally is ingested at a rate r. The mass of the drug in the gastrointestinal tract is denoted by m_1 and in the bloodstream by m_2. The rate of change of the mass of the drug in the gastrointestinal tract is equal to the rate at which the drug is ingested minus the rate at which the drug enters the blood stream, a rate that is taken to be proportional to the mass present. The rate of change of the mass in the blood stream is proportional to the amount coming from the gastrointestinal tract minus the rate at which mass is lost by metabolism, which is proportional to the mass present in the blood. Develop the state space representation of this system.

For the special case where the coefficients of A are equal to 1 (with the appropriate sign), determine the response when $m_1(0) = 1$ and $m_2(0) = 0$. Plot the state variables versus time and on the $x_1 - x_2$ state plane.

P9.35. The dynamics of a rocket are represented by

$$\frac{C(s)}{U(s)} = G(s) = \frac{1}{s^2},$$

and state variable feedback is used where $x_1 = c(t)$ and $u = -x_2 - 0.5x_1$. Determine the roots of the characteristic equation of this system and the response of the system when the initial conditions are $x_1(0) = 0$ and $x_2(0) = 1$.

Design Problems 

DP9.1. A spring-mass-damper system, as shown in Fig. 9.2, is used as a shock absorber for a large high-performance motorcycle. The original parameters selected are $m = 1$ kg, $f = 9\text{N}\cdot s/\text{m}$, and $k = 20\text{N/m}$. (a) Determine the system matrix, the characteristic roots, and the transition matrix, $\phi(t)$. The harsh initial conditions are assumed to be $y(0) = 1$ and $dy/dt|_{t=0} = 2$. (b) Plot the response of $y(t)$ and dy/dt for the first two seconds. (c) Redesign the shock absorber by changing the spring constant and the damping constant in order to reduce the effect of a high rate of acceleration force, d^2y/dt^2, on the rider. The mass must remain constant at 1 kg.

DP9.2. The motion control of a lightweight hospital transport vehicle can be represented by a system of two masses is shown in Fig. DP9.2, where $m_1 = m_2 = 1$ and $k_1 = k_2 = 1$ [24]. (a) Determine the state vector differential equation. (b) Find the roots of the characteristic equation. (c) We wish to stabilize the system by letting $u = -kx_i$, where u is the force on the lower mass and x_i is one of the state variables. Select an appropriate state variable x_i. (d) Choose a value for the gain k and sketch the root locus as k varies.

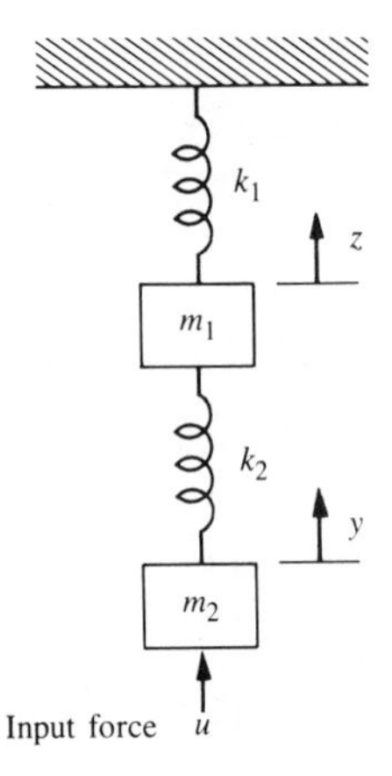

Figure DP9.2. Model of hospital vehicle.

DP9.3. Consider the inverted pendulum mounted to a motor, as shown in Fig. DP9.3. The motor and load are assumed to have no friction damping. The pendulum to be balanced is attached to the horizontal shaft of a servomotor. The servomotor carries a tach-

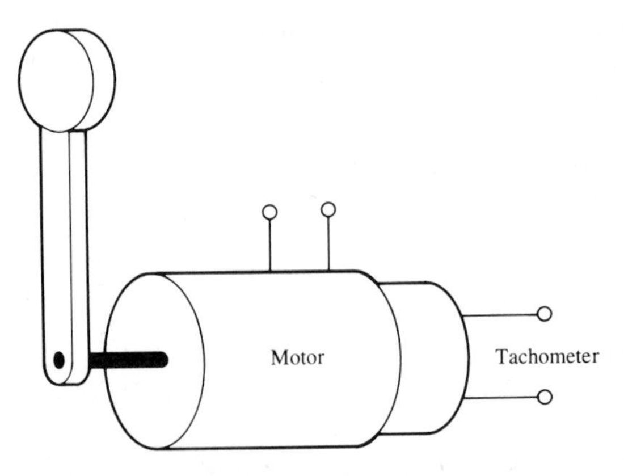

Figure DP9.3. Motor and inverted pendulum.

ogenerator, so that a velocity signal is available, but there is no position signal. When the motor is unpowered, the pendulum will hang vertically downward and if slightly disturbed will perform oscillations. If the pendulum is lifted to the top of its arc, it is unstable in that position. Devise a feedback compensator, $G_c(s)$, using only the velocity signal from the tachometer.

Terms and Concepts

Discrete-time approximation An approximation used to obtain the time response of a system based on the division of the time into small increments Δt.

Fundamental matrix See Transition matrix.

State of a system A set of numbers such that the knowledge of these numbers and the input function will, with the equations describing the dynamics, provide the future state of the system.

State-variable feedback The control signal for the process is a direct function of all the state variables.

State variables The set of variables that describe the system.

State vector The vector matrix containing all n state variables, $x_1, x_2, \ldots, x_n$.

State vector differential equation The differential equation for the state vector: $\dot{\mathbf{x}} = \mathbf{Ax} + \mathbf{Bu}$.

Time domain The mathematical domain that incorporates the time response and the description of a system in terms of time, t.

Time varying system A system for which one or more parameters may vary with time.

Transition matrix, $\phi(t)$ The matrix exponential function that describes the unforced response of the system.

References

1. R. C. Dorf, *Encyclopedia of Robotics,* John Wiley & Sons, New York, 1988.
2. L. B. Jackson, *Signals, Systems, and Transforms,* Addison-Wesley, Reading, Mass., 1991.
3. E. Kamen, *Introduction to Signals and Systems,* 2nd ed., Macmillan, New York, 1990.
4. R. C. Dorf, *Robotics and Automated Manufacturing,* Reston Publishing, Reston, Va., 1983.
5. R. E. Ziemer, *Signals and Systems,* 2nd ed., Macmillan, New York, 1989.
6. J. F. Engelberger, *Robotics in Service,* M.I.T. Press, Cambridge, Mass., 1989.
7. D. F. Delchamps, *State Space and Input-Output Linear Systems,* Springer-Verlag, New York, 1988.
8. K. J. Astrom, *Adaptive Control,* Addison-Wesley, Reading, Mass., 1989.
9. D. I. Lewin, "Orbital Mass Transit," *Mechanical Engineering,* July 1990; pp. 78–80.

10. T. Soderstrom, *System Identification,* Prentice Hall, Englewood Cliffs, N.J., 1989.
11. R. DeMeis, "Shuttling to the Space Station," *Aerospace America,* March 1990; pp. 44–47.
12. K. C. Cheok and N. K. Loh, "A Ball-Balancing Demonstration of Optimal Control," *IEEE Control Systems,* February 1987; pp. 54–57.
13. L. E. Ryan, "Closed-Loop Control of Impact Printer Hammer," *J. of Dynamic Systems,* March 1990; pp. 69–75.
14. R. C. Dorf and R. Jacquot, *Control System Design Program,* Addison-Wesley, Reading, Mass., 1988.
15. C. W. Anderson, "Learning to Control an Inverted Pendulum," *IEEE Control Systems,* April 1989; pp. 31–35.
16. F. Demeester, "Real-Time Optical Measurement of Robot Structural Deflections," *Mechatronics,* vol. 1, 1990; pp. 73–86.
17. J. C. Maciejowski, *Multivariable Feedback Design,* Addison-Wesley, Reading, Mass., 1990.
18. P. M. Leucht, "Active Four-Wheel Steering Design for an Advanced Vehicle," *Proceed. of American Automatic Control Conference,* 1990; pp. 2379–85.
19. I. Ha, "Feedback Linearizing Control of Vehicle Acceleration," *IEEE Trans. on Automatic Control,* July 1989; pp. 689–98.
20. E. K. Parsons, "An Experiment Demonstrating Pointing Control on a Flexible Structure," *IEEE Control Systems,* April 1989; pp. 79–83.
21. S. Boyd, "Linear Controller Design," *IEEE Proceed.,* March 1990; pp. 529–74.
22. H. C. Fowler, "Performance of a Direct Drive Manipulator," *Proceed. of 1991 IEEE Conference on Robotics,* April 1991; pp. 230–39.
23. H. Kazerooni, "Control of Robotic Systems Worn by Humans," *Proceed. of 1991 IEEE Conference on Robotics,* April 1991; pp. 2399–2403.
24. T. Skewis, "A Hospital Transport Robot," *Proceed. of 1991 IEEE Conference on Robotics,* April 1991; pp. 58–63.

✦

CHAPTER 10

The Design and Compensation of Feedback Control Systems

Preview

Thus far we have striven to achieve the desired performance of a system by adjusting one or two parameters. However, parameter adjustment may not result in the desired performance. Thus it may be necessary to introduce a new block within the feedback loop that will compensate for the original system's limitation. This block with a transfer function $G_c(s)$, is called a compensator. It is the purpose of this chapter to develop several design techniques in the frequency and time domain that enable us to achieve the desired system performance.

We will discuss various candidates for service as compensators and show how they help to achieve improved performance. We will also use the method of state variable feedback discussed in Chapter 9 to obtain a so-called optimum performance from a closed-loop control system.

10.1 Introduction

The performance of a feedback control system is of primary importance. This subject was discussed at length in Chapter 4 and quantitative measures of performance were developed. We have found that a suitable control system is stable and that it results in an acceptable response to input commands, is less sensitive to system parameter changes, results in a minimum steady-state error for input commands, and, finally, is able to eliminate the effect of undesirable disturbances. A feedback control system that provides an optimum performance without any necessary adjustments is rare indeed. Usually we find it necessary to compromise among the many conflicting and demanding specifications and to adjust the system parameters to provide a suitable and acceptable performance when it is not possible to obtain all the desired optimum specifications.

We have considered at several points in the preceding chapters the question of design and adjustment of the system parameters in order to provide a desirable response and performance. In Chapter 4, we defined and established several suitable measures of performance. Then, in Chapter 5, we determined a method of investigating the stability of a control system, since we recognized that a system is unacceptable unless it is stable. In Chapter 6, we utilized the root locus method to effect a design of a self-balancing scale (Section 6.4) and then illustrated a method of parameter design by using the root locus method (Section 6.5). Furthermore, in Chapters 7 and 8, we developed suitable measures of performance in terms of the frequency variable ω and utilized them to design several suitable control systems. Finally, using time-domain methods in Chapter 9, we investigated the selection of feedback parameters in order to stabilize a system. Thus we have been considering the problems of the design of feedback control systems as an integral part of the subjects of the preceding chapters. It is now our purpose to study the question somewhat further and to point out several significant design and compensation methods.

We have found in the preceding chapters that it is often possible to adjust the system parameters in order to provide the desired system response. However, we often find that we are not able simply to adjust a system parameter and thus obtain the desired performance. Rather we are forced to reconsider the structure of the system and redesign the system in order to obtain a suitable one. That is, we must examine the scheme or plan of the system and obtain a new design or plan that results in a suitable system. Thus *the design of a control system is concerned with the arrangement, or the plan, of the system structure and the selection of suitable components and parameters.* For example, if we desire a set of performance measures to be less than some specified values, we often encounter a conflicting set of requirements. Thus if we wish a system to have a percent overshoot less than 20% and $\omega_n T_p = 3.3$, we obtain a conflicting requirement on the system damping ratio, ζ, as can be seen by examining Fig. 4.8 again. Now, if we are unable to relax these two performance requirements, we must alter the system in some way. The alteration or adjustment of a control system in order to provide a suitable performance is called *compensation;* that is, compensation is the adjustment of a system in order to make up for deficiencies or inadequacies. It is the purpose of this chapter to consider briefly the issue of the design and compensation of control systems.

In redesigning a control system in order to alter the system response, an additional component is inserted within the structure of the feedback system. It is this additional component or device that equalizes or compensates for the performance deficiency. The compensating device may be an electric, mechanical, hydraulic, pneumatic, or other type of device or network, and is often called a *compensator.* Commonly, an electric circuit serves as a compensator in many control systems. The transfer function of the compensator is designated as $G_c(s) = E_{out}(s)/E_{in}(s)$ and the compensator can be placed in a suitable location within the structure of the system. Several types of compensation are shown in Fig. 10.1 for a simple single-loop feedback control system. The compensator placed in the feedforward path is called a *cascade* or series compensator. Similarly, the other

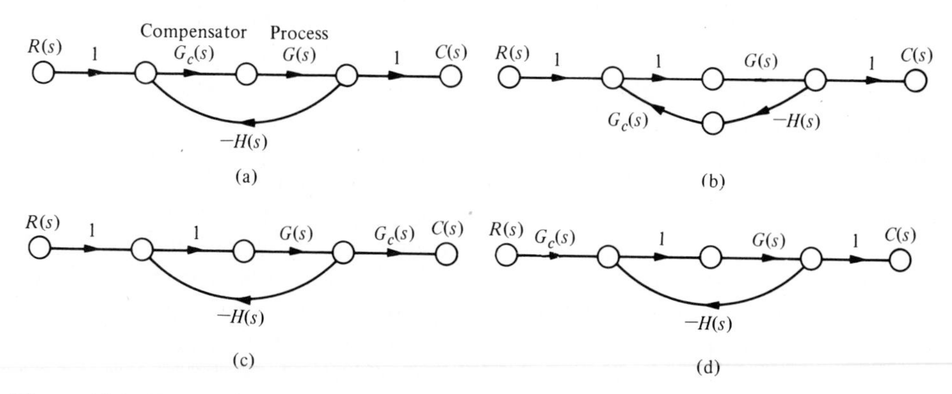

Figure 10.1. Types of compensation. (a) Cascade compensation. (b) Feedback compensation. (c) Output or load compensation. (d) Input compensation.

compensation schemes are called feedback, output or load, and input compensation, as shown in Fig. 10.1(b), (c), and (d), respectively. The selection of the compensation scheme depends upon a consideration of the specifications, the power levels at various signal nodes in the system, and the networks available for use. Usually the output $C(s)$ is a direct output of the process $G(s)$ and the output compensation of Fig. 10.1(c) is not physically realizable. It will not be possible for us to consider all the possibilities in this chapter, and the reader is referred to Chapters 11 and 12 following the introductory material of this chapter.

10.2 Approaches to Compensation

The performance of a control system can be described in terms of the time-domain performance measures or the frequency-domain performance measures. The performance of a system can be specified by requiring a certain peak time, T_p, maximum overshoot, and settling-time for a step input. Furthermore, it is usually necessary to specify the maximum allowable steady-state error for several test signal inputs and disturbance inputs. These performance specifications can be defined in terms of the desirable location of the poles and zeros of the closed-loop system transfer function, $T(s)$. Thus the location of the s-plane poles and zeros of $T(s)$ can be specified. As we found in Chapter 6, the locus of the roots of the closed-loop system can be readily obtained for the variation of one system parameter. However, when the locus of roots does not result in a suitable root configuration, we must add a compensating network (Fig. 10.1) in order to alter the locus of the roots as the parameter is varied. Therefore we can use the root locus method and determine a suitable compensator network transfer function so that the resultant root locus results in the desired closed-loop root configuration.

Alternatively, we can describe the performance of a feedback control system in terms of frequency performance measures. Then a system can be described in terms of the peak of the closed-loop frequency response, M_{p_ω}, the resonant frequency ω_r, the bandwidth, and the phase margin of the system. We can add a suitable compensation network, if necessary, in order to satisfy the system specifications. The design of the network $G_c(s)$, is developed in terms of the frequency response as portrayed on the polar plane, the Bode diagram, or the Nichols chart. Because a cascade transfer function is readily accounted for on a Bode plot by adding the frequency response of the network, we usually prefer to approach the frequency response methods by utilizing the Bode diagram.

Thus the compensation of a system is concerned with the alteration of the frequency response or the root locus of the system in order to obtain a suitable system performance. For frequency response methods we are concerned with altering the system so that the frequency response of the compensated system will satisfy the system specifications. Thus, in the case of the frequency response approach, we use compensation networks to alter and reshape the frequency characteristics represented on the Bode diagram and Nichols chart.

Alternatively, the compensation of a control system can be accomplished in the s-plane by root locus methods. For the case of the s-plane, the designer wishes to alter and reshape the root locus so that the roots of the system will lie in the desired position in the s-plane.

The time-domain method, expressed in terms of state variables, can also be utilized to design a suitable compensation scheme for a control system. Typically, we are interested in controlling the system with a control signal, $\mathbf{u}(t)$, which is a function of several measurable state variables. Then we develop a state-variable controller that operates on the information available in measured form. This type of system compensation is quite useful for system optimization and will be considered briefly in this chapter.

We have illustrated several of the aforementioned approaches in the preceding chapters. In Example 6.5, we utilized the root locus method in considering the design of a feedback network in order to obtain a satisfactory performance. In Chapter 8, we considered the selection of the gain in order to obtain a suitable phase margin and, therefore, a satisfactory relative stability. Also, in Example 9.6, we compensated for the unstable response of the pendulum by controlling the pendulum with a function of several of the state variables of the system.

Quite often, in practice, the best and simplest way to improve the performance of a control system is to alter, if possible, the process itself. That is, if the system designer is able to specify and alter the design of the process that is represented by the transfer function $G(s)$, then the performance of the system can be readily improved. For example, in order to improve the transient behavior of a servomechanism position controller, we can often choose a better motor for the system. In the case of an airplane control system we might be able to alter the aerodynamic design of the airplane and thus improve the flight transient characteristics. Thus a control-system designer should recognize that an alteration of the process may result in an improved system. However, often the process is fixed and unalterable or has been altered as much as possible and is still found to result in an unsatisfactory performance. Then the addition of compensation networks becomes useful for improving the performance of the system. In the following sections we will assume that the process has been improved as much as possible and the $G(s)$ representing the process is unalterable.

It is the purpose of this chapter to further describe the addition of several compensation networks to a feedback control system. First, we shall consider the addition of a so-called phase-lead compensation network and describe the design of the network by root locus and frequency response techniques. Then, using both the root locus and frequency response techniques, we shall describe the design of the integration compensation networks in order to obtain a suitable system performance. Finally, we shall determine an optimum controller for a system described in terms of state variables. While these three approaches to compensation are not intended to be discussed in a complete manner, the discussion that follows should serve as a worthwhile introduction to the design and compensation of feedback control systems.

10.3 Cascade Compensation Networks

In this section we shall consider the design of a cascade or feedback network as shown in Fig. 10.1(a) and Fig. 10.1(b), respectively. The compensation network, $G_c(s)$, is cascaded with the unalterable process $G(s)$ in order to provide a suitable loop transfer function $G_c(s)G(s)H(s)$. Clearly, the compensator $G_c(s)$ can be chosen to alter either the shape of the root locus or the frequency response. In either case the network may be chosen to have a transfer function

$$G_c(s) = \frac{K \prod_{i=1}^{M} (s + z_i)}{\prod_{j=1}^{N} (s + p_j)}. \tag{10.1}$$

Then the problem reduces to the judicious selection of the poles and zeros of the compensator. In order to illustrate the properties of the compensation network we shall consider a first-order compensator. The compensation approach developed on the basis of a first-order compensator can then be extended to higher-order compensators.

Consider the first-order compensator with the transfer function

$$G_c(s) = \frac{K(s + z)}{(s + p)}. \tag{10.2}$$

The design problem becomes, then, the selection of z, p, and K in order to provide a suitable performance. When $|z| < |p|$, the network is called a *phase-lead network* and has a pole–zero s-plane configuration as shown in Fig. 10.2. If the pole was negligible, that is, $|p| \gg |z|$, and the zero occurred at the origin of the s-plane, we would have a differentiator so that

$$G_c(s) \simeq \left(\frac{K}{p}\right) s. \tag{10.3}$$

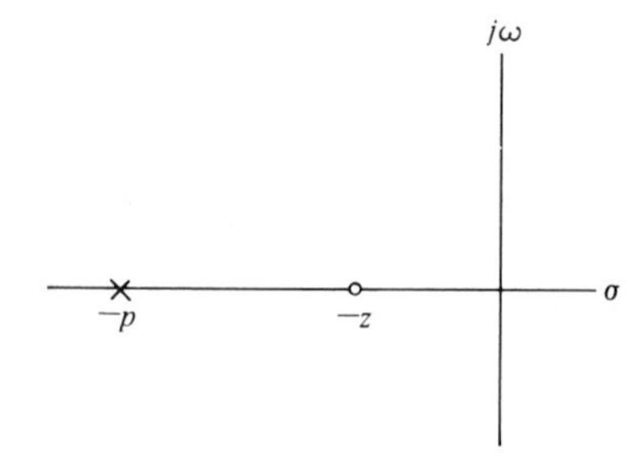

Figure 10.2. Pole–zero diagram of the phase-lead network.

Thus a compensation network of the form of Eq. (10.2) is a differentiator type network. The differentiator network of Eq. (10.3) has a frequency characteristic as

$$G_c(j\omega) = j\left(\frac{K}{p}\right)\omega = \left(\frac{K}{p}\omega\right)e^{+j90^\circ} \tag{10.4}$$

and phase angle of $+90°$, often called a phase-lead angle. Similarly, the frequency response of the differentiating network of Eq. (10.2) is

$$\begin{aligned} G_c(j\omega) &= \frac{K(j\omega + z)}{(j\omega + p)} \\ &= \frac{(Kz/p)(j(\omega/z) + 1)}{(j(\omega/p) + 1)} \\ &= \frac{K_1(1 + j\omega\alpha\tau)}{(1 + j\omega\tau)}, \end{aligned} \tag{10.5}$$

where $\tau = 1/p$, $p = \alpha z$, and $K_1 = K/\alpha$. The frequency response of this phase-lead network is shown in Fig. 10.3. The angle of the frequency characteristic is

$$\phi(\omega) = \tan^{-1}\alpha\omega\tau - \tan^{-1}\omega\tau. \tag{10.6}$$

Since the zero occurs first on the frequency axis, we obtain a phase-lead characteristic as shown in Fig. 10.3. The slope of the asymptotic magnitude curve is + 6 db/octave.

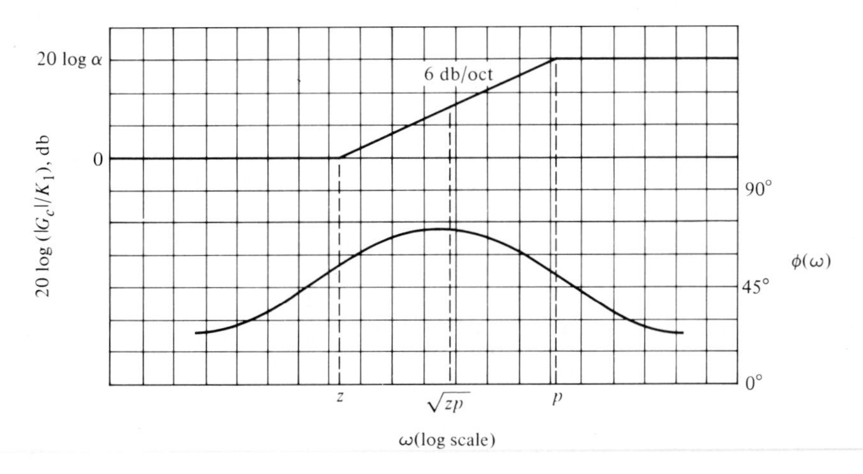

Figure 10.3. Bode diagram of the phase-lead network.

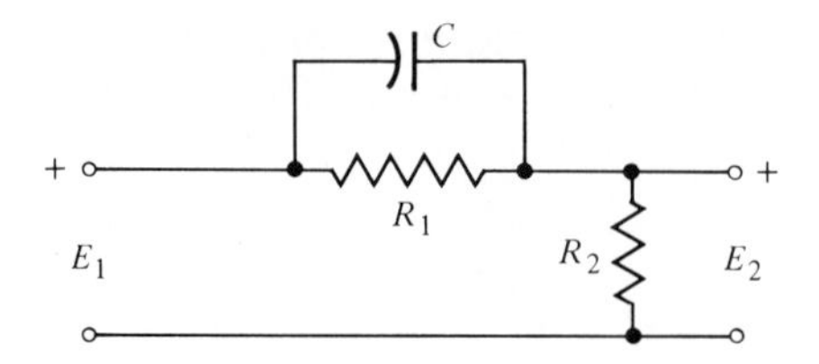

Figure 10.4. Phase-lead network.

The phase-lead compensation transfer function can be obtained with the network shown in Fig. 10.4. The transfer function of this network is

$$G_c(s) = \frac{E_2(s)}{E_1(s)} = \frac{R_2}{R_2 + \{R_1(1/Cs)/[R_1 + (1/Cs)]\}}$$
$$= \left(\frac{R_2}{R_1 + R_2}\right)\frac{(R_1Cs + 1)}{\{[R_1R_2/(R_1 + R_2)]Cs + 1\}}. \tag{10.7}$$

Therefore we let

$$\tau = \frac{R_1R_2}{R_1 + R_2}C \quad \text{and} \quad \alpha = \frac{R_1 + R_2}{R_2}$$

and obtain the transfer function

$$G_c(s) = \frac{(1 + \alpha\tau s)}{\alpha(1 + \tau s)}, \tag{10.8}$$

which is equal to Eq. (10.5) when an additional cascade gain K is inserted.

The maximum value of the phase lead occurs at a frequency ω_m, where ω_m is the geometric mean of $p = 1/\tau$ and $z = 1/\alpha\tau$; that is, the maximum phase lead occurs halfway between the pole and zero frequencies on the logarithmic frequency scale. Therefore

$$\omega_m = \sqrt{zp} = \frac{1}{\tau\sqrt{\alpha}}.$$

In order to obtain an equation for the maximum phase-lead angle, we rewrite the phase angle of Eq. (10.5) as

$$\phi = \tan^{-1}\frac{\alpha\omega\tau - \omega\tau}{1 + (\omega\tau)^2\alpha}. \tag{10.9}$$

Then, substituting the frequency for the maximum phase angle, $\omega_m = 1/\tau\sqrt{\alpha}$, we have

$$\tan\phi_m = \frac{(\alpha/\sqrt{\alpha}) - (1/\sqrt{\alpha})}{1 + 1}$$
$$= \frac{\alpha - 1}{2\sqrt{\alpha}}. \tag{10.10}$$

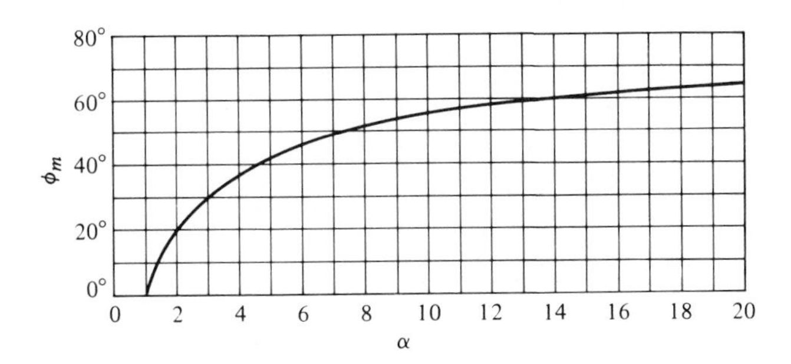

Figure 10.5. Maximum phase angle ϕ_m versus α for a lead network.

Because the tan ϕ_m equals $(\alpha - 1)/2\sqrt{\alpha}$, we utilize the triangular relationship and note that

$$\sin \phi_m = \frac{\alpha - 1}{\alpha + 1}. \tag{10.11}$$

Equation (10.11) is very useful for calculating a necessary α ratio between the pole and zero of a compensator in order to provide a required maximum phase lead. A plot of ϕ_m versus α is shown in Fig. 10.5. Clearly, the phase angle readily obtainable from this network is not much greater than 70°. Also, since $\alpha = (R_1 + R_2)/R_2$, there are practical limitations on the maximum value of α that one should attempt to obtain. Therefore, if one required a maximum angle of greater than 70°, two cascade compensation networks would be utilized. Then the equivalent compensation transfer function is $G_{c1}(s)G_{c2}(s)$ when the loading effect of $G_{c2}(s)$ on $G_{c1}(s)$ is negligible.

It is often useful to add a cascade compensation network that provides a phase-lag characteristic. The *phase-lag network* is shown in Fig. 10.6. The transfer function of the phase-lag network is

$$\begin{aligned} G_c(s) = \frac{E_2(s)}{E_1(s)} &= \frac{R_2 + (1/Cs)}{R_1 + R_2 + (1/Cs)} \\ &= \frac{R_2Cs + 1}{(R_1 + R_2)Cs + 1}. \end{aligned} \tag{10.12}$$

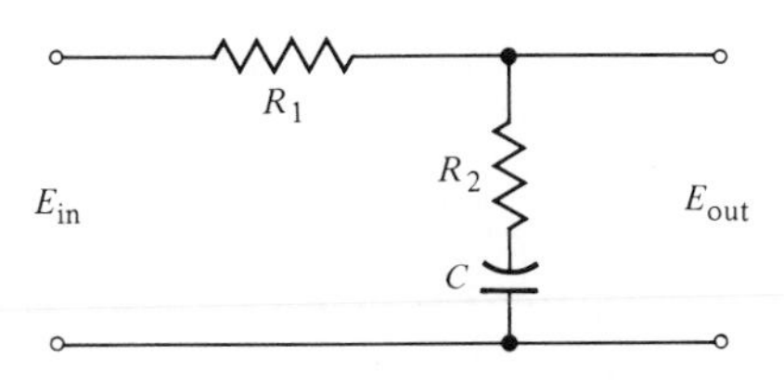

Figure 10.6. Phase-lag network.

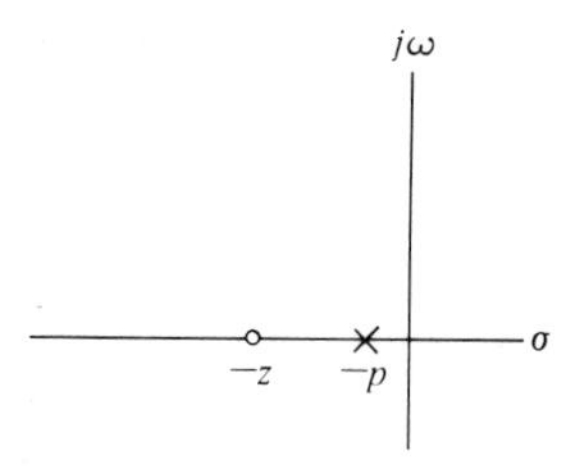

Figure 10.7. Pole-zero diagram of the phase-lag network.

When $\tau = R_2C$ and $\alpha = (R_1 + R_2)/R_2$, we have

$$\begin{aligned} G_c(s) &= \frac{1 + \tau s}{1 + \alpha\tau s} \\ &= \frac{1}{\alpha}\frac{(s + z)}{(s + p)}, \end{aligned} \tag{10.13}$$

where $z = 1/\tau$ and $p = 1/\alpha\tau$. In this case, because $\alpha > 1$, the pole lies closest to the origin of the s-plane as shown in Fig. 10.7. This type of compensation network

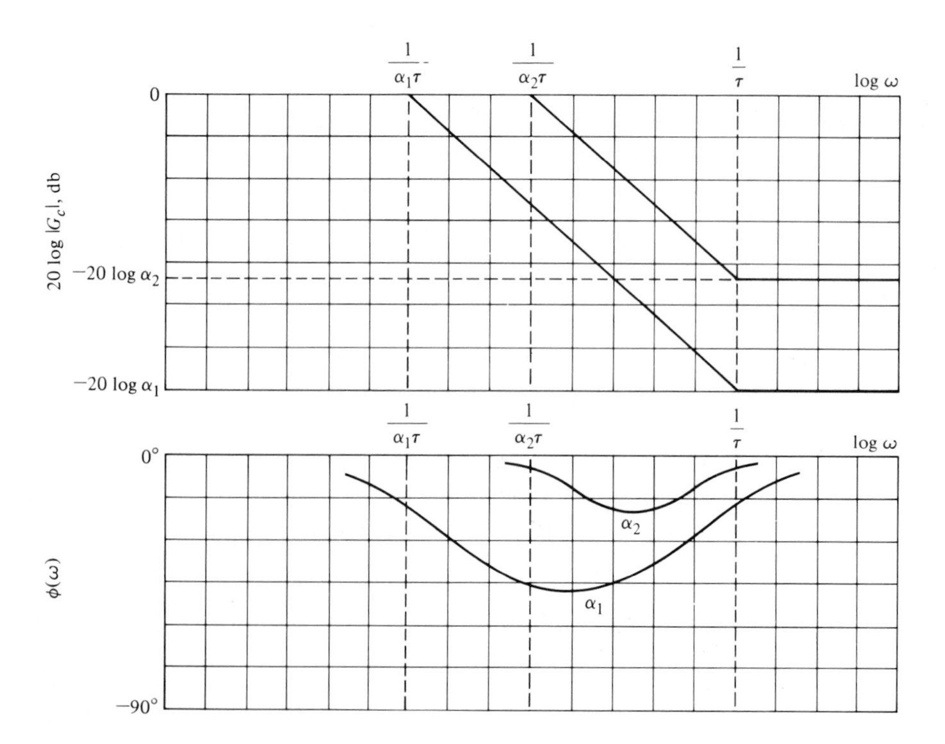

Figure 10.8. Bode diagram of the phase-lag network.

is often called an integrating network. The Bode diagram of the phase-lag network is obtained from the transfer function

$$G_c(j\omega) = \frac{1 + j\omega\tau}{1 + j\omega\alpha\tau} \tag{10.14}$$

and is shown in Fig. 10.8. The form of the Bode diagram of the lag network is similar to that of the phase-lead network; the difference is the resulting attenuation and phase-lag angle instead of amplification and phase-lead angle. However, note that the shape of the diagrams of Figs. 10.3 and 10.8 are similar. Therefore it can be shown that the maximum phase lag occurs at $\omega_m = \sqrt{zp}$.

In the succeeding sections, we wish to utilize these compensation networks in order to obtain a desired system frequency locus or s-plane root location. The lead network is utilized to provide a phase-lead angle and thus a satisfactory phase margin for a system. Alternatively, the use of the phase-lead network can be visualized on the s-plane as enabling us to reshape the root locus and thus provide the desired root locations. The phase-lag network is utilized not to provide a phase-lag angle, which is normally a destabilizing influence, but rather to provide an attenuation and increase the steady-state error constant [3]. These approaches to compensation utilizing the phase-lead and phase-lag networks will be the subject of the following four sections.

10.4 System Compensation on the Bode Diagram Using the Phase-Lead Network

The Bode diagram is used in order to design a suitable phase-lead network in preference to other frequency response plots. The frequency response of the cascade compensation network is added to the frequency response of the uncompensated system. That is, because the total loop transfer function of Fig. 10.1(a) is $G_c(j\omega)G(j\omega)H(j\omega)$, we will first plot the Bode diagram for $G(j\omega)H(j\omega)$. Then we can examine the plot for $G(j\omega)H(j\omega)$ and determine a suitable location for p and z of $G_c(j\omega)$ in order to satisfactorily reshape the frequency response. The uncompensated $G(j\omega)$ is plotted with the desired gain to allow an acceptable steady-state error. Then the phase margin and the expected M_p are examined to find whether they satisfy the specifications. If the phase margin is not sufficient, phase lead can be added to the phase angle curve of the system by placing the $G_c(j\omega)$ in a suitable location. Clearly, in order to obtain maximum additional phase lead, we desire to place the network such that the frequency ω_m is located at the frequency where the magnitude of the compensated magnitude curves crosses the 0-db axis. (Recall the definition of phase margin.) The value of the added phase lead required allows us to determine the necessary value for α from Eq. (10.11) or Fig. 10.5. The zero $\omega = 1/\alpha\tau$ is located by noting that the maximum phase lead should occur at $\omega_m = \sqrt{zp}$, halfway between the pole and the zero. Because the total magnitude gain for the network is 20 log α, we expect a gain of 10 log α at ω_m. Thus we determine the compensation network by completing the following steps:

1. Evaluate the uncompensated system phase margin when the error constants are satisfied.
2. Allowing for a small amount of safety, determine the necessary additional phase lead, ϕ_m.
3. Evaluate α from Eq. (10.11).
4. Evaluate 10 log α and determine the frequency where the uncompensated magnitude curve is equal to $-10 \log \alpha$ db. This frequency is the new 0-db crossover frequency and ω_m simultaneously, because the compensation network provides a gain of 10 log α at ω_m.
5. Draw the compensated frequency response, check the resulting phase margin, and repeat the steps if necessary. Finally, for an acceptable design, raise the gain of the amplifier in order to account for the attenuation $(1/\alpha)$.

■ Example 10.1 A lead compensator for a type 2 system

Let us consider a single-loop feedback control system as shown in Fig. 10.1(a), where

$$G(s) = \frac{K_1}{s^2} \tag{10.15}$$

and $H(s) = 1$. The uncompensated system is a type 2 system and at first appears to possess a satisfactory steady-state error for both step and ramp input signals. However, uncompensated, the response of the system is an undamped oscillation because

$$T(s) = \frac{C(s)}{R(s)} = \frac{K_1}{s^2 + K_1}. \tag{10.16}$$

Therefore the compensation network is added so that the loop transfer function is $G_c(s)G(s)H(s)$. The specifications for the system are

Settling time, $T_s \leq 4$ sec,

Percent overshoot for a step input $\leq 20\%$.

Using Fig. 4.8, we estimate that the damping ratio should be $\zeta \geq 0.45$. The settling time requirement is

$$T_s = \frac{4}{\zeta\omega_n} = 4, \tag{10.17}$$

and therefore

$$\omega_n = \frac{1}{\zeta} = \frac{1}{0.45} = 2.22.$$

Perhaps the simplest way to check the value of ω_n for the frequency response is to relate ω_n to the bandwidth, ω_B, and evaluate the -3db bandwidth of the closed-loop system. For a closed-loop system with $\zeta = 0.45$, we estimate from Fig. 7.9 that $\omega_B = 1.36\omega_n$. Therefore we require a closed-loop bandwidth $\omega_B = 1.36(2.22) = 3.02$. The bandwidth can be checked following compensation by utilizing the Nichols chart. For the uncompensated system, the bandwidth of the system is $\omega_B = 1.36\omega_n$ and $\omega_n = \sqrt{K}$. Therefore a loop gain equal to $K = \omega_n^2 \simeq 5$ would be sufficient. To provide a suitable margin for the settling time we will select $K = 10$ in order to draw the Bode diagram of

$$GH(j\omega) = \frac{K}{(j\omega)^2}.$$

The Bode diagram of the uncompensated system is shown as solid lines in Fig. 10.9.

By using Eq. (8.58), the phase margin of the system is required to be approximately

$$\phi_{pm} = \frac{\zeta}{0.01} = \frac{0.45}{0.01} = 45°. \tag{10.18}$$

The phase margin of the uncompensated system is 0° because the double integration results in a constant 180° phase lag. Therefore we must add a 45° phase-lead angle at the crossover (0-db) frequency of the compensated magnitude curve. Evaluating the value of α, we have

$$\begin{aligned}\frac{\alpha - 1}{\alpha + 1} &= \sin \phi_m \\ &= \sin 45°,\end{aligned} \tag{10.19}$$

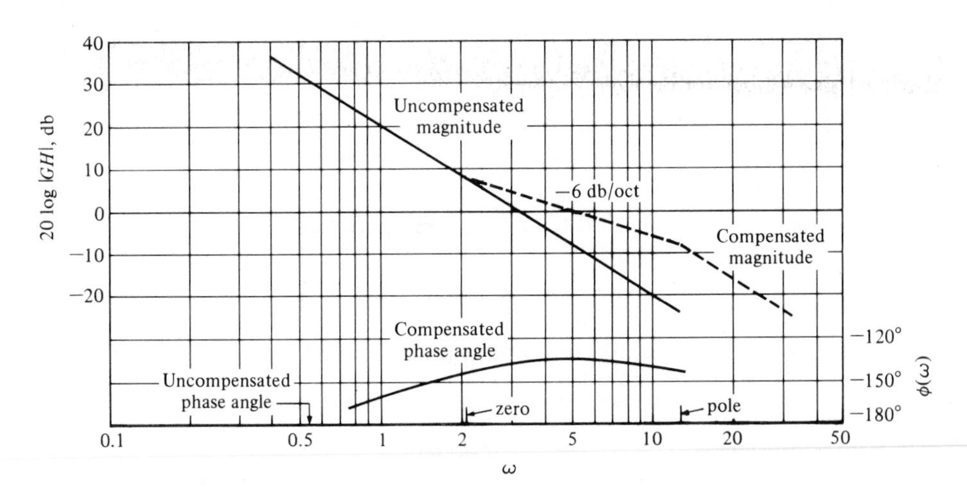

Figure 10.9. Bode diagram for Example 10.1.

and therefore $\alpha = 5.8$. In order to provide a margin of safety, we will use $\alpha = 6$. The value of 10 log α is then equal to 7.78 db. Then the lead network will add an additional gain of 7.78 db at the frequency ω_m, and it is desired to have ω_m equal to compensated slope near the 0-db axis (the dashed line) so that the new crossover is ω_m and the dashed magnitude curve is 7.78 db above the uncompensated curve at the crossover frequency. Thus the compensated crossover frequency is located by evaluating the frequency where the uncompensated magnitude curve is equal to -7.78 db, which in this case is $\omega = 4.9$. Then the maximum phase-lead angle is added to $\omega = \omega_m = 4.9$ as shown in Fig. 10.9. The bandwidth of the compensated system can be obtained from the Nichols chart. For estimating the bandwidth, we can simply examine Fig. 8.23 and note that the -3-db line for the closed-loop system occurs when the magnitude of $GH(j\omega)$ is -6-db and the phase shift of $GH(j\omega)$ is approximately $-140°$. Therefore, in order to estimate the bandwidth from the open-loop diagram, we will approximate the bandwidth as the frequency for which $20 \log |GH|$ is equal to -6 db. Thus the bandwidth of the uncompensated system is approximately equal to $\omega_B = 4.4$, while the bandwidth of the compensated system is equal to $\omega_B = 8.4$. The lead compensation doubles the bandwidth in this case and the specification that $\omega_B > 3.02$ is satisfied. Therefore the compensation of the system is completed and the system specifications are satisfied. The total compensated loop transfer function is

$$G_c(j\omega)G(j\omega)H(j\omega) = \frac{10[(j\omega/2.1) + 1]}{(j\omega)^2[(j\omega/12.6) + 1]}. \tag{10.20}$$

The transfer function of the compensator is

$$\begin{aligned} G_c(s) &= \frac{(1 + \alpha\tau s)}{\alpha(1 + \tau s)} \\ &= \frac{1}{6}\frac{[1 + (s/2.1)]}{[1 + (s/12.6)]} \end{aligned} \tag{10.21}$$

in the form of Eq. (10.8). Because an attenuation of $\frac{1}{6}$ results from the passive RC network, the gain of the amplifier in the loop must be raised by a factor of six so that the total dc loop gain is still equal to 10 as required in Eq. (10.20). When we add the compensation network Bode diagram to the uncompensated Bode diagram as in Fig. 10.9, we are assuming that we can raise the amplifier gain in order to account for this $1/\alpha$ attenuation. The pole and zero values are simply read from Fig. 10.9, noting that $p = \alpha z$.

■ Example 10.2 A lead compensator for a second-order system

A feedback control system has a loop transfer function

$$GH(s) = \frac{K}{s(s + 2)}. \tag{10.22}$$

It is desired to have a steady-state error for a ramp input less than 5% of the magnitude of the ramp. Therefore we require that

$$K_v = \frac{A}{e_{ss}} = \frac{A}{0.05A} = 20. \tag{10.23}$$

Furthermore, we desire that the phase margin of the system be at least 45°. The first step is to plot the Bode diagram of the uncompensated transfer function

$$\begin{aligned} GH(j\omega) &= \frac{K_v}{j\omega(0.5j\omega + 1)} \\ &= \frac{20}{j\omega(0.5j\omega + 1)} \end{aligned} \tag{10.24}$$

as shown in Fig. 10.10(a). The frequency at which the magnitude curve crosses the 0-db line is 6.2 rad/sec, and the phase margin at this frequency is determined readily from the equation of the phase of $GH(j\omega)$, which is

$$\underline{/GH(j\omega)} = \phi(\omega) = -90° - \tan^{-1}(0.5\omega). \tag{10.25}$$

At the crossover frequency, $\omega = \omega_c = 6.2$ rad/sec, we have

$$\phi(\omega) = -162°, \tag{10.26}$$

and therefore the phase margin is 18°. Using Eq. (10.25) to evaluate the phase margin is often easier than drawing the complete phase angle curve, which is shown in Fig. 10.10(a). Thus we need to add a phase-lead network so that the phase margin is raised to 45° at the new crossover (0-db) frequency. Because the

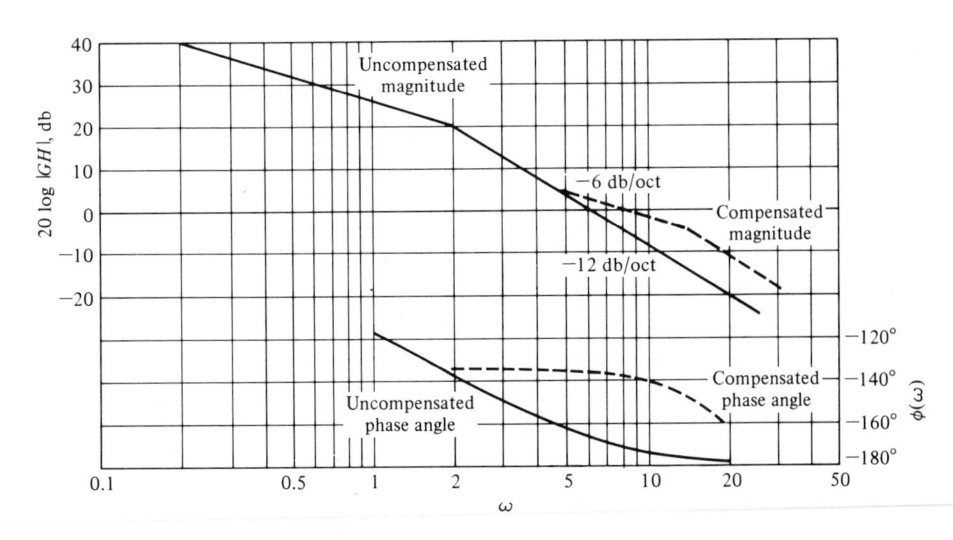

Figure 10.10(a). Bode diagram for Example 10.2.

compensation crossover frequency is greater than the uncompensated crossover frequency, the phase lag of the uncompensated system is greater also. We shall account for this additional phase lag by attempting to obtain a maximum phase lead of 45° − 18° = 27° plus a small increment (10%) of phase lead to account for the added lag. Thus we will design a compensation network with a maximum phase lead equal to 27° + 3° = 30°. Then, calculating α, we obtain

$$\frac{\alpha - 1}{\alpha + 1} = \sin 30° = 0.5, \tag{10.27}$$

and therefore $\alpha = 3$.

The maximum phase lead occurs at ω_m, and this frequency will be selected so that the new crossover frequency and ω_m coincide. The magnitude of the lead network at ω_m is $10 \log \alpha = 10 \log 3 = 4.8$ db. The compensated crossover frequency is then evaluated where the magnitude of $GH(j\omega)$ is -4.8 db and thus $\omega_m = \omega_c = 8.4$. Drawing the compensated magnitude line so that it intersects the 0-db axis at $\omega = \omega_c = 8.4$, we find that $z = 4.8$ and $p = \alpha z = 14.4$. Therefore the compensation network is

$$G_c(s) = \frac{1}{3}\frac{(1 + s/4.8)}{(1 + s/14.4)}. \tag{10.28}$$

The total dc loop gain must be raised by a factor of 3 in order to account for the factor $1/\alpha = 1/3$. Then the compensated loop transfer function is

$$G_c(s)GH(s) = \frac{20[(s/4.8) + 1]}{s(0.5s + 1)[(s/14.4) + 1]}. \tag{10.29}$$

In order to verify the final phase margin, we can evaluate the phase of $G_c(j\omega)GH(j\omega)$ at $\omega = \omega_c = 8.4$ and therefore obtain the phase margin. The phase angle is then

$$\begin{aligned} \phi(\omega_c) &= -90° - \tan^{-1} 0.5\omega_c - \tan^{-1}\frac{\omega_c}{14.4} + \tan^{-1}\frac{\omega_c}{4.8} \\ &= -90° - 76.5° - 30.0° + 60.2° \\ &= -136.3°. \end{aligned} \tag{10.30}$$

Therefore the phase margin for the compensated system is 43.7°. If we desire to have exactly 45° phase margin, we would repeat the steps with an increased value of α—for example, with $\alpha = 3.5$. In this case, the phase lag increased by 7° between $\omega = 6.2$ and $\omega = 8.4$, and therefore the allowance of 3° in the calculation of α was not sufficient. The step response of this system yields a 28% overshoot with a settling time of 0.75 sec.

The Nichols diagram for the compensated and uncompensated system is shown on Fig. 10.10(b). The reshaping of the frequency response locus is clear on this diagram. One notes the increased phase margin for the compensated system

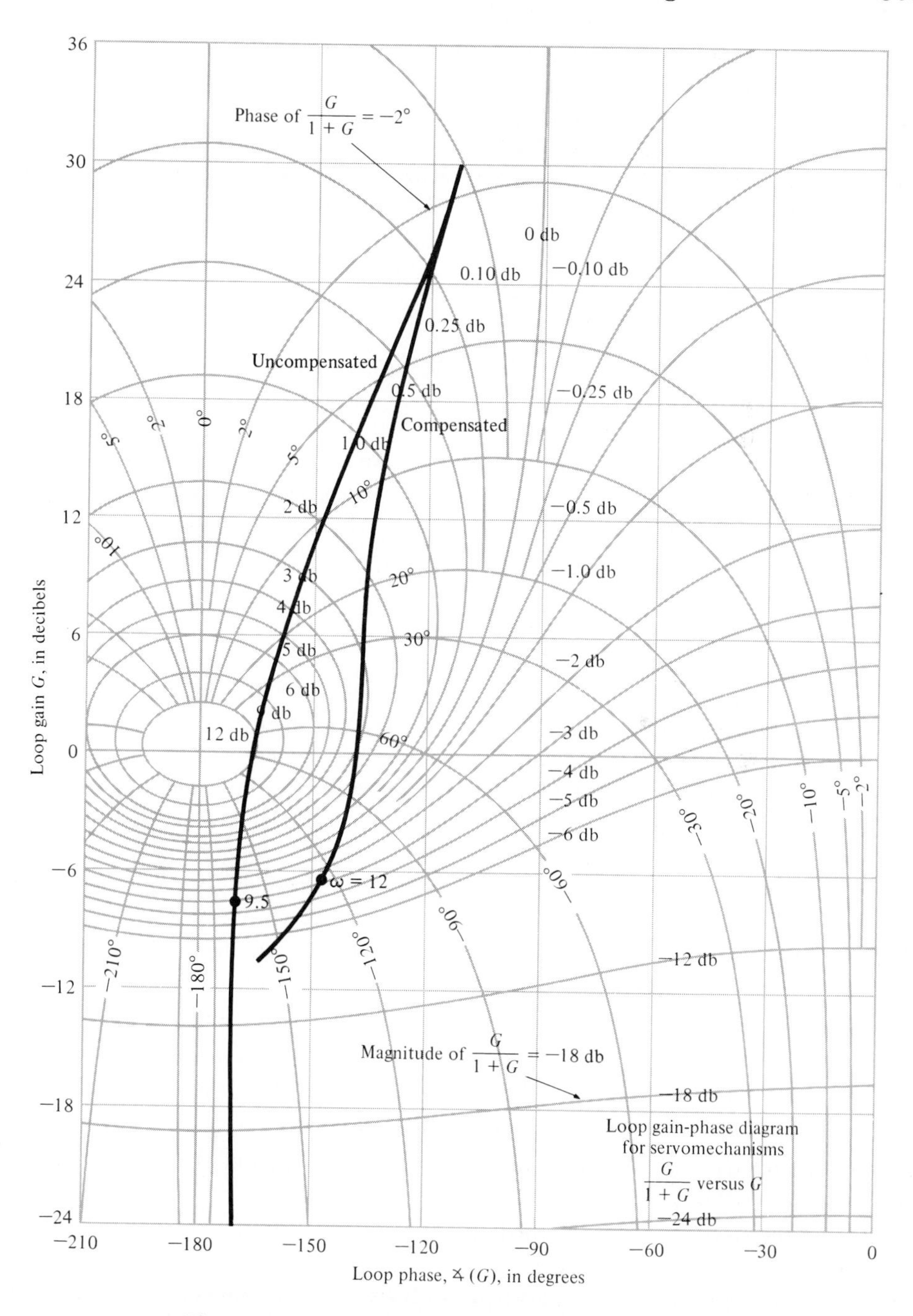

Figure 10.10(b). Nichols diagram for Example 10.2.

as well as the reduced magnitude of M_{p_ω}, the maximum magnitude of the closed-loop frequency response. In this case, M_{p_ω} has been reduced from an uncompensated value of +12 db to a compensated value of approximately +3.2 db. Also, we note that the closed-loop 3-db bandwidth of the compensated system is equal to 12 rad/sec compared with 9.5 rad/sec for the uncompensated system.

Looking again at Examples 10.1 and 10.2, we note that the system design is satisfactory when the asymptotic curve for the magnitude $20 \log|GG_c|$ crosses the 0 db line with a slope of −6 db/octave.

10.5 Compensation on the *s*-Plane Using the Phase-Lead Network

The design of the phase-lead compensation network can also be readily accomplished on the *s*-plane. The phase-lead network has a transfer function

$$\begin{aligned} G_c(s) &= \frac{[s + (1/\alpha\tau)]}{[s + (1/\tau)]} \\ &= \frac{(s + z)}{(s + p)}, \end{aligned} \tag{10.31}$$

where α and τ are defined for the *RC* network in Eq. (10.7). The locations of the zero and pole are selected in order to result in a satisfactory root locus for the compensated system. The specifications of the system are used to specify the desired location of the dominant roots of the system. The *s*-plane root locus method is as follows:

1. List the system specifications and translate these specifications into a desired root location for the dominant roots.
2. Sketch the uncompensated root locus and determine whether the desired root locations can be realized with an uncompensated system.
3. If the compensator is necessary, place the zero of the phase-lead network directly below the desired root location (or to the left of the second pole).
4. Determine the pole location so that the total angle at the desired root location is 180° and therefore is on the compensated root locus.
5. Evaluate the total system gain at the desired root location and then calculate the error constant.
6. Repeat the steps if the error constant is not satisfactory.

Therefore we first locate our desired dominant root locations so that the dominant roots satisfy the specifications in terms of ζ and ω_n as shown in Fig. 10.11(a). The root locus of the uncompensated system is sketched as illustrated in Fig. 10.11(b). Then the zero is added to provide a phase lead of +90° by placing it directly below the desired root location. Actually, some caution must be main-

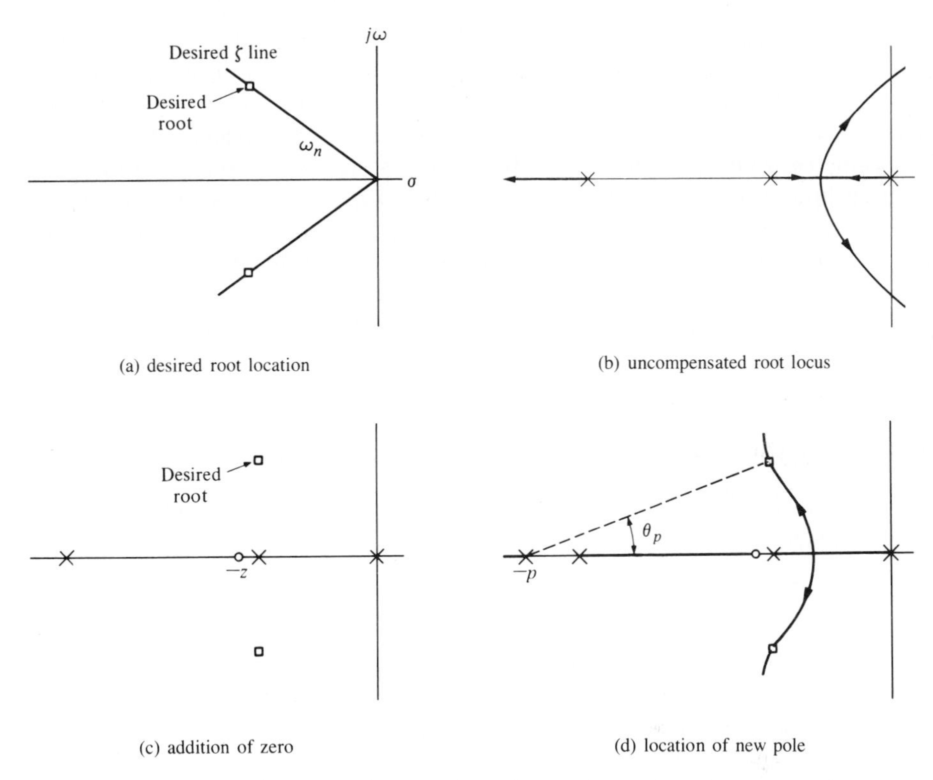

Figure 10.11. Compensation on the s-plane using a phase-lead network.

tained because the zero must not alter the dominance of the desired roots; that is, the zero should not be placed nearer the origin than the second pole on the real axis or a real root near the origin will result and will dominate the system response. Thus, in Fig. 10.11(c), we note that the desired root is directly above the second pole, and we place the zero z somewhat to the left of the pole.

Consequently the real root will be near the real zero and the coefficient of this term of the partial fraction expansion will be relatively small. Thus the response due to this real root will have very little effect on the overall system response. Nevertheless, the designer must be continually aware that the compensated system response will be influenced by the roots and zeros of the system and the dominant roots will not by themselves dictate the response. It is usually wise to allow for some margin of error in the design and to test the compensated system using a digital simulation.

Because the desired root is a point on the root locus when the final compensation is accomplished, we expect the algebraic sum of the vector angles to be 180° at that point. Thus we calculate the angle from the pole of compensator, θ_p, in order to result in a total angle of 180°. Then, locating a line at an angle θ_p intersecting the desired root, we are able to evaluate the compensator pole, p, as shown in Fig. 10.11(d).

The advantage of the s-plane method is the ability of the designer to specify the location of the dominant roots and, therefore, the dominant transient response. The disadvantage of the method is that one cannot directly specify an error constant (for example, K_v) as in the Bode diagram approach. After the design is completed, one evaluates the gain of the system at the root location, which depends upon p and z, and then calculates the error constant for the compensated system. If the error constant is not satisfactory, one must repeat the design steps and alter the location of the desired root as well as the location of the compensator pole and zero. We shall reconsider the two examples we completed in the preceding section and design a compensation network using the root locus (s-plane) approach.

■ Example 10.3 Lead compensator on the s-plane

Let us reconsider the system of Example 10.1 where the open-loop uncompensated transfer function is

$$GH(s) = \frac{K_1}{s^2}. \tag{10.32}$$

The characteristic equation of the uncompensated system is

$$1 + GH(s) = 1 + \frac{K_1}{s^2} = 0, \tag{10.33}$$

and the root locus is the $j\omega$-axis. Therefore we desire to compensate this system with a network, $G_c(s)$, where

$$G_c(s) = \frac{s + z}{s + p} \tag{10.34}$$

and $|z| < |p|$. The specifications for the system are

Settling time, $T_s \leq 4$ sec,

Percent overshoot for a step input $\leq 30\%$.

Therefore the damping ratio should be $\zeta \geq 0.35$. The settling time requirement is

$$T_s = \frac{4}{\zeta\omega_n} = 4,$$

and therefore $\zeta\omega_n = 1$. Thus we will choose a desired dominant root location as

$$r_1, \hat{r}_1 = -1 \pm j2 \tag{10.35}$$

as shown in Fig. 10.12 (thus $\zeta = 0.45$).

Now we place the zero of the compensator directly below the desired location

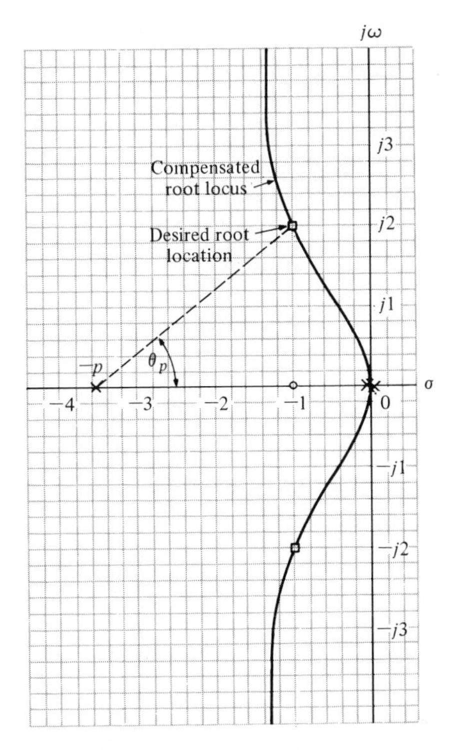

Figure 10.12. Phase-lead compensation for Example 10.3.

at $s = -z = -1$, as shown in Fig. 10.12. Then, measuring the angle at the desired root, we have

$$\phi = -2(116°) + 90° = -142°.$$

Therefore, in order to have a total of 180° at the desired root, we evaluate the angle from the undetermined pole, θ_p, as

$$-180° = -142° - \theta_p \tag{10.36}$$

or $\theta_p = 38°$. Then a line is drawn at an angle $\theta_p = 38°$ intersecting the desired root location and the real axis, as shown in Fig. 10.12. The point of intersection with the real axis is then $s = -p = -3.6$. Therefore the compensator is

$$G_c(s) = \frac{s + 1}{s + 3.6}, \tag{10.37}$$

and the compensated transfer function for the system is

$$GH(s)G_c(s) = \frac{K_1(s + 1)}{s^2(s + 3.6)}. \tag{10.38}$$

The gain K_1 is evaluated by measuring the vector lengths from the poles and zeros to the root location. Hence

$$K_1 = \frac{(2.23)^2(3.25)}{2} = 8.1. \tag{10.39}$$

Finally, the error constants of this system are evaluated. We find that this system with two open-loop integrations will result in a zero steady-state error for a step and ramp input signal. The acceleration constant is

$$K_a = \frac{8.1}{3.6} = 2.25. \tag{10.40}$$

The steady-state performance of this system is quite satisfactory, and therefore the compensation is complete. When we compare the compensation network evaluated by the s-plane method with the network obtained by using the Bode diagram approach, we find that the magnitudes of the poles and zeros are different. However, the resulting system will have the same performance and we need not be concerned with the difference. In fact, the difference arises from the arbitrary design step (Number 3), which places the zero directly below the desired root location. If we placed the zero at $s = -2.1$, we would find that the pole evaluated by the s-plane method is approximately equal to the pole evaluated by the Bode diagram approach.

The specifications for the transient response of this system were originally expressed in terms of the overshoot and the settling time of the system. These specifications were translated, on the basis of an approximation of the system by a second-order system, to an equivalent ζ and ω_n and therefore a desired root location. However, the original specifications will be satisfied only if the roots selected are dominant. The zero of the compensator and the root resulting from the addition of the compensator pole result in a third-order system with a zero. The validity of approximating this system with a second-order system without a zero is dependent upon the validity of the dominance assumption. Often the designer will simulate the final design by using an analog computer or a digital computer and obtain the actual transient response of the system. In this case, an analog computer simulation of the system resulted in an overshoot of 40% and a settling time of 3.8 sec for a step input. These values compare moderately well with the specified values of 30% and 4 sec and justify the utilization of the dominant root specifications. The difference in the overshoot from the specified value is due to the third root, which is not negligible. Thus again we find that the specification of dominant roots is a useful approach but must be utilized with caution and understanding. A second attempt to obtain a compensated system with an overshoot of 30% would utilize a compensator with a zero at -2 and then calculate the necessary pole location to yield the desired root locations for the dominant roots. This approach would move the third root farther to the left in the s-plane, reduce the effect of the third root on the transient response, and reduce the overshoot.

■ Example 10.4 Lead compensator for a type 1 system

Now let us reconsider the system of Example 10.2 and design a compensator based on the s-plane approach. The open-loop system transfer function is

$$GH(s) = \frac{K}{s(s+2)}. \tag{10.41}$$

It is desired that the damping ratio of the dominant roots of the system be $\zeta = 0.45$ and that the velocity error constant be equal to 20. In order to satisfy the error constant requirement, the gain of the uncompensated system must be $K = 40$. When $K = 40$, the roots of the uncompensated system are

$$s^2 + 2s + 40 = (s + 1 + j6.25)(s + 1 - j6.25). \tag{10.42}$$

The damping ratio of the uncompensated roots is approximately 0.16, and therefore a compensation network must be added. In order to achieve a rapid settling time, we will select the real part of the desired roots as $\zeta\omega_n = 4$ and therefore $T_s = 1$ sec. Also, the natural frequency of these roots is fairly large, $\omega_n = 9$; hence the velocity constant should be reasonably large. The location of the desired roots is shown on Fig. 10.13(a) for $\zeta\omega_n = 4$, $\zeta = 0.45$, and $\omega_n = 9$.

The zero of the compensator is placed at $s = -z = -4$, directly below the desired root location. Then the angle at the desired root location is

$$\phi = -116^\circ - 104^\circ + 90^\circ = -130^\circ. \tag{10.43}$$

Therefore the angle from the undetermined pole is determined from

$$-180^\circ = -130^\circ - \theta_p,$$

and thus $\theta_p = 50^\circ$. This angle is drawn to intersect the desired root location, and p is evaluated as $s = -p = -10.6$, as shown in Fig. 10.13(a). The gain of the compensated system is then

$$K = \frac{9(8.25)(10.4)}{8} = 96.5. \tag{10.44}$$

The compensated system is then

$$G_c(s)GH(s) = \frac{96.5(s+4)}{s(s+2)(s+10.6)}. \tag{10.45}$$

Therefore the velocity constant of the compensated system is

$$K_v = \lim_{s \to 0} s\{G(s)H(s)G_c(s)\} = \frac{96.5(4)}{2(10.6)} = 18.2. \tag{10.46}$$

The velocity constant of the compensated system is less than the desired value of 20. Therefore we must repeat the design procedure for a second choice of a desired root. If we choose $\omega_n = 10$, the process can be repeated and the resulting gain K will be increased. The compensator pole and zero location will also be

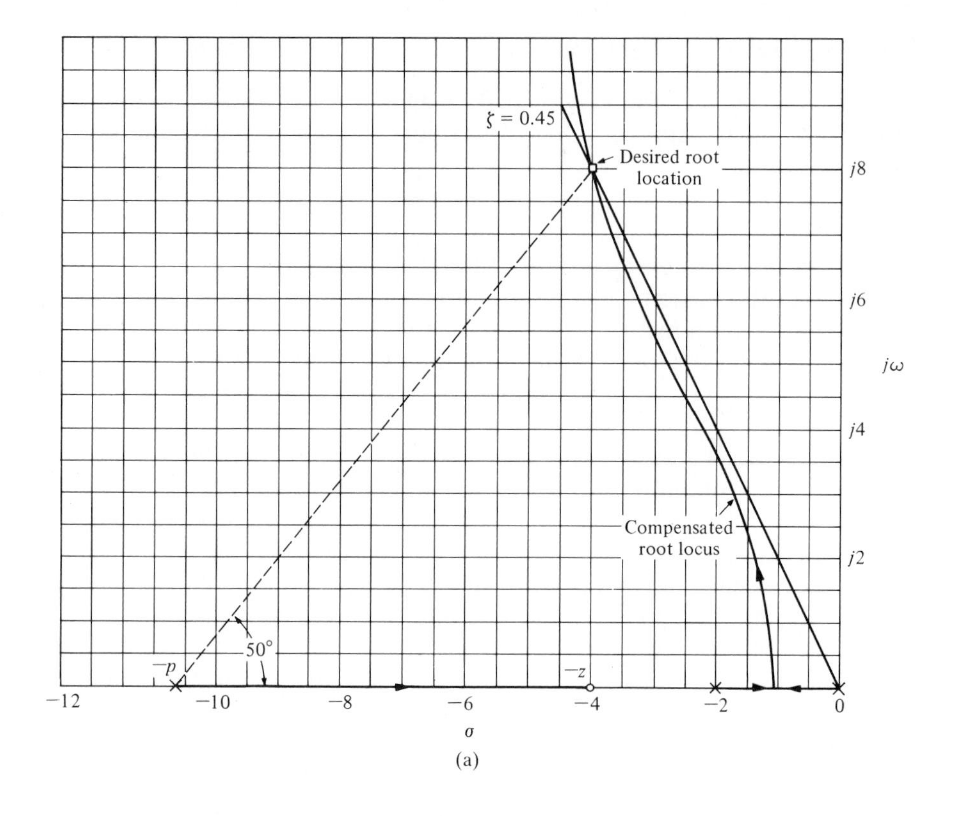

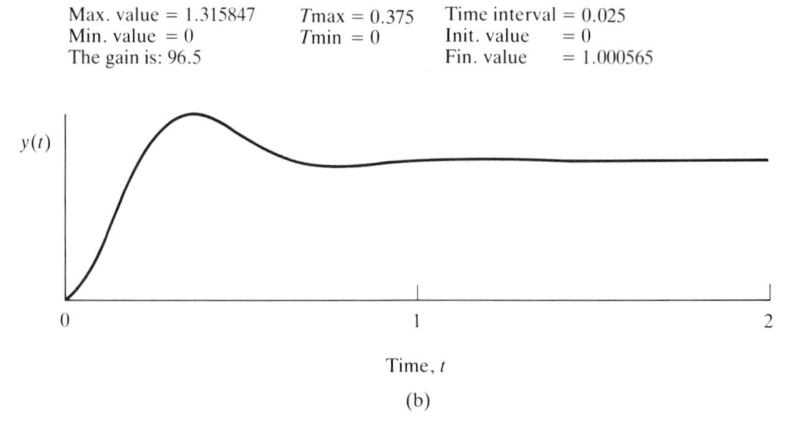

Figure 10.13. (a) Design of a phase-lead network on the s-plane for Example 10.4. (b) Step response of the compensated system of Example 10.4.

altered. Then the velocity constant can be again evaluated. We will leave it as an exercise for the reader to show that for $\omega_n = 10$, the velocity constant is $K_v = 22.7$ when $z = 4.5$ and $p = 11.6$.

Finally, for the compensation network of Eq. (10.45) we have

$$G_c(s) = \frac{s+4}{s+10.6} = \frac{(s+1/\alpha\tau)}{(s+1/\tau)}. \tag{10.47}$$

The design of an *RC*-lead network as shown in Fig. 10.4 follows directly from Eqs. (10.47) and (10.7), and is

$$G_c(s) = \left(\frac{R_2}{R_1+R_2}\right)\frac{(R_1Cs+1)}{(R_1R_2/(R_1+R_2)Cs+1)}. \tag{10.48}$$

Thus in this case we have

$$\frac{1}{R_1C} = 4$$

and

$$\alpha = \frac{R_1+R_2}{R_2} = \frac{10.6}{4}.$$

Then, choosing $C = 1\ \mu$f, we obtain $R_1 = 250{,}000$ ohms and $R_2 = 152{,}000$ ohms. The step response of the compensated system yields a 32% overshoot with a settling time of 0.8 sec as shown in Fig. 10.13(b). As shown here, we may use the Control System Design Program to verify the actual transient response.

The phase-lead compensation network is a useful compensator for altering the performance of a control system. The phase-lead network adds a phase-lead angle in order to provide adequate phase margin for feedback systems. Using an *s*-plane design approach, we can choose the phase-lead network in order to alter the system root locus and place the roots of the system in a desired position in the *s*-plane. When the design specifications include an error constant requirement, the Bode diagram method is more suitable, because the error constant of a system designed on the *s*-plane must be ascertained following the choice of a compensator pole and zero. Therefore the root locus method often results in an iterative design procedure when the error constant is specified. On the other hand, the root locus is a very satisfactory approach when the specifications are given in terms of overshoot and settling time, thus specifying the ζ and ω_n of the desired dominant roots in the *s*-plane. The use of a lead network compensator always extends the bandwidth of a feedback system, which may be objectionable for systems subjected to large amounts of noise. Also, lead networks are not suitable for providing high steady-state accuracy systems requiring very high error constants. In order to provide large error constants, typically K_p and K_v, we must consider the use of integration-type compensation networks, and therefore this will be the subject of concern in the following section.

10.6 System Compensation Using Integration Networks

For a large percentage of control systems, the primary objective is to obtain a high steady-state accuracy. Another goal is to maintain the transient performance of these systems within reasonable limits. As we found in Chapters 3 and 4, the steady-state accuracy of many feedback systems can be increased by increasing the amplifier gain in the forward channel. However, the resulting transient response may be totally unacceptable, if not even unstable. Therefore it is often necessary to introduce a compensation network in the forward path of a feedback control system in order to provide a sufficient steady-state accuracy.

Consider the single-loop control system shown in Fig. 10.14. The compensation network is to be chosen in order to provide a large error constant. The steady-state error of this system is

$$\lim_{t\to\infty} e(t) = \lim_{s\to 0} s\left[\frac{R(s)}{1 + G_c(s)G(s)H(s)}\right]. \tag{10.49}$$

We found in Section 3.5 that the steady-state error of a system depends upon the number of poles at the origin for $G_c(s)G(s)H(s)$. A pole at the origin can be considered an integration, and therefore the steady-state accuracy of a system ultimately depends upon the number of integrations in the transfer function $G_c(s)G(s)H(s)$. If the steady-state accuracy is not sufficient, we will introduce an *integration-type network* $G_c(s)$ in order to compensate for the lack of integration in the original transfer function $G(s)H(s)$.

One form of controller widely used is the proportional plus integral (PI) controller, which has a transfer function

$$G_c(s) = K_P + \frac{K_I}{s}. \tag{10.50}$$

For an example, let us consider a temperature control system where the transfer function $H(s) = 1$, and the transfer function of the heat process is

$$G(s) = \frac{K_1}{(\tau_1 s + 1)(\tau_2 s + 1)}.$$

The steady-state error of the uncompensated system is then

$$\lim_{t\to\infty} e(t) = \lim_{s\to 0} s\left\{\frac{A/s}{1 + G(s)H(s)}\right\} = \frac{A}{1 + K_1}, \tag{10.51}$$

where $R(s) = A/s$, a step input signal. Clearly, in order to obtain a small steady-state error (less than 0.05 A, for example), the magnitude of the gain K_1 must be quite large. However, when K_1 is quite large, the transient performance of the

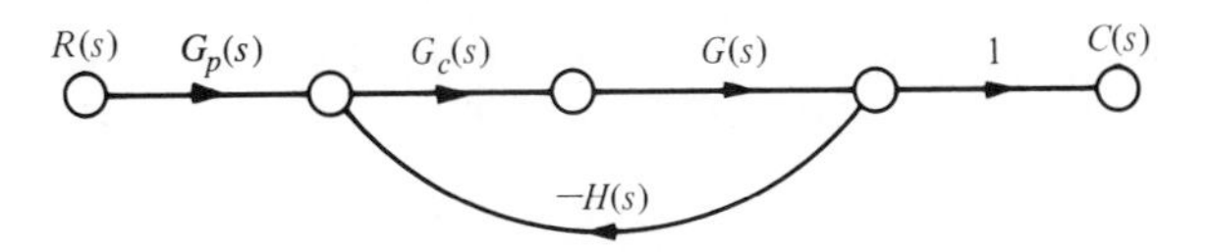

Figure 10.14. Single-loop feedback control system.

system will very likely be unacceptable. Therefore we must consider the addition of a compensation transfer function $G_c(s)$, as shown in Fig. 10.14. In order to eliminate the steady-state error of this system, we might choose the compensation as

$$G_c(s) = K_2 + \frac{K_3}{s} = \frac{K_2 s + K_3}{s}. \tag{10.52}$$

This PI compensation can be readily constructed by using an integrator and an amplifier and adding their output signals. Now, the steady-state error for a step input of the system is always zero, because

$$\begin{aligned} \lim_{t\to\infty} e(t) &= \lim_{s\to 0} s \frac{A/s}{1 + G_c(s)GH(s)} \\ &= \lim_{s\to 0} \frac{A}{1 + [(K_2 s + K_3)/s]\{K_1/[(\tau_1 s + 1)(\tau_2 s + 1)]\}} \\ &= 0. \end{aligned} \tag{10.53}$$

The transient performance can be adjusted to satisfy the system specifications by adjusting the constants K_1, K_2, and K_3. The adjustment of the transient response is perhaps best accomplished by using the root locus methods of Chapter 6 and drawing a root locus for the gain K_2K_1 after locating the zero $s = -K_3/K_2$ on the s-plane by the method outlined for the s-plane in the preceding section.

The addition of an integration as $G_c(s) = K_2 + (K_3/s)$ can also be used to reduce the steady-state error for a ramp input, $r(t) = t, t \geq 0$. For example, if the uncompensated system $GH(s)$ possessed one integration, the additional integration due to $G_c(s)$ would result in a zero steady-state error for a ramp input. In order to illustrate the design of this type of integration compensation, we will consider a temperature control system in some detail.

■ Example 10.5 Temperature control system

The uncompensated loop transfer function of a temperature control system is

$$GH(s) = \frac{K_1}{(2s + 1)(0.5s + 1)}, \tag{10.54}$$

where K_1 can be adjusted. In order to maintain zero steady-state error for a step input, we will add the compensation network

$$G_c(s) = K_2 + \frac{K_3}{s} = K_2\left(\frac{s + K_3/K_2}{s}\right). \tag{10.55}$$

Furthermore, the transient response of the system is required to have an overshoot less than or equal to 10%. Therefore the dominant complex roots must be on (or below) the $\zeta = 0.6$ line, as shown in Fig. 10.15. We will adjust the compensator zero so that the real part of the complex roots is $\zeta\omega_n = 0.75$ and thus the settling time is $T_s = 4/\zeta\omega_n = 16/3$ sec. Now, as in the preceding section, we will determine the location of the zero, $z = -K_3/K_2$, by ensuring that the angle at the desired root is $-180°$. Therefore the sum of the angles at the desired root is

$$-180° = -127° - 104° - 38° + \theta_z,$$

where θ_z is the angle from the undetermined zero. Therefore we find that $\theta_z = +89°$ and the location of the zero is $z = -0.75$. Finally, in order to determine the gain at the desired root, we evaluate the vector lengths from the poles and zeros and obtain

$$K = K_1K_2 = \frac{1.25(1.03)1.6}{1.0} = 2.08.$$

The compensated root locus and the location of the zero are shown in Fig. 10.15. It should be noted that the zero, $z = -K_3/K_2$, should be placed to the left of the

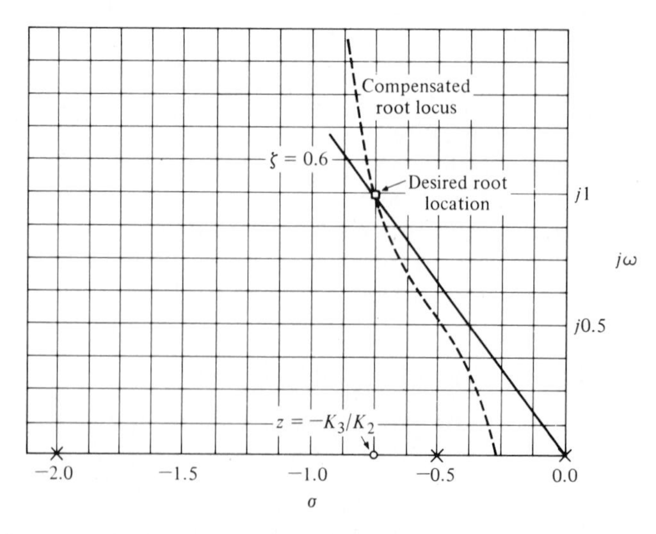

Figure 10.15. The s-plane design of an integration compensator.

pole at $s = -0.5$ in order to ensure that the complex roots dominate the transient response. In fact, the third root of the compensated system of Fig. 10.15 can be determined as $s = -1.0$, and therefore this real root is only ⅘ times the real part of the complex roots. Thus, although complex roots dominate the response of the system, the equivalent damping of the system is somewhat less than $\zeta = 0.60$ due to the real root and zero.

The closed-loop transfer function of the system of Fig. 10.14 is

$$T(s) = \frac{G_p G_c G(s)}{1 + G_c G(s)} = \frac{2.08(s + 0.75)G_p(s)}{(s + 1)(s + r_1)(s + \hat{r}_1)} \tag{10.56}$$

where $r_1 = -0.75 + j1$. The effect of the zero is to increase the overshoot to a step input (see Fig. 4.10). Thus, if we wish to attain an overshoot of 5%, we may use a prefilter $G_p(s)$ so that the zero is eliminated in $T(s)$ by setting

$$G_p(s) = \frac{0.75}{(s + 0.75)}. \tag{10.57}$$

Note that the overall *dc* gain (set $s = 0$) is $T(0) = 1.0$ when $G_p(s) = 1$ or if we use the prefilter of Eq. (10.57). The overshoot without the prefilter is 20% and with the prefilter it is 5%.

10.7 Compensation on the s-Plane Using a Phase-Lag Network

The phase-lag *RC* network of Fig. 10.6 is an integration-type network and can be used to increase the error constant of a feedback control system. We found in Section 10.3 that the transfer function of the *RC* phase-lag network is of the form

$$G_c(s) = \frac{1}{\alpha}\frac{(s + z)}{(s + p)}, \tag{10.58}$$

as given in Eq. (10.13), where

$$z = \frac{1}{\tau} = \frac{1}{R_2 C},$$

$$\alpha = \frac{R_1 + R_2}{R_2},$$

$$p = \frac{1}{\alpha\tau}.$$

The steady-state error of an uncompensated system is

$$\lim_{t\to 0} e(t) = \lim_{s\to 0} s\left\{\frac{R(s)}{1 + GH(s)}\right\}. \tag{10.59}$$

Then, for example, the velocity constant of a type-one system is

$$K_v = \lim_{s\to 0} s\{GH(s)\}, \tag{10.60}$$

as shown in Section 4.4. Therefore, if $GH(s)$ is written as

$$GH(s) = \frac{K\prod_{i=1}^{M}(s + z_i)}{s\prod_{j=1}^{Q}(s + p_j)}. \tag{10.61}$$

we obtain the velocity constant

$$K_v = \frac{K\prod_{i=1}^{M} z_i}{\prod_{j=1}^{Q} p_j}. \tag{10.62}$$

We will now add the integration type phase-lag network as a compensator and determine the compensated velocity constant. If the velocity constant of the uncompensated system (Eq. 10.62) is designated as $K_{v_{\text{uncomp}}}$, we have

$$\begin{aligned} K_{v_{\text{comp}}} &= \lim_{s\to 0} s\{G_c(s)GH(s)\} \\ &= \lim_{s\to 0} (G_c(s))K_{v_{\text{uncomp}}} \\ &= \left(\frac{z}{p}\right)\left(\frac{1}{\alpha}\right)K_{v_{\text{uncomp}}} \\ &= \left(\frac{z}{p}\right)\left(\frac{K}{\alpha}\right)\left(\frac{\prod z_i}{\prod p_j}\right). \end{aligned} \tag{10.63}$$

The gain on the compensated root locus at the desired root location will be (K/α). Now, if the pole and zero of the compensator are chosen so that $|z| = \alpha|p| < 1$, the resultant K_v will be increased at the desired root location by the ratio $z/p = \alpha$. Then, for example, if $z = 0.1$ and $p = 0.01$, the velocity constant of the desired root location will be increased by a factor of 10. However, if the compensator pole and zero appear relatively close together on the s-plane, their effect on the location of the desired root will be negligible. Therefore the compensator pole-zero combination near the origin can be used to increase the error constant of a feedback system by the factor α while altering the root location very slightly. The factor α does have an upper limit, typically about 100, because the required resistors and capacitors of the network become excessively large for a higher α. For example, when $z = 0.1$ and $\alpha = 100$, we find from Eq. (10.58) that

$$z = 0.1 = \frac{1}{R_2 C}$$

and

$$\alpha = 100 = \frac{R_1 + R_2}{R_2}.$$

If we let $C = 10\ \mu f$, then $R_2 = 1$ megohm and $R_1 = 99$ megohms. As we increase α, we increase the required magnitude of R_1. However, we should note that an attenuation, α, of 1000 or more may be obtained by utilizing pneumatic process controllers, which approximate a phase-lag characteristic (Fig. 10.8).

The steps necessary for the design of a phase-lag network on the s-plane are as follows:

1. Obtain the root locus of the uncompensated system.
2. Determine the transient performance specifications for the system and locate suitable dominant root locations on the uncompensated root locus that will satisfy the specifications.
3. Calculate the loop gain at the desired root location and, thus, the system error constant.
4. Compare the uncompensated error constant with the desired error constant and calculate the necessary increase that must result from the pole–zero ratio of the compensator, α.
5. With the known ratio of the pole–zero combination of the compensator, determine a suitable location of the pole and zero of the compensator so that the compensated root locus will still pass through the desired root location.

The fifth requirement can be satisfied if the magnitude of the pole and zero is less than one and they appear to merge as measured from the desired root location. The pole and zero will appear to merge at the root location if the angles from the compensator pole and zero are essentially equal as measured to the root location. One method of locating the zero and pole of the compensator is based on the requirement that the difference between the angle of the pole and the angle of the zero as measured at the desired root is less than 2°. An example will illustrate this approach to the design of a phase-lag compensator.

■ Example 10.6 Design of a phase lag compensator

Consider the uncompensated system of Example 10.2, where the uncompensated open-loop transfer function is

$$GH(s) = \frac{K}{s(s+2)}. \tag{10.64}$$

It is required that the damping ratio of the dominant complex roots is 0.45, while a system velocity constant equal to 20 is attained. The uncompensated root locus is a vertical line at $s = -1$ and results in a root on the $\zeta = 0.45$ line at $s = -1 \pm j2$, as shown in Fig. 10.16. Measuring the gain at this root, we have $K = (2.24)^2 = 5$. Therefore the velocity constant of the uncompensated system is

$$K_v = \frac{K}{2} = \frac{5}{2} = 2.5.$$

Thus the ratio of the zero to the pole of the compensator is

$$\left|\frac{z}{p}\right| = \alpha = \frac{K_{v_{\text{comp}}}}{K_{v_{\text{uncomp}}}} = \frac{20}{2.5} = 8. \tag{10.65}$$

Examining Fig. 10.17, we find that we might set $z = -0.1$ and then $p = -0.1/8$. The difference of the angles from p and z at the desired root is approximately 1°, and therefore $s = -1 \pm j2$ is still the location of the dominant roots. A sketch of the compensated root locus is shown as a heavy line in Fig. 10.17. Therefore the compensated system transfer function is

$$G_c(s)GH(s) = \frac{5(s+0.1)}{s(s+2)(s+0.0125)}, \tag{10.66}$$

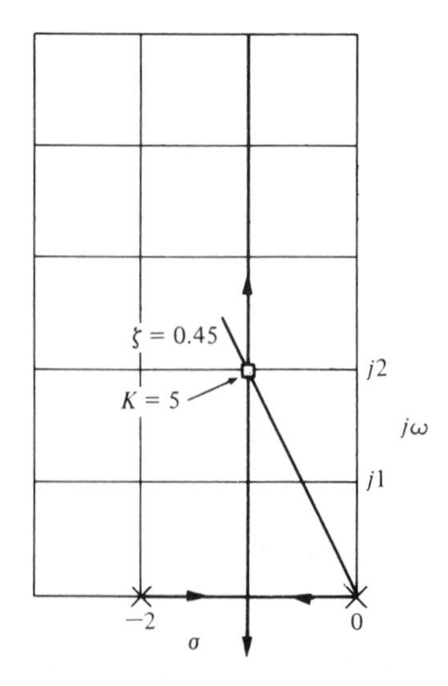

Figure 10.16. Root locus of the uncompensated system of Example 10.6.

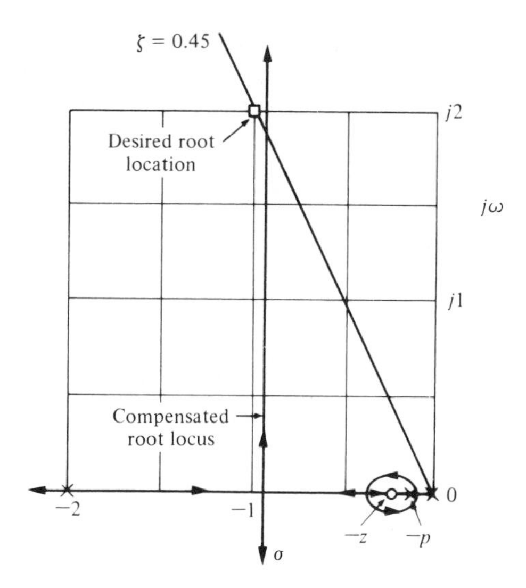

Figure 10.17. Root locus of the compensated system of Example 10.6. Note that the actual root will differ from the desired root by a slight amount. The vertical portion of the locus leaves the σ axis at $\sigma = -0.95$.

where $(K/\alpha) = 5$ or $K = 40$ in order to account for the attenuation of the lag network.

■ Example 10.7 Lag compensation of a third-order system

Let us now consider a system that is difficult to compensate by a phase-lead network. The open-loop transfer function of the uncompensated system is

$$GH(s) = \frac{K}{s(s + 10)^2}. \tag{10.67}$$

It is specified that the velocity constant of this system be equal to 20, while the damping ratio of the dominant roots be equal to 0.707. The gain necessary for a K_v of 20 is

$$K_v = 20 = \frac{K}{(10)^2}$$

or $K = 2000$. However, using Routh's criterion, we find that the roots of the characteristic equation lie on the $j\omega$-axis at $\pm j10$ when $K = 2000$. Clearly, the roots of the system when the K_v-requirement is satisfied are a long way from satisfying the damping ratio specification, and it would be difficult to bring the dominant roots from the $j\omega$-axis to the $\zeta = 0.707$ line by using a phase-lead compensator. Therefore we will attempt to satisfy the K_v and ζ-requirements by using a phase-lag network. The uncompensated root locus of this system is shown in Fig.

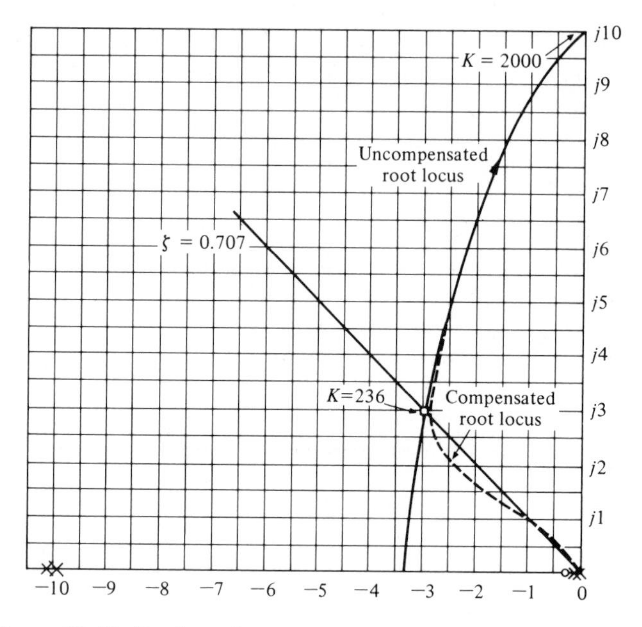

Figure 10.18. Design of a phase-lag compensator on the s-plane.

10.18 and the roots are shown when $\zeta = 0.707$ and $s = -2.9 \pm j2.9$. Measuring the gain at these roots, we find that $K = 236$. Therefore the necessary ratio of zero to pole of the compensator is

$$\alpha = \left|\frac{z}{p}\right| = \frac{2000}{236} = 8.5.$$

Thus we will choose $z = 0.1$ and $p = 0.1/9$ in order to allow a small margin of safety. Examining Fig. 10.18, we find that the difference between the angle from the pole and zero of $G_c(s)$ is negligible. Therefore the compensated system is

$$G_c(s)GH(s) = \frac{236(s + 0.1)}{s(s + 10)^2(s + 0.0111)}, \tag{10.68}$$

where $(K/\alpha) = 236$ and $\alpha = 9$.

The design of an integration compensator in order to increase the error constant of an uncompensated control system is particularly illustrative using s-plane and root locus methods. We shall now turn to similarly useful methods of designing integration compensation using Bode diagrams.

10.8 Compensation on the Bode Diagram Using a Phase-Lag Network

The design of a phase-lag RC network suitable for compensating a feedback control system can be readily accomplished on the Bode diagram. The advantage of the Bode diagram is again apparent for we will simply add the frequency response of the compensator to the Bode diagram of the uncompensated system in order to obtain a satisfactory system frequency response. The transfer function of the phase-lag network written in Bode diagram form is

$$G_c(j\omega) = \frac{1 + j\omega\tau}{1 + j\omega\alpha\tau}, \tag{10.69}$$

as we found in Eq. (10.14). The Bode diagram of the phase-lag network is shown in Fig. 10.8 for two values of α. On the Bode diagram, the pole and zero of the compensator have a magnitude much smaller than the smallest pole of the uncompensated system. Thus the phase lag is not the useful effect of the compensator, but rather it is the attenuation $-20 \log \alpha$ that is the useful effect for compensation. The phase-lag network is used to provide an attenuation and therefore to lower the 0-db (crossover) frequency of the system. However, at lower crossover frequencies we usually find that the phase margin of the system is increased and our specifications can be satisfied. The design procedure for a phase-lag network on the Bode diagram is as follows:

1. Draw the Bode diagram of the uncompensated system with the gain adjusted for the desired error constant.
2. Determine the phase margin of the uncompensated system and, if it is insufficient, proceed with the following steps.
3. Determine the frequency where the phase margin requirement would be satisfied if the magnitude curve crossed the 0-db line at this frequency, ω_c'. (Allow for 5° phase lag from the phase-lag network when determining the new crossover frequency.)
4. Place the zero of the compensator one decade below the new crossover frequency ω_c' and thus ensure only 5° of lag at ω_c' (see Fig. 10.8).
5. Measure the necessary attenuation at ω_c' in order to ensure that the magnitude curve crosses at this frequency.
6. Calculate α by noting that the attenuation is $-20 \log \alpha$.
7. Calculate the pole as $\omega_p = 1/\alpha\tau = \omega_z/\alpha$ and the design is completed.

An example of this design procedure will illustrate that the method is simple to carry out in practice.

■ Example 10.8 Design of a phase-lag network

Let us reconsider the system of Example 10.6 and design a phase-lag network so that the desired phase margin is obtained. The uncompensated transfer function is

$$\begin{aligned} GH(j\omega) &= \frac{K}{j\omega(j\omega + 2)} \\ &= \frac{K_v}{j\omega(0.5j\omega + 1)}, \end{aligned} \tag{10.70}$$

where $K_v = K/2$. It is desired that $K_v = 20$ while a phase margin of 45° is attained. The uncompensated Bode diagram is shown as a solid line in Fig. 10.19. The uncompensated system has a phase margin of 20°, and the phase margin must be increased. Allowing 5° for the phase-lag compensator, we locate the frequency ω where $\phi(\omega) = -130°$, which is to be our new crossover frequency ω_c'. In this case we find that $\omega_c' = 1.5$, which allows for a small margin of safety. The attenuation

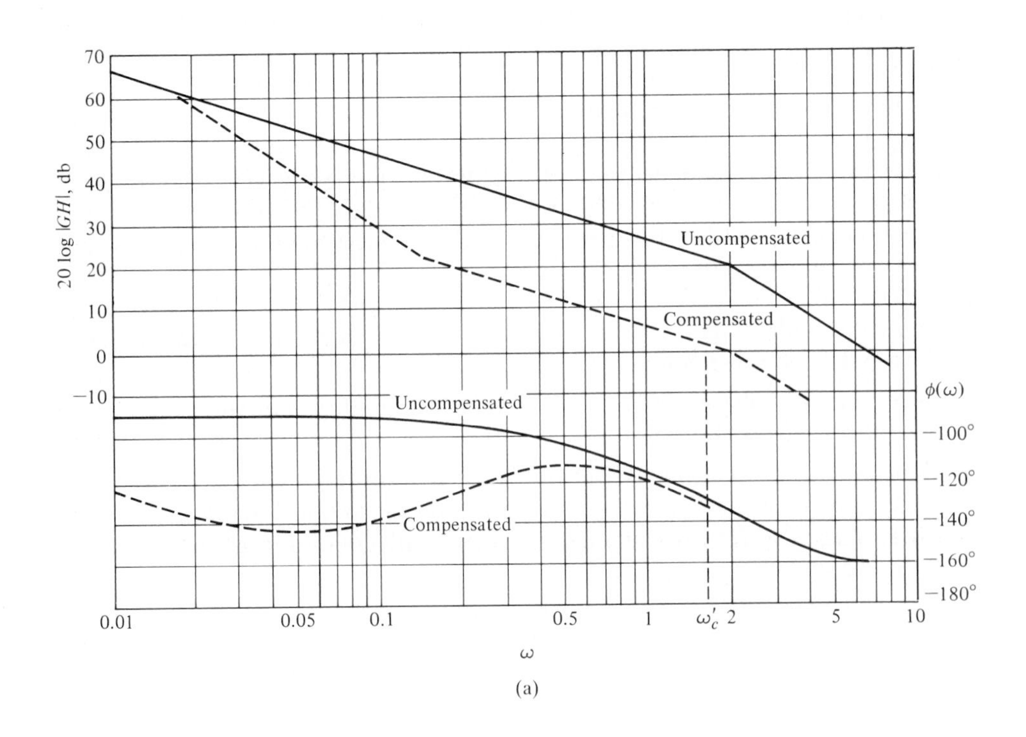

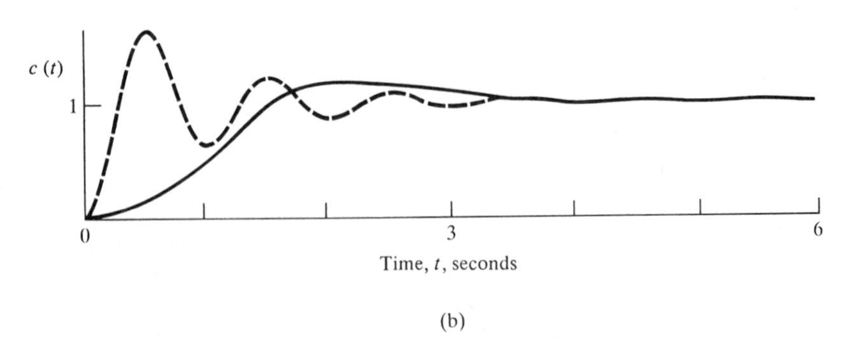

Figure 10.19. (a) Design of a phase-lag network on the Bode diagram for Example 10.8. (b) Time response to a step input for the uncompensated system (dashed line) and the compensated system (solid line) of Example 10.8.

necessary to cause ω_c' to be the new crossover frequency is equal to 20 db, accounting for a 2-db difference between the actual and asymptotic curves. Then we find that 20 db = 20 log α, or $\alpha = 10$. Therefore the zero is one decade below the crossover, or $\omega_z = \omega_c'/10 = 0.15$, and the pole is at $\omega_p = \omega_z/10 = 0.015$. The compensated system is then

$$G_c(j\omega)GH(j\omega) = \frac{20(6.66j\omega + 1)}{j\omega(0.5j\omega + 1)(66.6j\omega + 1)}. \tag{10.71}$$

The frequency response of the compensated system is shown in Fig. 10.19(a) with dashed lines. It is evident that the phase lag introduces an attenuation that lowers the crossover frequency and therefore increases the phase margin. Note that the phase angle of the lag network has almost totally disappeared at the crossover frequency ω_c'. As a final check, we numerically evaluate the phase margin at $\omega_c' = 1.5$ and find that $\phi_{pm} = 45°$, which is the desired result. Using the Nichols chart, we find that the closed-loop bandwidth of the system has been reduced from $\omega = 10$ rad/sec for the uncompensated system to $\omega = 2.5$ rad/sec for the compensated system.

The time response of the system is shown in Fig. 10.19(b). Note that the overshoot is 25% and the peak time is two seconds. Thus the response is within the specifications. The response is easy to check using the Control System Design Program.

■ Example 10.9 Lag compensation of a third-order system

Let us reconsider the system of Example 10.7, which is

$$GH(j\omega) = \frac{K}{j\omega(j\omega + 10)^2} = \frac{K_v}{j\omega(0.1j\omega + 1)^2} \tag{10.72}$$

where $K_v = K/100$. A velocity constant of K_v equal to 20 is specified. Furthermore, a damping ratio of 0.707 for the dominant roots is required. From Fig. 8.19 we estimate that a phase margin of 65° is required. The frequency response of the uncompensated system is shown in Fig. 10.20. The phase margin of the uncompensated system is zero degrees. Allowing 5° for the lag network, we locate the frequency where the phase is −110°. This frequency is equal to 1.74, and therefore we shall attempt to locate the new crossover frequency at $\omega_c' = 1.5$. Measuring the necessary attenuation at $\omega = \omega_{c_1}'$, we find that 23 db is required; 23 = 20 log α, or $\alpha = 14.2$. The zero of the compensator is located one decade below the crossover frequency, and thus

$$\omega_z = \frac{\omega_c'}{10} = 0.15.$$

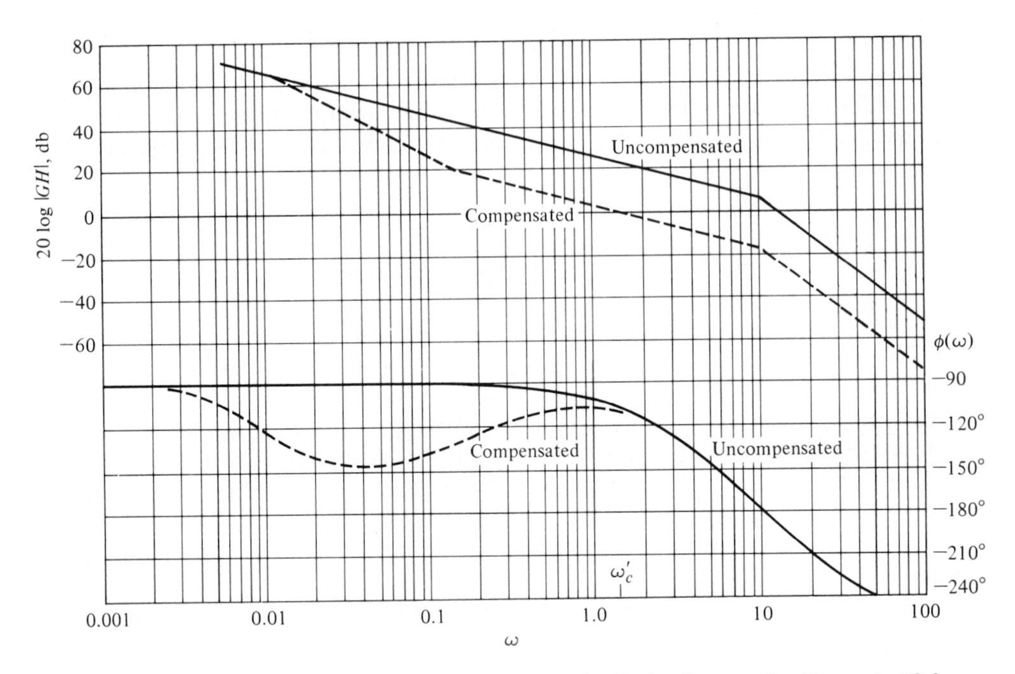

Figure 10.20. Design of a phase-lag network on the Bode diagram for Example 10.9.

The pole is then

$$\omega_p = \frac{\omega_z}{\alpha} = \frac{0.15}{14.2}.$$

Therefore the compensated system is

$$G_c(j\omega)GH(j\omega) = \frac{20(6.66j\omega + 1)}{j\omega(0.1j\omega + 1)^2(94.6j\omega + 1)}. \tag{10.73}$$

The compensated frequency response is shown in Fig. 10.20. As a final check, we numerically evaluate the phase margin at $\omega_c' = 1.5$ and find that $\phi_{pm} = 67°$, which is within the specifications.

Therefore a phase-lag compensation network can be used to alter the frequency response of a feedback control system in order to attain satisfactory system performance. Examining both Examples 10.8 and 10.9, we note again that the system design is satisfactory when the asymptotic curve for the magnitude of the compensated system crosses the 0-db line with a slope of -6 db/octave. The attenuation of the phase-lag network reduces the magnitude of the crossover (0-db) frequency to a point where the phase margin of the system is satisfactory. Thus, in contrast to the phase-lead network, the phase-lag network reduces the closed-loop bandwidth of the system as it maintains a suitable error constant.

One might ask, why do we not place the compensator zero more than one decade below the new crossover ω_c' (see item 4 of the design procedure) and thus ensure less than 5° of lag at ω_c' due to the compensator? This question can be answered by considering the requirements placed on the resistors and capacitors of the lag network by the values of the poles and zeros (see Eq. 10.12). As the magnitudes of the pole and zero of the lag network are decreased, the magnitudes of the resistors and the capacitor required increase proportionately. The zero of the lag compensator in terms of the circuit components is $z = 1/R_2C$, and the α of the network is $\alpha = (R_1 + R_2)/R_2$. Thus, considering Example 10.9, we require a zero at $z = 0.15$, which can be obtained with $C = 1\ \mu$f and $R_2 = 6.66$ megohms. However, for $\alpha = 14$ we require a resistance R_1 of $R_1 = R_2(\alpha - 1) = 88$ megohms. Clearly, a designer does not wish to place the zero z further than one decade below ω_c' and thus require larger values of R_1, R_2, and C.

The phase-lead compensation network alters the frequency response of a network by adding a positive (leading) phase angle and, therefore increases the phase margin at the crossover (0-db) frequency. It becomes evident that a designer might wish to consider using a compensation network that provided the attenuation of a phase-lag network and the lead-phase angle of a phase-lead network. Such a network does exist. It is called a *lead-lag network* and is shown in Fig. 10.21. The transfer function of this network is

$$\frac{E_2(s)}{E_1(s)} = \frac{(R_1C_1s + 1)(R_2C_2s + 1)}{R_1R_2C_1C_2s^2 + (R_1C_1 + R_1C_2 + R_2C_2)s + 1}. \tag{10.74}$$

When $\alpha\tau_1 = R_1C_1$, $\beta\tau_2 = R_2C_2$, $\tau_1\tau_2 = R_1R_2C_1C_2$, we note that $\alpha\beta = 1$ and then Eq. (10.74) is

$$\frac{E_2(s)}{E_1(s)} = \frac{(1 + \alpha\tau_1 s)(1 + \beta\tau_2 s)}{(1 + \tau_1 s)(1 + \tau_2 s)}, \tag{10.75}$$

where $\alpha > 1$ and $\beta < 1$. The first terms in the numerator and denominator, which are a function of τ_1, provide the phase-lead portion of the network. The second terms, which are a function of τ_2, provide the phase-lag portion of the compensation network. The parameter β is adjusted to provide suitable attenuation of the low frequency portion of the frequency response, and the parameter α is adjusted to provide an additional phase lead at the new crossover (0-db) frequency. Alternatively, the compensation can be designed on the s-plane by placing the lead pole and zero compensation in order to locate the dominant roots in

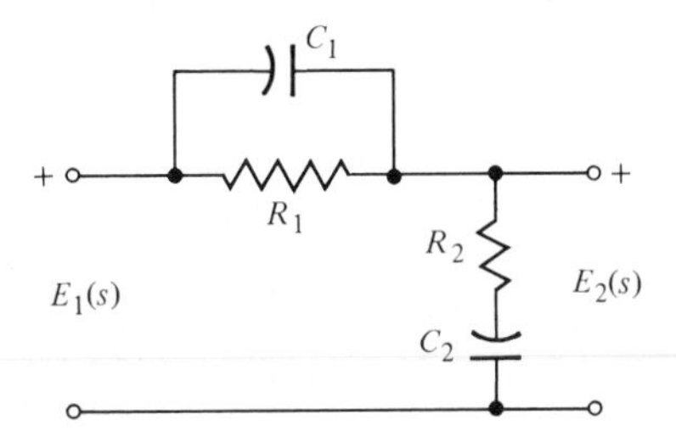

Figure 10.21. An *RC* lead-lag network.

a desired location. Then the phase-lag compensation is used to raise the error constant at the dominant root location by a suitable ratio, $1/\beta$. The design of a phase lead-lag compensator follows the procedures already discussed, and the reader is referred to further literature illustrating the utility of lead-lag compensation [2, 3].

10.9 Compensation on the Bode Diagram Using Analytical and Computer Methods

It is desirable to use computers, when appropriate, to assist the designer in the selection of the parameters of a compensator. The development of algorithms for computer-aided design is an important alternative approach to the trial-and-error methods considered in earlier sections. By the use of compensators, computer programs have been developed for the selection of suitable parameter values based on satisfaction of frequency response criteria such as phase margin [5, 6].

An analytical technique of selecting the parameters of a lead or lag network has been developed for Bode diagrams [6, 7]. For a single-stage compensator

$$G_c(s) = \frac{1 + \alpha\tau s}{1 + \tau s}, \tag{10.76}$$

where $\alpha < 1$ yields a lag compensator and $\alpha > 1$ yields a lead compensator. The phase contribution of the compensator at the desired crossover frequency ω_c (see Eq. 10.9) is

$$p = \tan\phi = \frac{\alpha\omega_c\tau - \omega_c\tau}{1 + (\omega_c\tau)^2\alpha}. \tag{10.77}$$

The magnitude M (in db) of the compensator at ω_c is

$$c = 10^{M/10} = \frac{1 + (\omega_c\alpha\tau)^2}{1 + (\omega_c\tau)^2}. \tag{10.78}$$

Eliminating $\omega_c\tau$ from Eqs. (10.77) and (10.78), we obtain the nontrivial solution equation for α as

$$(p^2 - c + 1)\alpha^2 + 2p^2c\alpha + p^2c^2 + c^2 - c = 0. \tag{10.79}$$

For a single-stage compensator it is necessary that $c > p^2 + 1$. If we solve for α from Eq. (10.79), we can obtain τ from

$$\tau = \frac{1}{\omega_c}\sqrt{\frac{1 - c}{c - \alpha^2}}. \tag{10.80}$$

The design steps for a lead compensator are:

1. Select the desired ω_c.
2. Determine the phase margin desired and therefore the required phase ϕ for Eq. (10.77).

3. Verify that the phase lead is applicable, $\phi > 0$ and $M > 0$.
4. Determine whether a single stage will be sufficient when $c > p^2 + 1$.
5. Determine α from Eq. (10.79).
6. Determine τ from Eq. (10.80).

If we need to design a single lag compensator, then $\phi < 0$ and $M < 0$ (step 3). Also, step 4 will require $c < [1/(1 + p^2)]$. Otherwise the method is the same.

■ Example 10.10 Design using an analytical technique

Let us reconsider the system of Example 10.1 and design a lead network by the analytical technique. Examine the uncompensated curves in Fig. 10.9. We select $\omega_c = 5$. Then, as before, we desire a phase margin of 45°. The compensator must yield this phase, so

$$p = \tan 45° = 1. \tag{10.81}$$

The required magnitude contribution is 8 db or $M = 8$, so that

$$c = 10^{8/10} = 6.31. \tag{10.82}$$

Using c and p, we obtain

$$-4.31\alpha^2 + 12.62\alpha + 73.32 = 0. \tag{10.83}$$

Solving for α we obtain $\alpha = 5.84$. Solving Eq. (10.80), we obtain $\tau = 0.087$. Therefore the compensator is

$$G_c(s) = \frac{1 + 0.515s}{1 + 0.087s}. \tag{10.84}$$

The pole is equal to 11.5 and the zero is 1.94. This design is similar to that obtained by the iteration technique of Section 10.4.

10.10 The Design of Control Systems in the Time Domain

The design of automatic control systems is an important function of control engineering. The purpose of design is to realize a system with practical components which will provide the desired operating performance. The desired performance can be readily stated in terms of time-domain performance indices. For example, the maximum overshoot and rise time for a step input are valuable time-domain indices. In the case of steady-state and transient performance, the performance indices are normally specified in the time domain and, therefore, it is natural that we wish to develop design procedures in the time domain.

The performance of a control system can be represented by integral performance measures, as we found in Section 4.5. Therefore the design of a system must be based on minimizing a performance index such as the integral of the squared error (ISE), as in Section 4.5. Systems that are adjusted to provide a minimum performance index are often called *optimum control systems.* In this section, we shall consider the design of an optimum control system where the system is described by a state variable formulation.

However, before proceeding to the specifics, we should note that we earlier designed a system in the time domain in Example 9.6. In this example, we considered the unstable portion of an inverted pendulum system and developed a suitable feedback control so that the system was stable. This design was based on measuring the state variables of the system and using them to form a suitable control signal $u(t)$ so that the system was stable. In this section, we shall again consider the measurement of the state variables and their use in developing a control signal $u(t)$ so that the performance of the system is optimized.

The performance of a control system, written in terms of the state variables of a system, can be expressed in general as

$$J = \int_0^{t_f} g(\mathbf{x}, \mathbf{u}, t)\, dt, \tag{10.85}$$

where $\mathbf{x}$ equals the state vector and $\mathbf{u}$ equals the control vector.*

We are interested in minimizing the error of the system and, therefore, when the desired state vector is represented as $\mathbf{x}_d = \mathbf{0}$, we are able to consider the error as identically equal to the value of the state vector. That is, we desire the system to be at equilibrium, $\mathbf{x} = \mathbf{x}_d = \mathbf{0}$, and in any deviation from equilibrium is considered an error. Therefore we will consider, in this section, the design of optimum control systems using *state-variable feedback* and error-squared performance indices [1, 2, 3].

The control system we will consider is shown in Fig. 10.22 and can be represented by the vector differential equation

$$\dot{\mathbf{x}} = \mathbf{A}\mathbf{x} + \mathbf{B}\mathbf{u}. \tag{10.86}$$

We will select a feedback controller so that $\mathbf{u}$ is some function of the measured state variables $\mathbf{x}$ and therefore

$$\mathbf{u} = \mathbf{h}(\mathbf{x}). \tag{10.87}$$

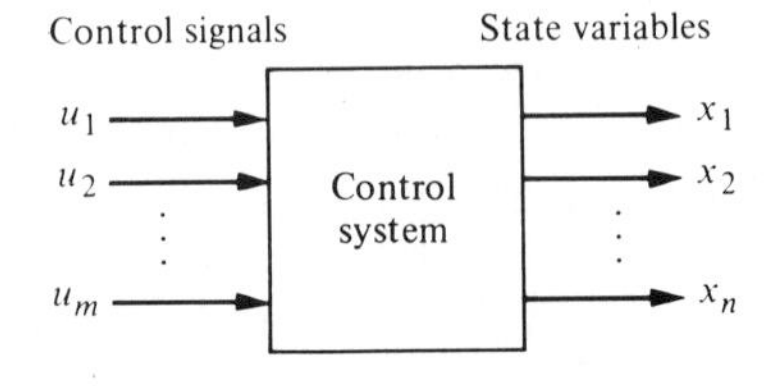

Figure 10.22. A control system in terms of x and u.

*Note that J is used to denote the performance index, instead of I, which was used in Chapter 4. This will enable the reader to distinguish readily the performance index from the identity matrix, which is represented by the boldfaced capital **I**.

For example, we might use

$$\begin{aligned} u_1 &= k_1 x_1, \\ u_2 &= k_2 x_2, \\ &\vdots \\ u_m &= k_m x_m. \end{aligned} \tag{10.88}$$

Alternatively, we might choose the control vector as

$$\begin{aligned} u_1 &= k_1(x_1 + x_2), \\ u_2 &= k_2(x_2 + x_3), \\ &\vdots \end{aligned} \tag{10.89}$$

The choice of the control signals is somewhat arbitrary and depends partially upon the actual desired performance and the complexity of the feedback structure allowable. Often we are limited in the number of state variables available for feedback, since we are only able to utilize measurable state variables.

Now, in our case we limit the feedback function to a linear function so that $\mathbf{u} = \mathbf{Hx}$, where $\mathbf{H}$ is an $m \times n$ matrix. Therefore, in expanded form, we have

$$\begin{bmatrix} u_1 \\ u_2 \\ \vdots \\ u_m \end{bmatrix} = \begin{bmatrix} h_{11} \cdots h_{1n} \\ \\ \vdots \qquad \vdots \\ h_{m1} \cdots h_{mn} \end{bmatrix} \begin{bmatrix} x_1 \\ x_2 \\ \vdots \\ x_n \end{bmatrix}. \tag{10.90}$$

Then, substituting Eq. (10.90) into Eq. (10.86), we obtain

$$\dot{\mathbf{x}} = \mathbf{Ax} + \mathbf{BHx} = \mathbf{Dx}, \tag{10.91}$$

where $\mathbf{D}$ is the $n \times n$ matrix resulting from the addition of the elements of $\mathbf{A}$ and $\mathbf{BH}$.

Now, returning to the error-squared performance index, we recall from Section 4.5 that the index for a single state variable, x_1, is written as

$$J = \int_0^{t_f} (x_1(t))^2 \, dt. \tag{10.92}$$

A performance index written in terms of two state variables would then be

$$J = \int_0^{t_f} (x_1^2 + x_2^2) \, dt. \tag{10.93}$$

Therefore, because we wish to define the performance index in terms of an integral of the sum of the state variables squared, we will utilize the matrix operation

$$\mathbf{x}^T\mathbf{x} = [x_1, x_2, x_3, \ldots, x_n] \begin{bmatrix} x_1 \\ x_2 \\ \vdots \\ x_n \end{bmatrix} = (x_1^2 + x_2^2 + x_3^2 + \ldots + x_n^2), \tag{10.94}$$

where $\mathbf{x}^T$ indicates the transpose of the $\mathbf{x}$ matrix.* Then the general form of the performance index, in terms of the state vector, is

$$J = \int_0^{t_f} (\mathbf{x}^T\mathbf{x})\, dt. \tag{10.95}$$

Again considering Eq. (10.95), we will let the final time of interest be $t_f = \infty$. In order to obtain the minimum value of J, we postulate the existence of an exact differential so that

$$\frac{d}{dt}(\mathbf{x}^T\mathbf{P}\mathbf{x}) = -\mathbf{x}^T\mathbf{x}, \tag{10.96}$$

where $\mathbf{P}$ is to be determined. A symmetric $\mathbf{P}$ matrix will be used in order to simplify the algebra without any loss of generality. Then, for a symmetric $\mathbf{P}$ matrix, $p_{ij} = p_{ji}$. Completing the differentiation indicated on the left-hand side of Eq. (10.96), we have

$$\frac{d}{dt}(\mathbf{x}^T\mathbf{P}\mathbf{x}) = \dot{\mathbf{x}}^T\mathbf{P}\mathbf{x} + \mathbf{x}^T\mathbf{P}\dot{\mathbf{x}}.$$

Then, substituting Eq. (10.91), we obtain

$$\begin{aligned}\frac{d}{dt}(\mathbf{x}^T\mathbf{P}\mathbf{x}) &= (\mathbf{D}\mathbf{x})^T\mathbf{P}\mathbf{x} + \mathbf{x}^T\mathbf{P}(\mathbf{D}\mathbf{x}) \\ &= \mathbf{x}^T\mathbf{D}^T\mathbf{P}\mathbf{x} + \mathbf{x}^T\mathbf{P}\mathbf{D}\mathbf{x} \\ &= \mathbf{x}^T(\mathbf{D}^T\mathbf{P} + \mathbf{P}\mathbf{D})\mathbf{x},\end{aligned} \tag{10.97}$$

where $(\mathbf{D}\mathbf{x})^T = \mathbf{x}^T\mathbf{D}^T$ by the definition of the transpose of a product. If we let $(\mathbf{D}^T\mathbf{P} + \mathbf{P}\mathbf{D}) = -\mathbf{I}$, then Eq. (10.97) becomes

$$\frac{d}{dt}(\mathbf{x}^T\mathbf{P}\mathbf{x}) = -\mathbf{x}^T\mathbf{x}, \tag{10.98}$$

which is the exact differential we are seeking. Substituting Eq. (10.98) into Eq. (10.95), we obtain

$$\begin{aligned}J &= \int_0^{\infty} -\frac{d}{dt}(\mathbf{x}^T\mathbf{P}\mathbf{x})\, dt \\ &= -\mathbf{x}^T\mathbf{P}\mathbf{x}\,\Big|_0^{\infty} \\ &= \mathbf{x}^T(0)\mathbf{P}\mathbf{x}(0).\end{aligned} \tag{10.99}$$

In the evaluation of the limit at $t = \infty$, we have assumed that the system is stable and hence $\mathbf{x}(\infty) = 0$ as desired. Therefore, in order to minimize the performance index J, we consider the two equations

$$J = \int_0^{\infty} \mathbf{x}^T\mathbf{x}\, dt = \mathbf{x}^T(0)\mathbf{P}\mathbf{x}(0) \tag{10.100}$$

---

*The matrix operation $\mathbf{x}^T\mathbf{x}$ is discussed in Appendix C, Section C.4.

and

$$\mathbf{D}^T\mathbf{P} + \mathbf{P}\mathbf{D} = -\mathbf{I}. \tag{10.101}$$

The design steps are then as follows:

1. Determine the matrix **P** which satisfies Eq. (10.101), where **D** is known.
2. Minimize J by determining the minimum of Eq. (10.100).

■ Example 10.11 State variable feedback

Consider the control system shown in Fig. 10.23 in signal-flow graph form. The state variables are identifed as x_1 and x_2. The performance of this system is quite unsatisfactory because an undamped response results for a step input or disturbance signal. The vector differential equation of this system is

$$\frac{d}{dt}\begin{bmatrix} x_1 \\ x_2 \end{bmatrix} = \begin{bmatrix} 0 & 1 \\ 0 & 0 \end{bmatrix}\begin{bmatrix} x_1 \\ x_2 \end{bmatrix} + \begin{bmatrix} 0 \\ 1 \end{bmatrix} u(t), \tag{10.102}$$

where

$$\mathbf{A} = \begin{bmatrix} 0 & 1 \\ 0 & 0 \end{bmatrix}.$$

We will choose a feedback control system so that

$$u(t) = -k_1x_1 - k_2x_2, \tag{10.103}$$

and therefore the control signal is a linear function of the two state variables. The sign of the feedback is negative in order to provide negative feedback. Then Eq. (10.102) becomes

$$\begin{aligned} \dot{x}_1 &= x_2, \\ \dot{x}_2 &= -k_1x_1 - k_2x_2, \end{aligned} \tag{10.104}$$

or, in matrix form, we have

$$\begin{aligned} \dot{\mathbf{x}} &= \mathbf{D}\mathbf{x} \\ &= \begin{bmatrix} 0 & 1 \\ -k_1 & -k_2 \end{bmatrix}\mathbf{x}. \end{aligned} \tag{10.105}$$

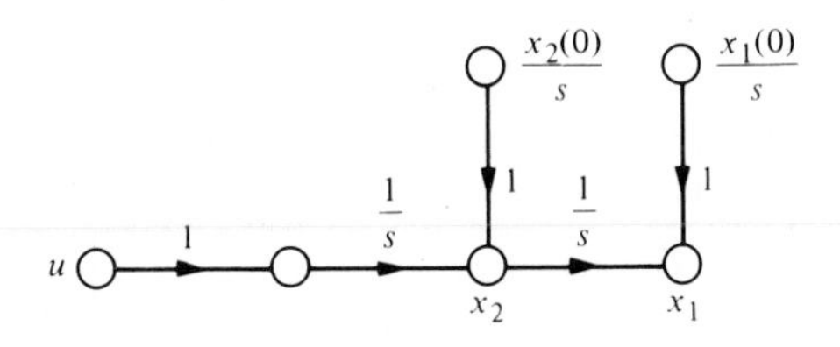

Figure 10.23. Signal-flow graph of the control system of Example 10.11.

We note that x_1 would represent the position of a position-control system and the transfer function of the system would be $G(s) = 1/Ms^2$, where $M = 1$ and the friction is negligible. In any case, in order to avoid needless algebraic manipulation, we will let $k_1 = 1$ and determine a suitable value for k_2 so that the performance index is minimized. Then, writing Eq. (10.101), we have

$$\mathbf{D}^T\mathbf{P} + \mathbf{P}\mathbf{D} = -\mathbf{I},$$

$$\begin{bmatrix} 0 & -1 \\ 1 & -k_2 \end{bmatrix}\begin{bmatrix} p_{11} & p_{12} \\ p_{12} & p_{22} \end{bmatrix} + \begin{bmatrix} p_{11} & p_{12} \\ p_{12} & p_{22} \end{bmatrix}\begin{bmatrix} 0 & 1 \\ -1 & -k_2 \end{bmatrix} = \begin{bmatrix} -1 & 0 \\ 0 & -1 \end{bmatrix}. \tag{10.106}$$

Completing the matrix multiplication and addition, we have

$$\begin{aligned} -p_{12} - p_{12} &= -1, \\ p_{11} - k_2 p_{12} - p_{22} &= 0, \\ p_{12} - k_2 p_{22} + p_{12} - k_2 p_{22} &= -1. \end{aligned} \tag{10.107}$$

Then, solving these simultaneous equations, we obtain

$$p_{12} = \frac{1}{2}, \qquad p_{22} = \frac{1}{k_2}, \qquad p_{11} = \frac{k_2^2 + 2}{2k_2}.$$

The integral performance index is then

$$J = \mathbf{x}^T(0)\mathbf{P}\mathbf{x}(0), \tag{10.108}$$

and we shall consider the case where each state is initially displaced one unit from equilibrium so that $\mathbf{x}^T(0) = [1, 1]$. Therefore Eq. (10.108) becomes

$$\begin{aligned} J &= [1, 1]\begin{bmatrix} p_{11} & p_{12} \\ p_{12} & p_{22} \end{bmatrix}\begin{bmatrix} 1 \\ 1 \end{bmatrix} \\ &= [1, 1]\begin{bmatrix} (p_{11} + p_{12}) \\ (p_{12} + p_{22}) \end{bmatrix} \\ &= (p_{11} + p_{12}) + (p_{12} + p_{22}) = p_{11} + 2p_{12} + p_{22}. \end{aligned} \tag{10.109}$$

Substituting the values of the elements of $\mathbf{P}$, we have

$$\begin{aligned} J &= \frac{k_2^2 + 2}{2k_2} + 1 + \frac{1}{k_2} \\ &= \frac{k_2^2 + 2k_2 + 4}{2k_2}. \end{aligned} \tag{10.110}$$

In order to minimize as a function of k_2, we take the derivative with respect to k_2 and set it equal to zero as follows:

$$\frac{\partial J}{\partial k_2} = \frac{2k_2(2k_2 + 2) - 2(k_2^2 + 2k_2 + 4)}{(2k_2)^2} = 0. \tag{10.111}$$

Therefore $k_2^2 = 4$ and $k_2 = 2$ when J is a minimum. The minimum value of J is obtained by substituting $k_2 = 2$ into Eq. (10.110). Thus we obtain

$$J_{\min} = 3.$$

The system matrix $\mathbf{D}$, obtained for the compensated system, is then

$$\mathbf{D} = \begin{bmatrix} 0 & 1 \\ -1 & -2 \end{bmatrix} \tag{10.112}$$

The characteristic equation of the compensated system is therefore

$$\begin{aligned} \det[\lambda \mathbf{I} - \mathbf{D}] &= \det\begin{bmatrix} \lambda & -1 \\ 1 & \lambda + 2 \end{bmatrix} \\ &= \lambda^2 + 2\lambda + 1. \end{aligned} \tag{10.113}$$

Because this is a second-order system, we note that the characteristic equation is of the form $(s^2 + 2\zeta\omega_n s + \omega_n^2)$, and therefore the damping ratio of the compensated system is $\zeta = 1.0$. This compensated system is considered to be an optimum system in that the compensated system results in a minimum value for the performance index. Of course, we recognize that this system is optimum only for the specific set of initial conditions that were assumed. The compensated system is shown in Fig. 10.24. A curve of the performance index as a function of k_2 is shown in Fig. 10.25. It is clear that this system is not very sensitive to changes in k_2 and will maintain a near minimum performance index if the k_2 is altered some percentage. We define the sensitivity of an optimum system as

$$S_k^{\text{opt}} = \frac{\Delta J/J}{\Delta k/k}, \tag{10.114}$$

where k is the design parameter. Then for this example we have $k = k_2$, and therefore

$$S_{k_2}^{\text{opt}} \simeq \frac{0.08/3}{0.5/2} = 0.107. \tag{10.115}$$

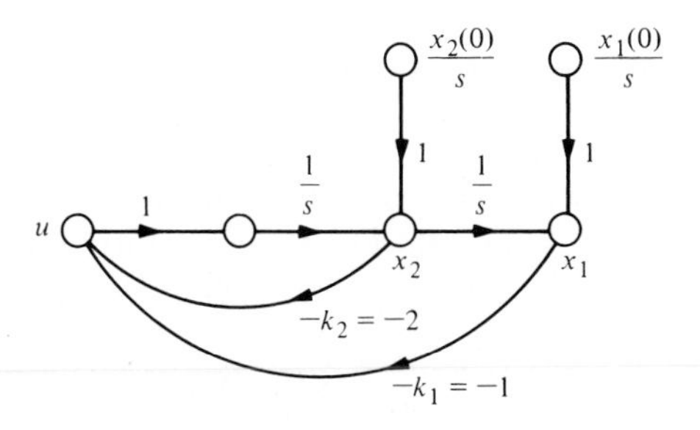

Figure 10.24. Compensated control system of Example 10.11.

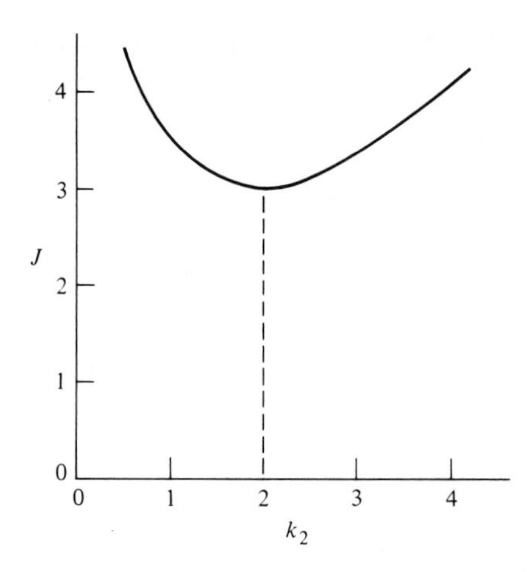

Figure 10.25. Performance index versus the parameter k_2.

■ Example 10.12 Determination of an optimum system

Now let us reconsider the system of Example 10.11, where both the feedback gains, k_1 and k_2, are unspecified. In order to simplify the algebra without any loss in insight into the problem, let us set $k_1 = k_2 = k$. The reader can prove that if k_1 and k_2 are unspecified then $k_1 = k_2$ when the minimum of the performance index (Eq. 10.100) is obtained. Then for the system of the previous example, Eq. (10.105) becomes

$$\begin{aligned}\dot{\mathbf{x}} &= \mathbf{D}\mathbf{x} \\ &= \begin{bmatrix} 0 & 1 \\ -k & -k \end{bmatrix} \mathbf{x}.\end{aligned} \tag{10.116}$$

In order to determine the **P** matrix we utilize Eq. (10.101), which is

$$\mathbf{D}^T\mathbf{P} + \mathbf{P}\mathbf{D} = -\mathbf{I}. \tag{10.117}$$

Solving the set of simultaneous equations resulting from Eq. (10.117), we find that

$$p_{12} = \frac{1}{2k}, \qquad p_{22} = \frac{(k+1)}{2k^2}, \qquad p_{11} = \frac{(1+2k)}{2k}.$$

Let us consider the case where the system is initially displaced one unit from equilibrium so that $\mathbf{x}^T(0) = [1, 0]$. Then the performance index (Eq. 10.100) becomes

$$J = \int_0^\infty \mathbf{x}^T\mathbf{x}\, dt = \mathbf{x}^T(0)\mathbf{P}\mathbf{x}(0) = p_{11}. \tag{10.118}$$

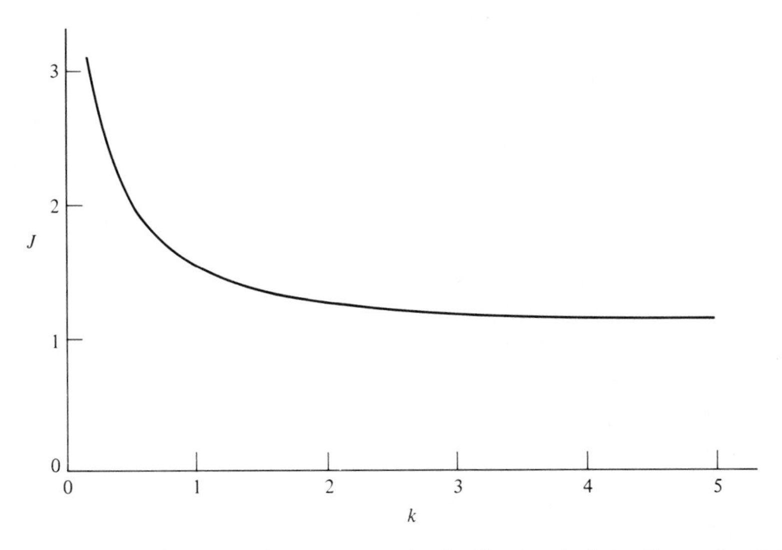

Figure 10.26. Performance index versus the feedback gain k for Example 10.12.

Thus the performance index to be minimized is

$$J = p_{11} = \frac{(1 + 2k)}{2k} = 1 + \frac{1}{2k}. \tag{10.119}$$

Clearly, the minimum value of J is obtained when k approaches infinity; the result is $J_{\text{min}} = 1$. A plot of J versus k is shown in Fig. 10.26. This plot illustrates that the performance index approaches a minimum asymptotically as k approaches an infinite value. Now we recognize that in providing a very large gain k, we can cause the feedback signal

$$u(t) = -k(x_1(t) + x_2(t))$$

to be very large. However, we are restricted to realizable magnitudes of the control signal $u(t)$. Therefore we must introduce a *constraint* on $u(t)$ so that the gain k is not made too large. Then, for example, if we establish a constraint on $u(t)$ so that

$$|u(t)| \leq 50, \tag{10.120}$$

we require that the maximum accceptable value of k in this case is

$$k_{\text{max}} = \frac{|u|_{\text{max}}}{x_1(0)} = 50. \tag{10.121}$$

Then the minimum value of J is

$$\begin{aligned} J_{\min} &= 1 + \frac{1}{2k_{\max}} \\ &= 1.01, \end{aligned} \tag{10.122}$$

which is sufficiently close to the absolute minimum of J in order to satisfy our requirements.

Upon the examination of the performance index (Eq. 10.95), we recognize that the reason the magnitude of the control signal is not accounted for in the original calculations is that $u(t)$ is not included within the expression for the performance index. However, there are many cases where we are concerned with the expenditure of the control signal energy. For example, in a space vehicle attitude control system, $[u(t)]^2$ represents the expenditure of jet fuel energy and must be restricted in order to conserve the fuel energy for long periods of flight. In order to account for the expenditure of the energy of the control signal, we will utilize the performance index

$$J = \int_0^\infty (\mathbf{x}^T\mathbf{I}\mathbf{x} + \lambda\mathbf{u}^T\mathbf{u})\,dt, \tag{10.123}$$

where λ is a scalar weighting factor and $\mathbf{I}$ = identity matrix. The weighting factor λ will be chosen so that the relative importance of the state variable performance is contrasted with the importance of the expenditure of the system energy resource which is represented by $\mathbf{u}^T\mathbf{u}$. As in the previous paragraphs we will represent the state variable feedback by the matrix equation.

$$\mathbf{u} = \mathbf{H}\mathbf{x} \tag{10.124}$$

and the system with this state variable feedback as

$$\begin{aligned} \dot{\mathbf{x}} &= \mathbf{A}\mathbf{x} + \mathbf{B}\mathbf{u} \\ &= \mathbf{D}\mathbf{x}. \end{aligned} \tag{10.125}$$

Now, substituting Eq. (10.124) into Eq. (10.123), we have

$$\begin{aligned} J &= \int_0^\infty (\mathbf{x}^T\mathbf{I}\mathbf{x} + \lambda(\mathbf{H}\mathbf{x})^T(\mathbf{H}\mathbf{x}))\,dt \\ &= \int_0^\infty [\mathbf{x}^T(\mathbf{I} + \lambda\mathbf{H}^T\mathbf{H})\mathbf{x}]\,dt \\ &= \int_0^\infty \mathbf{x}^T\mathbf{Q}\mathbf{x}\,dt, \end{aligned} \tag{10.126}$$

where $\mathbf{Q} = (\mathbf{I} + \lambda\mathbf{H}^T\mathbf{H})$ is an $n \times n$ matrix. Following the development of Eqs. (10.95) through (10.99), we postulate the existence of an exact differential so that

$$\frac{d}{dt}(\mathbf{x}^T\mathbf{P}\mathbf{x}) = -\mathbf{x}^T\mathbf{Q}\mathbf{x}. \tag{10.127}$$

Then in this case we require that

$$\mathbf{D}^T\mathbf{P} + \mathbf{P}\mathbf{D} = -\mathbf{Q}. \tag{10.128}$$

and thus we have as before (Eq. 10.99)

$$J = \mathbf{x}^T(0)\mathbf{P}\mathbf{x}(0). \tag{10.129}$$

Now the design steps are exactly as for Eqs. (10.100) and (10.101) with the exception that the left side of Eq. (10.128) equals $-\mathbf{Q}$ instead of $-\mathbf{I}$. Of course, if $\lambda = 0$, Eq. (10.128) reduces to Eq. (10.101). Now let us reconsider the previous example when λ is other than zero and account for the expenditure of control signal energy.

■ Example 10.13 Optimum system with energy constraint

Let us reconsider the system of Example 10.11, which is shown in Fig. 10.23. For this system we use a state variable feedback so that

$$\begin{aligned}\mathbf{u} &= \mathbf{H}\mathbf{x} \\ &= \begin{bmatrix} k & 0 \\ 0 & k \end{bmatrix}\begin{bmatrix} x_1 \\ x_2 \end{bmatrix} = k\mathbf{I}\mathbf{x}.\end{aligned} \tag{10.130}$$

Therefore the matrix $\mathbf{Q}$ is then

$$\begin{aligned}\mathbf{Q} &= (\mathbf{I} + \lambda\mathbf{H}^T\mathbf{H}) \\ &= (\mathbf{I} + \lambda k^2\mathbf{I}) \\ &= (1 + \lambda k^2)\mathbf{I}.\end{aligned} \tag{10.131}$$

As in Example 10.12 we will let $\mathbf{x}^T(0) = [1, 0]$ so that $J = p_{11}$. We evaluate p_{11} from Eq. (10.128) as

$$\begin{aligned}\mathbf{D}^T\mathbf{P} + \mathbf{P}\mathbf{D} &= -\mathbf{Q} \\ &= -(1 + \lambda k^2)\mathbf{I}.\end{aligned} \tag{10.132}$$

Thus we find that

$$J = p_{11} = (1 + \lambda k^2)\left(1 + \frac{1}{2k}\right), \tag{10.133}$$

and we note that the right-hand side of Eq. (10.133) reduces to Eq. (10.119) when $\lambda = 0$. Now the minimum of J is found by taking the derivative of J, which is

$$\begin{aligned}\frac{dJ}{dk} &= 2\lambda k + \frac{\lambda}{2} - \frac{1}{2k^2} \\ &= \frac{4\lambda k^3 + \lambda k^2 - 1}{2k^2} = 0.\end{aligned} \tag{10.134}$$

Therefore the minimum of the performance index occurs when $k = k_{\min}$, where $k_{\min}$ is the solution of Eq. (10.134).

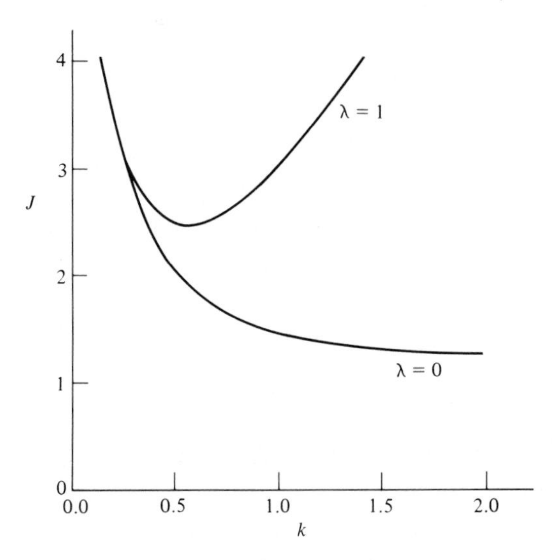

Figure 10.27. Performance index versus the feedback gain k for Example 10.13.

A simple method of solution for Eq. (10.134) is the Newton-Raphson method, illustrated in Section 5.4. Let us complete this example for the case where the control energy and the state variables squared are equally important so that $\lambda = 1$. Then Eq. (10.134) becomes $4k^3 + k^2 - 1 = 0$, and using the Newton-Raphson method, we find that $k_{\min} = 0.555$. The value of the performance index J obtained with $k_{\min}$ is considerably greater than that of the previous example, because the expenditure of energy is equally weighted as a cost. The plot of J versus k for this case is shown in Fig. 10.27. The plot of J versus k for Example 10.12 is also shown for comparison on Fig. 10.27. It has become clear from this and the previous examples that the actual minimum obtained depends upon the initial conditions, the definition of the performance index, and the value of the scalar factor λ.

The design of several parameters can be accomplished in a similar manner to that illustrated in the examples. Also, the design procedure can be carried out for higher-order systems. However, we must then consider the use of a digital computer to determine the solution of Eq. (10.101) in order to obtain the $\mathbf{P}$ matrix. Also, the computer would provide a suitable approach for evaluating the minimum value of J for the several parameters. The newly emerging field of adaptive and optimal control systems is based on the formulation of the time-domain equations and the determination of an optimum feedback control signal $\mathbf{u}(t)$ [1, 2, 9]. The design of control systems using time-domain methods will continue to develop in the future and will provide the control engineer with many interesting challenges and opportunities.

10.11 State-Variable Feedback

In Section 10.10 we considered the use of state-variable feedback in achieving optimization of a performance index. In this section we will use *state-variable feedback* in order to achieve the desired pole location of the closed-loop transfer function $T(s)$. The approach is based on the feedback of all the state variables, and therefore

$$\mathbf{u} = \mathbf{H}\mathbf{x}. \tag{10.135}$$

When using this state-variable feedback, the roots of the characteristic equation are placed where the transient performance meets the desired response. As an example of state-variable feedback, consider the feedback system shown in Fig. 10.28. This position control uses a field controlled motor, and the transfer function was obtained in Section 2.5 as

$$G(s) = \frac{K}{s(s + f/J)(s + R_f/L_f)}, \tag{10.136}$$

where $K = K_a K_m / J L_f$. For our purposes we will assume that $f/J = 1$ and $R_f/L_f = 5$. As shown in Fig. 10.28, the system has feedback of the three state variables: position, velocity, and field current. We will assume that the feedback constant for the position is equal to 1, as shown in Fig. 10.29, which provides a signal-flow

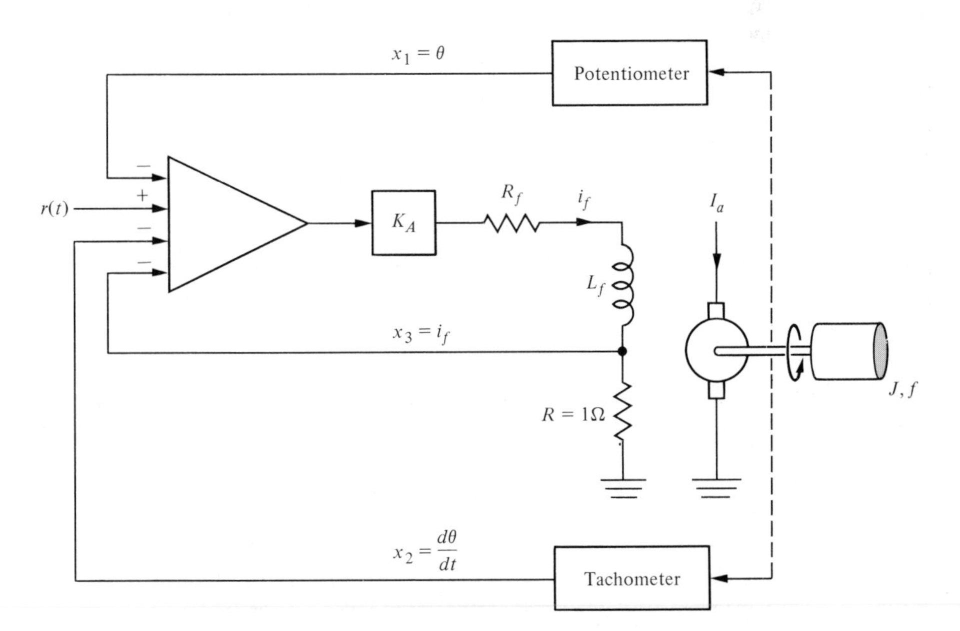

Figure 10.28. Position control system with state-variable feedback.

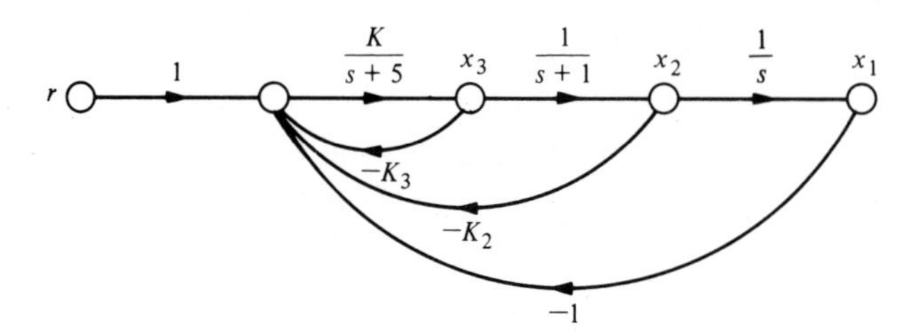

Figure 10.29. Signal-flow graph of the state-variable feedback system.

graph representation of the system. Without state-variable feedback of x_2 and x_3, we set $K_3 = K_2 = 0$ and we have

$$G(s) = \frac{K}{s(s + 1)(s + 5)}. \tag{10.137}$$

This system will become unstable when $K \geq 30$. However, with variable feedback of all the state variables, we can ensure that the system is stable and set the transient performance of the system to a desired performance.

In general, the state-variable feedback signal-flow graph can be converted to the block diagram form shown in Fig. 10.30. The transfer function $G(s)$ remains unaffected (as in Eq. 10.137) and the $H(s)$ accounts for the state-variable feedback. Therefore

$$H(s) = K_3\left[s^2 + \left(\frac{K_3 + K_2}{K_3}\right)s + \frac{1}{K_3}\right] \tag{10.138}$$

and

$$G(s)H(s) = \frac{M[s^2 + Qs + (1/K_3)]}{s(s + 1)(s + 5)}, \tag{10.139}$$

where $M = KK_3$ and $Q = (K_3 + K_2)/K_3$. Since K_3 and K_2 can be set independently, the designer can select the location of the zeros of $G(s)H(s)$.

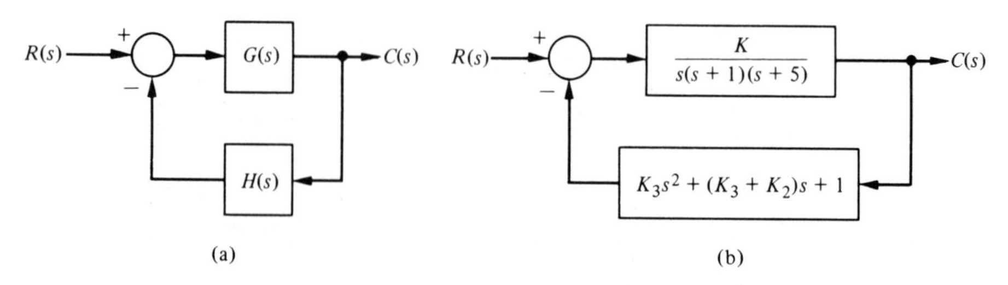

Figure 10.30. Equivalent block diagram representation of the state-variable feedback system.

As an illustration, let us choose the zeros of $GH(s)$ so that they cancel the real poles of $G(s)$. We set the numerator polynomial

$$H(s) = K_3\left(s^2 + Qs + \frac{1}{K_3}\right) = K_3(s+1)(s+5). \tag{10.140}$$

This requires $K_3 = 1/5$ and $Q = 6$, which sets $K_2 = 1$. Then

$$GH(s) = \frac{M(s+1)(s+5)}{s(s+1)(s+5)}, \tag{10.141}$$

where $M = KK_3$. The closed-loop transfer function is then

$$\frac{C(s)}{R(s)} = T(s) = \frac{G(s)}{1 + G(s)H(s)} = \frac{K}{(s+1)(s+5)(s+M)}. \tag{10.142}$$

Therefore, although we could choose $M = 10$, which would ensure the stability of the system, the closed-loop response of the system will be dictated by the poles at $s = -1$ and $s = -5$. Therefore we will usually choose the zeros of $GH(s)$ in order to achieve closed-loop roots in a desirable location in the left-hand plane and to ensure system stability.

■ Example 10.14 State variable feedback design

Let us again consider the system of Fig. 10.30(b) and set the zeros of $GH(s)$ at $s = -4 + j2$ and $s = -4 - j2$. Then the numerator of $GH(s)$ will be

$$\begin{aligned} H(s) &= K_3\left(s^2 + Qs + \frac{1}{K_3}\right) \\ &= K_3(s + 4 + j2)(s + 4 - j2) \\ &= K_3(s^2 + 8s + 20). \end{aligned} \tag{10.143}$$

Therefore $K_3 = 1/20$ and $Q = 8$ resulting in $K_2 = 7/20$. The resulting root locus for

$$G(s)H(s) = \frac{M(s^2 + 8s + 20)}{s(s+1)(s+5)} \tag{10.144}$$

is shown in Fig. 10.31. The system is stable for all values of gain $M = KK_3$. For $M = 10$ the complex roots have $\zeta = 0.73$, so that we might expect an overshoot for a step input of approximately 5%. The settling time will be approximately 1 second. The closed-loop transfer function is

$$\begin{aligned} \frac{C(s)}{R(s)} = T(s) &= \frac{G(s)}{1 + G(s)(H(s)} \\ &= \frac{200}{(s + 3.45 + j3.2)(s + 3.45 - j3.2)(s + 9.1)}. \end{aligned} \tag{10.145}$$

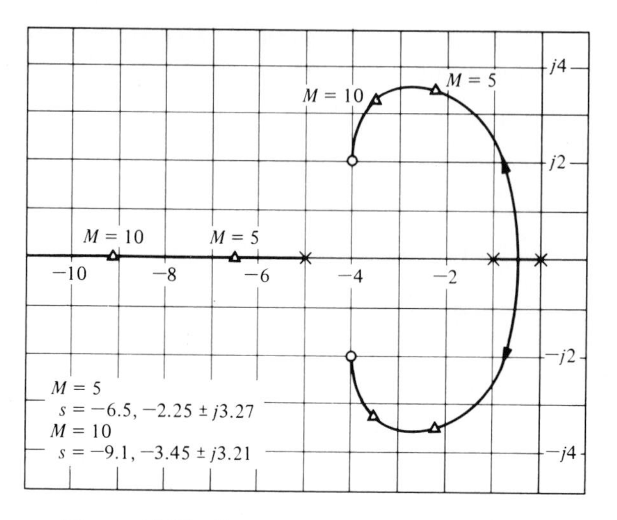

Figure 10.31. Compensated system root locus.

An alternative approach is to set the closed-loop roots of $1 + G(s)H(s) = 0$ at desired locations and then solve for the gain values of K, K_3, and K_2 that are required. For example, if we desire closed-loop roots at $s = -10$, $s = -5 + j$ and $s = -5 - j$, we have the characteristic equation

$$q(s) = (s + 10)(s^2 + 10s + 26) = s^3 + 20s^2 + 126s + 260 = 0. \tag{10.146}$$

Because

$$1 + G(s)H(s) = s(s + 1)(s + 5) + M\left(s^2 + Qs + \frac{1}{K_3}\right) = 0, \tag{10.147}$$

we equate Eq. (10.146) and Eq. (10.147), obtaining $M = 14$, $Q = 121/14$, $K_3 = 14/260$ and $K_2 = 0.41$.

In many cases the state variables are available and we can use state-variable feedback to obtain a stable, well-compensated system.

10.12 Design Example: Rotor Winder Control System

Our goal is to replace a manual operation with a machine to wind copper wire onto the rotors of small motors. Each motor has three separate windings of several hundred turns of wire. It is important that the windings are consistent, and that the throughput of the process is high. The operator simply inserts an unwound rotor, pushes a start button, and then removes the completely wound rotor. The dc motor is used to achieve accurate, rapid windings. Thus the goal is to achieve high steady-state accuracy for both position and velocity. The control

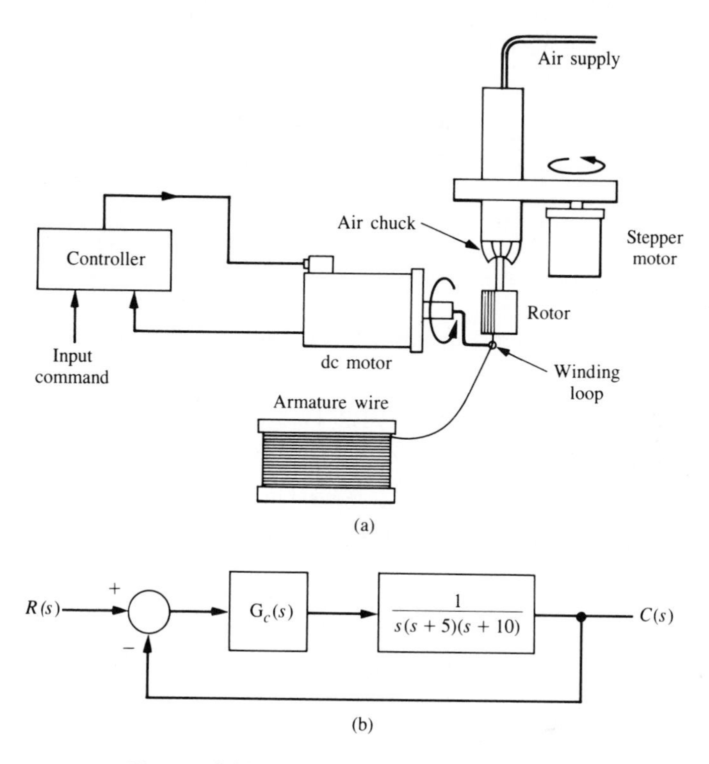

Figure 10.32. Rotor winder control system.

system is shown in Fig. 10.32(a) and the block diagram in Fig. 10.32(b). This system has zero steady-state error for a step input and the steady-state error for a ramp input is

$$e_{ss} = A/K_v,$$

where

$$K_v = \lim_{s \to 0} \frac{G_c(0)}{50}.$$

When $G_c(s) = K$, we have $K_v = K/50$. If we select $K = 500$, we will have $K_v = 10$, but the overshoot to a step is 70% and the settling time is 8 seconds.

First, let us try a lead compensator so that

$$G_c(s) = \frac{K(s + z_1)}{(s + p_1)}. \tag{10.148}$$

Selecting $z = 4$ and the pole p so that the complex roots have a ζ of 0.6, we have

$$G_c(s) = \frac{191.2(s + 4)}{(s + 7.3)}. \tag{10.149}$$

Table 10.1. Design Example Results

Controller	Gain, K	Lead Network	Lag Network	Lead-lag Network
Step overshoot	70%	3%	12%	5%
Settling time (seconds)	8	1.5	2.5	2.0
Steady-state error for ramp	10%	48%	2.6%	4.8%
K_v	10	2.1	38	21

Therefore the response to a step input has a 3% overshoot and a settling time of 1.5 seconds. However, the

$$K_v = \frac{191.2(4)}{7.3(50)} = 2.1,$$

which is inadequate.

If we use a phase-lag design, we select

$$G_c(s) = \frac{K(s + z_2)}{(s + p_2)}$$

in order to achieve $K_v = 38$. Thus the velocity constant of the lag-compensated system is

$$K_v = \frac{Kz}{50p}.$$

Using a root locus, we select $K = 105$ in order to achieve a reasonable uncompensated step response with an overshoot of less than or equal to 10%. We select $\alpha = z/p$ to achieve the desired K_v. We then have

$$\alpha = \frac{50K_v}{K} = \frac{50(38)}{105} = 18.1.$$

Selecting $z_1 = 0.1$ in order to not impact the uncompensated root locus, we have $p_z = 0.0055$. We then obtain a step response with a 12% overshoot and a settling time of 2.5 seconds.

The results for the simple gain, the lead network, and the lag network are summarized in Table 10.1. Let us return to the lead-network system and add a lag network so that the compensator is

$$G_c(s) = \frac{K(s + z_1)(s + z_2)}{(s + p_1)(s + p_2)}. \tag{10.150}$$

The lead compensator of Eq. (10.149) requires $K = 191.2$, $z_1 = 4$, and $p_1 = 7.3$. The root locus for the system is shown in Fig. 10.33. We recall that this lead

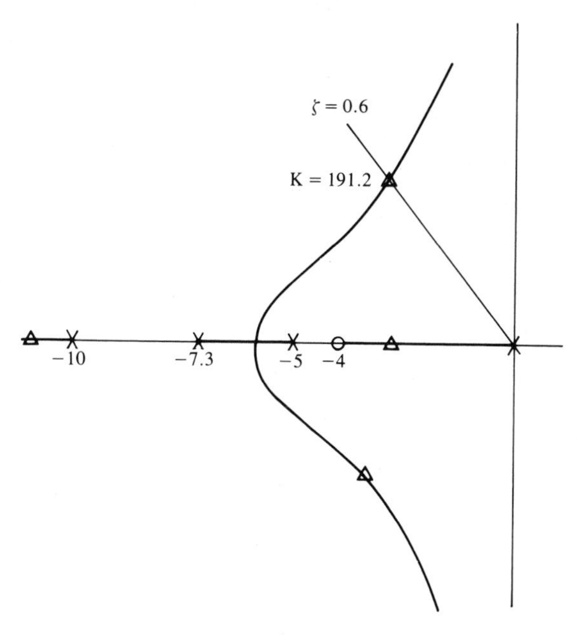

Figure 10.33. Root locus for lead compensator.

network resulted in $K_v = 2.1$ (see Table 10.1). In order to obtain $K_v = 21$, we use $\alpha = 10$ and select $z_2 = 0.1$ and $p_z = 0.01$. Then the total system is

$$G(s)G_c(s) = \frac{191.2(s + 4)(s + 0.1)}{s(s + 5)(s + 10)(s + 7.28)(s + 0.01)}. \tag{10.151}$$

The step response and ramp response of this system are shown in Fig. 10.34 in parts (a) and (b), respectively, and are summarized in Table 10.1. Clearly, the lead-lag design is suitable for satisfaction of the design goals.

10.13 Summary

In this chapter we have considered several alternative approaches to the design and compensation of feedback control systems. In the first two sections, we discussed the concepts of design and compensation and noted the several design cases which we completed in the preceding chapters. Then, the possibility of introducing cascade compensation networks within the feedback loops of control systems was examined. The cascade compensation networks are useful for altering the shape of the root locus or frequency response of a system. The phase-lead network and the phase-lag network were considered in detail as candidates for system compensators. Then system compensation was studied by using a phase-lead s-plane network on the Bode diagram and the root locus s-plane. We noted that the phase-lead compensator increases the phase margin of the system and

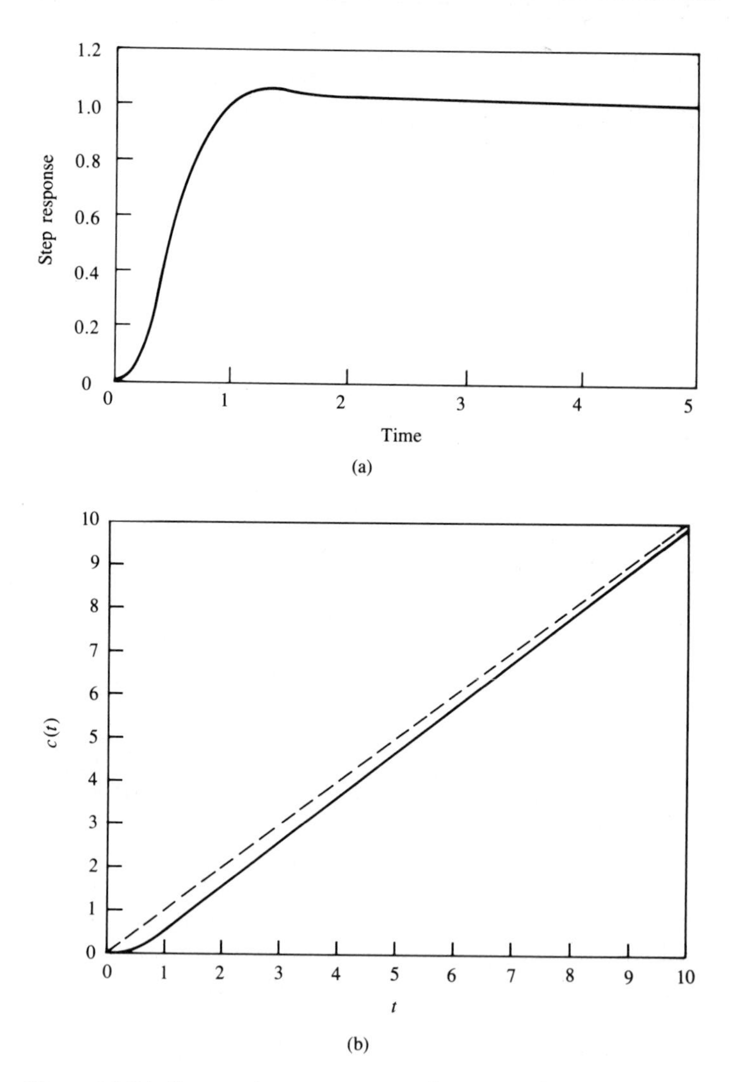

Figure 10.34. Step and ramp response for rotor winder system.

thus provides additional stability. When the design specifications include an error constant, the design of a phase-lead network is more readily accomplished on the Bode diagram. Alternatively, when an error constant is not specified, but the settling time and overshoot for a step input are specified, the design of a phase-lead network is more readily carried out on the s-plane. When large error constants are specified for a feedback system, it is usually easier to compensate the system by using integration (phase-lag) networks. We also noted that the phase-lead compensation increases the system bandwidth, whereas the phase-lag compensation decreases the system bandwidth. The bandwidth often may be an important factor when noise is present at the input and generated within the system. Also we noted that a satisfactory system is obtained when the asymptotic course for magnitude of the compensated system crosses the 0-db line with a slope of -6 db/octave. The characteristics of the phase-lead and phase-lag compensation networks are summarized in Table 10.2. Operational amplifier circuits for phase-lead and lag, and PI and PD compensators are summarized in Table 10.3.

The design of control systems in the time domain was briefly examined. Specifically, the optimum design of a system using state-variable feedback and an integral performance index was considered. Finally, the s-plane design of systems utilizing state-variable feedback was examined.

Table 10.2. A Summary of the Characteristics of Phase-Lead and Phase-Lag Compensation Networks

	Compensation	
	Phase-lead	**Phase-lag**
Approach	Addition of phase-lead angle near crossover frequency or to yield desired dominant roots in s-plane	Addition of phase-lag to yield an increased error constant while maintaining desired dominant roots in s-plane or phase margin on Bode diagram
Results	1. Increases system bandwidth 2. Increases gain at higher frequencies	1. Decreases system bandwidth
Advantages	1. Yields desired response 2. Speeds dynamic response	1. Suppresses high frequency noise 2. Reduces steady-state error
Disadvantages	1. Requires additional amplifier gain 2. Increases bandwidth and thus susceptibility to noise 3. May require large values of components for RC network	1. Slows down transient response 2. May require large values of components for RC network
Applications	1. When fast transient response is desired	1. When error constants are specified
Not applicable	1. When phase decreases rapidly near crossover frequency	1. When no low frequency range exists where phase is equal to desired phase margin

Table 10.3. Operational Amplifier Circuits for Compensators

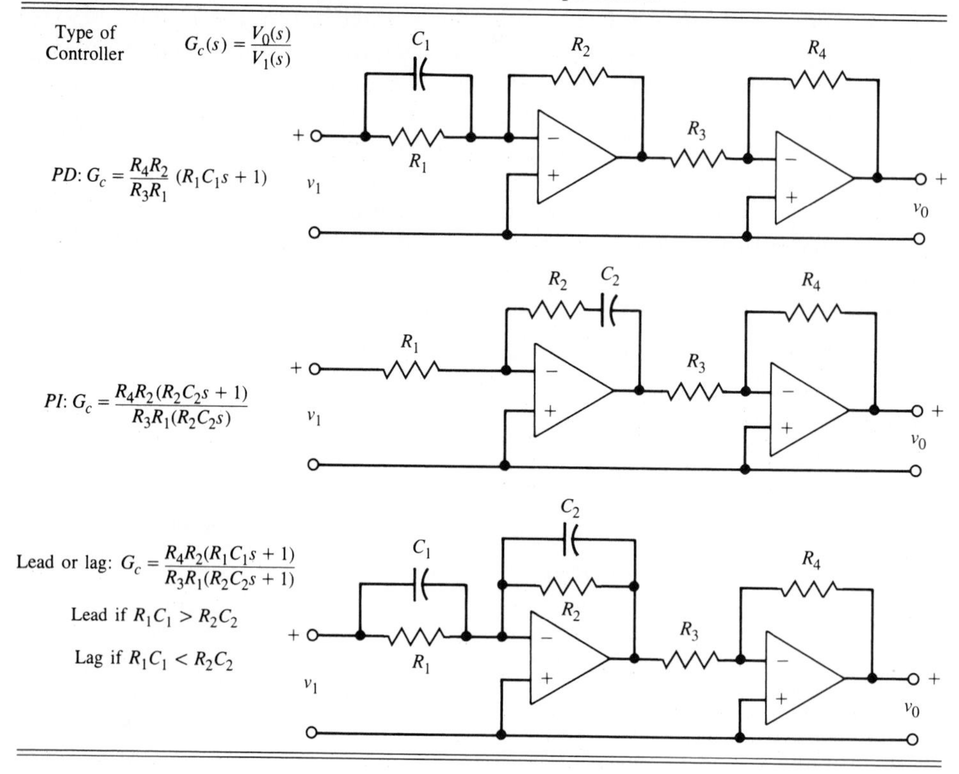

Type of Controller	$G_c(s) = \dfrac{V_0(s)}{V_1(s)}$
PD	$G_c = \dfrac{R_4R_2}{R_3R_1}(R_1C_1s + 1)$
PI	$G_c = \dfrac{R_4R_2(R_2C_2s + 1)}{R_3R_1(R_2C_2s)}$
Lead or lag	$G_c = \dfrac{R_4R_2(R_1C_1s + 1)}{R_3R_1(R_2C_2s + 1)}$ Lead if $R_1C_1 > R_2C_2$ Lag if $R_1C_1 < R_2C_2$

Exercises

E10.1. A negative feedback control system has a transfer function

$$G(s) = \frac{K}{(s + 3)}.$$

We select a compensator,

$$G_c(s) = \frac{s + a}{s},$$

in order to achieve zero steady-state error for a step input. Select a and K so that the overshoot to a step is approximately 5% and the settling time is approximately 1 sec.

Answer: K = 5, a = 6.4

E10.2. A control system with negative unity feedback has a process

$$G(s) = \frac{400}{s(s + 40)},$$

and we wish to use a proportional plus integral compensation where

$$G_c(s) = K_1 + \frac{K_2}{s}.$$

Note that the steady-state error of this system for a ramp input is zero. (a) Set $K_2 = 1$ and find a suitable value of K_1 so the step response will have an overshoot of approximately 20%. (b) What is the expected settling time of the compensated system?

E10.3. A unity negative feedback control system in a manufacturing system has a process transfer function

$$G(s) = \frac{e^{-s}}{s + 1},$$

and it is proposed to use a compensator to achieve a 5% overshoot to a step input. The compensator is [4]

$$G_c(s) = K\left(1 + \frac{1}{\tau s}\right),$$

which provides proportional plus integral control. Show that one solution is $K = 0.5$ and $\tau = 1$.

E10.4. Consider a unity negative feedback system with

$$G(s) = \frac{K}{s(s + 2)(s + 3)},$$

where K is set equal to 20 in order to achieve a specified $K_v = 3.33$. We wish to add a lead-lag compensator

$$G_c(s) = \frac{(s + 0.15)(s + 0.7)}{(s + 0.015)(s + 7)}.$$

Show that the gain margin of the compensated system is 24 db and the phase margin is 75°.

E10.5. Consider a unity feedback system with the transfer function

$$G(s) = \frac{K}{s(s + 2)(s + 4)}.$$

It is desired to obtain the dominant roots with $\omega_n = 3$ and $\zeta = 0.5$. We wish to obtain a $K_v = 2.7$. Show that we require a compensator

$$G_c(s) = \frac{7.53(s + 2.2)}{(s + 16.4)}.$$

Determine the value of K that should be selected.

E10.6. Reconsider the wind tunnel control system of Problem 6.31. When $K = 326$, find $T(s)$ and estimate the expected overshoot and settling time. Compare your estimates with the actual overshoot of 60% and a settling time of 4 sec. Explain the discrepancy in your estimates.

E10.7. NASA astronauts retrieved a satellite and brought it into the cargo bay of the space shuttle, as shown in Fig. E10.7(a). An astronaut within the shuttle controlled the robot arm

(a)

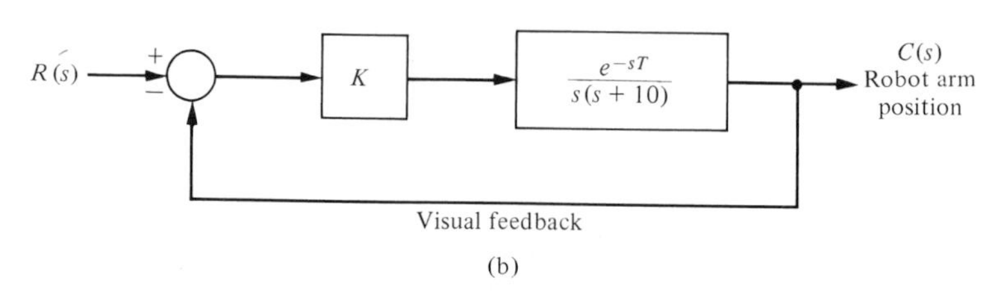

(b)

Figure E10.7. Retrieval of a satellite.

upon which Joseph Allen is seen standing. A model of the feedback control system is shown in Fig. E10.7(b). Determine the value of K that will result in a phase margin of 50° when $T = 0.1$ sec.

Answer: $K = 37.8$

E10.8. A negative-unity feedback system has a plant

$$G(s) = \frac{2257}{s(\tau s + 1)},$$

where $\tau = 2.8$ ms. Select a compensator

$$G_c(s) = K_1 + K_2/s$$

so that the dominant roots of the characteristic equation have ζ equal to $1/\sqrt{2}$. Plot $c(t)$ for a step input.

E10.9. A control system with a controller is shown in Fig. E10.9. Select K_1 and K_2 so that the overshoot to a step input is less than 1% and the velocity constant, K_v, is equal to 10. Use the Control System Design Program to verify the results of your design.

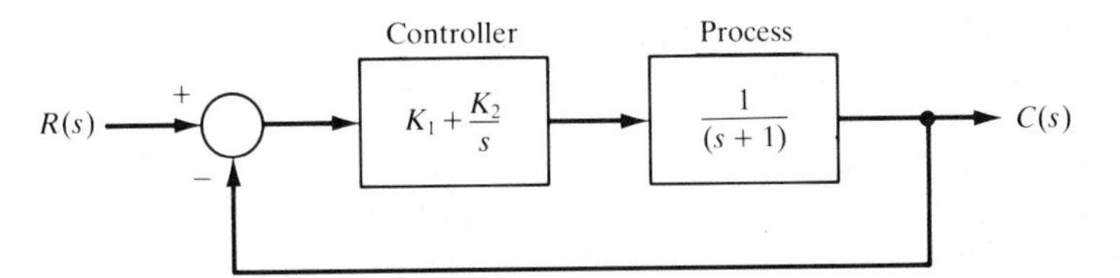

Figure E10.9. Design of a controller.

E10.10. A control system with a controller is shown in Fig. E10.10. We will select $K_2 = 4$ in order to provide a reasonable steady state error to a step [13]. Using the Control System Design Program, find K_1 to obtain a phase margin of 60°. Find the peak time and percent overshoot of this system.

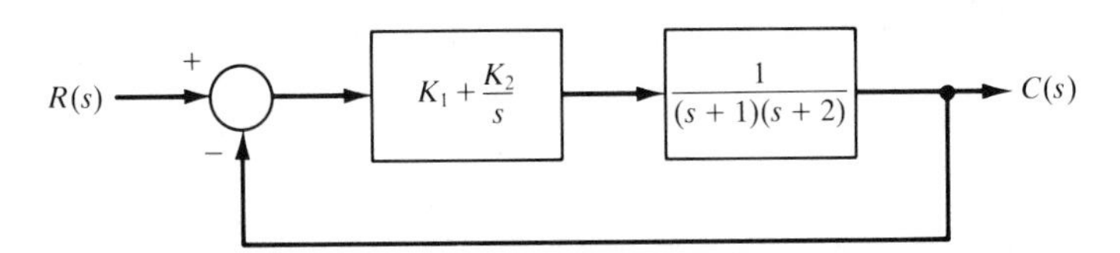

Figure E10.10. Design of a PI controller.

E10.11. A unity feedback system has

$$G(s) = \frac{1350}{s(s+2)(s+30)}.$$

A lead network is selected so that

$$G_c(s) = \frac{(1 + 0.25s)}{(1 + 0.025s)}$$

Determine the peak magnitude and the bandwidth of the closed-loop frequency response using the (a) Nichols chart, and (b) Control System Design Program plot of the closed-loop frequency response.

Answer: $M_{p_\omega} = 2.3$ db, $\omega_B = 22$

E10.12. The control of an automobile ignition system has unity negative feedback and a loop transfer function $G_c(s)G(s)$, where

$$G(s) = \frac{K}{s(s+10)}$$

and

$$G_c(s) = K_1 + K_2/s.$$

A designer selects $K_2/K_1 = 0.5$ and asks you to determine KK_1 so that the dominant roots have a ζ of $1/\sqrt{2}$ and the settling time is less than 2 sec.

E10.13. The design of Example 10.3 determined a lead network in order to obtain desirable dominant root locations using a cascade compensator $G_c(s)$ in the system configuration shown in Fig. 10.1(a). The same lead network would be obtained if we used the feedback compensation configuration of Fig. 10.1(b). Determine the closed-loop transfer function, $T(s) = C(s)/R(s)$, of both the cascade and feedback configuration and show how the transfer function of each configuration differs. Explain how the response to a step, $R(s)$, will be different for each system.

E10.14. A robot will be operated by NASA to build a permanent lunar station. The position control system for the gripper tool is shown in Fig. 10.1(a), where $H(s) = 1$ and

$$G(s) = \frac{3}{s(s+1)(0.5s+1)}.$$

Determine a compensator lag network, $G_c(s)$, that will provide a phase margin of 45 degrees.

Answer: $G_c(s) = \dfrac{1 + 20s}{1 + 106s}$

Problems

P10.1. The design of the lunar excursion module (LEM) is an interesting control problem. The *Apollo 11* lunar landing vehicle is shown in Fig. P10.1(a), on next page. The attitude control system for the lunar vehicle is shown in Fig. P10.1(b). The vehicle damping is negligible and the attitude is controlled by gas jets. The torque, as a first approximation, will be considered to be proportional to the signal $V(s)$ so that $T(s) = K_2V(s)$. The loop gain may be selected by the designer in order to provide a suitable damping. A damping ratio of $\zeta = 0.5$ with a settling time of less than 2 sec is required. Using a lead-network compensation, select the necessary compensator $G_c(s)$ by using (a) frequency response techniques and (b) root locus methods.

P10.2. A magnetic tape-recorder transport for modern computers requires a high-accuracy, rapid-response control system. The requirements for a specific transport are as follows: (1) the tape must stop or start in 10 msec; (2) it must be possible to read 45,000 characters per second. This system was discussed in Problem P6.11. It is desired to set $J = 5 \times 10^{-3}$, and K_1 is set on the basis of the maximum error allowable for a velocity input. In this case, it is desired to maintain a steady-state speed error of less than 5%. We will use a tachometer in this case and set $K_a = 50{,}000$ and $K_2 = 1$. In order to provide a suitable performance, a compensator $G_c(s)$ is inserted immediately following the photocell transducer. Select a compensator $G_c(s)$ so that the overshoot of the system for a step input is less than 30%. We will assume that $\tau_1 = \tau_a = 0$.

P10.3. A simplified version of the attitude rate control for the F-94 or X-15 type aircraft is shown in Fig. P10.3. When the vehicle is flying at four times the speed of sound (Mach 4) at an altitude of 100,000 ft, the parameters are

$$\frac{1}{\tau_a} = 1.0, \qquad K_1 = 0.5, \qquad \zeta\omega_a = 1.0, \qquad \omega_a = 2.$$

Design a compensator, $G_c(s)$, so that the response to a step input has an overshoot of less than 10% and a settling time of less than 5 sec.

(a)

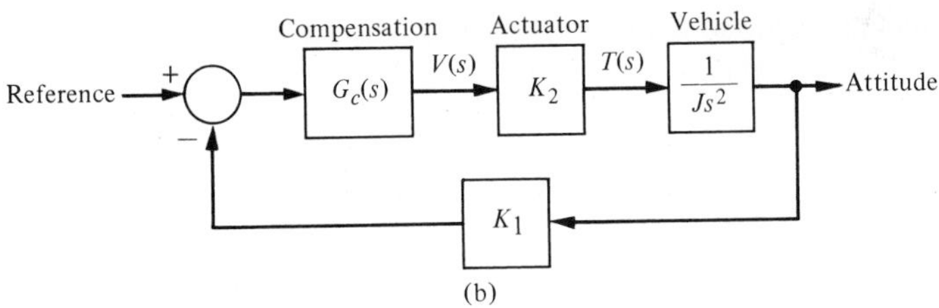

(b)

Figure P10.1. (a) *Apollo 11* lunar excursion module viewed from the command ship. Inside the LEM were astronauts Neil Armstrong and Edwin Aldrin, Jr. The LEM landed on the moon on July 20, 1969. (Photo courtesy of NASA Manned Spacecraft Center.)

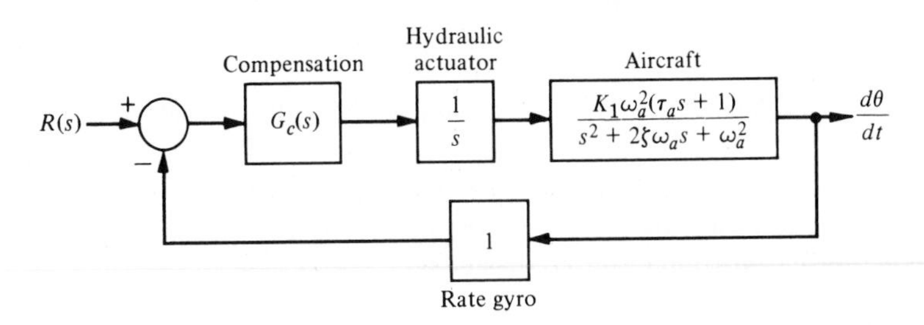

Figure P10.3. Aircraft attitude control.

P10.4. Magnetic particle clutches are useful actuator devices for high power requirements because they can typically provide a 200-w mechanical power output. The particle clutches provide a high torque-to-inertia ratio and fast time constant response. A particle-clutch positioning system for nuclear reactor rods is shown in Fig. P10.4. The motor drives two counter-rotating clutch housings. The clutch housings are geared through parallel gear trains, and the direction of the servo output is dependent on the clutch that is energized. The time constant of a 200-w clutch is $\tau = \frac{1}{40}$ sec. The constants are such that $K_T n/J = 1$. It is desired that the maximum overshoot for a step input is in the range of 10% to 20%. Design a compensating network so that the system is adequately stabilized. The settling time of the system should be less than or equal to 1 sec. This system requires two compensation networks in cascade.

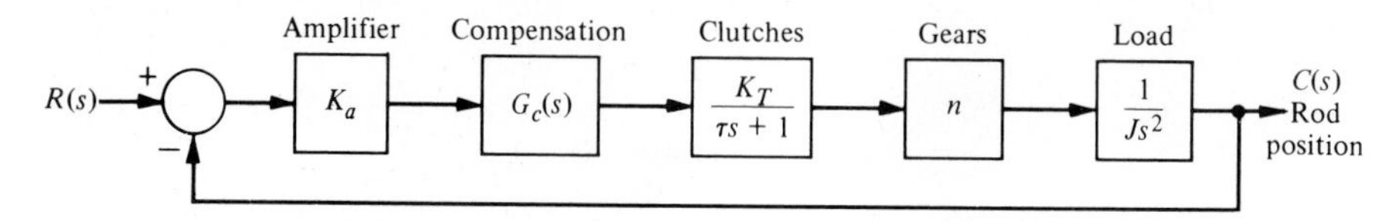

Figure P10.4. Nuclear reactor rod control.

P10.5. A stabilized precision rate table uses a precision tachometer and a dc direct-drive torque motor, as shown in Fig. P10.5. It is desired to maintain a high steady-state accuracy for the speed control. In order to obtain a zero steady-state error for a step command design, select a proportional plus integral compensator as discussed in Section 10.6. Select the appropriate gain constants so that the system has an overshoot of approximately 10% and a settling time in the range of 0.4 to 0.6 sec.

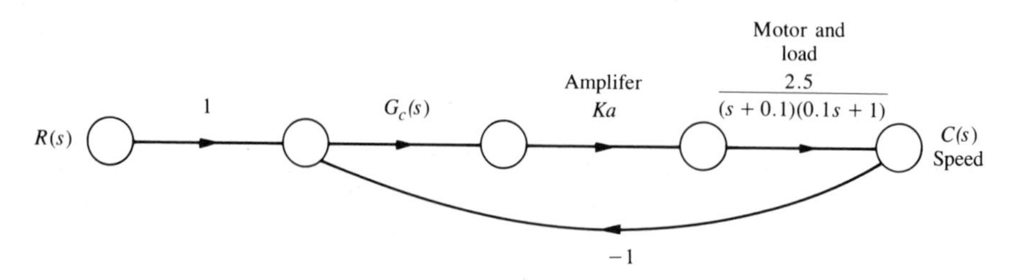

Figure P10.5. Stabilized rate table.

P10.6. Repeat Problem P10.5 by using a lead network compensator and compare the results.

P10.7. The primary control loop of a nuclear power plant includes a time delay due to the time necessary to transport the fluid from the reactor to the measurement point (see Fig. P10.7). The transfer function of the controller is

$$G_1(s) = \left(K_1 + \frac{K_2}{s}\right).$$

The transfer function of the reactor and time delay is

$$G(s) = \frac{e^{-sT}}{\tau s + 1},$$

where $T = 0.4$ sec and $\tau = 0.2$ sec. Using frequency response methods, design the controller so that the overshoot of the system is less than 20%. Estimate the settling time of the system designed. Use the Control System Design Program to determine the actual overshoot and settling time.

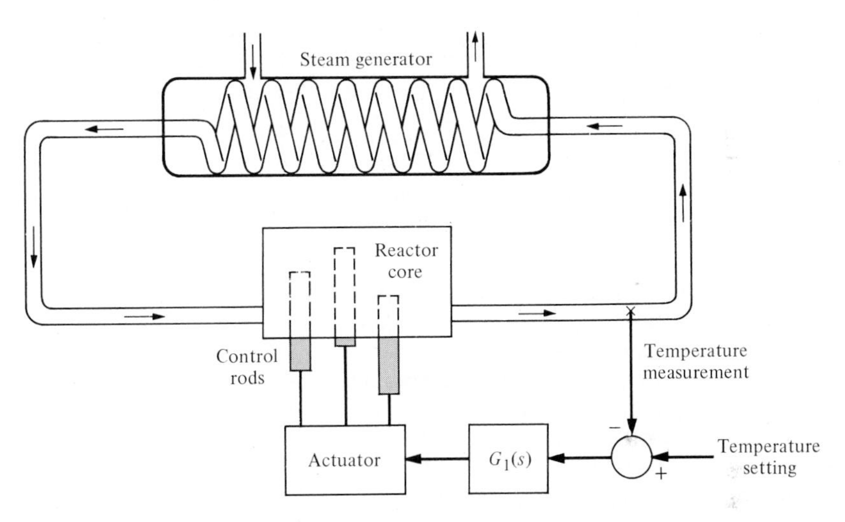

Figure P10.7. Nuclear reactor control.

P10.8. A chemical reactor process whose production rate is a function of catalyst addition is shown in block diagram form in Fig. P10.8 [4]. The time delay is $T = 10$ min and the time constant, τ, is approximately 10 min. The gain of the process is $K = 1$. Design a compensation by using Bode diagram methods in order to provide a suitable system response. It is desired to have a steady-state error for a step input, $R(s) = A/s$, less than 0.05 A. For the system with the compensation added, estimate the settling time of the system.

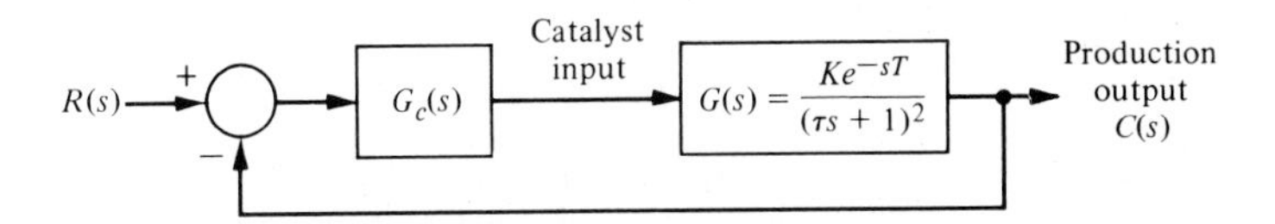

Figure P10.8. Chemical reactor control.

P10.9. A numerical path-controlled machine turret lathe is an interesting problem in attaining sufficient accuracy [2]. A block diagram of a turret lathe control system is shown in Fig. P10.9. The gear ratio is $n = 0.1$, $J = 10^{-3}$, and $f = 10^{-2}$. It is necessary to attain an accuracy of 5×10^{-4} in., and therefore a steady-state position accuracy of 3% is specified for a ramp input. Design a cascade compensator to be inserted before the silicon-controlled rectifiers in order to provide a response to a step command with an overshoot of less than 5%. A suitable damping ratio for this system is 0.7. The gain of the silicon-controlled rectifiers is $K_R = 5$. Design a suitable lag compensator by using the (a) Bode diagram method and (b) s-plane method.

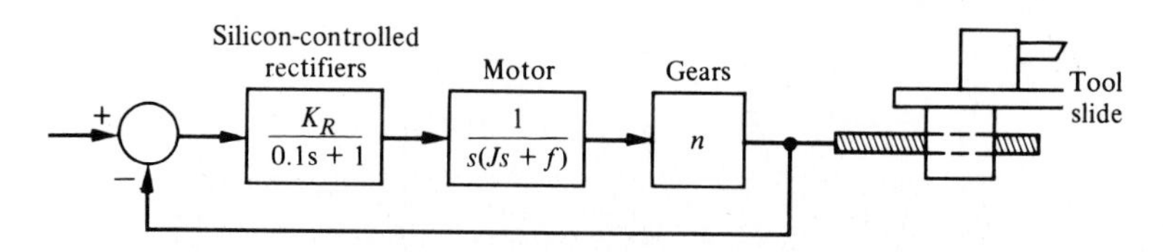

Figure P10.9. Path-controlled turret lathe.

P10.10. The HS *Denison,* shown in Fig. P10.10(a), on the next page, is a large hydrofoil seacraft built by Grumman Corp. for the U.S. Maritime Administration. The *Denison* is an 80-ton hydrofoil capable of operating in seas ranging to 9 ft in amplitude at a speed of 60 knots as a result of the utilization of an automatic stabilization control system. Stabilization is achieved by means of flaps on the main foils and the incidence of the aft foil. The stabilization control system maintains a level flight through rough seas. Thus a system that minimizes deviations from a constant lift force or, equivalently, that minimizes the pitch angle θ has been designed. A block diagram of the lift control system is shown in Fig. P10.10(b). The desired response of the system to wave disturbance is a constant-level travel of the craft. Establish a set of reasonable specifications and design a compensator $G_c(s)$ so that the performance of the system is suitable. Assume that the disturbance is due to waves with a frequency $\omega = 1$ rad/sec.

P10.11. A first-order system is represented by the time-domain differential equation

$$\dot{x} = 3x + 2u.$$

A feedback controller is to be designed where

$$u(t) = -kx$$

and the desired equilibrium condition is $x(t) = 0$ as $t \to \infty$. The performance integral is defined as

$$J = \int_0^\infty x^2\, dt,$$

and the initial value of the state variable is $x(0) = 1$. Obtain the value of k in order to make J a minimum. Is this k physically realizable? Select a practical value for the gain k and evaluate the performance index with that gain. Is the system stable without the feedback due to $u(t)$?

P10.12. In order to account for the expenditure of energy and resources, the control signal is often included in the performance integral. Then, the system may not utilize an unlim-

(a)

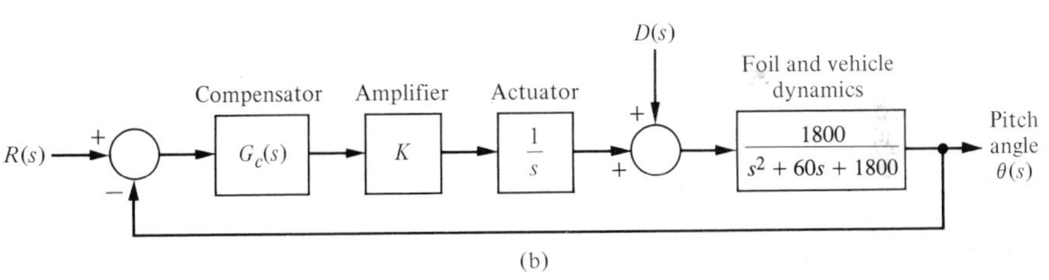

(b)

Figure P10.10. (a) The HS *Denison* hydrofoil. (Photo courtesy of Grumman Aircraft Engineering Corp.) (b) A block diagram of the lift control system.

ited control signal $u(t)$. One suitable performance index, which includes the effect of the magnitude of the control signal, is

$$J = \int_0^\infty (x^2(t) + \lambda u^2(t)\, dt.$$

(a) Repeat Problem P10.11 for the performance index. (b) If $\lambda = 1$, obtain the value of k that minimizes the performance index. Calculate the resulting minimum value of J.

P10.13. An unstable robot system is described by the vector differential equation [10]

$$\frac{d}{dt}\begin{bmatrix} x_1 \\ x_2 \end{bmatrix} = \begin{bmatrix} 1 & 0 \\ -1 & 2 \end{bmatrix}\begin{bmatrix} x_1 \\ x_2 \end{bmatrix} + \begin{bmatrix} 1 \\ 1 \end{bmatrix} u(t).$$

Both state variables are measurable, and so the control signal is set as $u(t) = -k(x_1 + x_2)$. Following the method of Section 10.10, design gain k so that the performance index is minimized. Evaluate the minimum value of the performance index. Determine the sensitivity of the performance to a change in k. Assume that the initial conditions are

$$\mathbf{x}(0) = \begin{bmatrix} 1 \\ 1 \end{bmatrix}.$$

Is the system stable without the feedback signals due to $u(t)$?

P10.14. Determine the feedback gain k of Example 10.12 that minimizes the performance index

$$J = \int_0^\infty \mathbf{x}^T\mathbf{x}\, dt$$

when $\mathbf{x}^T(0) = [1, 1]$. Plot the performance index J versus the gain k.

P10.15. Determine the feedback gain k of Example 10.13 that minimizes the performance index

$$J = \int_0^\infty (\mathbf{x}^T\mathbf{x} + \mathbf{u}^T\mathbf{u})\, dt$$

when $\mathbf{x}^T(0) = [1, 1]$. Plot the performance index J versus the gain k.

P10.16. For the solutions of Problems P10.13., P10.14, and P10.15 determine the roots of the closed-loop optimal control system. Note how the resulting closed-loop roots depend upon the performance index selected.

P10.17. A system has the vector differential equation as given in Eq. (10.101). It is desired that both state variables are used in the feedback so that $u(t) = -k_1x_1 - k_2x_2$. Also, it is desired to have a natural frequency, ω_n, for this system equal to 2. Find a set of gains k_1 and k_2 in order to achieve an optimal system when J is given by Eq. (10.95). Assume $\mathbf{x}^T(0) = [1, 0]$.

P10.18. A unity feedback control system for a robot submarine has a plant with a third-order transfer function:

$$G(s) = \frac{K}{s(s+10)(s+50)}.$$

It is desired that the overshoot be approximately 7.5% for a step input and the settling time of the system be 400 msec. Find a suitable phase-lead compensator by using root locus methods. Let the zero of the compensator be located at $s = -15$ and determine the compensator pole. Determine the resulting system K_v.

P10.19. Now that expenditures for manned space exploration have leveled off, NASA is developing remote manipulators. These manipulators can be used to extend the hand and the power of humankind through space by means of radio. A concept of a remote manipulator is shown in Fig. P10.19(a), on the next page. The closed loop control is shown schematically in Fig. P10.19(b). Assuming an average distance of 238,855 miles from earth to moon, the time delay in transmission of a communication signal T is 1.28 sec. The operator uses a control stick to control remotely the manipulator placed on the moon to assist in geological experiments, and the TV display to access the response of the manipulator. The time constant of the manipulator is ⅓ sec. (a) Set the gain K_1 so that the system has a phase margin of approximately 20°. Evaluate the percentage steady-state error for this system for a step input. (b) In order to reduce the steady-state error for a position command input to 5%, add a lag compensation network in cascade with K_1.

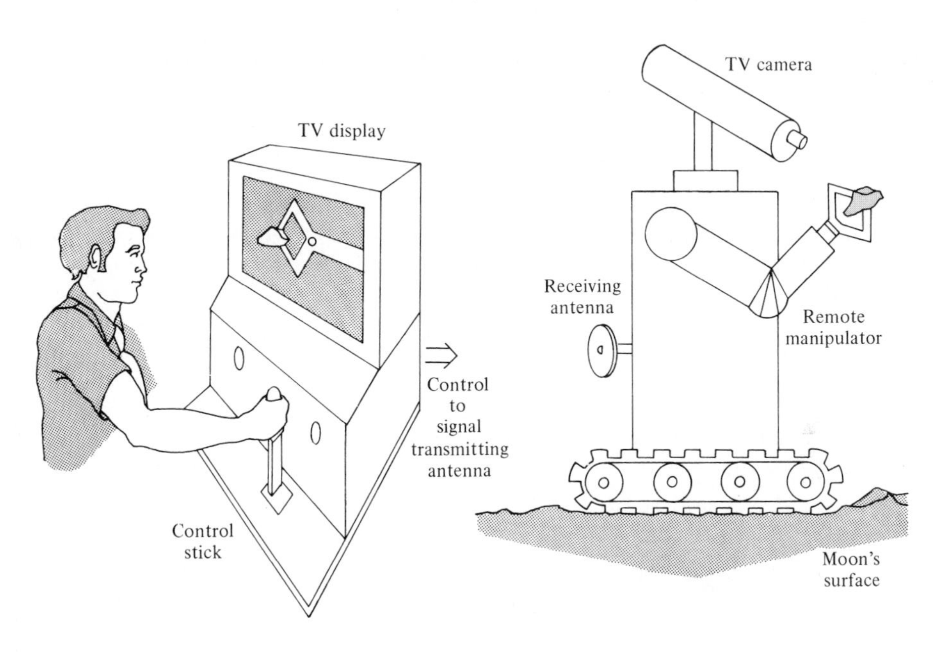

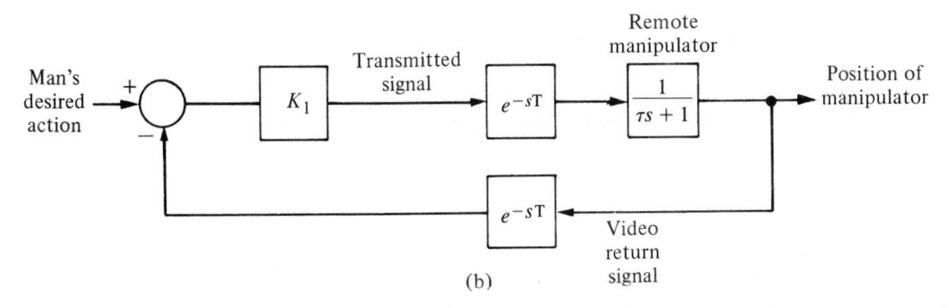

Figure P10.19. (a) Conceptual diagram of a remote manipulator on the moon controlled by a person on the earth. (b) Feedback diagram of the remote manipulator control system with τ = transmission time delay of the video signal.

P10.20. There have been significant developments in the application of robotics technology to nuclear power plant maintenance problems. Thus far, robotics technology in the nuclear industry has been used primarily on spent-fuel reprocessing and waste management. Today, the industry is beginning to apply the technology to such areas as primary containment inspection, reactor maintenance, facility decontamination, and accident recovery activities. These developments suggest that the application of remotely operated devices can significantly reduce radiation exposure to personnel and improve maintenance-program performance.

Currently, an operational robotic system is under development to address particular operational problems within a nuclear power plant. This device, IRIS (Industrial Remote Inspection System), is a general-purpose surveillance system that conducts particular inspection and handling tasks with the goal of significantly reducing personnel exposure to high radiation fields [16]. The device is shown in Fig. P10.20. The open loop transfer function is

$$G(s) = \frac{Ke^{-sT}}{(s+1)(s+3)}.$$

(a) Determine a suitable gain K for the system when T = 0.5 sec so that the overshoot to a step input is less than 30%. Determine the steady-state error. (b) Design a lead compensator

$$G_c(s) = \frac{s+2}{s+b}$$

to improve the step response for the system in part (a) so that the steady-state error is less than 12%. Assume the closed-loop system of Figure 10.1a.

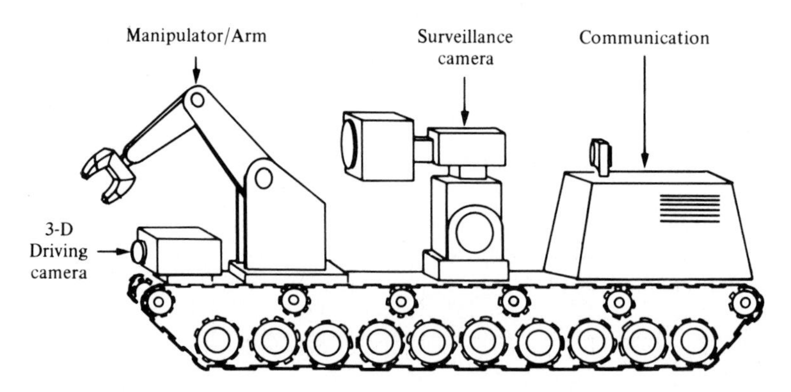

Figure P10.20. Remote controlled robot for nuclear plants.

P10.21. An uncompensated control system with unity feedback has a plant transfer function

$$G(s) = \frac{K}{s(s/2+1)(s/6+1)}.$$

It is desired to have a velocity error constant of $K_v = 20$. It is also desired to have a phase margin approximately to 45° and a closed-loop bandwidth greater than $\omega = 4$ rad/sec. Use two identical cascaded phase-lead networks to compensate the system.

P10.22. For the system of Problem P10.21, design a phase-lag network to yield the desired specifications, with the exception that a bandwidth equal to or greater than 2 rad/sec will be acceptable.

P10.23. For the system of Problem P10.21 we wish to achieve the same phase margin and K_v, but in addition we wish to limit the bandwidth to less than 10 rad/sec but greater than 2 rad/sec. Utilize a lead-lag compensation network to compensate the system. The lead-lag network could be of the form

$$G_c(s) = \frac{(1+s/10a)(1+s/b)}{(1+s/a)(1+s/10b)},$$

where a is to be selected for the lag portion of the compensator and b is to be selected for the lead portion of the compensator. The ratio α is chosen to be 10 for both the lead and lag portions.

P10.24. For the system of Example 10.10, determine the optimum value for k_2 when $k_1 = 1$ and $\mathbf{x}^T(0) = [1, 0]$.

P10.25. The stability and performance of the rotation of a robot (similar to waist rotation) is a challenging control problem. The system requires high gains in order to achieve high resolution; yet a large overshoot of the transient response cannot be tolerated. The block diagram of an electro-hydraulic system for rotation control is shown in Fig. P10.25 [16]. The arm-rotating dynamics are represented by

$$G(s) = \frac{360}{s(s^2/6400 + s/50 + 1)}.$$

It is desired to have $K_v = 10$ for the compensated system. Design a compensator that results in an overshoot to a step input of less than 15%.

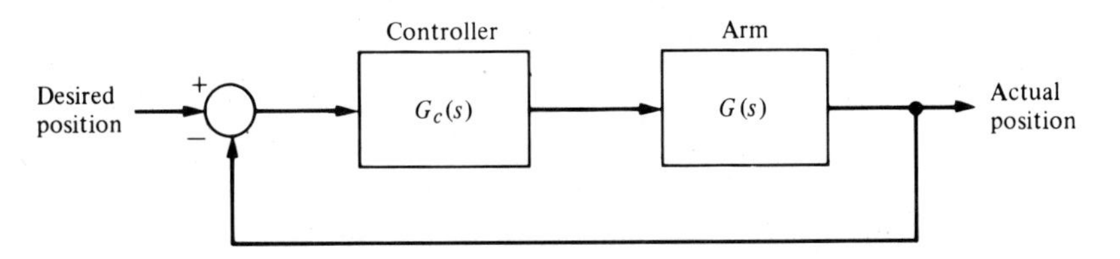

Figure P10.25. Robot position control.

P10.26. The possibility of overcoming wheel friction, wear, and vibration by contactless suspension for passenger-carrying mass-transit vehicles is being investigated throughout the world. One design uses a magnetic suspension with an attraction force between the vehicle and the guideway with an accurately controlled airgap. A system is shown in Fig. P10.26, which incorporates feedback compensation. Using root-locus methods, select a suitable value for K_1 and b so the system has a damping ratio for the underdamped roots of $\zeta = 0.50$. Assume, if appropriate, that the pole of the airgap feedback loop ($s = -200$) can be neglected.

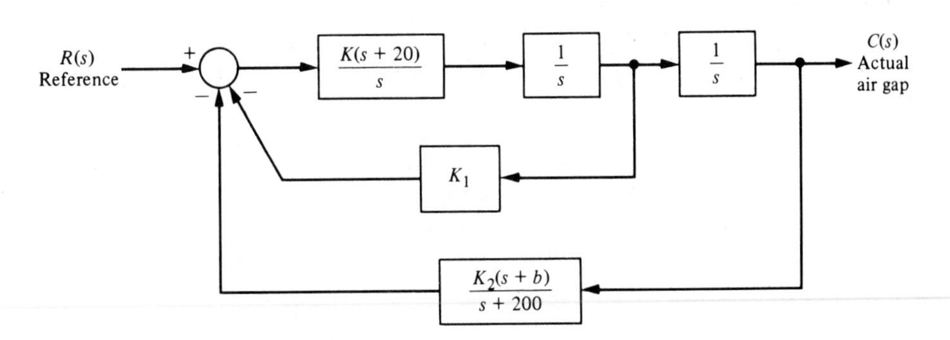

Figure P10.26. Airgap control of train.

P10.27. A computer uses a printer as a fast output device. It is desirable to maintain accurate position control while moving the paper rapidly through the printer. Consider a system with unity feedback and a transfer function for the motor and amplifier as

$$G(s) = \frac{0.15}{s(s+1)(5s+1)}.$$

Design a lead network compensator so that the system bandwidth is 0.75 rad/sec and the phase margin is 30°. Use a lead network with $\alpha = 10$.

P10.28. An engineering design team is attempting to control a process shown in Fig. P10.28. The system has a controller $D(s)$, but the design team is unable to appropriately select $D(s)$. It is agreed that a system with a phase margin of 50° is acceptable, but $D(s)$ is unknown. You are asked to determine $D(s)$.

First, let $D(s) = K$ and using the Control System Design Program (CSDP) or an equivalent program find (a) a value of K which yields a phase margin of 50° and the system's step response for this value of K. (b) Determine the settling time, percent overshoot, and the peak time. (c) Obtain the system's closed-loop frequency response and determine M_{p_ω} and the bandwidth.

The team has decided to let

$$D(s) = \frac{K(s+12)}{(s+20)}$$

and to repeat parts (a), (b), and (c) as above. Determine the gain K that results in a phase margin of 50° and then proceed to evaluate the time response and the closed-loop frequency response. Prepare a table contrasting the results of the two selected controllers for $D(s)$ by comparing settling time, percent overshoot, peak time, M_{p_ω}, and bandwidth.

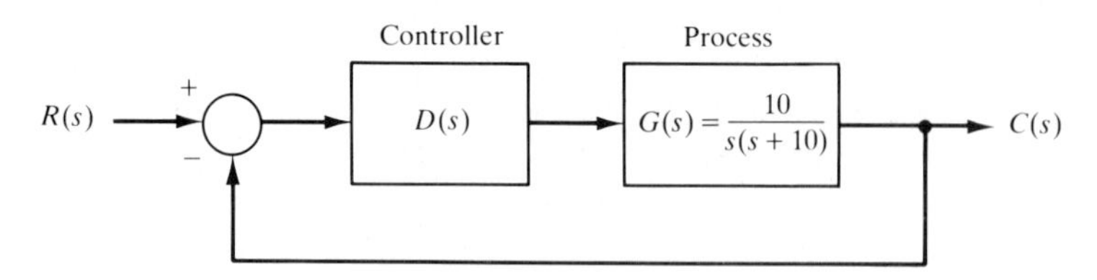

Figure P10.28. Controller design.

P10.29. An adaptive suspension vehicle uses a legged locomotion principle. The control of the leg can be represented by a unity feedback system with [10]

$$G(s) = \frac{K}{s(s+10)(s+5)}.$$

It is desired to achieve a steady-state error for a ramp input of 1% and a damping ratio of the dominant roots of 0.707. Determine a suitable lag compensator. Using the Control System Design Program, determine the actual overshoot and settling time of this system.

P10.30. A liquid level control system (see Fig. 8.26) has a loop transfer function

$$G_c(s)G(s)H(s)$$

where $H(s) = 1$, $G_c(s)$ is a compensator, and the plant is

$$G(s) = \frac{10e^{-sT}}{s^2(s + 10)}$$

where $T = 50$ ms. Design a compensator so that M_{p_ω} does not exceed 3.5 db and ω_r is approximately 1.4 rad/s. Predict the overshoot and settling time of the compensated system when the input is a step.

P10.31. An automated guided vehicle (AGV) can be considered as an automated mobile conveyor designed to transport materials. Most AGVs require some type of guide path. The steering stability of the guidance control system has not been fully solved. The slight "snaking" of the AGV about the track generally has been acceptable, although it indicates instability of the steering guidance control system [9].

Most AGVs have a specification of maximum speed of about 1 m/sec, although in practice they are usually operated at half that speed. In a fully automated manufacturing environment, there should be few personnel in the production area; therefore the AGV should be able to be run at full speed. As the speed of the AGV increases, so does the difficulty in designing stable and smooth tracking controls.

A steering system for an AGV is shown in Fig. P10.31, where $\tau_1 = 40$ ms and $\tau_2 = 1$ ms. It is required that the velocity constant, K_v, is 100 so that the steady-state error for a ramp input is 1% of the slope of the ramp. Neglect τ_2 and design a lead compensator so that the phase margin is

$$45° \leq \text{PM} \leq 65°.$$

Attempt to obtain the two limiting cases for phase margin and compare your results for the two designs by determining the actual percent overshoot and settling time for a step input.

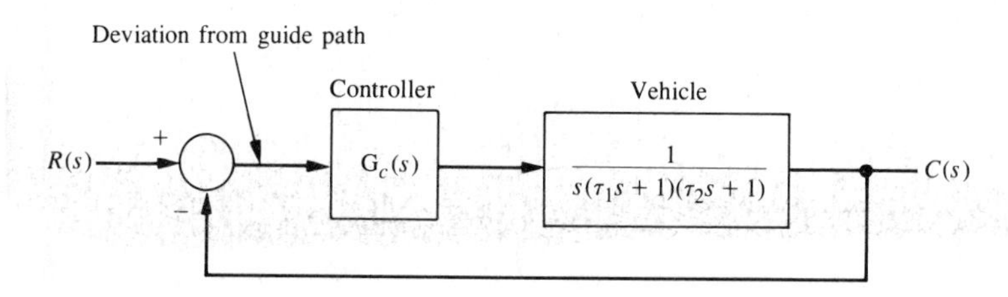

Figure P10.31. Steering control for vehicle.

P10.32. For the systems of Problem P10.31, use a phase-lag compensator and attempt to achieve a phase margin of approximately 50°. Determine the actual overshoot and peak time for the compensated system.

P10.33. When a motor drives a flexible structure, the structure's natural frequencies, as compared to the bandwidth of the servodrive, determine the contribution of the structural flexibility to the errors of the resulting motion. In current industrial robots, the drives are often relatively slow and the structures are relatively rigid, so that overshoots and other errors are caused mainly by the servodrive. However, depending on the accuracy required, the structural deflections of the driven members may become significant. Structural flexibility must be considered the major source of motion errors in space structures and manipulators. Because of weight restrictions in space, large arm lengths result in flexible structures. Furthermore, future industrial robots should require lighter and more flexible manipulators.

To investigate the effects of structural flexibility and how different control schemes can reduce unwanted oscillations, an experimental apparatus consisting of a dc motor driving a slender aluminum beam was constructed. The purpose of the experiments was to identify simple and effective control strategies to deal with the motion errors that occur when a servomotor is driving a very flexible structure [14].

The experimental apparatus is shown in Fig. P10.33(a) and the control system is shown in Fig. P10.33(b). The goal is that the system will have a K_v of 100. (a) When $G_c(s) = K$, determine K and plot the Bode diagram. Find the phase margin and gain margin. (b) Using the Nichols chart, find ω_r, M_{P_ω}, and ω_B. (c) Select a compensator so that the phase margin is greater than 35° and find ω_r, M_{P_ω}, and ω_B for the compensated system.

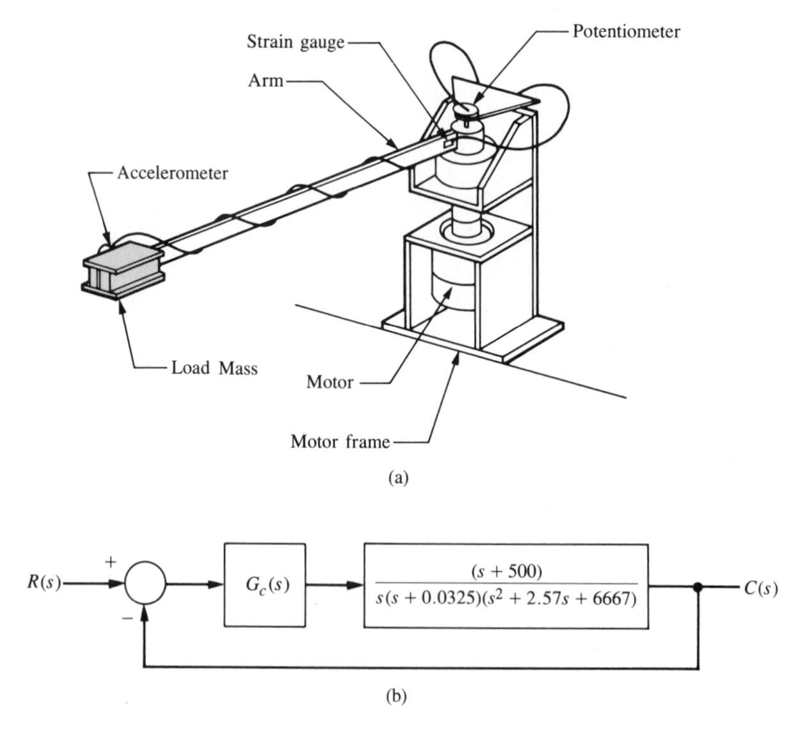

Figure P10.33. Flexible arm control.

P10.34. A human's ability to perform physical tasks is limited not by intellect but by physical strength. If in an appropriate environment a machine's mechanical power is closely integrated with a human arm's mechanical strength under the control of the human intellect, the resulting system will be superior to a loosely integrated combination of a human and a fully automated robot.

Extenders are defined as a class of robot manipulators that extend the strength of the human arm while maintaining human control of the task [15]. The defining characteristic of an extender is the transmission of both power and information signals. The extender is worn by the human; the physical contact between the extender and the human allows direct transfer of mechanical power and information signals. Because of this unique interface, control of the extender trajectory can be accomplished without any type of joystick, keyboard, or master-slave system. The human provides a control system for the extender, while the extender actuators provide most of the strength necessary for the task. The human becomes a part of the extender and "feels" a scaled-down version of the load that the extender is carrying. The extender is distinguished from a conventional master-slave system; in that type of system, the human operator is either at a remote location or close to the slave manipulator but is not in direct physical contact with the slave in the sense of transfer of power. An extender is shown in Fig. P10.34(a) [15]. The block diagram of the system is shown in Fig. P10.34(b). The goal is that the compensated system will have a velocity constant, K_v, equal to 80, that the settling time will be 1.6 sec, and that the overshoot will be 16% so that the dominant roots have a ζ of 0.5. Determine a lead-lag compensator using root locus methods.

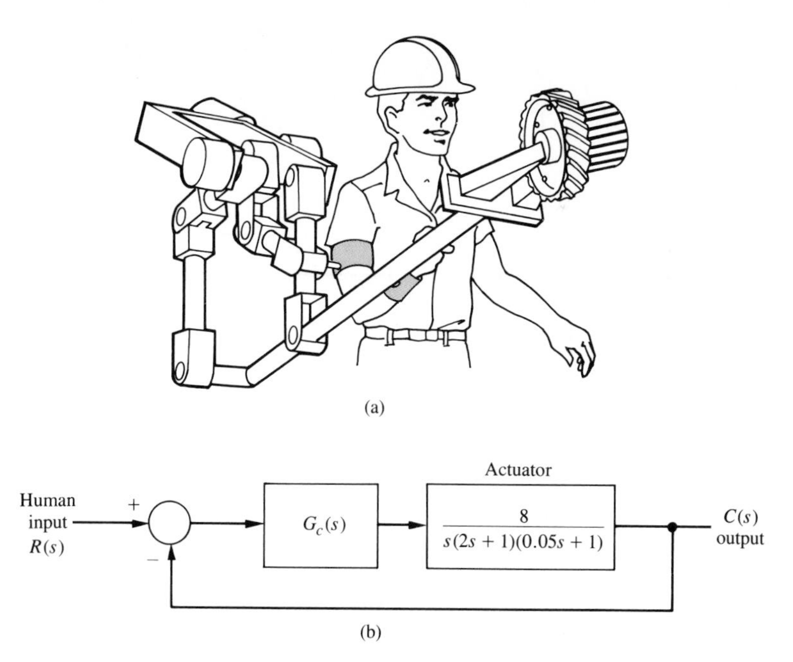

Figure P10.34. Extender robot control.

P10.35. A magnetically levitated train is operating in Berlin, Germany, as shown in Fig. P10.35(a). The M-Bahn 1600-m line represents the current state of worldwide systems. Fully automated trains can run at short intervals and operate with excellent energy efficiency. The control system for the levitation of the car is shown in Fig. P10.35(b). Select a compensator so that the phase margin of the system is $55° \leq \text{PM} \leq 65°$. Predict the response of the system to a step command and determine the actual step response for comparison.

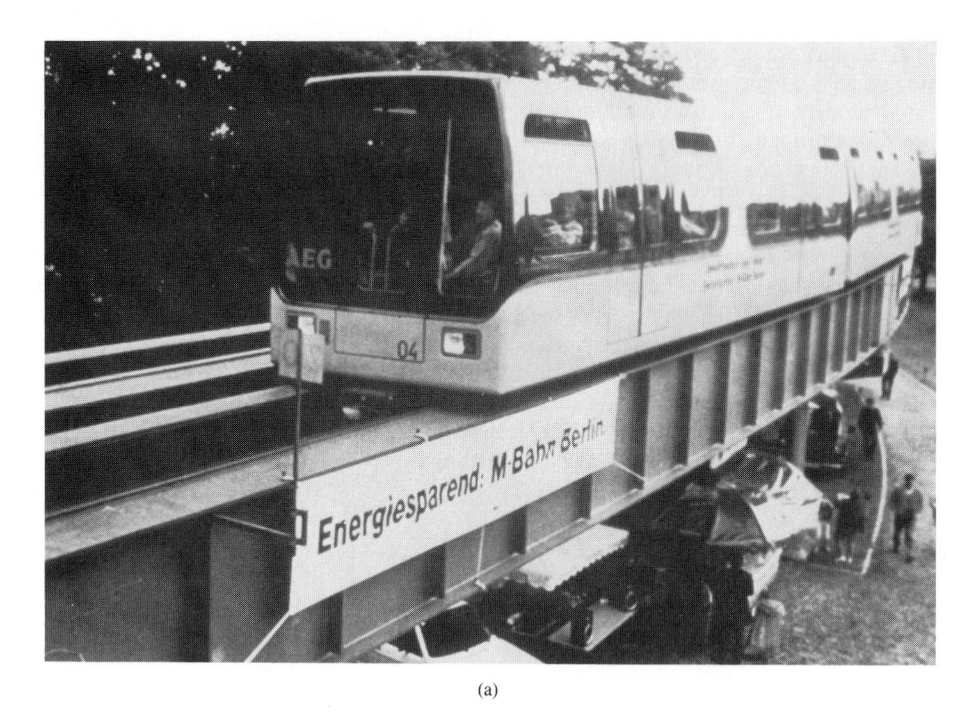

(a)

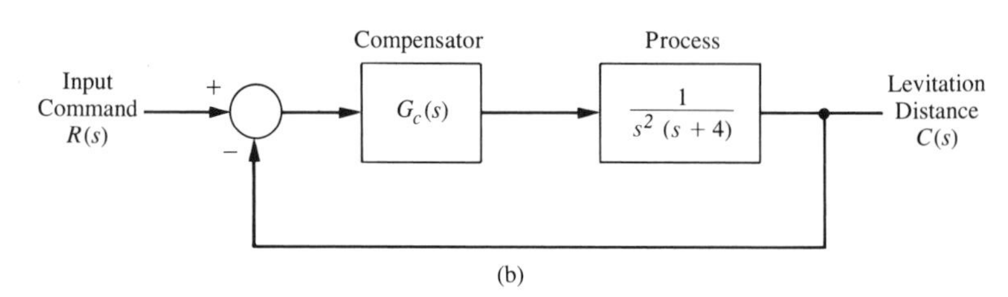

(b)

Figure P10.35. Magnetically levitated train. (Photo courtesy of AEG Westinghouse Transportation Systems.)

P10.36. Engineers are developing new sensors for machining and other manufacturing processes. A new sensing technique gleans information about the cutting process from acoustic emission (AE) signals. AE is a low-amplitude, high-frequency stress wave from a rapid release of strain energy in a continuous medium. AE is sensitive to material, tool geometry, tool wear, and cutting parameters such as feed and speed. AE sensors are commonly piezoelectric crystals sensitive in the 100 kHz to 1 MHz range. They are cost effective and can be mounted on most machine tools.

Current research has been extracting depth-of-cut information from AE signals. One study links sensitivity of the AE power signal to small depth-of-cut changes in diamond turning [18]. The power of the AE signals is sensitive to small, instantaneous changes in depth of cut during diamond turning and conventional milling. A milling machine using an AE sensor is shown in Fig. P10.36(a). The system for controlling the depth of cut of a milling machine is shown in Fig. P10.36(b). Design a compensator so that the percent overshoot to a step input is less than or equal to 20% while the velocity constant is greater than 8.

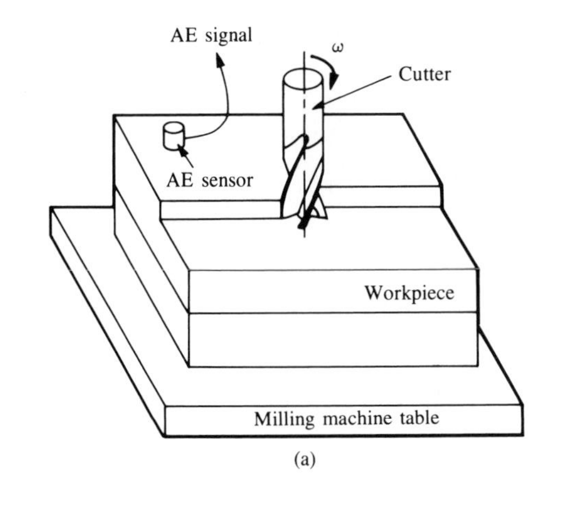

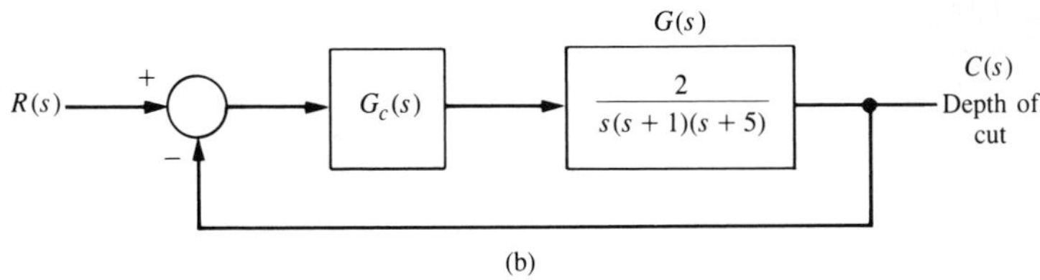

Figure P10.36. Milling machine control.

P10.37. A system's open-loop transfer function is a pure time delay of 1 sec so that $G(s) = e^{-s}$. Select a compensator $G_c(s)$ so that the steady-state error for a step input is less than 2% of the magnitude of the step and the phase margin is greater than 30°.

P10.38. A unity-negative feedback system has

$$G(s) = \frac{1}{s(s + 5)(s + 10)}.$$

The objective is that the dominant roots have a ζ equal to 0.707 while achieving zero steady-state error for a ramp input. Select a proportional plus integral (PI) controller so that the requirements are met. Determine the resulting peak time and settling time of the system.

Design Problems 

DP10.1. Two robots are shown in Fig. DP10.1 cooperating with each other to manipulate a long shaft prior to inserting it into the hole in the block resting on the table. Long part insertion is a good example of a task that can benefit from cooperative control. The feedback control system of one robot joint of the form shown in Fig. 10.1, has $H(s) = 1$ and

$$G(s) = \frac{4}{s(s + 0.5)}.$$

The specifications require a steady-state error for a unit ramp input of 0.0125, and the step response has an overshoot of less than 20% with a settling time of less than 2 sec. Determine a lead-lag compensator that will meet the specifications, and plot the compensated and uncompensated response for the ramp and step inputs.

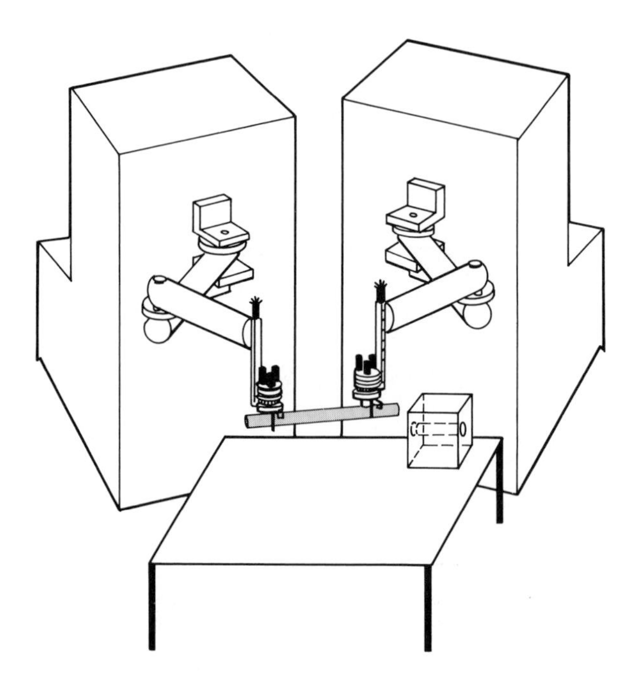

Figure DP10.1. Two robots cooperate to insert a shaft.

DP10.2. The heading control of the traditional bi-wing aircraft shown in Fig. DP10.2(a), on the next page is represented by the block diagram of Fig. DP10.2(b). (a) Determine the minimum value of the gain K when $G_c(s) = K$ so that the steady-state effect of a unit step disturbance, $D(s) = 1/s$, is less than or equal to 1% of the unit step $[c(\infty) = 0.01]$. (b) Determine if the system using the gain of part (a) is stable. (c) Design a compensator using one stage of lead compensation so that the phase margin is 30°. (d) Design a two-stage lead compensator so that the phase margin is 55°. (e) Compare the bandwidth of the systems of part (c) and (d). (f) Plot the step response $c(t)$ for the systems of part (c) and (d) and compare percent overshoot, settling time, and peak time.

DP10.3. NASA has identified the need for large deployable space structures, which will be constructed of lightweight materials and will contain large numbers of joints or structural connections. This need is evident for programs such as the space station. These deployable space structures may have precision shape requirements and a need for vibrations suppression during on-orbit operations [8].

One such structure is the mast flight system, which is shown in Fig. DP10.3(a). The intent of the system is to provide an experimental test bed for controls and dynamics. The basic element in the mast flight system is a 60.7-m-long truss beam structure, which is attached to the shuttle orbiter. Included at the tip of the truss structure are the primary

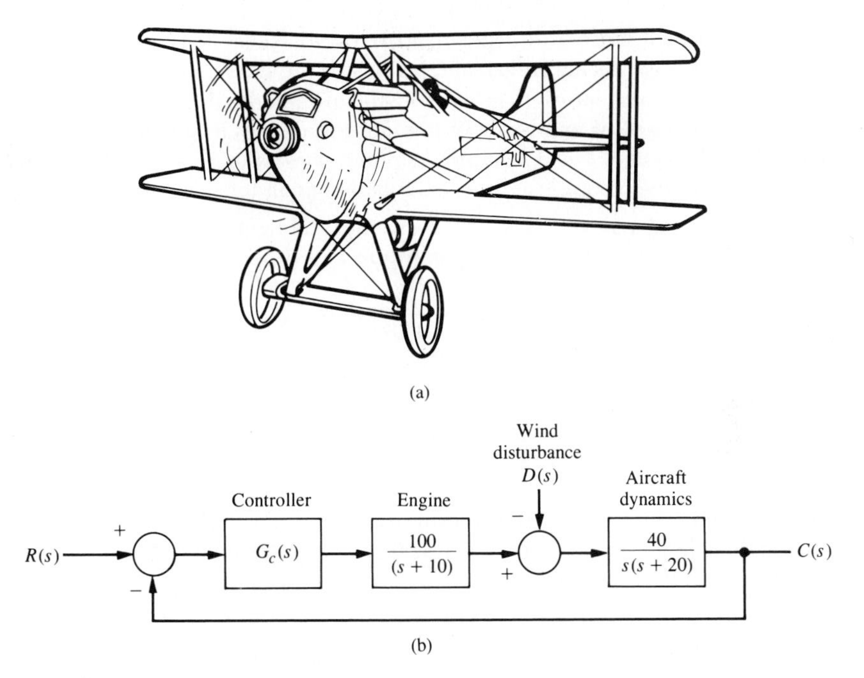

Figure DP10.2. (a) Bi-wing aircraft (Source: *The Illustrated London News,* Oct. 9, 1920.) (b) Control system.

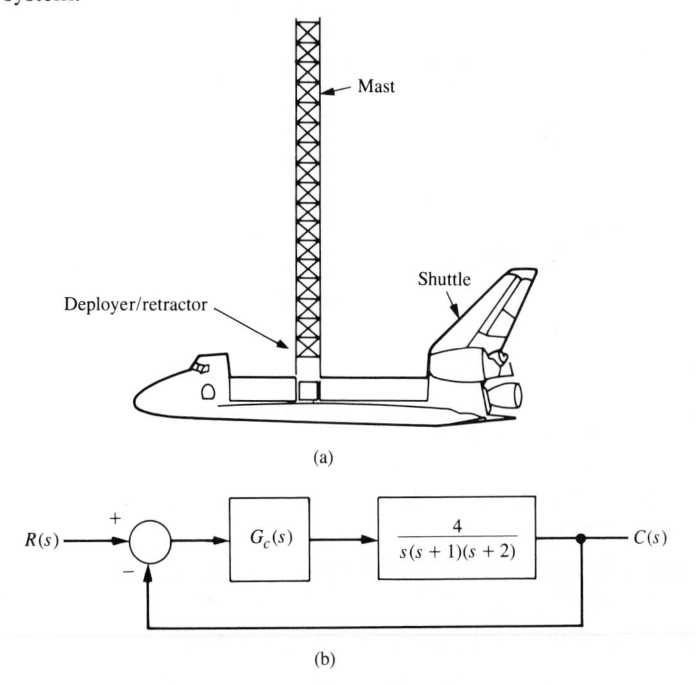

Figure DP10.3. Mast flight system.

actuators and collocated sensors. A deployment/retraction subsystem, which also secures the stowed beam package during launch and landing, is provided.

The system uses a large motor to move the structure and has the block diagram shown in Fig. DP10.3(b). The goal is an overshoot to a step response of less than or equal to 16%; thus we estimate the system ζ as 0.5 and the required phase margin as 50°. Design for $0.1 < K < 1$ and record overshoot, rise time, and phase margin for selected gains.

DP10.4. A robot using a vision system as the measurement device is shown in Fig. DP10.4. The control system is of the form shown in Fig. 10.14, where

$$G(s) = \frac{1}{(s + 1)(0.5s + 1)}$$

and $G_c(s)$ is selected as a PI controller so that the steady-state error for a step input is equal to zero. We then have

$$G_c(s) = K_1 + \frac{K_2}{s} = \frac{K_1 s + K_2}{s}.$$

Determine a suitable G_c so that (a) the percent overshoot for a step input is 5% or less; (b) the settling time is less than 6 sec; (c) the system K_v is greater than 0.9; and (d) the peak time for a step input is minimized.

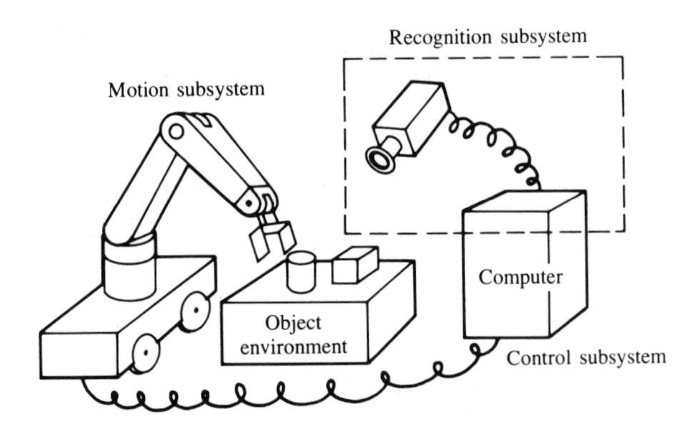

Figure DP10.4. A robot and vision system.

Terms and Concepts

Cascade compensation network A compensator network placed in cascade or series with the system process.

Compensation The alteration or adjustment of a control system in order to provide a suitable performance.

Compensator An additional component or circuit that is inserted into the system to equalize or compensate for the performance deficiency.

Design of a control system The arrangement or the plan of the system structure and the selection of suitable components and parameters.

Integration network A network that acts, in part, like an integrator.

Lag network See Phase lag network.

Lead-lag network A network with the characteristics of both a lead network and a lag network.

Lead network See Phase lead network.

Phase-lag network A network that provides a negative phase angle and a significant attenuation over the frequency range of interest.

Phase-lead network A network that provides a positive phase angle over the frequency range of interest. Thus phase lead can be used to cause a system to have an adequate phase margin.

PID controller See Three-mode controller.

Process controller See Three-mode controller.

Three-mode controller A controller with three terms so that the output of the controller is the sum of a proportional term, an integrating term, and a differentiating term, with an adjustable gain for each term.

References

1. R. C. Dorf and R. Jacquot, *Control System Design Program,* Addison-Wesley, Reading, Mass., 1988.
2. C. L. Phillips and R. D. Harbor, *Feedback Control Systems,* Prentice Hall, Englewood Cliffs, N.J., 1988.
3. D. F. Delchamps, *State-Space and Input-Output Linear Systems,* Springer-Verlag, New York, 1988.
4. F. H. Raven, *Automatic Control Engineering,* McGraw-Hill, New York, 1987.
5. J. R. Mitchell and W. L. McDaniel, Jr., "A Computerized Compensator Design Algorithm with Launch Vehicle Applications," *IEEE Trans. on Automatic Control,* June 1976; pp. 366–71.
6. W. R. Wakeland, "Bode Compensator Design," *IEEE Trans. on Automatic Control,* October 1976; pp. 771–73.
7. J. R. Mitchell, "Comments on Bode Compensator Design," *IEEE Trans. on Automatic Control,* October 1977; pp. 869–70.
8. F. Ham, "Active Vibration Suppression for the Mast Flight System," *IEEE Control Systems,* January 1989; pp. 85–89.
9. E. Sung, "Parallel Linkage Steering for an Automated Guided Vehicle," *IEEE Control Systems,* October 1989; pp. 3–8.
10. K. J. Waldron and R. B. McGhee, "The Adaptive Suspension Vehicle," *IEEE Control Systems,* December 1986; pp. 7–12.
11. B. Anderson, "Controller Reduction: Concepts and Approaches," *IEEE Trans. on Automatic Control,* August 1989; pp. 802–12.

12. H. Butler, "Model Reference Adaptive Control of a Direct-Drive dc Motor," *IEEE Control Systems,* January 1989; pp. 80–84.
13. C. L. Phillips, "Analytical Bode Design of Controllers," *IEEE Trans. on Education,* February 1985; pp. 43–44.
14. R. L. Wells, "Control of a Flexible Robot Arm," *IEEE Control Systems,* January 1990; pp. 9–15.
15. H. Kazerooni, "Force Augmentation in Human-Robot Interaction," *IEEE Proceed. of the American Control Conference,* May 1990; pp. 2821–26.
16. R. C. Dorf, *The Encyclopedia of Robotics,* John Wiley & Sons, New York, 1988.
17. S. Boyd, "Linear Controller Design," *Proceed. of the IEEE,* March 1990; pp. 529–74.
18. J. Koelsch, "Sensors—the Missing Link," *Manufacturing Engineering,* October 1990; pp. 53–56.

✦

CHAPTER 11

Robust Control Systems

Preview

We have described the benefits of using a compensator, $G_c(s)$, in order to achieve the desired performance of a feedback system. In this chapter we extend the concept of compensation and introduce a powerful controller, the proportional-integral-derivative (PID) device.

The design of highly accurate control systems in the presence of significant uncertainty requires the designer to seek a robust system. Thus, in this chapter we will utilize three methods for robust design.

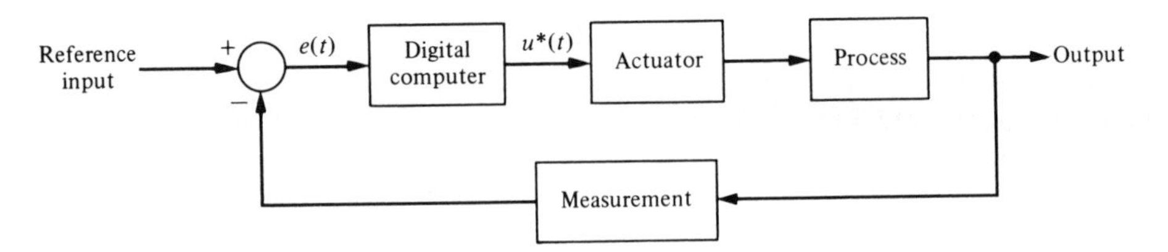

Figure 11.1. Block diagram of a computer control system.

11.1 Introduction

The use of a digital computer as a compensator device has grown during the past decade as the price and reliability of digital computers have improved dramatically [1,2]. A block diagram of a single-loop digital control system is shown in Fig. 11.1. The digital computer in this system configuration receives the error, $e(t)$, and performs calculations in order to provide an output $u^*(t)$. The computer may be programmed to provide an ouput, $u^*(t)$, so that the performance of the process is near or equal to the desired performance. Many computers are able to receive and manipulate several inputs, so a digital computer control system can often be a multivariable system.

11.2 Digital Computer Control System Applications

A more complete block diagram of a computer control system is shown in Fig. 11.2. This diagram recognizes that a digital computer receives and operates on signals in digital (numerical) form, as contrasted to continuous signals [3]. A *digital control system* uses digital signals and a digital computer to control a process. The measurement data are converted from analog form to digital form by means of the converter shown in Fig. 11.2. After the digital computer has processed the inputs, it provides an output in digital form. This output is then converted to analog form by the digital-to-analog converter shown in Fig. 11.2.

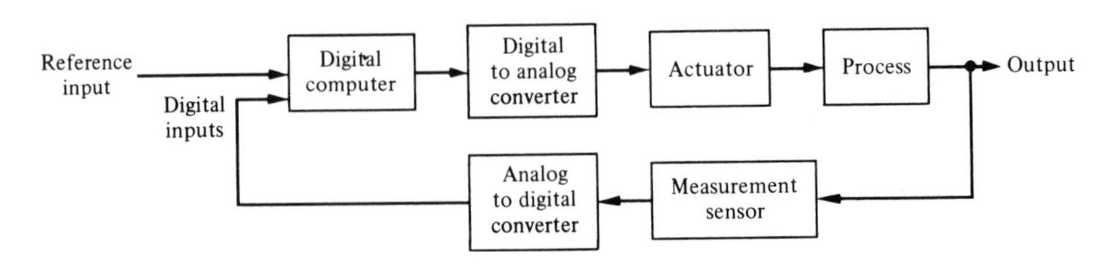

Figure 11.2. Block diagram of a computer control system, including the signal converters.

While digital computers or microprocessors require a sequence of discrete signals as inputs, they provide a series of output signals that are converted to a continuous signal by the digital-to-analog converter. Thus, if the time interval between each digital signal is small compared to the time constants of the actuator and process, the system essentially acts as a continuous system. The digital computer and the analog-to-digital and digital-to-analog converter can all be then represented by the summation node and $G_c(s)$, as shown in Fig. 11.3.

The period between digital data is called the sampling period, T. If the time constant τ of the system is much greater than T, then the system may be considered continuous. For example, many microprocessors accommodate data at a rate equal to 1 megahertz so that $T = 1\ \mu s$. Thus if $\tau > 100\ \mu s$, we may consider the system continuous. Most electromechanical and chemical control systems have time constants that exceed 1 ms.

Digital control systems are used in many applications: for machine tools, metal working processes, chemical processes, aircraft control, and automobile traffic control, among others [1,2,3]. An example of a computer control system used in the aircraft industry is shown in Fig. 11.4. Automatic computer controlled systems are used for purposes as diverse as measuring the objective refraction of the human eye and controlling the engine spark timing or air-fuel ratio of automobile engines. The latter innovations are necessary to reduce automobile emissions and increase fuel economy.

The advantage of using digital control includes (1) improved sensitivity, (2) digitally coded signals, (3) digital sensors and transducers, and (4) the use of microprocessors. Improved sensitivity results from the low energy signals required by digital sensors and devices. The use of digitally coded signals permits the wide application of digital devices and communications. Digital sensors and transducers can effectively measure, transmit, and couple signals and devices. In addition, many systems are inherently digital because they send out pulse signals. Examples of such a digital system are a radar tracking system and a space satellite. However, we must reemphasize that data rates are usually so high that most digital control systems may be treated as continuous systems.

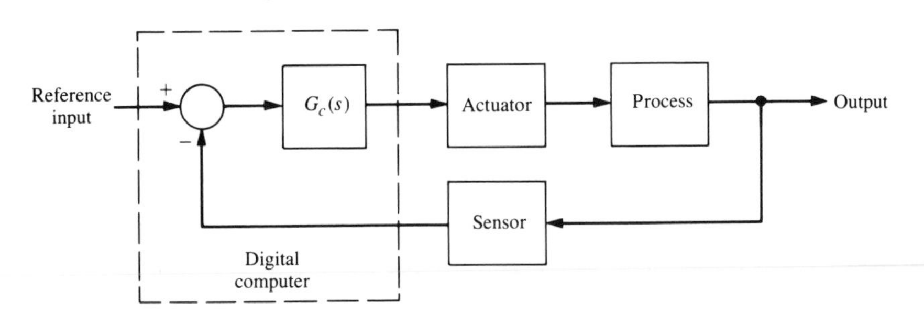

Figure 11.3. Continuous system model of the computer control system of Fig. 11.2.

Figure 11.4. The flight deck of the Boeing 757 and 767 features digital control electronics, including an engine indicating system and a crew alerting system. All systems controls are within reach of either pilot. The system includes an inertial reference system making use of laser gyroscopes and an electronic attitude director indicator. A flight-management computer system integrates navigation, guidance, and performance data functions. When coupled with the automatic flight-control system (automatic pilot), the flight-management system provides accurate engine thrust settings and flight-path guidance during all phases of flight from immediately after takeoff to final approach and landing. The system can predict the speeds and altitudes that will result in the best fuel economy and command the airplane to follow the most fuel-efficient, or the "least time," flight paths. (Courtesy of Boeing Airplane Co.)

11.3 Automatic Assembly and Robots

Automatic handling equipment for home, school, and industry is particularly useful for hazardous, repetitious, dull, or simple tasks. Machines that automatically load and unload, cut, weld, or cast are used by industry in order to obtain accuracy, safety, economy, and productivity [1,2,14,15]. The use of computers integrated with machines that perform tasks as a human worker does has been foreseen by several authors. In his famous 1923 play, entitled *R.U.R.* [23], Karel Capek called artificial workers *robots,* deriving the word from the Czech noun *robota,* meaning "work." Two modern "robots" that appeared in the films *Star Wars* and *The Empire Strikes Back* are shown in Fig. 11.5.

Figure 11.5. Two modern "robots" appeared in the films *Star Wars* and *The Empire Strikes Back.* R2-D2 (left) and C-3PO were the able assistants and supporters of Luke and Leia, the hero and heroine of the film. (Courtesy of Twentieth Century Fox-Film Corporation.)

Robots are programmable computers integrated with machines. They often substitute for human labor in specific repeated tasks. Some devices even have anthropomorphic mechanisms, including what we might recognize as mechanical arms, wrists, and hands. Robots can be used extensively in space exploration and assembly [1]. They can be flexible, accurate aids on assembly lines, as shown in Fig. 11.6.

A *robot* can be defined as a reprogrammable, multifunctional manipulator designed to move material, parts, tools, or specialized devices through variable

Figure 11.6. At its Mirafiori auto assembly plant, Fiat of Italy uses 190 Unimate robots for welding and loading and unloading tasks. The robots are shown here on both sides of the welding line. (Courtesy of Unimation, Inc.)

programmed motions for the performance of a variety of tasks. It is the ability of the robot to be reprogrammed that enables the user to use rapidly the robot in new tasks. Approximately 25,000 robots have been installed for industrial uses over the past decade, and about 3,000 new robots are now being installed annually.

11.4 Robust Control Systems

Designing highly accurate systems in the presence of significant plant uncertainty is a classical feedback design problem. The theoretical bases for the solution of this problem date back to the works of H. S. Black and H. W. Bode in the early 1930s, when this problem was referred to as the sensitivities design problem. However, a significant amount of literature has been published since then regarding the design of systems subject to large plant uncertainty. The designer seeks to obtain a system that performs adequately over a large range of uncertain parameters. A system is said to be *robust* when it is durable, hardy, and resilient.

A control system is robust when (1) it has low sensitivities, (2) it is stable over the range of parameter variations, and (3) the performance continues to meet the specifications in the presence of a set of changes in the system parameters [4,7]. Robustness is the sensitivity to effects that are not considered in the analysis and design phase—for example, disturbances, measurement noise, and unmodeled dynamics. The system should be able to withstand these neglected effects when performing the tasks of interest.

For small parameter perturbations, we may use, as a measure of robustness, the differential sensitivities discussed in Sections 3.2 (system sensitivity) and 6.6 (root sensitivity).

System sensitivity is defined as

$$S_\alpha^T = \frac{\partial T/T}{\partial \alpha/\alpha}, \tag{11.1}$$

where α is the parameter and T the transfer function of the system.

Root sensitivity is defined as

$$S_\alpha^{r_i} = \frac{\partial r_i}{\partial \alpha/\alpha}. \tag{11.2}$$

When the zeros of $T(s)$ are independent of the parameter α, we showed that

$$S_\alpha^T = -\sum_{i=1}^{n} S_\alpha^{r_i} \cdot \frac{1}{(s + r_i)} \tag{11.3}$$

for an nth order system. For example, if we have a closed-loop system, as shown in Fig. 11.7, where the variable parameter is α, then $T(s) = 1/[s + (\alpha + 1)]$ and

$$S_\alpha^T = \frac{-\alpha}{s + \alpha + 1}. \tag{11.4}$$

Furthermore, the root is $s = r_1 = (\alpha + 1)$ and

$$S_\alpha^{r_1} = \alpha. \tag{11.5}$$

$$S_\alpha^T = -S_\alpha^{r_1} \frac{1}{(s + \alpha + 1)}. \tag{11.6}$$

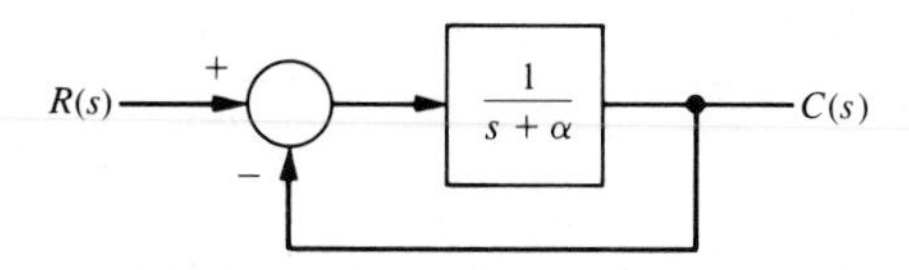

Figure 11.7. A first-order system.

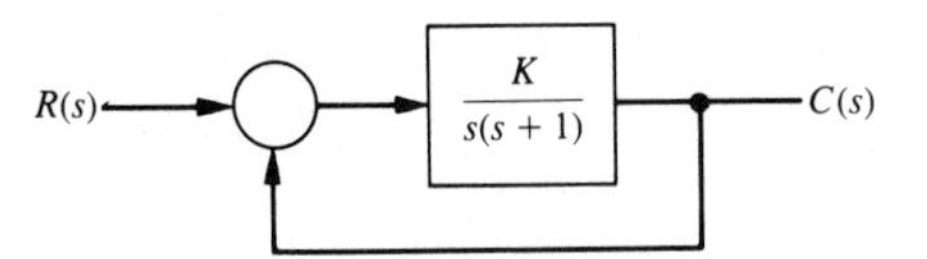

Figure 11.8. A second-order system.

Let us examine the sensitivity of the second-order system shown in Fig. 11.8. The transfer function of the closed-loop system is

$$T(s) = \frac{K}{s^2 + s + K}. \tag{11.7}$$

As we saw in Eq. (3.13), the system sensitivity for K is

$$S_K^T = \frac{1}{1 + GH(s)} = \frac{s(s + 1)}{s^2 + s + K}. \tag{11.8}$$

A Bode plot of the asymptotes of $20 \log|T(j\omega)|$ and $20 \log|S(j\omega)|$ are shown in Fig. 11.9 for $K = 1/4$ (critical damping). Note that the sensitivity is small for lower frequencies, while the transfer function primarily passes low frequencies. Also, note for this case, $T(s) = 1 - S(s)$.

Of course, the sensitivity S only represents robustness for small changes in gain. If K changes from $1/4$ within the range $K = 1/16$ to $K = 1$, the resulting range of step response is shown in Fig. 11.10. This system, with an expected wide range

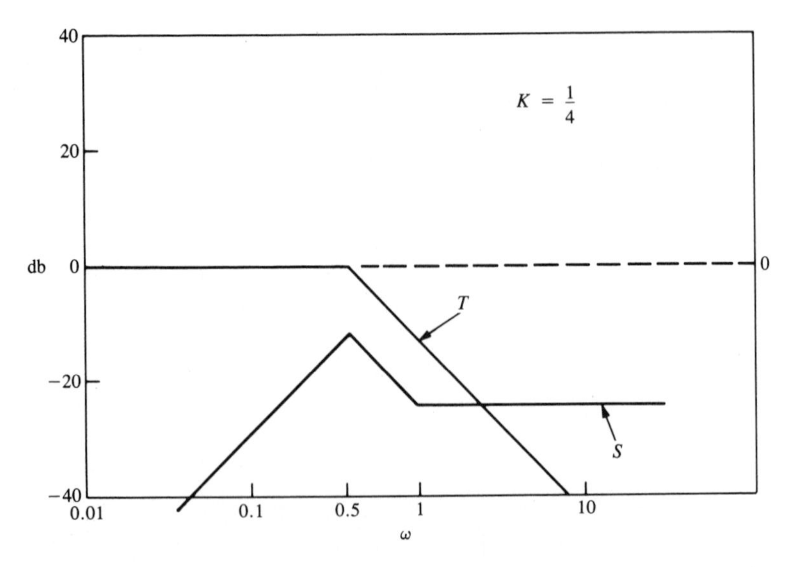

Figure 11.9. Sensitivity and 20 log T(jω) for the second-order system in Fig. 11.8.

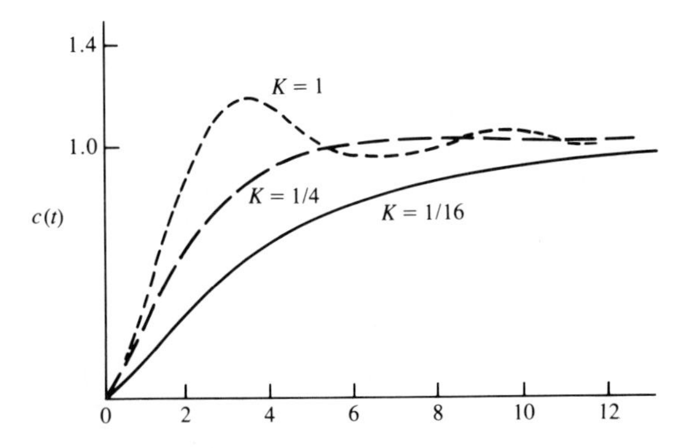

Figure 11.10. Step response for selected gain K.

of K, may not be considered adequately robust. A robust system would be expected to yield essentially the same (within an agreed-upon variation) response to a selected input.

■ Example 11.1 Sensitivity of a controlled system

Consider the system shown in Fig. 11.11, where $G(s) = 1/s^2$ and a PD controller $G_c(s) = b_1 + b_2 s$. Then, the sensitivity with respect to changes in $G(s)$ is

$$S_G^T = \frac{1}{1 + GG_c} = \frac{s^2}{s^2 + b_2 s + b_1} \tag{11.9}$$

and

$$T(s) = \frac{b_2 s + b_1}{s^2 + b_2 s + b_1}. \tag{11.10}$$

Consider the condition $\zeta = 1$ and $\omega_n = \sqrt{b_1}$. Then $b_2 = 2\omega_n$ to achieve $\zeta = 1$. Therefore we may plot $20 \log|S|$ and $20 \log|T|$ on a Bode diagram, as shown in Fig. 11.12. Note that the frequency ω_n is an indicator on the boundary between the frequency region in which the sensitivity is the important design criterion and the region in which the stability margin is important. Thus, if we specify ω_n properly to take into consideration the extent of modeling error and the frequency of

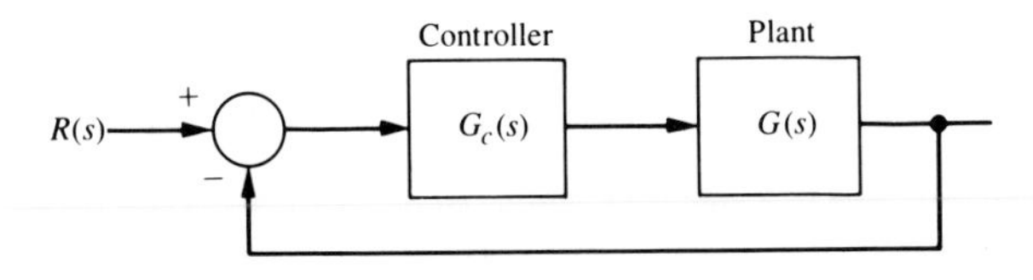

Figure 11.11. A system with a PD controller.

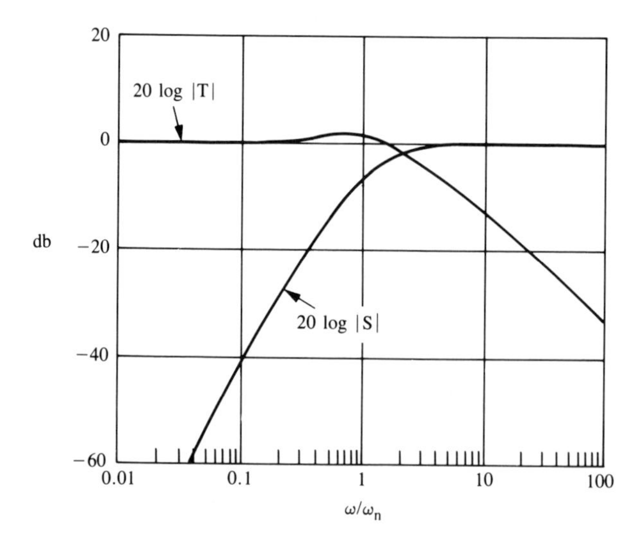

Figure 11.12. Sensitivity and $T(s)$ for the second-order system in Fig. 11.11.

external disturbance, we can expect that the system has an acceptable amount of robustness. Note $G_c(s)$ is a proportional-derivative (PD) controller.

■ Example 11.2 System with a right-hand plane zero

Consider the system shown in Fig. 11.13, where the plant has a zero in the right-hand plane. The closed-loop transfer function is

$$T(s) = \frac{K(s - 1)}{s^2 + (2 + K)s + (1 - K)}. \tag{11.11}$$

The system is stable for a gain $-2 < K < 1$. The steady-state error is

$$e_{ss} = \frac{1}{1 - K} \tag{11.12}$$

and $e_{ss} = 0$ when $K = \infty$. The response to a negative unit step input $R(s) = -1/s$ is shown in Fig. 11.14. Note the initial undershoot at $t = 1$ second. This system

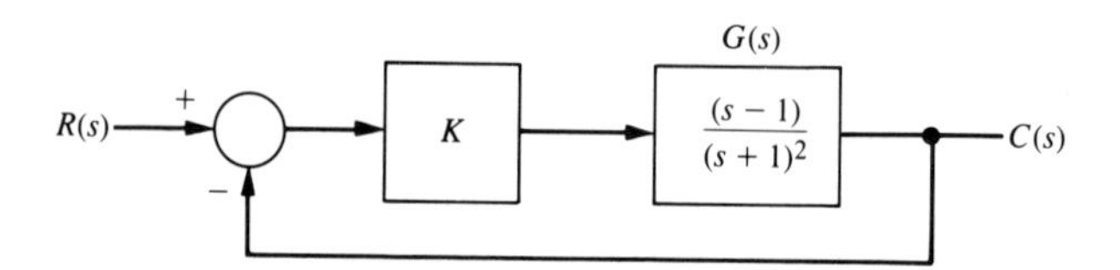

Figure 11.13. A second-order system.

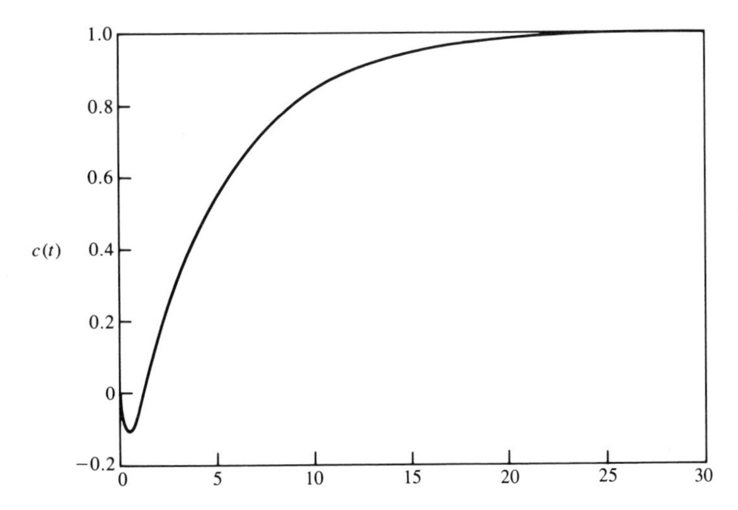

Figure 11.14. Step response of the system in Fig. 11.13 with $K = \frac{1}{2}$.

is sensitive to changes in K as recorded in Table 11.1. The performance of this system might be considered acceptable for a change of gain of only ± 10%. Thus this system would not be considered robust.

11.5 The Design of Robust Control Systems

The design of robust control systems is based on two tasks: determining the structure of the controller and adjusting the controller's parameters to give an "optimal" system performance. This design process is normally done with "assumed complete knowledge" of the plant. Furthermore, the plant is normally described by a linear time-invariant continuous model. The structure of the controller is chosen such that the system's response can meet certain performance criteria.

One possible objective in the design of a control system is that the controlled system's output should exactly and instantaneously reproduce its input. That is, the system's transfer function should be unity,

$$T(s) = C(s)/R(s) = 1. \tag{11.13}$$

In other words, the system should be presentable on a Bode gain versus frequency diagram with a 0-db gain of infinite bandwidth and zero phase shift. In practice,

Table 11.1. Results for Example 11.2.

K	0.25	0.45	0.50	0.55	0.75
$\lvert e_{ss} \rvert$	0.67	0.18	0	0.22	1.0
Undershoot	5%	9%	10%	11%	15%
settling time (seconds)	15	24	27	30	45

this is not possible since every system will contain inductive- and capacitive-type components that store energy in some form. It is these elements and their interconnections with energy dissipative components that produce the system's dynamic response characteristics. Such systems reproduce some inputs almost exactly, while other inputs are not reproduced at all, signifying that the system's bandwidth is less than infinite.

Once it is recognized that the system's dynamics cannot be ignored, a new design objective is needed. One possible design objective is to maintain the magnitude response curve as flat and as close to unity for a large bandwidth as possible for a given plant and controller combination [25].

Another important goal of a control system design is that the effect on the output of the system due to disturbances is minimized. Thus we wish to minimize $C(s)/D(s)$ over a range of frequency.

Consider the control system shown in Fig. 11.15, where $G(s) = G_1(s)G_2(s)$ is the plant and $D(s)$ is the disturbance. We then have

$$T(s) = \frac{C(s)}{R(s)} = \frac{G_cG_1G_2}{1 + G_cG_1G_2} \tag{11.14}$$

and

$$\frac{C(s)}{D(s)} = \frac{G_2(s)}{1 + G_cG_1G_2}. \tag{11.15}$$

Note that both the reference and disturbance transfer functions have the same denominator, or, in other words, they have the same characteristic equation—that is,

$$1 + G_c(s)G_1(s)G_2(s) = 1 + L(s) = 0. \tag{11.16}$$

Furthermore, we recall that the sensitivity of $T(s)$ with respect to $G(s)$ is

$$S_G^T = \frac{1}{1 + G_cG_1G_2(s)} \tag{11.17}$$

and the characteristic equation is the influencing factor on the sensitivity. Equation 11.17 shows that for low sensitivity S we require a high value of loop gain

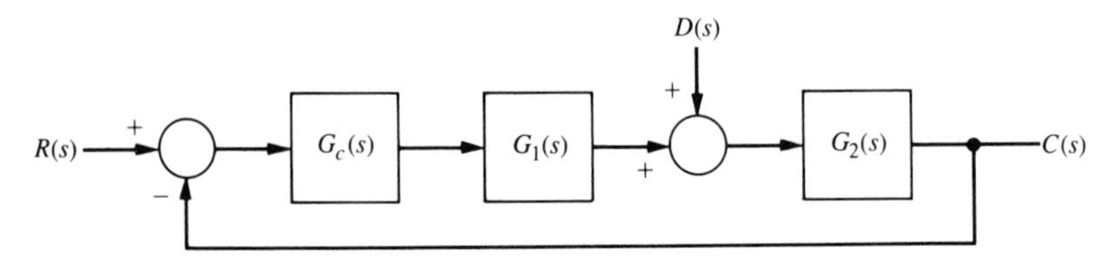

Figure 11.15. A system with a disturbance.

$L(j\omega)$, but it is known that high gain could cause instability or poor responsiveness of $T(s)$. Thus the designer seeks the following:

1. $T(s)$ with wide bandwidth and faithful reproduction of $R(s)$.
2. Large loop gain $L(s)$ in order to minimize sensitivity S.
3. Large loop gain $L(s)$ attained primarily by $G_c(s)G_1(s)$ since $C(s)/D(s) \approx 1/G_cG_1(s)$.

Setting the design of robust systems in frequency domain terms, we are required to find a proper compensator, $G_c(s)$, such that the closed-loop sensitivity is less than some tolerance value and sensitivity minimization involves finding a proper compensator such that the closed-loop sensitivity equals or is arbitrarily close to the minimal attainable sensitivity. Similarly, the gain margin problem is to find a proper compensator to achieve some prescribed gain margin, and gain margin maximization involves finding a proper compensator to achieve the maximal attainable gain margin. For the frequency domain specifications, we require the following for the Bode diagram of $G_cG(j\omega)$, shown in Fig. 11.16:

1. For relative stability, $G_cG(j\omega)$ must have, for an adequate range of ω, not more than -20 db/decade slope at or near the crossover frequency ω_c.
2. Steady-state accuracy achieved by the low frequency gain.
3. Accuracy over a bandwidth ω_B by not allowing $|G_cG(\omega)|$ to fall below a prescribed level,
4. Disturbance rejection by high gain for $G_c(j\omega)$ over the system bandwidth.

Using the root sensitivity concept, we can state that we require that S_α^r be minimized while attaining $T(s)$ with dominant roots that will provide the appro-

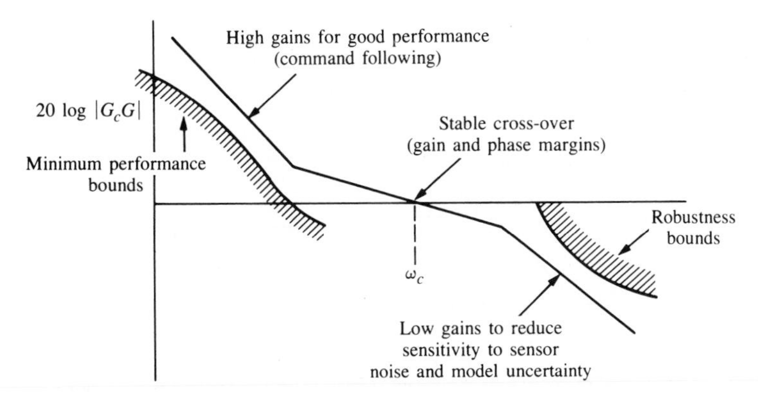

Figure 11.16. Bode diagram for 20 log $|G_cG(j\omega)|$.

priate response and minimize the effect of $D(s)$. Again we see that the goal is to have the gain of the loop primarily attained by $G_c(s)$. As an example, let $G_c(s) = K$, $G_1(s) = 1$, and $G_2(s) = 1/s(s+1)$ for the system in Fig. 11.15. This system has two roots and we select a gain K so that $C(s)/D(s)$ is minimized, S_K^r is minimized, and $T(s)$ has desirable dominant roots. The sensitivity is

$$S_K^r = \frac{dr}{dK} \cdot \frac{K}{r} = \left.\frac{ds}{dK}\right|_{s=r} \cdot \frac{K}{r} \tag{11.18}$$

and the characteristic equation is

$$s(s+1) + K = 0. \tag{11.19}$$

Therefore $dK/ds = -(2s+1)$ since $K = -s(s+1)$. We then obtain

$$S_K^r = \frac{-1}{(2s+1)} \cdot \left.\frac{(-s(s+1))}{s}\right|_{s=r}. \tag{11.20}$$

When $\zeta < 1$, the roots are complex and $r = -0.5 + j\omega$. Then,

$$|S_K^r| = \left(\frac{0.25 + \omega^2}{4\omega^2}\right)^{1/2}. \tag{11.21}$$

The magnitude of the sensitivity is plotted in Fig. 11.17 for $K = 0.2$ to $K = 5$. Also, the percent overshoot to a step is shown. Clearly, it is best to reduce the sensitivity but to limit K to 1.5 or less. We then attain the majority of the attain-

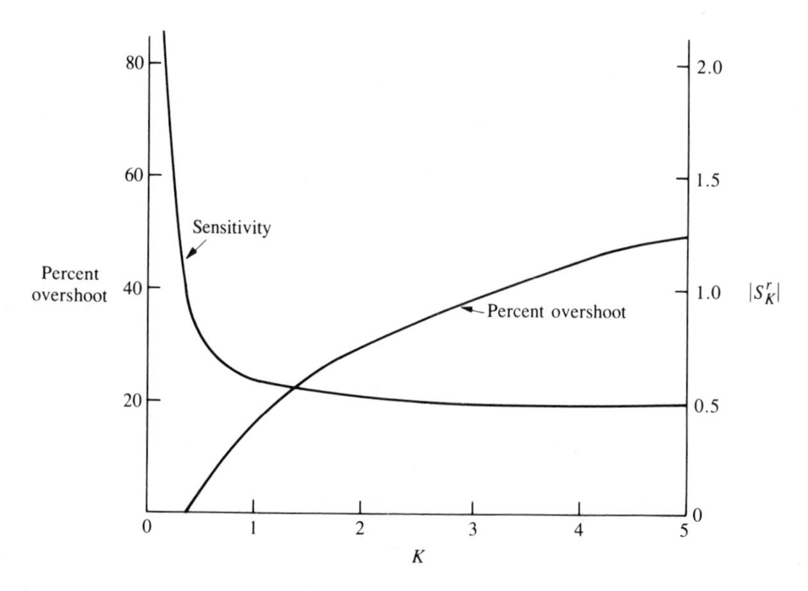

Figure 11.17. Sensitivity and percent overshoot for a second-order system.

able reduction in sensitivity while maintaining good performance for the step response. In general, we can use the design procedure as follows:

1. Draw the root locus of the compensated system with $G_c(s)$ chosen to attain the desired location for the dominant roots.
2. Maximize the gain of $G_c(s)$ so that the effect of the disturbance is reduced.
3. Determine S_α^r and attain the minimum value of the sensitivity consistent with the transient response required as described in step 1.

■ Example 11.3 Sensitivity and compensation

Let us reconsider the system in Example 10.1 when $G(s) = 1/s^2$, $G_1(s) = 1$, and $G_c(s)$ is to be selected by frequency response methods. Therefore the compensator is to be selected to achieve an appropriate gain and phase margin while minimizing sensitivity and the effect of the disturbance. Thus we choose

$$G_c(s) = \frac{K(s/z + 1)}{(s/p + 1)}. \tag{11.22}$$

As in Example 10.1, we choose $K = 10$ to reduce the effect of the disturbance. In order to attain a phase margin of 45° we select $z = 2.1$ and $p = 12.6$. We then attain the compensated diagram shown in Fig. 11.18. Recall that the closed-loop bandwidth is $\omega_B = 1.6\ \omega_c$. Thus we will increase the bandwidth by using the compensator and improve the fidelity of reproduction of the input signals.

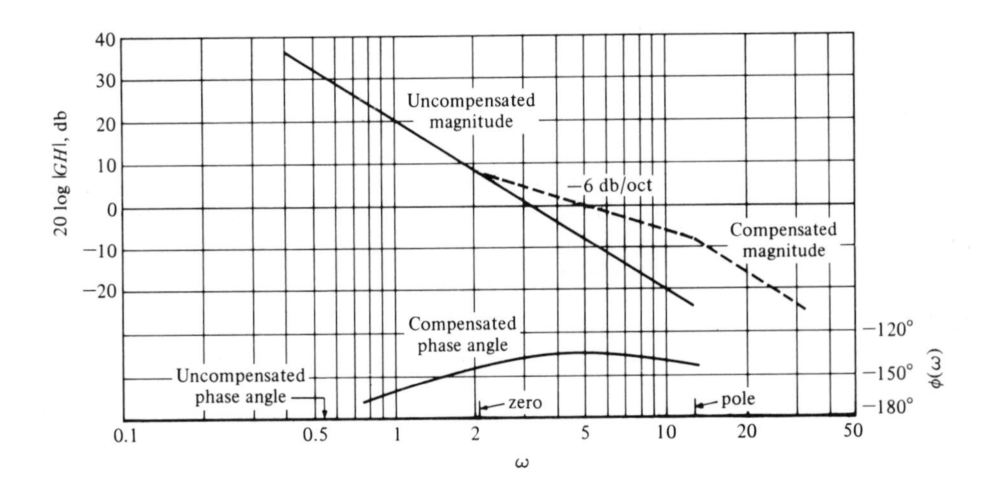

Figure 11.18. Compensated Bode diagram for Example 11.3.

The sensitivity may be ascertained at ω_c as

$$|S_G^T(\omega_c)| = \left|\frac{1}{1 + G_cG(j\omega)}\right|_{\omega_c}. \tag{11.23}$$

In order to estimate $|S_G^T|$, we recall that the Nichols chart enables us to obtain

$$|T(\omega)| = \left|\frac{G_cG(\omega)}{1 + G_cG(\omega)}\right|. \tag{11.24}$$

Thus we can plot a few points of $G_cG(j\omega)$ on the Nichols chart and read $T(\omega)$ from the Nichols chart. Then,

$$|S_G^T(\omega_1)| = \frac{|T(\omega_1)|}{|G_cG(\omega_1)|}, \tag{11.25}$$

where ω_1 is chosen arbitrarily as $\omega_c/2.5$. In general, we choose a frequency below ω_c to determine the value of $|S|$. Of course, we desire a low value of sensitivity. The Nichols chart for the compensated system is shown in Fig. 11.19. For $\omega_1 = \omega_c/2.5 = 2$, we have $20 \log T = 2.5$ db and $20 \log G = 9$ db. Therefore

$$|S(\omega_1)| = \frac{|T|}{|G_cG|} = \frac{1.33}{2.8} = 0.47.$$

■ Example 11.4 Sensitivity with a lead compensator

Let us again consider the system in Example 11.3, using the root locus design obtained in Example 10.3. The compensator was chosen as

$$G_c(s) = \frac{8.1(s + 1)}{(s + 3.6)}, \tag{11.26}$$

for the system of Fig. 11.20. The dominant roots are thus $s = -1 \pm j2$. Since the gain is 8.1, the effect of the disturbance is reduced and the time response meets the specifications. The sensitivity at r may be obtained by assuming that the system, with dominant roots, may be approximated by the second-order system

$$T(s) = \frac{K}{s^2 + 2\zeta\omega_n s + K}$$

$$= \frac{K}{s^2 + 2s + K},$$

since $\zeta\omega_n = 1$. The characteristic equation is thus

$$s^2 + 2s + K = 0.$$

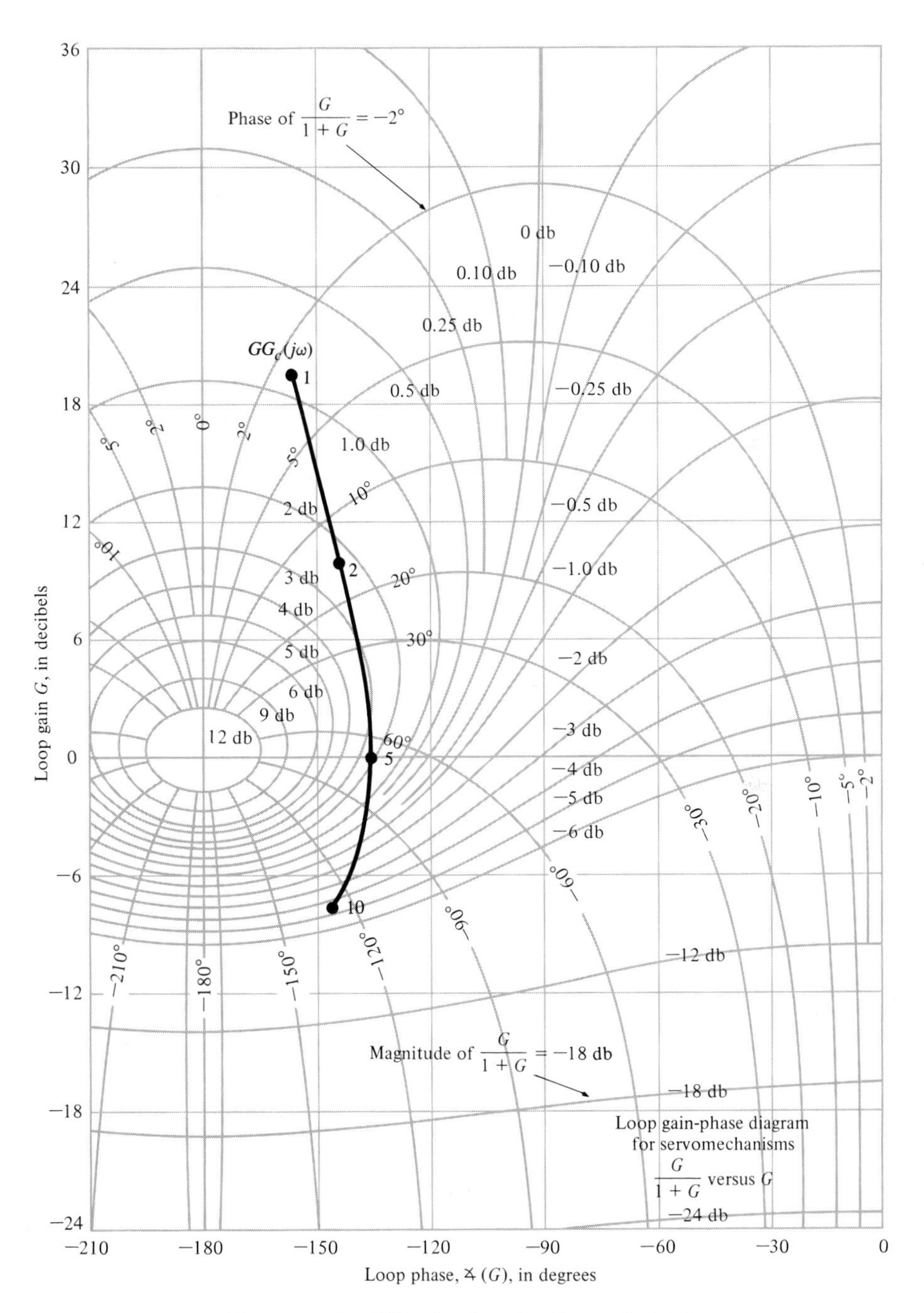

Figure 11.19. Nichols chart for Example 11.3

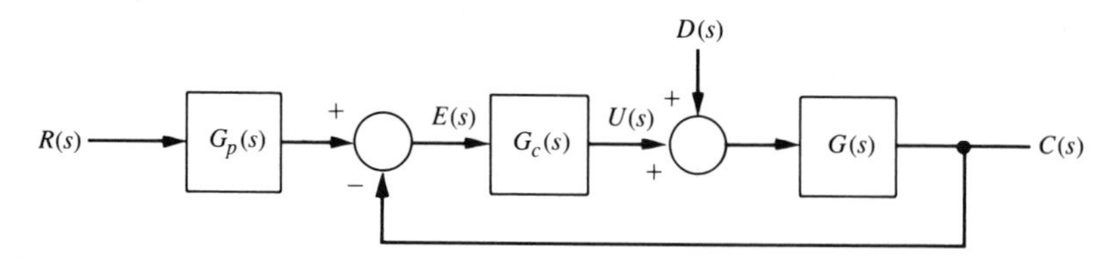

Figure 11.20. Feedback control system with a desired input $R(s)$ and an undesired input $D(s)$.

Then $dK/ds = -(2s + 2)$ and $K = -(s^2 + 2s)$. Therefore

$$\begin{aligned} S_K^r &= \frac{-1}{(2s+2)} \cdot \frac{-(s^2+2s)}{s}\bigg|_{s=r} \\ &= \frac{(s+2)}{2s+2)}\bigg|_{s=r} \end{aligned} \tag{11.27}$$

where $r = -1 + j2$. Then, substituting $s = r$, we obtain

$$|S_K^r| = 0.56.$$

If we raise the gain K to $K = 10$, we expect $r \cong -1.1 \pm j2.2$. Then the sensitivity is

$$|S_K^r| = 0.53.$$

Thus as K increases the sensitivity decreases but the transient performance deteriorates.

11.6 Three-Term (PID) Controllers

One form of controller available and widely used in industrial process control is called a *three-term controller* or *process controller.* This controller has a transfer function

$$\frac{U(s)}{E(s)} = G_c(s) = K_p + \frac{K_I}{s} + K_D s. \tag{11.28}$$

The controller provides a proportional term, an integration term, and a derivative term [26]. The equation for the output in the time domain is

$$u(t) = K_p e(t) + K_I \int e(t)\,dt + K_D \frac{de(t)}{dt}. \tag{11.29}$$

The three-mode controller is also called a PID controller because it contains a proportional, an integration, and a derivative term. The transfer function of the derivative term is actually

$$G_d(s) = \frac{K_D s}{\tau_d s + 1}, \tag{11.30}$$

but usually τ_d is much smaller than the time constants of the process itself and it may be neglected.

If we set $K_D = 0$, then we have the familiar PI controller discussed in Section 10.6. When $K_I = 0$, we have

$$G_c(s) = K_p + K_D s, \tag{11.31}$$

which is called a proportional plus derivative (PD) controller (Example 11.1).

Many industrial processes are controlled using proportional-integral-derivative (PID) controllers. The popularity of PID controllers can be attributed partly to their robust performance in a wide range of operating conditions and partly to their functional simplicity, which allows engineers to operate them in a simple and straightforward manner. To implement such a controller, three parameters must be determined for the given process: proportional gain, integral gain, and derivative gain.

Consider the PID controller

$$\begin{aligned} G_c(s) &= K_1 + \frac{K_2}{s} + K_3 s \\ &= \frac{K_3 s^2 + K_1 s + K_2}{s} \\ &= \frac{K_3(s^2 + as + b)}{s}, \end{aligned} \tag{11.32}$$

where $a = K_1/K_3$ and $b = K_2/K_3$. Therefore a PID controller introduces a transfer function with one pole at the origin and two zeros that can be located anywhere in the left-hand s-plane.

Recall that a root locus begins at the poles and ends at the zeros. If we have a system as shown in Fig. 11.20 with

$$G(s) = \frac{1}{(s + 2)(s + 3)}$$

and we use a PID controller with complex zeros, we can plot the root locus as shown in Fig. 11.21. As the gain, K_3, of the controller is increased, the complex

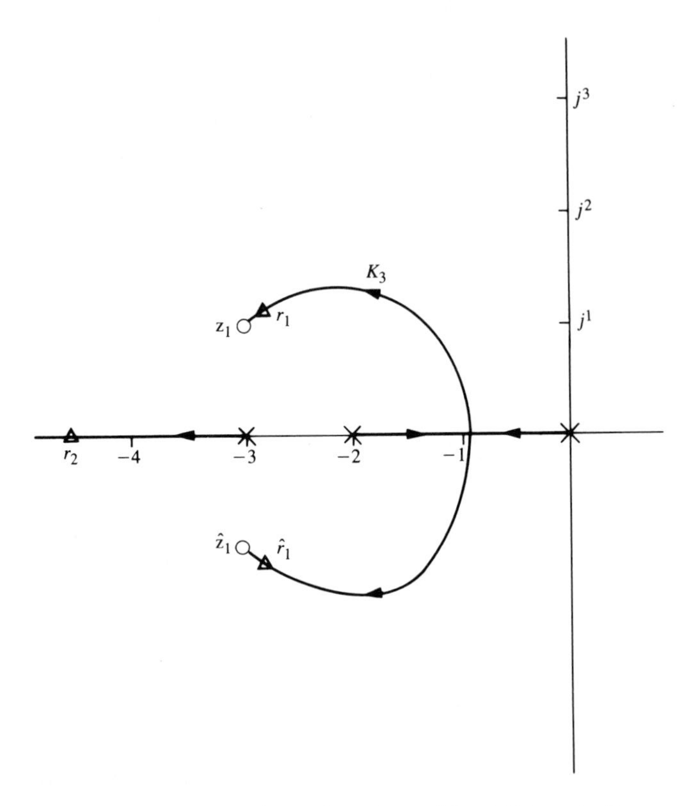

Figure 11.21. Root locus for plant with a PID controller.

roots approach the zeros. The closed-loop transfer function is

$$T(s) = \frac{G(s)G_c(s)G_p(s)}{1 + G(s)G_c(s)}$$

$$= \frac{K_3(s + z_1)(s + \hat{z}_1)}{(s + r_2)(s + r_1)(s + \hat{r}_1)} G_p(s) \tag{11.33}$$

$$\cong \frac{K_3 G_p(s)}{(s + r_2)}.$$

because the zeros and the complex roots are approximately equal ($r_1 \approx z_1$). Setting $G_p(s) = 1$, we have

$$T(s) = \frac{K_3}{s + r_2}$$

$$\approx \frac{K_3}{s + K_3} \tag{11.34}$$

when $K_3 \gg 1$. The only limiting factor is the allowable magnitude of $U(s)$ (Fig. 11.20) when K_3 is large. If K_3 is 100, the system has a fast response and zero steady-state error. Furthermore, the effect of the disturbance is reduced significantly.

In general, we note that PID controllers are particularly useful for reducing steady-state error and improving the transient response when $G(s)$ has one or two poles (or may be approximated by a second-order plant).

We may use frequency response methods to represent the addition of a PID controller. The PID controller, Eq. (11.32), may be rewritten as

$$G_c(s) = \frac{K_2\left(\frac{K_3}{K_2}s^2 + \frac{K_1}{K_2}s + 1\right)}{s} = \frac{K_2(\tau s + 1)\left(\frac{\tau}{\alpha}s + 1\right)}{s}. \quad (11.35)$$

The Bode diagram of Eq. (11.35) is shown in Fig. 11.22 for $\omega\tau$, $K_2 = 2$, and $\alpha = 10$. The PID controller is a form of a lag-lead compensator with a variable gain, K_2. Of course, it is possible that the controller will have complex zeros and a Bode diagram that will be dependent on the ζ of the complex zeros. The contribution by the zeros to the Bode chart may be visualized by reviewing Fig. 7.9 for complex poles and noting that the phase and magnitude change as ζ changes. The PID controller with complex zeros is

$$G_c(\omega) = \frac{K_2[1 + (2\zeta/\omega_n)j\omega - (\omega/\omega_n)^2]}{j\omega}. \quad (11.36)$$

Normally, we choose $0.9 < \zeta < 0.7$.

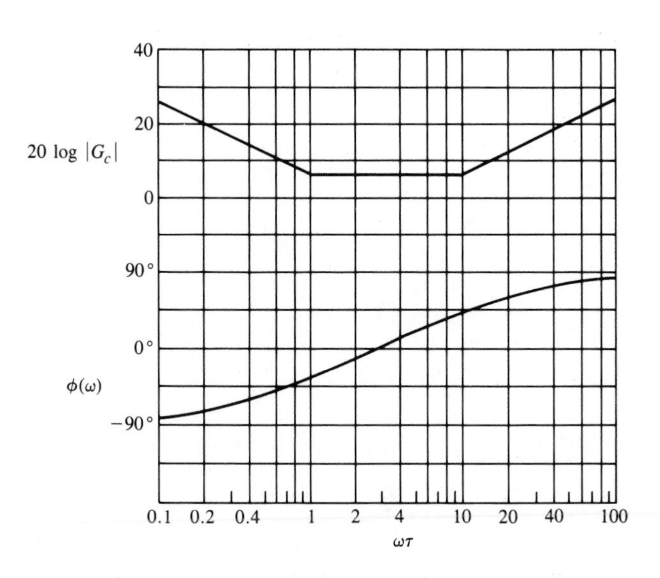

Figure 11.22. Bode diagram for a PID controller.

11.7 The Design of Robust PID Controlled Systems

The selection of the three coefficients of PID controllers is basically a search problem in a three-dimensional space. Points in the search space correspond to different selections of a PID controller's three parameters. By choosing different points of the space, we can produce, for example, different step responses for a step input. A PID controller can be determined by moving in this search space on a trial and error basis.

The main problem in the selection of the three coefficients is that these coefficients do not readily translate into the desired performance and robustness characteristics that the control system designer has in mind. Several rules and methods have been proposed to solve this problem. In this section we consider several design methods using root locus and performance indices.

The first design method uses the ITAE performance index of Section 4.5 and the optimum coefficients of Table 4.5 for a step input or Table 4.6 for a ramp input. Thus we select the three PID coefficients to minimize the ITAE performance index, which produces an excellent transient response to a step (Fig. 4.22c) or a ramp. The design procedure consists of three steps:

1. Select the ω_n of the closed-loop system by specifying the settling time.
2. Determine the three coefficients using the appropriate optimum equation (Table 4.5) and the ω_n of step 1 to obtain $G_c(s)$.
3. Determine a prefilter $G_p(s)$ so that the closed-loop system transfer function, $T(s)$, does not have any zeros as required by Eq. (4.47).

■ Example 11.5 Robust control of temperature

Consider a temperature controller with a control system as shown in Fig. 11.20 and a plant

$$G(s) = \frac{1}{(s+1)^2}. \tag{11.37}$$

If $G_c(s) = 1$, the steady-state error is 50% and the settling time is 3.2 seconds for a step input. We desire to obtain an optimum ITAE performance for a step input and a settling time of less than 0.5 seconds. Using a PID controller, we have

$$G_c(s) = \frac{K_3 s^2 + K_1 s + K_2}{s}. \tag{11.38}$$

Therefore the closed loop transfer function without prefiltering [$G_p(s) = 1$] is

$$\begin{aligned} T_1(s) = \frac{C(s)}{R(s)} &= \frac{G_c G(s)}{1 + G_c G(s)} \\ &= \frac{K_3 s^2 + K_1 s + K_2}{s^3 + (2 + K_3)s^2 + (1 + K_1)s + K_2}. \end{aligned} \tag{11.39}$$

The optimum coefficients of the characteristic equation for ITAE are obtained from Table 4.5 as

$$(s^3 + 1.75\,\omega_n s^2 + 2.15\,\omega_n^2 s + \omega_n^3). \tag{11.40}$$

We need to select ω_n in order to meet the settling time requirement. Since $T_s = 4/\zeta\omega_n$ and ζ is unknown, but near 0.8, we set $\omega_n = 10$. Then, equating the denominator of Eq. (11.39) to Eq. (11.40), we obtain the three coefficients as $K_1 = 214$, $K_3 = 15.5$, and $K_2 = 1000$.

Then Eq. (11.39) becomes

$$\begin{aligned} T_1(s) &= \frac{15.5s^2 + 214s + 1000}{s^3 + 17.5s^2 + 215s + 1000} \\ &= \frac{15.5(s + 6.9 + j4.1)(s + 6.9 - j4.1)}{s^3 + 17.5s^2 + 215s + 1000}. \end{aligned} \tag{11.41}$$

The response of this system to a step input has an overshoot of 32% as recorded in Table 11.2.

We select a prefilter $G_p(s)$ so that we achieve the desired ITAE response as

$$\begin{aligned} T(s) &= \frac{G_cGG_p(s)}{1 + G_cG(s)} \\ &= \frac{1000}{s^3 + 17.5s^2 + 215s + 1000}. \end{aligned} \tag{11.42}$$

Therefore we require

$$G_p(s) = \frac{64.5}{(s^2 + 13.8s + 64.5)} \tag{11.43}$$

in order to eliminate the zeros in Eq. (11.41) and bring the overall numerator to 1000. The response of the system, $T(s)$, to a step input is indicated in Table 11.2. The system has a small overshoot, a settling time of less than ½ second, and zero steady-state error. Furthermore, for a disturbance $D(s) = 1/s$, the maximum value of $c(t)$ due to the disturbance is 0.4% of the magnitude of the disturbance. This is a very favorable design.

Table 11.2. Results for Example 11.5

Controller	$G_c = 1$	PID and $G_p = 1$	PID with $G_p(s)$ prefilter
Percent overshoot	0	31.7%	1.9%
Settling time (seconds)	3.2	0.20	0.45
Steady-state error	50.1%	0.0%	0.00%
$\lvert c(t)/d(t)\rvert_{\text{maximum}}$	52%	0.4%	0.4%

■ Example 11.6 Robust system design

Let us reconsider the system in Example 11.5 when the plant varies significantly so that

$$G(s) = \frac{K}{(\tau s + 1)^2}, \tag{11.44}$$

where $0.5 \leq \tau \leq 1$, and $1 \leq K \leq 2$. It is desired to achieve robust behavior using an ITAE optimum system with a prefilter while attaining an overshoot of less than 4% and a settling time of less than 2 seconds while $G(s)$ can attain any value in the range indicated above. We select $\omega_n = 8$ in order to attain the settling time and determine the ITAE coefficients for $K = 1$ and $\tau = 1$. Then, completing the calculation, we obtain the system without a prefilter $[G_p(s) = 1]$ as

$$T_1(s) = \frac{12(s^2 + 11.38s + 42.67)}{s^3 + 14s^2 + 137.6s + 512} \tag{11.45}$$

and

$$G_c(s) = \frac{12(s^2 + 11.38s + 42.67)}{s}. \tag{11.46}$$

We select a prefilter

$$G_p(s) = \frac{42.67}{(s^2 + 11.38s + 42.67)} \tag{11.47}$$

to obtain the optimum ITAE transfer function

$$T(s) = \frac{512}{s^3 + 14s^2 + 137.6s + 512}. \tag{11.48}$$

We then obtain the step response for the four conditions: $\tau = 1$, $K = 1$; $\tau = 0.5$, $K = 1$; $\tau = 1$, $K = 2$; and $\tau = 0.5$, $K = 2$. The results are summarized in Table 11.3. Clearly, this is a very robust system.

The value of ω_n that can be chosen will be limited by considering the maximum allowable $u(t)$ where $u(t)$ is the output of the controller, as shown in Fig. 11.20. If the maximum value of $e(t)$ is 1, then $u(t)$ would normally be limited to 100 or less. As an example, consider the system in Fig. 11.20 with a PID controller, $G(s) = 1/s(s + 1)$, and the necessary prefilter $G_p(s)$ to achieve ITAE performance. If we select $\omega_n = 10$, 20, and 40, the maximum value of $u(t)$ is recorded

Table 11.3. Results for Example 11.6

Plant Conditions	$\tau = 1$ $K = 1$	$\tau = 0.5$ $K = 1$	$\tau = 1$ $K = 2$	$\tau = 0.5$ $K = 2$
Percent overshoot	2%	0%	0%	1%
Settling time (seconds)	1.25	0.8	0.8	0.9

Table 11.4. Maximum Value of Plant Input

ω_n	10	20	40
$u(t)$ maximum for $R(s) = 1/s$	35	135	550
Settling time (seconds)	0.9	0.5	0.3

in Table 11.4. If we wish to limit $u(t)$ to a maximum equal to 100, we need to limit ω_n to 16. Thus we are limited in the settling time we can achieve.

Let us consider the design of a PID compensator using frequency response techniques for a system with a time delay so that

$$G(s) = \frac{Ke^{-Ts}}{(\tau s + 1)}. \tag{11.49}$$

This type of plant represents many industrial processes which incorporate a time delay. We use a PID compensator to introduce two equal zeros so that

$$G_c(s) = \frac{K_2(\tau_1 s + 1)^2}{s}. \tag{11.50}$$

The design method is as follows:

1. Plot the uncompensated Bode diagram for $K_2\, G(s)/s$ with a gain K_2 that satisfies the steady-state error requirement.
2. Place the two equal zeros at or near the crossover frequency, ω_c.
3. Test the results and adjust K or the zero locations, if necessary.

■ Example 11.7 PID control of a system with a delay

Consider the system of Fig. 11.19 when

$$G(s) = \frac{20e^{-0.1s}}{(0.1s + 1)}, \tag{11.51}$$

where $K = 20$ is selected to achieve a small steady-state error for a step input. We want an overshoot to a step input of less than 5%.

Plotting the Bode diagram for $G(j\omega)$, we find that the uncompensated system has a negative phase margin and the system is unstable.

We will use a PID controller of the form of Eq. (11.50) to attain a desirable phase margin of 70°. Then the loop transfer function is

$$GG_c(s) = \frac{20e^{-0.1s}(\tau_1 s + 1)^2}{s(0.15s + 1)}, \tag{11.52}$$

where $K_2K = 20$. We plot the Bode diagram without the two zeros, as shown in Fig. 11.23. The phase margin is $-32°$ and the system is unstable prior to the introduction of the zeros.

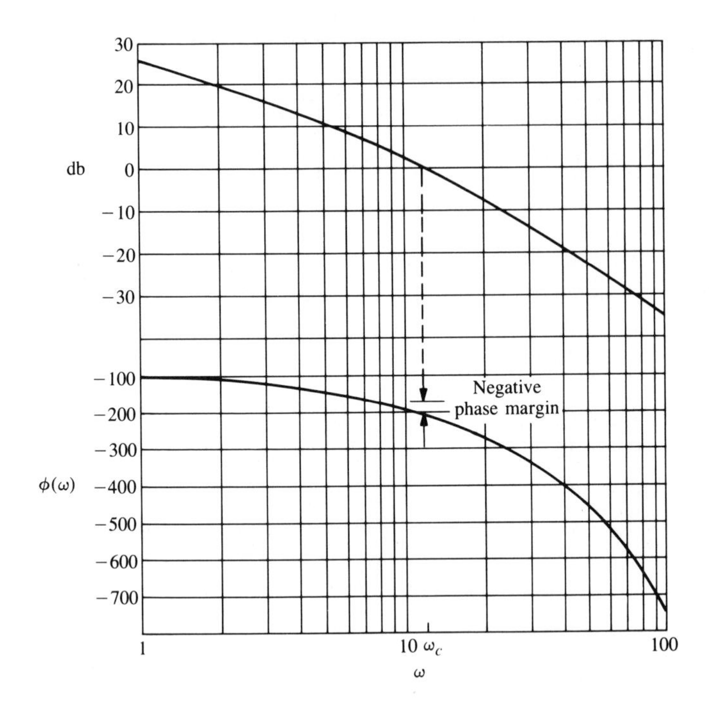

Figure 11.23. Bode diagram for $G(s)/s$ for Example 11.7.

Since we have introduced a pole at the origin because of the integral term in the PID compensator, we may reduce the gain K_2K because e_{ss} is now zero. We place the two zeros at or near the crossover, $\omega_c = 11$. We choose to set $\tau_1 = 0.06$ so that the two zeros are set at $\omega = 16.7$. Also, we reduce the gain to 4.5. Then we obtain the frequency response shown in Fig. 11.24, where

$$G_cG(s) = \frac{4.5\,(0.06s + 1)^2 e^{-0.1s}}{s(0.1s + 1)}. \tag{11.53}$$

The new crossover frequency is $\omega_c' = 4.5$ and the phase margin is 70°. The step response of this system has no overshoot and has a settling time of 0.80 seconds. This response satisfies the requirements. However, if the designer wanted to adjust the system further, we could raise K_2K to 10 and achieve a somewhat faster response with an overshoot of less than 5%.

As a final consideration of the design of robust control systems using a PID controller, we turn to an s-plane root locus method. This design approach may be simply stated as follows:

1. Place the poles and zeros of $G(s)/s$ on the s-plane.
2. Select a location for the zeros of $G_c(s)$ that will result in an acceptable root locus and suitable dominant roots.

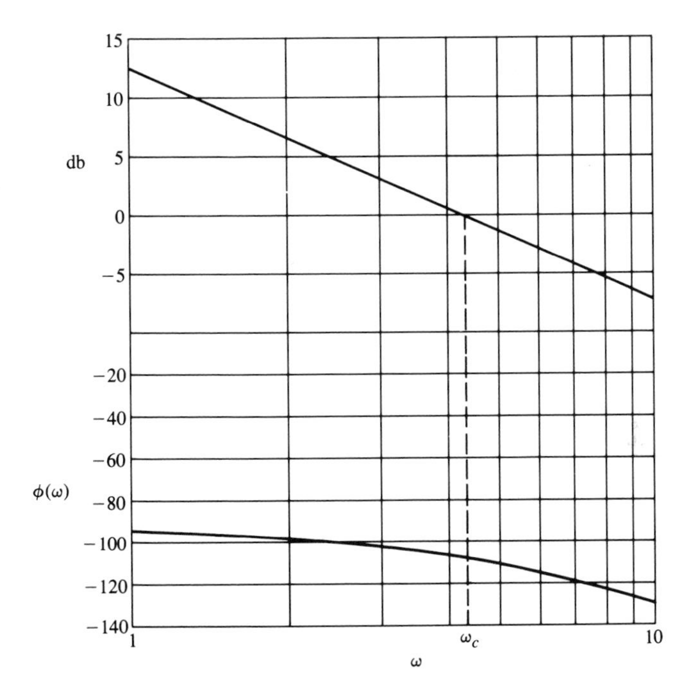

Figure 11.24. Bode diagram for $G_cG(s)$ for Example 11.7.

3. Test the transient response of the compensated system and iterate step 2, if necessary.

11.8 Design Example: Aircraft Autopilot

A typical aircraft autopilot control system consists of electrical, mechanical, and hydraulic devices that move the flaps, elevators, fuel-flow controllers, and other components that cause the aircraft to vary its flight. Sensors provide information on velocity, heading, rate of rotation, and other flight data. This information is combined with the desired flight characteristics (commands) electronically available to the autopilot. The autopilot should be able to fly the aircraft on a heading and under conditions set by the pilot. The command often consists of a predetermined heading. Design often focuses on a forward-moving aircraft that moves somewhat up or down without moving right or left and without rolling (rotating the wingtips). Such a study is called pitch plane design. The aircraft is represented by a plant

$$G(s) = \frac{K}{(s + 1/\tau)(s^2 + 2\zeta_1\omega_1 s + \omega_1^2)}, \tag{11.54}$$

where τ is the time constant of the actuator. Let $\tau = 1/4$, $\omega_1 = 2$, and ζ equal to $1/2$. Then, the s-plane plot has two complex poles—a pole at the origin and a pole at

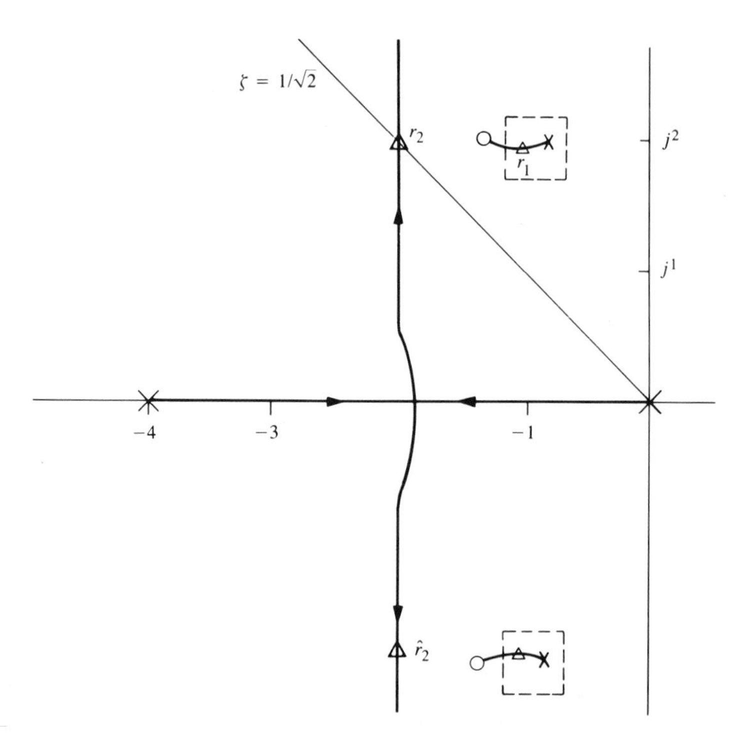

Figure 11.25. Root locus for Example 11.8. The complex poles can vary within the dashed-line box.

$s = -4$, as shown in Fig. 11.25. The complex poles, representing the aircraft dynamics, can vary within the dashed-line box shown in Fig. 11.25. We then choose the zeros of the controller as $s = -1.3 \pm j2$, as shown. We select the gain K so that the roots r_2 and $\hat{r}_2$ are complex with a ζ of $1/\sqrt{2}$. The other roots r_1 and $\hat{r}_1$ lie very near the zeros. Therefore the closed-loop transfer function is approximately

$$T(s) \cong \frac{\omega_n^2}{s^2 + 2\zeta\omega_n s + \omega_n^2} = \frac{5}{s^2 + 3.16s + 5}, \tag{11.55}$$

since $\omega_n = \sqrt{5}$ and $\zeta = 1/\sqrt{2}$. The resulting response to a step input has an overshoot of 4.5% and a settling time of 2.5 seconds.

11.9 The Future Evolution of Robotics and Control Systems

The continuing goal of robot and control systems is to provide extensive flexibility and a high level of autonomy. The two system concepts are approaching this goal by different evolutionary pathways, as illustrated in Fig. 11.26. Today's

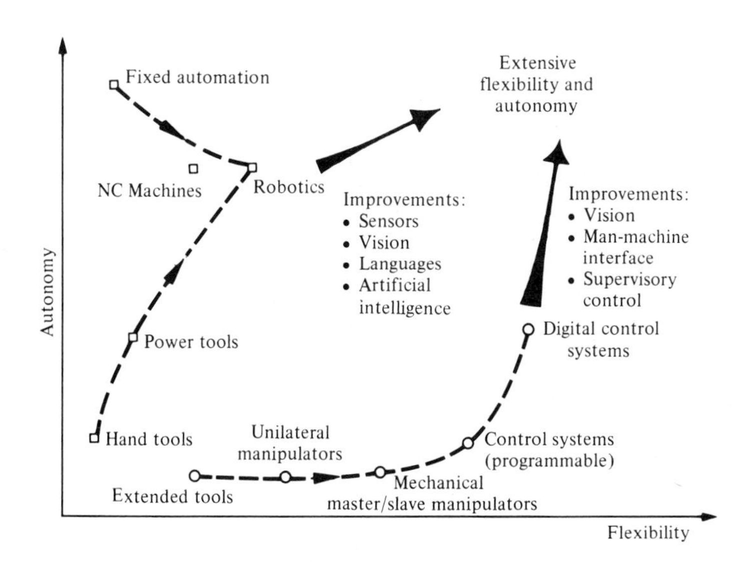

Figure 11.26. Future evolution of control systems and robotics.

industrial robot is perceived as quite autonomous—once it is programmed, further intervention is not normally required. Because of sensory limitations, these robot systems have limited flexibility in adapting to work environment changes, which is the motivation of computer vision research. The control system is very adaptable, but it relies on human supervision. Advanced robotic systems are striving for task adaptability through enhanced sensory feedback. Research areas concentrating on artificial intelligence, sensor integration, computer vision, and off-line CAD/CAM robot programming will make robots more universal and economical. Control systems are moving toward autonomous operation as an enhancement to human control. Research in supervisory control, human-machine interface methods to reduce operator burden, and computer data base management is intended to improve operator efficiency. Many research activities are common to both systems and are aimed toward reducing implementation cost and expanding the realm of application. These include improved communications methods and advanced programming languages.

11.10 Summary

The use of a digital computer as the compensation device for a closed-loop control system has grown during the past decade as the price and reliability of computers have improved dramatically. A computer can be used to complete many calculations during the sampling interval and to provide an output signal that is used to drive an actuator of a process. Computer control is used today for chemical processes, aircraft control, machine tools, and many common processes. Computers can also be used in devices commonly called robots (Fig. 11.27).

The design of highly accurate control systems in the presence of significant

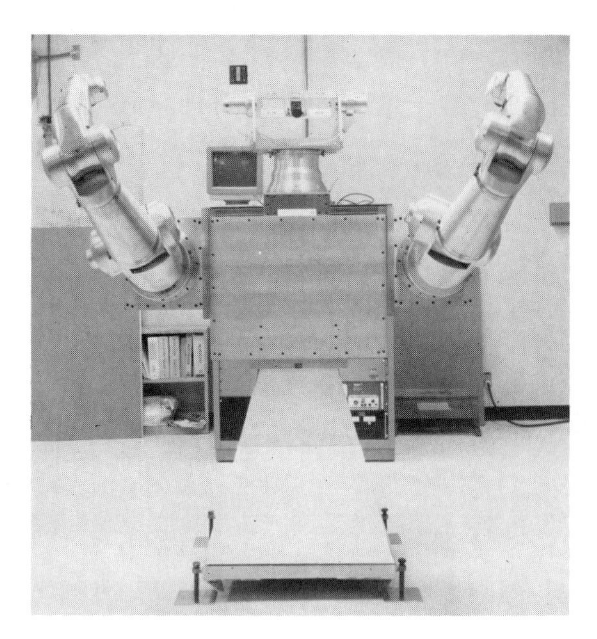

Figure 11.27. Anthropomorphic robot with two arms. (Courtesy of Lockheed Missiles and Space Co.)

plant uncertainty requires the designer to seek a robust control system. A robust control system exhibits low sensitivities to parameter change and is stable over a wide range of parameter variations.

The three-mode or PID controller was considered as a compensator that aids in the design of robust control systems. The design issue for a PID controller is the selection of the gain and two zeros of the controller transfer function. We utilized three design methods for the selection of the controller: (1) the root locus method, (2) the frequency response method, and (3) the ITAE performance index method. An operational amplifier circuit used for a PID controller is shown in Fig. 11.28.

In general, the use of a PID controller will enable the designer to attain a robust control system.

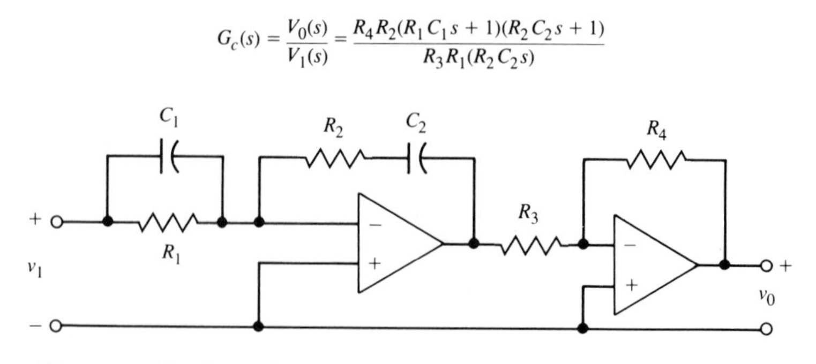

Figure 11.28. Operational amplifier circuit used for PID controller.

Exercises

E11.1. Consider a system of the form shown in Fig. 11.20 where

$$G(s) = \frac{1}{(s+1)}.$$

Using the ITAE performance method for a step input, determine the required $G_c(s)$. Assume $\omega_n = 20$ for Table 4.6. Determine the step response with and without a prefilter $G_p(s)$.

E11.2. For the ITAE design obtained in E11.1, determine the response due to a disturbance $D(s) = 1/s$.

E11.3. A closed-loop unity feedback system has

$$G(s) = \frac{9}{s(s+p)},$$

where p is normally equal to 3. Determine S_p^T and plot $|T(j\omega)|$ and $|S(j\omega)|$ on a Bode plot.

E11.4. A PID controller is used in the system in Fig. 11.20 where

$$G(s) = \frac{1}{(s+2)(s+8)}.$$

The gain K_3 of the controller (Eq. 11.32) is limited to 180. Select a set of compositor zeros so that the pair of closed-loop roots is approximately equal to the zeros. Find the step response for the approximation in Eq. (11.34) and the actual response and compare them.

E11.5. A system has a plant

$$G(s) = \frac{15{,}900}{s\left(\frac{s}{100}+1\right)\left(\frac{s}{200}+1\right)}$$

and negative unity feedback with a PD compensator

$$G_c(s) = K_1 + K_2 s.$$

The objective is to design $G_c(s)$ so that the overshoot to a step is less than 25% and the settling time is less than 60 ms.

E11.6 Consider the control system shown in Fig. E11.6 when $G(s) = \dfrac{1}{(s+4)}$, select a PI controller so that the settling time is less than 1 second for an ITAE step response. Plot $c(t)$ for a step input $r(t)$ with and without a prefilter. Determine and plot $c(t)$ for a step disturbance. Discuss the effectiveness of the system.

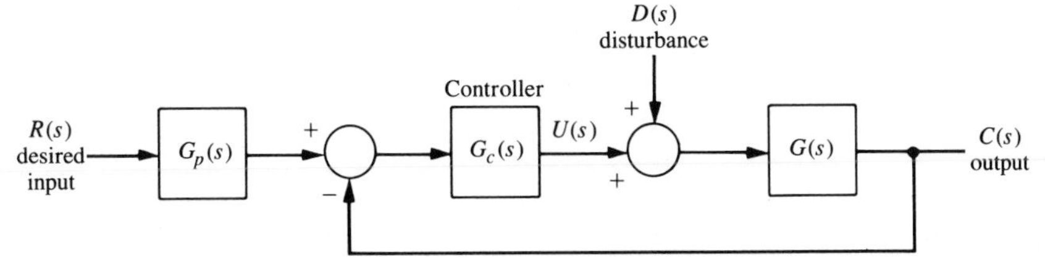

Figure E11.6. System with PI controller.

E11.7 For the control system of Figure E11.6 with $G(s) = \dfrac{1}{(s+4)^2}$, select a PID controller to achieve a settling time of less than 0.8 seconds for an ITAE step response. Plot $c(t)$ for a step input $r(t)$ with and without a prefilter. Determine and plot $c(t)$ for a step disturbance. Discuss the effectiveness of the system.

E11.8 Repeat Exercise 11.6 while striving to achieve a minimum settling time while adding the constraint that $|u(t)| \leq 100$ for $t > 0$ for a unit step input, $r(t) = 1$, $t \geq 0$.

Problems

P11.1. Interest in unmanned underwater vehicles (UUVs) has been increasing recently, with a large number of possible applications being considered. These include intelligence-gathering, mine detection, and surveillance. Regardless of the intended mission, a strong need exists for reliable and robust control of the vehicle. The proposed vehicle is shown in Fig. P11.1(a) [13].

It is desired to control the vehicle through a range of operating conditions. The vehicle is 30 feet long with a vertical sail near the front. The control inputs are sternplane, rudder, and shaft speed commands. In this case we wish to control the vehicle roll by using the stern planes. The control system is shown in Fig. P11.1(b), where $R(s) = 0$, the desired roll angle, and $D(s) = 1/s$. We select $G_c(s) = K(s + 1)$, where $K = 2$. (a) Plot $20 \log |T|$ and $20 \log |S_K^T|$ on a Bode diagram. (b) Evaluate $|S_K^T|$ at ω_B, $\omega_B/2$, and $\omega_B/4$. ($T = C/R$).

P11.2. A new suspended, mobile, remote-controlled video-camera system to bring three-dimensional mobility to professional NFL football is shown in Fig. P11.2(a) [24]. The cam-

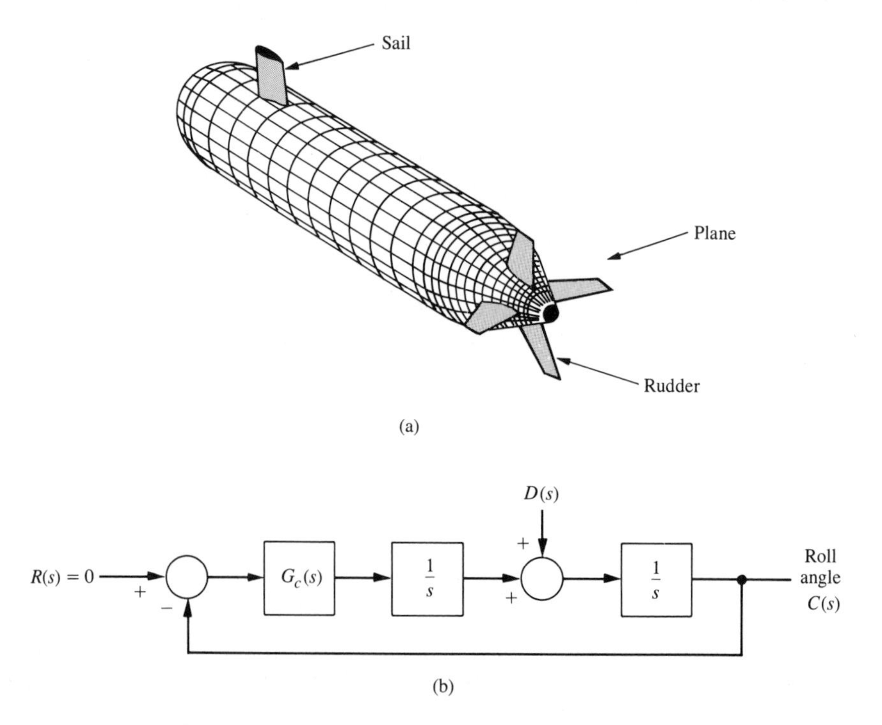

Figure P11.1. Control of an underwater vehicle.

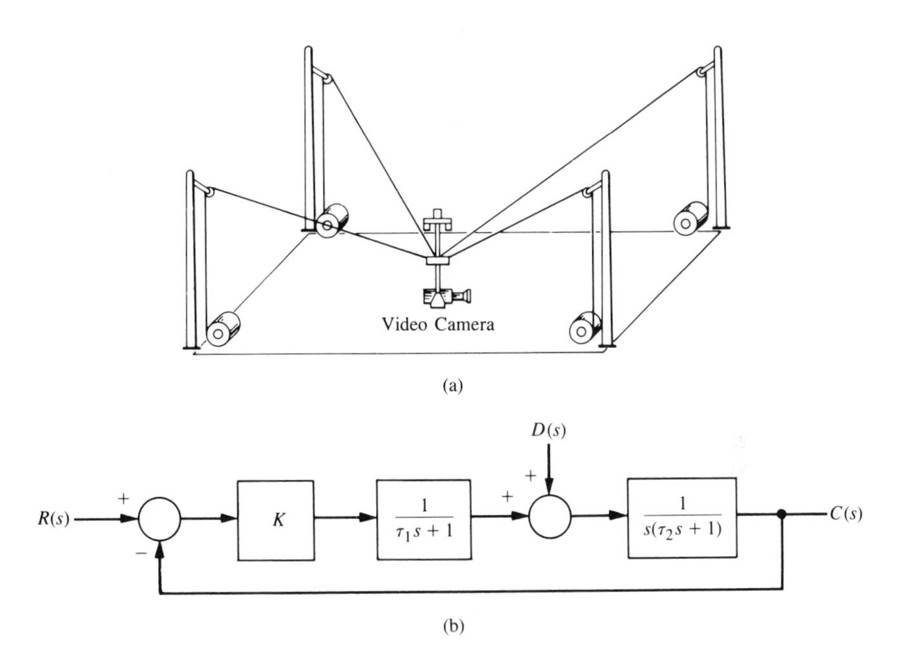

Figure P11.2. Remote controlled tv camera.

era can be moved over the field as well as up and down. The motor control on each pulley is represented by the system in Fig. P11.2(b), where $\tau_1 = 20$ ms and $\tau_2 = 2$ ms. (a) Select K so that $M_{p_\omega} = 1.84$. (b) Plot $20 \log |T|$ and $20 \log |S_K^T|$ on one Bode diagram. (c) Evaluate $|S_K^T|$ at ω_B, $\omega_{B/2}$, and $\omega_{B/4}$. (d) Let $R(s) = 0$ and determine the effect of $D(s) = 1/s$ for the gain K of part (a) by plotting $c(t)$.

P11.3. Magnetic levitation (maglev) trains may replace airplanes on routes shorter than 200 miles. The maglev train developed by a German firm uses electromagnetic attraction to propel and levitate heavy vehicles, carrying up to 400 passengers at 300-mph speeds. But the ¼-inch gap between car and track is hard to maintain [12].

The air-gap control system is shown in Fig. P11.3(a), on the next page. The block diagram of the air-gap control system is shown in Fig. P11.3(b). The compensator is

$$G_c(s) = \frac{K(s + 1)}{(s + 10)}.$$

(a) Find the range of K for a stable system. (b) Select a gain so that the steady-state error of the system is zero for a step input command. (c) Find $c(t)$ for the gain of part (b). (d) Find $c(t)$ when K varies $\pm 10\%$ from the gain of part (b).

P11.4. Computer control of a robot to spray-paint an automobile is shown by the system shown in Fig. P11.4(a), on the next page [1]. We wish to investigate the system when K = 2, 4, and 6. (a) For the three values of K, determine ζ, ω_n, percent overshoot, settling time, and steady-state error for a step input. Record your results in a table. (d) Determine the sensitivity $|S_K^T|$ for the three values of K. (c) Select the best of the three values of K. (d) For the value selected in part (c), determine $c(t)$ for a disturbance $D(s) = 1/s$ when $R(s)$ = 0.

P11.5. An automatically guided vehicle is shown in Fig. P11.5(a) and its control system is shown in Fig. P11.5(b). The goal is to accurately track the guide wire, to be insensitive to

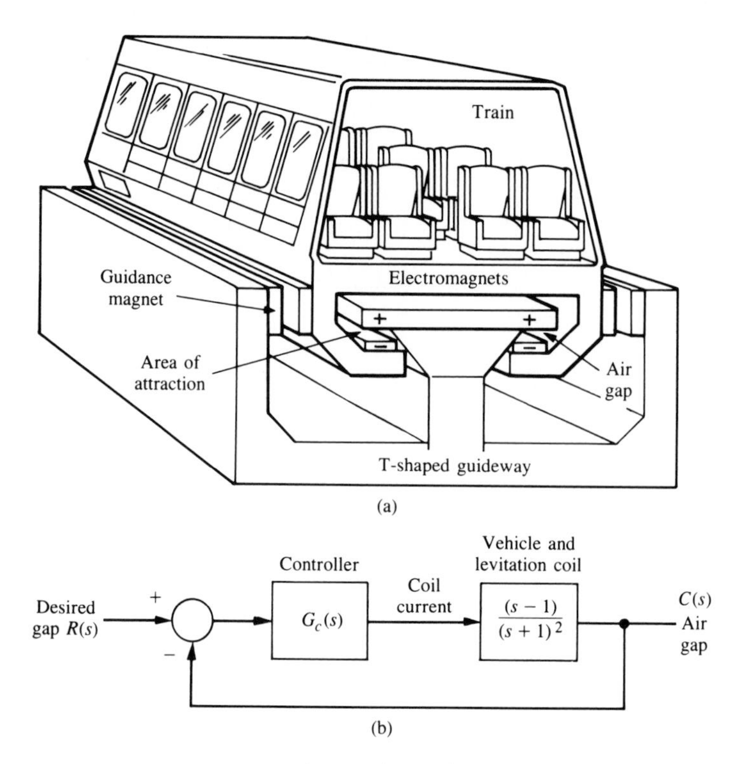

Figure P11.3. Maglev train control.

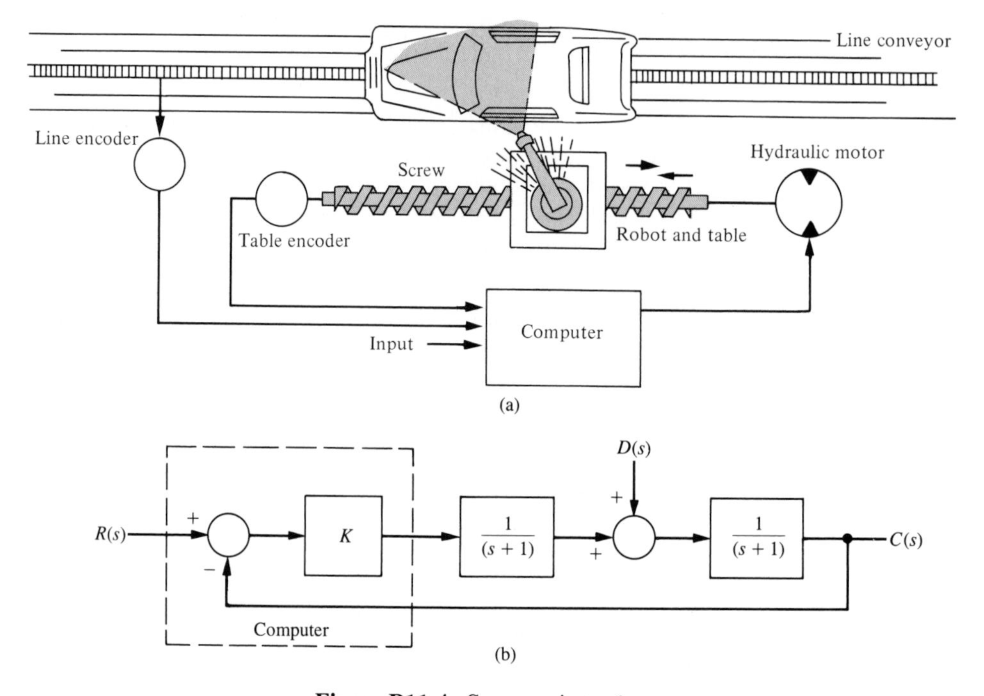

Figure P11.4. Spray paint robot.

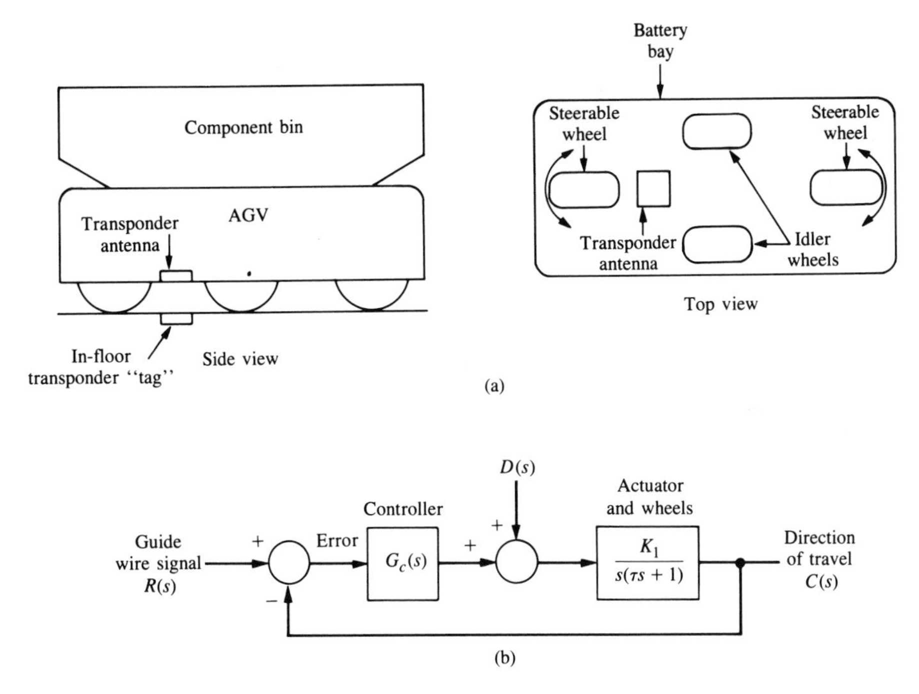

Figure 11.5. Automatically-guided vehicle.

changes in the gain K_1, and to reduce the effect of the disturbance. The gain K_1 is normally equal to 1 and $\tau_1 = 1/25$ sec. (a) Select a compensator, $G_c(s)$, so that the percent overshoot to a step input is less than or equal to 10%, the settling time is less than 100 ms, and the velocity constant, K_v, for a ramp input is 100. (b) For the compensator selected in part (a), determine the sensitivity of the system to small changes in K_1 by determining $S^r_{K_1}$ or $S^T_{K_1}$. (c) If K_1 changes to 2 while $G_c(s)$ of part (a) remains unchanged, find the step response of the system and compare selected performance figures with those obtained in part (a). (d) Determine the effect of $D(s) = 1/s$ by plotting $c(t)$ when $R(s) = 0$.

P11.6. A roll-wrapping machine (RWM) receives, wraps, and labels large paper rolls produced in a paper mill [15]. The RWM consists of several major stations: positioning station, waiting station, wrapping station, and so forth. We will focus on the positioning station shown in Fig. P11.6(a). The positioning station is the first station that sees a paper roll. This station is responsible for receiving and weighing the roll, measuring its diameter and width, determining the desired wrap for the roll, positioning it for down-stream processing, and finally ejecting it from the station.

Functionally, the RWM can be categorized as a complex operation because each functional step (e.g., measuring the width) involves a large number of field device actions and relies upon a number of accompanying sensors.

The control system for accurately positioning the width-measuring arm is shown in Fig. P11.6(b). The pole of the positioning arm, p, is normally equal to 2, but it is subject to change because of loading and misalignment of the machine. (a) For p = 2, design a compensator so that the complex roots are $s = -2 \pm j2\sqrt{3}$. (b) Plot $c(t)$ for a step input $R(s) = 1/s$. (c) Plot $c(t)$ for a disturbance $D(s) = 1/s$, with $R(s) = 0$. (d) Repeat parts (b)

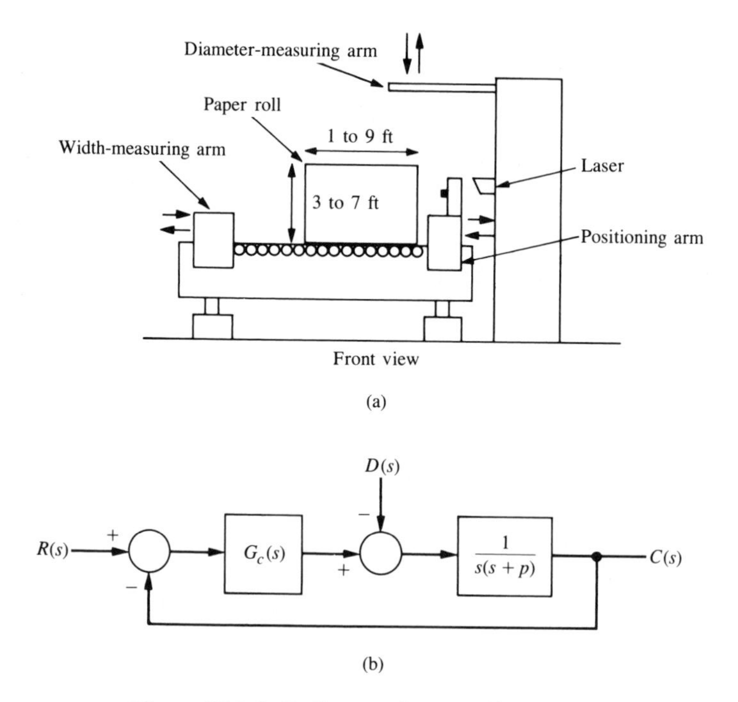

Figure P11.6. Roll-wrapping machine control.

and (c) when p changes to 1 and $G_c(s)$ remains as designed in part (a). Compare the results for the two values of the pole p.

P11.7. The function of a steel plate mill is to roll reheated slabs into plates of scheduled thickness and dimension [27]. The final products are of rectangular plane view shapes having a width of up to 3300 mm and a thickness of 180 mm.

A schematic layout of the mill is shown in Fig. P11.7(a). The mill has two major roll-

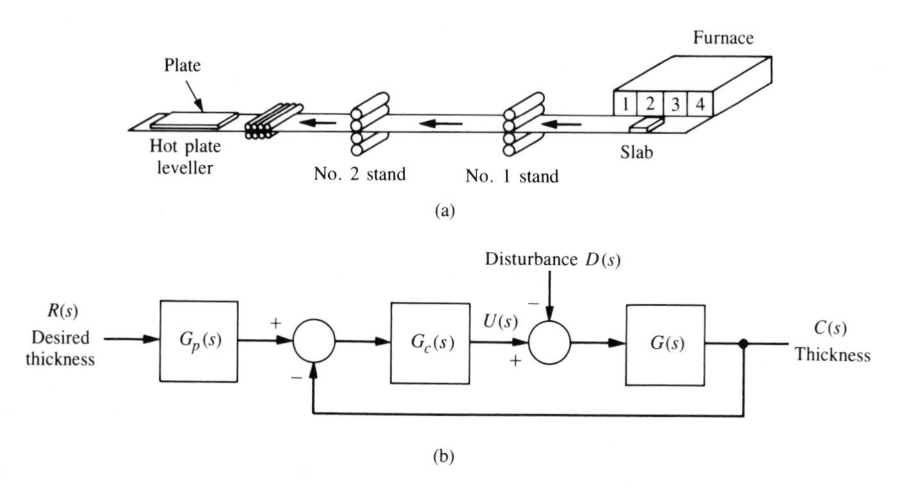

Figure P11.7. Steel rolling mill control.

ing stands, denoted No. 1 and No. 2. These are equipped with large rolls (up to 508 mm in diameter), which are driven by high-power electric motors (up to 4470 kW). Roll gaps and forces are maintained by large hydraulic cylinders.

Typical operation of the mill can be described as follows. Slabs coming from the reheating furnace initially go through the No. 1 stand, whose function is to reduce the slabs to the required width. The slabs proceed through the No. 2 stand, where finishing passes are carried out to produce required slab thickness, and finally through the hot plate leveller, which gives each plate a smooth finishing surface.

One of the key systems controls the thickness of the plates by adjusting the rolls. The block diagram of this control system is shown in Fig. P11.7(b). The plant is represented by

$$G(s) = \frac{1}{s(s^2 + 4s + 5)}.$$

The controller is a PID with two equal real zeros. (a) Select the PID zeros and the gains so that the real part of the four closed-loop roots are equal. (b) For the design of part (a), obtain the step response without a prefilter [$G_p(s) = 1$]. (c) Repeat part (b) for an appropriate prefilter. (d) For the system, determine the effect of a unit step disturbance by evaluating $c(t)$ with $r(t) = 0$.

P11.8. A motor and load with negligible friction and a voltage-to-current amplifier, K_a, is used in a feedback control system, as shown in Fig. P11.8. A designer selects a PID controller

$$G_c(s) = K_1 + \frac{K_2}{s} + K_3 s,$$

where $K_1 = 5$, $K_2 = 500$, and $K_3 = 0.0475$. (a) Determine the appropriate value of K_a so that the phase margin of the system is 42 degrees. (b) For the gain K_a, plot the root locus of the system and determine the roots of the system for the K_a of part (a). (c) Determine the maximum value of $c(t)$ when $D(s) = 1/s$ and $R(s) = 0$ for the K_a of part (a).

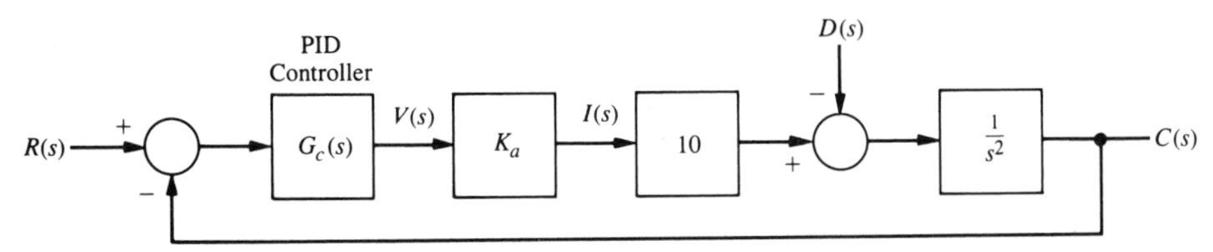

Figure P11.8.

Design Problems ✦

DP11.1. A position control system for a large turntable is shown in Fig. DP11.1(a) and the block diagram of the system is shown in Fig. DP11.1(b). This system uses a large torque motor with $K_m = 15$. The objective is to reduce the steady-state effect of a step change in the load disturbance to 5% of the magnitude of the step disturbance while maintaining a fast response to a step input command, $R(s)$, with less than 5% overshoot. Select K_1 and the compensator when (a) $G_c(s) = K$ and (b) $G_c(s) = K_2 + K_3 s$ (a PD compensator). Plot the step response for the disturbance and the input for both compensators.

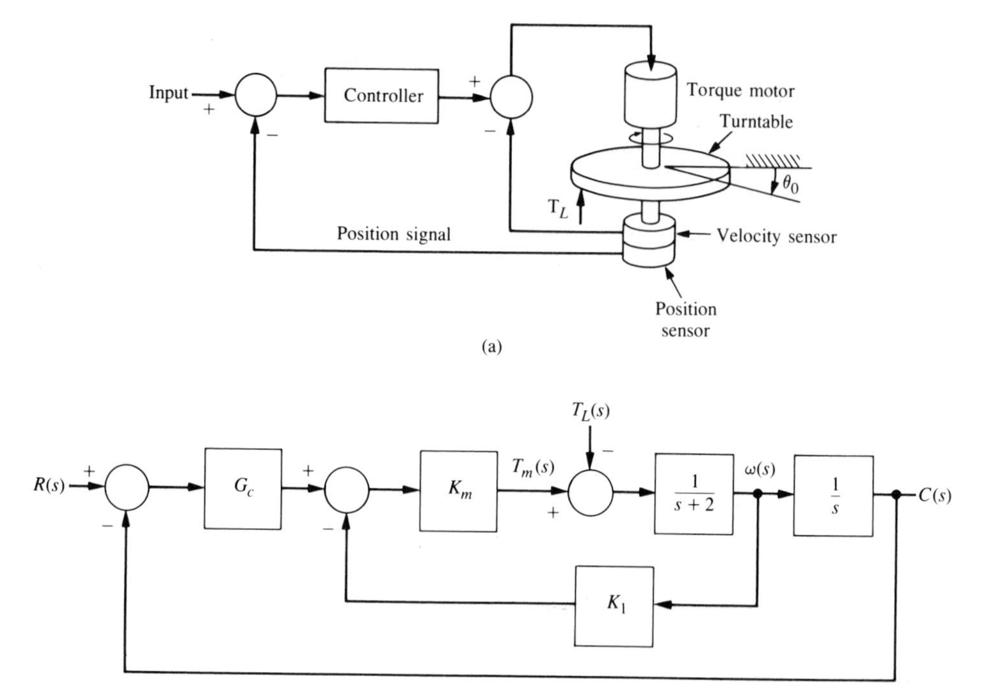

Figure DP11.1. Turntable control.

DP11.2. Digital audio tape (DAT) stores 1.3 gigabytes of data in a package the size of a credit card—roughly nine more times than half-inch-wide reel-to-reel tape or quarter-inch-wide cartridge tape. A DAT sells for a few dollars—about the same as a floppy disk even though it can store 1000 times more data. A DAT can record for two hours (longer than either reel-to-reel or cartridge tape), which means that it can run longer unattended and requires fewer tape changes and hence fewer interruptions of data transfer. DAT gives access to a given data file within 20 sec, on the average, compared with a few to several minutes for either cartridge or reel-to-reel tape [17].

The tape drive electronically controls the relative speeds of the drum and tape so that the heads follow the tracks on the tape, as shown in Fig. DP11.2(a). The control system is much more complex than that for a CD ROM because more motors have to be accurately controlled: capstan, take-up and supply reels, drum, and tension control.

Let us consider the speed control system shown in Fig. DP11.2(b). The motor and load transfer function varies because the tape moves from one reel to the other. The transfer function is

$$G(s) = \frac{K_m}{(s + p_1)(s + p_2)},$$

where the nominal values are $K_m = 4$, $p_1 = 1$, and $p_2 = 4$. However, the range of variation is: $3 \leq K_m \leq 5$; $0.5 \leq p_1 \leq 1.5$; $3.5 \leq p_2 \leq 4.5$. The system must respond rapidly with an overshoot to a step of less than 13% and a settling time of less than 0.5 sec. The system requires a fast peak time so the overdamped condition is not allowed. Determine a PID controller that results in an acceptable response over the range of parameter variation. The gain $K_m K_3$ cannot exceed 20 when the nominal $K_m = 4$ and K_3 is defined in Eq. (11.32).

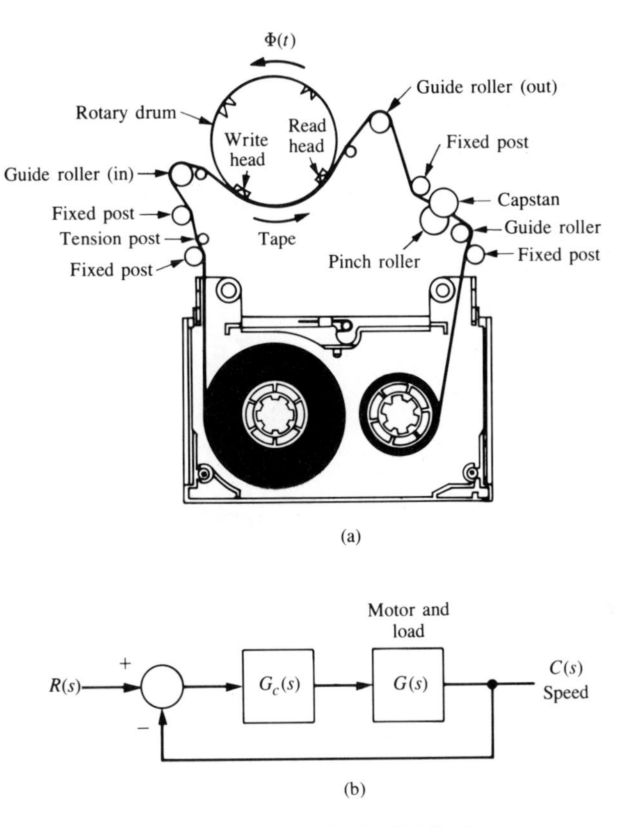

Figure DP11.2. Control of a DAT player.

DP11.3. The Gamma-Ray Imaging Device (GRID) is a NASA experiment to be flown on a long-duration, high-altitude balloon during the coming solar maximum. The GRID on a balloon is an instrument that will qualitatively improve hard X-ray imaging and carry out the first gamma-ray imaging for the study of solar high-energy phenomena in the next phase of peak solar activity. From its long-duration balloon platform, GRID will observe numerous hard X-ray bursts, coronal hard X-ray sources, "superhot" thermal events, and microflares [19]. Parts (a) and (b) of Fig. DP11.3 depict the system elements. The major components of the GRID experiment consist of a 5.2-meter canister and mounting gondola, a high-altitude balloon, and a cable connecting the gondola and balloon. The instrument-sun pointing requirements of the experiment are 0.1 degrees pointing accuracy and 0.2 arc second per 4 ms pointing stability.

An optical sun sensor provides a measure of the sun-instrument angle and is modeled as a first-order system with a dc gain and a pole at $s = -500$. A torque motor actuates the canister/gondola assembly. The azimuth angle control system is shown in Fig. DP11.3(c). (a) The PID controller is selected by the design team so that

$$G_c(s) = \frac{K_3(s^2 + as + b)}{s},$$

where $a = 6$, and b is to be selected. A prefilter is used as shown in Fig. DP11.3(c). Determine the value of K_3 and b so that the dominant roots have a ζ of 0.8 and the overshoot to a step input is less than 3%. Determine the actual step response percent overshoot, t_p, and T_s. (b) Design a PID filter using the ITAE performance criteria when $\omega_n = 8$. Compare the actual results with those obtained in part (a).

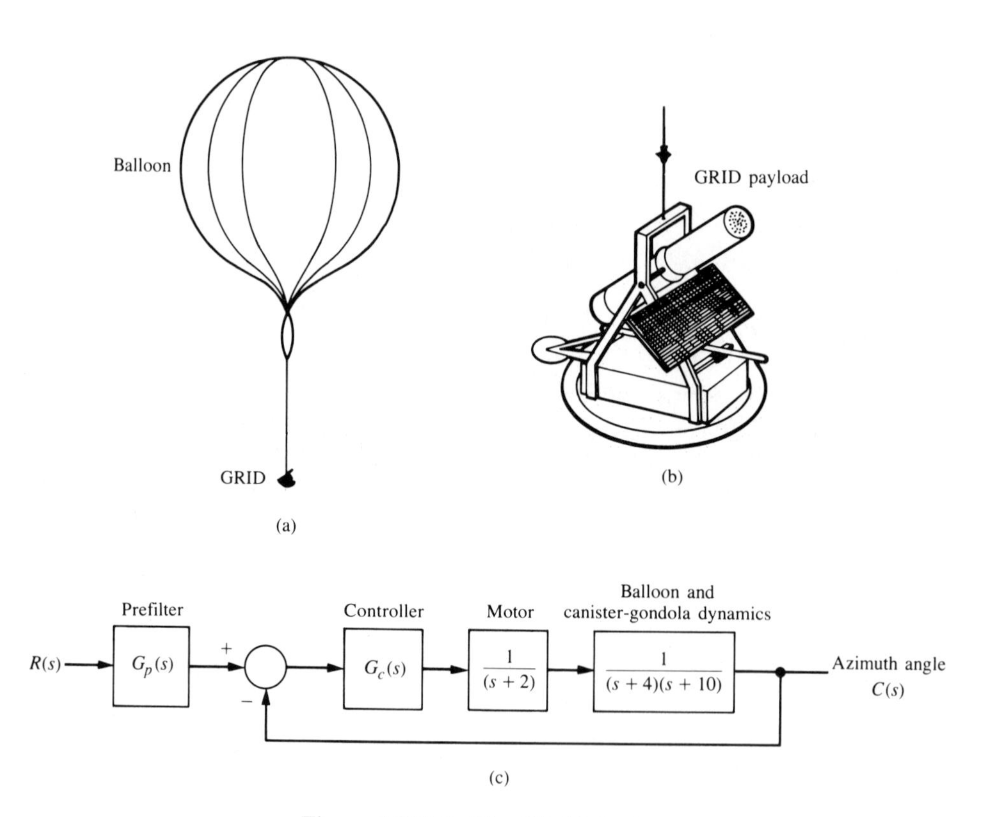

Figure DP11.3. The GRID device.

DP11.4. Many university and government laboratories have constructed robot hands capable of grasping and manipulating objects. But teaching the artificial devices to perform even simple tasks required formidable computer programming. Now, however, the Dexterous Hand Master (DAM) can be worn over a human hand to record the side-to-side and bending motions of finger joints. Each joint is fitted with a sensor that changes its signal depending on position. The signals from all the sensors are translated into computer data and used to operate robot hands.

The DAM is shown in parts (a) and (b) of Fig. DP11.4. The joint angle control system is shown in part (c). The normal value of K_m is 1.0. The goal is to design a PID controller so that the steady-state error for a ramp input is zero. Also, the settling time must be less than 3 sec for the ramp input. It is desired that the controller is

$$G_c(s) = \frac{K_3(s^2 + 6s + 18)}{s}.$$

(a) Select K_3 and obtain the ramp response. Plot the root locus as K_3 varies. (b) If K_m changes to one-half of its normal value, and $G_c(s)$ remains as designed in part (a), obtain the ramp response of the system. Compare the results of parts (a) and (b) and discuss the robustness of the system.

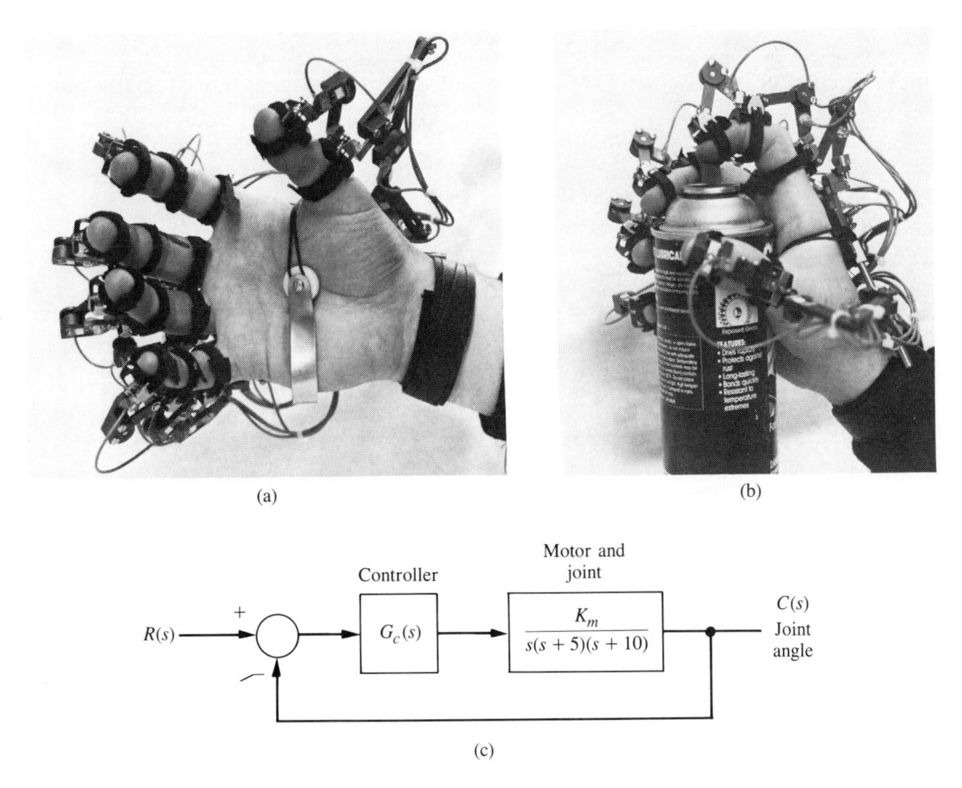

Figure DP11.4. Dexterous Hand Master.

DP11.5. Objects smaller than the wavelengths of visible light are a staple of contemporary science and technology. Biologists study single molecules of protein or DNA; materials scientists examine atomic-scale flaws in cyrstals; microelectronics engineers lay out circuit patterns only a few tens of atoms thick. Until recently, this minute world could be seen only by cumbersome, often destructive methods such as electron microscopy and X-ray diffraction. It lay beyond the reach of any instrument as simple and direct as the familiar light microscope. New microscopes, typified by the scanning tunneling microscope (STM), are now available [10].

The precision of position control required is in the order of nanometers. The STM relies on piezoelectric sensors that change size when an electric voltage across the material is changed. The "aperture" in the STM is a tiny tungsten probe, its tip ground so fine that it may consist of only a single atom and measure just 0.2 nanometer in width. Piezoelectric controls maneuver the tip to within a nanometer or two of the surface of a conducting specimen—so close that the electron clouds of the atom at the probe tip and of the nearest atom of the specimen overlap. A feedback mechanism senses the variations in tunneling current and varies the voltage applied to a third, z, control. The z-axis piezoelectric moves the probe vertically to stabilize the current and to maintain a constant gap between the microscope's tip and the surface. The control system is shown in Fig. DP11.5(a), and the block diagram is shown in Fig. DP11.5(b). The process is

$$G(s) = \frac{17{,}640}{s(s^2 + 59.4s + 1764)}$$

and the controller is chosen to have two real, unequal zeros so that (Eq. 11.35) we have:

$$G_c(s) = \frac{K_2(\tau_1 s + 1)(\tau_2 s + 1)}{s}.$$

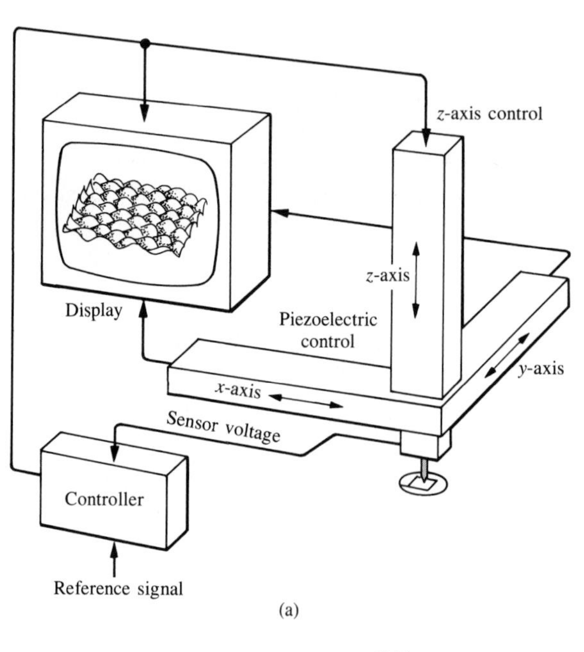

(a)

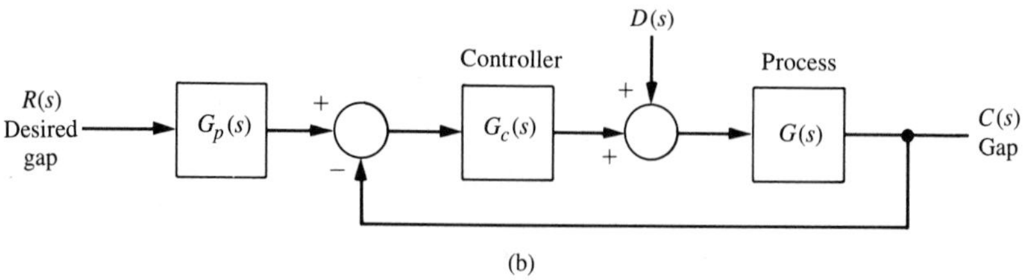

(b)

Figure DP11.5. Microscope control.

(a) Use the ITAE design method to determine $G_c(s)$. (b) Determine the step response of the system with and without a prefilter $G_p(s)$. (c) Determine the response of the system to a disturbance when $D(s) = 1/s$. (d) Using the prefilter and $G_c(s)$ of parts (a) and (b), determine the actual response when the process changes to

$$G(s) = \frac{16{,}000}{s(s^2 + 40s + 1600)}.$$

DP11.6. The system described in DP11.5 is to be designed using the frequency response techniques described in Section 11.7 with

$$G_c(s) = \frac{K_2(\tau_1 s + 1)(\tau_2 s + 1)}{s}.$$

Select the coefficients of $G_c(s)$ so that the phase margin is approximately 70°. Obtain the step response of the system with and without a prefilter, $G_p(s)$.

Terms and Concepts

Digital computer compensator A system that uses a digital computer as the compensator element.

Digital control system A control system that uses digital signals and a digital computer to control a process.

PID controller A controller with three terms in which the output of the controller is the sum of a proportional term, an integrating term, and a differentiating term, with an adjustable gain for each term.

Prefilter A transfer function, $G_p(s)$, that filters the input signal $R(s)$ prior to the calculation of the error signal.

Robot A programmable computer integrated with a manipulator; a reprogrammable, multifunctional manipulator used for a variety of tasks.

Robust control A system that exhibits the desired performance in the presence of significant plant uncertainty.

Sampling period The period when all the numbers leave or enter the computer; the period during which the sampled variable is held constant.

Three-term controller See PID controller.

References

1. R. C. Dorf, *Encyclopedia of Robotics,* John Wiley & Sons, New York, 1988.
2. G. Franklin, *Digital Control Systems,* 2nd ed., Addison-Wesley, Reading, Mass., 1991.
3. J. G. Bollinger and N. A. Duffie, *Computer Control of Machines and Processes,* Addison-Wesley, Reading, Mass., 1988.
4. P. Dorato, *Robust Control,* IEEE Press, New York, 1987.
5. B. Barmish, "The Robust Root Locus," *Automatica,* vol. 26, no. 2, 1990; pp. 283–92.
6. M. Morari, *Robust Process Control,* Prentice Hall, Englewood Cliffs, N.J., 1991.

7. P. Dorato, "Case Studies in Robust Control Design," *IEEE Proceed. of the Decision and Control Conference,* December 1990; pp. 2030–31.
8. P. Dorato, *Recent Advances in Robust Control,* IEEE Press, New York, 1990.
9. K. Lorell, "Control Technology Test Bed for Large Segmented Reflectors," *IEEE Control Systems,* October 1989; pp. 13–20.
10. H. K. Wickramasinghe, "Scanned-Probe Microscopes," *Scientific American,* October 1989; pp. 98–106.
11. K. Chew, "Digital Control of Errors in Disk Drive Systems," *IEEE Control Systems,* January 1990; pp. 16–19.
12. "The Flying Train Takes Off," *U.S. News and World Report,* July 23, 1990; pp. 52–53.
13. M. Ruth, "Robust Control of an Undersea Vehicle," *Proceed. of the American Automatic Control Conference,* June 1990; pp. 2374–79.
14. T. Yoshikawa, *Foundations of Robotics,* M.I.T. Press, Cambridge, Mass., 1990.
15. M. Mavrovouniotis, *Artificial Intelligence in Process Engineering,* Academic Press, New York, 1990.
16. K. Wright, "The Shape of Things to Go," *Scientific American,* May 1990; pp. 92–101.
17. E. Tan, "Digital Audio Tape for Data Storage," *IEEE Spectrum,* October 1989; pp. 34–38.
18. C. T. Chen, "The Inward Approach in the Design of Control Systems," *IEEE Trans. on Automatic Control,* August 1990; pp. 270–78.
19. J. Walls, "Active Control of the Gamma Ray Imaging Device Experiment," *Proceed. of the American Automatic Control Conference,* 1990; pp. 1082–83.
20. J. Slotine, *Applied Nonlinear Control,* Prentice Hall, Englewood Cliffs, N.J., 1991.
21. S. Boyd, "Linear Controller Design," *Proceed. of the IEEE,* March 1990; pp. 529–72.
22. M. Rimer, "Solutions of the Second Benchmark Control Problem," *IEEE Control Systems,* August 1990; pp. 33–39.
23. K. Capek, *Rossum's Universal Robots;* English edition by P. Selver and N. Playfair, Doubleday, Page, New York, 1923.
24. L. L. Cone, "Skycam: An Aerial Robotic Camera System," *Byte,* October 1985; pp. 122–28.
25. J. Umland, "Magnitude and Symmetric Optimum Criterion for the Design of Linear Control Systems," *IEEE Trans. on Industry Applications,* June 1990; pp. 489–97.
26. P. J. Gawthrop and P. Nomikos, "Automatic Tuning of Commercial PID Controllers," *IEEE Control Systems,* January 1990; pp. 34–41.
27. T. S. Ng, "An Expert System for Shape Diagnosis in a Plate Mill," *IEEE Transactions on Industry Applications,* December 1990; pp. 1057–64.

CHAPTER 12

Design Case Studies

Preview

Engineering design is the central task of the engineer. The design of control systems for practical applications requires the creative configuration and specification of the system and its performance. The control engineer attempts to achieve the best performance available and may be required to reconfigure the system in order to attain the best performance.

In this chapter we use the methods of the preceding chapters to achieve the design of control systems for actual systems. These case studies will illustrate the complexity and challenge of control-system design.

12.1 Engineering Design

Engineering design is the central task of the engineer. *Design is the process of conceiving or inventing the forms, parts, and details of a system to achieve a reasoned purpose.* It is a complex process in which both creativity and analysis play a central role.

Design activity can be thought of as planning for the emergence of a particular product or system. Design is a structured, innovative act whereby the engineer creatively uses knowledge and materials to specify the shape, function, and material content of a system. The design steps are (1) to determine a need arising from the values of various groups, covering the spectrum from public policy makers to the consumer; (2) to specify in detail what the solution to that need must be and to embody these values; (3) to develop and evaluate various alternative solutions to meet these specifications; and (4) to decide which one is to be designed in detail and fabricated.

Of course, the design process is not a linear, step-by-step activity, but rather an iterative, reactive process. It includes several test points and continuous examination of intermediate results. Another factor in realistic design is the limitation of time. Design takes place under imposed schedules, and we eventually settle for a design which may be less than ideal but considered "good enough."

A major challenge for the designer is to write the specifications for the technical product. *Specifications* are statements that explicitly state what the device or product is to be and do. The design of technical systems aims to achieve appropriate design specifications and rests on four characteristics: (1) complexity, (2) tradeoffs, (3) gaps, and (4) risk.

Complexity of design results from the wide range of tools, issues, and knowledge to be used in the process. The large number of factors to be considered illustrates the complexity of the design specification activity, not only in assigning these factors their relative importance in a particular design, but also in giving them substance either in numerical or written form, or both.

The concept of *tradeoff* involves the need to make a judgment about how much of a compromise is made between two conflicting criteria, both of which are desirable. The design process requires an efficient compromise between desirable but conflicting criteria.

In making a technical device there is frequently a design *gap* or void to the extent that the final product does not appear the same as it had been visualized. For example, our image of the problem we're solving is not what appears in written description and ultimately in the specifications. Such gaps are intrinsic in the progression from an abstract idea to its realization.

This inability to make absolutely sure predictions of the performance of a technological object leads to major uncertainties about the actual effects of the designed devices and products. These uncertainties are embodied in the idea of unintended consequences or *risk.* The result is that designing a system is a risk-taking activity.

Complexity, tradeoff, gaps, and risk are inherent in designing new systems and devices. While they can be minimized by considering all the effects of a given design, they are always present in the design process.

Within engineering design, there is a fundamental difference between the two major types of thinking that must take place: engineering analysis and synthesis. Attention is focused on models of the physical systems that are analyzed to provide insight and that point in directions for improvement. On the other hand, *synthesis* is the process by which these new physical configurations are created.

Design is an iterative process that may proceed in many directions before the desired one is found. It is a deliberate process by which a designer creates something new in response to a recognized need while recognizing realistic constraints.

The main approach to the most effective engineering design is parameter analysis and optimization. Parameter analysis is based on (1) identification of the key parameters; (2) generation of the system configuration; and (3) evaluation of how well the configuration meets the needs. These three steps form an iterative loop. Once the key parameters are identified and the configuration synthesized, the designer can optimize the parameters. Typically, the designer strives to identify a limited set of parameters—hopefully less than five—to be adjusted.

12.2 Control-System Design

The design of control systems is a specific example of engineering design. Again, the goal of control engineering design is to obtain the configuration, specifications, and identification of the key parameters of a proposed system to meet an actual need.

The first step in the design process consists of identifying the variables that we desire to control. For example, we may state that the goal is to accurately control the velocity of a motor. The second step is to write the specifications in terms of the accuracy we must attain. This required accuracy of control will then lead to the identification of a sensor to measure the controlled variable.

The designer then proceeds to the first attempt to configure a system that will result in the desired control performance. This system configuration will normally consist of a sensor, the process under control, an actuator, and a controller, as shown in Fig. 12.1. The next step consists of identifying a candidate for the

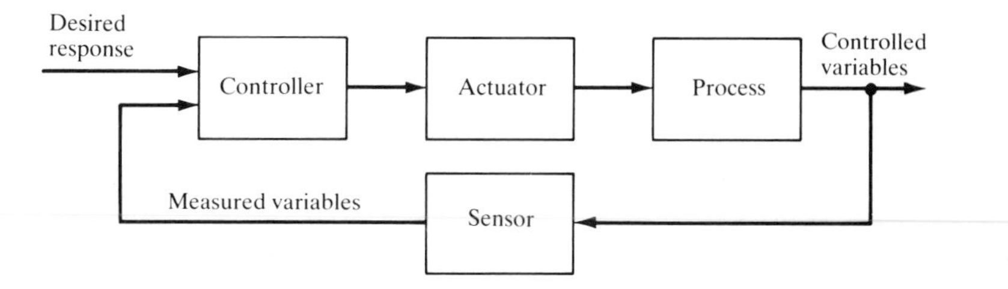

Figure 12.1. The control system configuration.

actuator. This will, of course, depend on the process, but the actuation chosen must be capable of effectively adjusting the performance of the process. For example, if we wish to control the speed of a rotating flywheel, we will select a motor as the actuator. The sensor, in this case, will need to be capable of accurately measuring the speed.

The next step is the selection of a controller, which will often consist of a summing amplifier that will compare the desired response and the actual response and then forward this error-measurement signal to a compensator.

The final step in the design process is the adjustment of the parameters of the system in order to achieve the desired performance. If we can achieve the desired performance by adjusting the parameters, we will finalize the design and proceed to document the results. If not, we will need to establish an improved system configuration and perhaps select an enhanced actuator and sensor. Then we will repeat the design steps until we are able to meet the specifications, or until we decide the specifications are too demanding and should be relaxed. The control system design process is summarized in Fig. 12.2.

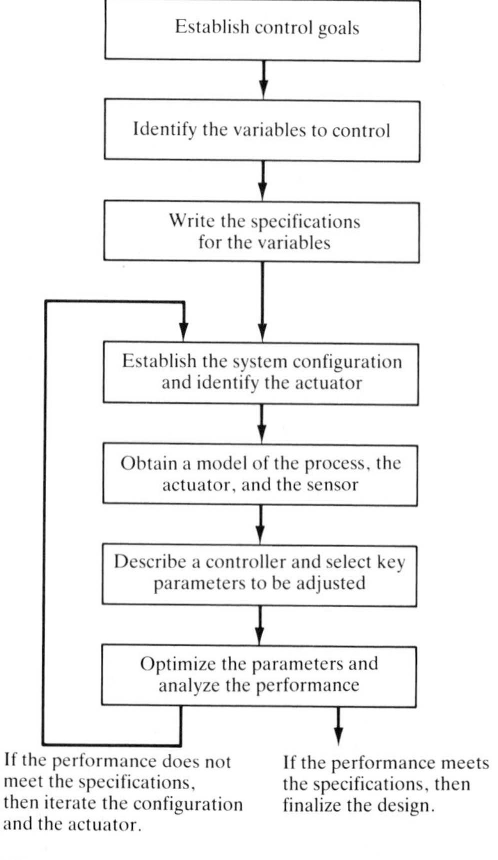

Figure 12.2. The control system design process.

The performance specifications will describe how the closed-loop system should perform and will include (1) good regulation against disturbances; (2) desirable responses to commands; (3) realistic actuator signals; (4) low sensitivities; and (5) robustness.

The controller design problem is as follows: Given a model of the system to be controlled (including its sensors and actuators) and a set of design goals, find a suitable controller, or determine that none exists.

12.3 The Design of an X-Y Plotter

Many physical phenomena are characterized by parameters that are transient or slowly varying. If these changes can be recorded, they can be examined at leisure and stored for future reference or comparison. To accomplish this recording, a number of electro-mechanical instruments have been developed, among them the X-Y recorder. In this instrument, the displacement along the X-axis represents a variable of interest or time, and the displacement along the Y-axis varies as a function of yet another variable [10].

Such recorders can be found in many laboratories recording experimental data such as changes in temperature, variations in transducer output levels, and stress versus applied strain, to name just a few. The HP 7090 plotting system is shown in Fig. 12.3.

The purpose of a plotter is to accurately follow the input signal as it varies. We will consider the design of the movement of one axis, since the movement

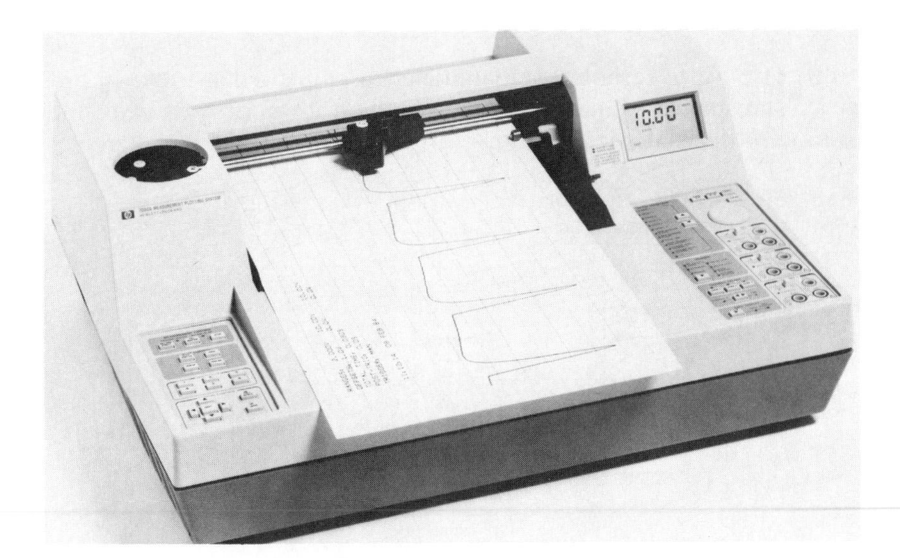

Figure 12.3. The HP 7090A Measurement Plotting System©. Copyright 1986, Hewlett-Packard Co. Reproduced with permission.

dynamics of both axes are identical. Thus we will strive to control very accurately the position and the movement of the pen as it follows the input signal.

In order to achieve accurate results, our goal is to achieve (1) a step response with an overshoot of less than 5% and a settling time less than 0.5 second, and (2) a percentage steady-state error for a step equal to zero. If we achieve these specifications, we will have a fast and accurate response.

Since we wish to move the pen, we select a dc motor as the actuator. The feedback sensor will be a 500-line optical encoder. By detecting all state changes of the two-channel quadrature output of the encoder, 2000 encoder counts per revolution of the motor shaft can be detected. This yields an encoder resolution of 0.001 inch at the pen tip. The encoder is mounted on the shaft of the motor. Since the encoder provides digital data, it is compared with the input signal by using a microprocessor. Next, we propose to use the difference signal calculated by the microprocessor as the error signal, and then use the microprocessor to calculate the necessary algorithm to obtain the designed compensator. The output of the compensator is then converted to an analog signal that will drive the motor.

The model of the feedback position-control system is shown in Fig. 12.4. Since the microprocessor calculation speed is very fast compared to the rate of change of the encoder and input signals, we assume that the continuous signal model is very accurate.

The model for the motor and pen carriage is

$$G(s) = \frac{1}{s(s+10)(s+1000)}, \tag{12.1}$$

and our initial attempt at a specification of a compensator is to use a simple gain so that

$$G_c(s) = K.$$

In this case we have only one parameter to adjust—that is, K. In order to achieve a fast response, we have to adjust K so that it will provide two dominant s-plane roots with a damping ratio of 0.707, which will result in a step response

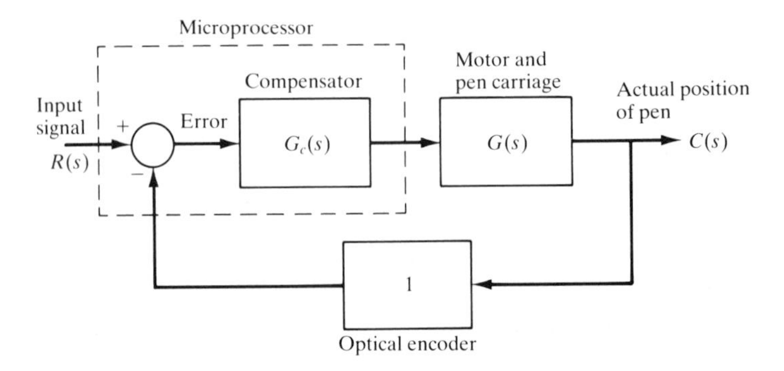

Figure 12.4. Model of the pen-plotter control system.

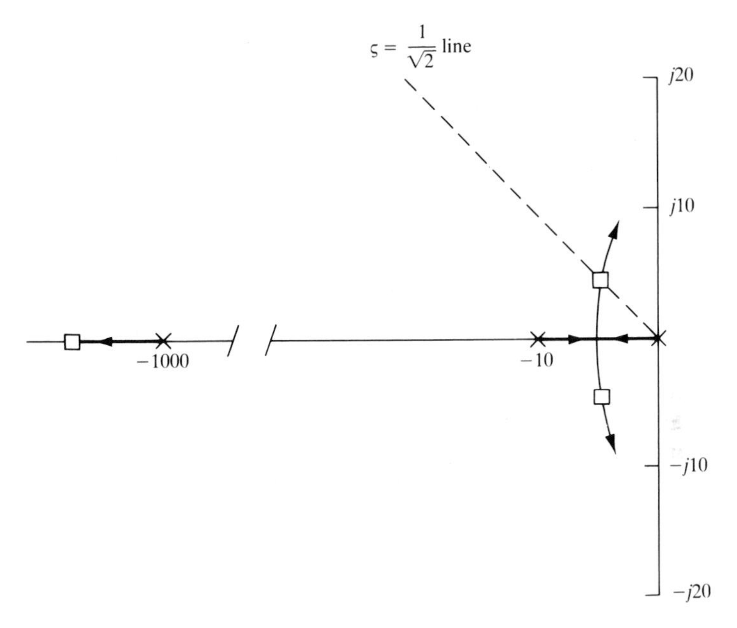

Figure 12.5. Root locus for the pen plotter, showing the roots with a damping ratio of $1\sqrt{2}$. The dominant roots are $s = -4.9 \pm j4.9$.

overshoot of about 4.5 percent. A sketch of the root locus (note the break in the real axis) is shown in Fig. 12.5.

Adjusting the gain to $K = 47{,}200$, we obtain a system with an overshoot of 3.6% to a step input and a settling time of 0.8 second. Since the transfer function has a pole at the origin, we have a steady-state error of zero for a step input.

This system does not meet our specifications, so we select a compensator that will reduce the settling time. Let us select a lead compensator so that

$$G_c(s) = \frac{K\alpha(s+z)}{(s+p)}, \tag{12.2}$$

where $p = \alpha z$. Let us use the method of Section 10.5, which selects the phase-lead compensator on the s-plane. Therefore we place the zero at $s = -20$ and determine the location of the pole, p, that will place the dominant roots on the line that has the damping ratio of $1\sqrt{2}$. Thus we find that $p = 60$ and that $\alpha = 3$, so that

$$G(s)G_c(s) = \frac{142{,}600(s+20)}{s(s+10)(s+60)(s+1000)}. \tag{12.3}$$

Obtaining the actual step response, we determined that the percent overshoot was 2% and that the settling time was 0.35 seconds, which meet the specifications.

The third design approach is to recognize that the encoder can be used to

Table 12.1. Results for Three Designs

System	Step Response: Percent Overshoot	Settling Time (milliseconds)
Gain adjustment	3.6	800
Gain and lead compensator adjustment	2.0	350
Gain adjustment plus velocity signal multiplied by gain K_2	4.3	8

generate a velocity signal by counting the rate at which encoder lines pass by a point. Thus we can use the microprocessor. Since the position signal and the velocity signal are available, we may describe the compensator as

$$G_c(s) = K_1 + K_2 s, \tag{12.4}$$

where K_1 is the gain for the error signal and K_2 is the gain for the velocity signal. Then we may write

$$G(s)G_c(s) = \frac{K_2(s + K_1/K_2)}{s(s + 10)(s + 1000)}.$$

If we set $K_1/K_2 = 10$, we cancel the pole at $s = -10$ and obtain

$$G(s)G_c(s) = \frac{K_2}{s(s + 1000)}.$$

The characteristic equation for this system is

$$s^2 + 1000s + K_2 = 0, \tag{12.5}$$

and we desire that ζ is $1/\sqrt{2}$. Noting that $2\zeta\omega_n = 1000$, we have $\omega_n = 707$ and $K_2 = \omega_n^2$. Therefore we obtain $K_2 = 5 \times 10^5$ and the compensated system is

$$G_c G(s) = \frac{5 \times 10^5}{s(s + 1000)}. \tag{12.6}$$

The response of this system will provide an overshoot of 4.3% and a settling time of 8 milliseconds.

The results for the three approaches to system design are compared in Table 12.1. Clearly, the best design uses the velocity feedback, and this is the actual design adopted for the HP 7090A.

12.4 The Design of a Space Telescope Control System

Scientists have proposed the operation of a space vehicle as a space-based research laboratory and test bed for equipment to be used on a manned space station. The industrial space facility (ISF) would remain in space and the astro-

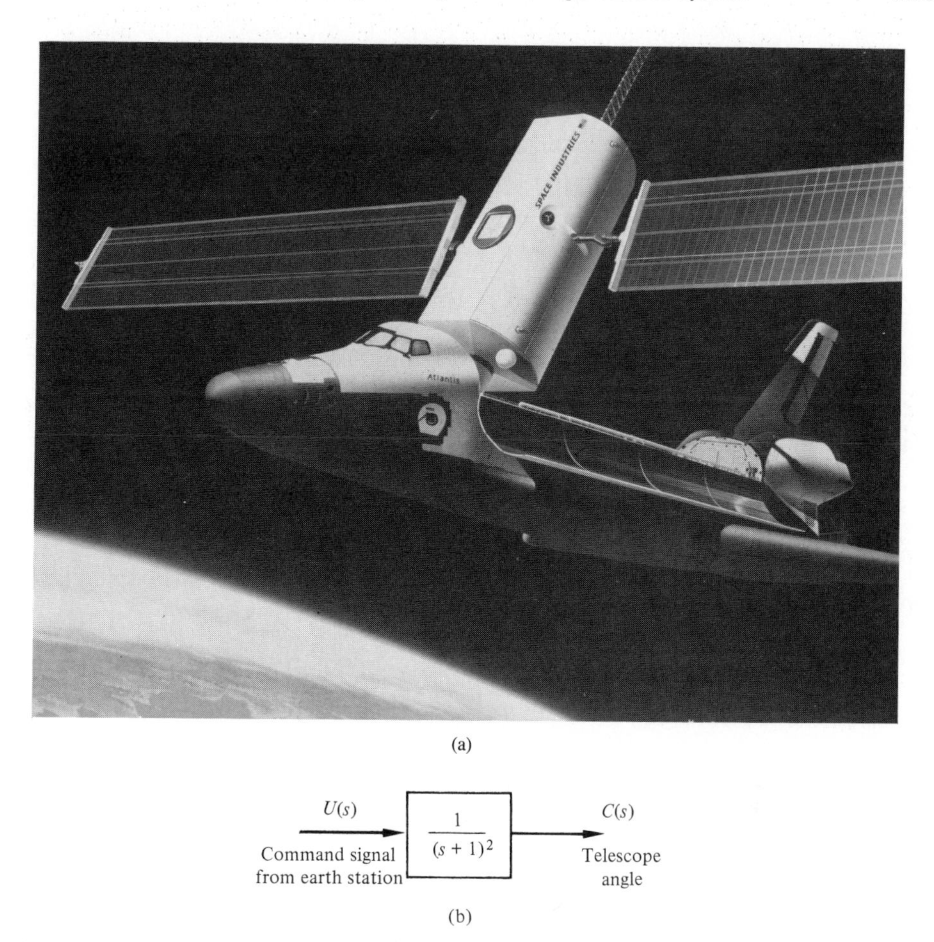

Figure 12.6. (a) The industrial space facility and the space shuttle shown in docking position. Courtesy of Space Industries, Inc. (b) The model of a low-power actuator and telescope.

nauts would be able to use it only when the shuttle is attached, as shown in Fig. 12.6(a) [14]. The ISF will be the first permanent, human-operated commercial space facility designed for R & D, testing, and, eventually, processing in the space environment.

We will consider an experiment operated in space but controlled from earth. The goal is to manipulate and position a small telescope to accurately point at a planet. The goal is to have a steady-state error equal to zero, while maintaining a fast response to a step with an overshoot of less than 5%. The actuator chosen is a low-power actuator, and the model of the combined actuator and telescope is shown in Fig. 12.6(b). The command signal is received from an earth station with a delay of $\pi/16$ seconds. A sensor will measure the pointing direction of the telescope accurately. However, this measurement is relayed back to earth with a delay

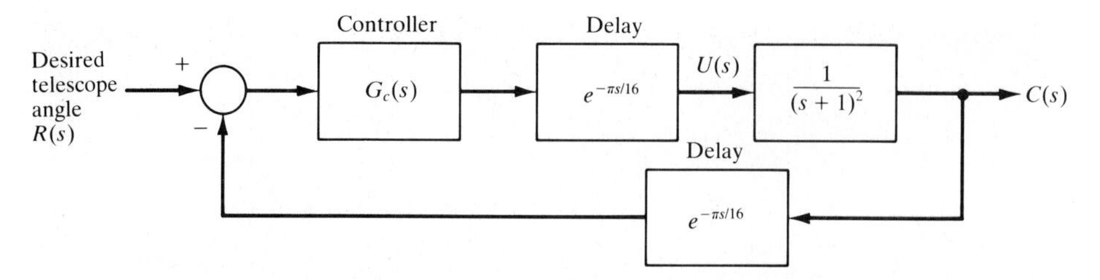

Figure 12.7. Feedback control system for the telescope experiment.

of $\pi/16$ seconds. Thus the total transfer function of the telescope, actuator, sensor, and round-trip delay (Fig. 12.7) is

$$G(s) = \frac{e^{-s\pi/8}}{(s + 1)^2}. \tag{12.7}$$

We propose a controller that is the three-term controller described in Section 10.6 and Section 11.5, where

$$\begin{aligned} G_c(s) &= K_1 + \frac{K_2}{s} + K_3 s \\ &= \frac{K_1 s + K_2 + K_3 s^2}{s}. \end{aligned} \tag{12.8}$$

Clearly, the use of only the proportional term will not be acceptable since we require a steady-state error of zero for a step input. Thus we must use a finite value of K_2 and thus we may elect to use either proportional plus integral control (PI) or proportional plus integral plus derivative control (PID).

We will first try PI control, so that

$$G_c(s) = K_1 + \frac{K_2}{s} = \frac{K_1 s + K_2}{s}. \tag{12.9}$$

Since we have a pure delay, e^{-sT}, we use the frequency response methods for the design process. Thus we will translate the overshoot specification to the frequency domain. If we have two dominant characteristic roots, the overshoot to a step is 5% when $\zeta = 0.7$ or the phase margin is about 70°.

If we choose $K_1 = 0.022$ and $K_2 = 0.22$, we have

$$G_c(s)G(s) = \frac{0.22(0.1s + 1)e^{-s\pi/8}}{s(s + 1)^2} \tag{12.10}$$

and the Bode diagram is shown in Fig. 12.8. The location of the zero at $s = 10$ was chosen to add a phase lead angle so that the desired phase margin is attained. An iterative procedure of using the Control System Design Program (CSDP) or equivalent yields a series of trials for K_1 and K_2 until the desired phase margin is achieved. Note that we have achieved a phase margin of about 63 degrees. The

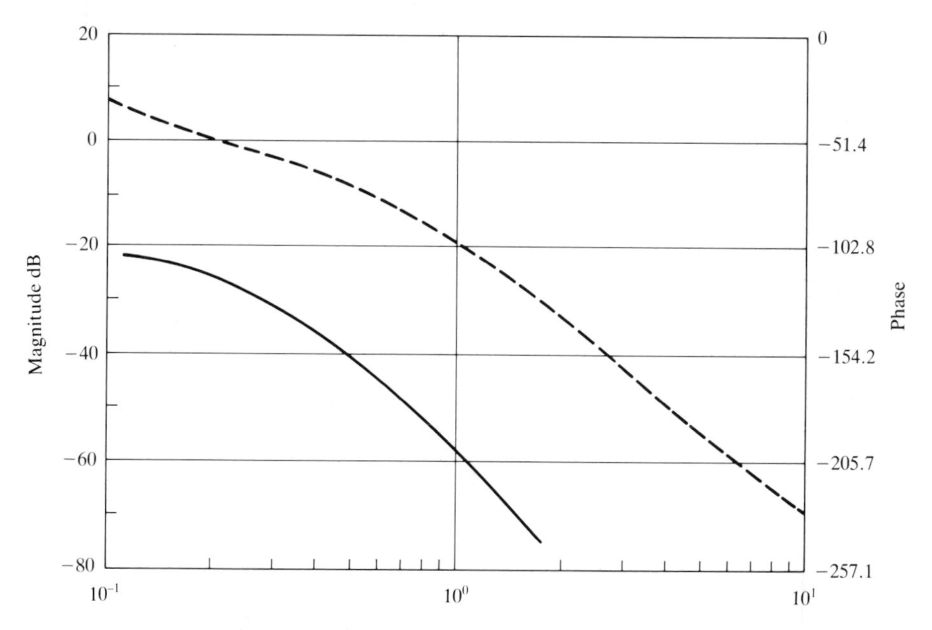

Figure 12.8. Bode diagram for the system with the PI controller.

actual step response was plotted using CSDP, and it is determined that the overshoot was 4.7% with a setting time of 16 seconds.

The proportional plus integral plus derivative controller is

$$G_c(s) = \frac{K_1 s + K_2 + K_3 s^2}{s}. \tag{12.11}$$

Now, we have three parameters to vary in order to achieve the desired phase margin. If we select, after some iteration, $K_1 = 0.8$, $K_2 = 0.5$, and $K_3 = 10^{-3}$, we obtain a phase margin of 64 degrees. The percentage overshoot is 3.7% and the setting time is 5.8 seconds. Perhaps the easiest way to select the gain constants is to initially let K_3 be a small, but nonzero, number and $K_1 = K_2 = 0$. Then plot the frequency response using CSDP. In this case, we choose $K_3 = 10^{-3}$ and obtain a Bode plot. We then use $K_1 \approx K_2$ and iterate to obtain the appropriate values of these unspecified gains.

The performance of the PI and the PID compensated systems is recorded in Table 12.2. Clearly, the PID controller is the most desirable.

Table 12.2. Step Response of the Space Telescope for Two Controllers

	Steady-State Error	**Percent Overshoot**	**Settling Time (seconds)**
PI controller	0	4.7	16.0
PID controller	0	3.7	5.8

12.5 The Design of a Robot Control System

The concept of robot replication is relatively easy to grasp. The central idea is that robots replicate themselves and develop a factory that automatically produces robots. An example of a robot replication facility is shown in Fig. 12.9. In order to achieve the rapid and accurate control of a robot, it is important to keep the robot arm stiff and yet lightweight.

The specifications for controlling the motion of a lightweight, flexible arm are (1) a settling time of less than two seconds; (2) a percent overshoot of less than 10% for a step input; and (3) a steady-state error of zero for a step input.

The block diagram of the proposed system with a controller is shown in Fig. 12.10. The configuration proposes the use of velocity feedback as well as the use of a controller $G_c(s)$. Since the robot is quite light and flexible, the transfer function of the arm is

$$\frac{C(s)}{U(s)} = \left(\frac{1}{s^2}\right) G(s)$$

and

$$G(s) = \frac{(s^2 + 4s + 10{,}004)(s^2 + 12s + 90{,}036)}{(s + 10)(s^2 + 2s + 2501)(s^2 + 6s + 22{,}509)}. \tag{12.12}$$

Therefore the complex zeros are located at

$$s = -2 \pm j100$$

and

$$s = -6 \pm j300.$$

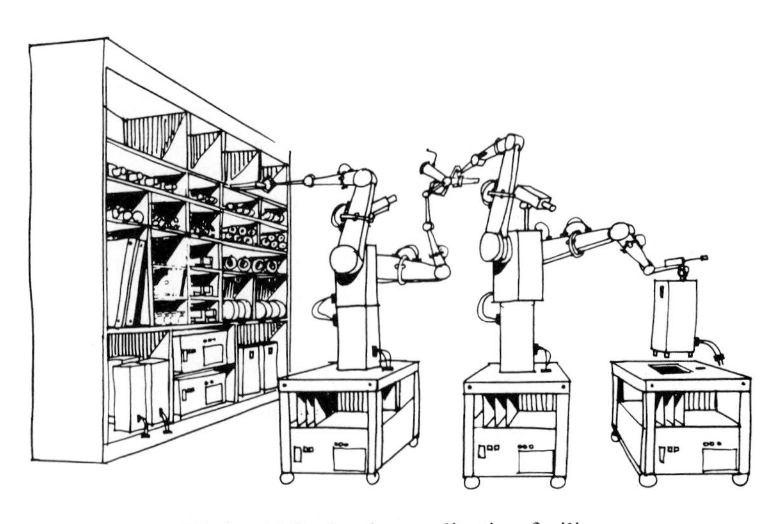

Figure 12.9. A robot replication facility.

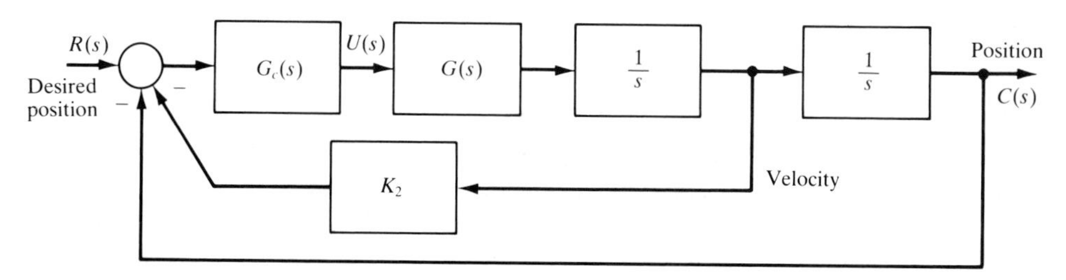

Figure 12.10. Proposed configuration for control of the lightweight robot arm.

The complex poles are located at

$$s = -1 \pm j50$$

and

$$s = -3 \pm j150.$$

A sketch of the root locus when $K_2 = 0$ and the controller is an adjustable gain, K_1, is shown in Fig. 12.11. Clearly, the system is unstable since two roots of the characteristic equation appear in the right-hand s-plane for $K_1 > 0$.

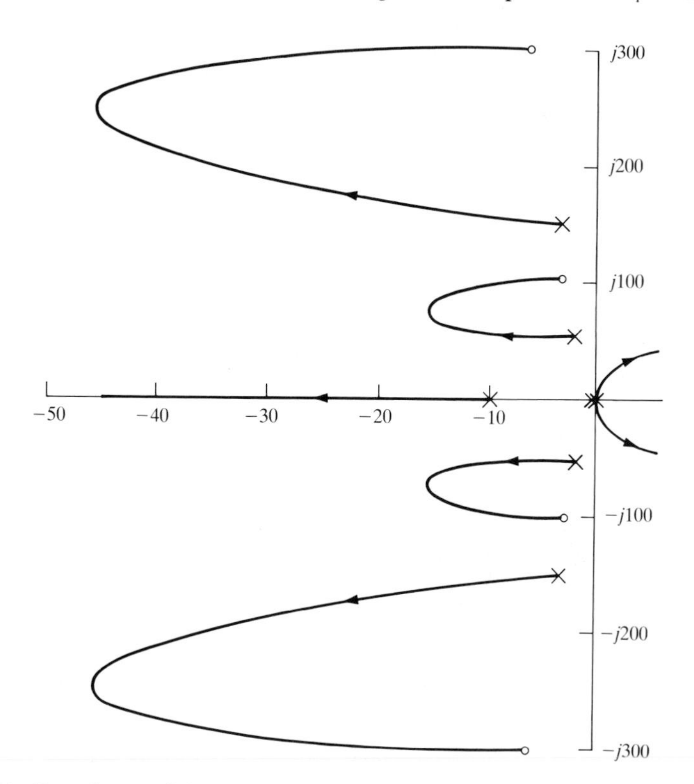

Figure 12.11. Root locus of the system if $K_2 = 0$ and K_1 is varied from $K_1 = 0$ to $K_1 = \infty$.

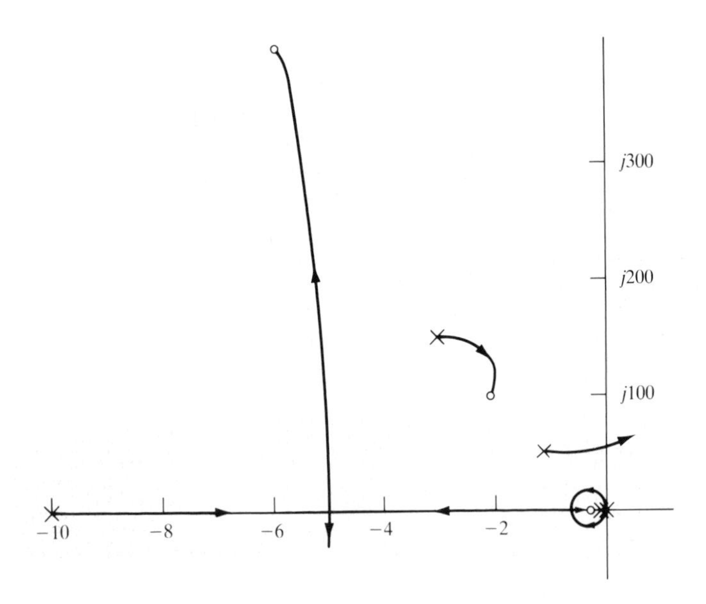

Figure 12.12. Root locus for the robot controller with a zero inserted at $s = -0.2$.

It is clear that we need to introduce the use of velocity feedback by setting K_2 to a positive magnitude. Then we have $H(s) = (1 + K_2 s)$ and therefore the loop transfer function is

$$\left(\frac{1}{s^2}\right) G(s)H(s) = \frac{K_1 K_2 \left(s + \dfrac{1}{K_2}\right)(s^2 + 4s + 10{,}004)(s^2 + 12s + 90{,}036)}{s^2(s + 10)(s^2 + 2s + 2501)(s^2 + 6s + 22{,}509)}$$

where K_1 is the gain of $G_c(s)$. We now have available two parameters, K_1 and K_2, that we may adjust. We select $5 < K_2 < 10$ in order to place the adjustable zero near the origin.

When $K_2 = 5$ and K_1 is varied, we obtain the root locus sketched in Fig. 12.12. When $K_1 = 0.8$ and $K_2 = 5$, we obtain a step response with a percent overshoot of 12% and a settling time of 1.8 seconds. This is the optimum achievable response. If one tries $K_2 = 7$ or $K_2 = 4$, the overshoot will be larger than desired. Therefore we have achieved the best performance with this system. If we desired to continue the design process, we would use a lead network for $G_c(s)$ in addition to retaining the velocity feedback with $K_2 = 5$.

One possible selection of a lead network is

$$G_c(s) = \frac{K_1(s + z)}{(s + p)}. \tag{12.13}$$

If one selects $z = 1$ and $p = 5$, then when $K_1 = 5$ we obtain a step response with an overshoot of 8% and a settling time of 1.6 seconds.

Figure 12.13. Hermies II, an upgraded version of an 18-degree-of-freedom robot developed at the Center for Engineering Systems Advanced Research (Oak Ridge National Laboratory). Both robotic manipulators are now operational and the unit has a phased array of 20 Polaroid transceivers for position and object sensing. (Courtesy of Oak Ridge National Laboratory, Oak Ridge, Tennessee.)

12.6 The Design of a Mobile Robot Control System

An 18-degree-of-freedom mobile robot is shown in Fig. 12.13. The objective is to accurately control the position of one of the wheels of the robot. Each wheel control will be identical. The goal is to obtain a fast response with a rapid rise time and settling time to a step command while not exceeding an overshoot of 5%.

The specifications are then (1) a percent overshoot equal to 5%; (2) minimum settling time and rise time. Rise time is defined as the time to reach the magnitude of the command and is illustrated in Fig. 4.7 by T_R.

In order to configure the system, we choose a power amplifier and motor so that the system is described by Fig. 12.14. Then, obtaining the transfer function of the motor and power amplifier, we have

$$G(s) = \frac{1}{s(s+10)(s+20)}. \tag{12.14}$$

First we select the controller as a simple gain, K, in order to determine the response that can be achieved without a compensator. Plotting the root locus, we find that when $K = 700$ the dominant complex roots have a damping ratio of 0.707 and we expect a 5% overshoot. Then using the Control System Design Program, we find that the overshoot is 5%, the rise time is 0.48 seconds, and the settling time is 1.12 seconds. These values are recorded as item 1 in Table 12.3.

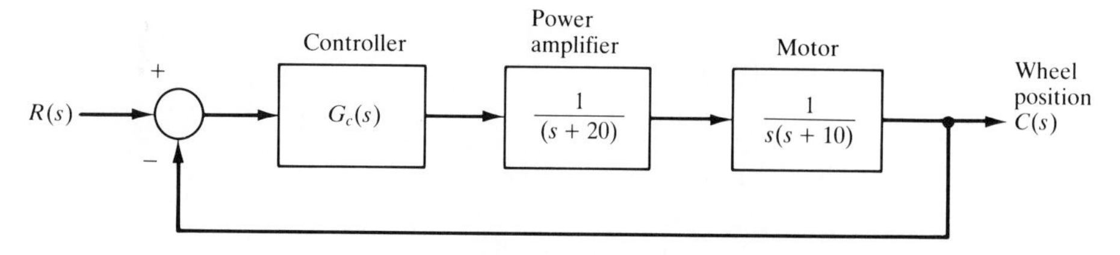

Figure 12.14. Model of the wheel control for a mobile robot.

The next step is to introduce a lead compensator, so that

$$G_c(s) = \frac{K(s + z)}{(s + p)}. \tag{12.15}$$

We will select the zero at $s = -11$ so that the complex roots near the origin dominate. Then using the method of Section 10.5, we find that we require the pole at $s = -62$. Evaluating the gain at the roots shown in Fig. 12.15, we find that $K = 8000$. Then the step response has a rise time of 0.25 seconds and a settling time of 0.60 seconds. This is an improved response and we could finalize this system as acceptable.

However, if we wish to try to speed up the response, we could select a compensator with two lead elements as follows:

$$G_c(s) = \frac{K(s + z)^2}{(s + p)^2}.$$

Again placing the zero at $s = -11$, we find that two poles will be required at $s = -24$, so that

$$G_c(s) = \frac{K(s + 11)^2}{(s + 24)^2} \tag{12.16}$$

and $K = 8200$ in order to obtain dominant roots with a ζ of 0.707. The sketch of the root locus is shown in Fig. 12.16. The dominant complex roots have a ζ of 0.707 when $K = 8200$. The overshoot is 5.1%, rise time is 0.20 seconds, and the settling time is 0.38 seconds as recorded as item 3 in Table 12.3. The double lead compensator gives the fastest response, and we will choose this compensator for the final system.

Table 12.3. Performance for Three Controllers

Compensator $G_c(s)$	**K**	**Percent Overshoot**	**Settling Time (seconds)**	**Rise Time (seconds)**
1. K	700	5.0	1.12	0.40
2. $K(s + 11)/(s + 62)$	8000	5.0	0.60	0.25
3. $K(s + 11)^2/(s + 24)^2$	8200	5.1	0.38	0.20

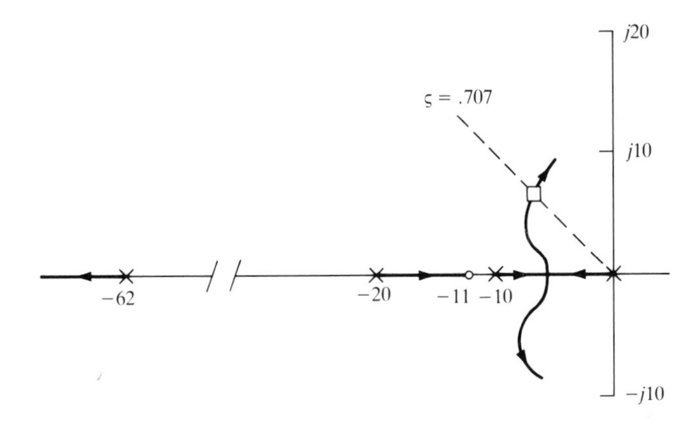

Figure 12.15. Root locus for a lead compensator $G_c(s)$.

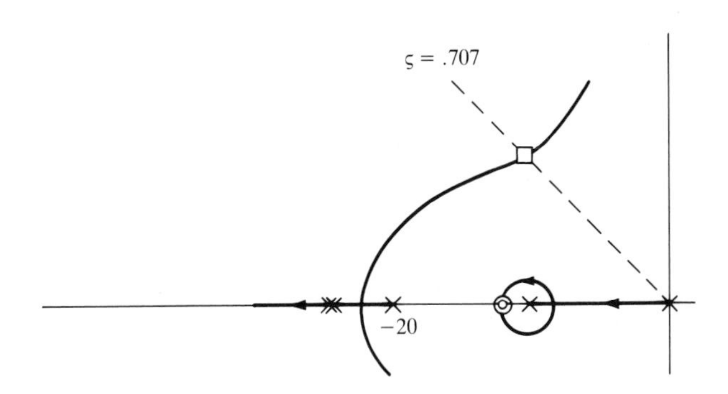

Figure 12.16. Root locus for two lead compensators.

12.7 The Design of an ROV Control System

The design of a remotely operated vehicle (ROV) for undersea exploration as shown in Fig. 12.17 requires the control of the heading (direction) of the vehicle. The system will use an electric motor and propeller to drive the ROV, and an electrically controlled rudder will provide the steering.

The configuration for the control system is shown in Fig. 12.18. Our goal is to obtain a rapid response to a step command with 25% overshoot or less and a steady-state error of less than 3.33% for a ramp command.

Therefore we desire a velocity constant, K_v, for the ramp command of

$$K_v = \frac{1}{0.0333} = 30.$$

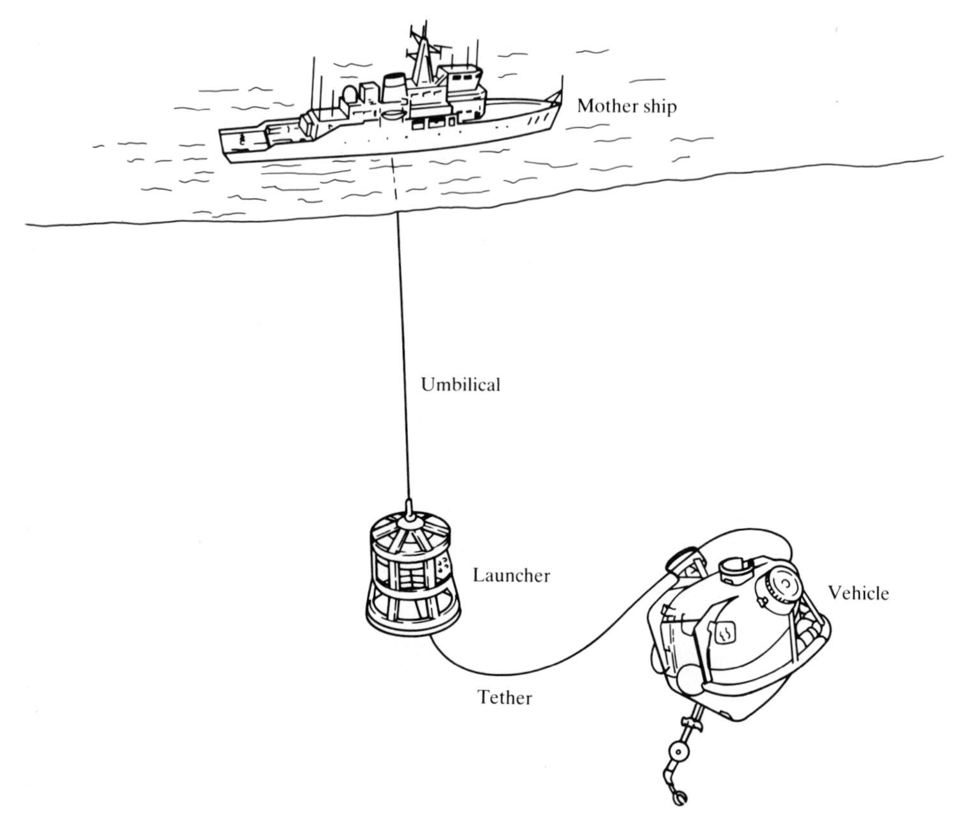

Figure 12.17. A remotely operated vehicle (ROV) for undersea exploration.

If we select a compensator as a simple gain K, we have

$$\begin{aligned} K_v &= \lim_{s \to 0} s\{G_c(s)G(s)\} \\ &= \lim_{s \to 0} s\left\{\frac{K}{s(s+5)^2}\right\} \\ &= \frac{K}{25}. \end{aligned}$$

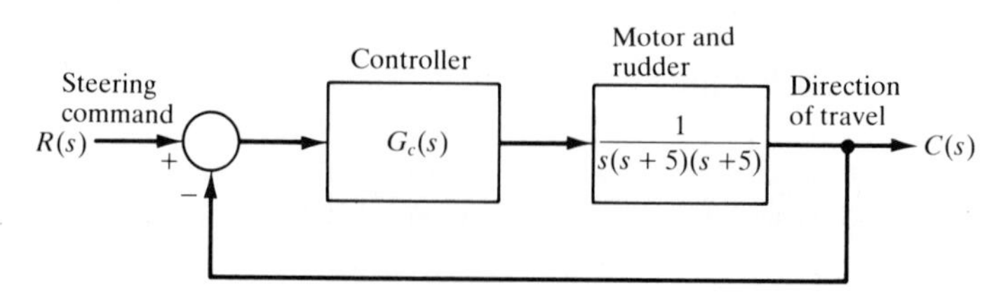

Figure 12.18. Steering control system for the ROV.

This would require $K = 750$ in order to achieve the desired K_v of 30. However, using the Routh criteria with the characteristic equation, it may be shown that the system has two roots on the imaginary axis when $K = 250$. Therefore we cannot attain the desired K_v with a simple gain K.

Let us try to design a lag network compensator so that we may achieve the desired K_v and the desired step response. Therefore we propose that

$$G_c(s) = \frac{K(s + z)}{(s + p)}, \tag{12.17}$$

where $z > p$. The specification for the overshoot of 25% or less will lead us to select a damping ratio for the complex roots of 0.45 (see Fig. 4.8). We expect the real root introduced by the compensator to provide some damping to the system, and we allow for that by selecting $\zeta = 0.45$ for the complex roots. We will use the root locus method of Section 10.7 to obtain the desired lag compensator. Later in this section we also use the frequency response method of Section 10.8 to obtain the desired compensator.

First we draw the root locus of the uncompensated system and determine that $K = 53.5$ when the complex roots have $\zeta = 0.45$. Since K_v is desired as 30, we have the actual K_v as

$$K_{v\text{uncomp.}} = \frac{53.5}{25} = 2.14.$$

Therefore

$$\alpha = \left|\frac{z}{p}\right| = \frac{K_{v\text{comp.}}}{K_{v\text{uncomp.}}} = \frac{30}{2.14} = 14.0.$$

We will use $\alpha = 14.5$ to allow for some margin of error. Then we choose $z = 0.10$ and therefore

$$p = \frac{z}{\alpha} = \frac{0.1}{14.5} = 0.0069.$$

Thus the compensated system is

$$G_c(s)G(s) = \frac{53.5(s + 0.1)}{s(s + 5)^2(s + 0.0069)}. \tag{12.18}$$

Using the Control System Design Program, we obtain the step response and determine that the percent overshoot is 25%, the rise time is 1.2 seconds, and the settling time is 7 seconds. The results are summarized as item 2 in Table 12.4.

Now let us use the frequency response method of design for the lag compensator discussed in Section 10.8. We will use the Bode diagram and attempt to obtain a phase margin of about 60 degrees, which should result in a step response with an overshoot of 25%. First we plot the Bode diagram with

$$G_cG(s) = \frac{K_v}{s(0.2s + 1)^2}, \tag{12.19}$$

Table 12.4. Performance Results

Compensator, $G_c(s)$	Percent Overshoot	Rise Time (seconds)	Settling Time (seconds)	K_v
1. $G_c = 53.5$	0	—	90	2.14
2. $G_c(s) = \dfrac{53.5(s+0.1)}{(s+0.0069)}$	25	1.2	7.0	30
3. $G_c(s) = \dfrac{30(3s+1)}{(170s+1)}$	25	2.0	12.0	30

where $K_v = 30$. Then we find that the system has a phase margin of $-15°$, as shown in Fig. 12.19. We examine the diagram and determine that the frequency where the phase is $-130°$ is $\omega_c' = 3.3$. This is the new desired crossover frequency. Then set the zero at $\omega_z = \omega_c'/10 = 0.33$. Measuring the attenuation required at ω_z, we find that 35 db attenuation is necessary. Then

$$20 \log \alpha = 35$$

and therefore $\alpha = 56.2$. Using this α, we have

$$\omega_p = \frac{\omega_z}{\alpha} = \frac{0.33}{56.2} = 0.0059.$$

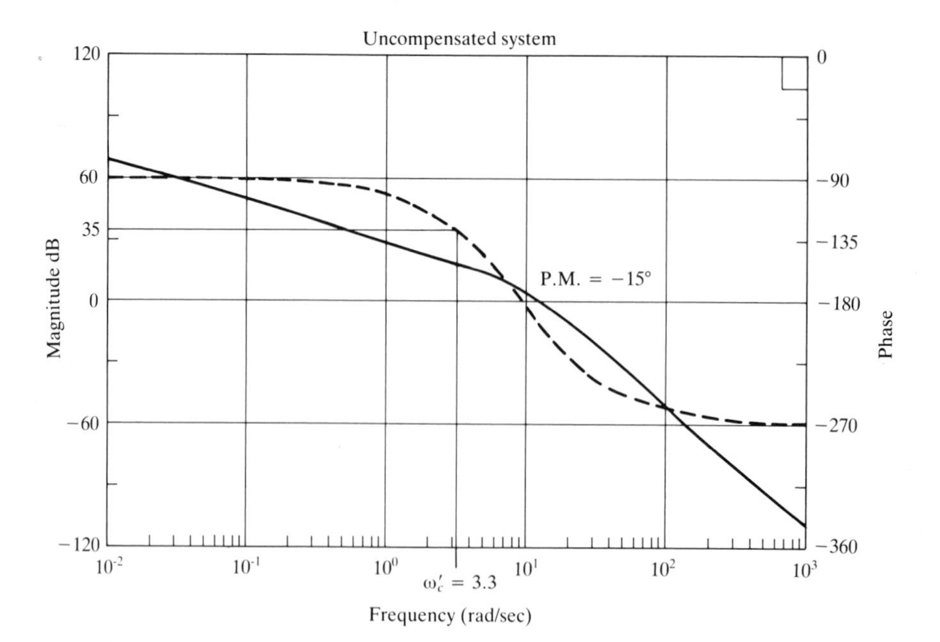

Figure 12.19. Bode diagram to determine lag compensator.

The compensated system is then

$$G_c(s)G(s) = \frac{30(3s + 1)}{s(0.2s + 1)^2(170s + 1)}. \quad (12.20)$$

Using CSDP to obtain the step response, we have an overshoot of 25% and a settling time of 12 seconds. The results are summarized as item 3 in Table 12.4.

It is clear that a lag network compensator will provide the desired K_v for the system as well as a reasonable step response. The design using the root locus procedure resulting in item 2 of Table 12.4 will be selected since it provides a shorter rise time.

12.8 The Design of a Solar-Powered Racing Car Motor Control System

A group of 14 cars completed the race of solar-powered cars in Australia in 1987. The *Sunraycer,* shown in Fig. 12.20, won the 1867-mile Pentax World Solar Challenge race [18]. Powered by 9500 solar cells, the electric auto completed the race with an average speed of 41.6 miles per hour.

Let us consider the design of the ac motor drive for a solar-powered vehicle. A block diagram of the drive system is shown in Fig. 12.21, where the disturbance incorporates the effects of wind, component change, and road variations. The goal is to maintain low steady-state error and low overshoot for step changes in the input command, $R(s)$, while minimizing the effects of a step disturbance $D(s) = 1/s$.

We will attempt to maintain an overshoot of less than 10% when

$$G(s) = \frac{1}{(s + 1)^2}. \quad (12.21)$$

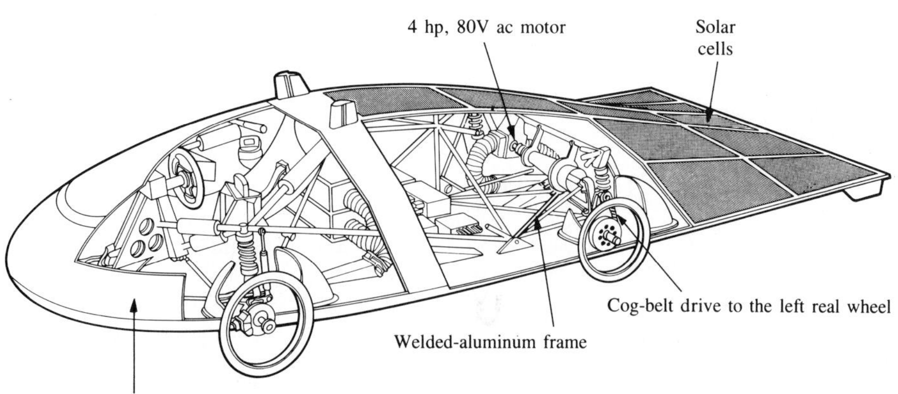

Figure 12.20. The *Sunraycer,* a solar-powered electric car.

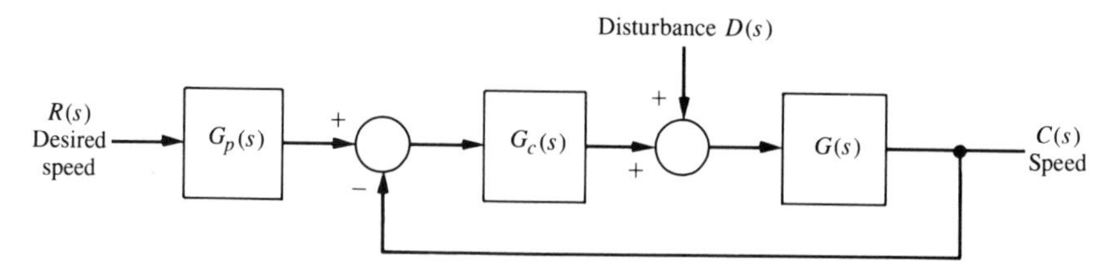

Figure 12.21. Speed control system for a solar-powered electric car.

We will try to design a system by considering $G_c(s)$ as a gain, a lead network, a lag network, and a PID controller.

First, we let $G_c(s) = K$ and determine $K = 6.11$ when the maximum allowable overshoot is obtained. The results of the response to $R(s)$ and $D(s)$ are summarized in Table 12.5. This simple system has a significant steady-state error for a step command and the effect of the disturbance is significant.

The next approach is to design a lead network with

$$G_c(s) = \frac{s + z}{s + p}, \tag{12.22}$$

where α cannot exceed 20. We select

$$G_c(s) = \frac{391(s + 1.1)}{(s + 20)}$$

where the zero of $G_c(s)$ is placed to the left of the poles of $G(s)$. We achieve significant improvement in terms of the steady-state error and the effect of the disturbance, as summarized in Table 12.5.

If we design a lag network, we obtain

$$G_c(s) = \frac{5.4(s + 0.1)}{(s + 0.005)}. \tag{12.23}$$

Table 12.5. Design Results for Solar-Powered Auto

$G_c(s)$		Step Input, $R(s) = 1/s$: Percent Over shoot (%)	Step Input: Steady-State Error (%)	Step Input: Settling Time (seconds)	Step Disturbance $D(s) = 1/s$: $c(\infty)$ Steady-State $c(t)$ (%)	Step Disturbance: $c(t)$ maximum (%)
K	$K = 6.11$	10	14.1	4.1	14	18
Lead	$K = 391$	10	4.1	0.4	4.4	4.4
Lag	$K = 5.4$	10	1.3	19.5	0.9	19
PID	$K = 86$	10	0	0.2	0	0.06
PID with filter	$K = 15.5$	2	0	0.75	0	0.4

The lag network reduces the steady-state error and the steady-state effect of the disturbance, as summarized in Table 12.5. Unfortunately, the settling time and the maximum value of $c(t)$ due to the disturbance are both increased significantly above those achieved with the lead network design.

Now, let us try to achieve the design of a PID controller. We will use the ITAE optimum coefficients provided in Table 4.5 for a step input. Then, the controller is

$$G_c(s) = \frac{K_2(s^2 + as + b)}{s}, \tag{12.24}$$

where $a = K_1/K_2$ and $b = K_3/K_2$. We select $\omega_n = 10$ for the third-order optimum equation

$$s^3 + 1.75\omega_n s^2 + 2.15\omega_n^2 s + \omega_n^3.$$

The optimum coefficients are $K_1 = 214$, $K_2 = 15.5$, and $K_3 = 1000$. Therefore we have

$$G_c(s) = \frac{15.5(s^2 + 13.8s + 64.5)}{s}. \tag{12.25}$$

Without a prefilter $[G_p(s) = 1]$, this system gives an overshoot to a step of 30%. In order to reduce the overshoot to 10%, we increase the gain of $G_c(s)$ to 86 so that

$$G_c(s) = \frac{86(s^2 + 13.8s + 64.5)}{s}. \tag{12.26}$$

The effect of increasing the gain to 86 can be illustrated by Fig. 12.22. The steady-state error is then zero and the settling time is reduced 0.2 seconds. The effect of the disturbance is reduced to 0.06% of the magnitude of the disturbance, as recorded in Table 12.5.

An alternative is to utilize a prefilter $G_p(s)$, as shown in Fig. 12.21, with $G_c(s)$ as given in Eq. (12.25). Using the prefilter

$$G_p(s) = \frac{64.5}{(s^2 + 13.8s + 64.5)}, \tag{12.27}$$

we have the optimum closed-loop transfer function for the ITAE index as

$$T(s) = \frac{1000}{s^3 + 17.5s^2 + 215s + 1000}. \tag{12.28}$$

The overshoot is then 2%, and the maximum value of $c(t)$ for a unit step disturbance is 0.4%. The PID controller with a prefilter provides an excellent system response.

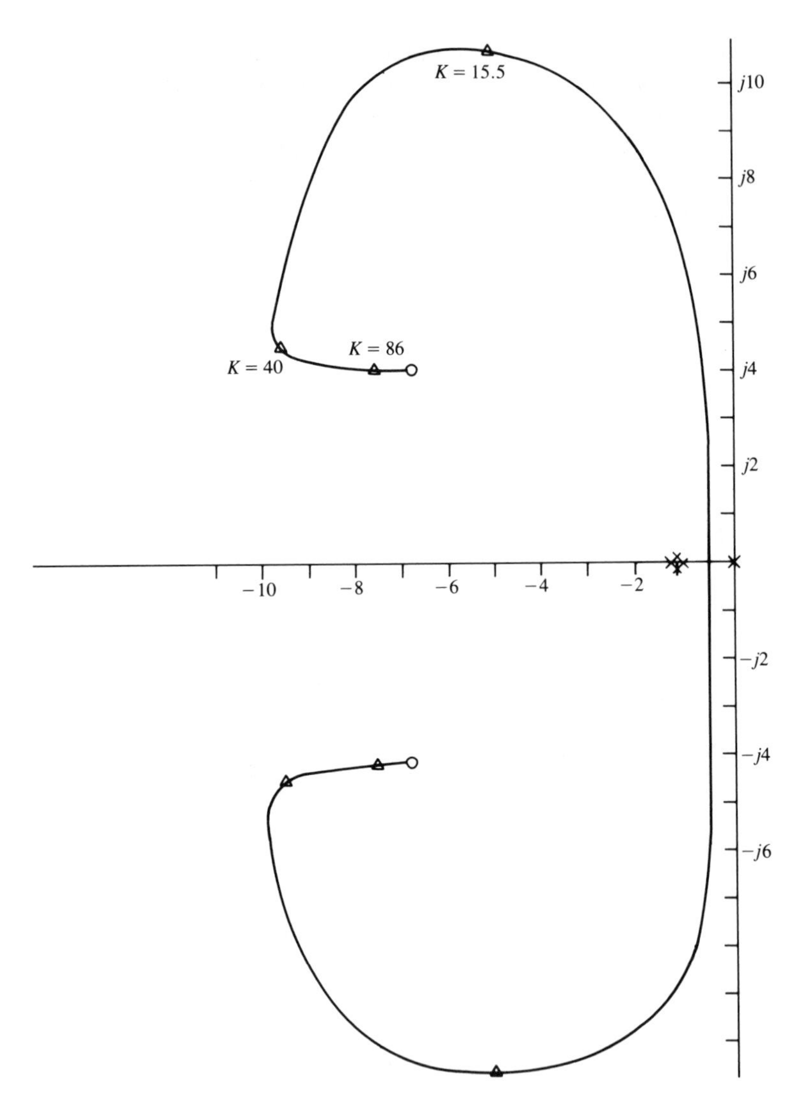

Figure 12.22. Root locus design for solar-powered racing car.

12.9 Summary

Engineering design is the central task of the engineer. The design of control systems incorporates obtaining the configuration, specifications, and identification of the key parameters of the proposed system to meet an actual need. Then the designer attempts to optimize the parameters to achieve the best system performance.

The designer of a control system needs to choose an actuator and a sensor and to determine a linear model of these devices as well as the process under control. The next step is to select a controller and to optimize the parameters of the system. If a suitable performance cannot be achieved, the designer will need to reconfigure the system and perhaps select an improved actuator and sensor.

In this chapter we have illustrated the design process by examining a number of case studies. The creativity of the designer will be challenged by the complexity of the required design. The utility of a number of control schemes was illustrated by a series of case studies.

Design Problems 

DP12.1. Consider the device for the magnetic levitation of a steel ball, as shown in Fig. DP12.1(a) and (b). Obtain a design that will provide a stable response where the ball will remain within 10% of its desired position.

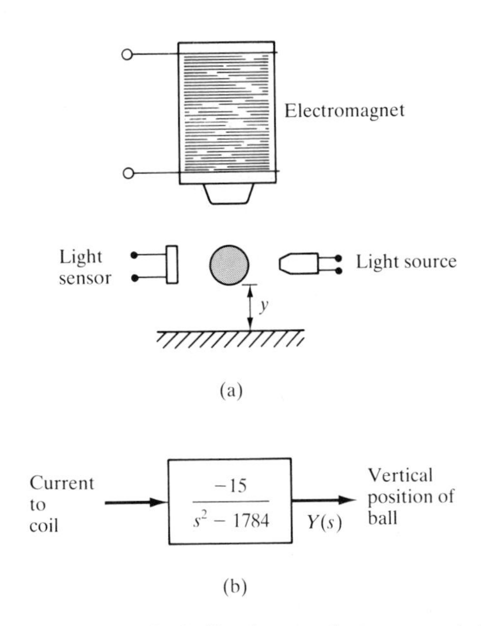

Figure DP12.1. (a) The levitation of a ball using an electromagnet. (b) The model of the electromagnet and the ball.

P12.2. A photovoltaic system is mounted on a space station in order to develop the power for the station. The photovoltaic panels should follow the sun with good accuracy in order to maximize the energy from the panels. The system uses a dc motor so that the transfer function of the panel mount and the motor is

$$G(s) = \frac{1}{s(s + 25)}.$$

We will select a controller $G_c(s)$ assuming that an optical sensor is available to accurately track the sun's position, and thus $H(s) = 1$.

The goal is to design $G_c(s)$ so that (1) the percent overshoot to a step is less than 7% and (2) the steady-state error to a ramp input should be less than or equal to 1%. Determine the best phase-lead controller.

DP12.3. The control of the fuel-to-air ratio in an automobile carburetor became of prime importance in the 1980s as automakers worked to reduce exhaust-pollution emissions. Thus auto engine designers turned to the feedback control of the fuel-to-air ratio. A sensor was placed in the exhaust stream and used as an input to a controller. The controller actually adjusts the orifice that controls the flow of fuel into the engine.

Select the devices and develop a linear model for the entire system. Assume that the sensor measures the actual fuel-to-air ratio with a delay. With this model, determine the optimum controller when a system is desired with a zero steady-state error to a step input and the overshoot for a step command is less than 10%.

DP12.4. High-performance tape transport systems are designed with a small capstan to pull the tape past the read/write heads and with take-up reels turned by dc motors. The tape is to be controlled at speeds up to 200 inches per second, with start-up as fast as possible, while preventing permanent distortion of the tape. Since we wish to control the speed and the tension of the tape, we will use a dc tachometer for the speed sensor and a potentiometer for the position sensor. We will use a dc motor for the actuator. Then the linear model for the system is a feedback system with $H(s) = 1$ and

$$\frac{C(s)}{E(s)} = G(s) = \frac{K(s + 4000)}{s(s + 1000)(s + 3000)(s + p_1)(s + p_1^*)},$$

where $p_1 = +2000 + j2000$ and $C(s)$ is position.

The specifications for the system are (1) settling time of less than 12 ms, (2) an overshoot to a step position command of less than 10%, and (3) a steady-state velocity error of less than .5%. Determine a compensator scheme to achieve these stringent specifications.

DP12.5. Electromagnetic suspension systems for aircushioned trains are known as magnetic levitation (maglev) trains. One maglev train uses a superconducting magnet system, as shown in Fig. DP12.5, on the next page [19]. It uses superconducting coils and the levitation distance $x(t)$ is inherently unstable. The model of the levitation is

$$\frac{X(s)}{V(s)} = \frac{K}{(s\tau_1 + 1)(s^2 - \omega_1^2)},$$

where $V(s)$ is the coil voltage; τ_1 is the magnet time constant; and ω_1 is the natural frequency. The system uses a position sensor with a negligible time constant. A train traveling at 300 km/h would have $\tau_1 = 1$ second and $\omega_1 = 100$ rad/s. Determine a controller that can maintain steady, accurate levitation when disturbances occur along the railway.

DP12.6. The four-wheel-steering automobile has several benefits. The system gives the driver a greater degree of control over the automobile. The driver gets a more forgiving vehicle over a wide variety of conditions. The system enables the driver to make sharp, smooth lane transitions. It also prevents yaw, which is the swaying of the rear end during sudden movements. Furthermore, the four-wheel-steering system gives a car increased maneuverability. This enables the driver to park the car in extremely tight quarters. Finally, with additional closed-loop computer operating systems, a car could be prevented from sliding out of control in abnormal icy or wet road conditions.

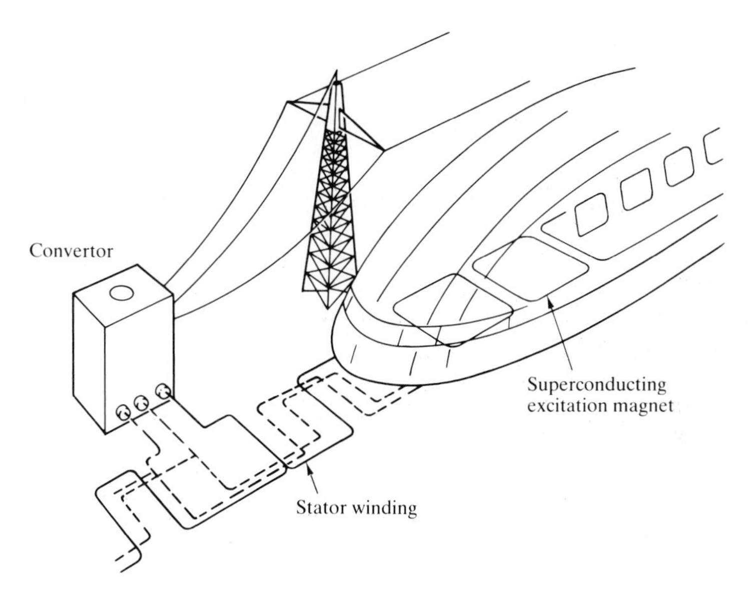

Figure DP12.5. Maglev train.

The system works by moving the rear wheels relative to the front-wheel-steering angle. The control system takes information about the front wheels' steering angle and passes it to the actuator in the back. This actuator then moves the rear wheels appropriately.

When the rear wheels are given a steering angle relative to the front ones, the vehicle can vary its lateral acceleration response according to the transfer function

$$G(s) = K\frac{1 + (1 + \lambda)T_1 s + (1 + \lambda)T_2 s^2}{1 + (2\zeta/\omega_n)s + (1/\omega_n^2)s^2},$$

where $\lambda = 2q/(1 - q)$ and q is the ratio of rear wheel angle to front wheel steering angle [17]. We will assume that $T_1 = T_2 = 1$ second and $\omega_n = 4$. Design a unity feedback system for $G(s)$, selecting an appropriate set of parameters (λ, K, ζ) so that the steering control response is rapid and yet will yield modest overshoot characteristics. In addition, q must be between 0 and 1.

DP12.7. A bridge played a leading role in the Broadway musical *Starlight Express* [12]. The biggest star of the show is an eight-ton automatically controlled bridge that hangs 90 feet above the stage and extends up to 60 feet across it. The bridge is under the control of a computer and is shown in Fig. DP12.7.

Starlight Express dramatizes a young boy's dream of racing railroad trains. Twenty-seven actors portray the trains by zooming around on roller skates. The skaters crisscross the stage by means of the bridge, which connects the sides of the three-tiered set. The bridge is in motion almost constantly throughout the show, operating like a gantry robot with six axes of motion. Its ascents and descents are controlled by hydraulics; 360-degree rotation, tilting, and extension are the work of dc drives. The entire bridge moves upstage and downstage on the tracks located just below the ceiling.

The movements of the bridge coincide precisely with musical cues in the score and the skaters' movements. When the bridge is lowered and its sides extended to connect with either side of the stage, gates open just in time for the skaters to get across.

Figure DP12.7.

Commands for moving the bridge require position as well as velocity accuracy. A computer is available for controlling all six axes of motion independently. Assume that the rotational motion of the bridge is controlled while the other axes are fixed. Then obtain a model of the bridge motion and its control. Design a three-term (PID) controller that will provide 1-mm position accuracy with a rapid response to both step and velocity commands. Establish a set of specifications and proceed to the determination of a suitable controller.

DP12.8. The automatic control of an airplane is one example where multiple-variable feedback methods are required. In this system, the attitude of an aircraft is controlled by three sets of surfaces: elevators, rudder, and ailerons, as shown in Fig. DP12.8(a). By manipulating these surfaces, a pilot can set the aircraft on a desired flight path.

An autopilot, which will be considered here, is an automatic control system that controls the roll angle ϕ by adjusting aileron surfaces. The deflection of the aileron surfaces by an angle Θ generates a torque due to air pressure on these surfaces. This causes a rolling motion of the aircraft. The aileron surfaces are controlled by a hydraulic actuator with a transfer function b/s.

The actual roll angle ϕ is measured and compared with the input. The difference between the desired roll angle ϕ_d and the actual angle ϕ will drive the hydraulic actuator, which in turn adjusts the deflection of the aileron surface.

A simplified model where the rolling motion can be considered independent of other motions is assumed, and its block diagram is shown in Fig. DP12.8(b). Assume that $K_1 = 1$ and that the roll rate ϕ is fed back using a rate gyro. Establish a set of specifications and select the two parameters K_a and K_2 for the best performance.

DP12.9. The goal is to design an elevator control system so that the elevator moves from floor to floor rapidly and stops accurately at the selected floor (Fig. DP12.9). The elevator can contain from one to three occupants. However the weight of the elevator is greater than the weight of the occupants and you may assume that the elevator weighs 1000

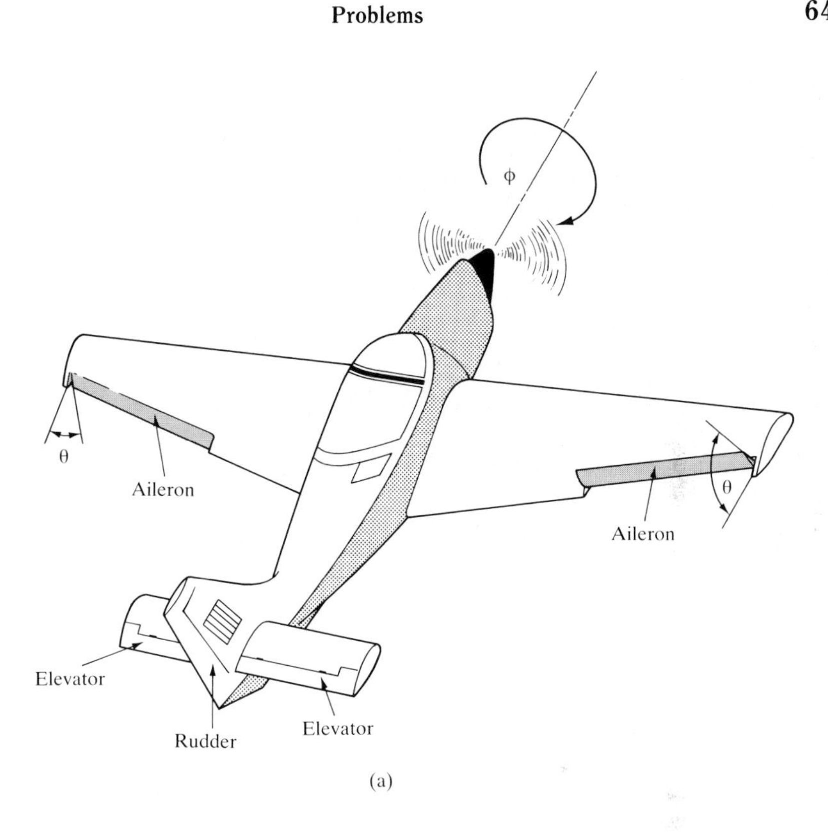

(a)

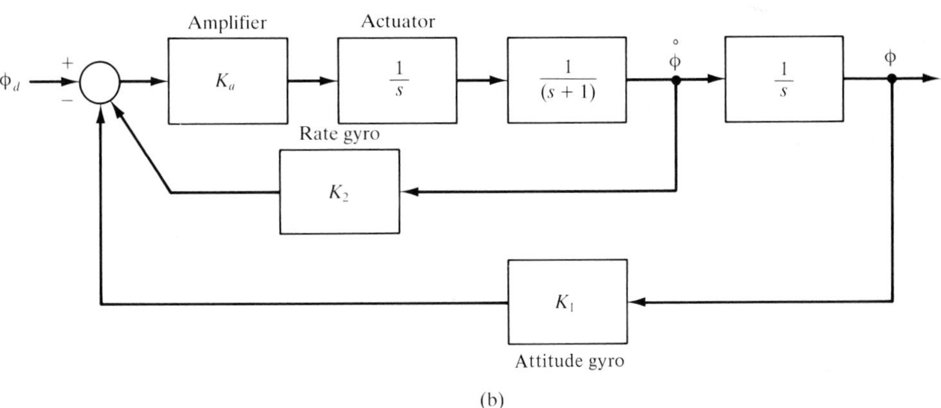

(b)

Figure DP12.8. (a) An airplane with a set of ailerons. (b) The block diagram for controlling the roll rate of the airplane.

pounds and each occupant weighs 150 pounds. Design a system to accurately control the elevator to within one centimeter. Assume that the large dc motor is field controlled. Assume that the time constant of the motor and load is one second and the time constant of the power amplifier driving the motor is one-half second.

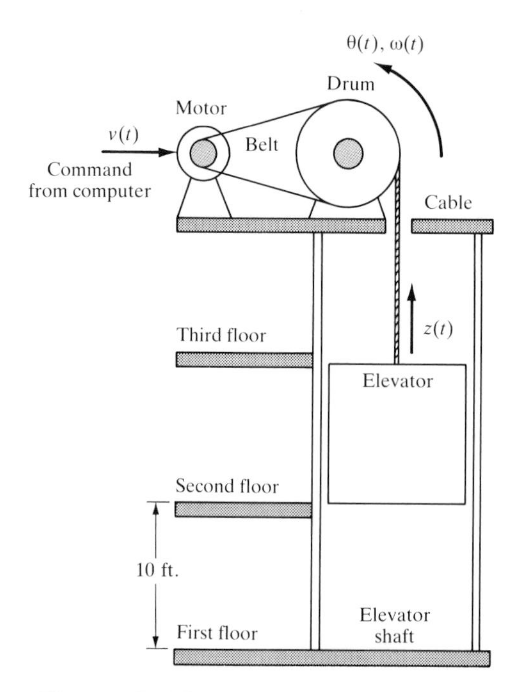

Figure DP12.9. Elevator position control.

DP12.10. The efficiency of a diesel engine is very sensitive to the speed of rotation. Thus we wish to control the speed of the diesel engine that drives the electric motors of a diesel-electric locomotive for large railroad trains. The locomotive is powered by dc motors located on each of the axles. Consider the model of the electric drive shown in Fig. DP12.10 where the throttle position is set by moving the input potentiometer.

The controlled speed ω_0 is sensed by a tachometer, which supplies a feedback voltage v_0. The tachometer may be belt driven from the motor shaft. An electronic amplifier amplifies the error $(v_r - v_0)$ between the reference and feedback voltage signals and provides a voltage v_f that is supplied to the field winding of a dc generator.

The generator is run at a constant speed ω_d by the diesel engine and generates a voltage v_g that is supplied to the armature of a dc motor. The motor is armature controlled with a fixed current i supplied to its field. As a result, the motor produces a torque T and drives the load connected to its shaft so that the controlled speed ω_0 tends to equal the command speed ω_r.

It is known that the generator constant k_g is 100 and the motor constant is $K_m = 10$. The back emf constant is $K_b = 31/50$. The constants for the motor are $J = 1$, $f = 1$, $L_a = 0.2$ and $R_a = 1$. The generator has a field resistance, $R_f = 1$ and a field inductance $L_f = 0.1$. Also the tachometer has a gain of $K_t = 1$ and the generator has $L_g = 0.1$ and $R_g = 1$. All three constants are in the appropriate SI units.

(a) Develop a linear model for the system and analyze the performance of the system as K is adjusted. What is the steady-state error for a fixed throttle position so that $v_r = A$? (b) Include the effect of a load torque disturbance, T_d, in the model of the system and determine the effect of a load disturbance torque on the speed, $\omega(s)$. For the design of part (a), what is the steady-state error introduced by the disturbance? (c) Select the three state

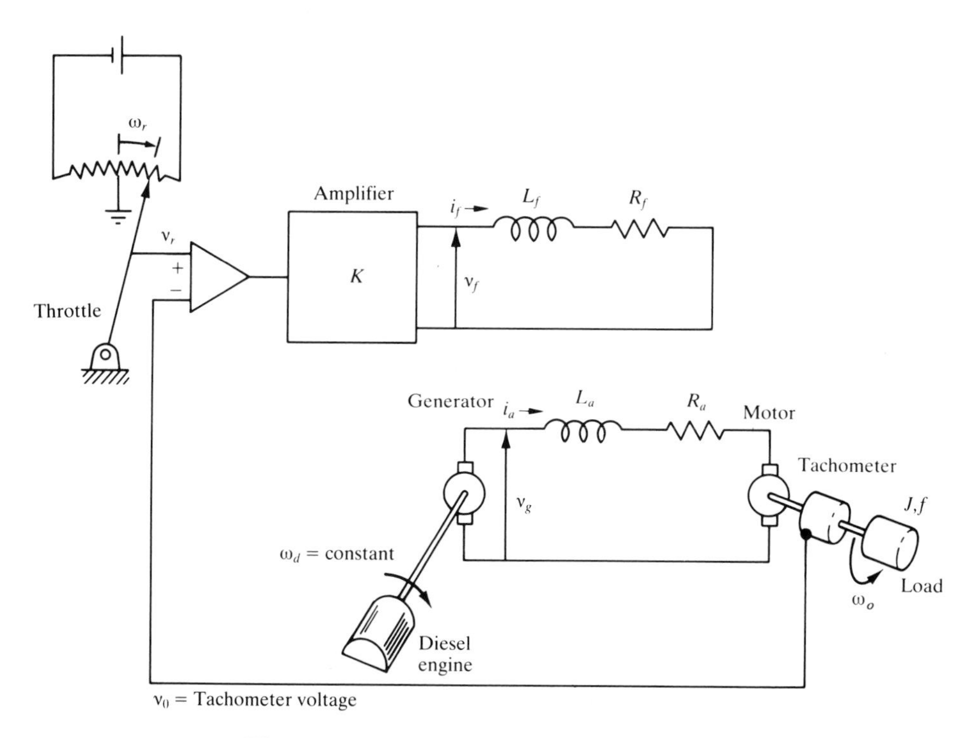

Figure DP12.10. Diesel-electric locomotive.

variables that are readily measured and obtain the state vector differential equation. Then design a state-variable feedback system to achieve favorable performance.

DP12.11. A high performance helicopter has a model shown in Fig. DP.12.11. The goal is to control the pitch angle, θ, of the helicopter by adjusting the rotor angle, δ.

The equations of motion of the helicopter are

$$\frac{d^2\theta}{dt^2} = -\sigma^1 \frac{d\theta}{dt} - \alpha^1 \frac{dx}{dt} + n\delta$$

$$\frac{d^2x}{dt^2} = g\theta - \alpha^2 \frac{d\theta}{dt} - \sigma^2 \frac{dx}{dt} + g\delta$$

where x is the translation in the horizontal direction. For a military, high-performance helicopter we find that

$$\begin{aligned} \sigma^1 &= 0.415 & \alpha^2 &= 1.43 \\ \sigma^2 &= 0.0198 & n &= 6.27 \\ \alpha^1 &= 0.0111 & g &= 9.8 \end{aligned}$$

all in appropriate SI units.

Find (a) the state variable representation of this system and (b) the transfer function representation for $\theta(s)/\delta(s)$. (c) Use state variable feedback to achieve adequate performances for the controlled system. (d) Use a lead-lag compensator and an appropriate amplifier gain to achieve an adequate response for the system in the transfer function form.

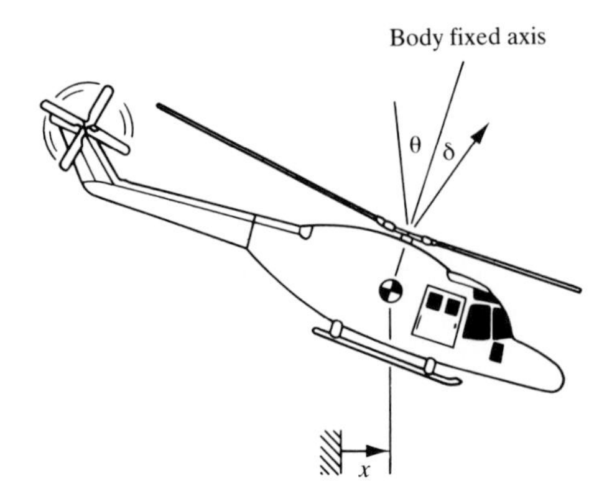

Figure DP12.11. Helicopter pitch angle, θ, control.

Desired specifications include (1) a steady-state for an input step command for $\theta_d(s)$, the desired pitch angle, less than 20% of the input step magnitude; (2) an overshoot for a step input command less than 20% and (3) a settling time for a step command of less than 1.5 sec.

DP12.12. A pilot crane control is shown in Fig. DP12.12(a). The trolley is moved by an input $F(t)$ in order to control $x(t)$ and $\phi(t)$ [13]. The model of the pilot crane control is shown in Fig. DP12.12(b). Design a controller that will achieve control of the desired variables. First use $G_c(s) = K$ and then use more advanced controllers.

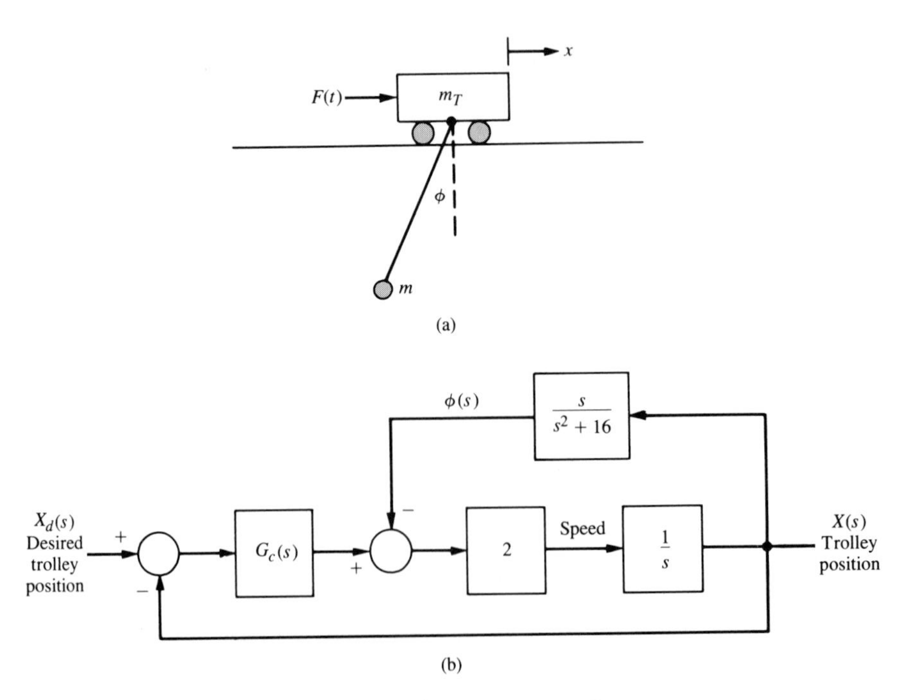

Figure DP12.12. Pilot crane control system.

DP12.13. Electronic systems currently make up about 6% of a car's value. That figure will climb to 20% by the year 2000 as antilock brakes, active suspensions, and other computer-dependent technologies move into full production. Much of the added computing power will be used for new technology for smart cars and smart roads, or IVHS (intelligent vehicle/highway systems) [11, 18]. The term refers to a varied assortment of electronics that provide real-time information on accidents, congestion, routing, and roadside services to drivers and traffic controllers. IVHS also encompasses devices that would make vehicles more autonomous: collision-avoidance systems and lane-tracking technology that alert drivers to impending disaster or allow a car to drive itself.

An example of an automated highway system is shown in Fig. DP12.13(a). A position control system for maintaining the distance between vehicles is shown in Fig. DP12.13(b). Design a controller so that the steady-state error for a ramp input is less than 25% of the input magnitude, A, of the ramp $R(s) = A/s^2$. The response to a step command should have an overshoot of less than 3% and a settling time of less than 1.5 sec.

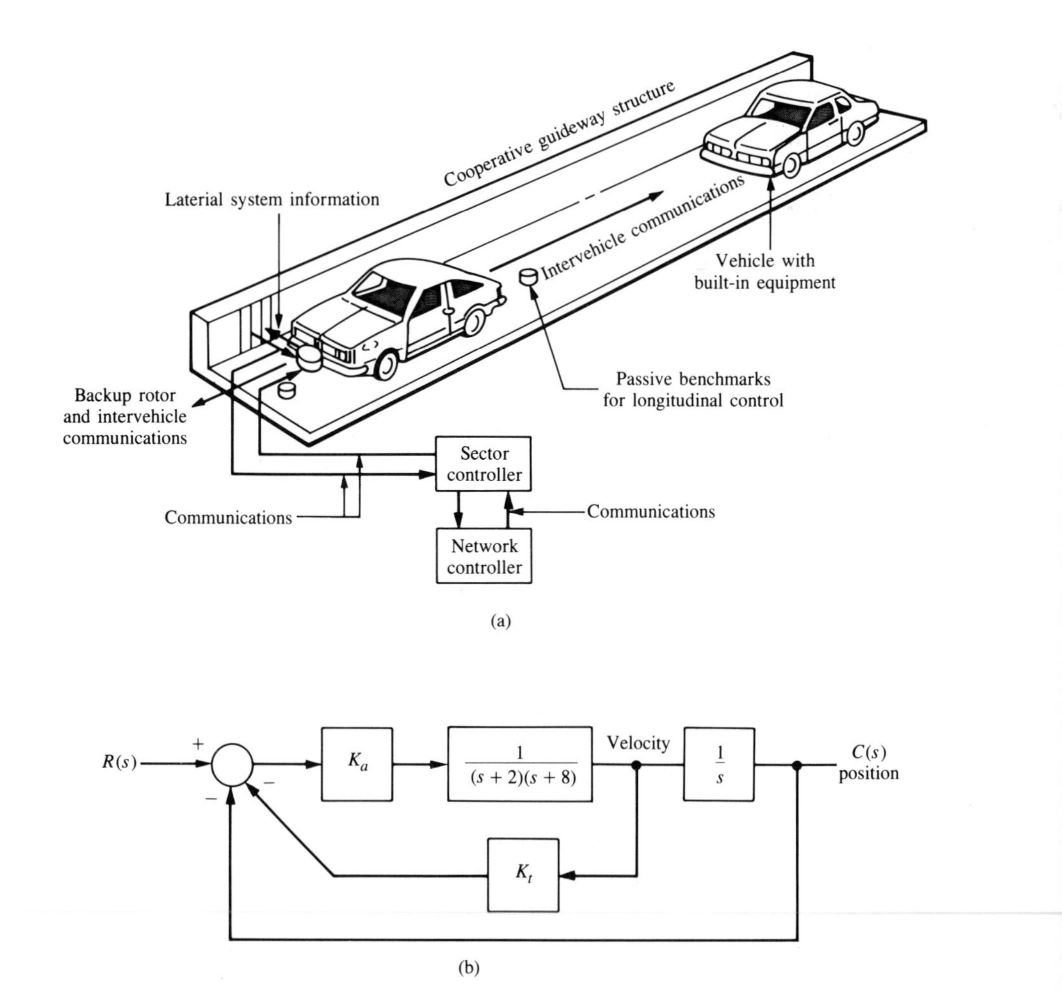

Figure DP12.13. Vehicle distance control system.

DP12.14. The past several years have witnessed a significant engine model building activity in the automotive industry in a category referred to as "control-oriented" or "control design" models. These models contain representations of the throttle body, engine pumping phenomena, induction process dynamics, fuel system, engine torque generation, and rotating inertia.

The control of the fuel-to-air ratio in an automobile carburetor became of prime importance in the 1980s as automakers worked to reduce exhaust-pollution emissions. Thus auto engine designers turned to the feedback control of the fuel-to-air ratio. Operation of an engine at or near a particular air-to-fuel ratio requires management of both air and fuel flow into the manifold system. The fuel command is considered the input and the engine speed as the output [9, 11].

The block diagram of the system is shown in Fig. DP12.14, where $T = 0.066$ seconds. A compensator is required to yield zero steady-state error for a step input, and an overshoot of less than 10%. It is also desired that the settling time not exceed 10 sec.

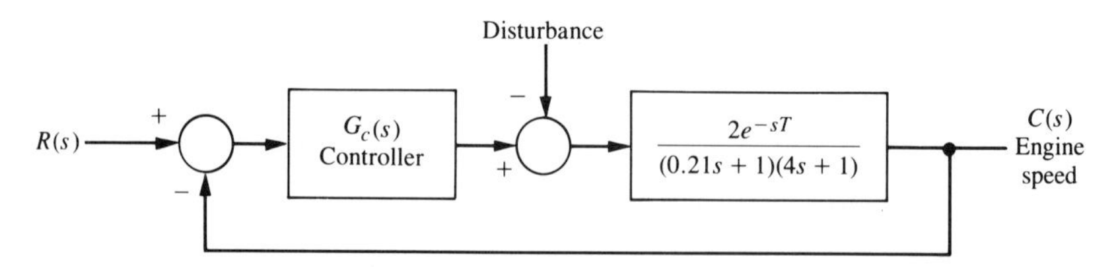

Figure DP12.14. Engine control system.

DP12.15. Plastic extrusion is a well-established method widely used in the polymer processing industry [7]. Such extruders typically consist of a large barrel divided into several temperature zones, with a hopper at one end and a die at the other. Polymer is fed into the barrel in raw and solid form from the hopper and is pushed forward by a powerful screw. Simultaneously, it is gradually heated while passing through the various temperature zones set in gradually increasing temperatures. The heat produced by the heaters in the barrel, together with the heat released from the friction between the raw polymer and the surfaces of the barrel and the screw, eventually causes the melting of the polymer, which is then pushed by the screw out from the die, to be processed further for various purposes.

The output variables are the outflow from the die and the polymer temperature. The main controlling variable is the screw speed, since the response of the process to it is rapid.

The control system for the output polymer temperature is shown in Fig. DP12.15, on the next page. Select a controller to provide zero steady-state error for a step input and an overshoot of less than 10%. Also, try to minimize the settling time of the system.

DP12.16. A computer controller for a robot that picks up hot ingots and places them in a quenching tank is shown in Fig. DP12.16(a). The robot places itself over the ingot and then moves down in the y axis. The control system is shown in Fig. DP12.16(b), where

$$G(s) = \frac{e^{-sT}}{(s + 1)^2}$$

and $T = \pi/4$ sec. Design a controller that reduces the steady-state error for a step input to 10% of the input magnitude while maintaining an overshoot of less than 10%.

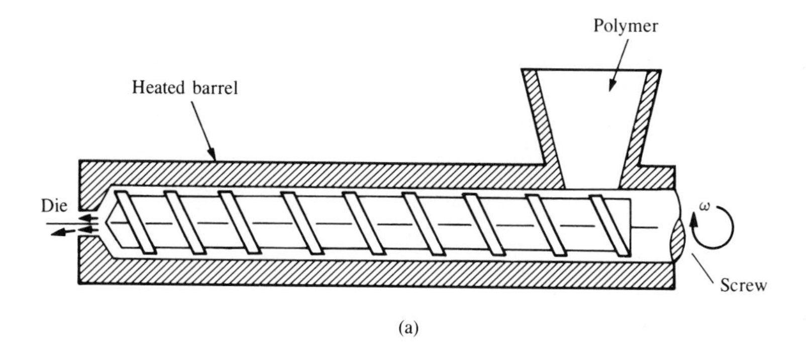

(a)

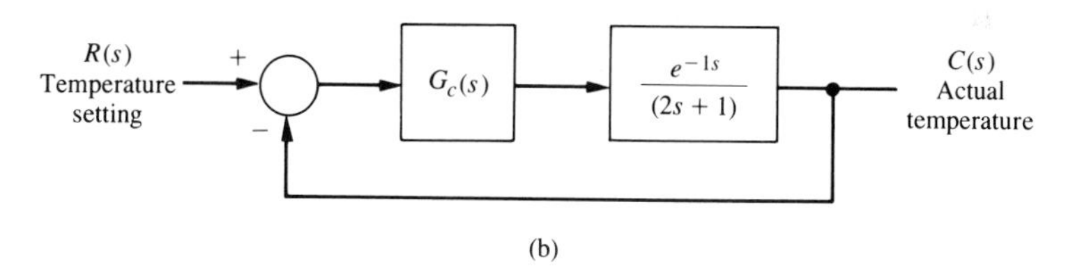

(b)

Figure DP12.15. Control system for an extruder.

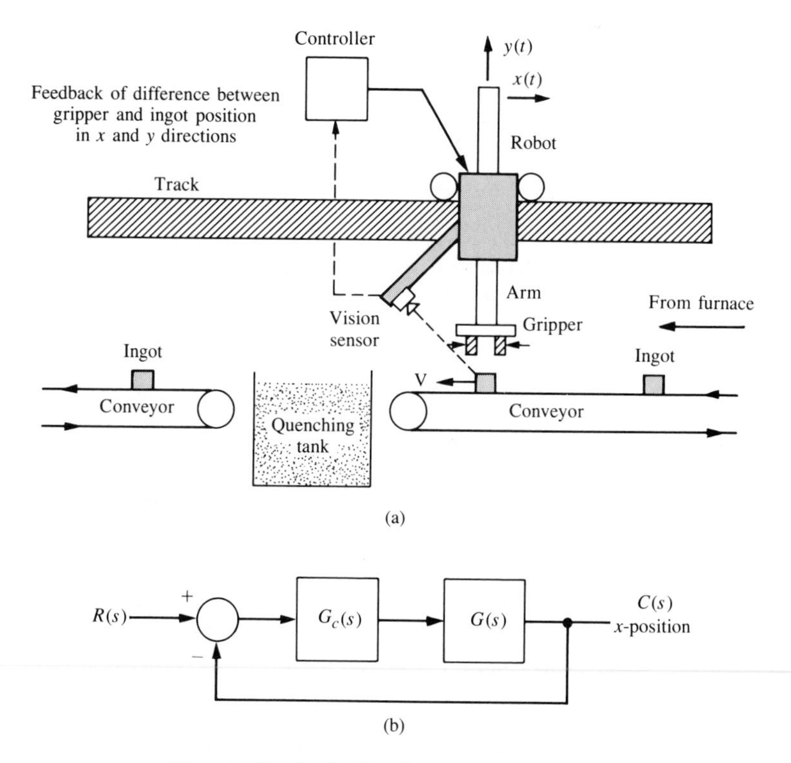

(b)

Figure DP12.16. Hot ingot robot control.

DP12.17. A high-performance jet airplane is shown in Fig. DP12.17(a), and the roll-angle control system is shown in Fig. DP12.17(b). Design a controller $G_c(s)$ so that the step response is well-behaved and the steady-state error is small.

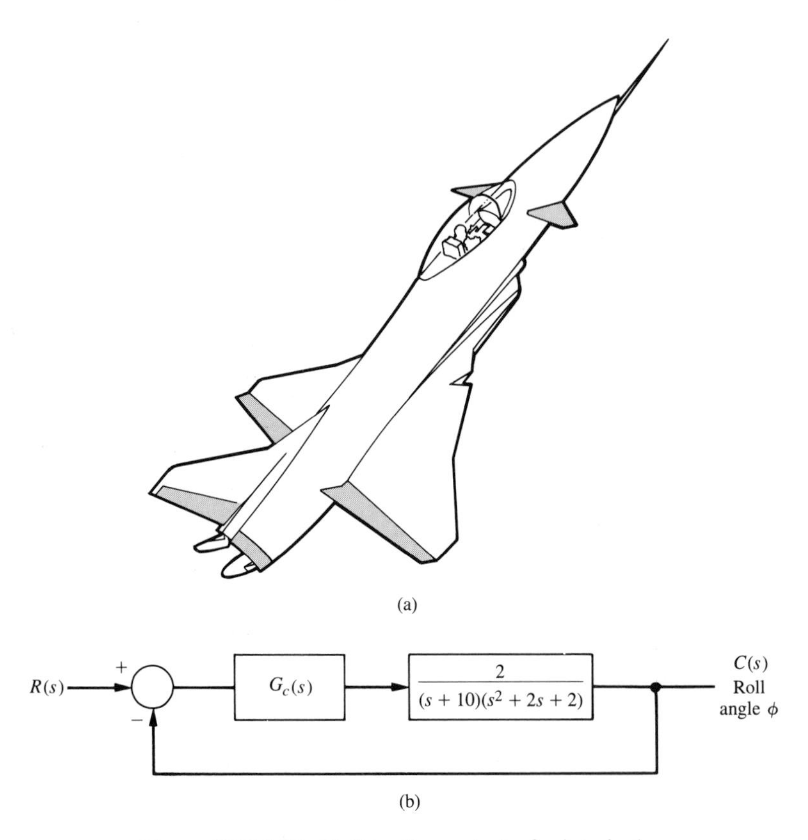

Figure DP12.17. Roll angle control of a jet airplane.

DP12.18. A two-tank system containing a heated liquid has the model shown in Fig. DP12.18(a), where T_0 is the temperature of the fluid flowing into the first tank and T_2 is the temperature of the liquid flowing out of the second tank. The block diagram model is shown in Fig. DP12.18(b). The system of the two tanks has a heater in tank 1 with a controllable heat input, Q. The time constants are $\tau_1 = 10$ sec and $\tau_2 = 50$ sec. (a) Determine $T_2(s)$ in terms of $T_0(s)$ and $T_{2d}(s)$. (b) If $T_{2d}(s)$, the desired output temperature, is changed instantaneously from $T_{2d}(s) = A/s$ to $T_{2d}(s) = 2A/s$, determine the transient response of $T_2(t)$ when $G_c(s) = K = 500$. Assume that prior to the abrupt temperature change the system is at steady state. (c) Find the steady-state error, e_{ss}, for the system of part (b), where $E(s) = T_{2d}(s) - T_2(s)$. (d) Let $G_c(s) = K/s$ and repeat parts (b) and (c). Use a gain, K, such that the percent overshoot is less than 10%. (e) Design a PI compensator that will result in a system with a settling time of $T_s < 150$ sec, and a percent overshoot of less than 10%, while maintaining a zero steady-state error. (f) Design a PID controller to achieve a settling time of less than 150 sec and an overshoot of less than 10%. (g) Prepare a table comparing the percent overshoot, settling time, and steady-state error for the designs of parts (b), through (f).

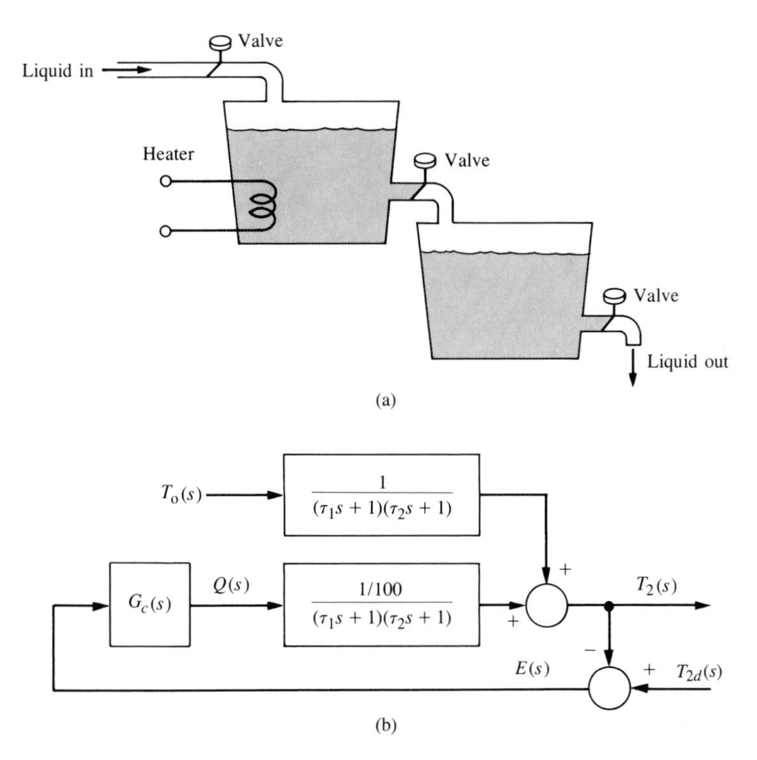

Figure DP12.18. Two-tank temperature control.

DP12.19. A rover vehicle designed for use on other planets and moons is shown in Fig. DP12.19(a), on the next page [14]. The block diagram of the steering control is shown in Fig. DP12.19(b), where

$$G(s) = \frac{(s + 1.5)}{(s + 1)(s + 2)(s + 4)(s + 10)}.$$

(a) When $G_c(s) = K$, draw the root locus as K varies from 0 to 1000. Find the roots for K equal to 100, 300, and 600. (b) Predict the overshoot, settling time, and steady-state error for a step input, assuming dominant roots. (c) Determine the actual time response for a step input for the three values of the gain K and compare the actual results with the predicted results. (d) Design a lead compensator so that the overshoot is less than 5%. The maximum gain K is 1500. Determine the actual percent overshoot, settling time, and steady-state error. (d) Design a PID controller so that the overshoot is less than 5%. Determine the actual performance of the system. (f) Prepare a table comparing the percent overshoot, settling time, and steady-state error for the designs of parts (a), (d), and (e).

DP12.20. One arm of a space robot is shown in Fig. DP12.20(a). The block diagram for the control of the arm is shown in Fig. DP12.1(b). The transfer function of the motor and arm is

$$G(s) = \frac{1}{s(s + 10)}.$$

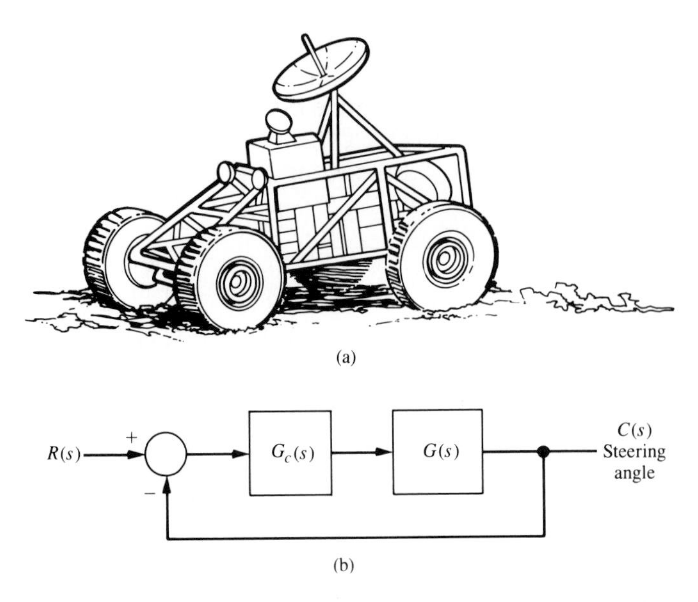

Figure DP12.19. Planetary rover vehicle steering control.

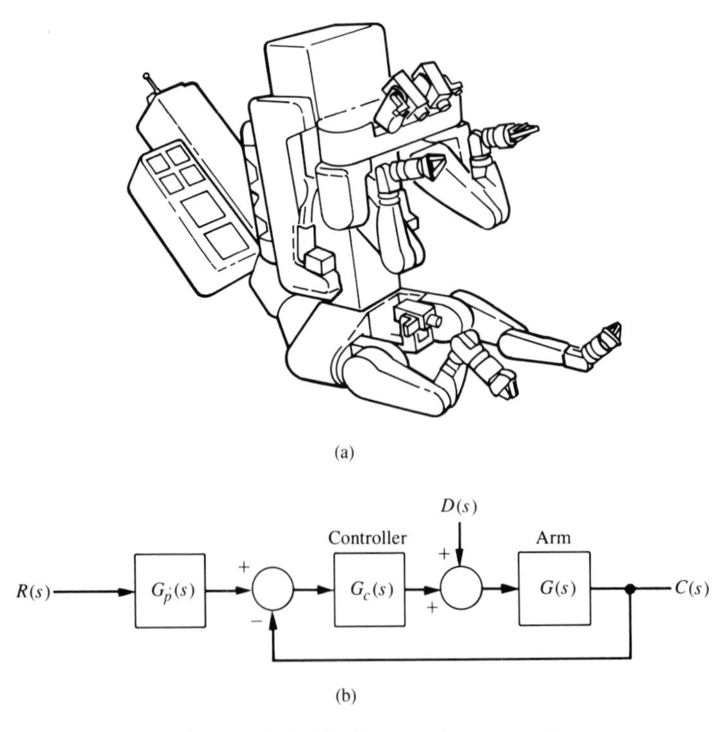

Figure DP12.20. Space robot control.

(a) If $G(s) = K$, determine the gain necessary for an overshoot of 4.5%, and plot the step response. (b) Design a proportional plus derivative (PD) controller using the ITAE method and $\omega_n = 10$. Determine the required prefilter $G_p(s)$. (c) Design a PI controller and a prefilter using the ITAE method. (d) Design a PID controller and a prefilter using the ITAE method with $\omega_n = 10$. (e) Determine the effect of a unit step disturbance for each design. Record the maximum value of $c(t)$ and the final value of $c(t)$ for the disturbance input. (f) Determine the overshoot, peak time, and settling time for an input step $R(s)$ for each design above. (g) The plant is subject to variation due to load changes. Find the magnitude of the sensitivity at $\omega = 5$, $|S_G^T(j5)|$ where

$$T = \frac{GG_c}{1 + GG_c}.$$

(h) Based on the results of parts (e), (f), and (g), select the best controller.

Terms and Concepts

Complexity of design The intricate pattern of interwoven parts and knowledge required.

Design The process of conceiving or inventing the forms, parts, and details of a system to achieve a reasoned purpose.

Engineering design The process of designing a technical system.

Gap The void between what is intended (or visualized) as the product or device and the actual, practical form of the final design.

Optimization The adjustment of the parameters to achieve the most favorable or advantageous design.

Risk Uncertainties embodied in the unintended consequences of the design.

Specifications Statements that explicitly state what the device or product is to be and to do. A set of prescribed performance criteria.

Synthesis The process by which new physical configurations are created. The combining of separate elements or devices to form a coherent whole.

Tradeoff The need to make a judgment about how much compromise is made between conflicting criteria.

References

1. R. C. Dorf and R. Jacquot, *Control Systems Design Program,* Addison-Wesley, Reading, Mass., 1988.
2. B. Wada, "Adaptive Structures," *Mechanical Engineering,* November 1990; pp. 41–46.
3. J. G. Bollinger and N. A. Duffie, *Computer Control of Machines and Processes,* Addison-Wesley, Reading, Mass., 1988.

4. J. P. Peatmen, *Design with Microcontrollers,* McGraw-Hill, New York, 1988.
5. M. J. Shah, *Engineering Simulation,* Prentice Hall, Englewood Cliffs, N.J., 1988.
6. D. R. Pugh, "Technical Description of the Adaptive Suspension Vehicle," *International Journal of Robotics Research,* vol. 9, no. 2, 1990; pp. 24–42.
7. P. J. Gawthrop, "Automatic Tuning of Commercial PID Controllers," *IEEE Control Systems,* January 1990; pp. 34–41.
8. R. C. Dorf, *Encyclopedia of Robotics,* John Wiley & Sons, New York, 1988.
9. J. Cook, "Modeling of an Internal Combustion Engine for Control Analysis," *IEEE Control Systems,* August 1988; pp. 20–25.
10. S. T. Van Voorhis, "Digital Control of Measurement Graphics," *Hewlett-Packard Journal,* January 1986; pp. 24–26.
11. R. K. Jurgen, "Putting Electronics to Work in the 1991 Car Models," *IEEE Spectrum,* December 1990; pp. 75–75.
12. "Choreographing a Bridge," *Mechanical Engineering,* November 1987; pp. 42–43.
13. A. Marttinen, "Control Study with Pilot Crane," *IEEE Trans. on Education,* August 1990; pp. 298–305.
14. W. Whittaker, "Japan Robotics Aim for Unmanned Space Exploration," *IEEE Spectrum,* December 1990; pp. 64–68.
15. K. Wright, "The Shape of Things to Go," *Scientific American,* May 1990; pp. 92–101.
16. S. Boyd, *Linear Controller Design,* Prentice Hall, Englewood Cliffs, N.J., 1990.
17. H. Nakaya and Y. Oguchi, "Characteristics of the Four-Wheel Steering Vehicle," *Int. J. of Vehicle Design,* vol. 8, no. 3, 1987; pp. 314–25.
18. H. G. Wilson and P. B. MacCready, "Lessons of Sunraycer," *Scientific American,* March 1989; pp. 90–97.
19. "The Flying Train Takes Off," *U.S. News and World Report,* July 23, 1990; pp. 52–53.

APPENDIXES

✦

APPENDIX A

Laplace Transform Pairs

Table A.1.

$F(s)$	$f(t),\ t \geq 0$
1. 1	$\delta(t_0)$, unit impulse at $t = t_0$
2. $1/s$	1, unit step
3. $\dfrac{n!}{s^{n+1}}$	t^n
4. $\dfrac{1}{(s+a)}$	e^{-at}
5. $\dfrac{1}{(s+a)^n}$	$\dfrac{1}{(n-1)!}\, t^{n-1} e^{-at}$
6. $\dfrac{a}{s(s+a)}$	$1 - e^{-at}$
7. $\dfrac{1}{(s+a)(s+b)}$	$\dfrac{1}{(b-a)}\,(e^{-at} - e^{-bt})$
8. $\dfrac{s+\alpha}{(s+a)(s+b)}$	$\dfrac{1}{(b-a)}\,[(\alpha - a)e^{-at} - (\alpha - b)e^{-bt}]$
9. $\dfrac{ab}{s(s+a)(s+b)}$	$1 - \dfrac{b}{(b-a)}\, e^{-at} + \dfrac{a}{(b-a)}\, e^{-bt}$

Table A.1.—*Continued*

	$F(s)$	$f(t),\ t \geq 0$
10.	$\dfrac{1}{(s+a)(s+b)(s+c)}$	$\dfrac{e^{-at}}{(b-a)(c-a)} + \dfrac{e^{-bt}}{(c-a)(a-b)} + \dfrac{e^{-ct}}{(a-c)(b-c)}$
11.	$\dfrac{s+\alpha}{(s+a)(s+b)(s+c)}$	$\dfrac{(\alpha-a)e^{-at}}{(b-a)(c-a)} + \dfrac{(\alpha-b)e^{-bt}}{(c-b)(a-b)} + \dfrac{(\alpha-c)e^{-ct}}{(a-c)(b-c)}$
12.	$\dfrac{ab(s+\alpha)}{s(s+a)(s+b)}$	$\alpha - \dfrac{b(\alpha-a)}{(b-a)} e^{-at} + \dfrac{a(\alpha-b)}{(b-a)} e^{-bt}$
13.	$\dfrac{\omega}{s^2+\omega^2}$	$\sin \omega t$
14.	$\dfrac{s}{s^2+\omega^2}$	$\cos \omega t$
15.	$\dfrac{s+\alpha}{s^2+\omega^2}$	$\dfrac{\sqrt{\alpha^2+\omega^2}}{\omega} \sin(\omega t + \phi),\ \phi = \tan^{-1} \omega/\alpha$
16.	$\dfrac{\omega}{(s+a)^2+\omega^2}$	$e^{-at} \sin \omega t$
17.	$\dfrac{(s+a)}{(s+a)^2+\omega^2}$	$e^{-at} \cos \omega t$
18.	$\dfrac{s+\alpha}{(s+a)^2+\omega^2}$	$\dfrac{1}{\omega} [(\alpha-a)^2+\omega^2]^{1/2} e^{-at} \sin(\omega t + \phi),\ \phi = \tan^{-1} \dfrac{\omega}{\alpha - a}$
19.	$\dfrac{\omega_n^2}{s^2+2\zeta\omega_n s+\omega_n^2}$	$\dfrac{\omega_n}{\sqrt{1-\zeta^2}} e^{-\zeta\omega_n t} \sin \omega_n \sqrt{1-\zeta^2}\, t, \quad \zeta < 1$
20.	$\dfrac{1}{s[(s+a)^2+\omega^2]}$	$\dfrac{1}{a^2+\omega^2} + \dfrac{1}{\omega\sqrt{a^2+\omega^2}} e^{-at} \sin(\omega t - \phi),$ $\phi = \tan^{-1} \omega/-a$
21.	$\dfrac{\omega_n^2}{s(s^2+2\zeta\omega_n s+\omega_n^2)}$	$1 - \dfrac{1}{\sqrt{1-\zeta^2}} e^{-\zeta\omega_n t} \sin(\omega_n \sqrt{1-\zeta^2}\, t + \phi),$ $\phi = \cos^{-1} \zeta,\ \zeta < 1$

Table A.1.—*Continued*

	$F(s)$	$f(t),\ t \geq 0$
22.	$\dfrac{(s+\alpha)}{s[(s+a)^2+\omega^2]}$	$\dfrac{\alpha}{a^2+\omega^2} + \dfrac{1}{\omega}\left[\dfrac{(\alpha-a)^2+\omega^2}{a^2+\omega^2}\right]^{1/2} e^{-at}\sin(\omega t+\phi),$ $\phi = \tan^{-1}\dfrac{\omega}{\alpha-a} - \tan^{-1}\dfrac{\omega}{-a}$
23.	$\dfrac{1}{(s+c)[(s+a)^2+\omega^2]}$	$\dfrac{e^{-ct}}{(c-a)^2+\omega^2} + \dfrac{e^{-at}\sin(\omega t+\phi)}{\omega[(c-a)^2+\omega^2]^{1/2}},\ \phi = \tan^{-1}\dfrac{\omega}{c-a}$

APPENDIX B

Symbols, Units and Conversion Factors

Table B.1. Symbols and Units

Parameter or Variable Name	Symbol	SI	English
Acceleration, angular	$\alpha(t)$	rad/sec^2	rad/sec^2
Acceleration, translational	$a(t)$	m/sec^2	ft/sec^2
Friction, rotational	f	$\frac{\text{n-m}}{\text{rad/sec}}$	$\frac{\text{ft-lb}}{\text{rad/sec}}$
Friction, translational	f	$\frac{\text{n}}{\text{m/sec}}$	$\frac{\text{lb}}{\text{ft/sec}}$
Inertia, rotational	J	$\frac{\text{n-m}}{\text{rad/sec}^2}$	$\frac{\text{ft-lb}}{\text{rad/sec}^2}$
Mass	M	kg	slugs
Position, rotational	$\Theta(t)$	rad	rad
Position, translational	$x(t)$	m	ft
Speed, rotational	$\omega(t)$	rad/sec	rad/sec
Speed, translational	$v(t)$	m/sec	ft/sec
Torque	$T(t)$	n-m	ft-lb

Table B.2. Conversion Factors

To Convert	Into	Multiply by
Btu	ft-lb	778.3
Btu	joules	1054.8
Btu/hr	ft-lb/sec	0.2162
Btu/hr	watts	0.2931
Btu/min	hp	0.02356
Btu/min	kw	0.01757
Btu/min	watts	17.57
cal	joules	4.182
cm	ft	3.281×10^{-2}
cm	in	0.3937
cm^3	ft^3	3.531×10^{-5}
deg (angle)	rad	0.01745
deg/sec	rpm	0.1667
dynes	gm	1.020×10^{-3}
dynes	lb	2.248×10^{-6}
dynes	newtons	10^{-5}
ft/sec	miles/hr	0.6818
ft/sec	miles/min	0.01136
ft-lb	gm-cm	1.383×10^4
ft-lb	oz-in	192
ft-lb/min	Btu/min	1.286×10^{-3}
ft-lb/sec	hp	1.818×10^{-3}
ft-lb/sec	kw	1.356×10^{-3}
$\frac{\text{ft-lb}}{\text{rad/sec}}$	$\frac{\text{oz-in}}{\text{rpm}}$	20.11
gm	dynes	980.7
gm	lb	2.205×10^{-3}
$gm\text{-}cm^2$	$oz\text{-}in^2$	5.468×10^{-3}
gm-cm	oz-in	1.389×10^{-2}
gm-cm	ft-lb	1.235×10^{-5}
hp	Btu/min	42.44
hp	ft-lb/min	33,000
hp	ft-lb/sec	550.0
hp	watts	745.7
in	meters	2.540×10^{-2}
in	cm	2.540
joules	Btu	9.480×10^{-4}
joules	ergs	10^7
joules	ft-lb	0.7376
joules	watt-hr	2.778×10^{-4}
kg	lb	2.205
kg	slugs	6.852×10^{-2}
kw	Btu/min	56.92
kw	ft-lb/min	4.462×10^4
kw	hp	1.341
miles (statute)	ft	5280
mph	ft/min	88
mph	ft/sec	1.467
mph	m/sec	0.44704
mils	cms	2.540×10^{-3}
mils	in	0.001
min (angles)	deg	0.01667
min (angles)	rad	2.909×10^{-4}
n-m	ft-lb	0.73756
n-m	dyne-cm	10^7
n-m-sec	watt	1.0
oz	gm	28.349527
oz-in	dyne-cm	70,615.7
$oz\text{-}in^2$	$gm\text{-}cm^2$	1.829×10^2
oz-in	ft-lb	5.208×10^{-3}
oz-in	gm-cm	72.01
lb(force)	newtons	4.4482
lb/ft^3	gm/cm^3	0.01602
$lb\text{-}ft\text{-}sec^2$	$oz\text{-}in^2$	7.419×10^4
rad	deg	57.30
rad	min	3438
rad	sec	2.063×10^5
rad/sec	deg/sec	57.30
rad/sec	rpm	9.549
rad/sec	rps	0.1592
rpm	deg/sec	6.0
rpm	rad/sec	0.1047
sec (angle)	deg	2.778×10^{-4}
sec (angle)	rad	4.848×10^{-6}
slugs (mass)	kg	14.594
$slug\text{-}ft^2$	km^2	1.3558
watts	Btu/hr	3.413
watts	Btu/min	0.05688
watts	ft-lb/min	44.27
watts	hp	1.341×10^{-3}
watts	n-m/sec	1.0
watts-hr	Btu	3.413

APPENDIX C

An Introduction to Matrix Algebra

C.1 Definitions

There are many situations in which we have to deal with rectangular arrays of numbers or functions. The rectangular array of numbers (or functions)

$$\mathbf{A} = \begin{bmatrix} a_{11} & a_{12} & \dots & a_{1n} \\ a_{21} & a_{22} & \dots & a_{2n} \\ \vdots & \vdots & & \vdots \\ a_{m1} & a_{m2} & \dots & a_{2mn} \end{bmatrix} \tag{C.1}$$

is known as a *matrix*. The numbers a_{ij} are called *elements* of the matrix, with the subscript i denoting the row and the subscript j denoting the column.

A matrix with m rows and n columns is said to be a matrix of *order* (m, n) or alternatively called an $m \times n$ (m by n) matrix. When the number of the columns equals the number of rows ($m = n$) the matrix is called a *square matrix* of order n. It is common to use boldfaced capital letters to denote an $m \times n$ matrix.

A matrix comprised of only one column, that is, an $m \times 1$ matrix, is known as a column matrix or, more commonly, a *column vector.* We shall represent a column vector with boldfaced lower-case letters as

$$\mathbf{y} = \begin{bmatrix} y_1 \\ y_2 \\ \vdots \\ y_m \end{bmatrix}. \tag{C.2}$$

Analogously, a *row vector* is an ordered collection of numbers written in a row—that is, a $1 \times n$ matrix. We will use boldfaced lowercase letters to represent vectors. Therefore a row vector will be written as

$$\mathbf{z} = [z_1, z_2, \ldots, z_n] \tag{C.3}$$

with n elements.

A few matrices with distinctive characteristics are given special names. A square matrix in which all the elements are zero except those on the principal diagonal, $a_{11}, a_{22}, \ldots, a_{nn}$, is called a *diagonal matrix*. Then, for example, a 3×3 diagonal matrix would be

$$\mathbf{B} = \begin{bmatrix} b_{11} & 0 & 0 \\ 0 & b_{22} & 0 \\ 0 & 0 & b_{33} \end{bmatrix}. \tag{C.4}$$

If all the elements of a diagonal matrix have the value 1, then the matrix is known as the *identity matrix* **I**, which is written as

$$\mathbf{I} = \begin{bmatrix} 1 & 0 & \ldots & 0 \\ 0 & 1 & \ldots & 0 \\ \vdots & \vdots & \ldots & \vdots \\ 0 & 0 & \ldots & 1 \end{bmatrix}. \tag{C.5}$$

When all the elements of a matrix are equal to zero, the matrix is called the *zero* or *null matrix*. When the elements of a matrix have a special relationship so that $a_{ij} = a_{ji}$, it is called a *symmetrical* matrix. Thus, for example, the matrix

$$\mathbf{H} = \begin{bmatrix} 3 & -2 & 1 \\ -2 & 6 & 4 \\ 1 & 4 & 8 \end{bmatrix} \tag{C.6}$$

is a symmetrical matrix of order (3, 3).

C.2 Addition and Subtraction of Matrices

The addition of two matrices is possible only for matrices of the same order. The sum of two matrices is obtained by adding the corresponding elements. Thus if the elements of **A** are a_{ij} and the elements of **B** are b_{ij}, and if

$$\mathbf{C} = \mathbf{A} + \mathbf{B}, \tag{C.7}$$

then the elements of **C** that are c_{ij} are obtained as

$$c_{ij} = a_{ij} + b_{ij}. \tag{C.8}$$

Then, for example, the matrix addition for two 3 × 3 matrices is as follows:

$$\mathbf{C} = \begin{bmatrix} 2 & 1 & 0 \\ 1 & -1 & 3 \\ 0 & 6 & 2 \end{bmatrix} + \begin{bmatrix} 8 & 2 & 1 \\ 1 & 3 & 0 \\ 4 & 2 & 1 \end{bmatrix} = \begin{bmatrix} 10 & 3 & 1 \\ 2 & 2 & 3 \\ 4 & 8 & 3 \end{bmatrix}. \quad \text{(C.9)}$$

From the operation used for performing the operation of addition, we note that the process is commutative; that is

$$\mathbf{A} + \mathbf{B} = \mathbf{B} + \mathbf{A}. \quad \text{(C.10)}$$

Also, we note that the addition operation is associative, so that

$$(\mathbf{A} + \mathbf{B}) + \mathbf{C} = \mathbf{A} + (\mathbf{B} + \mathbf{C}). \quad \text{(C.11)}$$

In order to perform the operation of subtraction we note that if a matrix **A** is multiplied by a constant α, then every element of the matrix is multiplied by this constant. Therefore we can write

$$\alpha\mathbf{A} = \begin{bmatrix} \alpha a_{11} & \alpha a_{12} & \dots & \alpha a_{1n} \\ \alpha a_{12} & \alpha a_{22} & \dots & \alpha a_{2n} \\ \vdots & \vdots & & \vdots \\ \alpha a_{m1} & \alpha a_{m2} & \dots & \alpha a_{mn} \end{bmatrix} \quad \text{(C.12)}$$

Then, in order to carry out a subtraction operation, we use $\alpha = -1$, and $-\mathbf{A}$ is obtained by multiplying each element of **A** by -1. Then, for example,

$$\mathbf{C} = \mathbf{B} - \mathbf{A} = \begin{bmatrix} 2 & 1 \\ 4 & 2 \end{bmatrix} - \begin{bmatrix} 6 & 1 \\ 3 & 1 \end{bmatrix} = \begin{bmatrix} -4 & 0 \\ 1 & 1 \end{bmatrix}. \quad \text{(C.13)}$$

C.3 Multiplication of Matrices

Matrix multiplication is defined in such a way as to assist in the solution of simultaneous linear equations. The multiplication of two matrices **AB** requires that the number of columns of **A** is equal to the number of rows of **B**. Thus if **A** is of order $m \times n$ and **B** is of order $n \times q$, then the product is of order $m \times q$. The elements of a product

$$\mathbf{C} = \mathbf{AB} \quad \text{(C.14)}$$

are found by multiplying the *i*th row of **A** and the *j*th column of **B** and summing these products to give the element c_{ij}. That is,

$$c_{ij} = a_{i1}b_{1j} + a_{i2}b_{2j} + \cdots + a_{iq}b_{qj} = \sum_{k=1}^{q} a_{ik}b_{kj}. \quad \text{(C.15)}$$

Thus we obtain c_{11}, the first element of **C**, by multiplying the first row of **A** by the first column of **B** and summing the products of the elements. We should note that, in general, matrix multiplication is not commutative, that is

$$\mathbf{AB} \neq \mathbf{BA}. \tag{C.16}$$

Also, we will note that the multiplication of a matrix of $m \times n$ by a column vector (order $n \times 1$) results in a column vector of order $m \times 1$.

A specific example of multiplication of a column vector by a matrix is

$$\mathbf{x} = \mathbf{Ay} = \begin{bmatrix} a_{11} & a_{12} & a_{13} \\ a_{21} & a_{22} & a_{23} \end{bmatrix} \begin{bmatrix} y_1 \\ y_2 \\ y_3 \end{bmatrix} = \begin{bmatrix} (a_{11}y_1 + a_{12}y_2 + a_{13}y_3) \\ (a_{21}y_1 + a_{22}y_2 + a_{23}y_3) \end{bmatrix}. \tag{C.17}$$

Note that **A** is of order 2×3 and **y** is of order 3×1. Therefore the resulting matrix **x** is of order 2×1, which is a column vector with two rows. There are two elements of **x**, and

$$x_1 = (a_{11}y_1 + a_{12}y_2 + a_{13}y_3) \tag{C.18}$$

is the first element obtained by multiplying the first row of **A** by the first (and only) column of **y**.

Another example, which the reader should verify, is

$$\mathbf{C} = \mathbf{AB} = \begin{bmatrix} 2 & -1 \\ -1 & 2 \end{bmatrix} \begin{bmatrix} 3 & 2 \\ -1 & -2 \end{bmatrix} = \begin{bmatrix} 7 & 6 \\ -5 & -6 \end{bmatrix}. \tag{C.19}$$

For example, the element c_{22} is obtained as $c_{22} = -1(2) + 2(-2) = -6$.

Now we are able to use this definition of multiplication in representing a set of simultaneous linear algebraic equations by a matrix equation. Consider the following set of algebraic equations:

$$\begin{aligned} 3x_1 + 2x_2 + x_3 &= u_1, \\ 2x_1 + x_2 + 6x_3 &= u_2, \\ 4x_1 - x_2 + 2x_3 &= u_3. \end{aligned} \tag{C.20}$$

We can identify two column vectors as

$$\mathbf{x} = \begin{bmatrix} x_1 \\ x_2 \\ x_3 \end{bmatrix} \quad \text{and} \quad \mathbf{u} = \begin{bmatrix} u_1 \\ u_2 \\ u_3 \end{bmatrix}. \tag{C.21}$$

Then we can write the matrix equation

$$\mathbf{Ax} = \mathbf{u}, \tag{C.22}$$

where

$$\mathbf{A} = \begin{bmatrix} 3 & 2 & 1 \\ 2 & 1 & 6 \\ 4 & -1 & 2 \end{bmatrix}.$$

We immediately note the utility of the matrix equation as a compact form of a set of simultaneous equations.

The multiplication of a row vector and a column vector can be written as

$$\mathbf{xy} = [x_1, x_2, \ldots, x_n] \begin{bmatrix} y_1 \\ y_2 \\ \vdots \\ y_n \end{bmatrix} = x_1y_1 + x_2y_2 + \cdots + x_ny_n. \tag{C.23}$$

Thus we note that the multiplication of a row vector and a column vector results in a number that is a sum of a product of specific elements of each vector.

As a final item in this section, we note that the multiplication of any matrix by the identity matrix results in the original matrix, that is $\mathbf{AI} = \mathbf{A}$.

C.4 Other Useful Matrix Operations and Definitions

The *transpose* of a matrix $\mathbf{A}$ is denoted in this text as $\mathbf{A}^T$. One will often find the notation $\mathbf{A}'$ for $\mathbf{A}^T$ in the literature. The transpose of a matrix $\mathbf{A}$ is obtained by interchanging the rows and columns of $\mathbf{A}$. Then, for example, if

$$\mathbf{A} = \begin{bmatrix} 6 & 0 & 2 \\ 1 & 4 & 1 \\ -2 & 3 & -1 \end{bmatrix},$$

then

$$\mathbf{A}^T = \begin{bmatrix} 6 & 1 & -2 \\ 0 & 4 & 3 \\ 2 & 1 & -1 \end{bmatrix}. \tag{C.24}$$

Therefore, we are able to denote a row vector as the transpose of a column vector and write

$$\mathbf{x}^T = [x_1, x_2, \ldots, x_n]. \tag{C.25}$$

Because $\mathbf{x}^T$ is a row vector, we obtain a matrix multiplication of $\mathbf{x}^T$ by $\mathbf{x}$ as follows:

$$\mathbf{x}^T\mathbf{x} = [x_1, x_2, \ldots, x_n] \begin{bmatrix} x_1 \\ x_2 \\ \vdots \\ x_n \end{bmatrix} = x_1^2 + x_2^2 + \cdots + x_n^2. \tag{C.26}$$

Thus the multiplication $\mathbf{x}^T\mathbf{x}$ results in the sum of the squares of each element of $\mathbf{x}$.

The transpose of the product of two matrices is the product in reverse order of their transposes, so that

$$(\mathbf{AB})^T = \mathbf{B}^T\mathbf{A}^T. \tag{C.27}$$

The sum of the main diagonal elements of a square matrix $\mathbf{A}$ is called the *trace* of $\mathbf{A}$, written as

$$\text{tr}\,\mathbf{A} = a_{11} + a_{22} + \cdots + a_{nn}. \tag{C.28}$$

The *determinant* of a square matrix is obtained by enclosing the elements of the matrix $\mathbf{A}$ within vertical bars as, for example,

$$\det \mathbf{A} = \begin{vmatrix} a_{11} & a_{12} \\ a_{21} & a_{22} \end{vmatrix}. \tag{C.29}$$

If the determinant of $\mathbf{A}$ is equal to zero, then the determinant is said to be singular. The value of a determinant is determined by obtaining the minors and cofactors of the determinants. The *minor* of an element a_{ij} of a determinant of order n is a determinant of order $(n - 1)$ obtained by removing the row i and the column j of the original determinant. The cofactor of a given element of a determinant is the minor of the element with either a plus or minus sign attached; hence

$$\text{cofactor of } a_{ij} = \alpha_{ij} = (-1)^{i+j}M_{ij},$$

where M_{ij} is the minor of a_{ij}. For example, the cofactor of the element a_{23} of

$$\det \mathbf{A} = \begin{vmatrix} a_{11} & a_{12} & a_{13} \\ a_{21} & a_{22} & a_{23} \\ a_{31} & a_{32} & a_{33} \end{vmatrix} \tag{C.30}$$

is

$$\alpha_{23} = (-1)^5 M_{23} = -\begin{vmatrix} a_{11} & a_{12} \\ a_{31} & a_{32} \end{vmatrix}. \tag{C.31}$$

The value of a determinant of second order (2×2) is

$$\begin{vmatrix} a_{11} & a_{12} \\ a_{21} & a_{22} \end{vmatrix} = (a_{11}a_{22} - a_{21}a_{12}). \tag{C.32}$$

The general nth-order determinant has a value given by

$$\begin{aligned} \det \mathbf{A} &= \sum_{j=1}^{n} a_{ij}\alpha_{ij} \quad \text{with } i \text{ chosen for one row, or} \\ \det \mathbf{A} &= \sum_{i=1}^{n} a_{ij}\alpha_{ij} \quad \text{with } j \text{ chosen for one column.} \end{aligned} \tag{C.33}$$

That is, the elements a_{ij} are chosen for a specific row (or column) and that entire row (or column) is expanded according to Eq. (C.33). For example, the value of a specific 3 × 3 determinant is

$$\det \mathbf{A} = \det \begin{bmatrix} 2 & 3 & 5 \\ 1 & 0 & 1 \\ 2 & 1 & 0 \end{bmatrix}$$

$$= 2 \begin{vmatrix} 0 & 1 \\ 1 & 0 \end{vmatrix} - 1 \begin{vmatrix} 3 & 5 \\ 1 & 0 \end{vmatrix} + 2 \begin{vmatrix} 3 & 5 \\ 0 & 1 \end{vmatrix} \qquad \text{(C.34)}$$

$$= 2(-1) - (-5) + 2(3) = 9,$$

where we have expanded in the first column.

The *adjoint matrix* of a square matrix **A** is formed by replacing each element a_{ij} by the cofactor α_{ij} and transposing. Therefore

$$\text{adjoint } \mathbf{A} = \begin{bmatrix} \alpha_{11} & \alpha_{12} & \cdots & \alpha_{1n} \\ \alpha_{21} & \alpha_{22} & \cdots & \alpha_{2n} \\ \vdots & \vdots & & \vdots \\ \alpha_{n1} & \alpha_{n2} & \cdots & \alpha_{nn} \end{bmatrix}^T = \begin{bmatrix} \alpha_{11} & \alpha_{21} & \cdots & \alpha_{n1} \\ \alpha_{12} & \alpha_{22} & \cdots & \alpha_{n2} \\ \vdots & \vdots & & \vdots \\ \alpha_{1n} & \alpha_{2n} & \cdots & \alpha_{nn} \end{bmatrix}. \qquad \text{(C.35)}$$

C.5 Matrix Inversion

The inverse of a square matrix **A** is written as $\mathbf{A}^{-1}$ and is defined as satisfying the relationship

$$\mathbf{A}^{-1}\mathbf{A} = \mathbf{A}\mathbf{A}^{-1} = \mathbf{I}. \qquad \text{(C.36)}$$

The inverse of a matrix **A** is

$$\mathbf{A}^{-1} = \frac{\text{adjoint of } \mathbf{A}}{\det \mathbf{A}} \qquad \text{(C.37)}$$

when the det **A** is not equal to zero. For a 2 × 2 matrix we have the adjoint matrix

$$\text{adjoint } \mathbf{A} = \begin{bmatrix} a_{22} & -a_{12} \\ -a_{21} & a_{11} \end{bmatrix} \qquad \text{(C.38)}$$

and the det $\mathbf{A} = a_{11}a_{22} - a_{12}a_{21}$. Consider the matrix

$$\mathbf{A} = \begin{bmatrix} 1 & 2 & 3 \\ 2 & -1 & 4 \\ 0 & -1 & 1 \end{bmatrix}. \qquad \text{(C.39)}$$

The determinant has a value det **A** = −7. The cofactor α_{11} is

$$\alpha_{11} = (-1)^2 \begin{vmatrix} -1 & 4 \\ -1 & 1 \end{vmatrix} = 3. \qquad \text{(C.40)}$$

In a similar manner we obtain

$$\mathbf{A}^{-1} = \frac{\text{adjoint } \mathbf{A}}{\det \mathbf{A}} = \left(-\frac{1}{7}\right)\begin{bmatrix} 3 & -5 & 11 \\ -2 & 1 & 2 \\ -2 & 1 & -5 \end{bmatrix}. \tag{C.41}$$

C.6 Matrices and Characteristic Roots

A set of simultaneous linear algebraic equations can be represented by the matrix equation

$$\mathbf{y} = \mathbf{Ax}, \tag{C.42}$$

where the $\mathbf{y}$ vector can be considered as a transformation of the vector $\mathbf{x}$. The question may be asked whether or not it may happen that a vector $\mathbf{y}$ may be a scalar multiple of $\mathbf{x}$. Trying $\mathbf{y} = \lambda\mathbf{x}$, where λ is a scalar, we have

$$\lambda\mathbf{x} = \mathbf{Ax}. \tag{C.43}$$

Alternatively, Eq. (C.43) can be written as

$$\lambda\mathbf{x} - \mathbf{Ax} = (\lambda\mathbf{I} - \mathbf{A})\mathbf{x} = \mathbf{0}, \tag{C.44}$$

where $\mathbf{I}$ = identity matrix. Thus the solution for $\mathbf{x}$ exists if and only if

$$\det(\lambda\mathbf{I} - \mathbf{A}) = 0. \tag{C.45}$$

This determinant is called the characteristic determinant of $\mathbf{A}$. Expansion of the determinant of Eq. (C.45) results in the *characteristic equation.* The characteristic equation is an nth-order polynomial in λ. The n roots of this characteristic equation are called the *characteristic roots.* For every possible value λ_i ($i = 1, 2, \ldots, n$) of the nth-order characteristic equation, we can write

$$(\lambda_i\mathbf{I} - \mathbf{A})\mathbf{x}_i = 0. \tag{C.46}$$

The vector $\mathbf{x}_i$ is the *characteristic vector* for the ith root. Let us consider the matrix

$$\mathbf{A} = \begin{bmatrix} 2 & 1 & 1 \\ 2 & 3 & 4 \\ -1 & -1 & -2 \end{bmatrix}. \tag{C.47}$$

The characteristic equation is found as follows:

$$\det\begin{bmatrix} (\lambda - 2) & -1 & -1 \\ -2 & (\lambda - 3) & -4 \\ 1 & 1 & (\lambda + 2) \end{bmatrix} = (-\lambda^3 + 3\lambda^2 + \lambda - 3) = 0. \tag{C.48}$$

The roots of the characteristic equation are $\lambda_1 = 1$, $\lambda_2 = -1$, $\lambda_3 = 3$. When $\lambda = \lambda_1 = 1$, we find the first characteristic vector from the equation

$$\mathbf{Ax}_1 = \lambda_1\mathbf{x}_1, \tag{C.49}$$

and we have $\mathbf{x}_1^T = k[1, -1, 0]$, where k is an arbitrary constant usually chosen equal to 1. Similarly, we find

$$\mathbf{x}_2^T = [0, 1, -1]$$

and

$$\mathbf{x}_3^T = [2, 3, -1]. \tag{C.50}$$

C.7 The Calculus of Matrices

The derivative of a matrix $\mathbf{A} = \mathbf{A}(t)$ is defined as

$$\frac{d}{dt}[\mathbf{A}(t)] = \begin{bmatrix} da_{11}(t)/dt & da_{12}(t)/dt & \ldots & da_{1n}(t)/dt \\ \vdots & \vdots & & \vdots \\ da_{n1}(t)/dt & da_{n2}(t)/dt & \ldots & da_{nn}(t)/dt \end{bmatrix}. \tag{C.51}$$

That is, the derivative of a matrix is simply the derivative of each element $a_{ij}(t)$ of the matrix.

The *matrix exponential function* is defined as the power series

$$\exp[\mathbf{A}] = e^{\mathbf{A}} = \mathbf{I} + \frac{\mathbf{A}}{1!} + \frac{\mathbf{A}^2}{2!} + \cdots + \frac{\mathbf{A}^k}{k!} + \cdots = \sum_{k=0}^{\infty} \frac{\mathbf{A}^k}{k!}, \tag{C.52}$$

where $\mathbf{A}^2 = \mathbf{AA}$ and, similarly, $\mathbf{A}^k$ implies $\mathbf{A}$ multiplied k times. This series can be shown to be convergent for all square matrices. Also, a matrix exponential that is a function of time is defined as

$$e^{\mathbf{A}t} = \sum_{k=0}^{\infty} \frac{\mathbf{A}^k t^k}{k!}. \tag{C.53}$$

If we differentiate with respect to time, then we have

$$\frac{d}{dt}(e^{\mathbf{A}t}) = \mathbf{A}e^{\mathbf{A}t}. \tag{C.54}$$

Therefore, for a differential equation

$$\frac{d\mathbf{x}}{dt} = \mathbf{Ax}, \tag{C.55}$$

we might postulate a solution $\mathbf{x} = e^{\mathbf{A}t}\mathbf{c} = \phi\mathbf{c}$, where the matrix ϕ is $\phi = e^{\mathbf{A}t}$ and $\mathbf{c}$ is an unknown column vector. Then we have

$$\frac{d\mathbf{x}}{dt} = \mathbf{Ax} \tag{C.56}$$

or

$$\mathbf{A}e^{\mathbf{A}t} = \mathbf{A}e^{\mathbf{A}t}, \tag{C.57}$$

and we have in fact satisfied the relationship, Eq. (C.55). Then, the value of $\mathbf{c}$ is simply $\mathbf{x}(0)$, the initial value of $\mathbf{x}$ because, when $t = 0$, we have $\mathbf{x}(0) = \mathbf{c}$. Therefore the solution to Eq. (C.55) is

$$\mathbf{x}(t) = e^{\mathbf{A}t}\mathbf{x}(0). \tag{C.58}$$

References

1. R. C. Dorf, *Matrix Algebra—A Programmed Introduction,* John Wiley & Sons, New York, 1969.
2. C. R. Wylie, Jr., *Advanced Engineering Mathematics,* 4th ed., McGraw-Hill, New York, 1975.

APPENDIX D

Decibel Conversion

M	0	1	2	3	4	5	6	7	8	9
0.0	$m =$	−40.00	−33.98	−30.46	−27.96	−26.02	−24.44	−23.10	−21.94	−20.92
0.1	−20.00	−19.17	−18.42	−17.72	−17.08	−16.48	−15.92	−15.39	−14.89	−14.42
0.2	−13.98	−13.56	−13.15	−12.77	−12.40	−12.04	−11.70	−11.37	−11.06	−10.75
0.3	−10.46	−10.17	−9.90	−9.63	−9.37	−9.12	−8.87	−8.64	−8.40	−8.18
0.4	−7.96	−7.74	−7.54	−7.33	−7.13	−6.94	−6.74	−6.56	−6.38	−6.20
0.5	−6.02	−5.85	−5.68	−5.51	−5.35	−5.19	−5.04	−4.88	−4.73	−4.58
0.6	−4.44	−4.29	−4.15	−4.01	−3.88	−3.74	−3.61	−3.48	−3.35	−3.22
0.7	−3.10	−2.97	−2.85	−2.73	−2.62	−2.50	−2.38	−2.27	−2.16	−2.05
0.8	−1.94	−1.83	−1.72	−1.62	−1.51	−1.41	−1.31	−1.21	−1.11	−1.01
0.9	−0.92	−0.82	−0.72	−0.63	−0.54	−0.45	−0.35	−0.26	−0.18	−0.09
1.0	0.00	0.09	0.17	0.26	0.34	0.42	0.51	0.59	0.67	0.75
1.1	0.83	0.91	0.98	1.06	1.14	1.21	1.29	1.36	1.44	1.51
1.2	1.58	1.66	1.73	1.80	1.87	1.94	2.01	2.08	2.14	2.21
1.3	2.28	2.35	2.41	2.48	2.54	2.61	2.67	2.73	2.80	2.86
1.4	2.92	2.98	3.05	3.11	3.17	3.23	3.29	3.35	3.41	3.46
1.5	3.52	3.58	3.64	3.69	3.75	3.81	3.86	3.92	3.97	4.03
1.6	4.08	4.14	4.19	4.24	4.30	4.35	4.40	4.45	4.51	4.56
1.7	4.61	4.66	4.71	4.76	4.81	4.86	4.91	4.96	5.01	5.06
1.8	5.11	5.15	5.20	5.25	5.30	5.34	5.39	5.44	5.48	5.53
1.9	5.58	5.62	5.67	5.71	5.76	5.80	5.85	5.89	5.93	5.98
2.	6.02	6.44	6.85	7.23	7.60	7.96	8.30	8.63	8.94	9.25
3.	9.54	9.83	10.10	10.37	10.63	10.88	11.13	11.36	11.60	11.82
4.	12.04	12.26	12.46	12.67	12.87	13.06	13.26	13.44	13.62	13.80
5.	13.98	14.15	14.32	14.49	14.65	14.81	14.96	15.12	15.27	15.42
6.	15.56	15.71	15.85	15.99	16.12	16.26	16.39	16.52	16.65	16.78
7.	16.90	17.03	17.15	17.27	17.38	17.50	17.62	17.73	17.84	17.95
8.	18.06	18.17	18.28	18.38	18.49	18.59	18.69	18.79	18.89	18.99
9.	19.08	19.18	19.28	19.37	19.46	19.55	19.65	19.74	19.82	19.91
	0.	**1.**	**2.**	**3.**	**4.**	**5.**	**6.**	**7.**	**8.**	**9.**

Decibels = $20 \log_{10} M$

APPENDIX E

Complex Numbers

E.1 A Complex Number

We all are familiar with the solution of the algebraic equation

$$x^2 - 1 = 0, \tag{E.1}$$

which is $x = 1$. However, we often encounter the equation

$$x^2 + 1 = 0. \tag{E.2}$$

A number that satisfies Eq. (E.2) is not a real number. We note that Eq. (E.2) may be written as

$$x^2 = -1, \tag{E.3}$$

and we denote the solution of Eq. (E.3) by the use of an imaginary number $j1$, so that

$$j^2 = -1 \tag{E.4}$$

and

$$j = \sqrt{-1}\,. \tag{E.5}$$

An *imaginary number* is defined as the product of the imaginary unit j with a real number. Thus we may, for example, write an imaginary number as jb. A complex number is the sum of a real number and an imaginary number, so that

$$c = a + jb, \tag{E.6}$$

where a and b are real numbers. We designate a as the real part of the complex number and b as the imaginary part and use the notation

$$Re\{c\} = a \tag{E.7}$$

and

$$\text{Im}\{c\} = b. \tag{E.8}$$

E.2 Rectangular, Exponential, and Polar Forms

The complex number $a + jb$ may be represented on a rectangular coordinate place called a *complex plane.* The complex plane has a real axis and an imaginary axis, as shown in Fig. E.1. The complex number c is the directed line identified as c with coordinates a, b. The *rectangular form* is expressed in Eq. (E.6) and pictured in Fig. E.1.

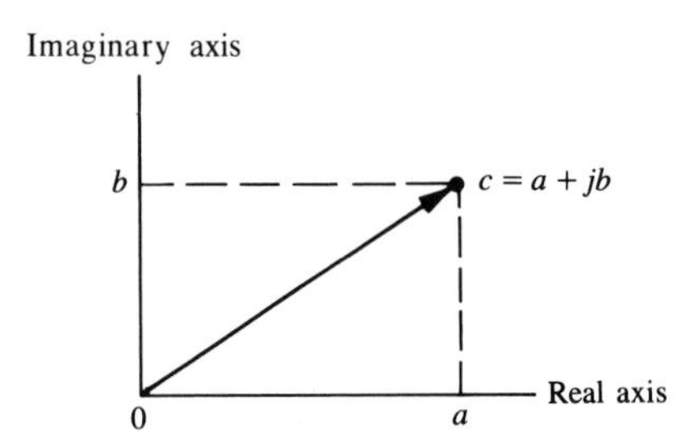

Figure E.1. Rectangular form of a complex number.

An alternative way to express the complex number c is to use the distance from the origin and the angle θ, as shown in Fig. E.2. The *exponential form* is written as

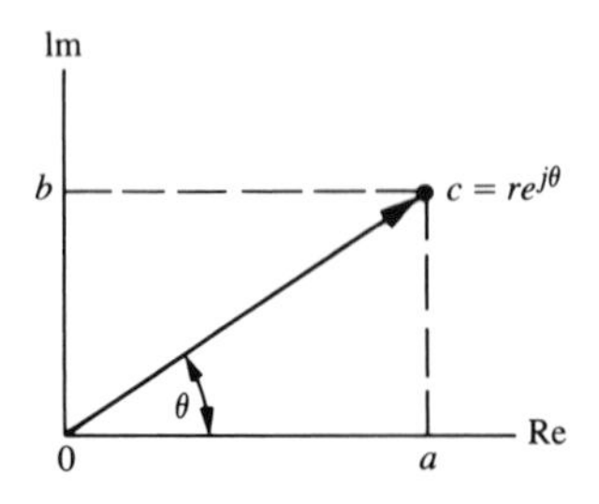

Figure E.2. Exponential form of a complex number.

$$c = re^{j\theta}, \tag{E.9}$$

where

$$r = (a^2 + b^2)^{1/2} \tag{E.10}$$

and

$$\theta = \tan^{-1}(b/a). \tag{E.11}$$

Note that $a = r \cos \theta$ and $b = r \sin \theta$.

The number r is also called the *magnitude* of c, denoted as $|c|$. The angle θ can also be denoted by the form $\underline{/\theta}$. Thus we may represent the complex number in *polar form* as

$$\begin{aligned} c &= |c|\underline{/\theta} \\ &= r\underline{/\theta}. \end{aligned} \tag{E.12}$$

■ Example E.1

Express $c = 4 + j3$ in exponential and polar form.

Solution First, draw the complex plane diagram as shown in Fig. E.3.

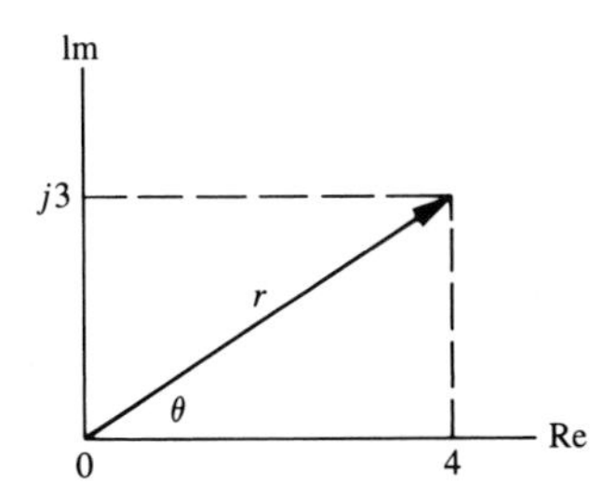

Figure E.3. Complex plane for Example E.1.

Then find r as

$$r = (4^2 + 3^2)^{1/2} = 5$$

and θ as

$$\theta = \tan^{-1}(3/4) = 36.9°.$$

The exponential form is then

$$c = 5e^{j36.9°}.$$

The polar form is

$$c = 5\underline{/36.9°}.$$

E.3 Mathematical Operations

The *conjugate* of the complex number $c = a + jb$ is called c^* and is defined as

$$c^* = a - jb \tag{E.13}$$

In polar form, we have

$$c^* = r\underline{/-\theta}. \tag{E.14}$$

To add or subtract two complex numbers, we add (or subtract) their real parts and their imaginary parts. Therefore, if $c = a + jb$ and $d = f + jg$, then

$$\begin{aligned} c + d &= (a + jb) + (f + jg) \\ &= (a + f) + j(b + g). \end{aligned} \tag{E.15}$$

The multiplication of two complex numbers is obtained as follows (note $j^2 = -1$):

$$\begin{aligned} cd &= (a + jb)(f + jg) \\ &= af + jag + jbf + j^2bg \\ &= (af - bg) + j(ag + bf). \end{aligned} \tag{E.16}$$

Alternatively, we use the polar form to obtain

$$\begin{aligned} cd &= (r_1\underline{/\theta_1})(r_2\underline{/\theta_2}) \\ &= r_1r_2\,\underline{/\theta_1 + \theta_2}, \end{aligned} \tag{E.17}$$

where

$$c = r_1\underline{/\theta_1} \quad \text{and} \quad d = r_2\underline{/\theta_2}.$$

Division of one complex number by another complex number is easily obtained using the polar form as follows:

$$\begin{aligned} \frac{c}{d} &= \frac{r_1\underline{/\theta_1}}{r_2\underline{/\theta_2}} \\ &= \frac{r_1}{r_2}\,\underline{/\theta_1 - \theta_2}. \end{aligned} \tag{E.18}$$

It is easiest to add and subtract complex numbers in rectangular form and to multiply and divide them in polar form.

A few useful relations for complex numbers are summarized in Table E.1.

Table E.1. Useful Relationships for Complex Numbers

(1) $\frac{1}{j} = -j$
(2) $(-j)(j) = 1$
(3) $j^2 = -1$
(4) $1\underline{/\pi/2} = j$
(5) $c^k = r^k\underline{/k\theta}$

■ **Example E.2**

Find $c + d$, $c - d$, cd, and c/d when $c = 4 + j3$ and $d = 1 - j$.

Solution First, we will express c and d in polar form as

$$c = 5\underline{/36.9^\circ}$$

and

$$d = \sqrt{2}\,\underline{/-45^\circ}.$$

Then, for addition, we have

$$\begin{aligned} c + d &= (4 + j3) + (1 - j) \\ &= 5 + j2. \end{aligned}$$

For subtraction, we have

$$\begin{aligned} c - d &= (4 + j3) - (1 - j) \\ &= 3 + j4. \end{aligned}$$

For multiplication, we use the polar form to obtain

$$\begin{aligned} cd &= (5\underline{/36.9^\circ})(\sqrt{2}\,\underline{/-45^\circ}) \\ &= 5\sqrt{2}\,\underline{/-8.1^\circ}. \end{aligned}$$

For division, we have

$$\begin{aligned} \frac{c}{d} &= \frac{5\underline{/36.9^\circ}}{\sqrt{2}\,\underline{/-45^\circ}} \\ &= \frac{5}{\sqrt{2}}\underline{/81.9^\circ}. \end{aligned}$$

✦

APPENDIX F

The Control System Design Program

F.1 Introduction

The Control System Design Program is a package of software tools organized to aid the engineer in designing feedback control systems. Given a single-input/single-output plant $G(s)$ and a controller $D(s)$ (Fig. F.1), the program provides the user with the following tools:

1. Open-loop time domain response to a step, ramp, or arbitrary input: data and graphical plot
2. Closed-loop time domain response to a step, ramp, impulse, or arbitrary input: data and graphical plot
3. Root locus data for a given gain
4. Open-loop frequency response (Bode plot): data and graphical plot
5. Closed-loop frequency response: data and graphical plot
6. State variable analysis: data and graphical plot

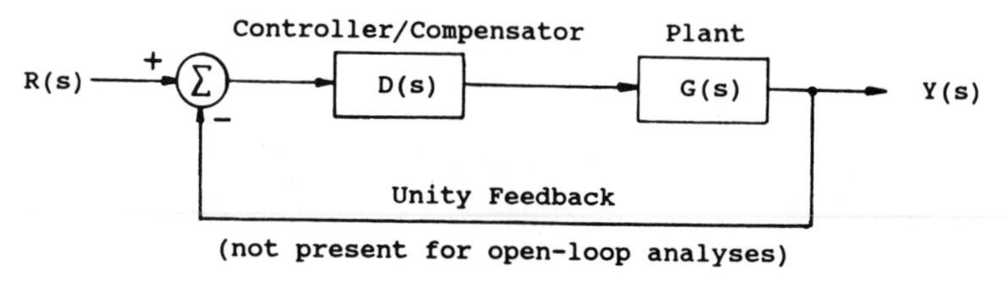

Figure F.1. Block diagram of a closed-loop system.

Using any or all of these techniques, it is possible for the designer to vary quickly and easily the gain and compensation to achieve the most desirable results.

F.2 Getting Started

F.2.1 Equipment Needed

To run the CSDP program, you need the following equipment, as a minimum:

1. IBM PC, XT, AT, or closely compatible system
2. One disk drive
3. 256K memory
4. CGA compatible graphics card
5. Color monitor or CGA compatible monochrome monitor
6. DOS 2.0 or later version

F.2.2 Customer Support

If you encounter technical problems in using this software, or find the diskette to be defective, you may obtain help by calling Addison-Wesley at one of the following numbers: Outside Massachusetts: (800)527-5210; Within Massachusetts: (617)944-3700 (Ask for the software hotline.)

F.2.3 Starting Up

The DOS operating system must be present prior to running CSDP and after quitting operation. To make a simple-to-use working copy, obtain a blank floppy diskette and follow these directions:

1. Insert the operating system (DOS) disk into drive A and turn the machine on. If the machine has a hard drive containing the operating system, then simply turn the computer on.
2. Format the blank disk using the "FORMAT A: /S" command followed by a carriage return ("/S" copies the DOS system files onto the blank disk). Following the instructions on the screen, remove the DOS disk, insert the blank disk, and press the Enter key.
3. When formatting is complete, remove the disk and insert the CSDP program disk. If you have a two-drive system, insert the newly formatted disk into drive B; otherwise, just continue. Copy the files CSDP.EXE and AUTOEXEC.BAT from the CSDP disk to the newly formatted disk by typing *COPY A:*. *B:*, followed by a carriage return. This will work with a one-drive system, but the disks will have to be swapped frequently.

   Note: For systems with a hard drive, the file CSDP.EXE can be copied to a directory on drive C by typing *COPY A:CSDP.EXE C:\ ⟨directory name⟩*.

4. Transfer the graphics printer driver program to your working disk, using the "COPY" command. On IBM machines the graphics program is called "graphics.com". If your system uses a different file name, then you must rename this file on your working disk.

The working disk is now ready. The file AUTOEXEC.BAT is provided to load the graphics printer driver and run CSDP as soon as you insert the disk and turn the machine on. Store the original CSDP disk someplace safe as a copy.

To run CSDP from the working disk, insert that disk into drive A. Turn on the computer. The program will automatically start after a few moments.

Note: Consult your DOS manual regarding utilities such as GRAPHICS.COM to allow output to be printed.

F.3 Main Task Menu Structure

The first choice you have to make occurs at the MAIN MENU. Here you are given three task sets to choose from: A, B, or C, depending upon which tool or tools you wish to use. The MAIN MENU screen looks like this:

```
          A                          B                          C

- OPEN-LOOP TIME  - CLOSED-LOOP TIME      - STATE VARIABLE
  RESPONSE          RESPONSE                ANALYSIS

                  - ROOT LOCUS
                    CALCULATION

                  - OPEN-LOOP
                    FREQUENCY RESPONSE

                  - CLOSED-LOOP
                    FREQUENCY RESPONSE

SELECT A TASK SET: A, B, OR C (OR TYPE 'Q' TO QUIT):
```

The tasks are broken up into these sets because each set has a different way of describing the system. The descriptions will be explained in the appropriate sections, and the differences should become clear.

The user must enter a system description the first time a particular task set is chosen. Once entered, the program will remember it (as long as you don't quit) and will give you the chance to modify the description every time you return to that task set. For task set B, you will enter the plant description and, if you choose, the compensator. You can then run any of the tools under task set B using this description.

In the following sections, each tool will be discussed. You will be told briefly what it does and what the system description looks like. A walk-through showing the inputs is then given, and the expected outputs are discussed. The tasks are arranged in the order encountered in this text.

F.4 Task A: Open-Loop Time Response

F.4.1 Introduction

The open-loop time response is a method of determining the output of a system in time given the input signal and the plant description. In this case, the "open-loop" means that there is no feedback involved in the total system (see Fig. F.2).

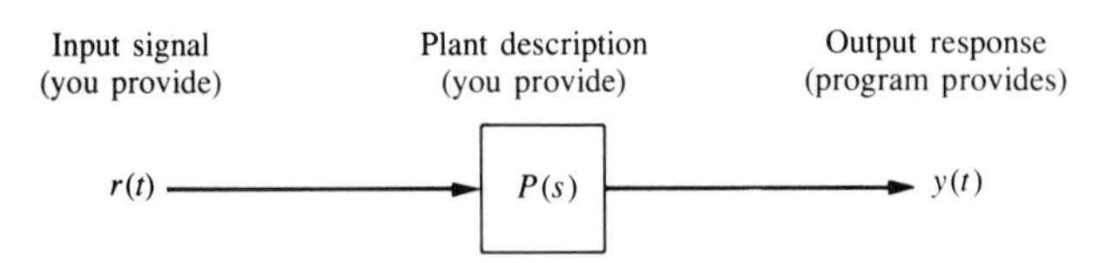

Figure F.2. Block diagram of open-loop system.

The plant description for Task A, the open-loop time response, is represented by the s-domain transfer function:

$$P(s) = \frac{K * [\, b_n s^n + b_{n-1} s^{n-1} + \cdots + b_1 s + b_0 \,]}{s^n + a_{n-1} s^{n-1} + \cdots + a_1 s + a_0} \tag{F.1}$$

The program shows it to you this way:

$$P(S) = \frac{K * [B(N)*S\hat{}N + B(N-1)*S\hat{}(N-1) + \cdots + B(1)*S + B(0)]}{S\hat{}N + A(N-1)*S\hat{}(N-1) + \cdots + A(1)*S + A(0)}. \tag{F.2}$$

Note that you are given the opportunity to specify equal order numerator and denominator polynomials. The numerator can be made a lower order by simply setting the higher order coefficients to zero. Also, because compensation does not usually occur in open-loop systems, it is assumed to be a simple gain, K.

F.4.2 Entering the Plant Description

To enter the coefficients, your plant description must be in the form shown above. This means that if a series of multiplicands are given, you must multiply them out to obtain the polynomial, making sure that the coefficient for s^n in the denominator is 1.

1. You are first asked for the order N of the system. Look at the denominator to figure this out; the largest order of s will tell you. Input this value. N has a minimum value of 1 and a maximum value of 10.

2. Next, enter the DENOMINATOR coefficients, starting with A(0). Make sure that each coefficient value is grouped with its proper order of s. For any orders of s not specified, enter a value of zero for that coefficient.
3. Then, enter the NUMERATOR coefficients, starting with B(0).
4. When you finish these steps, the values the computer received will be displayed for your review, and you will be asked if they are correct. Answer *Y* or *N*.

F.4.3 Adjustable Parameters

You are next asked to enter a couple of adjustable parameters. If you go to another task set and return to Task A, you will not need to reenter the plant description, but you will need to respecify these two parameters.

The first parameter is the gain, K. This can be any value except 0, which is not allowed. The second parameter is the simulation time increment, T, in seconds. The program calculates the output response incrementally, or in other words, at times 0, T, 2T, 3T, 4T, and so on, up to 80T. When all these points have been calculated, a "snapshot" of the response in time can be drawn on the screen. The duration of the response shown is given by the 80T. For example, if T = 0.1, the plot will range from t = 0 to t = 8.0 seconds.

T is a value (usually small and always greater than zero) that tells the program how often to calculate a new point. For responses with high frequency content, T should be small to resolve short duration events. For slowly changing responses, T can be longer to see a longer period of time. However, too large a value may make a stable system appear unstable. Trial and error may be required to find the best value of T, and this program is ideal for that. Start with a low value for T and increase it to obtain the total period 80T that you want while reviewing each response to verify it does not change due to changing T.

F.4.4 Input Function

The input specification is a general one for all three time-based response tools (open- and closed-loop response and state variable analysis). Thus this description will apply to all three tools.

The four choices of input function you can give to your system are IMPULSE, STEP, RAMP, and ARBITRARY. The IMPULSE function is described by the formula

$$\lim_{e \to 0} r(t) = \begin{cases} 1/e, \text{ if } 0 \leq t \leq e \\ 0, t > e \end{cases} \tag{F.3}$$

and looks like a spike occurring at time zero. The program approximates this by letting e = T, the time increment you select.

The STEP function looks like an abrupt jump from zero to some constant value and is described by the formula

$$r(t) = \begin{cases} A, t \geq 0 \\ 0, t < 0. \end{cases} \tag{F.4}$$

You simply specify the amplitude A (height) of the step. It can be positive or negative, but not zero.

The RAMP function is a straight line sloping up or down away from t = 0 and is described by the formula

$$r(t) = \begin{cases} At, & t \geq 0 \\ 0, & t < 0. \end{cases}$$

You simply specify the slope A of the ramp. It can be positive or negative, but not zero.

The ARBITRARY function allows you to specify an arbitrary input of up to 80 points. You are prompted for the quantity of nonzero consecutive points, and then you enter them.

After choosing which input you want, you will have to wait a short time for the computer to do its work. Refer to Section F.12, "Error Messages," if an "overflow" occurs. The program is still operating.

F.4.5 Outputs: Data and Graphics

Two types of data output are available for all three time-based tools (open- and closed-loop response and state variable analysis). Thus this description will apply to all three tools.

After waiting, you are prompted for the first output type: Do you want to see the list of tabular data? If you answer *Y*, the calculated output value for each time increment is given. This is useful in determining various amplitudes on the graphic plot. If you choose, you may enter *N* and go on to view the graphical plot. You will have another opportunity later to view the data table.

You are next prompted to see if you want a screen plot, even if you answered *N* to the tabular data. If you answer *N* here, the program will go on to its final set of queries. If you want to see the plot, enter *Y* to view it. In addition to the plot, you get a set of basic data:

MAX VALUE:	Maximum value on the graph
MIN VALUE:	Minimum value on the graph
Tmax:	Time at which MAX VALUE occurs
Tmin:	Time at which MIN VALUE occurs
TIME INTERVAL:	Your entered value for T
INIT. VALUE:	Value at time t = 0
FIN. VALUE:	Value for the last time point
GAIN:	Your entered value for gain

You are again asked if you want to see the list of tabular data. If you enter *Y*, you may view it. You can go back and forth between the data and the plot as often as you want. If you answer *N*, the program goes on to the final questions.

F.4.6 Final Questions

You have the chance to change the time increment T and recalculate. This is desirable if you are trying to find the best value for T for your system and input. You also have the chance to change the gain, K, and redo the calculations.

If you want to try a different input type (with or without the same T or K values), type *Y* to either final question. Enter the values for T (and K, if necessary), and you will then be prompted to enter the input type. When you are through with the open-loop time response, just say *N* to these questions. You will then be returned to the MAIN MENU.

F.4.7 Example

The system we want to examine for the open-loop response is

$$P(s) = \frac{(s + 2)(s + 3)}{(s + 10)(s^2 + 4s + 5)}. \tag{F.6}$$

The program will not accept it in this form, so we multiply the terms out to obtain polynomials in the numerator and denominator:

$$P(s) = \frac{s^2 + 5s + 6}{s^3 + 14s^2 + 45s + 50}. \tag{F.7}$$

Now, we choose *A* at the MAIN MENU and proceed. You are shown the form that the equation must be in, and as you are faced with the following prompts, respond with the italic numbers shown below (enter "Return" after each entry):

INPUT THE PLANT ORDER N: *3*

INPUT THE DENOMINATOR COEFFICIENTS
A(0) = *50*
A(1) = *45*
A(2) = *14*

INPUT THE NUMERATOR COEFFICIENTS
B(0) = *6*
B(1) = *5*
B(2) = *1*
B(3) = *0*

Note that there was no s^3 term in the numerator, so B(3) was set equal to zero.

The next questions ask you to enter the gain, K, and the time interval, T. Enter 10 for K and .05 for T. Wait until the input type prompt comes up. When it does, we want to do a step input of height 1. Enter the italic values at the prompts.

F.5 Task Set B: Four Tools

F.5.1 The Menu

When you choose task set *B*, you will have four tools at your disposal, all having access to the same plant and compensator descriptions of Fig. F.1. The first time you type *B* at the MAIN MENU, you will have to enter the plant description and, optionally, the compensator description. You will then be presented with the following menu:

MENU FOR TASK SET B

1 CHANGE PLANT DATA FOR THIS TASK SET
2 INPUT OR CHANGE CONTROLLER DATA
3 CLOSED-LOOP TIME DOMAIN RESPONSE
4 ROOT LOCUS CALCULATIONS
5 OPEN-LOOP FREQUENCY RESPONSE
6 CLOSED-LOOP FREQUENCY RESPONSE
7 RETURN TO MAIN MENU

INPUT A NUMBER AND RETURN:

This section will cover each element in the above menu in the order shown. You are returned to this menu after completing any task within it. You will also go straight to this menu if you temporarily go to Tasks A or C and then return to B.

F.5.2 Plant Description

The plant description for Task B is represented by the s-domain transfer function:

$$G(s) = \frac{c_{n-1}s^{n-1} + \cdots + c_1 s + c_0}{s^n + d_{n-1}s^{n-1} + \cdots + d_1 s + d_0}. \tag{F.8}$$

The program shows it to you this way:

$$G(S) = \frac{C(N-1)*S\hat{}(N-1) + \cdots + C(1)*S + C(0)}{S\hat{}N + D(N-1)*S\hat{}(N-1) + \cdots + D(1)*S + D(0)}. \tag{F.9}$$

This transfer function differs from the open-loop time response (Task A) only in that the order of the numerator is always less than the order of the denominator.

To enter the coefficients, your plant description must be in the form shown above. This means that if a series of multiplicands are given for the polynomials, you must multiply them out, making sure that the coefficient for s^n in the denominator is 1.

Follow the menu to enter the coefficients as shown in Section F4.7. Once you are satisfied with the coefficients, the program will ask you if the controller is a simple gain. If so, this means that the compensation is described by K, a value chosen when a tool is selected. Answer *Y*, and you will be returned to the menu for task set B. On the other hand, if your controller is also a transfer function, answer *N* and proceed to the next section. Note that a value for K is still expected for each tool.

F.5.3 Controller/Compensator Description

Before you enter the coefficients for the controller's transfer function, the program must first know if it has a higher order numerator than denominator. If not, answer *N*. The controller description will then be of the form:

$$D(s) = \frac{K * [\, s^m + p_{m-1}s^{m-1} + \cdots + p_1 s + p_0 \,]}{s^m + q_{m-1}s^{m-1} + \cdots + q_1 s + q_0}. \tag{F.10}$$

The program shows it as

$$D(S) = \frac{K * [S\hat{\ }M + P(M-1)*S\hat{\ }(M-1) + \cdots + P(1)*S + P(0)]}{S\hat{\ }M + Q(M-1)*S\hat{\ }(M-1) + \cdots + Q(1)*S + Q(0)}. \tag{F.11}$$

Note that here, the order of the numerator can equal that of the denominator. If the numerator order is less than the order of the denominator, set the higher coefficients of the numerator to zero. Note that K is included in the description but is specified while using the tools.

If the order of the numerator is greater than the order of the denominator, answer *Y* to the question. The controller transfer function will then be of the form

$$D(s) = \frac{K * [\, s^m + p_{n-1}s^{m-1} + \cdots + p_1 s + p_0 \,]}{s^{m-1} + q_{n-2}s^{m-2} + \cdots + q_1 s + q_0}. \tag{F.12}$$

The program shows:

$$D(S) = \frac{K * [S\hat{\ }M + P(M-1)*S\hat{\ }(M-1) + \cdots + P(1)*S + P(0)]}{S\hat{\ }(M-1) + Q(M-2)*S\hat{\ }(M-2) + \cdots + Q(1)*S + Q(0)}. \tag{F.13}$$

The data for either type are entered in the same fashion as for the plant description. The lower bound on the order M is 1. The upper bound depends on the order N of the plant. The sum of the two orders (N + M) cannot exceed 10.

F.6 Closed-Loop Time Response

F.6.1 Introduction

The closed-loop time response is a method of determining the time response given the reference input signal and the plant and controller description. In this case, the "closed-loop" means that there is a unity gain feedback element added to the system (Fig. F.1). It may be instructive to compare the open-loop responses to the closed-loop to become acquainted with the advantages of adding feedback control.

F.6.2 Inputs

The plant description has already been entered, along with the controller, if desired. Thus when you choose menu item #3, you are only requested to enter the gain, K, and the time increment used for analysis.

The gain, K, must be greater than zero for this case. The time increment, T, must also be greater than zero.

Next, you will be required to enter the desired input function, r(t). You may choose an IMPULSE, a STEP, a RAMP, or an ARBITRARY function.

F.6.3 Outputs: Data and Graphics

The program will present you with tabular data and a graphical plot, if desired. The nature of these outputs is identical to the outputs for the open-loop time response, and the reader is directed to that section for more information. You have the chance to change the time increment T and recalculate. This is desirable if you are trying to find the best value for T for your system and input.

You also have the chance to change the gain, K, and redo the calculations. If you want to try a different input type (with or without the same T or K values), type *Y* to either final question. Enter the values for T (and K, if necessary), and you will then be prompted to enter the input type. When you are through with the closed-loop time response, just say *N* to these questions. You will then be returned to the Menu for Task Set B.

F.7 Root Locus Calculations

F.7.1 Introduction

The root locus calculations will provide all the roots of the closed-loop system located in the s-plane for a particular gain. It also gives you the value of ZETA, the dimensionless damping ratio. By examining the locations of the roots of the system within the s-plane, it is possible to determine which gains will produce undesirable or unstable responses.

One general approach to using this tool is to input a variety of gain values, starting at $K = 0$, and build a table of root values. Determine which gain values cause breakaway (the condition where two or more roots meet on the real axis and split away off the real axis, causing nonzero imaginary components). Once you have obtained enough root locations, you can plot them by hand on graph paper to see the general trends of the system as the gain varies. This plot is known as the *root locus.* An initial qualitative sketch will aid in the selection of gains and interpretation of the results.

F.7.2 Input

Only one input is required here: the gain value, K. It can be any value, positive, negative, or zero.

F.7.3 Output

A short wait is required for the calculations, but the program soon returns with the real and imaginary parts for each root, along with its value of ZETA. A pole that is canceled by a zero will still show up in the list of roots unless you recalculate the system description to explicitly remove the pole before entering the coefficients.

The program will then ask you if you want to perform the calculations again for a different gain value. If you answer *Y*, you will return to the input. You can do this as often as you need. If you answer *N*, you will return to the Menu for Task Set B.

F.8 Open-Loop Frequency Response

F.8.1 Introduction

The open-loop frequency response is a tool for determining the amplitude and phase angle with respect to frequency for a particular system description. It is commonly referred to as a "Bode plot." The "open-loop" refers to the fact that there is no feedback component assumed in the calculations. The frequency response is dependent upon the gain value, K, and allows you to determine gain and phase margins in a system. These margins will give you an idea of the relative stability of the system in the closed-loop condition. You also have a general idea of which frequencies will be passed and which will be rejected, and you will be able to determine phase distortion values.

F.8.2 Inputs

The frequency response is calculated over a range of frequencies defined by the user by specifying the number of decades you want to view and the low-end frequency value in units of radians/sec. For example, if you want to calculate the frequency response from 0.1 rad/sec to 100 rad/sec, you would request *3* decades with a low-end frequency value of *0.1* rad/sec. The low-end frequency value is not limited to powers of 10 but can be any frequency you desire. The range of decades allowed is 1 to 6, and the frequency value must be greater than zero.

The desired gain value, K, is entered next. It must be greater than zero. You are then asked to wait a short time while the program calculates the numbers. When it is finished, it asks you if you would like to view the tabular data. If you answer *Y*, the data will be presented. If you answer *N*, the program then asks if you want to see the screen plot. This gives you an opportunity to view the Bode plot first; you are given a chance to see the tabular data after that if you desire.

F.8.3 Outputs

The program calculates 30 points of output data over the two-decade range. The tabular data will appear as:

THE GAIN IS 5

RAD/SEC	MAG	DB	PHASE
.1	1.007	.06	−5.769
.12	1.01	.087	−6.943
.14	1.013	.119	−8.128
.	.	.	.
.	.	.	.
.	.	.	.

Two screens of data are presented, one decade at a time. The frequency increases in a logarithmic fashion, as is characteristic of Bode plots.

The table headings have the following interpretations:

RAD/SEC: Radian frequency used for a particular calculation
MAG: Magnitude, or the (output/input) ratio
DB: Logarithmically scaled magnitude: 20 log(out/in)
PHASE: Phase angle, in degrees

After viewing the tabular data, you have a chance to see a graphical plot of the data. Should you want to review the tabular data again, you will have the chance.

The Bode plot you will see shows the magnitude (in decibels, db) and phase plots for the number of frequency decades you chose. If the zero db line is crossed in this interval, a horizontal line indicating "0 db" will be added to the graph. Similarly, if the −180 degree line is crossed by the phase plot, a "−180 DEG" line will be indicated. Other data are provided with the plot:

MAX.MAG: Maximum magnitude, in db
MIN.MAG: Minimum magnitude, in db
MAX.PHASE: Maximum phase, in degrees
MIN.PHASE: Minimum phase, in degrees
GAIN value: Gain value entered

A horizontal logarithmic scale of frequency values is also shown at the bottom of the plot. Vertical values of magnitude and phase are not provided but can be obtained from the tabular data.

After viewing the plot, you are given another chance to view the tabular data. Thus you can go back and forth between the tabular data and graphical plot as desired.

When you are through viewing the tabular and plotted data, you are presented with a couple of final questions allowing you to retry the system using different input parameters. The first question asks if you want to use a different number of decades. Answer *Y* if you want to change the number of decades and/ or the low-end frequency value.

If you want the same range of frequencies but a different gain value, answer *N* to the prior question. The following question will give you the chance to change the gain, K, and redo the calculations.

When you are through with the open-loop frequency response, just say *N* to these questions. You will then be returned to the Menu for Task Set B.

F.9 Closed-Loop Frequency Response

F.9.1 Introduction

The closed-loop frequency response is a tool for determining the amplitude and phase angle with respect to frequency of a particular system description. It is commonly referred to as a "Bode plot." The "closed-loop" refers to the fact that there is a unity gain feedback component assumed in the calculations (Fig. F.1). The frequency response indicates which frequencies will be passed and which will be rejected, allowing you to determine the closed-loop bandwidth as a measure of system fidelity. Altering the gain, K, will modify the system's response to certain frequencies and the system's relative stability.

F.9.2 Inputs

The frequency response is calculated over a range of frequencies defined by the user by specifying the number of decades you want to view and the low-end frequency value in units of radians/sec. For example, if you want to calculate the frequency response from 1 rad/sec to 100 rad/sec, you would request 2 decades with a low-end frequency value of 1 rad/sec to be evaluated. The low-end frequency value is not limited to powers of 10 but can be any frequency you desire. The range of decades allowed is 1 to 6, and the frequency value must be greater than zero.

The desired gain value, K, is entered next. It must be greater than zero. You are then asked to wait a short time while the program calculates the numbers. When the program is finished, it asks you if you would like to view the tabular data. If you answer *Y*, the data will be presented. If you answer *N*, the program then asks if you want to see the screen plot. This gives you an opportunity to view the Bode plot first; you are then given a chance to see the tabular data.

F.9.3 Outputs

The program will present you with tabular data and a graphical plot, if desired. The nature of these outputs is identical to the outputs for the open-loop frequency response, and the reader is directed to Section F.8 for more information. To review, the data at the top of the graph are interpreted as follows:

MAX.MAG: Maximum magnitude, in db
MIN.MAG: Minimum magnitude, in db

MAX.PHASE:	Maximum phase, in degrees
MIN.PHASE:	Minimum phase, in degrees
GAIN value:	Gain value entered

A horizontal logarithmic scale of frequency values is also shown at the bottom of the plot. Vertical values of magnitude and phase can be obtained from the tabular data.

When you are through viewing the tabular and plotted data, you are presented with a couple of final questions allowing you to retry the system using different input parameters. The first question asks if you want to use a different set of decades. Answer *Y* if you want to change the number of decades or the low-end frequency value.

If you want the same range of frequencies but a different gain value, answer *N* to the prior question. The following question will give you the chance to change the gain, K, and redo the calculations.

When you are through with the closed-loop frequency response, just say *N* to these questions. You will then be returned to the Menu for Task Set B.

F.10 State Variable Analysis

F.10.1 Introduction

State variable analysis provides a time-domain approach to solving a system. It is able to handle time-varying systems, and the calculations are facilitated by the use of the computer. This gives some distinct advantages over the more traditional frequency-domain approaches when systems have more than one variable or parameters that vary in time.

The state variable representation of a system consists of a series of independent, first-order differential equations—one equation for each distinct state variable. When the equations are rewritten in matrix form, it is then possible to enter the elements of the matrices and vectors into the program. Once this is done, an input type (as specified in the open-loop TIME response) is chosen, and the time-domain output response is calculated and presented.

F.10.2 Entering the Matrix Elements

The system is described by the state equations (given in matrix form)

$$\dot{\mathbf{x}} = \mathbf{A}\mathbf{x} + \mathbf{b}u \tag{F.14}$$

and

$$y(t) = [\, c_1\ c_2 \cdots c_n] \begin{bmatrix} x_1 \\ x_2 \\ \cdot \\ \cdot \\ x_n \end{bmatrix} . \tag{F.15}$$

Because the state variables $[x_1\ x_2 \ldots x_n]$ are solutions to first-order differential equations, it is also necessary to specify an initial state vector, meaning a series of initial values. This vector, $[x_1(0)\ x_2(0) \ldots x_n(0)]$, is entered later, when the type of input function is chosen. Note that the input, u(t), and the output, y(t), are single variables. This is because the program works with single-input, single-output systems.

The program expects you to enter the elements of matrix **A** and vectors **b** and **c**. It prompts you this way:

```
This task solves the STATE EQUATION: DX/DT=A*X+B*U   X(0)
                                                     KNOWN
        with OUTPUT EQUATION:          Y=C*X

MATRIX A IS NxN—INPUT THE SYSTEM ORDER N:
```

The system order N must not be less than 1 or exceed 10.

You are then asked to input the elements of the **A** matrix, starting with A(1,1), A(1,2), and so on, up to A(N,N). If you made a mistake typing in the value, don't worry. You'll have a chance to correct it later.

Next, enter the elements for **b** and **c** vectors as prompted. Once again, there will be an opportunity to correct typing mistakes.

Once you complete an analysis using this tool (Task C), you are returned to the main menu. If you choose Task C again later, you are prompted first with a question: Do you want to review or change the state equation previously entered? If you want to enter an entirely new system, answer yes and you will be presented with the menu for matrix **A** just described. Then typing *S* at the arrow prompt will allow you to enter a new system description.

F.10.3 Methods of Calculation: The Phi and Gamma Matrices

The program solves the *continuous-time* state equations by finding and solving the discrete-time equivalent system

$$x[(n + 1)T] = PHI(T)*x[nT] + GAMMA(T)*u[nT]. \tag{F.16}$$

For each point in discrete time, n, x_n is known and u_n is given. PHI and GAMMA are functions of matrix **A** and vector **b**, so it is possible to calculate the next point in time. T is a small interval of time that the user enters. Thus continuous time, t, is divided into equally spaced increments of nT along the time axis. The next state output (x[(n + 1)T]) is calculated on the basis of the previous one.

Hence, before the PHI and GAMMA matrices can be calculated, the value for T must be entered at the appropriate prompt. You will then be asked to enter the number of terms in the EXP(**A**t) series. For the calculations, the computer approximates the exponential exp(**A**t) using a series approximation. Usually four to seven terms are adequate for this approximation.

F.10.4 Entering the Input Type

Next, you will be required to enter the desired input function, u(t). You may choose an IMPULSE, a STEP, a RAMP, or an ARBITRARY function. Refer to Section F.4, "Open-Loop Time Response," for further details about these inputs.

Once you have chosen the input type, the program requires the initial state vector, $[x_1(0)\ x_2(0)\ \ldots\ x_n(0)]$, or $\mathbf{x}(0)$. These values are entered one at a time, as prompted.

F.10.5 Outputs: Data and Graphics

The program will present you with tabular data and a graphical plot, if desired. The nature of these outputs is identical to the outputs for the open-loop time response, and the reader is directed to Section F.4 for more information.

The data information at the top of the graph is the same as for other time response tools except for the lack of a GAIN value, which is not necessary for this task. The data are interpreted as follows:

MAX VALUE:	Maximum value on graph
MIN VALUE:	Minimum value on graph
Tmax:	Time at which MAX VALUE occurs
Tmin:	Time at which MIN VALUE occurs
TIME INTERVAL:	Value entered for T
INIT. VALUE:	Value at time t = 0
FIN. VALUE:	Value for the last time point

Recall that you can go back and forth between the tabular data and the graphical plot as often as you need, just by answering *Y* to the appropriate questions. When you are satisfied, answer *N*, and you will be returned to the MAIN MENU.

F.11 Compensation of Feedback Systems

Compensation can be implemented explicitly by the designer in CSDP only for the tools in Task Set B. Compensation is rare for open-loop time responses, and for state analysis it is best added prior to using CSDP. For Task Set B tools, however, it is quite easy to add and remove compensation networks by choosing Task *2* while at the "Menu for Task Set B." The controller/compensator description was described fully in the previous section introducing Task Set B. (See Eq. F.12.)

The advantage of CSDP is that you have the freedom to access a variety of tools for the same system. Each tool gives you a different perspective on a particular system. As this last example showed, it is wise not to limit yourself to just one tool when solving a problem, but rather to submit the system to other tools. It may provide insight, uncover problems, or reassure you that your methods were correct.

F.12 Error Messages

There are a few types of error messages you may encounter while using CSDP. To help you deal with these conditions, a review of the error types you may encounter is presented here.

Note: None of these errors will stop the program from running. They are considered to be warnings.

F.12.1 Input Errors

If you get any messages, the most likely occurrence is during an input. This is a list of some of the warnings you may encounter:

?Redo from start: This warning usually means that you entered a non-numeric value for an input expecting a number, such as a coefficient of the plant description. All you must do is just retry that input.

*> **Not a valid input** (or a variation): This means that your input was inappropriate. Maybe you entered a letter where a number should have gone. This message also occurs when you try to choose a menu selection that doesn't exist.

*> **Value out of range:** Some inputs, such as the system order, N, have a certain range of values they will accept. Only use these values.

Most other input-type errors should be self-explanatory.

F.12.2 Run-Time Errors

There is only one error message which may occur during calculations:

Overflow: This warning signals that the program has an intermediate result that exceeded its largest number. There is very little chance you will ever see this error because the program will usually anticipate this problem. In rare cases, however, it might arise and stop the program. Reevaluate your entries to make sure they are valid.

ANSWERS

ANSWERS TO SELECTED PROBLEMS

Chapter 1

1.1

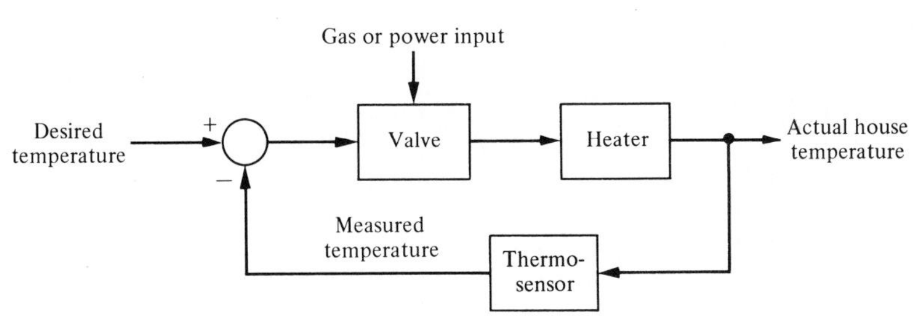

1.7 The feedback is positive.
Time lost per day = 5/3 minutes.
Total error after 15 days = 25 minutes.

Chapter 2

2.1 $R_1 i_1 + \frac{1}{C_1}\int i_1 dt + L_1 \frac{d(i_1 - i_2)}{dt} + R_2(i_1 - i_2) = v(t)$ loop 1.

$L_2 \frac{di_2}{dt} + \frac{1}{C_2}\int i_2 dt + R_2(i_2 - i_1) + L_1 \frac{d(i_2 - i_1)}{dt} = 0$ loop 2.

2.4 a) $v_o = \frac{v_{in}}{2}$ for $-0.5 \leq v_{in} \leq 0.5$.

b) $v_o = 2v_{in} - 1$ for $0.5 \leq v_{in} \leq 1.5$ or $\Delta v_o = 2\Delta v_{in}$

2.7 $T(s) = \frac{s + 1/R_1 C}{s + (R_1 + R_2)/R_1 R_2 C}.$

2.8 $T(s) = \dfrac{s^2 + 4s + 8}{s^2 + 8s + 8}.$

2.14 $\dfrac{\theta(s)}{V_f(s)} = \dfrac{0.0278}{s(s + 1.39)}.$

2.16 $x_1 = 2; x_2 = 3.$

2.18 $T(s) = \dfrac{V_2(s)}{V_1(s)} = \dfrac{Y_1Z_2Y_3Z_4}{1 + Y_1Z_2 + Y_3Z_2 + Y_3Z_4 + Y_1Z_2Z_4Y_3}.$

2.20 a) $\dfrac{e_o}{e_{in}} = \dfrac{g_mR_s}{1 + g_mR_s}.$

b) $\dfrac{e_o}{e_{in}} = \dfrac{20}{21}.$

2.25 $C(s) = G(s)R(s).$
With the effect of the disturbance eliminated.

Chapter 3

3.1 a) Open loop $S_R^T = \dfrac{1}{RCs + 1}.$

Closed loop $S_R^T = \dfrac{-KR}{RCs + 1 + KR},$

where

$$T(s) = \frac{G_1(s)}{1 + KG_1(s)}.$$

3.3 a) Closed loop $T(s) = \dfrac{KG_1(s)}{1 + KK_tG_1(s)}$; $G_1(s) = \dfrac{1}{\tau s + 1}$; $S_K^T = \dfrac{1}{1 + KK_tG_1(s)}.$

b) $\mathcal{T}(s) = \dfrac{G_1(s)}{1 + KK_tG_1(s)}\mathcal{T}_e(s) \approx \dfrac{\mathcal{T}_e(s)}{KK_t}.$

c) $e_{ss} = A/1 + KK_t$ closed-loop error.

3.7 a) $C(s) = \dfrac{G(s)T_L(s)}{1 + G_c(s)G(s)H(s)}$

b) $S_{K_2}^T = \dfrac{1}{1 + G_cGH(s)}$

c) $e_{ss} = 1 - \dfrac{1}{k_3}$

3.11. a) $T(s) = \dfrac{G_cG}{1 + G_cG} = \dfrac{K}{20s^2 + 12s + 1 + K}$

b) $S_K^T = \dfrac{(10s + 1)(2s + 1)}{20s^2 + 12s + 1 + K}$

3.12 c) $S_{K_1}^{T_1} = 0.01; S_{K_1}^{T_2} = 0.1.$

Chapter 4

4.1 a) $E(s) = \dfrac{R(s)}{1 + K_aK_mK_t/(s\tau_m + 1)}$

b) $K_aK_m \geq 24$

c) $K_aK_m \geq 39.$

4.5 a) $T(s) = \dfrac{2}{s^2 + 2s + 2}$; $\zeta = 1/\sqrt{2}$.

c) ITAE for step input

$$T(s) = \frac{\omega_n^2}{s^2 + 1.4\omega_n s + \omega_n^2}; \zeta = \frac{1}{\sqrt{2}}$$
$$= \frac{2}{s^2 + 2s + 2}.$$

4.7 a) A type one system, $K_p = \infty$, $K_v = 1/K_3$.

b) Set $\zeta = 0.6$, $\omega_n = 120$ and $K_1K_2 = 36 \times 10^4$.

4.9 $K_v = 2.05$.

4.12 $T(s) = \dfrac{3.25\omega_n^2 s + \omega_n^3}{s^3 + 1.75\omega_n s^2 + 3.25\omega_n^2 s + \omega_n^3}$; $\omega_n = 6$

Chapter 5

5.1 a) Stable

b) Stable

c) Unstable

d) Unstable

5.10 $0 \leq K \leq 28.1$.

5.13 a) $0 < K < 32.3$.

Chapter 6

6.1 a) $\phi_A = +60°, -60°, -180°$.

$\sigma_A = -2/3$.

Locus crosses imaginary axis at $K = 2$ and $s = \pm j1$.

Breakaway from real axis at $s = -1/3$.

6.5 a) $K_2 = 1.6$; complex roots: $s = -3.83 \pm j3.88$.
real roots: $s = -1.33, -0.045$.

6.8 Require a zero degree locus, $\angle GH = 0°, \pm 360°, \ldots$.

6.14 Stable for $320 \leq K \leq 27{,}000$.

6.26 b) Stable for $K \geq 2$.

c) $K = 2$, $s = \pm j1.4$.

d) Complex roots do not dominate.

6.29 $K = 7.35$.

6.30 $S_R^{r_1} = 5/6$; $S_R^{r_2} = -10/3$.

Chapter 7

7.1 a)

ω	0	1	5	∞
$\lvert GH \rvert$	1	0.4	0.037	0
ϕ	0°	−90°	−153°	−180°

7.2 a)

ω	0.5	1	2	8
db	−3.27	−8	−15.3	−36.4
ϕ	−59°	−90°	−121°	−162°

7.3 b) Evaluate at $\omega = 1.1\omega_n$.
Twin T: $|G| = 0.05$.
Bridged T: $|G| = 0.707$; A narrower band filter.

7.6 a) $GH(s) = \dfrac{0.8(5s + 1)}{s(0.25s + 1)^2}$.

7.14 $G(s) = \dfrac{809.7}{s(s^2 + 6.35s + 161.3)}$.

Chapter 8

8.2 a) $K < 4$ for stability.

8.6 a) Phase margin = +20°.
Gain margin = +6 db.
b) $Mp_\omega = 10.5$ db at $\omega_\gamma = 0.0075$.

8.8 a) Phase margin = −9°, unstable.
$\omega_c = 49$.

8.9 a) $\omega_c = 8$ and phase margin = 83°.

8.12 a) Phase margin = 27°, stable.
b) Phase margin = 13°, stable.

8.17 $K = 3$ yields phase margin = 30°.

8.18 a) $K_1 = 95$.

Chapter 9

9.1 c) $\dot{\mathbf{x}} = \begin{bmatrix} -R/L & -1/L \\ 1/C & 0 \end{bmatrix} \mathbf{x} + \begin{bmatrix} 1/L \\ 0 \end{bmatrix} u(t)$, where $x_1 = i$, $x_2 = v_c$.

9.6 a) For stability; $h > k$, $ab > kh$.
b) When $h > k$, then rabbits grow in number.

9.10 b) Unstable system.

9.11 a) $\phi(t) = \begin{bmatrix} (2t + 1)e^{-t} & -2te^{-t} \\ 2te^{-t} & (-2t + 1)e^{-t} \end{bmatrix}$.
b) $x_1(t) = x_2(t) = 10e^{-t}$.

9.14 a) System is unstable.

b) Add $r(t) = -Kx_2(t)$; then stable for $K > 1$.

Chapter 10

10.1 a) Lead network compensation:

$$G(s) = \frac{20}{s^2}.$$

$$G_c(s) = \frac{(s/2.8) + 1}{(s/22.4) + 1}.$$

Phase margin $\simeq 50°$.

b) Root locus, choose $\zeta = 0.5$, $\omega_n = 4.5$.

$$G_c(s)G(s) = \frac{40(s + 2.28)}{s^2(s + 9.0)}.$$

10.5 $G_c(s) = K_2 + \frac{K_3}{s} = \frac{K_2 s + K_3}{s}$.

$e_{ss} = 0$.

One solution: Let $K_3/K_2 = 0.1$ and cancel one pole of $G(s)$, set $\zeta = 0.6$, $\zeta\omega_n = 5$, $\omega_n = 8.333$, roots $s = -5 \pm j6.67$.

10.6 Set phase margin $= 60°$.

$$G_c(s)G(s) = \frac{1000(s/10 + 1)}{(s/70 + 1)(0.1s + 1)(10s + 1)}.$$

10.9 Desire $K_v = 33.3$, use root locus.

Desire complex roots at $s = -3.2 \pm j3.2$ ($\zeta = 0.7$).

10.11 $J = p_{11} = \frac{1}{2(2k - 3)}$, so desire k large.

10.12 a) $J = \frac{(1 + \lambda k^2)}{2(2k - 3)}$.

b) $\lambda = 1$; $k = 3.3$.

10.13 $J = 1/(2k - 1)$.

10.15 $k = 0.9$ for minimum J.

10.16 c) For Problem 10.15: $k = 0.9$.

Roots $s = -0.45 \pm j0.835$; $\zeta = 0.47$.

10.17 $k_2 = \sqrt{20}$ for minimum J; $J = \frac{k_2^2 - 20}{8k_2}$.

10.18 Set zero at $s = -15$; then find pole at $s = -63$. Then $K_v = 15$.

Chapter 11

P11.2 (b) $|S| = 0.3$ at $\omega_B/4$

P11.3 (a) Stable for $K < 10$

P11.8 (a) $K_a = 383$

Chapter 12

P12.1 $G_c(s) = -\dfrac{200(s^2 + 160s + 8900)}{s}$

PID controller

INDEX / GLOSSARY

Modern History of Control Systems

1769 James Watt's flyball governor for engine speed control
1868 James C. Maxwell's analysis of the flyball governor
1880 E. J. Routh's stability analysis
1910 Elmer A. Sperry develops the gyroscope and autopilot
1927 H. S. Black's feedback amplifier
1932 H. Nyquist's stability criterion
1938 H. W. Bode demonstrates the logarithmic frequency diagram
1947 N. B. Nichols's chart is provided for frequency analysis
1948 Walter R. Evans develops the root locus method
1958 J. Engelberger and G. Devol build the first modern industrial robot
1969 W. Hoff develops the microprocessor
1970 State variable feedback control widely used
1980 Robust control system design widely studied
1990 Export-oriented manufacturing companies emphasize automation

Standard Variables and Parameters

$C(s)$	output
$D(s)$	disturbance
$E(s)$	error
$G(s)$	plant or process
$G_c(s)$	compensator
$H(s)$	feedback block transfer function
K	gain constant
$q(s)$	polynomial in s, the characteristic equation
S_G^T	sensitivity function
$R(s)$	input or command
$T(s)$	closed-loop transfer function
$T_d(s)$	torque disturbance
$x(t)$	state variable
ω_n	natural frequency
ζ, zeta	damping ratio

ISBN 0-201-54343-5